Tropical Forest Diversity and Dynamism

Tropical Forest Diversity and Dynamism

Findings from a Large-Scale Plot Network

Edited by Elizabeth C. Losos and Egbert G. Leigh, Jr.

The University of Chicago Press   Chicago and London

ELIZABETH C. LOSOS is director of the Center for Tropical Forest Science at the Smithsonian Tropical Research Institute in Washington, DC.

EGBERT G. LEIGH, JR., is a biologist with the Smithsonian Tropical Research Institute in Balboa, Panama.

The University of Chicago Press, Chicago 60637
The University of Chicago Press, Ltd., London
© 2004 by The University of Chicago
All rights reserved. Published 2004
Printed in the United States of America
13 12 11 10 09 08 07 06 05 04   1 2 3 4 5
ISBN: 0-226-49345-8 (cloth)
ISBN: 0-226-49346-6 (paper)

Library of Congress Cataloging-in-Publication Data

Tropical forest diversity and dynamism : findings from a large-scale plot network / edited by Elizabeth C. Losos and Egbert G. Leigh, Jr.
    p. cm.
    Includes bibliographical references and index.
    ISBN 0-226-49345-8 (cloth : alk. paper)—ISBN 0-226-49346-6 (pbk. : alk. paper)
    1. Forest ecology—Tropics. 2. Forest dynamics—Tropics. 3. Biological diversity—Tropics.
I. Losos, Elizabeth Claire. II. Leigh, Egbert Giles.
    QK938.F6T725 2004
    577.34—dc22
                                                                                2004006904

♾ The paper used in this publication meets the minimum requirements of the American National Standard for Information Sciences—Permanence of Paper for Printed Library Materials, ANSI Z39.48-1992.

Contents

Preface

Elizabeth C. Losos and Egbert G. Leigh, Jr.

This book grew out of a symposium organized by the Center for Tropical Forest Science (CTFS) and held in August 1998 at the Smithsonian Institution in Washington, DC. The gathering was a landmark event for CTFS. For the first time, scientists involved in the network of Forest Dynamics Plots—large-scale (typically 25 or 50 ha) permanent plots in which all free-standing trees and large shrubs over 1 cm in trunk diameter are marked, measured, mapped, and identified—met to present their findings. Talks were given on tropical forest diversity, tree distribution patterns, species' habitat preferences, and other topics related to Forest Dynamics Plots. Roughly 200 biologists, forest managers, policy makers, and students attended this gathering. A symposium proceedings seemed like a natural outcome. In fact, all the chapters in the middle sections of this book are based on talks presented at this symposium.

When the manuscript began to arrive, however, we realized that we had an opportunity for something more than the traditional symposium volume. By then, nearly all of the Forest Dynamics Plots had already completed or were soon to complete their first census; quite a few were about to complete their first recensus. These newly emerging datasets were providing a wealth of new opportunities for comparing tropical tree demography in forests around the world. Yet no single source existed for obtaining information on Forest Dynamics Plots or making comparisons among them. After discussions with symposium participants, host-country partners within CTFS, and the University of Chicago Press, it became clear that the book should be expanded to provide standardized quantitative and qualitative information for every plot, as well as a comparative analyses of the plots. And so this volume blossomed.

Literally thousands of people have made this network of large plots, and therefore this book, possible. Individuals ranging from directors of major research institutions to undergraduate students all deserve recognition for their labors. We gratefully acknowledge the enormous efforts and commitments of each of these participants. Although space limitations do not allow us to mention everyone here, the names of many of the individuals that made each of 15 Forest Dynamics Plots a reality can be found in part 7 of the book as co-authors of plot summary chapters.

A few individuals must be singled out for their help with this book and support of the CTFS network. Suzanne Loo de Lao, CTFS Data Manager, played an essential role in the book's production. She calculated, recalculated, and recalculated again analysis after analysis. Her accuracy, efficiency, and professionalism made this book possible; indeed, for many years she has been an important part of the glue holding the CTFS network together. Three CTFS program managers also worked tirelessly on this volume and deserve recognition for their valuable efforts here and in support of other network activities: Shallin Busch, Christie Young, and Marie Massa. It is hard to imagine CTFS functioning as a network without them. We are also grateful to Patricia Ojeda, who assisted with literature reviews and bibliographic assistance.

Of course this book would not be possible if not for the visionaries that saw the power of a simple but grand idea and worked doggedly to pursue it. Stephen Hubbell and Robin Foster created the first Forest Dynamics Plot; Peter Ashton had the farsightedness to see it replicated and spread into Asia; and Ira Rubinoff had the determination to turn it into a truly global venture. Many others have joined them and played key roles in this pursuit, most notably James LaFrankie, Richard Condit, and Stuart Davies. Lisa Barnett has provided critical support in making the CTFS network sustainable over the long term.

We gratefully acknowledge input from dozens of reviewers of the individual chapters, and two very thorough and constructive anonymous reviewers of the entire text. Our thanks also to Christie Henry, Jennifer Howard, Zachary Dorsey, Anita Samen, and the University of Chicago Press, who have all lavished endless care and patience in assisting us with this volume.

Finally, the network of Forest Dynamics Plots could not have flourished without the support of donors willing to invest in projects whose payoff is best measured in decades. We especially appreciate general CTFS support from the U.S. National Science Foundation, The John D. and Catherine T. MacArthur Foundation, Frank H. Levinson and the Agora Foundation at the Peninsula Community Foundation, Andrew W. Mellon Foundation, U.S. Agency for International Development, John Merck Fund, the Rockefeller Foundation, W. Alton Jones Foundation, The Southern Company, AVINA Foundation, Conservation, Food & Health Foundation, United Nations Environment Programme, and several anonymous donors. Additional donors to each of the plots, though too numerous to mention here, are acknowledged in the last section of this volume. Institutional backing has been equally critical to the success of this venture, and so we recognize the invaluable support of the Smithsonian Tropical Research Institute, Arnold Arboretum of Harvard University, National Institute of Education/Nanyang Technological University (Singapore), National Institute for Environmental Studies (Japan), University of Puerto Rico, Pontificia Universidad Católica del Ecuador, Instituto de Investigación de Recursos Biológicos "Alexander von Humboldt" (Colombia), Forest Research Institute Malaysia, Center for Ecological Sciences of the Indian Institute of Science, Royal Forest Department (Thailand), Kasetsart University

(Thailand), Mahidol University (Thailand), University of Peradeniya (Sri Lanka), Sri Lanka Department of Forestry, Sarawak Forest Department (Malaysia), Osaka City University (Japan), Ehime University (Japan), Kyoto University (Japan), Singapore National Parks Board, Conservation International–Philippines, Isabela State University (Philippines), Federal Department of Environment and Natural Resources (Philippines), Tunghai University, National Taiwan University, Centre de Formation et de Recherche en Conservation Forestiere (Democratic Republic of Congo), Wildlife Conservation Society, BioResources Development and Conservation Programme–Cameroon, and Limbe Botanical Garden (Cameroon). We also thank the Centre for International Forestry Research, which helped support the publication of this book.

Tropical Forest Diversity and Dynamism

PART 1: Introduction

1 The Growth of a Tree Plot Network

Elizabeth C. Losos and Egbert G. Leigh, Jr.

On the Malay Peninsula, mature tropical forests reach 60 and even 70 m in height and shelter a dizzying array of mammals, birds, reptiles, and insects. These southeast Asian forests are also host to some of the most profitable timber stands in the world. Throughout the last century and up to the present, foresters have grappled with the dilemma of extracting timber without diminishing the productivity of these lush forests. This pursuit has drawn attention to several fundamental questions: What factors allowed these diverse, productive tropical forests to arise in the first place? How does human intervention—such as logging or conversion to agriculture—upset a forest's balance? Do the natural mechanisms by which tropical forests maintain their productivity and diversity offer us lessons on how to conserve or manage tropical forests more effectively?

On the other side of the world, researchers have been struggling with similar questions in neotropical forests. On Barro Colorado Island (BCI), Panama, biologists working at the Smithsonian Tropical Research Institute have long been seeking to understand what factors maintain the diversity and productivity of flora and fauna in that Central American forest. How did diversity arise in that tropical forest? Has specialization to the island's physical features allowed the coexistence of these diverse plants and animals? What role have history and chance played in that forest's development?

In the last three decades, real strides have been made toward addressing such questions, with Barro Colorado Island playing a key role in this advance. To discern more clearly what processes maintain the diversity of the island's trees, Stephen Hubbell and Robin Foster initiated a study of a 50-ha permanent plot within the island's mature forest. In this "Forest Dynamics Plot," all free-standing, woody trees and large shrubs over 1 cm diameter at breast height (dbh) were tagged, measured, mapped, identified to species, and repeatedly censused. Two decades, almost 350,000 tagged trees, and hundreds of publications later, findings from this 50-ha Forest Dynamics Plot, together with other research on BCI, have provided perhaps the best foundation for understanding the structure, dynamics, diversity, species distribution, and interactions of plants, animals, and fungi in the lowland tropics. In chapter 2, Hubbell describes how BCI's 50-ha plot has transformed ecologists' thinking on topics such as the stability of species composition within tropical forests, the influence of adult abundance and distribution on conspecific sapling recruitment, and how quickly tree population sizes change in response to the changing climate.

How well do findings from BCI's 50-ha plot apply to tropical forests elsewhere in the world? Certainly, this Panamanian forest has some unique features: The island's biota reflects the interchange of plants and animals over a land bridge that connected the Americas 3 million years ago; no such event influenced the biotas of the other tropical areas in Africa or southeast Asia. Twelve thousand years ago, the isthmus lost its megafauna. More recently, during the construction of the Panama Canal in 1910, Barro Colorado was severed from the adjacent mainland and transformed into a 1600-ha island. Climatically, BCI has a formidable dry season; its median rainfall for the year's first quarter is only 100 mm—far less rain than everwet tropical forests experience. Barro Colorado Island also tends to be far less diverse than everwet forests. On BCI, the trees over 10 cm dbh in a 25-ha subplot of the Forest Dynamics Plot account for 210 species, whereas the 25-ha subplots of everwet forest at Yasuni in Amazonian Ecuador and Lambir in Sarawak, Malaysia, respectively, contain 822 and 920 species of tree over 10 cm dbh. What causes these differences and how much do these differences matter when researchers analyze the factors attributable to tropical diversity?

To find out, in 1986 Peter S. Ashton and Salleh Mohammed Nor convinced the Forest Research Institute of Malaysia, Harvard University, and the Smithsonian Tropical Research Institute to establish a comparable 50-ha Forest Dynamics Plot in the Pasoh Forest Reserve, peninsular Malaysia. To create comparable datasets, the trees in the Pasoh plot were tagged, mapped, measured, and recensused using an identical protocol to that developed by Hubbell and Foster for the BCI plot. While both plots lie on relatively flat terrain, Pasoh has only 1 month per year that averages less than 100 mm rainfall, compared to BCI's 3; Pasoh's 50-ha plot contains 678 species of tree over 10 cm dbh, compared to BCI's 226. Do the same processes maintain tree diversity on both plots? Comparisons of the two Forest Dynamics Plots revealed that Pasoh has a far smaller number and far fewer species of pioneer trees specialized for colonizing large treefall gaps than does BCI. Thus specialization to treefall gaps of different sizes—a popular explanation for the diversity of tropical trees—could not explain why Pasoh's tree diversity is so much higher than BCI's. Increasing evidence from Pasoh indicates that sapling recruitment is diminished near conspecific adults, though no more so, apparently, than on BCI. And so the pursuit to explain why Pasoh's tree diversity is higher than BCI's continues.

The third 50-ha Forest Dynamics Plot provided a drier contrast to BCI. In 1987, R. Sukumar of the Indian Institute of Science established a large plot at Mudumalai Wildlife Sanctuary in southern India. The driving force behind this plot was the newly created Nilgiri Biosphere Reserve and World Heritage Site—of which Mudumalai was a part—and the need for rigorous information on forest regeneration to develop appropriate management plans for the reserve. Mudumalai averages only about 1200 mm of rain per year and its dry season lasts 4 months, while Barro Colorado Island averages 2600 mm per year and has a 3-month dry season. Mudumalai's 50-ha plot has only 63 species of tree over 10 cm dbh, compared to BCI's 226. Past logging allowed grass to take over Mudumalai's understory, providing fuel for frequent anthropogenic fires. Recent evidence has shown that many species recruit only when several years have passed without fire. BCI, by contrast, has suffered no fires in at least the past 80 years. Moreover, Mudumalai has elephants and wild cattle, which are far larger and more common than BCI's biggest herbivore, the tapir. Animal seed dispersers and seed predators play a less active role in Mudumalai, where most of the common species do not produce fleshy fruits attractive to frugivorous mammals or birds. Given these differences, can the processes shaping forest structure, dynamics, and diversity at Mudumalai bear any resemblance to those at work on BCI? Tree species composition is less stable and tree density is lower at Mudumalai than BCI, in part because of heavy and selective browsing by elephants and other mammalian herbivores at Mudumalai. An even more striking difference involves patterns of tree mortality. Barro Colorado Island resembles many, if not most, moist and wet tropical forests in that tree mortality is the same in nearly all size classes between 7 and 70 cm in trunk diameter. In the dry deciduous forest of Mudumalai, however, mortality is higher for smaller trees: a tree must attain a trunk diameter of 20 cm or more to be reasonably safe from fire and elephants. Moreover, unlike BCI, Mudumalai offers little evidence to link habitat specificity or density dependence to the regulation of its tree populations.

Both the similarities and the differences among these first three plots aroused widespread interest in establishing large-scale Forest Dynamics Plots in other tropical forests. As a result, in the late 1980s and early 1990s, plots were established in Sarawak, Thailand, Puerto Rico, Sri Lanka, Democratic Republic of Congo, Cameroon, Ecuador, Colombia, Philippines, Taiwan, and Singapore. In 1992, under the guidance of its director, Ira Rubinoff, the Smithsonian Tropical Research Institute (STRI) took the lead in establishing the Center for Tropical Forest Science (CTFS) to coordinate the rapidly growing network of Forest Dynamics Plots and widely disseminate plot findings. Since that time, CTFS has represented a voluntary global consortium of forestry agencies, research institutions, universities, and nongovernmental organizations, each involved in one or more Forest Dynamics Plot. The mission of the center is to promote and coordinate long-term research in the natural and social sciences based on standardized data from Forest Dynamics Plots and to translate these findings into information relevant to tropical forest conservation, management, and natural resource policy. In pursuit of these goals, over the last decade CTFS has developed a standardized forest censusing protocol; provided assistance in field training, data management, and data analysis to Forest Dynamics Plot programs; promoted communication within and outside the network; and catalyzed conservation and management applications of the plot research. By 2003, CTFS included more than three dozen research institutions and hundreds of scientists at 16 Forest Dynamics Plots in tropical forests spanning the globe. The network is now monitoring nearly 3 million trees of about 6000 species.

Due to the intensive sampling protocol of Forest Dynamics Plots, each individual plot supports detailed investigations that address the maintenance of tree diversity within its local setting. Researchers at each plot can examine, for example, whether tree composition is due to the specialization of individual tree species to habitats or light gaps, whether it is related to the concentrated effect of pests on seeds and saplings near their parents, or whether it is more attributable to the site's biogeographic history. Many plots are well suited for asking additional questions particular to their own forests: The Forest Dynamics Plots in Puerto Rico, Taiwan, and the Philippines are good places to investigate the effect of hurricanes or typhoons on the structure, dynamics, and diversity of these forests. The plots in the Congo's Ituri Forest were chosen to determine what factors allow a single tree species to dominate the canopy and why this happens in some areas but not others. Elsewhere in Africa, Cameroon's Korup National Park is distinctive for its nutrient-poor white sandy soils. Thailand's Huai Kha Khaeng and India's Mudumalai plots are well placed to investigate the influence of large herbivores and fire on tropical forests. With its high endemism, Sri Lanka's Sinharaja provides an opportunity to evaluate the dynamics of an island flora. The plot in Sarawak's Lambir Hills National Park is ideal for assessing the effect of heterogeneity in soil and topography on the distributions of different tree species. Montane forests can be explored in Thailand's Doi Inthanon and Colombia's La Planada.

Together, these plots can do even more. The network of Forest Dynamics Plots provides opportunities for global comparisons and the synthesis of research that could not be accomplished through any individual plot. For example, what climatic feature best predicts tree diversity in tropical forests? One consistent pattern that arises throughout tropical forests, whether in America, Africa, or southeast Asia, is that diversity is greater in more aseasonal climates. Yet, despite their rainy everwet conditions, tree diversity is much lower in islands such as Puerto Rico or depauperate areas such as the Deccan plate of India and Sri Lanka. What conditions must be satisfied for a tropical forest to attain the diversity appropriate to its climate? Similarly, how does soil quality influence the ways trees die, the role of pioneer species in forest dynamics, the forest's turnover rate, and the diversity of tree species? What factors influence species distributions? To facilitate global analyses of the features thought to most likely influence species composition and diversity, CTFS has selected a core set of Forest Dynamics Plots that represent a range of climatic conditions, soil topography and geology, and natural disturbance regimes. The introduction to part 2 describes the rationale behind the plot selection in greater detail.

This book represents the first attempt to assemble comparable data from the individual Forest Dynamics Plots. After reviewing the two-decade history and findings of the "mother plot" on Barro Colorado Island (chap. 2), this volume presents what we now know about the entire network of plots. It surveys basic features of tropical forests within the plots—forest structure, diversity, species accumulation, floristics—relating them to global variation in climate, biogeographic history, natural disturbances, and soil quality and topography (part 2). The volume then turns toward more detailed analyses of individual plots, illustrating the depth and range of information generated by Forest Dynamics Plots on tree diversity, seed dispersal limitation, canopy disturbance, fire response, canopy monodominance, pollination-mediated population dynamics, and forest fragmentation (parts 3–6). With the exception of a handful of chapters, most of these studies focus on an individual plot. All, however, present methods, analyses, and findings that can be readily transferred to other Forest Dynamics Plots and sites around the world. Finally, this book documents standardized qualitative and quantitative baseline information for each plot (part 7). These data, for use within this volume and beyond, provide the framework for assessing and comparing how biotic, abiotic, and stochastic factors affect the dynamics and diversity of tropical forests worldwide.

2

Two Decades of Research on the BCI Forest Dynamics Plot

Where We Have Been and Where We Are Going

Stephen P. Hubbell

Introduction

In community ecology research, there is perhaps no greater challenge than explaining the origin and maintenance of biodiversity in tropical rainforests, which are arguably the most species-rich ecosystems in the world. In particular, the tree component in the richest of these forests can be remarkably diverse, sometimes exceeding 1100 tree species in a single 25- or 50-ha plot, such as the 25-ha Yasuní Forest Dynamics Plot in Amazonian Ecuador or the 52-ha Lambir Forest Dynamics Plot in Sarawak, Malaysia. To put these plots' species richness in a global perspective, 1100 species is approximately double the entire tree flora native to continental North America north of Mexico, an area 40 million times larger. In this chapter, I summarize a few of the most salient results from 23 years of research on these questions in one particular tropical forest, Barro Colorado Island (BCI), Panama. Then, based on this experience, I offer a few suggestions for future research directions for the BCI Forest Dynamics Plot, as well as more generally for the whole set of Forest Dynamics Plots coordinated by the Smithsonian Tropical Research Institute's Center for Tropical Forest Science (CTFS).

Robin Foster and I began the first 50-ha Forest Dynamics Plot of the CTFS network on BCI in 1980 (Hubbell and Foster 1983). At that time tropical forest ecology was still using methods designed for the temperate zone that we considered inadequate for studying community-level questions about tropical forests. Tropical moist forest tree communities are extraordinary not only in species richness but also in the extreme rarity of many of their tree species. In 1980, the standard tropical forest study plot was 1 ha or smaller. We assumed that we would need a larger plot if we wanted to obtain demographic information at the individual species level rather than pooling together, as is necessary when using small plots. The final choice of 50 ha was an estimate, however, because the 50-ha BCI plot was the first of its kind, and of course we did not know BCI tree species abundances in advance. Retrospectively, we now know that although a single hectare contains about half the species found in 50 ha, most are too rare for species-level analysis. For example, suppose we take 100 individuals as a reasonable minimum number of trees for statistical analysis of survival and growth. Then 10 or fewer species meet this abundance criterion in a single hectare on BCI, even including small trees down to 1 cm diameter at breast height (dbh). Approximately half of the species in the 50-ha BCI Forest Dynamics Plot, however, meet this abundance criterion. In fact, this is true for five of the six Forest Dynamics Plots that are 50 ha or more in size.

The second assumption underlying our design was that the processes controlling tree diversity were likely to take considerable time to play out and reveal themselves. Tropical trees have lifespans ranging from 40 to 300+ years, depending upon life history differences. Thus we anticipated that we would have to commit to a long-term study, minimally lasting several decades and potentially much longer. We also appreciated the fact that the longer the study continued, the more valuable the data would become.

The final assumption was that whatever processes were governing tree diversity in the BCI forest, these processes were likely to operate in a spatial context and to be stronger in the younger life history stages of trees. This meant that we would have to collect our demographic information on individually tagged and mapped plants. We were among the first researchers in community ecology to take what later became known as the "individual-based" approach, with explicitly mapped populations of individual plants (e.g., Pacala and Silander 1998). It also meant that we would have to measure plants considerably smaller than the conventional diameter cutoff. Previous studies in 1-ha plots of tropical forest had generally used minimum diameter cutoffs of 10 cm or a 30-cm girth. We chose a much lower diameter cutoff, including all free-standing woody saplings with a stem of 1 cm dbh. Of course at the outset we had no idea of how many trees ≥ 1 cm dbh to expect in a 50-ha plot. We chose the lower cutoff diameter of 1 cm as an a priori compromise between the impossibility of a total seedling census of the entire 50 ha and the expected loss of power to detect the mortality factors in the young stages.

In 1980, Foster and I had no conception that a sustained, globally extensive, collaborative research program in tropical forest science would grow out of our initial BCI project. In fact, in the early years, it was touch and go whether the BCI project itself was sustainable. We had to overcome considerable resistance to the idea that such large plots were a good idea and worth the money, not only in the ecology community at large, but also at the U.S. National Science Foundation (NSF), our primary source for funding. In fact, NSF initially rejected the proposal for the second census in 1985, largely on the grounds that data from a single plot were "pseudoreplicated." Ironically, had we proposed instead to study 10 separate 1-ha plots across the same 50-ha area on BCI, I suspect that we would have had no trouble getting funded. In the end, partial funding for the second census was obtained from NSF, but most of the second census had to be completed with a series of small grants from individuals and private foundations.

Now, however, all these objections have been forgotten, largely, I believe, because of four main factors: (1) the creation of the CTFS in 1992 to coordinate the establishment of Forest Dynamics Plots across the tropics, censused with a standardized methodology (Condit 1998), (2) the depth and global breadth of international collaboration that CTFS has built, (3) the accelerating research productivity of the CTFS network, proving without question the phenomenal power of large plots to answer many previously intractable questions about tropical forest structure and dynamics, and (4) a sea change in the ecological community that has belatedly come to appreciate the significance of space and spatially dependent processes in ecology (Tilman and Kareiva 1999). Since 1980, spatial statistics has also come of age, providing the theory and statistical tools to properly analyze spatially autocorrelated ecological patterns and processes (e.g., Cressie 1995); these technical advances have largely eliminated the old issue of pseudoreplication, at least within the 50-ha plot. We not only have proven the concept but also are ahead of the curve on tropical forest science, and our leadership and global partnership has not gone unnoticed and unrecognized. Indeed, E. O. Wilson has recently described it as among "the most extensive field research ever conducted in biology" (Hubbell 2001).

We can be justly proud of our accomplishments, but at the same time, we should never allow the CTFS research program to ossify into an intellectually sterile routine of data collection. We need to ensure that we continue to ask questions that challenge our most cherished assumptions in community ecology about how tropical forests work and about the implications of our findings for applied ecology, particularly tropical forestry. For this reason, it is important to think ahead about old questions in new ways as well as brand new questions that can guide and invigorate the CTFS research enterprise over the coming decades. In particular, we should begin to imagine a time when the censuses are relatively routine and providing ongoing baseline data on forest structure and dynamics but the most penetrating scientific questions depend not only on the census results but also on other data of very different types. I do not presume to know what these questions will be, and indeed it is probably impossible to predict the questions of the future with any certainty. However, trying to think ahead is well worth the effort. Each CTFS scientist has his or her own list of plot-specific research ideas and favorite questions, but in this particular endeavor, it is especially important that we work together to decide what questions really need to be addressed across the entire CTFS network and what additional standardized data and protocols we will require to answer these questions. It is also important that the CTFS network welcome new scientists to work in the plots and bring fresh perspectives and questions to bear on the databases and on new data that they add to the databases.

My recommendations for future research directions are strongly colored by my own experiences in the BCI forest and based on certain assumptions and biases that should be stated at the outset. The most important bias is that I am a scientist interested in the basic ecology of tropical forests for their own sake. In my lifetime I hope to understand, far better than I do now, the fundamental mechanisms underlying the natural origin, maintenance, and loss of tropical tree diversity on local to global scales. The challenge in reaching this understanding is both theoretical and empirical (Hubbell 2001). In my opinion, however, the greater challenge is not theoretical. Currently we do not lack for theories capable of explaining such extraordinary diversity (Hubbell 1998). Today at least a dozen sufficient theories exist. Rather, the challenge lies in figuring out which of these mechanisms and in what quantitative mix (since they are not mutually exclusive) is responsible for maintaining tree diversity in which particular tropical forests. It is largely this combined theoretical and empirical challenge that motivates my questions for future research in the CTFS. I strongly believe, however, that this basic science will also be of great value to applied tropical forestry.

Before discussing my thoughts for future research directions, it is perhaps appropriate to summarize briefly the history of major findings from the BCI project that led to my research recommendations.

The Main Starting Hypotheses

When we began the BCI project, there were many fewer hypotheses for the mainte-nance of tree species richness in tropical forests than exist today. At that time, these hypotheses fell into two main groups: (1) the enemies hypothesis championed by Janzen (1970) and Connell (1971) and (2) the regeneration niche/gap partition-ing hypothesis (e.g., Grubb 1977; Ricklefs 1977; Denslow 1980; Hartshorn 1980; Orians 1994). The enemies hypothesis proposes that tree diversity is maintained in a tropical forest by an interaction of seed dispersal and density-dependent seed and seedling mortality. Most seeds fall beneath maternal parents where the major-ity are killed by predators and pathogens. A smaller number of seeds escape this fate by dispersing away from the mother tree. Assuming the near-parent mortality is sufficiently strong to overcome the greater local rain of seeds, then surviving seedlings (recruitment) of the given tree species will tend to be displaced away from existing maternal adults. This in turn will decrease the probability that the given tree species will replace itself at the same location in the next generation rel-ative to other species. If all species experience these effects, then a greater diversity of trees can be maintained locally because individual tree species are prevented from becoming locally monodominant and displacing other species (Chave et al. 2002).

The regeneration niche/gap partitioning hypothesis states that treefalls create heterogeneous microenvironments of light, nutrients, and other resources to which individual tree species specialize in their regeneration requirements. There is a greater range of light intensities in light gaps in tropical versus temperate forests because of greater insolation at tropical latitudes. Ricklefs (1977) suggested that this greater range of microsite conditions would support more regeneration niche specialists than in temperate forests. Related to the gap hypothesis is the intermediate disturbance hypothesis (Connell 1978), which is built on older ideas of r-K life history strategies (MacArthur 1972) and the ideas of Hutchinson (1951) about "fugitive" or pioneer species that coexist with superior competitors by virtue of being better dispersers. This is now called the competition-colonization tradeoff hypothesis (Tilman 1994). Connell (1978) argued that tree diversity would be maximal at intermediate rates of forest disturbance. At very low rates of disturbance, there would be few light gaps for pioneers to colonize, and so competitive species with high shade tolerance would dominate. Conversely, at high rates of disturbance, diversity would again be low because only good dispersers would manage to colonize the disturbances. However, at intermediate disturbance rates, both shade tolerant species and gap specialists would be present, and diversity would be higher.

Foster and I set out to test these ideas in the BCI plot. In testing the Janzen–Connell hypothesis, we reasoned that whatever effect enemies had on seedlings, the spatial signature of that effect would have to pass through the small sapling size classes on its way to producing the spatial pattern of the adult tree population. We predicted that we could pick up this signature as a density- and distance-dependent pattern of sapling mortality, with sapling mortality rates increasing in sites closer to seed-bearing adults. To test the gap hypothesis, we decided to collect plot-wide data on the creation and subsequent closure of gaps, so that we could overlay gap disturbances on maps of growth, mortality, and recruitment of individual species and on the entire tree community. We therefore began in 1983 an annual canopy height census over the entire plot on a 5-m grid. Canopy heights were classified into six canopy height categories (Hubbell and Foster 1986a), with finer categories at low heights to enable better resolution of the early stages of gap regeneration.

The Main BCI Findings

Forest Dynamics

We have now completed five censuses (through the year 2000) and have an 18-year record of change in the BCI forest. The most remarkable finding of all is how dynamic the BCI forest is, a result for which we were completely unprepared. In just 18 years, 40% of all trees and saplings over 1 cm dbh in the plot have turned over. For trees over 10 cm dbh, 34% died during the same period. At the very start

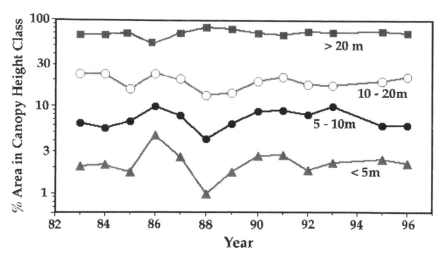

Fig. 2.1. Changes in canopy height in the BCI 50-ha plot from 1982 through 1996, by canopy height class. Low canopy height classes represent gap sites. There was a delayed increase in gap frequency that peaked in 1986, accompanying elevated canopy tree mortality several years after the severe El Niño drought of 1982–83. This pulse event had effects on the structure of the BCI forest that lasted for nearly a decade.

of the project, BCI experienced a very severe El Niño drought (1982–83). This event caused delayed increases in tree mortality, which was particularly evident in the 1986 census (Hubbell and Foster 1990; Condit et al. 1995). This mortality caused increases in gap frequencies that lasted for several years before falling back to "background" gap levels (fig. 2.1). Increasing gap frequencies stimulated a transient increase in the total abundance of the sapling size classes (table 2.1). Thus the effects of this El Niño event lasted much longer—nearly a decade—than the pulse event itself, and they propagated through the entire BCI tree community. A second smaller El Niño related drought event occurred in 1991–92, but it had no detectable effect on the forest's stand structure. This suggests that there is a threshold El Niño drought severity that must be crossed before it can have a large demographic (mortality) effect on the forest. However, lesser El Niño events are not without important effects on tree growth and reproduction, and are now believed to drive multiyear, forest-wide fluctuations in fruit production that have a major impact on food availability for mammals and birds (Wright and Calderon 1999).

In addition to these transient El Niño related changes in forest structure, large changes occurred in the species composition of the BCI forest. Three quarters (75.3%) of the tree and shrub species changed by >10% in total abundance, and a third (34.1%) of all species changed by >50% in total abundance in just under two decades. These large percentage changes in abundance were not limited to

Table 2.1. Changes in the Relative Abundances of All Trees

Size Class (cm)	1982	1985	1990	1995	2000
1–2	98,783	106,177	99,810	89,531	77,881
2–4	74,136	71,287	77,463	73,229	71,223
4–8	34,432	36,652	37,955	37,357	36,257
8–16	17,166	17,534	18,233	18,258	17,856
16–32	6,972	6,748	6,891	7,017	6,837
32+	3,852	3,690	3,707	3,657	3,748
Total	235,341	242,088	244,059	229,049	213,802

Note: An increase in tree number was seen in the 1–2 cm dbh class in 1985, which moved into the next size class (2–4 cm dbh) in 1990. At a slower rate, these plants moved into the 4–8 cm dbh class in 1995 and 2000.

Table 2.2. Contingency Analysis of the Relationship between Original 1982 Abundance and the Number of Changes in Population Size during the Four Intercensus Intervals through the 2000 Census

Abundance Class	0+ 4−		1+ 3−		2+ 2−		3+ 1−		4+ 0−		Row totals
	Obs	Exp	Obs	Exp	Obs	Exp	Obs	Exp	Obs	Exp	
0–9	31	23.5	22	15.7	15	22.1	5	6.2	0**	5.5	73
10–99	25	27.7	20	18.4	23	26.0	8	7.3	10	6.5	86
100–999	39	36.7	23	24.5	33	34.5	10	9.7	9	8.7	114
1000+	7*	14.2	3*	9.4	25**	13.3	4	3.7	5	3.3	44
Total	102		68		96		27		24		317

Note: For example, the column 0+ 4− means that these species exhibited four consecutive declines in abundance (monotonic decline), whereas 4+ 0− refers to species that exhibited four consecutive increases in abundance (monotonic increase). The remaining columns are for nonmonotonically changing species; the changes are counts of the number of negative or positive changes in abundance, without regard to their sequence or order.
$*p < 0.05$; $**p < −.025$; overall chi-square: 40.8, 12 DF, $p < 0.0001$.

rare species. More than a quarter (27.3%) of the 44 species with > 1000 individuals apiece changed by > 25% in total abundance over this relatively short time period. Of these changes in common species, roughly half were increases (52.3%) and half were decreases (47.7%).

There is mounting evidence not only that the old-growth BCI forest is changing in species composition but also that it is changing directionally (table 2.2). We now have four intercensus intervals over which to compute the directionality of change of individual species over 18 years. The relative monotonicity of change can be quantified by the number of increases or decreases among the intercensus intervals for particular tree species. By this measure, about a third (30.3%) of the species exhibited no population trend (two increases and two decreases). However, nearly 40% of the species exhibited either monotonic increases (24 species, 7.6%) or monotonic decreases (102 species, 32.1%). There was a highly significant interaction between the initial 1982 abundance of a species and the probability of increasing or decreasing ($p < 0.0064$). Of the 73 extremely

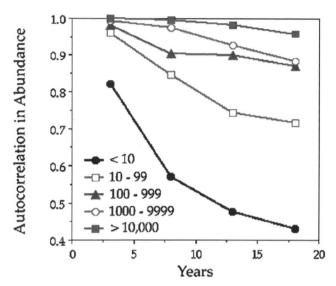

Fig. 2.2. Temporal autocorrelation of species abundances over the four census intervals, 1982–85, 1982–90, 1982–95, and 1982–2000. Mean autocorrelation coefficients were computed for five abundance classes. Species in progressively rarer abundance classes showed progressively faster rates of temporal decay in autocorrelation of abundance.

rare species with fewer than 10 individuals in 1982, 31 species (42.5%) exhibited monotonic declines, whereas among the 44 most abundant species with >1000 individuals in 1982, only 7 species (15.9%) declined monotonically. Conversely, no extremely rare species enjoyed monotonic increases, whereas 5 (11.4%) of the most abundant species did so. These trends can be summarized by examining the temporal autocorrelation of abundances for species grouped into orders-of-magnitude abundance classes (fig. 2.2). All abundance classes exhibit downward trends in temporal autocorrelation of abundances, as must be so, but the changes are progressively more rapid in the rarer abundance classes.

We do not know for certain what is causing these directional changes in the BCI forest, but by examining which species are increasing and which are decreasing, we can make some educated guesses. For example, many understory shrubs, particularly members of the genera *Piper* (Piperaceae) and *Psychotria* (Rubiaceae) and understory palms such as *Bactris* spp. (Palmae), are in decline. These are generally moisture-loving species. Conversely, a number of tree species on the increase are also found in drier forests to the south toward the Pacific Ocean. Over the last half century there has been a drying trend in central Panama (Windsor 1990), and it is possible that these changes are being driven by a long-term change in the climate of central Panama. Whether this is indicative of larger global changes in tropical precipitation patterns is not yet clear. However, several record wet years have occurred in the last decades; so if there is a drying trend, it has high year-to-year variance. Through the 1990s, the declines in these species have continued unabated, regardless of wet or dry years.

Is the BCI forest undergoing succession? The dynamism of the BCI forest does not fit the classical theory of secondary succession because the old-growth forest has never been cleared for agriculture (Piperno 1990). However, it was probably used for extractive purposes by nearby farms in pre-Canal days and by hunters and gatherers in pre-Columbian times. Because there was cleared land in farms immediately adjacent to the plot, there was likely an influx of ruderal and pioneer species into the old forest (e.g., Janzen 1986). The secondary forest northeast of the plot is now about 90 years old (Knight 1975) and is nearly indistinguishable in species composition from the old-growth forest (Lang and Knight 1983). We may now be witnessing the gradual die-off of these species because their immigration subsidy has been cut off (Hubbell and Foster 1986b). This may explain the loss of the rare species, but it cannot explain the rapid changes also occurring in the abundant tree species in the BCI plot. Is such continual change the rule or the exception in old-growth tropical forests? This is a question that the CTFS network should be able to answer relatively easily in the near future. Already it appears that at least some of the CTFS forests, such as the Pasoh Forest Reserve in Peninsular Malaysia, are much less dynamic than BCI (Condit et al. 1999). In any case, the idea that old-growth tropical forests are in equilibrium is severely challenged by the BCI results.

The Enemies Hypothesis

For a time, the data coming in from the early recensuses seemed to provide little support for the enemies hypothesis (Hubbell and Foster 1986c). We were able to detect density-dependent sapling mortality in only a few of the most common species (Hubbell et al. 1990; Condit et al. 1992), so the effects did not appear to be a community-wide phenomenon. Moreover, in many species the patterns of sapling mortality were not concentrated around adults. Indeed, new sapling recruitment was found to be statistically spatially attracted to the vicinity of adult trees in many species (Condit et al. 1992). Then the evidence began to change. First, we discovered that the spatial signature of Janzen–Connell effects was unlikely to persist through the 1 cm dbh and larger sapling stages, or only weakly in many cases. Our assumption was wrong because saplings turned out to be much older than we originally thought. The estimated median sapling age upon arrival at 1 cm dbh in shade tolerant species is on the order of two decades, if not older (Hubbell 1998). When the age of a 1 cm dbh sapling is considered in light of the adult tree mortality rate, which varies between 1.5 and 5% annually, depending upon the species, there is a problem: A large fraction ($>30\%$) of the original parent population will have died before their surviving 1 cm dbh saplings enter the main census. Simulations show that this considerably weakens the ability to detect spacing effects (Hubbell unpublished). Thus our analyses conducted with a focus on the nearest surviving adult were weak tests. Meanwhile, Schupp (1992) argued that it was more appropriate to analyze sapling mortality and recruitment data not in relation to distance to nearest adults but in relation to conspecific adult density in some local area. He reasoned that most tree species are patchily distributed, and seed predators and pathogens are more likely to respond to clumps of trees than to isolated single trees. This suggestion inspired us to conduct quadrat-based analyses. After the completion of the 1995 census, we performed such an analysis and found that per capita rates of sapling mortality and recruitment were indeed positive and negative functions of local adult density, respectively (Wills et al. 1997). Moreover, these effects were found across a large fraction of the BCI tree community, as required by the Janzen–Connell hypothesis (chap. 22). Wright (2002) pointed out statistical problems with the analysis and conclusions, so recently we completed new analyses that confirm the importance of local conspecific density on focal plant survival. In this volume we consider these effects in fixed tree neighborhood sizes of 20 neighbors (chap. 23). In other papers we consider the problem in continuous space, taking statistical account of the fact that mortality is spatially autocorrelated. These analyses once again support the Janzen–Connell hypothesis (Hubbell et al. 2001).

The most convincing support for the enemies hypothesis has come from a 13-year seed rain/seedling germination study. The origin of this study was in the late 1980s, when Joe Wright suggested following the fate of seedlings and small saplings <1 cm dbh in a complementary study to the main census in the BCI plot. Seed traps and associated seedling plots were established at 200 sites throughout the BCI Forest Dynamics Plot. When the data were analyzed recently, a remarkable result emerged (Harms et al. 2000). Tree species whose seeds were relatively more abundant in a given trap were much less relatively abundant as seedlings—or were actually less abundant than other species—in the seedling germination plots at the same location. Density dependence was evaluated for each species by the slope of the regression of log numbers of new seedlings on log numbers of seeds of the species in the adjacent trap. Regression slopes less than unity indicate negative density dependence. All species tested showed negative density dependence (fig. 2.3). Quite a few species exhibited such strong negative density dependence that they actually had negative regression slopes, indicating that they recruited absolutely smaller numbers of seedlings when they had more seeds in a given trap. The pattern of mortality was only explicable if the agents killing the seeds were host-specific. Otherwise, traps containing more total seeds of all species would have suffered higher total mortality across all species, but this was not the case (Harms et al. 2000). Moreover, the effect was pervasive through-out the tree community, a requirement for the Janzen–Connell mechanism to work.

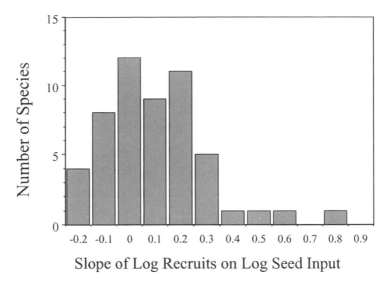

Slope of Log Recruits on Log Seed Input

Fig. 2.3. Histogram of the individual species' slopes of the regression of log seedlings germinated against log number of seeds of a given species collected in the adjacent traps, for 53 tree species. All species had regressions with slopes less than unity, indicating negative density dependence. However, note that 24 (45%) of the species had zero or negative slopes, so that the number of seedlings germinating was independent of the number of seeds falling into the traps, or was actually lower absolutely when there were more seeds in a given trap, indicating extreme density dependence (redrawn from Harms et al. 2000).

There are still many unanswered questions about the quantitative importance of all our findings on the Janzen–Connell effects for the maintenance of tree diversity in the BCI forest. First, the main criticism of the finding of similar diversities in the seedlings and in the main census is that seedling diversity was measured across all 600 seedling plots over the whole 50-ha plot. This spatial averaging will tend to converge on the average diversity in the plot simply by the law of large numbers. We still do not know quantitatively and on what spatial scales the density dependence operates in each life history stage. In saplings at least, the negative effect of conspecific density on survival is highly significant (chap. 23), but the effect is also extremely local, becoming indistinguishable from the "mean field" background mortality rate in as little as 5–10 m. Whether the spatial scales are similarly very small for the density-dependent mortality in the seed-to-seedling transition remains to be determined. Second, if the effects are local and operate only in the very earliest life history stages, it is not clear whether they can regulate the adult tree population across large landscapes. It is an open theoretical question as to what constraints such local Janzen–Connell effects impose on population growth rates and abundances of tropical trees.

The seed rain/seedling germination experiment showed that the local diversity of surviving seedlings was actually higher than the diversity of the seed rain. However, this does not answer the question of the importance of Janzen–Connell for maintaining adult tree diversity in the BCI forest. There are two questions. One is how the very local increase in seedling diversity translates into adult tree diversity on large spatial scales. A second question is how to estimate the tree species diversity that would result from the *absence* of Janzen–Connell effects. Seven years ago we started a large exclosure experiment on BCI and the adjacent mainland to remove ground-foraging mammals larger than spiny rats (Carson and Hubbell unpublished). These mammals are seed and seedling predators, but some of them are also scatter-hoarders that bury seeds, so their effects on seedling recruitment can be positive as well as negative. The results so far indicate that fencing increases seedling densities, at least in the near term. Seedling diversity also increased but less dramatically, in part because of huge seedling inputs from the few adult trees whose crowns happen to overlap or are inside the exclosures. Eventually mortality will have to take more of these seedlings because packing constraints will drive compensatory thinning mortality as the seedlings grow into saplings. Whether this mortality occurs randomly or exhibits species-specific density dependence is the next question to answer.

Finally, the role of Janzen–Connell effects in maintaining tropical tree diversity may be severely challenged by the rapid dynamics of the forest. Large changes in species composition in the BCI forest are not being prevented by Janzen–Connell effects, though we do not know to what extent Janzen–Connell effects are retarding these changes. Whatever the diversifying effects of conspecific density dependence are on very local spatial scales, they are not able to prevent species turnover on larger spatial (50 ha) and temporal (decadal) scales. In the final analysis, the Janzen–Connell mechanism does not explain tropical tree diversity because it does not uniquely determine or predict what local species diversity to expect. Its only role is to increase the fraction of the regional tree species pool that can coexist in a given local forest stand, but the mechanism itself imposes no cap on local diversity. Obviously the Janzen–Connell mechanism cannot maintain more species locally than are in the regional species pool. If the regional pool is species-poor, the local community will also be species-poor. For example, recent experimental evidence suggests that such Janzen–Connell effects are also present in relatively species-poor temperate forests (Packer and Clay 2000; Hill-Ris-Lambers et al. 2002).

Intermediate Disturbance and the Gap Hypothesis

Because gap species cannot persist in a closed-canopy forest without light gaps, this absolute gap requirement means that some disturbance is necessary for the pioneer guild of the tree community to be present in a tropical forest. This self-evident

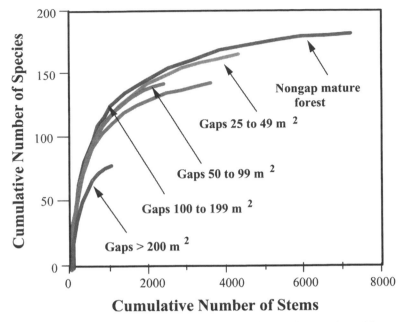

Fig. 2.4. Species accumulation curves for samples of mature forest in the BCI plot and for gaps of different sizes, showing that larger gaps have slower species accumulation curves than small gaps collectively, and that accumulation is lower in all gaps than in mature forest. This result is contrary to predictions of the intermediate disturbance hypothesis (redrawn from Hubbell 1999).

fact delayed us from testing the intermediate disturbance hypothesis in the BCI forest until quite recently (Hubbell et al. 1999). It was therefore with considerable surprise that we discovered absolutely no relationship between the rate of gap disturbance in different parts of the BCI plot over the past 18 years and local tree diversity. To our great surprise, gaps were actually poorer in species than the mature forest (Hubbell et al. 1999). Moreover, larger gaps had slower species accumulation curves (with increasing samples of trees) than did smaller gaps (fig. 2.4). How could we explain these results when gap species cannot be present except in gaps?

Our explanation is threefold (Hubbell 1999). First, most gaps suitable for individual pioneer species are not successfully colonized by the species—or at least saplings of the given pioneer species are not present in them. We attribute this to severe dispersal and recruitment limitation. There is mounting evidence for such limitation from germinating seeds of pioneer species out of the soil seed bank (Dalling et al. 1998). Even though pioneers are excellent dispersers, they fail to colonize most of the gaps suitable for their germination and growth. This dispersal

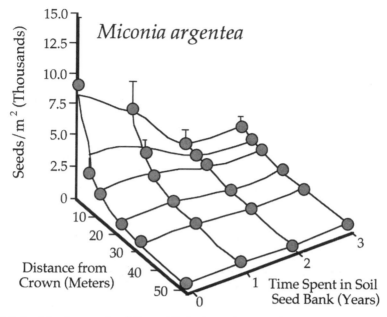

Fig. 2.5. Spatial and temporal seed dispersal limitation in the gap pioneer, *Miconia argentea* (Melastomataceae). Data were collected by germinating *Miconia* seeds in the soil seed bank at different distances from an isolated maternal adult and different times since deposition (seeds were placed in nylon bags in the soil and later recovered and germinated). Few seeds disperse farther than 20 m from the maternal parent and viably persist longer than 2 years. (Data courtesy of James Dalling.)

limitation is both spatial and temporal. There is a rapid decay in seed densities in the soil away from focal adults. Also, at least for some pioneer species, viable seeds are less persistent in the soil than had been previously thought. For example, seeds of the gap pioneer *Miconia argentea* (Melastomataceae) last only about 2 years in the soil and disperse generally to less than 20 m (fig. 2.5). Based on experiments on seeds placed in bags in the soil, the median seed survival time of all pioneers is only 2 years. However, some seeds, particularly larger seeded pioneers, do occasionally persist in a viable state for decades, as revealed by recent C^{14} measurements (J. Dalling, personal communication). Therefore, only a fraction of pioneer species can persist for decades in the soil—many fewer than previously thought—waiting for a gap to open overhead.

The second factor explaining lower diversity in gaps is the reduced species richness of shade tolerant species. Many of the saplings of shade tolerant species that were present in the understory before the gap was created managed to survive the treefall. However, a large proportion subsequently died relatively rapidly in the high-light environment of the gap. Gap abundance and distribution, the third factor, explains why larger gaps have slower species accumulation curves. There are many more small gaps than large, and the small gaps are distributed much more widely over the entire 50-ha plot. Therefore, small gaps are collectively less affected by dispersal limitation, whereas large gaps receive seeds dispersed from a much smaller fraction of the forest. The combined result of these three factors explains the lack of correlation of diversity with gap disturbance regimes in the BCI forest on local spatial scales <50 ha, in spite of the fact that pioneers require gaps.

We also used a quadrat-based analysis to demonstrate the failure of the intermediate disturbance hypothesis to explain diversity at small spatial scales in the BCI forest. Following a suggestion of Sean Thomas (personal communication), we can compute the change in basal area in a given 20×20 m quadrat over each intercensus interval in relation to changes in diversity. Quadrats with sudden large drops in basal area are sites where large trees died and sites of new light gaps. If gaps are enriching very local diversity in the BCI forest, then these sites should have steeper species accumulation curves. These curves are well characterized in the BCI forest by Fisher's α (Condit et al. 1996; chap. 7).

Fisher's α can be estimated from the number of species S and individuals N in each quadrat by solving the equation, $S = \alpha \ln(1 + N/\alpha)$ (Fisher et al. 1943). Fisher's α is a more stable measure of quadrat diversity than species per tree, which was used in our previous paper (Hubbell et al. 1999); unlike species per tree, α depends relatively little on sample size N (see also chap. 7). Figure 2.6 shows changes in basal area over the four census intervals (1982–2000). If sites having large drops in basal area show increases in diversity, these sites should show increases in Fisher's α. They do not (fig. 2.7). The same is true for all other intercensus intervals. We tested for a statistical relationship between changes in basal area and changes in Fisher's α for all census intervals and all temporal lags after the census in which the change in basal area occurred. The average r^2 of these correlations was 0.000017 ± 0.000028, completely insignificant. Presumably the reasons for this lack of correlation are the same as the ones discussed above.

Recently, Sheil and Burslem (2002) criticized the analysis done by Hubbell et al. (1999). They argued that perhaps BCI was anomalous in several ways (an artificial island, impacted by humans, lack of top predators, etc.). There is little evidence in support of any of these contentions. However, one criticism has some validity. They noted that we standardized all our gaps to be 2 years of age (by considering only gaps created 2 years before each 5-year census). By taking gaps that were only 2 years old, we may have measured gaps that were still largely dominated by their mortality phase rather than by their recruitment phase. Studying slightly older gaps may help clarify this question. However, whatever the outcome of such a reanalysis, it is clear that areas of greater long-term disturbance do not show enhanced tree diversity or any pattern consistent with the intermediate disturbance hypothesis (fig. 2.7).

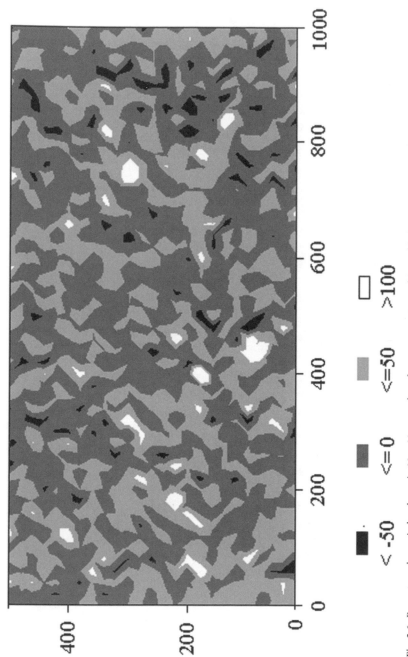

Fig. 2.6. Percentage change in basal area in 20 × 20 m quadrats from 1982 to 2000. White and light gray areas show increases in basal area, while dark gray and black areas show decreases. There is no correlation of changes in basal area and changes in Fisher's α over the entire 18-year study (see fig. 2.7), as might have been expected if treefall gaps were a major diversifying force. There is also no correlation between any of the changes in basal area by census interval and any concurrent or lagged change in Fisher's α (not shown).

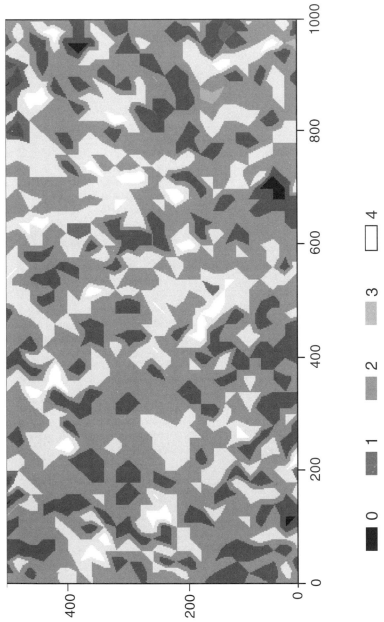

Fig. 2.7. Changes in Fisher's α shown as the number of positive changes over the four census intervals. Changes that monotonically lost diversity over the 18-year study are shown in black; areas that consistently increased in diversity are shown in white. Note the spatially fine-grained nature of diversity changes in the BCI 50-ha plot. Diversity decreases were weakly correlated with slopes, particularly in the east and south, where diversity had been greatest, and may be associated with declines in moisture-loving species.

We do not know how general these results are in relation to other forests in the CTFS network. Hubbell et al. (1999) predicted that the degree of generality will depend on how strongly the pioneer tree guild is dispersal limited in other forests. Preliminary tests for changes in basal area and Fisher's α over one inter-census interval suggest that Pasoh may provide some support for the intermediate disturbance hypothesis; but the correlations, although significant, are very weak (Sean Thomas, personal communication).

Lessons from the BCI Experience

We have learned several important lessons from the BCI experience. First, we were correct in assuming that a large permanent plot of individually tagged, mapped, identified, and monitored plants is essential to understanding the dynamics of the BCI forest. Without the large dataset we would have been unable to evaluate the life history differences of the trees at the species level. The long-term repeated recensuses have also been essential to understanding the maintenance of diversity in the BCI forest. The second lesson is that during the conduct of a long-term study such as ours, the theoretical ground is likely to shift. A whole spate of new hypotheses for the maintenance of tropical tree diversity appeared in the 1980s and 1990s that had not been considered when we designed our plot. Included among these new hypotheses, for example, was the idea that dispersal limitation per se could have a dramatic effect on diversity (Tilman 1994; Hurtt and Pacala 1995; Hubbell et al. 1999). Nevertheless, because we had given considerable forethought to the generic data that would be needed for testing arbitrary hypotheses that required spatially explicit data, we were in a reasonably good position to test many of these newer ideas. The third lesson, however, is less self-congratulatory. We were wrong in certain key assumptions about what information we would need to test our original hypotheses. In particular, we needed—and still need—much more information about seed dispersal, seed rain, seed-to-seedling germination, and mortality patterns. We should have developed a seed and seedling sampling regime to accompany the main census from the very outset.

Recommendations for Future Research Directions in CTFS

Before giving my specific recommendations for research directions, I would like to make a general comment about the strengths and limitations of the census data for testing hypotheses about the maintenance of tree diversity in tropical forests. At one level, the BCI census information has been sufficient for testing many hypotheses, but most of these tests are still what I would call phenomenological. They are tests for the existence of a process, and if it exists, they assess the strength of the process for maintaining local tree diversity, using data on changes in diversity as the phenomenological response variable. However, at a deeper level, the census results are only descriptive of patterns, both static and dynamic, and the correlations of these patterns with presumed causal phenomena. This is not to devalue such work. Long-term data of this sort are extremely important for generating hypotheses. Moreover, many hypotheses can be eliminated because they are inconsistent with observed patterns of forest dynamics.

Nevertheless, we should now seek the deeper mechanistic explanations for the phenomenology that we have so thoroughly documented in the Forest Dynamics Plots. For example, if the phenomenology says that there must be host-specific seed and seedling mortality agents operating to diversify the local seedling community in a particular forest, then we need to identify these mortality agents and then conduct experiments to determine, for example, how host-specific they are. If there is strong phenomenological evidence of dispersal and recruitment limitation in a particular forest, then we need (1) better data on the seed and seedling shadows (the "dispersal kernels") and (2) experiments to test how dispersal mechanistically interacts with mortality caused by seed predators and pathogens. How does the probability of a seed being killed by a predator or pathogen depend on the local variation in the density of adult or juvenile conspecifics?

Many of these and other mechanistic questions will require supplementary datasets on additional factors that are currently being collected at few, if any, CTFS sites. My highest priorities for desirable basic science datasets and experiments to accompany every Forest Dynamics Plot include those in the following list. Some of these priorities follow directly from what I have said above, whereas others are long-term monitoring data that are needed for testing all hypotheses.

1. Long-term weather records for each site, minimally including temperature, barometric pressure, wind, rainfall, and insolation
2. Detailed soil maps over the Forest Dynamics Plot at each CTFS site
3. Community-wide phenological studies over multiple years to capture inter-annual variation in patterns of leaf flushing, flowering, and seed set (ideally also done in different habitat types)
4. Long-term seed trap/seedling germination experiments at each site
5. Regular (e.g., annual or every 2 years) mapping of canopy height and gaps, particularly in the closed-canopy sites
6. Development of genetic markers for quantifying actual seed dispersal distances
7. Isolation and identification of plant and fungal pathogens, particularly those attacking seed and seedling stages, and their level of host specificity
8. Studies of the mycorrhizal associations of the species, their host specificity, and their interactions with plant and other fungal pathogens

9. Exclosure experiments to test the effects of removing ground-foraging mammals and other site-specific disturbance agents (e.g., fire)
10. Ecophysiological and life historical studies of plant adaptation to habitat variation in the plots, and reaction-norm experiments to test the habitat specificity of tree species
11. Studies of the long-term dynamics of animal populations in these forests, particularly those important to the reproductive biology and recruitment success of tree populations
12. Systematic and molecular studies of the phylogenetic relationships and phylogeography of tree taxa in the plots and their implications for speciation rates in different taxa and regions
13. Addition of networks of complementary smaller plots to understand the generality of the results obtained from the Forest Dynamics Plots in a broader geographic context

Clearly this is not an exhaustive list, but it encompasses an ambitious agenda of new research that will require a much larger level of staffing, infrastructure, and funding than the institutions comprising CTFS currently possess. Nevertheless I believe that these studies should be included if we are really to understand how natural tropical forests work and how tree species diversity is maintained in them. It is also extremely important to remain flexible; no such list should be cast in stone. A new list should be constructed from time to time as new questions and hypotheses come along. Also new technologies are likely to revolutionize approaches to many of these new and old questions and make many formerly intractable problems amenable to attack. For example, personal computers were primitive when the BCI project began. We now have more computing power in our laptops than we had in 1980-model supercomputers, and they are equipped with incredibly powerful software such as "R" for data analysis. New genetic technologies in genomics research will likely revolutionize the study of ecology in the near future by enabling us to ask fundamental but unsolved questions about such things as the genetic basis of ecological generalization and specialization, the frequency of long-distance dispersal events, and processes of speciation and detailed patterns of phylogeny. This technology will rapidly become inexpensive and will soon be available to scientists working at all Forest Dynamics Plots. It is possible to imagine the impact of these technologies because they are already here or in development. What technologies lie ahead that we cannot even imagine?

Another unknown factor is global climate change. Quite apart from the direct effects of humans on tropical forests, indirect effects will be imposed on them by changing climate. Temperature regimes at tropical latitudes are not expected to change significantly over the next 100 years, but precipitation regimes are. If the drying trend in central Panama is indicative of larger global changes in tropical precipitation, in part driven by changing frequencies of El Niño events, then we can expect potentially large changes in the composition of tropical forests, not just in Panama but also throughout the CTFS network. How will we recognize such exogenous forces and separate them from endogenous forces of change in tropical forests? There are no clear answers to this important but challenging question.

Whatever research directions CTFS takes in the future, it must remain true to its core commitment to fostering high-quality basic science on tropical forest ecology and evolutionary biology. I have deliberately emphasized the basic science research themes and needs as I see them in this paper because basic science is always under threat in a world that increasingly demands practical solutions to tropical forest conservation and management needs. As tropical forests dwindle, however, we are going to need far better scientific information about natural forests, not less. By all means CTFS should help supply the basic science needs of applied tropical forestry, but actually managing reforestation or conservation projects should never become the primary mission of CTFS.

References

Chave, J., H. C. Muller-Landau, and S. A. Levin. 2002. Comparing classical community models: Theoretical consequences for patterns of diversity. *American Naturalist* 159:1–23.

Condit, R. 1998. *Tropical Forest Census Plots*. Springer-Verlag, Berlin.

Condit, R., S. P. Hubbell, and R. B. Foster. 1992. Recruitment near conspecific adults and the maintenance of tree and shrub diversity in a neotropical forest. *American Naturalist* 140:261–86.

———. 1995. Mortality rates of 205 neotropical trees and shrub species and the impact of a severe drought. *Ecological Monographs* 65:419–39.

Condit, R., S. P. Hubbell, J. V. LaFrankie, R. Sukumar, N. Manokaran, R. B. Foster, and P. Ashton. 1996. Species-area and species-individual relationships for tropical trees: A comparison of three 50-ha plots. *Journal of Ecology* 84:549–62.

Condit, R., P. S. Ashton, N. Manokaran, J. V. LaFrankie, S. P. Hubbell, and R. B. Foster. 1999. Dynamics of the forest communities at Pasoh and Barro Colorado: Comparing two 50-ha plots. *Philosophical Transactions of the Royal Society of London* 354:1739–48.

Connell, J. H. 1971. On the role of natural enemies in preventing competitive exclusion in some marine animals and in rain forest trees. Pages 298–312 in P. J. den Boer and G. R. Gradwell, editors. *Dynamics of Populations*. Center for Agricultural Publication and Documentation, Wageningen, Netherlands.

———. 1978. Diversity in tropical rain forests and coral reefs. *Science* 199:1302–10.

Cressie, N. A. C. 1995. *Statistics for Spatial Data*. Wiley Interscience, New York.

Dalling, J. W., S. P. Hubbell, and K. Silvera. 1998. Seed dispersal, seedling establishment and gap partitioning among tropical pioneer trees. *Journal of Ecology* 86:674–89.

Denslow, J. 1980. Gap partitioning among tropical rain forest trees. *Biotropica (Suppl.)* 12:47–55.

Fisher, R. A., A. S. Corbet, and C. B. Williams. 1943. The relation between the number of species and the number of individuals in a random sample of an animal population. *Journal of Animal Ecology* 12:42–58.

Grubb, P. 1977. The maintenance of species richness in plant communities: The importance of the regeneration niche. *Biological Reviews of the Cambridge Philosophical Society* 52:107–45.

Harms, K. E., S. J. Wright, O. Calderón, A. Hernández, and E. A. Herre. 2000. Pervasive density-dependent recruitment enhances seedling diversity in a tropical forest. *Nature* 404:493–95.

Hartshorn, G. S. 1980. Neotropical forest dynamics. *Biotropica (Suppl.)* 12:23–30.

Hill-Ris-Lambers, J., J. S. Clark, and B. Beckage. 2002. Density-dependent mortality and the latitudinal gradient in species diversity. *Nature* 417:732–35.

Hubbell, S. P. 1997. A unified theory of biogeography and relative species abundance and its application to tropical rain forests and coral reefs. *Coral Reefs* 16(Suppl.):S9–21.

———. 1998. The maintenance of diversity in a neotropical tree community: Conceptual issues, current evidence, and challenges ahead. Pages 17–44 in F. Dallmeier and J. A. Comiskey, editors. *Forest Biodiversity Research, Monitoring and Modeling: Conceptual Background and Old World Case Studies.* Man and the Biosphere Series, Volume 20. Parthenon Publishing, Pearl River, NY.

———. 1999. Tropical tree species richness and resource-based niches. *Science* 285:1459–61.

———. 2001. *The Unified Neutral Theory of Biodiversity and Biogeography.* Princeton University Press, Princeton, NJ.

Hubbell, S. P., J. A. Ahumada, R. Condit, and R. B. Foster. 2001. Local neighborhood effects on long-term survival of individual trees in a neotropical forest. *Ecological Research* 16:S45–61.

Hubbell, S. P., R. Condit, and R. B. Foster. 1990. Presence and absence of density dependence in a neotropical tree community. *Philosophical Transactions of the Royal Society of London B* 330:269–81.

Hubbell, S. P., and R. B. Foster. 1983. Diversity of canopy trees in a neotropical forest and implications for the conservation of tropical trees. Pages 25–41 in S. J. Sutton, T. C. Whitmore, A. C. Chadwick, editors. *Tropical Rainforest Ecology and Management.* Blackwell, Oxford.

———. 1986a. Canopy gaps and the dynamics of a neotropical forest. Pages 77–96 in M. J. Crawley, editor. *Plant Ecology.* Blackwell Scientific, Oxford, U.K.

———. 1986b. Biology, chance and history and the structure of tropical rain forest tree communities. Pages 314–29 in J. Diamond and T. J. Case, editors. *Community Ecology.* Harper and Row, New York.

———. 1986c. Commonness and rarity in a neotropical forest and implications for conservation. Pages 205–31 in M. E. Soule, editor. *Conservation Biology.* Sinauer Associates, Sunderland, MA.

———. 1990. Structure, dynamics, and equilibrium status of old-growth forest on Barro Colorado Island. Pages 522–41 in A. H. Gentry, editor. *Four Neotropical Rainforests.* Yale University Press, New Haven, CT.

Hubbell, S. P., R. B. Foster, S. T. O'Brien, K. E. Harms, R. Condit, B. Weschler, S. J. Wright, and S. Loo de Lao. 1999. Light gap disturbances, recruitment limitation, and tree diversity in a neotropical forest. *Science* 283:554–57.

Hurtt, G. C., and S. W. Pacala. 1995. The consequences of recruitment limitation: Reconciling chance, history, and competitive differences between plants. *Journal of Theoretical Biology* 176:1–12.

Hutchinson, G. E. 1951. Copepodology for the ornithologist. *Ecology* 32:571–77.

Janzen, D. H. 1970. Herbivores and the number of tree species in tropical forests. *American Naturalist* 104:501–28.

———. 1986. Does a hectare of cropland equal a hectare of wild host plant? *American Naturalist* 128:147–49.

Knight, D. H. 1975. A phytosociological analysis of species-rich tropical forest on Barro Colorado Island, Panama. *Ecological Monographs* 45:259–84.

Lang, G. E., and D. H. Knight. 1983. Tree growth, mortality, recruitment, and canopy gap formation during a 10-year period in a tropical moist forest. *Ecology* 64:1075–80.

MacArthur, R. H. 1972. *Geographical Ecology*. Harper and Row, New York.

Orians, G. H. 1994. Prospects for a comparative tropical ecology. Pages 329–40 in L. A. McDade, K. S. Bawa, H. A. Hespenheide, and G. S. Hartshorn, editors. *La Selva: Ecology and Natural History of a Neotropical Rain Forest.* University of Chicago Press, Chicago.

Pacala, S. W., and J. A. Silander, Jr. 1998. Field tests of neighborhood population dynamic models of two annual weed species. *Ecological Monographs* 60:113–34.

Packer, A., and K. Clay. 2000. Soil pathogens and spatial patterns of seedling mortality in a temperate tree. *Nature* 404:278–81.

Piperno, D. R. 1990. Fitolitos, arqueologia y cambios prehistoricos de la vegetacion en un lote de cincuenta hectareas de la Isla de Barro Colorado. Pages 153–56 in E. G. Leigh, Jr., A. S. Rand, and D. M. Windsor, editors. *Ecologia de un Bosque Tropical.* Smithsonian Tropical Research Institute, Balboa, Panama.

Ricklefs, R. E. 1977. Environmental heterogeneity and plant species diversity: A hypothesis. *American Naturalist* 978:376–81.

Schupp, E. W. 1992. The Janzen–Connell model for tropical tree diversity: Population implications and the importance of spatial scale. *American Naturalist* 140:526–30.

Sheil, D., and D. F. Burslem. 2002. Disturbing hypotheses in tropical forests. *Trends in Ecology and Evolution* 18:18–26.

Tilman, D. 1994. Competition and biodiversity in spatially structured habitats. *Ecology* 75:2–16.

Tilman. D., and P. Kareiva. 1999. *Spatial Ecology: The Role of Space in Population Dynamics and Interspecific Interactions.* Princeton University Press, Princeton, NJ.

Wills, C., R. Condit, R. B. Foster, and S. P. Hubbell. 1997. Strong density- and diversity-related effects help to maintain tree species diversity in a neotropical forest. *Proceedings of the National Academy of Science* 94:1252–57.

Windsor, D. M. 1990. Climate and moisture availability in a tropical forest: Long-term records from Barro Colorado Island, Panama. *Smithsonian Contributions to the Earth Sciences* 29:1–145.

Wright, S. J., and O. Calderon. 1999. Plant diversity in tropical forests. Pages 449–72 in F. I. Pugnaire and F. Valladares, editors. *Handbook of Functional Plant Ecology.* Dekker, New York, NY.

Wright, S. J. 2002. Plant diversity in tropical forests: A review of mechanisms of species coexistence. *Oecologia* 130:1–14.

PART 2: The Whole Is Greater Than the Sum of the Plots

Introduction

Elizabeth C. Losos

Because major biogeographic realms have long been isolated from each other, their forests share almost no species in common. Many forests of south and southeast Asia are dominated by a striking and distinctive subfamily of dipterocarps, which are entirely absent from Africa and virtually so from the neotropics. Similarly, bromeliads, a characteristic feature of the wetter New World forests, are rarely found in Africa and never in Asia. Despite such differences, studies of one tropical forest can tell us something about the physiognomy, trophic organization, and the mechanisms maintaining tree diversity of another.

When P. W. Richards (1952) first attempted to categorize rainforests at a global scale, he demonstrated consistent trends in stand structure, morphology, species richness, and dominance on different soils among the three continents with equatorial forests. But his samples were at most 1 ha in size and had been censused but once. Since then, the diversity and structure of tropical rainforests have been studied over an increasing range of spatial and temporal scales. Ecologists have extended Richards' work by comparing forests—typically utilizing 0.1- or 1.0-ha plots or transects—to identify local, regional, and global patterns of floristics, diversity, and forest structure (e.g., Gentry 1982, 1988a, b; Rollet 1974; Swaine et al. 1987a; Basnet 1992; Primack and Hall 1992; Phillips et al. 1994; Valencia et al. 1998; Potts et al. 2002; Phillips and Miller 2002). With this evidence, we can begin to ask how similarities in forest structure, floristics, and diversity across biogeographic realms are related to similarities in climate and soils. Yet despite these advances, our capabilities to generalize broadly are still limited by the scope and scale of these comparative studies.

The network of Forest Dynamics Plots within the Center for Tropical Forest Science (CTFS) builds upon this foundation while adding further dimensions: sample plots big enough to characterize the demography of individual species, monitored long enough to understand their dynamics, and inclusive of smaller individuals in order to understand forest regeneration. As the number of Forest Dynamics Plots grow—and with it, geographical coverage—regional and global trends are beginning to emerge with greater clarity and precision. Clear discernment of these patterns, however, depends on the location of the plots. One goal of the CTFS network has been to strategically site core Forest Dynamics Plots so that

Table 2I.1. Abiotic Conditions within All Forest Dynamics Plots

Forest Dynamics Plot	No. Dry Months < 100 mm Rain	Soil and Topography Heterogeneity	Natural Disturbance Regime	Altitude Above Sea Level (m)
LATIN AMERICA				
Barro Colorado Island Nature Monument, Panamá*	3	Low	Gap-dominated	120–160
Luquillo Experimental Forest, Puerto Rico*	0	Moderate	Hurricane-dominated	333–428
Yasuní National Park, Ecuador*	0	Moderate	Gap-dominated	215–245
La Planada Nature Reserve, Colombia	0	Moderate	Gap-dominated	1718–1844
ASIA				
Pasoh Forest Reserve, Peninsular Malaysia*	1	Low	Gap-dominated	70–90
Mudumalai Wildlife Sanctuary, India	4	Low	Fire/Elephant-dominated	980–1120
Huai Kha Khaeng Wildlife Sanctuary, Thailand*	6	Moderate	Gap/Fire-dominated	549–638
Sinharaja World Heritage Site, Sri Lanka*	0	Moderate	Gap-dominated	424–575
Lambir Hills National Park, Sarawak, Malaysia*	0	High	Landslide/Gap-dominated	104–244
Bukit Timah Nature Reserve, Singapore	0	Moderate	Gap-dominated	150
Palanan Wilderness Area, Philippines*	0	Moderate	Typhoon-dominated	100–180
Doi Inthanon National Park, Thailand	6	Moderate	Gap-dominated	1660–1740
Nanjenshan Nature Reserve, Taiwan	0	Moderate	Typhoon-dominated	300–340
Khao Chong Wildlife Refuge, Thailand*	2–3	Moderate	Gap-dominated	50–300
AFRICA				
Ituri Forest/Okapi Wildlife Reserve, Democratic Republic of Congo*	3–4	Low	Gap-dominated	700–850
Korup National Park, Cameroon*	3	Moderate	Gap-dominated	300–390

*Denotes core Forest Dynamics Plot in CTFS network.

they encompass the extremes and the center of several gradients—rainfall seasonality, soils and topography, natural disturbance regimes, and biogeography—that are predicted, from empirical research and theory, to be the major influences on tree diversity, species composition and relative abundance, and forest structure and dynamics (table 2I.1). The core plots in the CTFS network are therefore confined to the range of abiotic conditions in which the most species-rich tropical forests occur: climates with 0 to 6 dry months, altitudes from 0 to 1000 m, and yellow-red zonal soils derived from acid and basic substrates but excluding ultramafics, limestone, podsols, peats, and saline soils. At present, six core plots in Asia represent a balanced sample from these gradients. It is intended that plots established in the neotropics and Africa will eventually replicate the combination of abiotic conditions among the plots in Asia.

In part 2, we use data from the network of Forest Dynamics Plots to compare aspects of the structure, diversity, floristic, composition, and dynamics among the plots. We ask what patterns emerge from the global network of plots. How do similar habitats across different biogeographic realms differ? To what degree can phenomena be generalized, defying local mechanistic explanations?

Before tackling these questions, it is necessary to construct the global context in which the network of Forest Dynamics Plots resides. In the first chapter of this section, Leigh sets the stage by comparing the biogeographic history of tropical forests in different regions. He then considers, in the following chapter, tropical climates and how they influence different forests and their soils. Ashton provides a comprehensive comparison of the soils of the regions where the Forest Dynamics Plots occur.

With the setting defined, Losos examines how forest structure—using the surrogates of tree density and basal area—varies with climate, natural disturbances, and topography across the network of Forest Dynamics Plots. No single factor explains all variations, although under extreme conditions, such as in cyclone-prone forests, natural disturbance clearly dominates.

Next, Condit, Leigh, and Lao explore how tree species are assembled in a forest. Can the law relating the number of species encountered in a Forest Dynamics Plot be used to predict species diversity over larger areas? They explore how Fisher's α—an index for comparing diversity across samples—can be used to describe the form of a plot's species-accumulation curve. Using nine Forest Dynamics Plots with differing levels of species diversity, plot sizes, species abundances, and patterns of aggregation, they find that α estimates diversity relatively well for subplots that include over 500 trees within most of the CTFS plots outside of Africa, especially at sites with low habitat variation. At more heterogeneous sites, deviation is more substantial. This failure of α to predict species accumulation is most marked where there is a surplus of very abundant and very rare species, or where trees belonging to common species tend to have strongly clumped distributions.

Finally, Ashton relates broad-scale patterns of species diversity with floristic differences among the Forest Dynamics Plots. Despite enormous differences in species numbers among the plots, the degree to which they share the same common families, both in terms of dominance of tree number and basal areas, is striking. Most notably, two ectomycorrhyzal subfamilies—Dipterocarpaceae subfamily Dipterocarpoideae and Leguminosae subfamily Caesalpinoideae—comprise around a quarter to a third of the basal area of lowland Forest Dynamics Plots in

Asia and Africa, though substantially less in the neotropics. The relationship of trees, fungi, climate, soils, and disturbance will clearly reveal more about species diversity as these factors are further studied within the CTFS network.

Notes

Each CTFS site utilizes its own taxonomic convention, complicating comparisons across sites. In an effort to provide taxonomic consistency among the Forest Dynamics Plots located throughout Asia, Africa, and Latin America, all plant genus and family names in this volume are consistent with the dictionary of vascular plants by D. J. Mabberley (1997), developed primarily from Cronquist's *An Integrated System of Classification of Flowering Plants* (1981). While no one definitive classification system exists for flowering plants, the editors selected Mabberley's dictionary because it is a comprehensive, respected, commonly used, and easily obtained reference. In addition, the *Index Nominum Genericorum* (Farr and Zijlstra 2001)—a collaborative project of the International Association for Plant Taxonomy (IAPT) (http://www.botanik.univie.ac.at/iapt/) and the Smithsonian Institution—and the Missouri Botanical Garden's VAST (Vascular Tropicos) (http://mobot.mobot.org/W3T/Search/vast.html) database were used when plants were not listed in the second edition of the Mabberley text. Most individual Forest Dynamics Plots continue to use different classification and nomenclatural systems for research and publications outside of this book. Even with this standardization of nomenclature, it should be noted that species identifications are not always consistent across sites because they are identified by different botanists. CTFS is also currently standardizing taxonomic nomenclature using the framework of the Angiosperm Phylogeny Group 2003 (www.mobot.org/MOBOT/research/APweb/), though this framework has not been used in this volume.

Unless otherwise stated, Forest Dynamics Plot census data used throughout this section are from the fourth census of Barro Colorado Island, the fourth census of Bukit Timah, the second of Luquillo, the first of Yasuní (25 ha only), the first of Ituri (four 10-ha plots), the first of Korup, the first of the Pasoh, the first of Lambir, the first of Sinharaja, the second of Huai Kha Khaeng, the fourth of Mudumalai (1996), the first of Doi Inthanon, the first of Palanan, and the first of Nanjenshan. Enumeration of the plot in Khao Chong, Thailand, has only recently been completed and is not yet available for analyses in this volume.

Throughout this book, the term "tree" will refer to all woody, free-standing individuals greater than 1 cm dbh, thus including trees, treelets, and large shrubs.

References

Ashton, P. S., and P. Hall. 1992. Comparisons of structure among mixed dipterocarp forests of north-western Borneo. *Journal of Ecology* 80:459–81.

Aymard, G., N. Cuello, and R. Schargel. 1998. Floristic composition, structure, and diversity in moist forest communities along the Casiquiare channel, Amazonas state, Venezuela. Pages 495–506 in F. Dallmeier and J. A. Comiskey, editors. *Forest Biodiversity in North, Central and South America, and the Caribbean.* Parthenon Publishing Group, New York.

Balslev, H., R. Valencia, G. Paz y Miño, H. Christensen, and I. Nielsen. 1998. Species count of vascular plants in one hectare of humid lowland forest in Amazonian Ecuador. Pages 585–94 in F. Dallmeier and J. A. Comiskey, editors. *Forest Biodiversity in North, Central and South America, and the Caribbean.* Parthenon Publishing Group, New York.

Basnet, K. 1992. Effect of Topography on the Pattern of Trees in Tabonuco (*Dacryodes excelsa*) dominated rain forest of Puerto Rico. *Biotropica* 24(1):31–42.

Cerón, C. E., and C. Montalvo. 1997. Composición y estructura de una hectárea de bosque en la Amazonía Ecuatoriana—con información etnobotánica de los Huaorani. Pages 153–72 in R. Valencia, and H. Balslev, editors. *Estudios Sobre Diversidad y Ecología de Plantas.* PUCE, Proyecto ENRECA, Proyecto DIVA, Quito, Ecuador.

Clark, D. B., & D. A. Clark. 2000. Landscape-scale variation in forest structure and biomass in a tropical forest. *Forest Ecology and Management* 137:185–98.

Clinebell II, R. R., O. L. Phillips, A. H. Gentry, N. Stark, and H. Zuuring. 1995. Prediction of neotropical tree and liana species richness from soil and climatic data. *Biodiversity and Conservation* 4:56–90.

Cronquist, A. 1982. *An Integrated System of Classification of Flowering Plants.* Columbia University Press, New York.

Crow, T. R. 1980. A Rainforest Chronicle: A 30 year record of change in structure and composition at El Verde, Puerto Rico. *Biotropica* 12(1):42–55.

Faber-Langendoen, D., and A. H. Gentry. 1991. The structure and diversity of rain forests at Bajo Calima, Chocó Region, Western Colombia. *Biotropica* 23(1):2–11.

Farr, E., and G. Zijlstra, editors. 2001. *Index Nominum Genericorum.* http://rathbun.si.edu/botany/ing/INDEX.HTM

Galeano, G., J. Cediel, and M. Pardo. 1998. Structure and floristic composition of a one-hectare plot of wet forest at the Pacific Coast of Chocó, Colombia. Pages 551–68 in F. Dallmeier and J. A. Comiskey, editors. *Forest Biodiversity in North, Central and South America, and the Caribbean.* Parthenon Publishing Group, New York.

Gentry, A. H. 1982. Patterns of neotropical plant species diversity. *Evolutionary Biology* 15:1–84.

———. 1988a. Changes in plant community diversity and floristic composition on environmental and geographical gradients. *Annals of the Missouri Botanical Garden* 75:1–34.

———. 1988b. Tree species richness of upper Amazonian forests. *Proceedings of the National Academy of Sciences* 85:156–59.

———. 1993. A field guide to the families and genera of woody plants of Northwest South America (Colombia, Ecuador, Peru). University of Chicago Press, Chicago.

Kartawinata, K., R. Abdulhadi, and T. Partomihardjo. 1981. Composition and structure of a lowland dipterocarp forest at Wanariset, east Kalimantan. *Malaysian Forester* 44:397–406.

Korning, J., and H. Balslev. 1994. Growth and mortality of trees in Amazonian tropical rain forest in Ecuador. *Journal of Vegetation Science* 4:77–86.

Korning, J., K. Thomsen, and B. Ollgaard. 1991. Composition and structure of a species rich Amazonian rain forest obtained by two different sample methods. *Nordic Journal of Botany* 11:103–10.

Lieberman, D., and M. Lieberman. 1987. Forest tree growth and dynamics at La Selva, Costa Rica (1969–1982). *Journal of Tropical Ecology* 3:347–58.

Mabberley, D. J. 1997. *The Plant-Book: A Portable Dictionary of the Vascular Plants.* Cambridge University Press, Cambridge, U.K.

Makana, J., T. B. Hart, and J. A. Hart. 1998. Forest structure and diversity of lianas and understory treelets in monodominant and mixed stands in the Ituri forest, Democratic Republic of the Congo. Pages 429–46 in F. Dallmeier and J. A. Comiskey, editors. Forest Biodiversity Research, *Monitoring and Modeling: Conceptual Background and Old World Case Studies.* Parthenon Publishing Group, New York.

Manokaran, N., and K. M. Kochummen. 1987. Recruitment, growth and mortality of tree species in a lowland dipterocarp forest in peninsular Malaysia. *Journal of Tropical Ecology* 3:315–30.

Milliken, W. 1998. Structure and composition of one hectare of central Amazonian tierra firme forest. *Biotropica* 30(4):530–37.

Phillips, O. L., and J. S. Miller. 2002. Global Patterns of Plant Diversity: Alwyn H. Gentry's Forest Transect Data Set. Missouri Botanical Garden Press, St. Louis, MO.

Phillips, O. L., P. Hall, A. H. Gentry, S. A. Sawyer, and R. Vásquez. 1994. Dynamics and species richness of tropical rain forests. *Proceedings of the National Academy of Sciences* 91:2805–09.

Potts, M. D., P. S. Ashton, L. S. Kaufman, and J. B. Plotkin. 2002. The effect of habitat and distance on tropical tree species: A floristic comparison of 105 plots in Northwest Borneo. *Ecology* 83:2782–97.

Primack, R. B. and P. Hall. 1992. Biodiversity and Forest Change in Malaysian Borneo. *Bioscience* 42(11):829–37.

Richards, P. W. 1952. *The Tropical Rain Forest.* Cambridge University Press, Cambridge, U.K.

Rollet, B. 1974. L'Architecture des Forêts Denses Humides Sempervirentes de Plaine. Centre Technique Forestier Tropical, Nogent sur Marne, France.

Swaine, M. D., D. Lieberman, and F. Putz. 1987a. The dynamics of tree populations in tropical forest: A review. *Journal of Tropical Ecology* 3:359–66.

Swaine, M. D., J. B. Hall, and I. J. Alexander. 1987b. Tree population dynamics at Kade, Ghana (1968–1982). *Journal of Tropical Ecology* 3:331–45.

Uhl, C., and P. G. Murphy. 1981. Composition, structure, and regeneration of a tierra firme forest in the Amazon Basin of Venezuela. *Tropical Ecology* 22(2):217–37.

Valencia, R., H. Balslev, and G. Paz y Miño. 1994. High tree alpha-diversity in Amazonian Ecuador. *Biodiversity and Conservation* 2:21–28.

———. 1997. Tamaño y distribución vertical de los árboles en una hectárea de un bosque muy diverso de la Amazonía ecuatoriana. Pages 173–87 in R. Valencia and H. Balslev, editors. *Estudios Sobre Diversidad y Ecología de Plantas.* PUCE, Proyecto ENRECA, Proyecto DIVA, Quito, Ecuador.

Valencia, R., H. Balslev, W. Palacios, D. Neill, C. Josse, M. Tirado, and F. Skov. 1998. Diversity and family composition of trees in different regions of Ecuador: A sample of 18 one-hectare plots. Pages 569–84 in F. Dallmeier and J. A. Comiskey, editors. *Forest Biodiversity in North, Central and South America, and the Caribbean.* Man and the Biosphere Series. Parthenon Publishing Group, New York.

Valle Ferreira, L., and J. M. Rankin-de-Mérona. 1998. Floristic composition and structure of a one–hectare plot in tierra firme forest in Central Amazonia. Pages 649–62 in F. Dallmeier and J. A. Comiskey, editors. *Forest Biodiversity in North, Central and South America, and the Caribbean.* Man and the Biosphere Series. Parthenon Publishing Group, New York.

3

The Dance of the Continents

Egbert G. Leigh, Jr.

Continental Drift and Angiosperm Evolution

The continual collisions and divisions of land masses during the past few hundred million years have greatly influenced the species composition of today's Forest Dynamics Plots. A region's geologic and climatic history, and changes in this region's accessibility from other regions, ultimately governed what groups colonized and prospered there. To understand this history, we must first learn, at least in outline, how the continents have broken up and moved about since angiosperms came into the world.

In the early Mesozoic, 200 million years ago, nearly all the world's land was joined together in one great supercontinent, Pangaea (Hallam 1994). About 140 million years ago, in the early Cretaceous, the angiosperms first entered the fossil record as tropical herbaceous weeds (Crane et al. 1995; Wing and Boucher 1998). At that time, the 20-million-year process was completed that divided Pangaea into the northern Laurasia, including North America and most of Eurasia, and the southern Gondwana, including South America and Africa plus Arabia, India, Madagascar, and Antarctica (Pitman et al. 1993; Hallam 1994). At about the same time, Madagascar, which was still joined to India and Australia, became completely separated from east Africa (Rabinowitz et al. 1993; Hallam 1994), although one could still walk dryshod from Madagascar eastward to India, and southward to Antarctica (Hay et al. 1999), westward to Antarctica's junction with South America, and northeastward from South America to adjoining parts of Africa.

In the earliest Cretaceous, Africa (plus Arabia) also began splitting from South America from the south northward. The split was complete 120 million years ago (Hay et al. 1999), 20 million years earlier than was previously thought (Pitman et al. 1993); Africa remained separate until it reconnected to Europe in the early Miocene (Hallam 1994). Gondwana was now split in two. The other, larger fragment was a great crescent consisting of South America, Antarctica, Madagascar, India, and Australia (Hay et al. 1999). By 110 million years ago, magnoliids had spread as far south as Antarctica and as far north as Alberta (Wing and Boucher 1998).

Over 100 million years ago, the earth entered a "greenhouse state" in which tropical climates often extended to 50° latitude to both sides of the equator. This condition lasted for about 60 million years (Frakes et al. 1992). Although primitive angiosperms were insect-pollinated, only as the greenhouse state approached did angiosperms begin to evolve specializations to attract particular kinds of insects. Perhaps as a result, angiosperm diversity increased rapidly (Wing and Boucher 1998). Around this time, the epicontinental Turgai Sea separated eastern from western Asia, and another epicontinental sea separated eastern from western North America (Hay et al. 1999). The Turgai Sea closed in the early Oligocene (Hallam 1994). Eastern North America was connected to Europe via Greenland until the mid-Eocene, while Asia was intermittently connected to western North America by the Beringia land corridor until the present time (Hallam 1994). Until the late Eocene, Beringia was often warm enough to support subtropical plants.

One hundred million years ago, a diversity of angiosperms had spread to higher latitudes, and angiosperms already accounted for a majority of the world's diversity of vascular plants (Crane et al. 1995; Morley 2000). Eighty-five million years ago, soon after figs had evolved their remarkable pollination system (Machado et al. 2001), the Deccan plate, carrying India and Sri Lanka, split from Madagascar and began drifting away (Storey et al. 1995). India was still connected to Antarctica, probably by the Kerguelen plateau, which was then above sea level (Hay et al. 1999). It was previously thought that India separated from Antarctica 130 million years ago (Storey 1995), but India and Patagonia shared the same genera of late Cretaceous dinosaurs, and during the late Cretaceous and early Cenozoic India did not have an endemic flora (Hay et al. 1999, p. 2).

About 70 million years ago, near the end of the Cretaceous, the proportion of angiosperm species with animal-dispersed seeds, and the size of average seed and average fruit of angiosperms, were rising (Eriksson et al. 2000), as if angiosperms were invading mature forest (Wing and Boucher 1998). By this time, India had separated from Antarctica, and movements of the Americas were creating a land bridge, or at least a row of islands, connecting them, allowing the exchange of a few migrants (Hallam 1994). This connection was broken when the bridge was fragmented and thrust eastward into the Caribbean. Puerto Rico began as an oceanic island in a volcanic arc associated with this land bridge.

Sixty-five million years ago, the Cretaceous ended when a comet or asteroid struck the Yucatan, causing a "nuclear winter" and other catastrophes that extinguished the dinosaurs (Powell 1998). Afterward, angiosperms came to dominate mature forests (Tiffney 1984; Tiffney and Mazer 1995; Wing and Boucher 1998). Fifty million years ago, a Laurasian rainforest, many of whose genera now occur in Malaysia, extended from Hokkaido and Oregon to the London Clay of south England and southeastward along the north shore of the Tethys (Wolfe 1975). At that time, rainforest resembling that in Queensland today occurred near 50° south paleolatitude in Australia (Christophel and Greenwood 1989).

Fifty million years ago, in the Eocene, Australia separated from Antarctica (Crook 1981). Soon thereafter, the earth began a cooling and drying trend, ending 60 million years of greenhouse climate. The cooling and drying have continued to this day, with occasional warmer interludes such as that in the early Miocene. The range of tropical rainforest has consequently shrunk; 10 million years ago it disappeared from North America (Wolfe 1975) and occupied only a small part of Australia (Christophel and Greenwood 1989).

Forty or 50 million years ago, the Deccan plate collided with Laurasia, eventually raising the Himalayan massif. At this time, the Deccan plate had very diverse tropical forest, which inoculated southeast Asia with dipterocarps and other plant lineages while resisting Eurasian invaders (Morley 2000). No more than 10 million years later, South America was finally separated from Antarctica (Pitman et al. 1993) and entered upon a long period of splendid isolation.

Fifteen million years ago, Australia and New Guinea, drifting northward from Antarctica, collided with the Asian plate (Hallam 1994), allowing rodents to hop from island to island all the way to Australia and dipterocarps and other plants to invade New Guinea.

Only 3 million years ago, the Americas were rejoined by the Panama land bridge, allowing an extensive interchange of faunas and permitting South American rainforest to spread northward to Mexico (Gentry 1982).

Biogeography

In Asia, the Forest Dynamics Plots are all west of Wallace's line. The region including these plots forms a relatively homogeneous biogeographic entity. This is due in part to (1) the almost complete disappearance of the local biogeographic imprint of the collision between India and Laurasia, (2) the enormous influence on regional climate of the Himalayas, which extend east-west along the northern border of the Asian tropics, and (3) the dominant influence of the Pacific's oceanic climate on the Malesian archipelago in the Far East (P. S. Ashton, personal communication).

South America now has the most extensive block of tropical rainforest in the world. During the past million years Amazonian climates have followed a 100,000-year cycle, in which they were drier and cooler than now during those 90,000 years of every 100,000 when glaciers were marching forth from the poles. These cyclic dry periods, however, never shattered this rainforest into scattered remnant fragments. At least the western half of Amazonia always remained a solid block of forest (Piperno 1997), of which Ecuador's Yasuní plot is a part.

Three million years ago, the South American flora began to spread out along the new land bridge into Central America. Central American forests are less diverse than Amazonian forests of similar climate, and diversity declines progressively with increased distance from South America, even when latitude and climate are factored out. Most of Panama's trees and lianas belong to species that have spread from Amazonia, while most of its shrubs and epiphytes belong to lineages that are now spreading from the lower slopes of the Andes (Gentry 1982).

Africa suffered far more from drought in the Pleistocene than the other rainforest regions (Morley 2000). When glaciers spread from the poles into the temperate zones, Africa's forest was reduced to isolated tracts. The site of the Korup Forest Dynamics Plot was probably included in one of these "refuges" (Maley 1987). Even so, the species composition of Korup's forest today appears to be different from what it was in the late Pleistocene, when many montane species were present. Africa currently harbors no everwet lowland forest, even though some lowland forests, such as Korup, have very high annual rainfall.

Within each biogeographic realm, there are striking contrasts between mainland and island settings. Within southeast Asia, the vast forests of the Sunda shelf and its hinterland—represented by plots at Lambir (Sarawak, Malaysia), Pasoh (peninsular Malaysia), Khao Chong (Thailand), and Huai Kha Khaeng (Thailand)—can be compared with those on ancient continental islands such as the Philippines. The plot at Palanan in the Philippines is on the windward (eastern) side of Luzon, an island whose forests are typical of those down the eastern side of the Philippines as far as Mindanao. Forested islands of the Philippines were joined together at times during the Pleistocene, though they have not been linked to the Asian continent since the late Tertiary. During the past 2 million years, the area of the island including Palanan never greatly exceeded 100,000 km^2, an area far less than that of Asia's continental forest block. The forests of Nanjenshan on the island of Taiwan were connected by a land bridge to mainland China until about 10,000 years ago. In an ecological sense, the plot at Sinharaja in Sri Lanka is on an even more isolated island than Palanan or Nanjenshan. The rainforests of southwestern Sri Lanka are in an island of everwet climate of roughly 10,000 km^2. The closest analogues to these forests are the mixed dipterocarp rainforests of Sumatra, 1500 km to the east (P. S. Ashton, personal communication). During the Pliocene, the Deccan plate, which includes Sri Lanka, suffered severe drought (Morley 2000), and the diversity of its forests has not yet recovered. In the neotropics, we may compare mainland plots with the Luquillo everwet forest plot at El Verde, Puerto Rico, an island isolated for many millions of years, whose flora may all be descended from propagules that were dispersed over water.

Now artificial fragmentation is also influencing our plots. Bukit Timah (Singapore) is part of a small fragment of primary forest, isolated a century ago, which has lost half its bird species and more of its mammals (Turner and Corlett 1996). Its tree diversity is lower than Pasoh's, but we do not yet know how fragmentation is affecting regeneration in this forest. The 1600-ha Barro Colorado Island has been isolated from the mainland since 1914; it has lost many of its birds and some mammals, and the ecological impact of Barro Colorado Island's isolation from the mainland is hotly debated (Terborgh 1988; Glanz 1990; Wright et al. 1994). Plantations and shifting agriculture expand along one side of Lambir Hills National Park. Elsewhere in Borneo, the isolation of parks from surrounding forest has impaired the regeneration of dipterocarp forest (Curran et al. 1999).

References

Christophel, D. C., and D. R. Greenwood. 1989. Changes in climate and vegetation in Australia during the Tertiary. *Review of Paleobotany and Palynology* 58:95–109.

Crane, P. R., E. M. Friis, and K. R. Pedersen. 1995. The origin and early diversification of angiosperms. *Nature* 374:27–33.

Crook, K. W. A. 1981. The break-up of the Australian–Antarctic segment of Gondwanaland. Pages 1–14 in A. Keast, editor. *Ecology and Biogeography of Australia.* W. Junk, The Hague.

Curran, L. M., I. Caniago, G. D. Paoli, D. Astianti, M. Kusneti, M. Leighton, C. E. Nirarita, and H. Haeruman. 1999. Impact of El Niño and logging on canopy tree recruitment in Borneo. *Science* 286:2184–88.

Eriksson, O., E. M. Friis, and P. Löfgren. 2000. Seed size, fruit size, and dispersal systems in angiosperms from the early Cretaceous to the late Tertiary. *American Naturalist* 156:47–58.

Frakes, L. A., J. E. Francis, and J. L. Syktus. 1992. *Climate Modes of the Phanerozoic.* Cambridge University Press, Cambridge, U.K.

Gentry, A. H. 1982. Neotropical floristic diversity: Phytogeographical connections between Central and South America—Pleistocene climatic fluctuations, or an accident of the Andean orogeny? *Annals of the Missouri Botanical Garden* 69:557–93.

Glanz, W. E. 1990. Neotropical mammal densities: How unusual is the community on Barro Colorado Island, Panama? Pages 287–311 in A. H. Gentry, editor. *Four Neotropical Rainforests.* Yale University Press, New Haven, CT.

Hallam, A. 1994. *An Outline of Phanerozoic Biogeography.* Oxford University Press, Oxford, U.K.

Hay, W. W., R. M. DeConto, C. N. Wold, K. M. Wilson, S. Voigt, M. Schulz, A. R. Wold, W. C. Dullo, A. B. Ronov, A. N. Balukhovsky, and E. Söding. 1999. Alternative global Cretaceous paleogeography. Pages 1–47 in E. Barrera and C. C. Johnson, editors. *Evolution of the Cretaceous Ocean-Climate System.* Geological Society of America. Boulder, CO.

Machado, C. A., E. Jousselin, S. G. Compton, and E. A. Herre. 2001. Phylogenetic relationships, historical biogeography, and character evolution of fig-pollinating wasps. *Proceedings of the Royal Society of London* B 268:685–94.

Maley, J. 1987. Fragmentation de la forêt dense humide ouest-africaine et extension des biotopes montagnards au quaternaire récent: Nouvelles données polliniques et chronologiques: implications paléoclimatiques et biogéographiques. *Palaeoecology of Africa* 18:307–34.

Morley, R. J. 2000. Origin and Evolution of Tropical Rain Forests. Wiley, Chichester, U.K.

Piperno, D. R. 1997. Phytoliths and microscopic charcoal from leg 155: A vegetational and fire history of the Amazon basin during the last 75 K.Y. *Proceedings of the Ocean Drilling Program, Scientific Results* 155:411–18.

Pitman, W. C. III., S. Cande, J. LaBrecque, and J. Pindell. 1993. Fragmentation of Gondwana: The separation of Africa from South America. Pages 15–34 in P. Goldblatt, editor. *Biological Relationships Between Africa and South America.* Yale University Press, New Haven, CT.

Powell, J. L. 1998. *Night Comes to the Cretaceous.* Harcourt, Brace, San Diego, CA.

Rabinowitz, P. D., M. F. Coffin, and D. Falvey. 1983. The separation of Madagascar and Africa. *Science* 220:67–69.

Storey, B. C. 1995. The role of mantle plumes in continental breakup: case histories from Gondwanaland. *Nature* 377:301–08.

Storey, M., J. J. Mahoney, A. D. Saunders, R. A. Duncan, S. P. Kelley, and M. F. Coffin. 1995. Timing of hot-spot-related volcanism and the breakup of Madagascar and India. *Science* 267:852–55.

Terborgh, J. 1988. The big things that run the world—A sequel to E. O. Wilson. *Conservation Biology* 2:402–03.

Tiffney, B. H. 1984. Seed size, dispersal syndrome, and the rise of the angiosperms: Evidence and hypothesis. *Annals of the Missouri Botanical Garden* 71:551–76.

Tiffney, B. H., and S. J. Mazer. 1995. Angiosperm growth habit, dispersal and diversification reconsidered. *Evolutionary Ecology* 9:93–117.

Turner, I. M., and R. T. Corlett. 1996. The conservation value of small, isolated fragments of lowland tropical rain forest. *Trends in Ecology and Evolution* 11:330–33.

Wing, S. L., and L. D. Boucher. 1998. Ecological aspects of the Cretaceous flowering plant radiation. *Annual Review of Earth and Planetary Sciences* 26:379–421.

Wolfe, J. A. 1975. Some aspects of plant geography of the northern hemisphere during the late Cretaceous and Tertiary. *Annals of the Missouri Botanical Garden* 62:264–79.

Wright, S. J., M. E. Gompper, and B. de Leon. 1994. Are large predators keystone species in Neotropical forests? The evidence from Barro Colorado Island. *Oikos* 71:279–94.

4

How Wet Are the Wet Tropics?

Egbert G. Leigh, Jr.

The forests portrayed in this volume span a variety of climates (table 4.1). At the dry extreme, Mudumalai in south India averages about 1250 mm of rain per year. At the wet extreme, Korup, Cameroon, averages more than 5250 mm of rain per year. These forests differ not only in total annual rainfall but also in its seasonal distribution. At Huai Kha Khaeng, Thailand, the total rainfall over the 3 driest months averages 46 mm, and at Mudumalai, the total rainfall over the 3 driest months averages 67 mm. By comparison, at Luquillo, Puerto Rico, the year's driest month averages 203 mm. Cameroon's Korup and Sri Lanka's Sinharaja both average over 5000 mm of rain per year, but at Korup, the total rainfall for the 3 driest months averages 172 mm, while Sinharaja's driest month, February, averages 171 mm. What effects do these differences in climate have on tropical forest?

Total Rainfall

Heavy rainfall is a defining characteristic of tropical rainforest, as its very name suggests. The flow of water "through the stem from root to leaf replacing that lost by evaporation . . . is called the transpiration-stream. . . . [Evapotranspiration] becomes the overwhelming activity of the forest, which through plenteous evaporation engenders its own storms" (Corner 1964, p. 107; see also Salati and Vose 1984). When enough water is available, brighter sun and warmer weather increase transpiration, creating more rain clouds. On a large scale, the role of heat in spawning rainclouds is illustrated by the "intertropical convergence zone," a circumtropical belt of rainclouds that follows the latitude of the sun's zenith with a 6-week lag (Rand and Rand 1982; Hallé 1993).

By regulating cloudiness, rainforests regulate their water use. In watertight catchments of tropical lowlands where the forest receives as much rain as it can use, rainfall exceeds runoff by 1384 ± 129 mm per year (mean $\pm$ SD among 19 lowland tropical catchments), regardless of how much more rain falls (Bruijnzeel 1989). This approximate 1400-mm difference between rainfall and runoff is the water the forest uses. Closed forests occupy tropical lowlands, as a rule, only where at least 1600 mm of rain falls a year (Walsh 1996) and where at least 7 or 8 months average over 150 mm apiece. Among the Forest Dynamics Plots, only Mudumalai

Table 4.1. Monthly Rainfall P, Average Temperature T, and Average Diurnal Temperature Range T at Selected Forest Dynamics Plot

	Jan	Feb	Mar	Apr	May	Jun	Jul	Aug	Sep	Oct	Nov	Dec	Total/Avg
Luquillo, Puerto Rico 18°19′ N, 65°49′ W													
P	233	227	203	232	351	242	307	361	350	288	401	353	3548
T	20.9	21.0	21.6	22.5	23.4	24.4	24.3	24.4	24.1	23.5	22.5	21.5	22.8
ΔT	4.2	4.5	5.2	5.6	5.3	5.4	4.8	4.8	4.8	4.6	3.8	3.8	4.7
Huai Kha Khaeng, Thailand 15°38′ N, 99°13′ E													
P	6	30	39	82	226	120	123	155	278	360	47	10	1476
T	24.3	22.1	23.9	25.8	27.7	25.5	26.4	25.6	24.2	23.6	20.8	19.1	24.1
ΔT	15.0	14.1	17.7	16.7	15.0	10.1	9.1	8.2	9.3	9.8	11.7	14.9	12.7
Mudumalai, India 11°35′ N, 76°31′ E													
P	7	9	88	119	129	160	146	130	126	170	115	51	1250
T	21.4	22.6	26.9	24.2	22.9	22.6	24.2	24.8	18.9	22.5	21.2	21.0	22.8
ΔT	10.6	12.5	14.2	12.4	13.3	8.8	8.2	9.3	8.4	8.1	12.0	13.6	10.9
Barro Colorado, Panama 09°09′ N, 79°51′ W													
P	71	37	23	106	245	275	237	322	309	364	360	202	2551
T	26.9	27.1	27.5	27.9	27.8	27.2	27.1	27.0	27.0	26.8	26.8	26.8	26.9
ΔT	8.2	8.4	8.8	8.8	8.1	7.4	7.3	7.5	7.8	7.9	7.5	7.8	8.2
Sinharaja, Sri Lanka 6°24′ N, 80°24′ E													
P	191	171	237	434	695	610	390	424	553	562	471	274	5016
T	21.9	22.4	23.2	23.5	23.4	22.7	22.4	22.4	22.5	22.2	22.1	22.0	22.6
ΔT	5.1	6.2	7.2	5.3	3.3	2.6	2.6	3.1	3.3	3.5	4.0	5.0	4.3

Korup, Cameroon 5°04′ N, 8°51′ E

P	38	58	221	329	459	564	913	914	691	668	341	76	5272
T	26.8	27.9	27.9	27.5	27.2	26.6	25.5	25.1	25.6	26.1	26.8	27.1	26.7
ΔT	9.7	9.8	9.5	9.0	8.7	7.6	6.5	5.4	6.5	7.2	7.9	8.0	7.9

Lambir, Sarawak, Malaysia 04°11′ N, 114°01′ E

P	229	200	182	222	204	153	165	180	198	290	322	319	2664
T	25.4	26.0	27.0	27.2	27.4	27.4	26.9	26.8	26.6	26.3	26.4	26.0	26.6
ΔT	6.6	6.9	7.2	7.8	8.0	8.0	7.9	7.7	7.3	7.3	7.4	6.8	7.4

Pasoh, Malaysia 2°58′ N, 102°18′ E

P	94	109	153	167	162	125	115	120	162	189	224	168	1788
T	27.0	27.7	28.3	28.9	28.9	28.5	27.7	28.1	28.0	28.3	27.4	27.2	28.0
ΔT	10.1	10.9	11.3	11.5	11.0	10.7	10.2	10.5	10.4	10.8	9.3	8.8	10.5

Ituri, Congo (monodominant forest) 01°19′ N, 28°39′ E

P	51	41	90	195	168	152	165	165	161	226	172	88	1674
T	22.8	22.3	23.3	23.9	24.1	24.4	23.1	22.3	22.8	22.8	22.5	22.6	23.1
ΔT	10.2	10.0	10.9	10.2	10.4	11.1	9.2	7.7	9.1	8.5	7.7	8.8	9.5

Yasuní, Ecuador 00°41′ S, 76°24′ W

P	226	344	200	253	412	374	227	193	174	253	196	229	3081
T	27.7	29.4	28.4	28.3	27.5	27.6	27.8	28.2	28.9	29.0	28.8	28.4	28.4
ΔT	12.0	12.0	12.8	13.1	12.1	12.6	12.9	14.0	15.4	15.2	14.6	13.4	13.3

Table 4.2. Annual Rainfall P, and Annual Averages of Temperature T, Diurnal Temperature Range ΔT, Number of Hours of Sunshine per Day S, and Solar Radiation Q (W/m^2)

	Singapore	Trivandrum, S. India	Yangambi, Congo	Abidjan, Ivory Coast	Manaus, Brazil	Cristóbal, Panama
P	2413	1696	1828	2095	1897	3285
T	27	27	25	26	27	27
ΔT	7	5	10	8	8	5
S	5.6	6.3	5.6	4.7	5.8	6.4
Q	224	232	199	180	181	194

Notes: Radiation data for Cristobal from Galeta, 7 km NE (Cubit et al. 1988), radiation data for Abidjan from Bernhard-Reversat et al. (1978), other data from Müller (1982).

in south India and Huai Kha Khaeng in Thailand receive less than 1600 mm of rain per year. These two forests and the montane forest at Doi Inthanon in Thailand are the only CTFS sites with fewer than 7 months averaging over 150 mm of rain.

Rainforests do not benefit from excess rainfall. Prolonged high rainfall usually entails more clouds and less sunlight (table 4.2). This circumstance can lower forest productivity. In most forests, roughly 1% of the incident sunlight reaches the forest floor (Leigh 1999). These rainforests are limited by shortages of both light and water; many produce more dry matter in years that are somewhat drier and sunnier than usual (Wright et al. 1999).

In typical lowland equatorial forest, average temperature, defined as the mean of the average daily minimum and the average daily maximum, is about 25–27°C, and the average diurnal temperature range is about 7–9°C (table 4.1). This is also true during the rainy season in more seasonal climates (table 4.1), as illustrated by the wettest months in Huai Kha Khaeng and Mudumalai. Rainfall and cloudiness allow the forest to control temperature variation. In forests close to oceans, such as Luquillo and Sinharaja (table 4.1), the nearby water also restricts the diurnal temperature range, making it lower than in rainforests far from the sea. Insofar as they regulate cloudiness, lowland tropical forests share not only similar temperature regimes but also similar inputs of sunshine and radiation (table 4.2).

The Seasonal Distribution of Rainfall

Rainfall varies seasonally. During a severe dry season, some canopy trees benefit from dropping their leaves to save water. Dry season can kill seedlings and depress the abundance of insect pollinators. Where the dry season is severe, insect-pollinated plants are most readily pollinated if they flower at the beginning of the rainy season (Foster 1982), and it is best for plants to disperse seeds as close as possible to the beginning of the rainy season, to give seedlings as much time as possible to prepare for the next dry season (Garwood 1983). If sufficient water is available all through the year, plants tend to flower and to flush their leaves during the season with the most solar radiation, while in places where the dry season is severe, plants, at least those with shallow roots, wait until the rains begin before flowering and flushing leaves (Wright and van Schaik 1994).

A dry season, however, appears to be good for the soil. On Panama's Barro Colorado Island (BCI), seasonal changes in the chemistry of streamwater flowing from the steep-sided Lutz catchment indicate that the dry season favors the formation of smectites, whereas the wettest conditions favor the formation of kaolinites. Smectites greatly enhance the soil's cation exchange capacity, whereas kaolinitic clays do so far less. Soils that retain cations are more fertile. As the soils of Lutz catchment on BCI contain primarily smectites (Robert Stallard, personal communication), it appears that, at least in this catchment, dry season favors more fertile soils. Moreover, irrigating two 2.25-ha plots in BCI's mature forest for five successive dry seasons (6 mm/day, 5 days/week) reduced soil porosity relative to unirrigated controls (Kursar et al. 1995). This circumstance impaired gas exchange somewhat. By the end of the watering experiment, mean oxygen content of the air in the soil at the end of the rainy season averaged 15% for irrigated plots and 17% for unwatered controls. This difference increased year by year. Moreover, at the end of the watering experiment, water infiltrated the soil twice as fast on unwatered control plots as on the irrigated plots, whatever the time of year (Kursar et al. 1995). Both in West Africa and in Amazonia, the tallest, biggest trees and the greatest standing timber per hectare are found not in the wettest forests but in more seasonal ones (Martin 1991). Do more seasonal forests have better soils? Or are tall forests intermediate stages of succession after widespread disturbances occurring a few centuries back, regardless of seasonality, as is true for the tall Douglas-fir forests of Oregon? Probably both are true.

Seasonality also affects a forest's antiherbivore defenses. Where dry season is severe, plants can reduce losses of young leaves to herbivores by flushing leaves in the dry season, when pest populations are low (Aide 1988, 1992; Murali and Sukumar 1993). In everwet climates, plants are exposed to pest pressure all through the year (Leigh 1999).

How does one measure seasonality or integrate the annual drought stress? Traditionally, months averaging less than 100 mm of rain are considered dry because evaporation plus transpiration from lowland tropical forests usually exceeds 100 mm per month. Such a crude measure, however, fails to distinguish a month averaging 5 mm of rainfall from one averaging 95 mm. A more instructive measure of the severity of the dry season is the sum of the rainfall for the driest series of 3 consecutive months during the year. Even this measure of seasonality, however, fails to account for differences in carryover of soil moisture from preceding rainy months. For example, the rainfall for the driest 3 consecutive months averages 192 mm at Ituri and 172 mm at Korup. Considering how

Table 4.3. Rainfall Data in Forest Dynamics Plots

Site	Mean annual rainfall (mm)	Walsh's score	Rain (mm), 3 driest months	No. mos. <100 mm rain
Luquillo	3548	24	662	0
La Planada	4415	22	578	0
Sinharaja	5016	22	599	0
Yasuní	3081	21	563	0
Palanan	3379	21	447	0
Bukit Timah	2473	19	506	0
Lambir	2664	19	498	0
Nanjenshan	3582	19	334	0
Korup	5272	14.5	172	3
BCI	2551	12.5	131	3
Pasoh	1788	11.5	318	1
Ituri (Mixed Forest)	1785	6.5	192	3
Ituri (Monodominant)	1674	4.5	180	4
Doi Inthanon	1908	2.5	25	6
Mudumalai	1250	2.5	67	4
HKK	1476	−1.5	46	6

much rain Korup receives during the other 9 months—5100 mm versus Ituri's 1593 mm—can one really believe that Korup's dry season is more stressful than Ituri's? Walsh (1996) devised an index to integrate the effects of annual rainfall and its seasonality: score +2 for each month averaging over 200 mm of rainfall, +1 for each month averaging over 100 mm, −1 for each month averaging between 51 and 100 mm, and −2 for each month averaging 50 mm or less. Add a bonus of 0.5 each time a month averaging less than 100 mm follows a month averaging more than 100 mm, and sum these scores for the year. Walsh predicts that tropical lowlands scoring 10 or more for the year support rainforest, whereas those scoring between 5 and 9.5 support evergreen seasonal forest (table 4.3). Although none of these methods is entirely satisfactory, both Walsh's index and the 3 consecutive dry months provide useful information, and we shall use them both throughout this section.

What factors affect the seasonality of rainfall? Theoretically, rainfall should peak about 6 weeks after the sun stands directly overhead at midday (Rand and Rand 1982), when the earth is warmest and vapor-laden air rises highest and most abundantly (Lauer 1989). As the air rises, it cools and its vapor condenses as rainfall. The now-dry air subsides to the north and south of the cloudy intertropical convergence zone, which might span 20° of latitude (Lauer 1989). Much of this air blows back toward the intertropical zone, replacing the warm, rising air that fuels its clouds. These trade winds prevent air away from this cloud belt from rising high enough to produce rain, except when the winds encounter steeply rising coasts, such as the east end of Puerto Rico. Thus the intertropical convergence zone tends to be flanked on both sides by a belt of drought.

Other things being equal, rainfall should be least seasonal at the equator, with two-rainfall peaks per year in May and November and two intervening drier spells centered on January and July. As latitude increases to the north, January's dry season becomes stronger and July's weaker, while the opposite happens south of the equator (tables 4.4 to 4.6). This pattern appears in every biogeographic realm, and the Forest Dynamics Plots conform crudely to it, but local climate is influenced by many confounding factors

Extreme Events

The two most familiar types of seriously destructive weather are hurricanes (known as typhoons in east Asia and cyclones in northeast Australia, Madagascar, and nearby islands of the eastern Indian Ocean) and El Niño droughts.

A hurricane is a vortex of storm clouds hundreds of kilometers in diameter surrounding a central cloudless eye. Especially near this eye, rainfall is intense— a single hurricane can drop a half meter of rain—and winds blow at hundreds of kilometers per hour. Such vortices are amplifications of "tropical depressions" moving eastward over the ocean, usually 10° or more from the equator. Hurricanes tend to occur at the end of summer when the top 60 m of ocean water are warmest (Emanuel 1988). Among the Forest Dynamics Plots, Luquillo in Puerto Rico, Nanjenshan in Taiwan, and Palanan in the Philippines are most frequently visited by hurricanes. A hurricane crosses Luquillo about twice a century (Scatena and Larsen 1991). The winds of a severe hurricane can strip a forest of its leaves and topple or smash most of its big trees (Lauer 1989; Walker 1991). Moreover, the extremely heavy rainfall that hurricanes bring can flood valleys and cause abundant landslides in hill country. In the mountains of Puerto Rico, 104 mm of rain in 2 hours, 164 mm in 24 hours, or 200 mm in 72 hours is enough to trigger a rash of landslides. These thresholds appear to be typical of other tropical regions as well (Larsen and Simon 1993).

The El Niño is a warm current that flows south along the desert coast of Peru every few years, bringing rains, and sometimes catastrophic floods, to that desert shore and snuffing out the coastal upwellings that nourish a fishery on which many people and a whole host of seabirds depend (Hutchinson 1950, pp. 13–14). El Niño tends to flow during extreme swings of the "Southern Oscillation," which is reflected by the seesaw between atmospheric pressure at Tahiti and that at Darwin, Australia; when one is high, the other is low, and vice versa (Rasmussen and Wallace 1983). When atmospheric pressure is high over Darwin and low over Tahiti, the equatorial eastern Pacific warms, the El Niño flows, and Central America, Malesia, and tropical Australia have a dry year. An extreme El Niño brings severe drought to eastern Borneo, the Malay Peninsula, and southern Africa, warms seawater enough to bleach corals along the Pacific shore of Central

Table 4.4. Average Monthly Rainfall, in mm, at Inland Stations and Forest Dynamics Plots in Africa

	Latitude	Jan	Feb	Mar	Apr	May	Jun	Jul	Aug	Sep	Oct	Nov	Dec	Total
Largeau, Chad	18°00′ N	0	0	0	0	1	2	3	11	1	0	0	0	17
Abéché, Chad	13°51′ N	0	0	0	1	24	26	141	232	67	14	0	0	505
Birao, C. Afr. Rep.	10°17′ N	0	0	2	19	97	112	217	204	171	37	1	0	860
Bria, C. Afr. Rep.	6°32′ N	10	10	104	117	206	173	251	277	208	211	66	3	1636
Korup, Cameroon	**4°54′ N**	**38**	**58**	**221**	**329**	**459**	**564**	**913**	**914**	**691**	**668**	**341**	**76**	**5272**
Bongabo, Congo	3°8′ N	38	63	135	167	189	180	186	250	207	209	130	56	1810
Ituri, D.R. Congo (Mixed Forest)	**1°33′ N**	**32**	**80**	**103**	**183**	**186**	**152**	**195**	**183**	**192**	**205**	**194**	**80**	**1785**
Ituri, D.R. Congo (Monodominant Forest)	**1°19′ N**	**51**	**41**	**90**	**195**	**168**	**152**	**165**	**165**	**161**	**226**	**172**	**88**	**1674**
Eala, Congo	0°3′ N	84	107	127	178	157	145	71	178	178	216	193	170	1794
Kisozi, Burundi	3°33′ S	167	190	196	228	120	12	6	16	64	115	174	189	1447
Kalami, Congo	5°53′ S	110	77	137	207	96	9	1	7	43	51	146	181	1064
Kamina, Congo	8°44′ S	201	193	202	119	18	1	1	5	38	121	191	253	1343
Lumumbashi, Congo	11°39′ S	256	264	210	53	3	0	0	0	3	27	166	262	1244
Lusaka, Burundi	15°25′ S	217	196	106	21	4	0	0	0	0	15	91	186	837
Wankie, Zimbabwe	18°22′ S	147	147	79	18	5	3	0	3	3	18	58	119	592

Notes: Months with greater than 100 mm rain are shaded; Forest Dynamic Plots are in bold. Data for non-CTFS sites from Müller (1982).

Table 4.5. Average Monthly Rainfall, in mm, at Inland Stations and Forest Dynamics Plots in Southeast Asia

	Latitude	Jan	Feb	Mar	Apr	May	Jun	Jul	Aug	Sep	Oct	Nov	Dec	Total
Mandalay, Burma	21°59' N	3	3	5	30	147	160	89	104	137	109	51	10	828
Chiang Mai, Thailand	18°47' N	7	7	16	45	146	137	169	223	270	143	40	14	1217
HKK, Thailand	**15°40' N**	**6**	**30**	**39**	**82**	**226**	**120**	**123**	**155**	**278**	**360**	**47**	**10**	**1476**
Bangkok, Thailand	13°45' N	9	30	36	82	165	153	168	183	310	239	55	8	1438
Surat Thani, Thailand	9°7' N	53	18	41	114	256	140	114	137	201	236	320	216	1848
Pinang, Malaysia	5°25' N	94	79	142	188	272	196	191	295	401	429	302	147	2736
Lambir, Swk, Malaysia	**4°20' N**	**229**	**200**	**182**	**222**	**204**	**153**	**165**	**180**	**198**	**290**	**322**	**319**	**2664**
Pasoh, Malaysia	**2°58' N**	**94**	**109**	**153**	**167**	**162**	**125**	**115**	**120**	**162**	**189**	**224**	**168**	**1788**
Bukit Timah, Singapore	**1° 15' N**	**256**	**184**	**212**	**200**	**200**	**171**	**158**	**179**	**177**	**216**	**249**	**271**	**2473**
Tandjungpandan, Indonesia	2°45' S	277	165	193	267	256	191	170	140	163	274	371	404	2871
Surakarta, Indonesia	7°45' S	307	287	239	185	119	79	33	36	33	91	203	236	1949
Darwin, Australia	12°28' S	411	314	284	76	8	2	0	1	15	49	110	218	1330

Notes: Months with greater than 100 mm rain are shaded; Forest Dynamic Plots are in bold. Data for non-CTFS sites from Müller (1982).

Table 4.6. Average Monthly Rainfall at Inland Stations and the Yasuní Forest Dynamics Plot in South America

	Latitude	Jan	Feb	Mar	Apr	May	Jun	Jul	Aug	Sep	Oct	Nov	Dec	Total
Apure, Venezuela	7°53' N	1	4	14	71	186	277	303	262	171	129	42	1	1491
Puerto Ayacucho, Venezuela	5°41' N	14	17	66	156	337	437	436	292	175	189	110	40	2249
Santa Elena, Venezuela	4°31' N	68	69	78	145	221	248	229	182	109	108	130	115	1700
San Carlos de Rio Negro, Venezuela	1°54' N	222	229	206	395	381	390	330	328	249	257	314	220	3521
Uaupes, Brazil	0°8' S	284	261	284	263	329	244	234	186	160	164	190	270	2869
Yasuní, Ecuador	**0°41' S**	**226**	**344**	**200**	**253**	**412**	**374**	**227**	**193**	**174**	**253**	**196**	**229**	**3081**
Manaus, Brazil	3°8' S	266	247	269	267	194	100	64	38	60	124	152	216	1897
Alto Tapajos, Brazil	7°20' S	398	416	379	302	112	20	22	51	144	247	315	335	2741
Cuiaba, Brazil	15°35' S	213	200	222	106	46	14	9	27	48	124	162	208	1376
Corumbá, Brazil	19°00' S	176	147	119	83	67	32	18	24	66	103	122	163	1120

Notes: Months with greater than 100 mm rain are shaded; Forest Dynamic Plots are in bold. Data for non-CTFS sites from Müller (1982).

America, flings violent storms against the west coast of North America, brings abundant rain to southern California, diminishes the Nile's flood and weakens the summer monsoon in India (Rasmussen and Wallace 1983; Paine 1986; Glynn 1988, 1990; Diaz and Kiladis 1992; Quinn 1992).

Extreme El Niños can wreak havoc among tropical forests. During El Niño years, BCI suffers drought. The 1982–83 El Niño killed about 1.3% of the trees <20 cm dbh and 4% of the trees >20 cm dbh on BCI's 50-ha plot (Condit et al. 1995; Leigh 1999, Table 6.2: compare tree death rates 1982–85 with normal death rates 1985–90). This same El Niño so dried out eastern Borneo that fires there burned millions of hectares, mostly of clumsily and carelessly logged forest (Leighton and Wirawan 1986). The 1991–92 El Niño was less severe in Panama, but it hit western Borneo harder than did the 1982–83 event (Salafsky 1994). The 1997–98 El Niño paved the way for catastrophic fires in Amazonia (Laurance 1998) and Sumatra (Davies and Unam 1999), although it affected BCI less severely than did its predecessor of 1982–83. On Barro Colorado Island, at Pasoh, and in the Lambir Hills, flowering is copious and widespread and fruit falls in extraordinary abundance during El Niño years (Wright et al. 1999; Chan and Appanah 1980; Chan 1980; Ashton et al. 1988; Sakai et al. 1999).

References

Aide, T. M. 1988. Herbivory as a selective agent on the timing of leaf production in a tropical understory community. *Nature* 336:574–75.

———. 1992. Dry season leaf production: An escape from herbivory. *Biotropica* 24:532–37.

Ashton, P. S., T. J. Givnish, and S. Appanah. 1988. Staggered flowering in the Dipterocarpaceae: New insights into floral induction and the evolution of mast fruiting in the aseasonal tropics. *American Naturalist* 132:44–66.

Bernhard-Reversat, F., C. Huttel, and G. Lemée. 1978. Structure and functioning of evergreen rain forest ecosystems of the Ivory Coast. Pages 557–74 in UNESCO, UNEP, FAO, Tropical Forest Ecosystems. UNESCO, Paris.

Bruijnzeel, L. A. 1989. Nutrient cycling in moist tropical forests: the hydrologic framework. Pages 383–415 in J. Proctor, editor. *Mineral Nutrients in Tropical Forest and Savanna Ecosystems.* Blackwell Scientific, Oxford, U.K.

Chan, H. T. 1980. Reproductive biology of some Malaysian dipterocarps II, Fruiting biology and seedling studies. *Malaysian Forester* 43:438–51.

Chan, H. T., and S. Appanah. 1980. Reproductive biology of some Malaysian dipterocarps I, Flowering biology. *Malaysian Forester* 43:132–43.

Condit, R., S. P. Hubbell, and R. B. Foster. 1995. Mortality rates of 205 neotropical tree and shrub species and the impact of a severe drought. *Ecological Monographs* 65:419–39.

Corner, E. J. H. 1964. *The Life of Plants.* World Press, Cleveland, OH.

Cubit, J. D., R. C. Thompson, H. M. Caffey, and D. M. Windsor. 1988. Hydrographic and meteorological studies of a Caribbean fringing reef at Punta Galeta, Panamá: Hourly and daily variations for 1977–1985. *Smithsonian Contributions to the Marine Sciences* 32:1–220.

Davies, S. J., and L. Unam. 1999. Smoke-haze from the 1997 Indonesian forest fires: Effects on pollution levels, local climate, atmospheric CO_2 concentrations, and tree photosynthesis. *Forest Ecology and Management* 124:137–44.

Diaz, H. F., and G. N. Kiladis. 1992. Atmospheric teleconnections associated with the extreme phases of the Southern Oscillation. Pages 7–28 in H. F. Diaz and V. Markgraf, editors. *El Niño: Historical and Paleoclimatic Aspects of the Southern Oscillation.* Cambridge University Press, Cambridge, U.K.

Emanuel, K. A. 1988. Toward a general theory of hurricanes. *American Scientist* 76:370–79.

Foster, R. B. 1982. The seasonal rhythm of fruit fall on Barro Colorado Island. Pages 151–72 in E. G. Leigh, Jr., A. S. Rand, and D. M. Windsor, editors. *The Ecology of a Tropical Forest.* Smithsonian Institution Press, Washington, DC.

Garwood, N. C. 1983. Seed germination in a seasonal tropical forest: A community study. *Ecological Monographs* 53:159–81.

Glynn, P. W. 1988. El Niño–Southern Oscillation 1982–83: Nearshore population, community and ecosystem responses. *Annual Review of Ecology and Systematics* 19:309–45.

———. 1990. *Global Ecological Consequences of the El Niño–Southern Oscillation.* Elsevier, Amsterdam.

Hallé, F. 1993. *Un Monde Sans Hiver: Les Tropiques, Nature et Sociétés.* Éditions du Seuil, Paris.

Hutchinson, G. E. 1950. Survey of contemporary knowledge of geochemistry. 3. The biogeochemistry of vertebrate excretion. *Bulletin of the American Museum of Natural History* 96:1–554 + 16 plates.

Kursar, T. A., S. J. Wright, and R. Radulovich. 1995. The effects of the rainy season and irrigation on soil water and oxygen in a seasonal forest in Panama. *Journal of Tropical Ecology* 11:497–516.

Larsen, M. C., and A. Simon. 1993. A rainfall intensity-duration threshold for landslides in a humid-tropical environment, Puerto Rico. *Geografiska Annaler* 75A:13–23.

Lauer, W. 1989. Climate and weather. Pages 7–53 in H. Leith and M. J. A. Werger, editors. *Ecosystems of the World, Vol. 14B: Tropical Rain Forest Ecosystems, Biogeographical and Ecological Studies.* Elsevier, Amsterdam.

Laurance, W. F. 1998. A crisis in the making: Responses of Amazonian forests to land use and climate change. *Trends in Ecology and Evolution* 13:411–15.

Leigh, E. G., Jr. 1999. *Tropical Forest Ecology.* Oxford University Press, New York.

Leighton, M., and N. Wirawan. 1986. Catastrophic drought and fire in Borneo tropical rain forest associated with the 1982–1983 El Niño–Southern Oscillation event. Pages 75–102 in G. T. Prance, editor. *Tropical Rain Forests and the World Atmosphere.* Westview Press, Boulder, CO.

Martin, C. 1991. *The Rainforests of West Africa.* Birkhäuser, Basel, Switzerland.

Müller, M. J. 1982. *Selected Climatic Data for a Global Set of Standard Stations for Vegetation Science.* W. Junk, The Hague.

Murali, K. S., and R. Sukumar. 1993. Leaf flushing phenology and herbivory in a tropical dry deciduous forest, Southern India. *Oecologia* 94:114–19.

Paine, R. T. 1986. Benthic community–water column coupling during the 1982–83 El Niño. Are community changes at high latitudes attributable to cause or coincidence? *Limnology and Oceanography* 31:351–60.

Quinn, W. H. 1992. A study of Southern Oscillation-related activity for A. D. 622–1990 incorporating Nile River flood data. Pages 119–49 in H. F. Diaz and V. Markgraf, editors. *El Niño: Historical and Paleoclimatic Aspects of the Southern Oscillation.* Cambridge University Press, Cambridge, U.K.

Rand, A. S., and W. M. Rand. 1982. Variation in rainfall on Barro Colorado Island. Pages 47–59 in E. G. Leigh, Jr., A. S. Rand, and D. M. Windsor, editors. *The Ecology of a Tropical Forest.* Smithsonian Institution Press, Washington DC.

Rasmussen, E. M., and J. M. Wallace. 1983. Meteorological aspects of the El Niño/Southern Oscillation. *Science* 222:1195–202.

Sakai, S., K. Momose, T. Yumoto, T. Nagamitsu, H. Nagamasu, A. A. Hamid, T. Nakashizuka, and T. Inoue. 1999. Plant reproductive phenology over four years including an episode of general flowering in a lowland dipterocarp forest, Sarawak, Malaysia. *American Journal of Botany* 86:1414–36.

Salafsky, N. 1994. Drought in the rain forest: Effects of the 1991 El Niño–Southern Oscillation event on a rural economy in west Kalimantan, Indonesia. *Climatic Change* 27:373–96.

Salati, E., and P. Vose. 1984. Amazon basin: A system in equilibrium. *Science* 225:129–38.

Scatena, F. N., and M. C. Larsen. 1991. Physical aspects of Hurricane Hugo in Puerto Rico. *Biotropica* 23:317–23.

Walker, L. R. 1991. Tree damage and recovery from Hurricane Hugo in Luquillo Experimental Forest, Puerto Rico. *Biotropica* 23:379–85.

Walsh, R. P. D. 1996. Climate. Pages 159–205 in P. W. Richards. *The Tropical Rain Forest.* Cambridge University Press, Cambridge, U.K.

Wright, S. J., and C. P. van Schaik. 1994. Light and the phenology of tropical trees. *American Naturalist* 143:192–99.

Wright, S. J., C. Carrasco, O. Calderón, and S. Paton. 1999. The El Niño Southern Oscillation, variable fruit production, and famine in a tropical forest. *Ecology* 80:1632–47.

5

Soils in the Tropics

Peter S. Ashton

The tropics exhibit great diversity not only in their forests but also in their soils. Contrary to common perception, the diversity of tropical soils extends much further than the yellow-red soils of the Oxisol order. While this prototypical intensely leached, highly weathered soil is found in 22.5% of the tropics (Van Wambeke 1992), tropical soils can range from highly leached infertile soils to fertile, less weathered alluvial soils. Within the CTFS network, the Forest Dynamics Plots contain a number of different soil types (table 5.1). On all three tropical continents, the predominant soils in the plots are highly weathered Ultisols and Oxisols. Yet, less developed Entisols and Inceptisols—with their weak to incipient horizonation, geologically recent deposits, and often moderate to high fertility—can be found both in the bottomlands and waterways and on steeper surfaces of various plots.

This chapter provides a general overview of soils in the tropics with special reference to the soils of the Forest Dynamics Plots of the CTFS network. The first section briefly introduces soil development in the tropics with particular emphasis on the effect of seasonality and parent material on soil formation. The second section describes the major soil orders of the CTFS network along a gradient of soil development. Dominant soil orders of the CTFS plots such as Ultisols and Oxisols are discussed in the greatest detail.

For more detailed information about the soils of the humid tropics, Baillie (1996) provides a useful introduction to these subjects for ecologists. Brady and Weil (1999) have written the standard text in the U.S. on soils and their classification, with emphasis on agriculture, and Sanchez (1976) has written probably the most cited textbook on tropical soils.

Soil Development

Mineral soil is derived in part from dust and other sources of impurity in rainfall, decomposing organic matter, and the breakdown of rock. In tropical soils, this breakdown may occur through physical mechanisms (temperature changes, erosion by landslips, abrasion by water and wind, the action of roots and animals, etc.) and chemical weathering (hydration, hydrolysis, dissolution, carbonation and other acid reactions, oxidation–reduction, and complexation with organics

56

Table 5.1. Forest Type and Soil Orders of Forest Dynamics Plots

Forest Dynamics Plot	Forest Type	Soil Order(s)*
LATIN AMERICA		
BCI, Panamá	Lowland semievergreen moist forest	Oxisols
Luquillo, Puerto Rico	Subtropical wet tabonuco forest	Oxisols, Ultisols, Inceptisols
Yasuní, Ecuador	Evergreen lowland wet forest	Udult Ultisols, Inceptisols
La Planada, Colombia	Pluvial premontane forest	Humult Ultisols (temperate)
ASIA		
Pasoh, Peninsular Malaysia	Mixed dipterocarp forest	Udult Ultisols, Histisols
Mudumalai, India	Dry deciduous forest	Oxisols, Alfisols
Huai Kha Khaeng, Thailand	Seasonal dry evergreen dipterocarp forest	Oxisols, Inceptisols
Sinharaja, Sri Lanka	Mixed dipterocarp forest	Udult and humult Ultisols, Inceptisols
Lambir, Sarawak Malaysia	Mixed dipterocarp forest	Udult and humult Ultisols, Inceptisols
Bukit Timah, Singapore	Coastal hill dipterocarp forest	Udult and humult Ultisols
Doi Inthanon, Thailand	Lower montane oak–laurel forest	Udult Ultisols (temperate)
Nanjenshan, Taiwan	Evergreen monsoon forest	Udult Ultisols
Palanan, Philippines	Mixed dipterocarp forest	Udult Ultisols
Khao Chong, Thailand	Seasonal wet evergreen dipterocarp forest	Udult and humult Ultisols, Inceptisols
AFRICA		
Ituri, Democratic Republic of Congo	Lowland mixed canopy semievergreen forest and single-canopy dominant evergreen "mbau" forest	Oxisols
Korup FDP, Cameroon	Lowland evergreen forest	Ultisols, Oxisols

*Dominant soil orders, but not limited to the following.

and their interactions). Soil genesis occurs through additions, translocations, transformations, and losses in the soil profile over time.

Most of the nutrients in the system may be organically bound and located near the soil surface as surficial detritus and soil organic matter and within the living components of the forest. The organic layer on the soil generally varies, especially topographically but also with substrate. Nutrients may enter into the system from rock decomposition, floodwater, downslope water movement, and rainwater; they may be lost by erosion and through drainage in solution, called leaching. This variation contributes to spatial variations in pH levels and available nutrients to plants, which may then affect the distribution of plant communities.

Rainfall Seasonality

Of the five soil-forming factors—parent material, living organisms, topography, time, and climate (Jenny 1941)—climate is the fundamental causative factor that separates the development of temperate and tropical soils. Even within the tropics, climatic differences result in the development of varying soils. Seasons, where they exist, generally differ in precipitation rather than temperature because all altitudes experience relatively small seasonal variation in mean daily temperatures. Because the lapse rate—the decline in temperature with altitude—changes little throughout the year, there is more marked altitudinal zonation of soils than in temperate and boreal climates. In temperate and tropical zones, but of greater importance in the latter, "young" and "mature" soils are determined more by the soil formation processes (rates of additions, losses, translocations, and transformations) of organic and inorganic materials than by the actual "age" of the soil. In particular, climate (especially seasonality) plays a critical role in the development of soils in the tropics.

Within the tropics, the wettest lowlands have the greatest diversity of soils due to the continuous impact of high rainfall, which influences soil formation, nutrient leaching, organic matter distribution, and other processes. High rainfall is also associated with high surface erosion rates. In many mountainous regions of the humid tropics, soils are continuously truncated through landslips and thereby rejuvenated. This process can be even more extreme in seasonal climates, where heavy rain following desiccation during the dry season may lead to massive slips, exposing lower soil horizons and even bedrock surfaces. Dramatic examples are the characteristic granite and metamorphic inselbergs of west Africa, coastal Brazil, and south Asia.

At the same time, massive surface erosion and land slippage in the wet tropics lead to the rapid accumulation of sediments in floodplains and estuaries. There, very young soils called Entisols contain profiles that reveal little discernable change since deposition or exposure to weathering. Their structure shows little translocation of minerals or sediments within the profile. These soils represent the initial stage of soil development and are widespread.

Over time, weathering can lead to gradual impoverishment of the soil under certain conditions. In the wet tropics, weathering can impoverish soils on gentle topography and ridgetops or where substrate mineralogy yields porous mineral soils. In the latter category, leaching, the continuous or net downward movement and loss of solutes and clay minerals from the soil profile, is prevalent. Leaching contributes to the translocation of clays and sesquioxides (iron and aluminum oxides). The soils thereafter have less ability to absorb nutrient cations and anions from the soil solution thereby increasing chances for nutrient losses.

Under anoxic, waterlogged conditions, sesquioxides can become reduced to soluble hydroxides and thereby mobilized. In regions with a pronounced annual dry season, they may concentrate at a certain depth in the soil, due to alternating downward and upward movement of soil water and alternating solution and precipitation. When precipitation is less, evaporation from the soil surface leads to capillary upward movement of solutes. The surface horizons may dry out and some minerals, notably oxides of iron and aluminum, may be oxidized and precipitated. Alternate annual wet and dry seasons may even lead to the gradual concentration of iron and aluminum oxides at a certain depth that can eventually form an impervious, cheesy-textured horizon called laterite. Its exposure to air and drying leads to oxidation and irreversible hardening into a solid, brick-like layer (*lateros* is Greek for brick). The material may be sawn when still soft then dried and used, as it was by the Khmers in the construction of Angkor Wat. Because laterite, once oxidized and dry, is extremely resistant to weathering, it may even be found as a fossil soil horizon in present day climates unsuited to its formation. Laterite fragments occur on hilltops within the 50-ha Pasoh Forest Dynamics Plot.

It is doubtful whether the pull of surface evaporation is ever intense enough within an evergreen canopy to form laterite. Deciduousness during the dry season, or opening of the canopy in savanna, greatly intensifies water loss directly from the soil surface. Laterites are associated with deciduous forests and savannas worldwide. They prevent deep rooting and are associated with mixed deciduous forests in contrast to the tardily deciduous or evergreen, often deeply rooting, woodlands and savannas on freely draining siliceous Oxisols low in sesquioxides, such as the cerrado of Brazil and the sal (*Shorea robusta* [Dipterocarpaceae]) forests of south Asia.

Parent Material

The soil parent materials also play an important role in the development of tropical soils. Most of the tropics are occupied either by ancient continental rocks rich in silica—granite and its metamorphosed derivatives—or by the sediments derived from them, which have often been eroded and redeposited. In these soils, sand or short-lattice (1:1 type) clays, which include the mineral kaolinite, become predominant, forming generally nutrient-deficient soils because they lack sufficient cation exchange capacity (CEC, negatively charged sites that attract positively charged cations). In association with high precipitation, nutrient leaching occurs. These soils are acidic because the few available exchange sites for electrical bonding are generally occupied by hydrogen ions. Surface pH may descend below 4.0, creating conditions that favor slow fungal rather than faster bacterial decomposition of organic litter. Partially decomposed acid-reacting "raw humus" then begins to accumulate (humult, see p. 64) even under the hot moist environment of the tropical forest understory. Where raw humus accumulates, tree species are favored, which themselves yield litter, particularly leaves, which are leathery and rich in lignin, acid in reaction, and rich in protein-precipitating phenolic compounds, all of which favor fungal over bacterial decomposition. Surface raw humus is prone to drying out during periodic droughts, further impeding decomposition. Where these soils are freely draining, the surface organic horizon prevents surface erosion, and the process of soil development is accompanied by eventual leaching of all nutrients from the mineral soil; nutrients are instead bonded increasingly in the raw humus, which is characteristically densely matted with fine roots. Due to the dense surface root mat and low surface erosion, these soils are associated with a lower frequency of group windthrows of canopy trees. Instead mortality is usually through die-back of individual trees, leaving small gaps in the canopy. Fast-growing pioneer plant species are poorly represented in such forests where the range of growth rates is lower.

In the extreme case of some sandstones and sea beaches marooned by past surface uplift or changes in sea level, the mineral soil may contain no sesquioxides and little clay. Organic compounds in solution are leached from the surface to the water table, often sitting above an impervious substrate where they are redeposited to form a black, impervious horizon of cemented sand. In such soils, tree roots are confined to the surface organic horizon (Spodosols, see p. 65). Rejuvenation of these soils is solely initiated by catastrophes, especially by means of landslips, which in many landscapes are rare.

Often, these low-nutrient mineral soils are mostly comprised of short-lattice (1:1 type) clays or silts. The porosity of these finer textured soils is lower and generally consists of smaller pore diameters. Rainwater cannot easily penetrate, and soils on slopes may be prone to slippage. Here, rooting is shallow owing in part to anoxic conditions at depth. Such soils are widespread in the Far East and the Amazon Basin, including the Biological Dynamics of Forest Fragments Project near Manaus where these soils predominate.

In these soils, the organic horizon is highly combustible. Immediately following burning, they are productive; but in the long term, once the released nutrients are leached, these upland soils are among the least fertile in the world. It is very difficult to restore these soils to productivity. These humic lowland tropical soils are therefore most widespread in lowland regions lacking a dry season. They hardly occur in Africa and are concentrated in the Malesian archipelago and ancient sandstone landscapes of the Guyana Highlands in the neotropics, where they have been widely burned and now support savanna.

Where the substrate is rich in open-lattice (2:1 type) clay minerals such as those in the smectite group, sites of negatively charged cations (bases) are plentiful and create relatively high pH conditions of 4.5–7.0. These conditions often occur where soils in the humid tropics are juvenile; in such udult soils (see p. 64), bacteria as well as fungi are active in humus formation and litter breakdown, and nutrient release is rapid. Tree rooting may be deep in such soils if not limited by high bulk densities, bedrock, or high water tables; parsnip-like "sinkers" may descend several meters from the main surface roots. Such soils support the fastest growing trees, most of which are pioneers that may produce massive taproots. Some fast-growing species are confined to these soils. Based on these criteria, these soils may be termed fertile. However, such soils are not widespread on a global scale. They are concentrated where volcanic chains such as "the ring of fire" penetrate the humid tropics. In these locations, most of the forest has long since disappeared over these soils because this land is in the highest demand for agriculture.

The parent materials of the Forest Dynamics Plots of the CTFS network have been formed and shaped by the geological processes that created the land surface of their particular regions. In all of the neotropical sites (Barro Colorado Island, Panama; La Planada, Colombia; Yasuní, Ecuador; and Luquillo, Puerto Rico), some or all of the parent materials are volcanic in origin, although at BCI and Yasuní, where the parent material is sedimentary, the volcanic influence is in the form of periodic ash deposits. In Asia and Africa, many sites have igneous rock such as granite as their parent soil material (Ituri, Democratic Republic of Congo; Doi Inthanon, Thailand; Khao Chong, Thailand; Huai Kha Khaeng, Thailand; Mudumalai, India; Pasoh, Malaysia; Korup, Cameroon; Bukit Timah, Singapore), metamorphic rock, including gneiss (Sinharaja, Sri Lanka; Doi Inthanon; Mudumalai; and Korup), and sedimentary rock, including sandstones and shales (Lambir and Pasoh, Malaysia; Nanjenshan, Taiwan; Huai Kha Khaeng, Thailand; Palanan, Philippines; and Mudumalai).

Soil Orders of the Tropics

The U.S. Department of Agriculture has classified tropical soils in a manner that has become widely used around the world (Soil Survey Staff 1975). Soils of the lowland wet tropics are classified according to their level of development, as summarized in figure 5.1. Once developed, the soils of the tropics are dominated by soils of the Oxisol, Ultisol, Alfisol, and Aridisol soil orders. These soil orders will be discussed by their ranking along a gradient of rainfall seasonality (fig. 5.2). In addition, other soil orders found in the tropics will be briefly described in this section.

Ultisols

Found in areas with the least rainfall seasonality and the greatest annual rainfall, Ultisols develop in warm, humid climates on all three of the tropical continents and in temperate regions. This type dominates in southeast Asia and is also abundant in Africa and the neotropics. Ultisols are the predominant soils of the CTFS plot network (table 5.1). Formed from clay mineral weathering in a sandy matrix, they contain a concentration of clays at depth translocated from the upper horizons. Ultisols are often deep, well-drained soils that are red or yellow in color due to an accumulation of iron oxides in the soil profile. These soils possess recognizable horizons and a marked increase in color intensity with depth. Characterized by a low cation exchange capacity, Ultisols are highly leached and generally shallower than Oxisols, because of more rapid surface erosion and

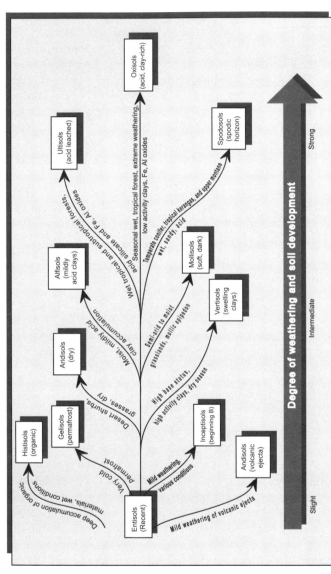

Fig. 5.1. The general degree of weathering and soil development in the different orders of mineral soils and the general climatic and vegetative conditions under which soils in each order are formed. Modified from Brady and Weil (1999). Reprinted with permission of Pearson Education, Inc., Upper Saddle River, NJ.

Ultisols	Oxisols		Alfisols	Aridisols
Water movement more or less constantly downward	Alternate seasonal downward and upward movement of water		Prevailing upward movement of water, with periodic intense reversals,	Intermittent upward water movement
Recognizable horizons; Marked increase in color intensity with depth; leaching of clay minerals and sesquioxides, and accumulation at depth. More or less accumulation of surface humus.	(Weakly seasonal climates) Deep soils with a uniform appearance with depth. Paler at the surface due to the leaching of sesquioxides. Intense yellow to red-brown color. Rapid litter decomposition shallow rooting.	(Strongly seasonal climates.) Accumulation of sesquioxides at a certain depth sometimes forming an impervious horizon; pale in the anoxic zone beneath shallow or deep rooting.	Higher nutrient concentration near surface; organic matter often-mixed in upper mineral horizons. Deep rooting.	Unsorted comminuted substrate; sometimes with sesquioxides or salts accumulate and precipitated at surface.
Variable depth of rooting.				Little or no vegetation.

Rainfall Seasonality

Fig. 5.2. The relationship between seasonality and mean annual rainfall on soil development in the tropics.

truncation, and are therefore often less developed. Nonetheless, Ultisols are the highly weathered, leached soils once mapped as Latosols with the Oxisols and Alfisols. They possess relatively low fertility without artificial inputs.

Within the Ultisols, there are two main suborders found in the tropics—udults and humults. Udult Ultisols have moderate to high clay content and demonstrate rapid litter decomposition and surface pH greater than 4.2. These soils are also often associated with soils of the Oxisol order. In contrast, humult Ultisols are produced by an accumulation of surface acid raw humus that is associated with more intense leaching, either on substrates high in silica and low in clay, or in areas of exceptionally high rainfall. Surface pH is less than 4.2. They possess recognizable horizons and are often found in the highlands.

Oxisols

Increased seasonality, usually where there are at least 4 contiguous months in which evapotranspiration exceeds precipitation, is accompanied by a major shift in the balance between leaching and capillary upward movement of soil solutes. Intensely leached soils become restricted to the most extreme sandy substrates. The predominant soils in these conditions are in the Oxisol order. Oxisols are the soils of the ancient surfaces of all three tropical continents, overlying the same rocks as the leached Ultisols of the more continuously wet regions. Ancient granites and metamorphic rocks are the prevalent parent material. They are the classical tropical soils most often referred to in the general texts: often red in color, but on siliceous soils, often pale, with low organic matter content associated with the high decomposition rates. In these deeply weathered soils, surface pH is nevertheless generally above 4.5, and litter decomposition rates are high. Tree rooting is frequently shallow due to high bulk densities and anoxic conditions during the wet growing season, so the ratio of biomass aboveground to that belowground is high.

Oxisols are highly weathered, mature soils that can reach 20 m in depth with a relatively uniform profile due to indistinct subsurface horizons. The spatial diversity of Oxisols is low compared with the soils of the everwet tropics; their variation is mainly induced locally by differences in drainage and regionally by rainfall seasonality. Oxisols are the predominant soils at the BCI, Mudumalai, Huai Kha Khaeng, and Ituri CTFS sites.

Alfisols and Aridisols

As the length of the dry season increases, the accumulation of soluble ions toward the soil surface also increases. Moisture becomes inadequate to translocate sesquioxides to depth. Such soils of semiarid regions, called Alfisols, support grassland and savanna and are gray-brown due to frequent burning and accumulation of charcoal in the mineral soil below the surface. These soils possess moderate to high native fertility. Eventually, in desert climates where rain falls intermittently and perhaps not annually—though still usually in cloudbursts—soil moisture may have been sufficient to mobilize clays and sesquioxides at the surface to accumulate at a lower depth (e.g., argillic horizon), forming soils called Aridisols. Where present, iron oxides coat exposed sand and stone surfaces creating the red and brown deserts found in Africa, the neotropics, and Asia. Forests are not supported in these climates.

Soils of Unusual Substrates in the Tropics

Soil classifications generally recognize several additional major soil types in the tropics:

- Soils of wet climates, in which all sesquioxides have been leached or where the substrate lacks sesquioxides, leaving a colorless, generally white, mineral profile, are termed Spodosols. Acid-reacting spongy raw humus may accumulate to as much as 50 cm on the surface, where nearly all roots are concentrated. In these soils, the sole source of additional nutrients is from precipitation. Organic acids percolate through the soil and are partially redeposited as insoluble humates that concrete the soil sand to form an impervious horizon that roots cannot penetrate. Sometimes, vertical organic channels found below this horizon, in soil still yellow with sesquioxides, betray former deeper rooting. In the mountains, this concreted pan often contains reoxidized sesquioxides. Lowland tropical Spodosols are widespread on siliceous substrates, especially sandstone and Pleistocene raised marine beaches, in regions with less than 3 dry months including southern Indochina, throughout Malesia, and in Guyana. They carry a forest formation known as "kerangas" in Borneo—occurring at the Lambir site although not within the Forest Dynamics Plot—with a characteristic structure lacking an emergent canopy, except sometimes where conifers are present, and with a characteristic flora including many species confined to them.
- Peats are Histisols in the USDA classification. On extensive floodplains and raised coastal alluvium of the everwet tropics, without seasons of massive flooding, sedimentation at the limits of the more limited flooding is followed by accumulation of organic litter above flood level, where it becomes densely rooted acid peat supporting tall forest. Once initiated, these peat hummocks become domes, eventually as much as 100 km in diameter and 20 m deep, accumulating at rates up to a meter in three centuries. Peat swamps are known only from the aseasonal wet lowlands of the Far East, from Sumatra to New Guinea, where they occupy many thousands of square kilometers in the coastal lowlands. There are

shallow Histosols in the swamp that penetrates the northern edge of the Pasoh Forest Dynamics Plot. The peat swamp flora includes many species endemic to it toward the margins, but becomes increasingly similar to the *kerangas* flora toward the center.

- Soils derived from volcanic ash are called Andisols. These are generally characterized by the open structure of their substrate (2:1 type clays), often imparting extreme through-drainage. In some subhumid tropical regions a horizon may be found that is kandic (i.e., iron and aluminum oxides with clays of low CEC). Ash can vary greatly in silica and clay mineral content and in pH. Soil development varies greatly according to ash composition, and forest species composition resembles that on soils derived from rock substrates of similar chemical composition.

- Montmorillonitic soils or Vertisols are soils dominated by the extraordinary clay molecule montmorillonite, which imparts a black color giving a false impression of high organic content. Sometimes known as black cotton soils, they are the most fertile soils of the tropics. But the open lattice of montmorillonite (2:1 type clays) results in immense expansion during the wet season when the soil has an extraordinary capacity to take up water and solutes; in the dry season, the soil shrinks to such an extent that large cracks may appear at the soil surface into which the sides may slip (self-mulching), while minerals in solution, particularly calcium bicarbonate, may be oxidized and precipitated within the profile. Other than the frequent presence of streaks of limestone at depth, the profile is typically uniform. Derived from basic volcanic rock and lava and prevalent on undulating land and plains adjacent to them, these soils support a forest flora typical of those substrates. In the everwet tropics of Asia, where they are rare, these soils once supported the tallest of all rainforests, with emergents exceeding 75 m. Now, they have been converted to supporting tobacco, bananas, and other nutrient-demanding crops. Windthrows in such forests are frequent, and the gaps are rapidly occupied by fast-growing, often tall, pioneer tree species.

- Soils of limestone are called Mollisols. Pure limestone in the humid tropics dissolves with rain to leave the characteristic karst formations favored by Chinese artists. The jagged tops may accumulate no mineral soil and, though prone to extreme drought and fire, they support an acid organic mantle in undisturbed very wet sites. On the lower slopes and around the base, though, the eroding limestone boulders and outwash create conditions for soils to develop that resemble Vertisols, but in which the clay minerals possess fewer of the expanding/contracting type clays found in the Vertisols and therefore do not self-mulch or mix. These soils are mostly dark and highly productive, where both cation exchange capacity and organic matter content are high. They support a flora similar to that in Vertisols though usually with a distinctive calcicole element.

Soils of the CTFS Network

Within the aseasonal wet tropics, the CTFS network of Forest Dynamics Plots are therefore overwhelmingly concentrated on Ultisols, their truncated form as Inceptisols, and their juvenile floodplain form as Entisols (table 5.1). The Yasuní, Nanjenshan, Palanan, Korup, Khao Chong, and Pasoh plots are sited on low hills bearing leached soils of moderate clay and sesquioxide content and therefore moderate fertility (due to a moderate cation exchange capacity), which are components of soils classified as udult Ultisols. These soils are all set in a landscape in which Entisols and Ultisols prevail at varying stages of development. Inceptisols predominate on the steeper slopes of the hills at Lambir, Sinharaja, and Yasuní where fully developed Ultisols are confined to convex surfaces. Humult Ultisols are found on the low nutrient sandy soils in much of the Lambir plot, on an incipient scale on the granite ridges at Bukit Timah and Khao Chong, and as skeletal soils on the siliceous metamorphic rocks underlying the ridge within the Sinharaja plot.

One of two montane tropical forest plots, the 15-ha Forest Dynamics Plot at Doi Inthanon in Thailand, is set in lower montane (Fagaceae-Lauraceae) forest and is associated with loamy, warm temperate Ultisols. The other montane plot, La Planada above the Colombian Choco, is in an area of exceptional, continuous rainfall, and the humult Ultisols predominate.

A representative range of Oxisols occur in the CTFS plots within the seasonal tropics. Classic Oxisols, with their high clay content, strong granular structure, and deep uniform bright rust-red profile, are the soils of the Forest Dynamics Plots in BCI and at lower altitudes at Luquillo, while the soils at Mudumalai resemble Oxisols morphologically, although no chemical confirmation is yet available. BCI and Mudumalai represent climates with 3 and 6 dry months respectively. Oxisols also predominate on siliceous continental basement and sandstone rocks found in Korup, Ituri, and Huai Kha Khaeng, and occur at Manaus. These soils are low sesquioxide, high silica soils with pale grey-brown profiles, which contrast with those of the typical red-brown Oxisol. Despite their atypical color, they are classified as Oxisols because they lack horizons or visual evidence of leaching and they lack raw humus, which are characteristics of these soils in aseasonal wet climates. These sites also experience dry seasons between 3 (Manaus, Korup, and Ituri) and 6 months (Huai Kha Khaeng).

Acknowledgments

The chapter was reorganized and edited for inclusion in this book by Christie Young. We have greatly benefited from the thorough review and extensive suggestions made by Dan Vogt and from the early advice of Mark Ashton.

References

Baillie, I. H. 1996. Soils of the humid tropics. Pages 256–386 in P. W. Richards. *The Tropical Rain Forest.* Cambridge University Press, Cambridge, U.K.

Brady, N., and R. R. Weil. 1999. *The Nature and Properties of Soils.* Prentice Hall, Upper Saddle River, NJ.

Jenny, H. 1941. *Factors of Soil Formation: A System of Quantitative Pedology.* McGraw-Hill, Mineola, NY.

Sanchez, P. A. 1976. *Properties and Management of Soils in the Tropics.* John Wiley and Sons, New York.

Soil Survey Staff. 1975. *Soil Taxonomy: A Basic System of Soil Classification for Making and Interpreting Soil Surveys.* Handbook 436. U.S. Department of Agriculture, Washington, DC.

Van Wambeke, A. 1992. *Soils of the Tropics. Their Properties and Appraisal.* McGraw-Hill, New York.

6

The Structure of Tropical Forests

Elizabeth C. Losos and CTFS Working Group*

Over half a century ago, H. C. Dawkins noted that despite floristic differences around the world, tropical forests were remarkably similar in structure. Most undisturbed forests, he pointed out, had a basal area of about 32 m^2/ha (cf. Hall and Swaine 1981). This is not the impression one gets when visiting Forest Dynamics Plots across the globe. Rather, visitors are initially struck by the structural differences, not the floristic ones, among the plots. Upon entering the Lambir Hills forest in Sarawak, Malaysia, for example, visitors are awed by the cathedral-like atmosphere created by its towering trees. In contrast, the forest of Luquillo, Puerto Rico, appears diminutive—less than half the height of Lambir, with many tangles and downed logs. The forest on Barro Colorado Island, Panama has enormously stout trees that can surpass 2 m in diameter. In the Amazonian forests of Yasuní, the understory appears crowded, filled with palms. The understory at Sinharaja feels much the same way, although monopodial treelets, not palms, fill the understory.

Which structural differences reflect the climate of the area? Which differences reflect the extremes of weather condition, such as tropical cyclones? Which differences reflect soil or topography? Which ones represent characteristics of the dominant groups—dipterocarps, for example—that happen to prevail in a given region? How might these patterns be distorted by recent or historical human disturbance? In this chapter, we consider two of these questions. What aspects of forest structure are correlated with climate and topography across the network of Forest Dynamics Plots?

Many ecologists have used forest plots to investigate the structure of tropical forests. Although studies vary considerably in plot size and minimum tree diameter cutoff, the forest plots consistently show a tree diameter size-class distribution

---

*For this book, the CTFS Working Group includes: P. S. Ashton, N. Brokaw, S. Bunyavejchewin, R. Condit, G. Chuyong, L. Co, H. S. Dattaraja, S. Davies, S. Esufali, C. Ewango, R. Foster, N. Gunatilleke, S. Gunatilleke, T. Hart, C. Hernandez, S. Hubbell, A. Itoh, R. John, M. Kanzaki, D. Kenfack, S. Kiratiprayoon, J. LaFrankie, H. S. Lee, I. Liengola, S. Lao, E. Losos, J. R. Makana, N. Manokaran, H. Navarrete, T. Ohkubo, R. Pérez, N. Pongpattananurak, C. Samper, Kriangsak Sri-ngernyuang, R. Sukumar, I.-F. Sun, H. S. Suresh, S. Tan, D. Thomas, J. Thompson, M. Vallejo, G. Villa Muñoz, R. Valencia, T. Yamakura, J. Zimmerman.

Table 6.1. Number of Trees per Hectare in Different Size Classes

Plot	P	W	P3	≥1 cm	≥2 cm	≥5 cm	≥10 cm	≥20 cm	≥30 cm	≥60 cm
Luquillo	3548	24	662	4198	2646	1440	874	218	109	11
Sinharaja	5016	22	599	8214	4498	1427	677	284	143	23
Yasuní	3081	21	563	6094	3883	1611	702	219	81	8
Palanan	3379	21	447	4124	2816	1075	537	221	110	28
Bukit Timah	2473	19	506	5959	3283	1005	422	185	102	20
Lambir	2664	19	498	6119	4034	1469	636	234	119	26
Nanjenshan	3582	19	334	12209	7978	3203	1052	187	46	1
Korup	5272	14.5	172	6581	3996	1319	492	192	84	11
BCI	2551	12.5	131	4581	2790	1024	429	156	82	17
Pasoh	1788	11.5	318	6707	3884	1375	531	169	76	15
Ituri-Mixed Forest*	1785	6.5	192	8112	4592	1301	438	127	77	22
Ituri-Monodominant Forest*	1674	4.5	180	6843	3524	940	358	151	98	33
Mudumalai	1250	2.5	67	353	340	315	281	180	102	13
HKK	1476	−1.5	46	1450	1130	738	438	171	82	19

Notes: P is average annual precipitation (mm); W is Walsh's index; P3 is the 3 consecutive months with the lowest average rainfall (mm).

*For Ituri, the two 10-ha Forest Dynamics Plots in similar forest types were combined into a 20-ha dataset.

that is J-shaped. Trees with the smallest diameters are the most numerous and large-girth trees are rare; otherwise they show a range of stem densities, basal area, biomass, and canopy heights. For example, across tropical America, Africa, Asia, and Australia, tree densities range from approximately 350 to 950 trees/ha and basal areas range from about 20 to 70 m²/ha for trees ≥10 cm dbh in mature tropical forests (e.g., Rollet 1974 and 1979; Crow 1980; Uhl and Murphy 1981; Hall and Swaine 1981; Kartawinata et al. 1981; Swaine et al. 1987, Manokaran and Kochummen 1987; Faber-Langendoen and Gentry 1991; Ashton and Hall 1992; Manokaran and Swaine 1994; Phillips et al. 1994; Nadkarni et al. 1995). For Forest Dynamics Plots (excluding subtropical Nanjenshan), tree densities for individuals ≥10 cm dbh range from 281 to 874 trees/ha (table 6.1). Basal areas for trees ≥10 cm dbh range from 24.7 to 40.1 m²/ha (table 6.2).

Climate

Climatic factors may affect some aspects of forest structure (Richards 1952; Gentry 1988; Clinebell et al. 1995; Pitman et al. 2002; Malhi et al. 2002, ter Steege et al. 2003). Malhi and colleagues (2002) speculated, for example, that basal area and stem density tend to vary with the intensity of the dry season. Using data from 32 Amazonian tree plots, they showed that aseasonal forests (fewer than 2 consecutive months with <100 mm precipitation) tend to have higher densities of tree ≥10 cm dbh than more seasonal forests. Basal area, by comparison, remains relatively constant among plots. They hypothesized that this condition occurs because

Table 6.2. Basal Area per Hectare in Different Size Classes

Plot	≥1 cm	≥2 cm	≥5 cm	≥10 cm	≥20 cm	≥30 cm	≥60 cm
Luquillo	38.2	37.9	36.7	34.3	23.3	18.1	5.2
Sinhajara	46.1	45.5	43.1	40.1	33.8	27.1	10.7
Yasuní	33.0	32.6	30.8	27.3	19.5	13.4	4.1
Palanan	39.8	39.6	38.2	36.1	31.0	26.0	15.6
Bukit Timah	34.6	34.2	32.5	30.3	26.7	22.8	11.5
Lambir	43.4	43.0	41.1	37.8	31.6	26.3	13.9
Nanjenshan	36.3	35.7	32.2	23.9	11.4	5.2	0.3
Korup	32.0	31.6	29.3	26.1	21.3	16.1	6.8
BCI	32.1	31.8	30.4	27.8	23.2	19.7	10.8
Pasoh	31.0	30.7	28.9	25.7	20.2	15.9	7.8
Ituri-Mixed Forest*	33.2	32.6	29.9	26.3	21.8	19.4	11.4
Ituri-Monodominant Forest*	37.5	36.9	34.8	32.6	29.5	27.0	17.5
HKK	30.8	30.8	30.4	29.2	24.8	20.7	12.1
Mudumalai	24.9	24.9	24.8	24.7	22.7	18.8	5.8

*For Ituri, the two 10-ha Forest Dynamics Plots in similar forest types were combined into a 20-ha dataset.

the most abundant trees are in small size classes, which make a disproportionately small contribution to basal area.

In this chapter, we test whether climate is a good predictor of forest structure within different size classes. Using linear regression analysis, we examine the correlation between climate and forest structure among saplings, juveniles, small trees, and shrubs (trees ≥1 cm dbh and <20 cm dbh), midstory trees (trees ≥20 cm dbh and <60 cm dbh), and canopy trees (≥60 cm dbh). We test 12 Forest Dynamics Plots in tropical climates ranging from strongly seasonal to aseasonal. "Forest structure" is represented by tree density and basal area within each of these size classes. We also compare three different climate parameters: average annual precipitation (P) to represent total rainfall, 3 consecutive driest months ($P3$) to represent the intensity of the dry season, and Walsh's index (W) as a weighted index of both seasonality and total rainfall. (See chap. 4 for further explanation of climate indices.)

Across the 12 Forest Dynamics Plots, saplings, juveniles, small trees, and shrubs (trees ≥1 cm dbh and <20 cm dbh) tended to be more abundant in wetter and less seasonal forests, but this relationship was not quite significant at the $p = 0.05$ level (table 6.3). For basal area in this same size class, the correlation was significant for $P3$ and Walsh's index but not for average annual precipitation (table 6.3). Apparently, variation in seasonal drought and, to a lesser extent, total rainfall can explain some of the variation in the abundance and volume in this smallest size class. One can imagine that many factors, including pest pressure, edaphic conditions, and plant–plant competition, would dilute the impact of climate during this establishment stage.

Table 6.3. Effect of Climate on Tree Density and Basal Area

Size Class (cm dbh)	Climate Variable	Slope	R^2(%)	F	DF
		Number of Trees			
$\geq$1 and <20	P	0.8779	22.4	2.88	10
$\geq$1 and <20	W	144.9492	24.7	3.28	10
$\geq$1 and <20	P3	4.8246	19.7	2.45	10
$\geq$20 and <60	P	0.0270	63.4	***17.32	10
$\geq$20 and <60	W	4.7024	78.6	***36.64	10
$\geq$20 and <60	P3	0.1563	62.4	***16.62	10
$\geq$60	P	−0.0002	0.1	0.01	10
$\geq$60	W	0.0515	0.4	0.04	10
$\geq$60	P3	0.0027	0.9	0.09	10
		Basal Area			
$\geq$1 and <20	P	0.0014	29.0	4.08	10
$\geq$1 and <20	W	0.2987	50.7	*10.27	10
$\geq$1 and <20	P3	0.0111	50.1	*10.05	10
$\geq$20 and <60	P	0.0015	27.6	3.81	10
$\geq$20 and <60	W	0.3305	57.7	**13.65	10
$\geq$20 and <60	P3	0.0132	65.8	***19.27	10
$\geq$60	P	−0.0002	0.6	0.06	10
$\geq$60	W	−0.0109	0.1	0.01	10
$\geq$60	P3	−0.0008	0.2	0.02	10

*Significant at 0.05 level.
**Significant at 0.01 level.
***Significant at 0.005 level.

Notes: Results of a least-squared simple linear regression of number of trees and basal area as a function of climate parameters for 12 Forest Dynamics Plots. (See plots in Table 6.1, excluding subtropical forest at Nanjenshan and the Ituri-Monodominant plot). W represents Walsh's index, P is the average annual precipitation (mm), and P3 is the average precipitation (mm) for 3 consecutive driest months.

The relationship between climate and midsize trees (trees $\geq$20 cm dbh and <60 cm dbh) for both tree density and basal area was markedly stronger among the 12 Forest Dynamics Plots (table 6.3). For tree abundance, the correlation was unambiguous. Three-quarters of the variation in tree density could be explained through Walsh's index. Annual precipitation was somewhat less correlated, and 3 consecutive driest months less still, but both were strongly significant. For basal area, seasonal drought showed a strong correlation, followed by Walsh's index. Annual precipitation did not significantly correlate with basal area. Thus, for midstory trees, the quantity and consistency of year-round moisture appear to be distinctly related to how many trees the forest can support. Seasonality shows a more pronounced relationship with tree density.

For canopy trees ($\geq$60 cm dbh), both tree density and basal area in the plots appeared to be completely uncorrelated with climate (table 6.3). For example, Mudumalai, with its intense annual drought, had a slightly higher basal area than

Luquillo, with its heavy year-long rains and frequent hurricanes. Meanwhile, Sinharaja experienced essentially the same seasonal rainfall regime as Yasuní, yet it had more than twice the basal area of its Amazonian counterpart. Apparently, once a tree attains about 60 cm in girth—roughly when it reaches the canopy—it escapes the vagaries of rainfall. This lack of correlation should not be surprising. Regardless of the climate in which they occur, very large trees are common in old-growth lowland tropical rainforest (Foster and Brokaw 1982; Clark 1996). Other factors appear to dictate the abundance and size of trees as they grow larger. For example, large trees are rare in Mudumalai, Lambir, Palanan, and Luquillo because the largest trees were harvested during the last half century in Mudumalai and Luquillo while blowdowns and hurricanes have taken their toll on large trees in Lambir, Palanan, and Luquillo.

An examination of size-dependent mortality and recruitment rates of five Forest Dynamics Plots reveals that the dynamics differed among the drier forests (Huai Kha Khaeng and Mudumalai), the semideciduous forest (BCI), and everwet forests (Lambir and Pasoh) (table 6.4). For all forests except BCI, recruitment rates were higher for larger size classes. For BCI, recruitment increased sharply until approximately 60 cm dbh, at which point it declined. However, the drier forests had substantially higher levels of recruitment among the small size classes. Mortality, in contrast, was fairly constant for all size classes among the wetter forests (BCI, Lambir, and Pasoh), ranging between about 1% and 3%. However, for the drier forests, mortality was dramatically higher in the small size classes, declining to the levels of wetter forests around 10 cm dbh. This imbalance between recruitment and mortality rates appears to indicate that these forests, especially Mudumalai, are recovering from past disturbances such as El Niño events, fire, blowdowns, and elephant damage.

Climate may affect forest structure more dramatically through periodic, large-scale events than annual cycles. Tropical cyclones and El Niño events, for instance, both leave their imprints on some forests. Three Forest Dynamics Plots are situated along tropical cyclone paths in the Atlantic and west Pacific Oceans: The Caribbean forest of Luquillo is struck by Category-4 hurricanes approximately twice a century. Nanjenshan forest along the Strait of Taiwan is hit annually by multiple hurricanes (regionally known as typhoons), with Category-4 hurricanes striking almost once a year. Palanan forest, on the other side of the same strait, is less exposed, with Category-3 typhoons hitting every several years. (See also chaps. 13, 32, 34, and 35.) Despite the plots' vulnerability to tropical cyclones, tree density and basal area within Luquillo and Palanan do not diverge notably from other Forest Dynamics Plots (tables 6.1 and 6.2). Nanjenshan, however, is strikingly different: It boasts an exceedingly dense understory, especially on its windward slope (chap. 34). Small-tree abundance is 50% higher than the next most abundant plot. Large trees are

Table 6.4. Number of Trees, Growth, and Mortality by Size Class

Trees	Pasoh			Lambir			BCI			HKK			Mudumalai		
Size Class	N	r	m	N	r	m	N	r	m	N	r	m	N	r	m
≥1 and <2 cm	141,160	0.69	1.73	140,745	0.40	2.10	98,783	0.57	2.55	21,354	1.91	7.44	2100	3.82	21.30
≥2 and <5 cm	125,430	0.78	1.26	130,049	0.51	1.50	89,147	0.84	2.52	20,124	1.72	4.46	2938	3.66	17.40
≥5 and <10 cm	42,208	1.15	1.27	42,605	0.92	1.30	26,530	1.48	2.18	15,975	1.91	2.29	5478	3.23	15.46
≥10 and <20 cm	18,097	1.79	1.50	20,707	1.50	1.32	12,938	2.15	2.08	13,340	2.41	1.81	6162	2.23	4.18
≥20 and <30 cm	4,654	2.56	1.75	5,817	2.11	1.23	3,677	4.17	2.36	4,380	2.77	2.13	3986	2.77	1.05
≥30 and <40 cm	1,764	3.02	1.58	2,676	2.24	1.18	1,633	4.97	2.43	1,723	3.01	2.30	2193	3.24	0.62
≥40 and <50 cm	813	3.33	1.56	1,358	2.61	0.94	919	5.87	2.69	946	3.11	1.77	1370	3.47	0.61
≥50 and <60 cm	495	3.10	1.42	748	3.14	0.95	548	6.23	2.54	583	3.26	1.86	758	3.62	0.62
≥60 cm and <80 cm	441	3.65	2.12	850	3.22	0.78	576	6.07	2.46	492	3.16	1.55	467	4.00	0.43
≥80 and <100 cm	189	3.37	1.79	341	3.77	0.66	265	5.95	1.77	230	3.45	0.94	73	4.47	0.56
≥100 cm	101	3.87	2.19	165	4.03	1.11	325	4.30	1.87	198	3.33	1.76	28	7.29	0.90

Notes: N represents the number of trees in the initial Forest Dynamics Plot census. Growth rate r and mortality rate m are average rates for all census intervals. (See part 7 Introduction for explanation of mortality and recruitment calculations.) Number of censuses used were 4 for Pasoh, 2 for Lambir, 4 for BCI, 2 for HKK, and 3 for Mudumalai. For Mudumalai, trees > 60 cm dbh are only included for the 1988–92 and 1996–2000 census intervals.

almost nonexistent. Canopy height is low on the protected leeward slopes (15–20 m) and a fraction of that (3–5 m) on the windward side. Whether these unusual structural features at Nanjenshan are due to the more frequent typhoons, continual monsoons, its subtropical location, or other aspects of the site will require further study and comparison among these hurricane-exposed forests.

Topography

Forest ecologists have long suspected that the topography of the substrate underlying a forest can influence its aboveground structure. The inclination or slope of forest floor, for example, affects water drainage and the leaching of nutrients (Ashton and Hall 1992; Chen et al. 1997, chap. 5). A sloping plane also provides more surface area on which trees can grow. Thus, it is not surprising that differences in tree density and basal area have been shown across slopes, plateaus, ridges, and valleys, including in Forest Dynamics Plots (e.g., Pasoh: Manokaran and LaFrankie 1987; Nanjenshan: Chen et al. 1997; Lambir: Lee et al. 2002; Sinharaja: chap. 10). Manokaran and LaFrankie (1987) proposed that the impact of topographic variation may differ across size classes, though the form of this relationship is unclear. Rollet (1974) found that Amazonian forests with low slopes had a higher density of trees with small diameters than did forests on plateaus, though this pattern did not hold for tropical forests in Venezuela and Malaysia. Does topography have a greater influence on young trees than old?

Here we test whether topography is a good predictor of forest structure among different size classes. We examine the correlation between topography and forest structure among saplings, juveniles, small trees, and shrubs (trees ≥1 cm dbh and <20 cm dbh), midstory trees (trees ≥20 cm dbh and <60 cm dbh), and canopy trees (≥60 cm dbh) across 10 Forest Dynamics Plots with homogeneous to heterogeneous topography (table 6.5). As a surrogate for forest structure, we use tree density within each size class. We also conducted identical analyses using basal area but do not include the data here because the results are very similar. As a proxy for topography, we use the slope of each 20 × 20 m quadrat in the Forest Dynamics Plots. The slope degree is calculated using the relative elevation of the four corners of each quadrat. (See chap. 15 for further explanation of how to calculate slope degree.) Linear regression analysis is used to estimate the correlation between slope and tree density across the 20 × 20 m quadrats for each plot.

The goodness of fit of the simple linear regression analyses, presented in table 6.5, demonstrate that slope accounted for virtually none of the variation in density of canopy trees, and only slightly more among smaller trees. At best—among small trees in Sinharaja—variation in the slope degree explained just over 10% of tree density variation. In most cases, slope accounted for less than 1% of

Table 6.5. Correlation between Topography and Tree Density

Slope	Pasoh	Ituri-Mixed[†]	BCI	Korup	HKK	La Planada	Yasuni	Palanan	Lambir	Sinharaja
Mean	2.94	3.58	4.96	8.45	10.83	12.28	12.46	17.94	21.53	24.54
Standard Deviation	2.41	2.42	3.70	7.36	6.09	6.19	7.11	8.25	9.64	7.69
Maximum	12.87	14.11	19.40	38.69	45.86	40.31	32.48	48.48	54.02	40.87
Size Class (cm dbh)						R-squared (%)				
≥1 and <20	1.7*	6.7*	0.1	5.6*	0.1	0.0	5.9*	0.2	0.2	10.8*
≥20 and <60	1.4*	1.5*	0.1	2.0*	2.0*	3.0*	6.8*	0.0	0.2	7.3*
≥60	0.7*	0.1	0.6*	0.1	0.8*	0.1	0.8*	0.1	0.9*	2.3*

*Significant at $p = 0.05$ level
[†]Two 10-ha plots averaged.

Notes: Goodness of fit (R-squared) given for simple linear regression analysis of tree density as a function of slope degree. Simple linear regressions calculated for three nonoverlapping size classes for each of 10 Forest Dynamics Plots.

the variation in density. Moreover, to the degree that a small amount of variation was explained, it did not vary consistently with topographic heterogeneity. The flattest, most homogeneous plots (Ituri, Pasoh, and BCI) showed roughly the same patterns as the most undulating (Sinharaja, Lambir, and Palanan). Thus, it appears that slope is not a strong predictor of tree density or basal area, especially among canopy trees, across 10 Forest Dynamics Plots. Future research will reveal whether other factors related to topography, including convexity, aspect, bedrock materials, and soil nutrients, show stronger correlations.

References

Ashton, P. S., and P. Hall. 1992. Comparisons of structure among mixed dipterocarp forests of north-western Borneo. *Journal of Ecology* 80:459–81.

Chen, Z., C. Hsieh, F. Jiang, T. Hsieh, and I. Sun. 1997. Relations of soil properties to topography and vegetation in a subtropical rain forest in southern Taiwan. *Plant Ecology* 132:229–41.

Clark, D. B. 1996. Abolishing virginity. *Journal of Tropical Ecology* 12:735–39.

Clinebell, R. R., II, O. L. Phillips, A. H. Gentry, N. Stark, and H. Zuuring. 1995. Prediction of neotropical tree and liana species richness from soil and climatic data. *Biodiversity and Conservation* 4:56–90.

Crow, T. R. 1980. A Rainforest chronicle: A 30 year record of change in structure and composition at El Verde, Puerto Rico. *Biotropica* 12:42–55.

Faber-Langendoen, D., and A. H. Gentry. 1991. The structure and diversity of rain forests at Bajo Calima, Chocó region, Western Colombia. *Biotropica* 23:2–11.

Foster, R. B., and N. V. L. Brokaw. 1982. Structure and history of vegetation of Barro Colorado Island. Pages 67–82 in E. G. Leigh, Jr., A. S. Rand, and D. M. Windsor, editors. *The Ecology of a Tropical Forest.* Smithsonian Institution Press, Washington, DC.

Gentry, A. H. 1988. Changes in plant community diversity and floristic composition on environmental and geographical gradients. *Annals of the Missouri Botanical Garden* 75:1–34.

Hall, J. B., and M. D. Swaine. 1981. *Distribution and Ecology of Vascular Plants in a Tropical Rain Forest: Forest Vegetation of Ghana.* Dr. W. Junk Publishers, The Hague.

Kartawinata, K., R. Abdulhadi, and T. Partomihardjo. 1981. Composition and structure of a lowland dipterocarp forest at Wanariset, east Kalimantan. *Malaysian Forester* 44:397–406.

Lee, H. S., S. J. Davies, J. V. LaFrankie, S. Tan, T. Yamakura, A. Itoh, T. Ohkubo, and P. S. Ashton. 2002. Floristic and structural diversity of mixed dipterocarp forest in Lambir Hills National Park, Sarawak, Malaysia. *Journal of Tropical Forest Science* 14:379–400.

Malhi, Y., O. L. Phillips, T. Baker, S. Almeida, T. Frederiksen, J. Grace, N. Higuchi, T. Killeen, W. F. Laurance, C. Leaño, J. Lloyd, P. Meier, A. Monteagudo, D. Neill, P. Nuñez Vargas, S. N. Panfil, N. Pitman, A. Rudas-Ll., R. Salomao, S. Saleska, N. Silva, M. Silveira, W. G. Sombroek, R. Valencia, R. Vásquez Martínez, I. Vieira, B. Vincenti. 2002. An international network to understand the biomass and dynamics of Amazonian forests (RAINFOR). *Journal of Vegetation Science* 13:439–50.

Manokaran, N., and K. M. Kochummen. 1987. Recruitment, growth, mortality of tree species in a lowland dipterocarp forest in peninsular Malaysia. *Journal of Tropical Ecology* 3:315–30.

Manokaran, N., and J. V. LaFrankie, Jr. 1987. Stand structure of Pasoh Forest Reserve, a lowland rain forest in peninsular Malaysia. *Journal of Tropical Ecology* 3:14–24.

Manokaran, N., and M. D. Swaine. 1994. *Population Dynamics of Trees in Dipterocarp Forests of Peninsular Malaysia.* Malayan Forest Records No. 40. Forest Research Institute Malaysia, Kepong, Malaysia.

Nadkarni, N. M., T. J. Matelson, and W. A. Haber. 1995. Structural characteristics and floristic composition of a neotropical cloud forest, Monteverde, Costa Rica. *Journal of Tropical Ecology* 11:481–95.

Phillips, O. L., P. Hall, A. H. Gentry, S. A. Sawyer, and R. Vásquez. 1994. Dynamics and species richness of tropical rain forests. *Proceedings of the National Academy of Sciences* 91:2805–09.

Pitman, N. C. A., J. W. Terborgh, M. R. Silman, P. Núñez V., D. A. Neill, C. E. Cerón, W. A. Palacios, and M. Aulestia. 2002. A comparison of tree species diversity in two upper Amazonian forests. *Ecology* 83:3210–24.

Richards, P. W. 1952. *The Tropical Rain Forest.* Cambridge University Press, Cambridge, U.K.

Rollet, B. 1974. *L'Architecture des Forêts Denses Humides Sempervirentes de Plaine.* Centre Technique Forestier Tropical, Nogent sur Marne, France.

Rollet, B. 1979. Application de diverses methodes d'analyse de données à des inventaires forestiers détaillés levés en forêt tropicale. *Oecologia plantarum* 14:319–44.

Swaine, M. D., D. Lieberman, and F. Putz. 1987. The dynamics of tree populations in tropical forest: A review. *Journal of Tropical Ecology* 3:359–66.

Ter Steege, H., N. Pitman, D. Sabatier, H. Castellanos, P. Van Der Hout, D. C. Daly, M. Silveira, O. L. Phillips, R. Vasquez, T. Van Andel, J. Duivenvoorden, A. Adalardo De Oliveira, R. Ek, R. Lilwah, R. Thomas, J. Van Essen, C. Baider, P. Maas, S. Mori, J. Terborgh, P. Núñez Vargas, H. Mogollón, and P. J. Horchler, 2003. A spatial model of tree alpha-diversity and density for the Amazon region. *Biodiversity and Conservation.*

Uhl, C., and P. G. Murphy. 1981. Composition, structure, and regeneration of a tierra firme forest in the Amazon basin of Venezuela. *Tropical Ecology* 22:217–37.

7

Species–Area Relationships and Diversity Measures in the Forest Dynamics Plots

Richard Condit, Egbert G. Leigh, Jr., Suzanne Loo de Lao, and CTFS Working Group*

Species–area and species–individual curves show how species accumulate with increased sample size (Rosenzweig 1995). If the curves have a specific form, they provide a means for predicting the number of species in larger areas with data from small samples. For example, if curves follow a power function, meaning they are linear on a log-log scale, then extrapolation to large areas is straightforward. Condit et al. (1996, 1998) examined species–individual curves in three large Forest Dynamics Plots and showed that the power function does not hold in samples less than about 10,000 individuals. For the three plots, the form of the species–individual curve is much better predicted by the equation for Fisher's α:

$$S = \alpha \ln \left(1 + \frac{N}{\alpha} \right), \tag{7.1}$$

where S is the number of species in a sample of N individuals, and α is a constant, Fisher's α, which is independent of N (see part 5, Introduction). On a log-log scale, the species–individual curve ascends steeply initially, but its slope diminishes progressively thereafter.

Fisher's α is frequently used as a diversity parameter (Fisher et al. 1943; Rosenzweig 1995; Condit et al. 1998). This parameter has a curious history. Fisher et al. (1943) found that in catches of moths at light traps, the distribution of individuals over species followed the log series: The number $S(m)$ of moth species with m sampled individuals apiece is $\alpha x^m / m$, where α is Fisher's α and x depends on N. This same log series also applies in a neutral community of N trees where, at each time-step, a tree chosen at random dies, another is chosen at random to be the

---

*For this book, the CTFS Working Group includes: P. S. Ashton, N. Brokaw, S. Bunyavejchewin, R. Condit, G. Chuyong, L. Co, H. S. Dattaraja, S. Davies, S. Esufali, C. Ewango, R. Foster, N. Gunatilleke, S. Gunatilleke, T. Hart, C. Hernandez, S. Hubbell, A. Itoh, R. John, M. Kanzaki, D. Kenfack, S. Kiratiprayoon, J. LaFrankie, H. S. Lee, I. Liengola, S. Lao, E. Losos, J. R. Makana, N. Manokaran, H. Navarrete, T. Ohkubo, R. Pérez, N. Pongpattananurak, C. Samper, Kriangsak Sri-ngernyuang, R. Sukumar, I.-F. Sun, H. S. Suresh, S. Tan, D. Thomas, J. Thompson, M. Vallejo, G. Villa Muñoz, R. Valencia, T. Yamakura, J. Zimmerman.

seed-parent of the immediately maturing young that replaces the dead tree, and this young has probability v of being an entirely new species (Watterson 1974, Hubbell 2001).

Where the log series applies, the number of trees belonging to a species with m trees apiece is $mS(m) = \alpha x^m$, the total number of trees sampled is

$$N = \sum_{m=1}^{\infty} mS(m) = \sum_{m=1}^{\infty} \alpha x^m = \frac{\alpha x}{1 - x},$$

and $x = N/(N + \alpha)$. Similarly, the number S of species in the sample is

$$S = \sum_{m=1}^{\infty} S(m) = \alpha \sum_{m=1}^{\infty} \frac{x^m}{m} = \alpha \ln \left[\frac{1}{(1 - x)} \right] = \alpha \ln \left(1 + \frac{N}{\alpha} \right).$$

Thus, where the log series is obeyed, so is equation 7.1.

The number $S(m)$ of species on a small plot with m trees ≥ 10 cm dbh apiece often approximates a log series (Williams 1964). For example, the distribution of trees in hectare 7.0 of the Barro Colorado Island (BCI) Forest Dynamics Plot, with 92 species among 424 trees ≥ 10 cm dbh in 1990 approximated a log series with $\alpha = 36.17$, $x = 0.9214$ (fig. 7.1). On the other hand, the distribution of trees ≥ 10 cm dbh by species on the whole 50-hectares plot does not obey the log series: there are too few rare species. Nonetheless, in 1990 this plot contained $S = 229$ species among $N = 21,233$ trees ≥ 10 cm dbh. Solving $S = \alpha \ln(1 + N/\alpha)$ for α by successive approximations (Condit et al. 1998; box 7.1) yields $\alpha = 35.87$.

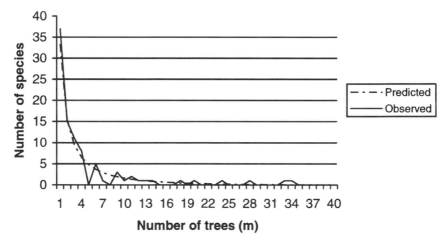

Fig. 7.1. Number of species represented on 1 ha (hectare 7.0) of the BCI Forest Dynamics Plot by m trees apiece, $S(m)$, compared with the predicted $S(m) = \alpha x^m/m$, where $\alpha = 36.17$, $x = 0.9214$.

The equation $S = \alpha \ln (1 + N/ \alpha)$ does not yield a formula for α in terms of N and S. However, we may use this equation to find α by successive approximations. Let $N = 21233$, $S = 229$. Choose $\alpha_0 = 100$. Then
$$\alpha_1 = S/\ln (1 + N/ \alpha_0) = 42.70; \quad \alpha_2 = S/\ln (1 + N/ \alpha_1) = 36.87$$
$$\alpha_3 = S/\ln (1 + N/ \alpha_2) = 36.02; \quad \alpha_4 = S/\ln (1 + N/ \alpha_3) = 35.89$$
Because the sequence of approximations never overshoots the real value, we can check whether $\alpha > \alpha^* = 35.85$ by evaluating $S/\ln (1 + N/ \alpha^*)$, which is 35.86. Thus, α has been shown to lie between 35.86 and 35.89 for $N = 21233$, $S = 229$.

Box 7.1. Finding α from N and S.

Thus the validity of equation 7.1 (as indicated by the constancy of α) transcends the validity of the log series.

In nine different Forest Dynamics Plots, S is predicted well by equation 7.1 (fig. 7.2). That is, α varies rather little with sample size (table 7.1). Fisher's α also depends relatively little on the lower diameter limit of trees sampled, as long as the sample includes over 500 trees (table 7.1). For trees over 20 cm dbh on a single hectare, Fisher's α falls below the values for larger plots or lower diameter limits. Fisher's α is much less sensitive to plot size than either the number of species S or the number S/N of species per individual on a plot (table 7.2).

For those Forest Dynamics Plots where equation 7.1 accurately describes the relation between N and S, curves for different plots should not intersect. Curves for the three most diverse plots, Lambir, Yasuní, and Pasoh, are much higher than those for other sites at all subplot sizes, while the curve for the least diverse site, Mudumalai, falls below all the others for all subplot sizes. Similarly, the curves for Barro Colorado Island, Sinharaja, and Huai Kha Khaeng do not intersect.

Equation 7.1, however, is misleading for some sites. At Mudumalai and Huai Kha Khaeng, Fisher's α increases with plot size even though, for any given plot size, it depends little on the lower diameter limit of trees sampled (table 7.1). Fisher's α is least useful at Korup and Ituri, where it increases rapidly with plot size and decreases markedly with increase in the lower diameter limit of trees sampled. Moreover, the species–individual curves for the two 10-ha mixed-forest plots and the two 10-ha monodominant forest plots at Ituri intersect the curves for Barro Colorado Island, Sinharaja, and Huai Kha Khaeng. Likewise, the Lambir curve crosses those for Pasoh and Yasuní. What causes these failures in equation 7.1?

The relationship between S and N is affected by two factors: the abundance of all the species, and the degree of spatial aggregation of individuals within each species. Species accumulation depends on abundances as follows. If all species are equally abundant, sampling is extremely efficient, and small samples will have many of the species present. In contrast, consider the extreme where one

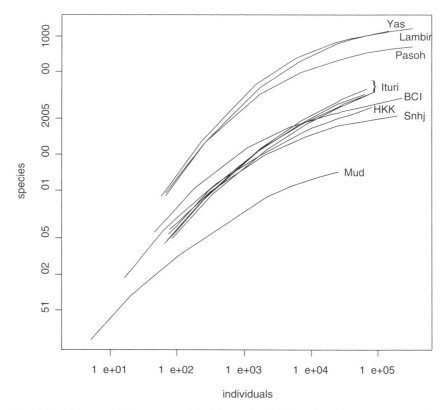

Fig. 7.2. Species accumulation curves at eight different sites, including four different plots at Ituri in Africa. The curves were calculated as described in Condit et al. (1996), using square quadrats of 10, 20, 50, 100, 200, 250, and 500 m on a side (the last two could not be done in the Ituri plots, though, which are 200 × 500 m rectangles). The rightmost point in all lines represents the entire plot, whether square or rectangle. For BCI, the 1995 census was used; for all other plots, the initial census was used.

species is extremely abundant and all others are rare. Sampling is then extremely inefficient—small samples have few species, and the rare species appear slowly. This contrast suggests a simple way to assess the form of the species accumulation curve: calculate the ratio of the number of species found in a small sample to the number found in a large sample. With an even abundance distribution, this ratio will be a relatively high number; but with very uneven abundances, it will be much lower.

We will use this efficiency ratio below to characterize the form of the species accumulation curve at all plots and then to judge which factors are responsible for causing the accumulation curves in figure 7.2 to deviate from the prediction of

Table 7.1. Different Diameter Thresholds on Different-Sized Subplots of CTFS Forest Dynamics Plots

Plot	1 ha			6.25 Ha			25 Ha		
	N	S	α	N	S	α	N	S	α
Sinharaja									
Trees ≥1 cm	8,214	141	24.5	51,337	186	24.3	205,373	205	22.5
≥5 cm	1,427	94	22.7	8,918	151	25.8	35,681	187	25.9
≥10 cm	677	69	19.5	4,234	126	24.3	16,937	167	25.7
≥20 cm	284	48	16.7	1,774	97	22.0	7096	138	24.3
	N	S	α	N	S	α	N	S	α
Yasuní									
Trees ≥1 cm	6,094	655	187	38,089	950	177	152,360	1,104	161
≥5 cm	1,611	413	181	10,071	759	190	40,284	954	175
≥10 cm	702	251	142	4,387	579	179	17,546	820	178
≥20 cm	219	115	101	1,369	347	150	5,476	590	168
	N	S	α	N	S	α	N	S	α
Lambir									
Trees ≥1 cm	6,907	618	165	43,170	955	173	172,679	1,120	160
≥5 cm	1,471	387	174	9,196	759	197	36,782	985	186
≥10 cm	637	247	154	3,979	591	194	15,916	851	193
≥20 cm	234	120	109	1,462	359	157	5,848	620	179
	N	S	α	N	S	α	N	S	α
Pasoh									
Trees ≥1 cm	6,707	495	124	41,918	681	116	167,673	781	106
≥5 cm	1,375	327	136	8,595	556	133	34,380	694	123
≥10 cm	531	206	125	3,319	440	136	13,276	604	130
≥20 cm	169	92	86	1057	260	110	4,229	439	123
	N	S	α	N	S	α	N	S	α
BCI									
Trees ≥1 cm	4,581	169	34.6	28,631	230	34.2	114,524	275	33.8
≥5 cm	1,024	128	39.0	6,401	199	39.0	25,603	242	36.9
≥10 cm	429	91	35.6	2,682	162	37.9	10,728	206	36.1
≥20 cm	156	53	28.5	978	114	33.4	3911	162	34.1
	N	S	α	N	S	α	N	S	α
HKK									
Trees ≥1 cm	1,450	96	23.3	9,064	166	28.9	36,255	218	30.8
≥5 cm	738	81	23.2	4,610	150	29.8	18,441	202	31.7
≥10 cm	438	65	21.3	2,734	130	28.5	10,938	185	31.5
≥20 cm	171	44	19.7	1,070	99	26.5	4,281	150	30.3
	N	S	α	N	S	α	N	S	α
Mudumalai									
Trees ≥1 cm	353	23.9	5.9	2,204	46.8	8.4	8,815	63	9.2
≥5 cm	315	22.9	5.9	1,970	44.9	8.2	7,882	60	8.9
≥10 cm	281	21.4	5.5	1,756	43.1	8.0	7,024	58	8.7
≥20 cm	180	18.0	5.1	1,123	37.1	7.4	4,494	53	8.5
	N	S	α	N	S	α	N	S	α
Korup									
Trees ≥1 cm	6,581	236	48.0	41,128	353	53.0	164,513	441	55.1
≥5 cm	1,319	134	37.6	8,243	241	46.5	32,970	332	51.3
≥10 cm	492	87	30.8	3,074	181	42.1	12,296	261	46.7
≥20 cm	192	54	25.3	1199	126	35.6	4796	200	42.1

(Continued)

Table 7.1. (Continued)

Plot	1 ha			6.25 Ha			25 Ha		
	N	S	α	N^*	S^*	α^*	N^{**}	S^{**}	α^{**}
Ituri (Monodominant Forest)									
Trees ≥1 cm	6,843	159	29.2	26,943	218	32.4	68,431	272	36.0
≥5 cm	940	90	24.8	3,676	145	30.1	9,403	208	37.6
≥10 cm	358	53	17.7	1,382	91	22.1	3,576	159	34.0
≥20 cm	151	22	7.9	605	50	13.5	1,511	100	24.0
	N	S	α	N^*	S^*	α^*	N^{**}	S^{**}	α^{**}
Ituri (Mixed Forest)									
Trees ≥1 cm	8,112	149	26.0	33,275	214	30.6	81,115	252	32.2
≥5 cm	1,301	94	23.2	5,289	154	29.8	13,010	194	32.3
≥10 cm	438	64	20.7	1,751	114	27.2	4,381	151	30.2
≥20 cm	127	36	17.3	508	79	26.1	1,272	115	30.6

Note: Average number of trees N, Number of Species S, and Fisher's α above.

*Subplots are 4.0 ha (200 × 200 m) and are averaged across both 10-ha Ituri Forest Dynamics Plot in each forest type.

**Subplots are 10.0 ha (200 × 500 m) and the average of both 10-ha Ituri Forest Dynamics Plot in each forest type.

equation 7.1 (which we refer to as the neutral prediction; see below). Specifically, we define the efficiency ratio as

$$R = \frac{S_{100}}{S_{25000}} \tag{7.2}$$

where S_{100} is the number of species found in a square quadrat with 100 individuals, and S_{25000} is the number found in a square quadrat with 25,000 individuals. The number 25,000 was chosen because the Mudumalai plot has just more than 25,000 individuals—it is the largest sample available in all plots; 100 was chosen arbitrarily to represent a small sample. Efficiency ratios were quite low (<10%) in all four Ituri plots and higher at BCI (16%; see table 7.3).

The second factor affecting the accumulation curve is spatial aggregation. If all species are uniformly spaced across a plot, then sampling is most efficient. Small samples will include many of the species, and the efficiency ratio will be high. But if species are highly aggregated, sampling is least efficient. In an extreme case—for example, where species occur in clumps that do not overlap—a sample smaller than the clump size would seldom have more than three species.

There are two different mechanisms that can cause aggregation: dispersal limitation and habitat preference (Condit et al. 2000). Both have a similar effect on the species accumulation curve, although habitat preference should have a more pronounced impact because it can lead to less intermingling of species distributions. In the current analysis, though, we cannot distinguish between the two mechanisms, so we only consider the importance of aggregated distributions in general.

Table 7.2. Number N of Trees ≥ 10 cm dbh, Number S of Species among Them, and Fisher's α in the Average 500 × 500 m Subplot, or the Whole Plot if Smaller, and the Average 100 × 100 m Subplot at Each Forest Dynamics Plot

	N	S	S/N	N	S	S/N
		500 × 500 m			100 × 100 m	
Lambir, Sarawak[†]	15,916	851	0.053	637	247	0.388
Pasoh, Malay Peninsula	13,276	604	0.046	531	206	0.390
HKK, Thailand	10,938	185	0.017	438	65	0.148
Mudumalai, South India	7,024	58	0.008	281	21	0.075
Sinharaja, Sri Lanka	16,937	167	0.010	677	69	0.102
Korup, Cameroon	12,296	261	0.021	492	87	0.177
Yasuní, Ecuador	17,546	820	0.047	702	251	0.358
La Planada, Colombia	14,650	179	0.012	586	88	0.150
Barro Colorado, Panama	10,728	206	0.019	429	91	0.212
		500 × 320 m			100 × 100 m	
Luquillo, Puerto Rico	13,988	86	0.006	876	42	0.048
		500 × 200 m			100 × 100 m	
Ituri (Mixed Forest), D.R. Congo	4,381	151	0.034	438	64	0.146

[†]The Lambir tree counts include only identified trees.

How can we separate the impact of the abundance distribution and spatial aggregation on the species accumulation curve? First, we use Hubbell's theory to predict a species–individual curve, which is very nearly equation 7.1 (see above). For each plot, the neutral species accumulation curve was calculated from equation 7.1, after finding α for the full plot (table 7.3), following the method given in Condit et al. (1998). Then the expected efficiency ratio (eq. 7.2) under the neutral model can be calculated with the use of equation 7.1, plugging in $N = 100$ and then $N = 25,000$ (actually, $N = 101$ and 25,344 at BCI, and likewise for the other plots; see table 7.3 notes).

The expected efficiency ratios under the neutral model vary with total species richness. In the three very diverse plots (Lambir, Pasoh, Yasuní), $R < 0.12$; it is about 0.2 in the middiversity plots; and >0.3 at Mudumalai (table 7.3, fig. 7.3). Thus, the efficiency of sampling species varies with species diversity even under the null model. Species are encountered more efficiently in small samples when diversity is low. This is an intuitive result, since species will be more abundant in a less diverse forest, assuming that abundance distributions have the shape that the neutral model predicts.

In all forests, observed species accumulation curves are less efficient—that is, the efficiency ratio is lower—than expected under the neutral model (fig. 7.3). For example, the efficiency ratio of the species accumulation curve at Yasuní would be 9.5% if a forest of the same species richness obeyed the predictions of the null model. The observed ratio is 7.4%. Is this because the abundance distribution differs from the prediction of the null theory, or because species are aggregated

Table 7.3. Number of Species in Square Plots Holding 100 Individuals and 25,000 Individuals in 11 Different Forest Dynamics Plots

	N = 100 Individuals			N = 25,000 Individuals			α	Top 3	Efficiency Ratio		
	Dimen.	Species	Individ.	Dimen.	Species	Individ.	Full Plot	Species	Neutral	Abund.	Observ.
BCI	14.8	35.4	100.7	233.6	224.9	25,343.5	33.9	0.332	0.209	0.179	0.158
Yasuni	13.1	67.2	100.7	207.3	903.5	25,733.3	162.4	0.070	0.095	0.085	0.074
Ituri-Mixed Forest 1	11.5	25.2	100.6	181.1	257.0	26,449.0	43.4	0.540	0.190	0.121	0.098
Ituri-Mixed Forest 2	10.8	21.9	100.0	170.5	253.5	26,975.0	43.7	0.633	0.188	0.104	0.086
Ituri-Mono Forest 1	11.9	23.5	99.9	187.7	280.5	25,203.0	49.0	0.608	0.179	0.097	0.084
Ituri-Mono Forest 2	12.3	22.9	99.2	194.8	268.5	24,765.5	44.3	0.580	0.185	0.102	0.085
HKK	25.0	29.9	101.2	395.8	203.5	25,728.0	31.7	0.258	0.213	0.188	0.147
Lambir	12.2	60.8	100.2	193.1	867.2	24,331.6	152.6	0.070	0.099	0.081	0.070
Mudumalai	44.5	13.9	100.5	1000 × 500	71.0	25,250.0	8.9	0.465	0.312	0.227	0.195
Pasoh	12.2	62.9	100.5	193.1	631.8	25,816.3	100.3	0.074	0.125	0.100	0.100
Sinharaja	11.3	26.7	101.2	178.6	172.8	23,952.8	23.2	0.269	0.240	0.239	0.154

Notes: We found the observed number of species by choosing quadrat dimensions that would include just 100 or just 25,000 trees. At BCI, for example, quadrats of 233.6 × 233.6 m have close to 25,000 individuals. These dimensions were found as the square area that would include 25000 individuals given the density of trees in all 50 ha. The actual number of individuals in the four nonoverlapping quadrats of 233.6 × 233.6 m that can be placed adjacent to one another within the plot, starting in the southwest corner, turned out to average 25,343.5, slightly more than 25,000 because the four quadrats do not encompass the entire plot. It was not necessary to be more precise than this, and in estimates of the expected efficiency ratio (see above) at BCI, N = 25,344 was always used instead of N = 25,000. These are given for each plot. Fisher's α is given for each plot, and the fraction of trees belonging to the three most abundant species. The final three columns give the efficiency ratio from the species individual curve—species in 100 individuals divided by species in 25,000 individuals. First is the efficiency ratio under the neutral assumption, then the ratio of species among 25,000 trees, sampled at random from the whole plot, and finally the observed ratio. Plots are listed geographically, first the two in the New World, then four plots in Africa, and finally five in Asia. The four African plots are all at a single location in the Ituri forest.

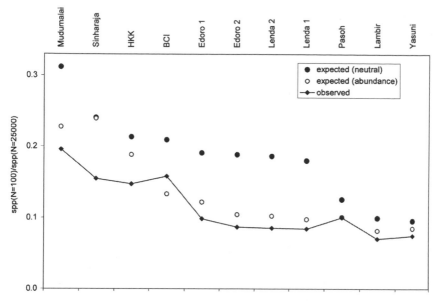

Fig. 7.3. The efficiency ratio of the species–individual curve, observed and predicted from two sets of assumptions. The first prediction is from the neutral theory and is thus based on equation 7.1. The second prediction is from the abundance distribution, following equation 7.3.

spatially? One more calculation allows us to answer this question. The impact of aggregation can be separated from the impact of the abundance distribution by calculating the species–individual curve using random subsets of individuals drawn from the observed species' abundances. This removes any impact of spatial arrangements, and it can be done analytically using the derivation presented in Hurlbert (1971),

$$S_n = S - \sum_{i=1}^{S} \prod_{j=0}^{n-1} \left(\frac{N - a_i - j}{N - j} \right), \tag{7.3}$$

where S_n is the number of species in n individuals, S is the total species in the full sample of N individuals, and a_i is the abundance of the ith species. Equation 7.3 was used to calculate S_{100} and S_{25000} in all plots. (Again, the exact numbers differed slightly from 100 and 25000 and are given in table 7.3).

Figure 7.3 gives the observed efficiency ratios for all plots, as well as the expected values under the neutral theory (eq. 7.1, and as calculated from the observed distribution of abundances for the whole plot using the rarefaction equation 7.3). The impact of abundance can be gleaned from figure 7.3 by comparing the neutral prediction (black circle) with the prediction from the observed abundance distribution (open circle). In most plots, the abundance distribution causes the greatest deviation from neutral, with the largest impact at the four plots in Ituri (Edoro 1 and 2 and Lenda 1 and 2), at Barro Colorado, and at Mudumalai. Table 7.3 shows that these plots also have the most dominated abundance distributions, with a few species occupying a high percentage of the forest (and thus deviating most from the abundance distribution predicted by the null model). At a few sites, the abundance distributions must be close to the prediction of the neutral model, especially at Sinharaja, where changing the abundance distribution had no effect on the efficiency ratio (fig. 7.3).

The effect of species aggregation was generally less important than that of the abundance distribution (fig. 7.3). At Pasoh, aggregation had no impact on the efficiency ratio, and indeed, this forest had a very low measure of species aggregation relative to the other plots. In contrast, at Sinharaja, Huai Kha Khaeng, and Lambir, aggregation had a greater impact than the abundance distribution in determining the efficiency ratio, and sure enough, these three forests had the highest indices of aggregation (Condit et al. 2000).

In summary, Fisher's α would be a perfect diversity index if the species accumulation curve were predicted by equation 7.1; then α would be independent of sample size and could be used to extrapolate species richness to large areas. The deviation between this prediction and the observed accumulation curve is indicated by the efficiency ratio in figure 7.3. The deviation is fairly small for BCI, Pasoh, and Yasuní, but substantial at other sites. In most forests, the failure of α to predict species accumulation can be attributed mostly to the abundance distribution, which deviates from the prediction of the neutral model in being more uneven (more very abundant species and more very rare species). Species aggregation, including dispersal clumps or habitat preferences, produces further deviations from expected, although with a smaller impact than the abundance distribution in most forests.

References

Condit, R., P. S. Ashton, P. Baker, S. Bunyavejchewin, S. Gunatilleke, N. Gunatilleke, S. P. Hubbell, R. B. Foster, L. Hua Seng, A. Itoh, J. V. LaFrankie, E. Losos, N. Manokaran, R. Sukumar, and T. Yamakura. 2000. Spatial patterns in the distribution of tropical tree species. *Science* 288:1414–18.

Condit, R., R. B. Foster, S. P. Hubbell, R. Sukumar, E. G. Leigh, N. Manokaran, and S. Loo de Lao. 1998. Assessing forest diversity on small plots: Calibration using species–individual curves from 50 ha plots. Pages 247–68 in F. Dallmeier and J. A. Comiskey, editors. *Forest Biodiversity Research, Monitoring, and Modeling.* Man and the Biosphere Series. Parthenon Publishing, Pearl River, NY.

Condit, R., S. P. Hubbell, J. V. LaFrankie, R. Sukumar, N. Manokaran, R. B. Foster, and P. S. Ashton. 1996. Species–area and species–individual relationships for tropical trees: A comparison of three 50 ha plots. *Journal of Ecology* 84:549–62.

Fisher, R. A., A. S. Corbet, and C. B. Williams. 1943. The relation between the number of species and the number of individuals in a random sample of an animal population. *Journal of Animal Ecology* 12:42–58.

Hubbell, S. P. 1997. A unified theory of biogeography and relative species abundance and its application to tropical rain forests and coral reefs. *Coral Reefs* 16(Suppl.):S9–S21.

———. 2001. *The Unified Neutral Theory of Biodiversity and Biogeography.* Princeton University Press, Princeton, NJ.

Hurlbert, S. H. 1971. The nonconcept of species diversity: A critique and alternative parameters. *Ecology* 52:577–86.

Rosenzweig, M. L. 1995. *Species Diversity in Space and Time.* Cambridge University Press, New York.

Watterson, G. A. 1974. Models for the logarithmic species abundance distribution. *Theoretical Population Biology* 6:217–50.

Williams, C. B. 1964. *Patterns in the in the Balance of Nature.* Academic Press, London.

8

Floristics and Vegetation of the Forest Dynamics Plots

Peter S. Ashton and CTFS Working Group*

The similarities and differences among plots in species diversity demonstrated in the previous chapter beget further questions: What abiotic and biotic factors contribute to the similarity in tree diversity of Lambir in southeast Asia and Yasuní in the Amazon? Why do species accumulate at a faster rate in the Congolese forest of the Ituri than in the Central American forest of BCI? Fundamentally, however, the signature of each forest depends not just on how many species are packed into an area but also on which ones. The floristic composition of a forest can reveal much about the abiotic and biotic factors influencing the dynamics of that forest, as well as its historical relationships with other forests in the region. In this chapter, we summarize and compare the floristic characteristics of the plots in the CTFS network so far as current data allow, identifying apparent correlates.

Floristic Richness and Climate

In continental settings of the lowland tropics, tree species diversity declines with increasing length and severity of the dry season, as judged by Fisher's α for trees ≥ 1 cm dbh in Forest Dynamics Plots (also see Clinebell et al. 1995). Severity is measured here by the average total rainfall for the driest 3 consecutive months of the year (table 8.1). Yasuní and Lambir, with 563 and 498 mm during their driest quarter, are the most diverse. Pasoh, with 318 mm during its driest quarter, follows. Of the core sites (see table 21.1, p. 32), Mudumalai and Huai Kha Khaeng (HKK) are the least diverse, as one would expect, although HKK has surprisingly high diversity for its climate while Mudumalai's tree diversity is surprisingly low—being less diverse than a deciduous forest in Madagascar with lower rainfall and a

---

*For this book, the CTFS Working Group includes: P. S. Ashton, N. Brokaw, S. Bunyavejchewin, R. Condit, G. Chuyong, L. Co, H. S. Dattaraja, S. Davies, S. Esufali, C. Ewango, R. Foster, N. Gunatilleke, S. Gunatilleke, T. Hart, C. Hernandez, S. Hubbell, A. Itoh, R. John, M. Kanzaki, D. Kenfack, S. Kiratiprayoon, J. LaFrankie, H. S. Lee, I. Liengola, S. Lao, E. Losos, J. R. Makana, N. Manokaran, H. Navarrete, T. Ohkubo, R. Pérez, N. Pongpattananurak, C. Samper, Kriangsak Sri-ngernyuang, R. Sukumar, I.-F. Sun, H. S. Suresh, S. Tan, D. Thomas, J. Thompson, M. Vallejo, G. Villa Muñoz, R. Valencia, T. Yamakura, J. Zimmerman.

Table 8.1. Annual Rainfall (P), Walsh's Index (W), Average Rainfall for Three Driest Months ($P3$), and the Average Fisher's α per ha for Trees ≥ 1 cm dbh (α_1) and for trees ≥ 10 cm dbh (α_{10})

Site	P (mm)	W	$P3$ (mm)	α_1	α_{10}
Yasuní, Ecuador	3081	21	563	187	142
Lambir, Malaysia	2664	19	498	165	154
Pasoh, Malaysia	1788	11.5	318	124	125
Bukit Timah, Singapore	2473	19	506	63	59
Korup, Cameroon	5272	14.5	172	48	31
Palanan, Philippines	3379	21	447	43	37
La Planada, Colombia	4415	22	578	31	29
BCI, Panama	2551	12.5	131	35	36
Ituri-Monodominant Forest*, Dem. Rep. Congo	1674	4.5	180	29	18
Ituri-Mixed Forest*, Dem. Rep. Congo	1785	6.5	192	26	21
Sinharaja, Sri Lanka	5016	22	599	24	20
HKK, Thailand	1476	−1.5	46	23	21
Doi Inthanon, Thailand	1908	2.5	25	19	21
Nanjenshan, Taiwan	3582	19	334	16	14
Luquillo, Puerto Rico	3548	24	662	14	9
Mudumalai, India	1250	2.5	67	6	6

*For Ituri, the two 10-ha Forest Dynamics Plots in similar forest types were combined into a 20-ha dataset.

Notes: For all tables in this chapter, Forest Dynamics Plots are arranged in descending order of Fisher's α per hectare for trees ≥ 1 cm dbh.

longer dry season (Abraham et al. 1996). Walsh's index is a much poorer predictor of tree diversity than total rainfall during the driest quarter. According to Walsh's index, Pasoh should be less diverse than Korup, and Ituri far less diverse than BCI, yet neither proposition is true. (See chap. 4 for description of Walsh's index.) Tree species diversity also appears to rise with increasing climatic unpredictability, notably in variable length of the dry season and the onset of the rains, but this still awaits detailed analysis. In contrast to the belief of Gentry (1982), total annual rainfall has little influence on tree diversity among the set of plots we consider. The highest species diversity scores of Fisher's α for a given seasonality are similar irrespective of continent.

Nonetheless, Fisher's α is surprisingly low in some everwet forests. Luquillo and Sinharaja do not fit the continental relationship between tree diversity and the severity of the dry season. Both plots are in geographical and ecological islands. Luquillo's diversity is low because it is located on an island, Puerto Rico, because all its tree species may have had to cross open sea to get there. Like Sinharaja, the forests nearby in continental India, including Mudumalai, are poor compared with those in similar climates in Far Eastern Asia. There is evidence of a massive reduction in area of the Deccan plate rainforest during the last Ice Age (Ashton 1997), when the Tibetan plateau was a high-pressure area covered by an ice sheet and the monsoon is thought to have ceased. Floristic poverty in Sinharaja, at least, could be due as much to Pleistocene extinctions as to current isolation. The greater richness of Lambir, on the continental island of Borneo, than Pasoh in peninsular Malaysia may likewise be due to Pleistocene change in Pasoh but not Lambir (Morley 2000), though greater habitat diversity at Lambir may account largely or fully for the difference (Lee et al. 2002).

The strikingly similar richness of the hyperdiverse Lambir to that of the Yasuní, Ecuadorian Amazon plot is surprising in view of Yasuní's position within a putative Pleistocene refugium in the world's most extensive rainforest (Whitmore and Prance 1987). Lambir may also be in a refugium (Ashton 1992; Morley 2000), but it was on an island during Pleistocene pluvial periods of extensive rainforest. Although Fisher's α for trees ≥ 1 cm dbh on a 25-ha square at Lambir is virtually the same as that for trees ≥ 1 cm dbh on the 25-ha plot at Yasuní, diversity on a 25-ha square at Lambir of trees ≥ 10 cm dbh is 196, compared to 178 at Yasuní. Gentry (1982) already remarked on the high richness of the understory of Andean forests. At Lambir, diversity of trees ≥ 1 cm dbh and ≤ 10 cm dbh may be much lower than the diversity of trees ≥ 10 cm dbh because understory specialists are squeezed out by the abundance of young dipterocarps (as shown later in table 8.3a). This higher diversity at Lambir of trees ≥ 10 cm dbh may reflect the circumstance that Lambir's forest, which is twice as tall as Yasuní's, offers more volume within which trees can diversify.

In the seasonal wet tropics, dry season severity is an unexpectedly good predictor of species richness. The African plots, contrary to predictions from the relatively impoverished flora as a whole (Richards 1973), are as rich for their climate as the Asian and neotropical plots. The extraordinary dominance of *Gilbertiodendron dewevrei* in the monodominant forest of Congo's Ituri does not reduce tree diversity in a 10-ha Forest Dynamics Plot; tree diversity in the plots is roughly as diverse as that in the nearby mixed plots (see also chap. 12). Huai Kha Khaeng, in Indo-Burma, however, has a climate just slightly wetter than Mudumalai's in peninsular India, but its Fisher's α is three times higher. The HKK plot has no history of human disturbance but includes patches at different stages of long-term succession, presumably following natural catastrophe. At least in the Old World, earlier successional stages in its climate were dominated by deciduous species characteristic of deciduous forest formations of drier climates, but later stages were predominantly evergreen. Huai Kha Khaeng is therefore heterogeneous while Mudumalai contains moist mixed deciduous forest dominated by teak (*Tectona grandis*) with evergreen species infrequent and confined to swales. Mudumalai also experiences more frequent and widespread (anthropogenic) fires than HKK, has higher densities of browsing mammals, has been influenced by logging operations for at least the last two centuries, and is on the site of an ancient agrarian culture. Therefore Mudumalai is probably secondary forest several centuries of age. It is on red lateritic dark Alfisol soils of higher nutrient status than the pale yellow-grey siliceous, highly weathered Oxisol soils of HKK. Mudumalai is also in a region with a poorer flora than that of Indo-Burma (Ashton 1997).

In Asia, it appears to be a general rule that a 50-ha Forest Dynamics Plot captures about half the species known from its community type(s), and about one quarter of the total flora of its climatic and geographic region. This implies that history is important in determining species richness. Pleistocene extinctions, however, appear to be more influential than biogeographic area.

Dominance

Broadly, dominance, here expressed as the relative dominance of the five most common species (tables 8.2a and b), is inversely correlated with species richness in the plots, but there are exceptions. The lowest relative dominance, with respect to either basal area or density, occurs at the aseasonal wet, Yasuní, Pasoh, and Lambir sites. There is, again, surprisingly little difference between Yasuní and the Asian (Pasoh and Lambir) plots despite canopy dominance by ectomycorrhizal Dipterocarpaceae in the latter. The small plot at Bukit Timah has lower richness and higher relative dominance than these three plots. This may be an artifact of its small size, though the peninsular Malaysian coastal hill dipterocarp forest community, in which the Bukit Timah plot is sited, is prone to drought and is thought to be less species rich. Sinharaja is an interesting case: Although it has among the highest annual rainfall of all CTFS sites and is aseasonal, it has comparable relative dominance, whether expressed as basal area or density, to HKK with 6 dry months. It would appear that Sinharaja's relative dominance is somehow a consequence of a poor available flora, there being less than 900 tree species in Sri Lanka's wet zone.

The degree of relative dominance attained may be influenced by the presence of ectomycorrhizal taxa (Malloch et al. 1980). The leading rainforest tree taxa with ectomycorrhizal associations taxa are Dipterocarpaceae subfamily Dipterocarpoideae, which attain family dominance in Asia (tables 8.3 and 8.4), and Leguminosae subfamily Caesalpinoideae (table 8.3a and b), which attain dominance elsewhere, as Davis and Richards (1933–34) first documented. Species of both families attain greatest dominance as expressed by basal area of trees reaching the canopy (table 8.3b), being frequently surpassed in relative density beneath the canopy by species of Rubiaceae, Violaceae, Euphorbiaceae, Annonaceae, Sapindaceae, and sometimes other families. Unlike Dipterocarpaceae, which achieve multiple-species family dominance, as at Pasoh and Lambir, Caesalpinoideae gains family dominance through the overwhelming canopy dominance of single species such as *Gilbertiodendron dewevrei* in the monodominant forest plot at Ituri (see chap. 12). Such single species dominance is confined among dipterocarps to limiting habitats, as the sal (*Shorea robusta*) deciduous

Table 8.2a. Dominance, Relative and Absolute, as Measured by the Basal Area of the Five Leading Species in Forest Dynamics Plots

Forest Dynamics Plot	Basal Area/ha	Basal Area (%)	# Trees (%)
Yasuní	5.08	16.9	3.3
Lambir	8.55	20.0	6.3
Pasoh	4.66	15.0	4.6
Bukit Timah	15.1	44.9	15.1
Korup	10.09	31.7	15.6
Palanan	19.8	51.7	11.2
BCI	7.83	24.4	10.8
Ituri-Monodominant Forest	29.00	78.1	60.4
Ituri-Mixed Forest	15.16	46.3	63.0
La Planada	10.40	35.4	24.5
Sinharaja	19.12	41.9	18.0
HKK	9.94	33.2	14.5
Doi Inthanon	14.36	35.3	11.4
Nanjenshan	12.57	34.6	21.6
Luquillo	18.58	48.6	14.8
Mudumalai	18.97	77.5	65.3

Notes: The number of trees (%) represents the number for the five species with the highest basal area.

Table 8.2b. Dominance, Relative and Absolute, as Measured by Tree Number of the Five Most Common Species in Forest Dynamics Plots

Forest Dynamics Plot	# Trees/ha	Relative density (%)
Yasuní	585	10.1
Lambir	612	9.2
Pasoh	769	11.5
Bukit Timah	1587	27.5
Korup	1978	30.1
Palanan	730	18.7
BCI	1919	41.9
Ituri-Monodominant Forest	4725	70.0
Ituri-Mixed Forest	5313	66.7
La Planada	1531	33.3
Sinharaja	2836	34.5
HKK	576	39.8
Doi Inthanon	810	27.4
Nanjenshan	3811	31.2
Luquillo	2111	50.3
Mudumalai	252	76.7

fire-climax forests on siliceous soils in south Asia and the alan (*S. albida*) forests of the Borneo peat swamps. Caesalpinoideae are better represented in African and neotropical than Asian forests, both in relative density and number of taxa (tables 8.3a and b). At Sinharaja, family dominance of Dipterocarpaceae, especially in basal area, is lower than in other aseasonal wet forests such as Pasoh and Lambir. Sinharaja's dominant species by density is the subcanopy Caesalpinoid, *Humboldtia laurifolia*.

Table 8.3a. Representation of Leguminosae Subfamily Caesalpinoideae and Dipterocarpaceae Subfamily Diptero-carpoideae, as Measured by Tree Density in Nine CTFS Sites

Forest Dynamics Plot	Trees/ha 1–10 cm		Trees/ha ≥10 cm dbh		Trees/ha ≥1 cm dbh		% Tree Density ≥1 cm dbh	
	Caes.	Dipt.	Caes.	Dipt.	Caes.	Dipt.	Caes.	Dipt.
Yasuní	169.5	0	31.1	0	200.6	0	3.4	0
Lambir	32.3	913.5	6.9	119.1	39.3	1032.6	0.6	15.4
Pasoh	103.3	553.3	19.8	64.9	123.0	618.3	1.8	9.2
Bukit Timah	17.0	276.5	2.0	69.0	19.0	345.5	0.3	6.0
Palanan	21.3	335.7	6.4	121.2	27.7	456.9	0.7	11.7
Korup	118.8	0	28.6	0	147.4	0	2.2	0
BCI	183.1	0	15.6	0	198.7	0	4.3	0
Ituri-Monodominant Forest	512.1	0	211.9	0	724.0	0	10.7	0
Ituri-Mixed Forest	1067.3	0	126.7	0	1194.0	0	15.0	0
Sinharaja	897.0	1033.9	1.4	125.5	898.4	1159.4	11.4	14.8
HKK	35.9	26.4	3.9	28.0	39.8	54.4	2.8	3.8
Luquillo	0	0	0	0	0	0	0	0
Mudumalai	16.8	0.3	5.5	0.1	22.3	0.4	6.4	0.1

Table 8.3b. Representation of Leguminosae Subfamily Caesalpinoideae and Dipterocarpaceae Subfamily Dipterocarpoideae for all Trees ≥1 cm dbh, as Measured by Basal Area in Nine CTFS Sites

Forest Dynamics Plots	Basal Area/Ha		Percent Basal Area	
	Caes.	Dipt.	Caes.	Dipt.
Yasuní	0.87	0	2.87	0
Lambir	0.65	17.46	1.52	40.95
Pasoh	2.02	8.74	6.54	28.22
Bukit Timah	0.95	12.98	2.81	38.41
Palanan	0.27	20.21	0.71	52.83
Korup	2.13	0	6.68	0
BCI	1.72	0	5.35	0
Ituri-Monodominant Forest	27.25	0	73.39	0
Ituri-Mixed Forest	13.51	0	41.25	0
Sinharaja	0.35	9.89	0.76	21.84
HKK	0.53	6.3	1.77	21.18
Luquillo	0	0	0	0
Mudumalai	0.19	0	0.76	0.01

Although many Caesalpinoids and Dipterocarpoids are subcanopy trees, the majority (albeit not all) of the tallest trees in lowland forests belong to one of these two ectomycorrhizal taxa. Thus, the tallest documented trees in Asian rainforests are an 84-m *Koompassia excelsa* (Caesalpinoideae) in Sarawak, an 81-m *K. excelsa* in the Malay peninsula (Foxworthy 1927, p. 84) and a 75-m *Dryobalanops lanceolata* (Dipterocarpaceae) at Lambir. In Africa and Latin America the tallest individual trees may not be Caesalpinoids, but the tallest stands are dominated by *G. dewevrei* in Africa and *Mora excelsa* (Caesalpinoideae) in Guyana, while *Dinizia excelsa* (Caesalpinoideae) is the tallest tree in much of Amazonia.

Table 8.4. Abundant Families: Percentage of Trees Comprising 24 Families in 13 CTFS Sites

Family	Yasuní	Lambir	Pasoh	Bukit Timah	Palanan	Korup	BCI	Ituri-Mono.*	Ituri-Mixed*	La Planada	Sinharaja	HKK	Luquillo	Mudumalai
Anacardiaceae	0.4	5.6	2.3	4.1	1.2	1.4	0.2	0.1	0	0	3.1	0.9	0.1	0.2
Annonaceae	3.1	4.3	7.1	3.69	5.1	3.3	7.3	2.1	2.8	0.6	1.3	21.5	0.3	0
Bombacaceae	6.9	0.5	0.4	0.1	0.8	0	1.1	0	0	1.6	3.0	0	0	0.2
Burseraceae	2.2	6.5	5.4	14.4	0	0.1	4.8	0.1	0.1	0	0	0.2	3.3	0.2
Combretaceae	0.2	0	0	0	0	0	0.1	0	0	0	0	0	0.3	28.3
Cyatheaceae	0.2	0	0	0	0	0	0	0	0	6.0	0	0	0	0
Dipterocarpaceae	0	15.4	9.2	6.0	11.7	0	0	3.3	0	0	14.1	3.8	0	0.2
Ebenaceae	0.1	2.7	4.2	7.0	1.0	5.8	0	14.8	2.7	0	1.6	4.8	0	0.7
Euphorbiaceae	4.7	14.3	13.1	10.8	9.9	13.8	1.8	14.8	9.7	4.4	14.8	21.6	2.6	3.2
Flacourtiaceae	2.9	1.9	1.8	0	0.1	9.5	1.7	0.9	2.1	0.2	0.6	1.8	8.8	0.3
Guttiferae	0.5	2.8	3.3	6.7	1.4	1.2	2.8	2.6	0.7	0.7	15.5	5.3	0.2	0
Lauraceae	5.5	3.5	2.2	2.7	8.2	1.0	1.9	0.1	0.1	5.4	2.2	3.1	1.5	0
Leguminosae	12.9	2.1	3.3	0.9	1.7	5.9	7.5	11.3	15.6	5.7	11.0	1.3	2.7	7.7
Lythraceae	0	0	0	0	0	0	0	0	0	0	0	0	0	22.1
Malvaceae	0	0	0	0	0	0	0	0	0	0	0	0	0	6.2
Melastomataceae	3.0	1.5	2.1	1.7	1.3	1.2	3.9	0.2	0.4	8.5	2.0	0.7	2.4	0
Meliaceae	5.3	1.9	2.7	0.9	12.1	0.5	7.3	0.7	0.7	1.5	0.6	2.2	2.6	0
Moraceae	4.2	1.5	0.7	8.9	1.6	0.1	3.5	0.3	0.1	1.6	0.6	0.2	0.4	0.2
Palmae	3.6	0	0.3	0.1	0	0	1.3	0	0	8.4	0	0	8.8	0
Rubiaceae	4.5	4.5	5.8	2.9	3.5	7.0	20.4	2.4	2.8	34.9	7.9	5.3	34.1	4.4
Sapindaceae	0.5	1.1	4.9	0.4	9.2	1.0	1.3	6.6	8.6	0.3	1.2	11.8	0.4	0.5
Sterculiaceae	1.3	1.5	0.8	0.4	2.7	22.0	0.3	46.7	44.3	0	0	1.6	0	5.6
Verbenaceae	0.3	0	0	0	0	0	0	0	0	0	0	0	0	10.9
Violaceae	5.0	0.2	2.5	0	0	9.6	16.8	0.4	0.3	0	0	0	0	0

Notes: The families included in the list are those that comprise more than 5% of trees that comprise more than 5% of trees ≥1 cm dbh in one or more Forest Dynamics Plots. Shaded cells represent more than 5% of the trees for a plot.

*For Ituri, the two 10-ha Forest Dynamics Plots in similar forest types were combined into a 20-ha dataset.

Mudumalai represents a special case among the plots, being exceptionally poor in species and high in relative dominance of leading species, yet without a single dominant species. None of Mudumalai's five leading species are Dipterocarpoids or Caesalpinoids. The low species richness there is correlated with exceptionally low subcanopy density (chaps. 22 and 33), which reflects the impact of fire and browsing by elephants.

Floristics

It is well known that a substantial number of tree families unique to the tropics are pantropical, occurring in Latin America, Asia, and Africa (Richards 1952). Less well known is that a much smaller number of families universally rank high in order of abundance or species richness (table 8.4). The 14 already censused Forest Dynamics Plots at 13 CTFS sites together include 135 families of trees. Only 24 of these families are represented in at least one plot by >5% of trees on that plot ≥1 cm dbh; only 17 are represented in at least one plot by >5% of that plot's species. (For Ituri, the four 10-ha Forest Dynamics Plots in mixed evergreen and monodominant forests are each combined into 20-ha datasets because patterns of family dominance are very similar.) Euphorbiaceae, Leguminosae, and Rubiaceae each comprise more than 5% of individuals (and often >10%) in 9, 8, and 7 out of 14 Forest Dynamics Plots, respectively. In all three continents, Leguminosae and Rubiaceae each include >5% of the trees in at least 1 plot.

As for species composition, Rubiaceae includes >5% of the species in 11 of 13 plots, Euphorbiaceae does so in 10, Leguminosae in 7, and Lauraceae in 6. Dipterocarpaceae are exceptional in ranking consistently high in relative abundance (though much less consistently rich in species) in the evergreen forest plots of one continent, while being absent from the others. The highest relative abundance of a family tends to range from 10 to 20%, though a few plots have particularly high abundance of certain families. Rubiaceae includes 34% of the trees ≥1 cm dbh in hurricane-prone Luquillo; Sterculiaceae includes 46.7% of the trees at monodominant stands of Ituri; Combretaceae includes 28% and Lythraceae 22% of the trees at Mudumalai.

Comparison of family rankings on a plot-by-plot basis (tables 8.4–8.6) identifies historic, geographic, and habitat-related differences. The three Far Eastern mixed dipterocarp forest plots (Pasoh, Lambir, and Bukit Timah) share high similarity in family representation, in both relative abundance and basal area; Pasoh and Lambir are broadly similar in family species richness (the Bukit Timah data are not comparable owing to small plot size), though Dipterocarpaceae ranks higher at the more heterogenous Lambir site. Sinharaja, in the same climate but ecologically and geographically isolated, is distinguished by the relatively high abundance of Guttiferae and *Humboldtia laurifolia* (Leguminosae–Caesalpinoideae);

Table 8.5. Dominant Families: Percentage of Basal Area Comprising 27 Families in 13 CTFS Sites

Family	Yasuní	Lambir*	Pasoh	Bukit Timah	Palanan	Korup	BCI	Ituri-Mono.**	Ituri-Mixed**	La Planada	Sinharaja	HKK	Luquillo	Mudumalai
Anacardiaceae	1.0	6.1	3.0	3.8	1.2	0.7	3.2	0.2	0.4	0	3.9	1.8	0.2	0.4
Annonaceae	1.6	1.6	3.3	1.4	0.8	2.6	2.3	1.0	1.7	0.7	1.5	19.5	0.3	0
Bombacaceae	4.9	0.9	0.6	0.1	0	0	11.4	0	0	4.1	8.8	0	0	0.9
Burseraceae	3.0	6.5	6.2	4.0	0.7	0.9	3.7	0.5	1.7	0	0.1	1.2	13.7	0.7
Cecropiaceae	5.5	0	0	0	0.1	0.4	1.3	0.2	0.5	2.2	0	0	5.2	0
Combretaceae	0.6	0.1	0.1	0	0.1	0.9	0.6	0.1	0.2	0	0	0.2	7.3	28.2
Dipterocarpaceae	0	41.0	28.2	38.4	52.8	0	0	0	0	0	21.6	21.2	0	0
Euphorbiaceae	5.3	6.9	6.9	6.0	6.1	16.1	6.7	3.5	7.8	14.9	5.9	6.7	5.0	1.6
Flacourtiaceae	1.1	0.8	1.0	0.0	0	3.4	1.6	0.4	1.8	0.4	1.2	0.2	8.3	0
Guttiferae	0.6	2.4	2.1	1.3	0.9	1.6	1.0	0.6	0.3	3.2	26.7	2.4	0.0	0
Lauraceae	5.2	3.4	1.2	4.2	2.5	1.6	3.9	0.1	0.1	5.6	3.3	7.4	1.6	0.1
Lecythidaceae	6.0	0.3	0.8	0	0.2	0	1.3	0	0	1.2	0.9	0	0	2.4
Leguminosae	14.9	2.1	8.6	3.5	2.3	9.0	9.9	74.4	42.4	6.3	0	2.5	6.5	15.8
Lythraceae	0	0	0	0	0.1	0	0	0	0	0	0	3.8	0	0
Melastomataceae	1.3	0.8	0.8	1.2	0.3	0.4	0.4	0	0.1	5.6	0.5	0.1	0.4	0
Meliaceae	5.4	0.7	1.3	1.6	5.4	1.1	7.7	1.2	2.2	1.5	2.0	3.3	5.9	0
Moraceae	5.6	1.6	1.6	19.5	1.2	0.1	6.1	0.2	0.5	3.4	0.1	4.9	2.3	1.5
Myristicaceae	4.7	2.3	2.7	1.5	0.6	1.3	3.0	0.4	0	11.4	3.2	0	0	0
Olacaceae	0.6	0.5	0.9	0.4	2.6	6.6	0.6	0.5	1.8	0	0.4	0	0	0
Palmae	8.4	0	0.1	0.1	0.4	0	3.9	0	0	4.5	0	0	17.3	0
Rubiaceae	2.7	1.2	2	2.6	0.4	2.2	9.1	1.3	4.8	14.0	1.7	0.3	2.6	2.8
Sapindaceae	0.7	0.5	2.8	0.1	4.6	0.4	0.3	3.1	5.4	0.1	1.0	5.9	2.2	2.4
Sapotaceae	2.2	1.8	1.6	1.5	1.2	6.7	1.7	1.7	2.5	0.3	3.9	0	6.6	0
Scytopetalaceae	0	0	0	0	0	14.3	0	0	0	0	0	0	0	0
Sterculiaceae	2.0	1.2	2.2	0.3	0.8	8.7	0.8	5.5	7.3	0	0	1.0	0	0.7
Tiliaceae	1.9	1.3	1.0	0.1	5.9	0.2	3.2	0.4	2.5	0	0	0.2	0	6.0
Verbenaceae	0	0	0	0	0	0	0	0	0	0	0	0	0	28.5

Notes: The families included in the list are those that comprise more than 5% of the basal area for trees ≥ 1 cm dbh in one or more Forest Dynamics Plots. Shaded cells represent more than 5% of basal area.

*Palmae was not included in Lambir.

**For Ituri, the two 10-ha Forest Dynamics Plots in similar forest types were combined into a 20-ha dataset.

Table 8.6. Speciose Families: Percentage of Species Comprising 18 Families in 13 CTFS Sites

Family	Yasuní	Lambir*	Pasoh	Bukit Timah	Palanan	Korup	BCI	Ituri-Mono.**	Ituri-Mixed**	La Planada	Sinharaja	HKK	Luquillo	Mudumalai
Annonaceae	3.6	4.6	5.2	7.3	4.2	4.5	3.0	4.3	4.8	0.4	2.0	3.6	1.5	0
Combretaceae	0.8	0.3	0.5	0	0.3	0.2	0.7	0.3	0.3	0	0	0.4	0.7	6.2
Dipterocarpaceae	0	7.4	3.7	2.7	3.6	0	0	0	0	0	6.3	2.8	0	1.5
Euphorbiaceae	3.1	10.6	10.4	10.6	11.0	7.5	4.0	10.1	12.2	3.9	10.2	12.8	5.1	7.7
Flacourtiaceae	2.7	1.7	1.5	0.6	1.2	2.7	5.0	3.3	3.4	2.2	2.4	2.8	5.1	3.1
Guttiferae	1.7	4.3	4.2	4.6	3.9	3.3	3.0	2.0	2.0	1.8	4.9	1.2	0	0
Lauraceae	7.3	6.5	5.9	4.0	5.4	1.6	3.3	1.3	1.4	9.2	5.4	4.0	5.8	0
Leguminosae	9.8	2.0	3.4	3.3	3.0	7.8	12.3	9.8	9.8	4.4	1.5	6.8	4.4	16.9
Melastomataceae	5.3	1.8	1.8	2.7	3.0	2.2	4.0	1.3	1.0	11.4	7.8	0.8	5.8	0
Meliaceae	3.5	4.7	5.3	2.7	6.0	2.2	2.0	4.3	3.1	2.2	2.9	4.0	2.9	1.5
Moraceae	4.6	3.2	2.8	3.0	4.8	1.0	7.0	2.9	3.1	4.4	1.5	6.8	3.6	9.2
Myrtaceae	5.1	4.8	5.9	6.7	6.9	1.0	2.3	0.7	0.7	3.9	6.8	2.8	5.1	1.5
Piperaceae	2	0	0	0	0	0	2.7	0	0	2.6	0	0	5.1	0
Rubiaceae	7.3	4.9	5.5	5.5	4.8	17.6	10.3	11.1	12.2	12.2	8.8	4.0	11.6	6.2
Sapindaceae	1.6	1.6	2.5	1.2	3.9	3.7	2.3	5.9	5.8	1.3	2.4	4.4	1.5	3.1
Sapotaceae	4.9	2.8	1.7	4.0	3.6	1.8	1.7	6.9	5.8	0.4	4.4	0	2.9	0
Sterculiaceae	0.8	1.7	1.7	1.5	1.5	5.7	1.3	3.6	3.4	0	0	2.8	0.7	3.1
Verbenaceae	0.2	0	0	0	0	0	0	0	0	0	0	0	2.2	6.2

Notes: The families included in the list are those that comprise more than 5% of the species of trees ≥1 cm dbh in one or more Forest Dynamics Plots. Shaded cells represent more than 5% of species in a plot.

*Palmae was not included in Lambir.

**For Ituri, the two 10-ha Forest Dynamics Plots in similar forest types were combined into a 20-ha dataset.

seasonally dry Huai Kha Khaeng by abundance of Annonaceae and species richness of Moraceae (especially *Ficus* trees and stranglers) and Leguminosae. The species-poor, low-density Mudumalai plot stands alone.

Data for family basal area are also available for CTFS Forest Dynamics Plots at Nanjenshan, Taiwan, at the margin of the tropics, and for Doi Inthanon, Thailand, in lower montane forest. Both share high rankings of families characteristic of warm temperate Asia including evergreen Fagaceae (dominant in both cases), Lauraceae, and Myrtaceae. Nanjenshan, which experiences more frequent and intense typhoons than other CTFS plots in Asia and which replicates Luquillo in Puerto Rico, gives little indication of disturbance in its tree flora, which is dominated by climax taxa.

The African plots boast high relative density of Euphorbiaceae, Sterculiaceae, and Caesalpinoideae. The forests of Ituri (both the mixed and mono-dominant stands) are distinguished by high dominance of their most abundant species, *Scaphopetalum dewevrei* (Sterculiaceae), which comprise 43% of the trees in the monodominant forest and 46% of the trees in mixed evergreen stands.

The neotropical plots share high relative representation, by both abundance and basal area (for which data from the Biological Dynamics of Forest Fragments Project at Manaus, Brazil, are also available), of Rubiaceae, Leguminosae, and Moraceae, and relatively low representation of Euphorbiaceae. Yasuní and Manaus, representative of Amazonian rainforests, score high also for palms and Lecythidaceae.

Currently available data permit some comparisons of familial and generic similarity using Sorenson's index $C_s = 2j/(a + b)$, where a is the number of species in the first plot, b is the number of species in the second plot, and j is the number of species shared between the first and second plots. This index was used in 12 of the plots at 11 sites (table 8.7), although it was distorted by the variability in sample size. Nevertheless, the 83% and 67% familial and generic similarities between the Pasoh 50-ha and Bukit Timah 2-ha plots, versus 90% and 77% for Pasoh and Lambir, 52-ha plots, suggest that the distortion is manageable. Overall, these comparisons stress the high relative similarity of plots sharing the same historical biogeography, but similarity is notably strengthened when they also share the same climate. The great range of family and genus richness, a real phenomenon, also influences results such that Mudumalai, a mixed deciduous forest with but 26 families and 53 genera, cannot score a close relationship with plots in its own region though they share many of the same taxa. Yet Mudumalai does score 55% familial and 31% generic similarity with Huai Kha Khaeng, a plot in which mixed deciduous and seasonal dry evergreen dipterocarp forests are both present; meanwhile, HKK also scores 74% and 44% similarity with the everwet dipterocarp forest at Pasoh.

Table 8.7a. Sorenson's Similarity Index for Shared Families among 11 CTFS Sites

	No. Families	Luquillo	Sinharaja	Yasuní	Bukit Timah	Lambir	Korup	Pasoh	BCI	Ituri-Mixed	Ituri-Mono.	HKK	Mudumalai
Luquillo	47	1.00											
Sinharaja	46	0.60	1.00										
Yasuní	81	0.67	0.61	1.00									
Bukit Timah	60	0.58	0.73	0.62	1.00								
Lambir	83	0.55	0.68	0.68	0.81	1.00							
Korup	62	0.54	0.57	0.67	0.63	0.64	1.00						
Pasoh	82	0.58	0.69	0.71	0.83	0.90	0.66	1.00					
BCI	58	0.76	0.62	0.79	0.64	0.65	0.69	0.65	1.00				
Ituri-Mixed Forest	48	0.61	0.62	0.62	0.62	0.64	0.76	0.66	0.70	1.00			
Ituri-Monodominant Forest	53	0.57	0.61	0.64	0.65	0.66	0.79	0.67	0.70	0.95	1.00		
HKK	58	0.62	0.59	0.64	0.70	0.70	0.56	0.74	0.63	0.58	0.57	1.00	
Mudumalai	26	0.52	0.52	0.46	0.44	0.41	0.53	0.45	0.57	0.51	0.53	0.55	1.00

Table 8.7b. Sorenson's Similarity Index for Shared Genera among 11 CTFS Sites

	No. Genera	Luquillo	Sinharaja	Yasuní	Bukit Timah	Lambir	Korup	Pasoh	BCI	Ituri-Mixed	Ituri-Mono.	HKK	Mudumalai
Luquillo	102	1.00											
Sinharaja	116	0.17	1.00										
Yasuní	328	0.30	0.12	1.00									
Bukit Timah	169	0.13	0.41	0.10	1.00								
Lambir	287	0.13	0.42	0.12	0.65	1.00							
Korup	235	0.15	0.17	0.13	0.16	0.17	1.00						
Pasoh	288	0.13	0.40	0.15	0.67	0.77	0.18	1.00					
BCI	180	0.40	0.13	0.53	0.09	0.12	0.16	0.13	1.00				
Ituri-Mixed Forest	189	0.18	0.18	0.15	0.21	0.20	0.57	0.23	0.17	1.00			
Ituri-Monodominant Forest	190	0.16	0.19	0.14	0.21	0.21	0.56	0.22	0.17	0.90	1.00		
HKK	161	0.21	0.32	0.14	0.36	0.40	0.16	0.44	0.15	0.23	0.22	1.00	
Mudumalai	53	0.10	0.18	0.07	0.07	0.10	0.09	0.11	0.07	0.13	0.12	0.31	1.00

Overall, it appears that similar trends in community species richness in continental sites, attained under similar rainfall seasonality and soils, reflect similar trends in structure and composition. Whether these apparently universal trends in maximum species richness share a causal explanation remains to be discovered by research within the CTFS plot network.

References

Abraham, J. P., B. Rakotonirina, M. Randrianasolo, J. U. Ganzhorn, V. Jeannoda, and E. G. Leigh Jr. 1996. Tree diversity on small plots in Madagascar: A preliminary review. *Rev. Ecol. (Terre Vie)*, 51:93–116.

Ashton, P. S. 1992. Plant conservation in the Malaysian region. Pages 86–93 in S. K. Yap and S. W. Lee, editors. *In Harmony with Nature. Proceedings of the International Conference on Conservation of Tropical Biodiversity.* Malayan Nature Society, Kuala Lumpur.

———. 1997. South Asian evergreen forests: Some thoughts towards biogeographic reevaluation. *Tropical Ecology* 2:71–180.

Beard, J. S. 1944. Climax vegetation in tropical America. *Ecology* 25:127–58.

Clinebell, R. R., O. L. Phillips, A. H. Gentry, N. Stark, and H. Zuuring. 1995. Prediction of neotropical tree and liana species richness from soil and climatic data. *Biodiversity and Conservation* 4:56–90.

Davis, T. A. W., and P. W. Richards. 1933–34. The vegetation of Moraballi Creek, British Guiana: An ecological study of a limited area of tropical rain forest. Parts I and II. *Journal of Ecology* 21:350–84; 22:106–55.

Foxworthy, F. W. 1927. Commercial timber trees of the Malay peninsula. *Malayan Forest Records 3.* Forest Research Institute, Kepong, Malaya.

Gentry, A. H. 1982. Patterns of neotropical plant species diversity. *Evolutionary Ecology* 15:1–84.

———. 1988. Changes in plant community diversity and floristic composition on environmental and geographical gradients. *Annals of the Missouri Botanical Garden* 75:1–34.

Lee, H. S., S. J. Davies, J. V. LaFrankie, S. Tan, T. Yamakura, A. Itoh, and P. S. Ashton.2002. Floristic and structural diversity of 52 hectare of mixed dipterocarp forest in Lambir National Park, Sarawak, Malaysia. *Journal of Tropical Forest Science* 14:379–400.

Malloch, D. W., K. A. Pirozynski, and P. H. Raven. 1980. Ecological and evolutionary significance of mycorrhizal symbioses in vascular plants (A review). *Proceedings of the National Academy of Sciences* 77:2113–18.

Morley, R. J. 2000. *Origin and Evolution of Tropical Rain Forests.* Wiley, Chichester, U.K.

Richards, P. W. 1952. *The Tropical Rain Forest: An Ecological Study.* Cambridge University Press, Cambridge, U.K.

———. 1973. Africa, the 'odd man out'. Pages 21–26 in B. J. Meggers, E. S. Ayensu, and W. D. Duckworth, editors. *Tropical Forest Ecosystems in Africa and South America: A Comparative Review.* Smithsonian Institution Press, Washington, DC.

Strong, D. R., Jr. 1977. Epiphyte loads, tree falls and perennial disruption: A mechanism for maintaining higher tree species richness in the tropics without animals. *Journal of Biogeography* 14:215–18.

Whitmore, T. C., and G. T. Prance. 1987. *Biogeography and Quaternary History in Tropical America.* Clarendon Press, Oxford, U.K.

PART 3: Habitat Specialization and Species Rarity in Forest Dynamics Plots

Introduction

Elizabeth C. Losos

Theories concerning the origin and maintenance of species diversity in tropical forests have multiplied many-fold over the last three decades (Wright 2002; chap 2). One of the original explanations for the maintenance of tree species diversity—specialization to different habitats or light levels—still occupies a key position among competing theories, though it is no longer considered a comprehensive explanation. This theory predicts that individual species are associated with different habitat types or regeneration niches and as such, species abundance is tied to the degree of specialization to habitat or regeneration conditions. According to this theory, rare species are more specialized to locally rare habitats or infrequent regeneration niches, whereas common species can regenerate under a wider range of conditions.

For two decades now, researchers have examined the habitat associations of species in the 50-ha Barro Colorado Island Forest Dynamics Plot, in part to test the degree of niche specialization. Plot data have revealed that certain groups of species show a strong preference for particular habitat features such as topography and light levels (chap. 2). Pioneers with their affinity to light gaps provide the most unambiguous case. Indeed, in the BCI plot, almost 60% of the species showed an association with slope, flatland, stream, or swamp. Moreover, niche specialists were found more frequently among rare species than common ones (Hubbell and Foster 1986). But researchers have not found that species segregate across finely divided habitats or regeneration space. Rather, broad guilds of species appear to compete more or less equivalently among each other in a particular habitat type (e.g., Welden et al. 1991; chap. 2). Furthermore, evidence from BCI has not supported the prediction of a rare-species advantage. In fact, for trees greater than 1cm dbh, rare species in the Forest Dynamics Plot are at a per capita disadvantage (Hubbell and Foster 1990; Hubbell 2001; chap. 2). How are rare species—comprising about one-third of all species in the BCI plot—sustained in the absence of frequency dependence? Data suggest that immigration may be the key in the BCI plot, and that many rare species are probably represented by sink populations (Hubbell and Foster 1986, 1990). Thus, the niche specialization theory has been only partially supported by evidence from the BCI Forest Dynamics

Plot. (For all chapters in this section, rare species have been defined as those with ≤ 1 tree/ha, cf. Hubbell and Foster 1986.)

Are other tropical forests exposed to similar pressures? One might expect to see a greater niche specialization in forests with more extreme and a greater range of habitats and regeneration conditions. Compared to most other Forest Dynamics Plots, BCI's habitat is relatively homogenous. BCI contains no dramatic soil ecotones as those found in Lambir Hills, no great variation in soil moisture as in Huai Kha Khaeng, no sharp topographic variation as in Sinharaja, and no large windthrows as found occasionally in Pasoh. Comparative data from other Forest Dynamics Plots can shed light on the more general role of niche specialization.

In this part, four chapters provide insights on habitat specialization. Each uses a different analytical technique to tease out the associations among habitat types, species abundance, diversity, and distributions. In part 4, three additional chapters turn toward the contribution of regeneration specialization and diversity distribution.

In the first two chapters of part 2, Valencia et al. and Gunatilleke et al. ask to what degree are species' distributions influenced by habitat features. Data from the first two hectares in the western Amazonian plot at Yasuní reveal that differences do exist between the ridge and bottomland habitats, but that the differences in species composition, species abundance, density, and forest structure are not sharp. For example, at any given distance, the species compositions of two quadrats are only slightly more similar (by a quarter) if they share the same habitat than if they do not. For the common species that could be analyzed, less than 4% were restricted to one of the two habitat types. Thus, Valencia and colleagues concluded that niche specialization could not by itself account for the enormously high species diversity of Yasuní.

Sinharaja along the southern tip of Sri Lanka provides an interesting contrast to Yasuní. Sinharaja has less than a quarter of the number of species found in Yasuní, despite their similar everwet climates and heterogeneous habitats. The low species diversity of Sinharaja appears to be a historical byproduct of the biogeography of the region (Ashton and Gunatilleke 1987, chap. 3). Might there be a greater role for niche specificity in a low-diversity, high-endemism system? Using principal component analysis of the 25-ha Forest Dynamics Plot data, Gunatilleke et al. found that species are clustered by elevation, aspect, and drainage within the Sinharaja site. Dominant species tend to be restricted to one cluster or the other, suggesting microtopography specialization. While direct comparison with Yasuní is not possible due to the different analytical techniques used for the two sites, it appears that Sinharaja may accommodate greater habitat segregation among its species than Yasuní. Clearly, further standardized analyses between the two sites need to be carried out to compare the relative strength of niche specialization.

Bunyavejchewin et al. address the relationship of rarity with niche specialization. In the seasonal dipterocarp forest of Huai Kha Khaeng (HKK) in Thailand, more than 60% of the species in the plot are considered rare, a higher percentage than found at virtually any other Forest Dynamics Plot. The distribution of most of these rare species shows no significant association with slopes, flat land, streams, or swamp. Common species comprise most of the habitat specialists in the 50-ha plot in HKK (in striking contrast to Barro Colorado Island, Hubbell and Foster 1986). In this case, niche specialization is linked to species success, as measured through abundance. The authors suggest that rarity at HKK arises not as a result of niche specialization but as a consequence of historical events.

Monodominant forests in the Congo Basin, where the mbau (*Gilbertiodendron dewevrei* [Leguminosae]) dominates the canopy, provide an excellent opportunity to better understand why species are rare and the role that they play contributing to species diversity. Makana et al. show that, in the monodominant Ituri Forest Dynamics Plots, about 55% of the species are rare, just slightly lower than that found in HKK. This high level of rarity is to be expected, given the crowding out by mbau. What is surprising is that in the complementary Forest Dynamics Plots in the nearby mixed evergreen forest—which is not dominated by one canopy species—an equally high proportion of rare species are found. Indeed, the authors showed that for areas larger than 1 ha, species richness was comparable between these two forest types. Though habitat associations have not yet been examined, it is worth noting that the Ituri is relatively homogenous, even more so than BCI. Only swamps break up the flat terrain. Given the contrast between the mixed and monodominant stands, this forest will provide a fascinating test of the role of niche specialization and species diversity.

Does high species diversity result, at least in part, from habitat heterogeneity? Evidence presented in this section, from the hyperdiverse Amazonian forest of Yasuní to the relatively low-diversity forests in Sinharaja and HKK, suggest that a significant portion of species are habitat specialists and this does contribute to the diversity levels found in these forests. Yet this niche differentiation is clearly just one contributing factor that competes with other forces such as regeneration specialization, recruitment limitation, sink populations, species drift, and history. Understanding habitat associations is just the first step for each of these Forest Dynamics Plots. For the network of plots as a whole, the important next step—currently underway—is to compare habitat associations across sites to assess the contribution of niche specialization at a broader scale.

References

Ashton, P. S., and C. V. S. Gunatilleke. 1987. New light on the plant geography of Ceylon I. Historical plant geography. *Journal of Biogeography* 14:249–85.

Hubbell, S. P. 2001. *The Unified Neutral Theory of Biodiversity and Biogeography.* Princeton University Press, Princeton, NJ.

Hubbell, S. P., and R. B. Foster. 1986. Commonness and rarity in a neotropical forest: Implications for tropical tree conservation. Pages 205–31 in M. E. Soule, editor. *Conservation Biology: The Science of Scarcity and Diversity.* Sinauer Associates, Sunderland, MA.

———. 1990. Structure, dynamics, and equilibrium status of old-growth forest on Barro Colorado Island. Pages 522–41 in A. H. Gentry, editor. *Four Neotropical Rainforests.* Yale University Press, New Haven, CT.

Welden, C. W., S. W. Hewett, S. P. Hubbell, and R. B. Foster. 1991. Sapling survival, growth, and recruitment: Relationship to canopy height in a neotropical forest. *Ecology* 72:35–50.

Wright, S. J. 2002. Plant diversity in tropical forests: A review of mechanisms of species coexistence. *Oecologia* 130:1–14.

9

Tree Species Diversity and Distribution in a Forest Plot at Yasuní National Park, Amazonian Ecuador

Renato Valencia, Richard Condit, Katya Romoleroux, Robin B. Foster, Gorky Villa Muñoz, Elizabeth C. Losos, Henrik Balslev, Jens-Christian Svenning, and Else Magård

Introduction

The local species richness of woody plants is extremely high in northwest Amazonia (Pitman 2000; Pitman et al. 2001). In a single hectare, trees with a diameter at breast height (dbh) ≥ 10 cm may number over 300 species (Gentry 1988; Valencia et al. 1998) and shrubs and treelets may contribute greater than 50% of the vascular plant species in 0.1- and 1.0-ha plots in western Amazonia (Duivenvoorden 1994; Balslev et al. 1998). In this chapter, we present results from a 2-ha plot that comprises 787 species of trees and shrubs ≥ 1 cm dbh.

How so many species can coexist in a small area is still an open question. Many theories explain the high diversity found in small portions of tropical rainforests (reviewed in Wright 1999, 2002). One of these, the niche differentiation hypothesis, suggests that because each species uses resources in a different way, coexistence occurs when resources vary spatially, and each species occurs where it is a superior competitor (Wright 1999). Consistent with the niche differentiation hypothesis, Tuomisto and Ruokolainen (1994) found that fern and Melastome species have distributions restricted to different topographic habitats in Amazonian forests. In this paper, we analyze the patterns of shrub and tree diversity in 2 ha that contain two main topographic habitats: ridge and bottomland. We then asked two questions related to niche diversification: 1) How much diversity can be attributed to niche-partitioning across the habitat boundary? 2) How much diversity can be attributed to the partitioning of forest by different life forms (shrubs, treelets, canopy trees)?

Study Area

Our study site is situated near the Yasuní Biological Research Station in a 50-ha Forest Dynamics Plot in eastern Ecuador (see chap. 38). The 50-ha plot is located within a mosaic of bottomland—including small streams, adjacent flat sections, and in one area a small swamp—and ridges.

Methods

The Yasuní 50-ha plot follows the standardized methodology of the Center for Tropical Forest Science (CTFS; Condit 1998; part 7, introduction). In Yasuní, a 50-ha Forest Dynamics Plot (500 × 1000 m) was topographically gridded, and all free-standing woody plants ≥1 cm dbh were mapped. Data entry and initial identifications have been completed for all trees in the western 25 ha of the plot and for trees ≥10 cm dbh in the eastern 25 ha of the plot.

For this paper, just two of the 50 ha are used, though since this paper was written, the tree database has become available for the first 25 ha. The two contiguous 1-ha study plots are located between 100–200 m east and 80–180 m north of the southwestern corner of the 50-ha plot. They include most of the entire plot's topographic variation: one plot is mostly bottomland, and the adjacent plot is a ridge with slopes and a hilltop (see fig. 38.1 in chap. 38).

The taxonomy of the 50-ha plot was far more difficult than first anticipated and took more than 5 years for 25 ha to be identified. Three of us (RF, KR, and GV) identified the species and morphospecies and standardized the taxonomy. We collected voucher specimens of all species and morphospecies and all forms of variation within them. A complete set of the collections is deposited in the Yasuní Biological Research Station and the Herbarium of the Pontifical Catholic University of Ecuador in Quito. In addition, duplicates of most of the specimens can be found in the Field Museum of Natural History in Chicago. Around 220 species were identified or confirmed by specialists (see acknowledgments in Romeroux et al. 1997), and about 2500 collections have been sent for identification to herbaria abroad. In the 2 ha, 14 trees have not been identified and 135 trees died before identification in a period of about 18 months. These trees are excluded from the 11,514 reported here.

Forest structure and species composition between the ridge and the bottomland were compared using 20 × 20 m quadrats. Ridge quadrats were those crossed by more than one 1-m contour line (the topographic map of fig. 9.2 shows 1-m contour lines), and bottomland quadrats were crossed by no more than one 1-m contour line. Nine quadrats including both slopes and bottomland were eliminated. We used the similarity index of Sorensen, weighted by individuals, and compared all possible pairs of quadrats within and across habitat types. The Sorensen index with cover (Babour et al. 1986) is defined as $\frac{2}{N} \sum_{i=1}^{S} n_{i_{min}}$, where $n_{i_{min}}$ is the number of individuals of a species i in the quadrat where it is less abundant, S is the total number of species i in a quadrat, the summation is across all species in the two quadrats, and N is the total number of individuals in the two quadrats. We used Fisher's α as a diversity index, where α is defined by $S = \alpha \ln(1 + \frac{N}{\alpha})$, where $S =$ the number of species and N the number of individuals (Fisher et al. 1943).

To see which life form contributes most to species diversity, we grouped the species into four life forms, defined by the maximum height they usually attain: shrubs (<5 m), treelets (5 to <10 m), midcanopy trees (10 to <20 m), and tall canopy trees (≥20 m). This follows the growth forms used in publications from the Barro Colorado Island (BCI) Forest Dynamics Plot (e.g., Hubbell and Foster 1986). The typical maximum height of 273 Yasuní species was obtained from florulas and taxonomic treatments (i.e., Berg et al. 1990; Brako and Zarucchi 1993; Croat 1978; Pennington 1990; Pennington et al. 1981; Prance 1979; Sleumer 1980; Rohwer 1993; Vásquez 1997); that information was confirmed by our own field experience. Life form information for an additional 461 species was assessed from field experience alone. In total, we assigned a life form category to 741 species and morphospecies (94.3% of the total). Only 46 species, mainly those with few individuals of small diameter (<5 cm), were not included in this analysis.

Results and Discussion

Exceptional Tree Diversity

The entire 25-ha dataset is still being revised, yet nearly every one of the 152,353 individual trees and shrubs has been assigned a morphospecies; another 4215 trees died before their identification over a 55-month period. There are a preliminary total of 1104 taxa represented in these 25 ha, which is higher than all the other Forest Dynamics Plot in the CTFS network except at Lambir Hills National Park in Sarawak, Malaysia, which has 1115 species in 25 ha (average of two 25-ha sections of the 52-ha plot). This is considerably more than the number of tree and shrub species in the entire North American flora north of Mexico. The concentration of species at Yasuní is remarkable in another way as well: Nearly half the previously recorded 2488 freestanding woody species (i.e., shrubs, treelets, and trees) of the Amazon region of Ecuador (Jørgensen and León-Yánez 1999) are in the plot. Currently, a careful analysis of this diversity by habitat and life form is restricted to the 2-ha subset of data that has been assembled and verified in the field.

The 2-ha study plot had 11,514 trees and 787 species ≥1 cm dbh, and 1329 trees and 351 species ≥10 cm dbh per hectare. Previously, Romoleroux et al. (1997) reported 638 individuals that are not considered in this study because they are lianas, regrowths of other trees, or below the diameter cutoff considered in this study. The earlier publication also reported a slightly higher number of species; taxonomic work since that paper was finished led us to collapse some of the variants into single species or morphospecies.

Habitat Differences

Forest structure differed between ridge and bottomland (table 9.1) In all dbh categories, tree density was higher on the ridge; this pattern was most conspicuous

Table 9.1. Structure and Species Diversity in Two Hectares of the Yasuní Forest Dynamics Plot

	Ridge				Bottomland			
dbh	Number Trees	Basal Area (m^2)	Number Species	Fisher's α	Number Trees	Basal Area (m^2)	Number Species	Fisher's α
$\geq$1 cm	6,304	36.8	653	183.0	5,210	27.6	645	193.8
$\geq$10 cm	725	31.2	255	140.0	604	22.2	234	140.2
$\geq$30 cm	105	16.6	64	69.6	62	9.1	46	80.7

Notes: Each hectare represents one of the main topographic habitats: ridge or bottomland.

for large trees ($\geq$30 cm dbh), which had nearly twice the density on the ridge. Basal area was also substantially higher on the ridge. This is consistent with the observation that the canopy height tends to be lower in the bottomland of the plot (Svenning 1999).

Species richness was also higher on the ridge, again especially in the largest dbh class (table 9.1). Since tree density was considerably higher on the ridge, we tested whether the larger number of species was due solely to the larger sample there. Fisher's α is effective for this test, since it accurately predicts the accumulation of species as a function of the number of individuals sampled (Condit et al. 1996). It turns out that α is actually higher on the bottomland for dbh $\geq$1 and dbh $\geq$30 cm (table 9.1), suggesting that equal samples in the two habitats would lead to slightly more species on the bottomland than the ridge. The full 25-ha dataset will allow us to test this assertion more fully.

Species composition also differed between ridge and bottomland. The similarity between pairs of 20 × 20 m quadrats declined with distance, by about 50% within 200 m (fig. 9.1). At a given distance, quadrats in different habitats were about 25% less similar than two quadrats in the same habitat (fig. 9.1). Svenning (1999) also found differences between the palm communities of the two habitats.

However, most species grew in both habitats. There were 649 species with $\geq$2 individuals in the 2 ha, and 137 were restricted to one habitat; but of course many of the rare species would be restricted by chance alone. Of those species with fairly large samples ($\geq$10 individuals), just 10 (3.5% of 288) grew exclusively in one of the two habitats: 6 on the ridge and 4 in the bottomland. Thus, although there are quantitative differences between the habitats, most species occur in both.

We illustrate distributions of individual species in figure 9.2. *Iriartea deltoidea* (Palmae), a canopy palm, occurred abundantly in both habitats. *Geonoma aspidifolia* (Palmae) and *Rinorea lindeniana* (Violaceae) are two abundant understory species associated with the ridge habitat, and *Rinorea viridifolia* (Violaceae) was only on the bottomland and transitional slopes. Rigorous statistical tests of habitat preference are currently being carried out on many more species as data from 25-ha and 50-ha plots become available.

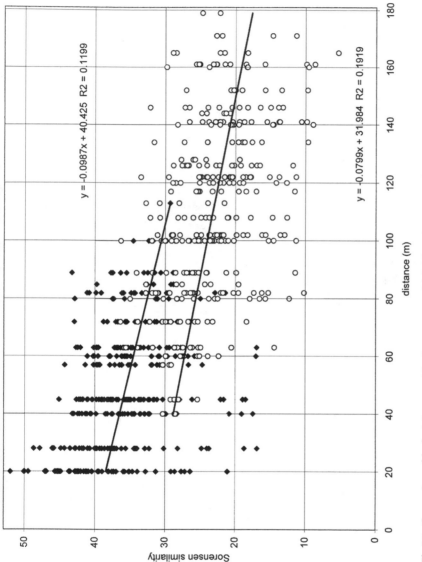

Fig. 9.1. Comparison of the floristic composition in two topographic habitats. Sorensen similarity calculated between pairs of quadrats is shown as a function of the distance between the quadrats. The similarity between all pairs of quadrats on the ridge is indicated by circles, and the similarity between all pairs of quadrats with one on the ridge and one in the bottomland is indicated by diamonds. Linear regression lines are given for each comparison, highlighting how pairs of quadrats in the same habitat are more similar than pairs on opposite habitats, when distance is controlled.

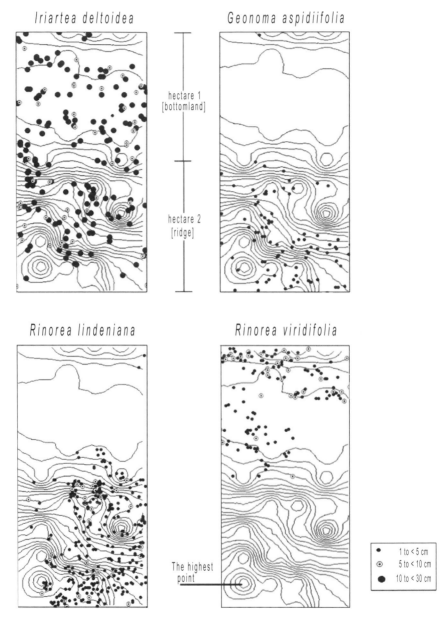

Fig. 9.2. Distribution of four species in a section of 2 ha of the Yasuní Forest Dynamics Plot. The circles represent individuals of different diameter size classes.

Apportionment of Diversity Among Life Forms

Is the high diversity at Yasuní concentrated in particular life forms? Are there particularly many species of shrubs and treelets as Duivenvoorden (1994) suggested for the Colombian Amazon? To evaluate this, we compared the Yasuní forest to the closest lowland sample available where complete censuses include shrubs and treelets—the Barro Colorado Island (BCI) Forest Dynamics Plot in Panama. Both Yasuní and BCI are lowland moist neotropical forests. The BCI plot, however, has 4 dry months per year, whereas Yasuní has no dry months and rains are evenly distributed throughout the year (see chaps. 24 and 38). Yasuní has more than twice as many species in 2 ha as BCI's plot has in 50 ha (see chap. 38). To make a precise comparison, though, we extracted two subsamples from BCI that match the size and shape of the 2-ha sample at Yasuní. Each subsample was 100 × 200 m and consisted of 1 ha (approximately) on the slope habitat, and one of the plateau; we averaged the results from the two 2-ha subsamples.

Most of the species in both plots were encountered in small diameter classes. Of 787 species in 2 ha at Yasuní, 93.8% were found in the 1.0–9.9 cm dbh class, compared to 92.1% of 195 species at BCI. In both plots, the proportion of species found in a dbh class decreased with increasing dbh, but more drastically at Yasuní (fig. 9.3). Thus, BCI has proportionally more large-sized species than Yasuní.

This is seen clearly by classifying species by life form (fig. 9.4). Nearly half the species at Yasuní are midsized trees, whose maximum height is 10–20 m, a much higher fraction than at BCI (Fig. 9.4a). Conversely, BCI is proportionally much richer in large trees (fig. 9.4a). Total tree density shows the same pattern, with Yasuní showing a much higher density of individuals of midcanopy trees compared to BCI (fig. 9.4b).

However, in absolute numbers, all life forms are far richer in species at the Yasuní plot. The Yasuní plot has 4 times the diversity of shrubs and treelets than BCI (259 vs. 63 species in 2 ha), nearly 6 times as many midcanopy trees (327 vs. 59 species), and about twice as many tall canopy trees (155 vs. 73 species). Thus, although species richness at Yasuní peaks among midsized trees, it is high in all life forms. The greater number of midstory trees could reflect (1) a generally greater shade tolerance in a more humid aseasonal climate, which allows more species to adapt to this niche (Wright 1992), (2) a higher mortality rate of tall canopy trees at Yasuní, which opens the high canopy so the midcanopy gets more light (though data are not yet available on this), and (3) a more heterogeneous topography at Yasuní, which allows more light to reach the understory and midcanopy compared to BCI.

Gentry (1982, 1990) believed that there is a higher diversity of shrubs in Central America than in the Amazon region and a higher diversity of trees in the Amazon region than in Central America. Our comparison partly bears this out. In contrast,

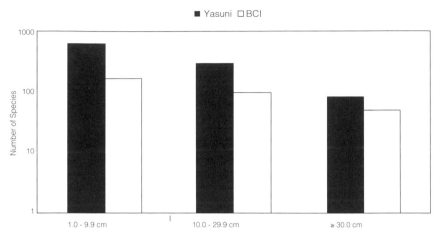

Fig. 9.3. Species richness by dbh class at a 2-ha plot in Yasuní National Park and a 2-ha sample from the Forest Dynamics Plot in Barro Colorado Island, Panama. Each bar gives the number of all species found in the plot that occur in the given dbh class. Nearly all species at both plots occur as saplings (<10 cm dbh), since this class includes juveniles of the largest species as well as adults of the small shrubs and treelets. The latter do not reach 10 or 30 cm dbh, and thus many fewer species occur in the larger size classes. The numbers are given on a logarithmic scale to facilitate direct comparison between the two plots.

our results contradict the idea that the forest understory contributes most to the great species richness found in the Amazon region (e.g., Balslev et al. 1998; Duivenvoorden 1994; Valencia et al. 1994). Shrubs and treelets are relatively just as important in the BCI flora as they are at Yasuní. Diversity comparisons such as this should be based on the maximum size a species attains, not just diameter as measured in a plot. Most species are represented in the smallest dbh classes, from shrubs to tall trees, and saplings of large tree species can live for decades in the forest understory (e.g., *Trichilia tuberculata* [Meliaceae]), a canopy tree at BCI, had individuals with an estimated age of 21.9 years at 1 cm dbh (Hubbell et al. 1998).

We should mention some caveats about our life form data, though. The Yasuní flora is poorly known, and we had to assemble information from a variety of taxonomic treatments, floras, and particularly our own field experience. In many cases we have not adopted the categories from taxonomic treatments and floras because specialists judge the life form based on collections from roads or rivers, where species tend to flower earlier and develop taller trunks than the average maximum inside mature forests. On the other hand, for rare species in the plot, we were unable to assign a life form based on our own experience. In addition,

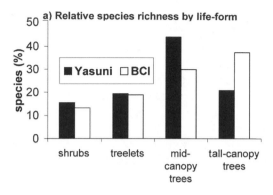

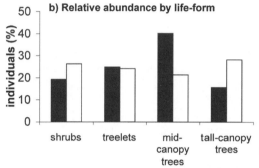

Fig. 9.4. Relative species richness (a) and relative abundance (b) by life form in a 2-ha plot in Yasuní National Park and a 2-ha sample from the Forest Dynamics Plot in Barro Colorado Island, Panama. Each bar gives the percentage of all species (a) or individuals (b) found in the plots that were classified in the given life form. Each species is only classified in a single category.

much taxonomic work in the Amazon region has concentrated on large trees (e.g., Neill and Palacios 1989; Pennington 1990), and the results presented here could reflect that in part. In any case, defining life forms of woody plants is a difficult task and observations in the long-term monitoring plot at Yasuní should greatly improve our knowledge. Our results are preliminary and will be refined in the coming years.

General Implications

Species richness is extraordinarily high in Yasuní National Park, Ecuador. The total number of species in 25 ha, 1104, is nearly double the entire North American tree flora north of Mexico. The ridge hectare had 255 species ≥10 cm, which is among the highest known in the lowland tropics (typically between 100 and 300 are found in a hectare; see Campbell et al. 1986; Gentry 1988b; Kochummen et al. 1990; Valencia et al. 1994).

This species richness cannot be attributed to habitat segregation. Although there were quantitative habitat distinctions, most tree species occurred in both of the major habitats in the plot, and total richness was high in both (255 species ≥10 cm dbh on the ridge hectare, 234 in the bottomland hectare). Diversity was also exceptionally high in all life forms included in the plot—shrubs, treelets, midcanopy trees, and tall canopy trees.

It would be shocking if this exceptional diversity in all areas and among all species groups does not result in a major impetus for conservation of large blocks of this forest.

Acknowledgments

We are indebted to the field workers who mapped, tagged, and identified thousands of trees since 1995. We thank especially those who worked more than 2 years in the project: Anellio Loor, Jairo Zambrano, Milton Zambrano, Gabriel Grefa, and Eleodan Velez. Margot Bass and Hugo Mogollón contributed greatly to the identification of the species. Suzanne Loo de Lao helped with data management and calculations. Nigel Pitman offered suggestions for the data analysis and worked as volunteer with the project. The Andrew W. Mellon Foundation supported the bulk of the research, and the National Science Foundation (U.S.) and the Tupper Family Foundation provided additional support. The Project Andean Biological and Cultural Diversity of the Pluvial Andean Forests (DIVA) provided financial support to the first author. The CTFS of the Smithsonian Institution advised in the methods for establishing large, permanent plots and assisted with the project.

References

Babour M., J. Burk, and W. D. Pitts. 1986. *Terrestrial Plant Ecology*. Benjamin/Cummings Publishing, Menlo Park, CA.
Balslev, H., J. Luteyn, B. Øllgaard, and L. B. Holm-Nielsen. 1987. Composition and structure of adjacent unflooded and floodplain forest in Amazonian Ecuador. *Opera Botanica* 92:37–57.
Balslev, H., R. Valencia, G. Paz y Miño, H. Christensen, and I. Nielsen. 1998. Species count of vascular plants in one hectare of humid lowland forest in Amazonian Ecuador. Pages 585–94 in F. Dallmeier and J. Comiskey, editors. *Forest Biodiversity in North, Central and South America, and the Caribbean: Research and Monitoring*. Man and Biosphere Series 21. Parthenon Publishing Group, Paris.
Berg, C. C., R. W. Akkermans, and E. C. van Heusden. 1990. Cecropiaceae: Coussapoa and Pourouma, With an introduction to the family. *Flora Neotropica* 51:1–208.

Brako, L. and J. L Zarucchi. 1993. Catálogo de angiospermas y gimnospermas del Perú. *Monographs in Systematic Botany from the Missouri Botanical Garden* 45:1–1286.

Campbell, D. G., D. C. Daly, G. T. Prance, and U. N. Maciel. 1986. Quantitative ecological inventory of terra firme and varzea tropical forest on Rio Xingú, Brazilian Amazon. *Brittonia* 38:369–93.

Condit, R. 1998. *Tropical Forest Census Plots.* Springer-Verlag, Berlin; RG Landes, Austin, TX.

Condit, R., S. P. Hubbell, and R. B. Foster. 1996. Changes in a tropical forest with a shifting climate: Results from a 50 ha permanent census plot in Panama. *Journal of Tropical Ecology* 12:231–56.

Condit, R., S. P. Hubbell, J. LaFrankie, R. Sukumar, N. Manokaran, R. Foster, and P. Ashton. 1996. Species-area and species-individual relationships for tropical trees: A comparison of three 50-ha plots. *Journal of Ecology* 84:549–62.

Croat, T. B. 1978. *Flora of Barro Colorado Island.* Stanford University Press, Stanford, CA.

Duivenvoorden, J. F. 1994. Vascular plant species counts in the rain forests of the middle Caqueta area, Colombian Amazonia. *Biodiversity and Conservation* 3:685–715.

Fisher, R. A., A. S. Corbet, and C. B. Williams. 1943. The relation between the number of species and the number of individuals in a random sample of an animal population. *Journal of Ecology* 12:42–58.

Gentry, A. 1982. Patterns of neotropical plant species diversity. Pages 1–84 in B. Wallace and G. T. Prance, editors. *Evolutionary Biology.* Plenum Press, New York.

———. 1988a. Changes in plant community diversity and floristic composition on environmental and geographical gradients. *Annals of the Missouri Botanical Garden* 75: 1–34.

———. 1988b. Tree species richness of upper Amazonian forest. *Proceedings of the National Academy of Sciences* 85:156–59.

———. 1990. Floristic similarities and differences between southern Central America and upper and central Amazonia. Pages 141–58 in A. Gentry, editor. *Four Neotropical Forests.* Yale University Press, New Haven, CT.

Harms, K. E., R. Condit, S. P. Hubbell, and R. B. Foster. 2001. Habitat associations of trees and shrubs in a 50-ha neotropical forest plot. *Journal of Tropical Ecology* 89:947–59.

Hubbell, S. P. 1998. The maintenance of diversity in a neotropical tree community: Conceptual issues, current evidence, and challenges ahead. Pages 17–44 in F. Dallmeier and J. Comiskey, editors. *Forest Biodiversity in North, Central and South America, and the Caribbean: Research and Monitoring.* Man and Biosphere Series 21. Parthenon Publishing Group, Paris.

Hubbell, S. P., and R. B. Foster. 1986. Commonness and rarity in a neotropical forest: Implications for tropical tree conservation. Pages 205–31 in M. Soule, editor. *Conservation Biology: Science of Scarcity and Diversity.* Sinauer Associates, Sunderland, MA.

Jørgensen, P. M., and S. León-Yánez. 1999. *Catalogue of the Vascular Plants of Ecuador.* Missouri Botanical Garden Press, St. Louis, MO.

Kochummen, K. M., J. V. LaFrankie, and N. Manokaran. 1990. Floristic composition of Pasoh Forest Reserve, A lowland rain forest in Peninsular Malaysia. *Journal of Tropical Forest Science* 3:1–13.

Neill, D., and W. Palacios. 1989. *Arboles de la Amazonía Ecuatoriana.* Ministerio de Agricultura y Ganadería, Quito, Ecuador.

Pennington, T. D. 1990. Sapotaceae. *Flora Neotropica* 52:1–770.

Pennington, T. D., D. Terence, and B. T. Styles. 1981. Meliaceae. *Flora Neotropica* 28:1–470.

Pitman, N. C. A. 2000. A large-scale inventory of two Amazonian tree communities. Ph.D. dissertation. Duke University, Durham, NC.

Pitman, N. C. A., J. Terborgh, M.R. Silman, P. Núñez V., D. A. Neill, C. E. Cerón, W. A. Palacios, and M. Aulestia. 2001. Dominance and distribution of tree species in upper Amazonian terra firme forests. *Ecology* 82:2101–17.

Prance, G. T. 1979. Chrysobalanaceae. *Flora of Ecuador* 10:1–24.

Renner, S. S., H. Balslev, and L. B. Holm-Nielsen. 1990. Flowering plants of Amazonian Ecuador—A checklist. *AAU Reports* 24:1–241.

Rohwer, J. G. 1993. Lauraceae: Nectandra. *Flora Neotropica* 60:1–332.

Romoleroux, K., R. Foster, R. Valencia, R. Condit, H. Balslev, and E. Losos. 1997. Pages 189–215 in R. Valencia and H. Balslev, editors. *Estudios Sobre Diversidad y Ecología de Plantas 1997*. Publicaciones de la Pontificia Universidad Católica del Ecuador, Quito, Ecuador.

Sleumer, H. O. 1980. Flacourtiaceae. *Flora Neotropica* 22:1–499.

Sukumar, R., H. S. Dattaraja, H. S. Suresh, J. Radhakrishnan, R. Vasudeva, S. Nirmala, and N. V. Joshi. 1992. Long-term monitoring of vegetation in a tropical deciduous forest in Mudumalai, Southern India. *Current Science* 62:608–16.

Svenning, J. C. 1999. Microhabitat specialization in a species-rich palm community in Amazonian Ecuador. *Journal of Tropical Ecology* 87:55–65.

Tuomisto, H., and K. Ruokolainen. 1994. Distribution of Pteridophyta and Melastomataceae along an edaphic gradient in an Amazonian rain forest. *Journal of Vegetation Science* 5:25–34.

Valencia, R., H. Balslev, and G. Paz y Miño. 1994. High tree alpha-diversity in Amazonian Ecuador. *Biodiversity and Conservation* 3:21–28.

Valencia, R., H. Balslev, W. Palacios, D. Neill, C. Josse, M. Tirado, and F. Skov. 1998. Species diversity and family composition in different regions of Ecuador: A sample of 18 1-ha plots. Pages 569–84 in F. Dallmeier and J. Comiskey, editors. *Forest biodiversity in North, Central and South America, and the Caribbean: Research and Monitoring*. Man and Biosphere Series 21. Parthenon Publishing Group, Paris.

Vásquez, R. 1997. *Flórula de las reservas biológicas de Iquitos, Perú*. Missouri Botanical Garden Press, St. Louis, MO.

Wright, S. J. 1992. Seasonal drought, soil fertility, and the species density of tropical forest plant communities. *Trends in Ecology and Evolution* 7:260–63.

———. 1999. Plant diversity in tropical forests. Pages 449–72 in F. I. Pugnaire and F. Valladares, editors. *Handbook of Functional Plant Ecology*. Marcel Dekker, New York.

———. 2002. Plant diversity in tropical forests: A review of mechanisms of species coexistence. *Oecologia* 130:1–14.

10

Community Ecology in an Everwet Forest in Sri Lanka

C. V. S. Gunatilleke, I. A. U. N. Gunatilleke, A. U. K. Ethugala,
N. S. Weerasekara, Shameema Esufali, Peter S. Ashton,
P. Mark S. Ashton, and D. S. A. Wijesundara

Introduction

Tropical forests are diverse and heterogeneous. Only recently have studies begun on a scale commensurate with their diversity. Since the late 1970s many studies have documented the diversity in Sri Lanka's forest types, in particular its rainforests (Gunatilleke and Ashton 1987; Jayasuriya 1999; Gunatilleke et al. 1995; Gunatilleke and Gunatilleke 1991; Seneviratne et al. 1999). The ancestry of these primeval rainforests dates back to the Deccan flora and hence its evolutionary and biogeographic significance (Ashton and Gunatilleke 1987). In geological times, these forest types were common to both peninsular India and Sri Lanka. Today they are restricted to a small area of 750 km^2 in southwest Sri Lanka, where the climate continues to remain aseasonal and perhumid. Elsewhere on the island, where the everwet climate changed to a more monsoonal one with many dry months, the rainforests that once prevailed in the rest of Sri Lanka and peninsular India have given way to the present day seasonal forests. Sri Lanka's rainforests, now climatically and geographically isolated from other similar forests in the region, harbor 94 % of the island's endemic flora and most of its endemic fauna including all 25 of its freshwater crabs (IUCN Sri Lanka 2000). Much is yet being discovered in the island as shown recently by Meegaskumbura et al. (2002). These authors report on an endemic radiation of over 100 species of Old World frogs (Rhacophorinae), primarily in the rainforests, where only 18 were previously known.

There is also an urgent need to understand the dynamics of the island's ecosystems, with respect to different types of disturbances and global climate change, if they are to be sustainably managed. It is this need that prompted Sri Lanka to become a partner in the Center for Tropical Forest Science (CTFS) program. In choosing to participate in the CTFS program, Sri Lanka selected Sinharaja, the largest block of relatively undisturbed lowland evergreen rainforest, as the country's study site. Since the 1970s when part of the forest was selectively logged, the forest has been used primarily for research, education, and recreation. This

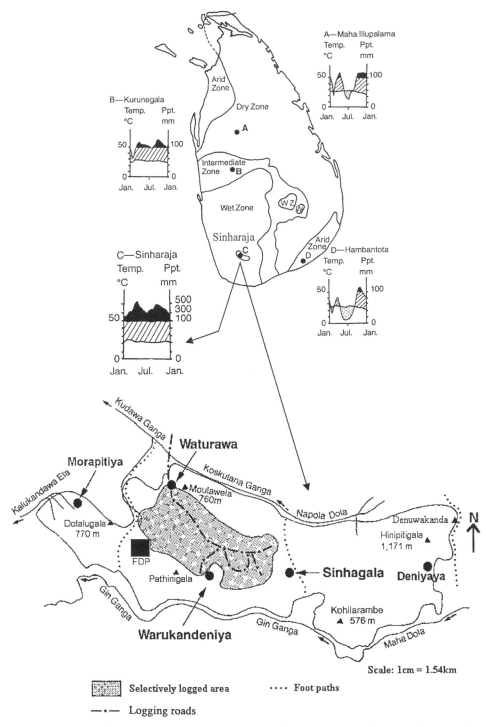

Fig. 10.1. Locations of Sinharaja (C) in the lowland wet zone of Sri Lanka and the 25-ha Forest Dynamics Plot (FDP, solid square) in the southwestern part of the reserve. Note the perhumid aseasonal climate in Sinharaja in the wet zone of the island compared to that of the other areas, and study sites (larger solid circles) of the phytosociological survey carried out in the late 1970s.

forest has been declared an International Man and Biosphere Reserve, a National Heritage Wilderness Area, and a Natural World Heritage Site (see chap. 37).

Research at Sinharaja since 1980 has been summarized by Gunatilleke et al. (1995) and Hadley and Ishwaran (1997). Floristic information on trees ≥10 cm dbh in 100 quarter-hectare plots distributed over five sites (fig. 10.1) provided the foundation for research at Sinharaja (Gunatilleke and Gunatilleke 1985). Due to the small sizes of the plots used, however, this initial effort could not determine the spatial distribution of different tree species along the microtopographic catena. A large plot was needed to assess spatial distribution patterns and study long-term forest dynamics in the relatively undisturbed part of Sinharaja. In this paper, we present the first findings related to the structure and floristics of the forest in relation to elevation and the contribution of the endemic species that form a dominant component to different aspects of the vegetation.

Methods

Study Area

The Sinharaja Forest Dynamics Plot (FDP) is located in an undisturbed part of the forest near the southwestern edge of the reserve, adjacent to a selectively logged area (fig. 10.1). The plot is 25 ha, 500 × 500 m. It has a central valley lying between two slopes, a steeper, higher slope facing southwest, and a less steep slope facing northeast. Seepage ways, spurs, and small hillocks cut across these slopes. This plot has two perennial streams and several seasonal streamlets (fig. 10.2, chap. 37).

Surveying, Enumeration, and Plant Identification

The plot was leveled and surveyed according to the methods of Hubbell and Foster (1983) and Manokaran et al. (1990). Plants were censused according to the methods of Manokaran et al. (1990).

When the census first started, herbarium material was collected for every plant censused, but after the census-takers learned the plants, specimens were taken only from those plants that could not be identified with certainty. Collected specimens were identified to species with the help of the floras of Trimen (1891–1900) and Dassanayake and Fosberg (1980–96).

Analyses

We constructed species–area and species–individual curves. For the species–area curve, we divided the Sinharaja Forest Dynamics Plot into 5 × 5, 10 × 10, 20 × 20, 25 × 25, 50 × 50, 100 × 100, 125 × 125 and 250 × 250 m quadrats. For each plot size, we calculated the mean number of species per plot and its standard error. For the species–individual curve, all trees censused in the 25-ha plot were

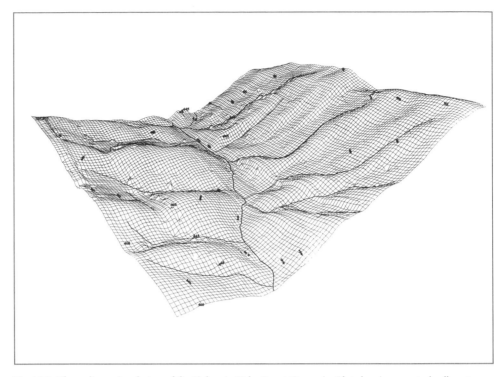

Fig. 10.2. Three-dimensional view of the Sinharaja 25-ha Forest Dynamics Plot showing a central valley, steeper southwest facing slope (right), less steep northwest facing slope (left) and streams (indicated by lines) flowing through the plot. Figure created by GIS Works by Enviromental & Forest Conservation Division, Mahaweli Authority, Polgolla, Sri Lanka, 1998.

divided into groups of contiguous trees containing 100, 200, 400, 800, . . . 102,400 and 204,800 stems apiece, chosen to overlap minimally. For each group size, we calculated the number of species per group and its standard error.

We partitioned the censused species on the plot into five groups:

(1) Canopy species, which reach the uppermost level of the forest. These trees have umbrella-shaped crowns that are directly exposed to light. They correspond to the emergents of mixed dipterocarp forests in Malesia, but the everwet midelevation lowland forests of Sri Lanka lack an emergent layer, perhaps because of the strong winds to which they are subject.

(2) Subcanopy species, found immediately below the uppermost stratum. Trees of these species can also grow in bright light.

(3) Understory tree species, with narrow, deep crowns, are always found under shade.

(4) Treelets and shrubs, most of whose adults are less than 10 cm dbh.

(5) Lianas.

Fisher's α, the Shannon–Wiener index (H'), and the evenness index (J) were used to compare species diversity among plants of these different life forms, for the 25-ha plot as a whole and for different ranges of elevation within it (Magurran 1988; Kent and Coker 1992). This was done to understand the variation in diversity among the different elevation ranges and different plant groups.

Data on the relative abundance of species in the 20 × 20 m quadrats of the Sinharaja Forest Dynamics Plot were classified by a cluster analysis using SAS software. Minimum variance clustering (Ward 1963) was used as the clustering algorithm. In this method the distance between two clusters is the sum of squares between the two clusters added up over all variables. At each step the two clusters to be united were those when fused yielded the least increase in within-cluster sum of squares (Pielou 1984). While computations were done using the PROC CLUSTER procedure, dendrograms were constructed using the PROC TREE procedure, both in SAS (SAS Institute 1989, 1993).

Results

Physical Features

We divided the Sinharaja FDP into five elevation ranges: 424–429 m, 430–459 m, 460–489 m, 490–519 m, and 520–575 m. We chose these elevation ranges because the habitats within each were relatively homogeneous. The approximate areas within each range were 1, 11, 7, 3, and 3 ha respectively. In the lowest range, 424–429 m, a permanent stream bisects the plot. A relatively flat valley lies between 430 and 459 m. The three upper ranges represent the middle slopes, the upper slopes, and the ridgetops. The contours of the FDP and the location of features such as spurs, seepage ways, and ridges are shown in figure 10.2.

Vegetation: Density and Basal Area

Within the elevation range of the plot, tree density increased with elevation from 4870 to 11,278/ha (fig. 10.3). The density of both small trees and big trees increased with elevation. The basal area also increased with elevation from 21.58 to 59.44 m^2/ha (fig. 10.3).

Floristic Richness

The Sinharaja plot contained 28 species of canopy tree, 47 species of subcanopy tree, 51 species of understory tree, 79 species of shrubs and treelets, and 10 species of liana (fig. 10.4). Above 430 m, the number of species of each free-standing growth form declined with increase of elevation. The number of species of shrub

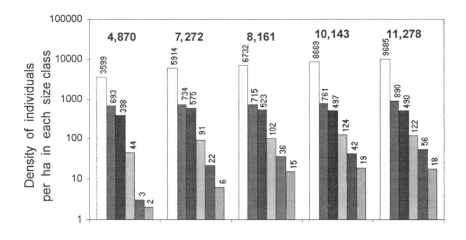

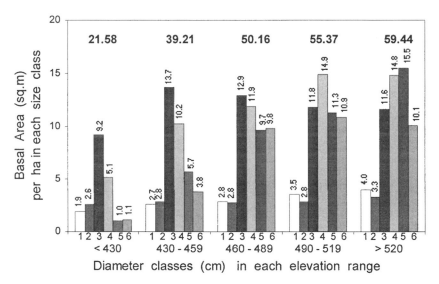

Fig. 10.3. Diameter distribution of individuals in different elevation (m) ranges in the Sinharaja 25-ha Forest Dynamics Plot. Mean density/ha and mean basal area in m²/ha at each elevation class are also shown in bold above the respective set of histograms. Key to size classes: 1–4 cm dbh (column 1), 5–9 (column 2), 10–29 (column 3), 30–49 (column 4), 50–69 (column 5), and >70 cm (column 6).

and treelet decreased most rapidly with increased elevation, while the number of canopy and subcanopy tree species decreased most slowly (fig. 10.5). This difference may also be partly due to the progressively decreasing extents of forests in the different elevation ranges. However, as reported above under density and basal area, the number of individuals per hectare showed a marked increase with elevation.

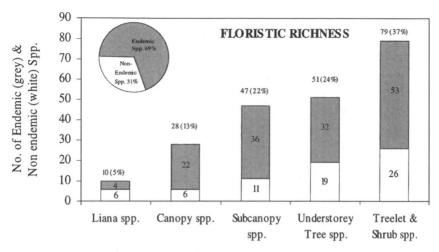

Fig. 10.4. Floristic richness and the proportion of endemic/nonendemic species (excluding the five unidentified species) in the Sinharaja 25-ha Forest Dynamics Plot. Inset shows the proportion of endemic to nonendemic species in the whole plot.

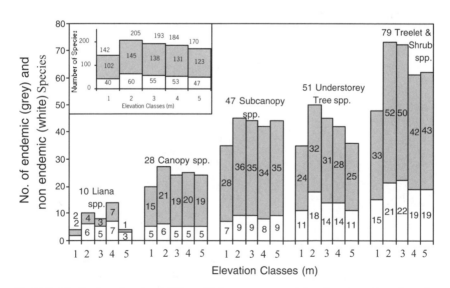

Fig. 10.5. Distribution of species of different life form groups across elevation classes. Inset shows the proportion of endemic and nonendemic species in each elevation class, the ranges of which are given in figure 10.3.

Table 10.1. Dominant Species Based on Density in Each Life Form Group in the Different Elevation Classes and in the 25-ha Sinharaja Forest Dynamics Plot

Species	Density (Trees/ha) by Elevation Range (m)					
	<430	430–459	460–489	490–519	>520	424–574
Canopy Species						
Shorea megistophylla (E)	**518**	142	156	29	2	129
Shorea trapezifolia (E)	**74**	**156**	100	60	10	108
Mesua ferrea (E)	**315**	145	**264**	380	365	**239**
Mesua nagassarium (NE)	2	**191**	**652**	**1482**	**1242**	**596**
Shorea disticha (E)	27	66	**250**	595	**1022**	**296**
Shorea worthingtonii (E)	0	15	106	**569**	649	185
Subcanopy Species						
Semecarpus gardneri (E)	141	83	33	11	25	56
Palaquium canaliculatus (E)	**253**	**193**	90	87	48	**138**
Semecarpus walkeri (E)	116	115	76	56	48	89
Cullenia ceylanica (E)	71	**165**	**177**	133	92	**152**
Myristica dactyloides (E)	35	106	117	113	111	108
Shorea cordifolia (E)	30	72	**156**	**305**	**333**	**153**
Cullenia rosayroana (E)	31	34	79	**174**	**289**	94
Understory tree species						
Dillenia retusa	**43**	31	10	0	0	18
Garcinia hermonii (E)	**93**	**414**	338	220	152	**325**
Humboldtia laurifolia (E)	7	**453**	**1143**	**1642**	**1534**	**898**
Diospyros acuminata (E)	18	50	77	**97**	59	63
Xylopia championii (E)	15	87	**101**	68	**117**	**89**
Shrub and Treelet Species						
Agrostistachys hookeri (E)	**770**	**522**	101	10	1	**293**
Schumacheria castanaefolia (E)	**268**	227	109	5	2	142
Urophyllum ellipticum (E)	**163**	253	**145**	28	20	**164**
Psychotria nigra (E)	119	**380**	**247**	15	6	**243**
Agrostistachys intramarginalis (E)	12	**254**	**876**	**1449**	**1585**	**721**
Memecylon arnottianum (E)	1	31	60	33	**202**	59
Lianas						
Uncaria thwaitesii (E)	**5**	**7**	**6**	**3**	**2**	**6**
Dalbergia pseudo-sissoo (NE)	**5**	**7**	4	2	1	5

Note: All but two species are endemic to Sri Lanka. E and NE represent endemic and nonendemic species respectvely. Numbers in bold indicate the three dominant species in each category.

Dominance

The most common trees (per hectare) in the plot were the understory tree *Humboldtia laurifolia* (Leguminosae), with 898 trees ≥ 1 cm dbh, the treelet *Agrostistachys intramarginalis* (Euphorbiaceae), with 721 trees, and the canopy tree *Mesua nagassarium* (Guttiferae) with 596 trees. These three species were the most common of their life form group (table 10.1), and the two latter species had the highest basal area within their group (table 10.2). Among subcanopy trees, *Shorea cordifolia* (Dipterocarpaceae) with 153 trees and *Cullenia ceylanica* (Bombacaceae) with

Table 10.2. Dominant Species Based on Basal Area in Each Life Form Group in the Different Elevation Classes in the 25-ha Sinharaja Forest Dynamics Plot

Species	Basal Area per Hectare by Elevation Range (m)					
	<430	430–459	460–489	490–519	>520	424–574
Canopy Species						
Mesua ferrea (E)	**1.525**	0.356	0.694	0.815	0.636	0.583
Shorea megistophylla (E)	**1.900**	1.103	0.990	0.151	0.064	0.859
Shorea trapezifolia (E)	**1.336**	**4.048**	**2.654**	0.747	0.517	**2.730**
Mesua nagassarium (NE)	0.001	**2.165**	**10.850**	**22.291**	**21.377**	**9.237**
Litsea gardneri (E)	0.914	**1.255**	1.736	0.227	0.254	1.118
Shorea stipularis (E)	0.064	0.667	**2.378**	1.219	1.215	1.233
Shorea disticha (E)	0.104	0.471	2.158	**2.909**	**4.593**	**1.710**
Palaquium petiolare (E)	0.000	0.268	0.725	**2.068**	1.981	0.812
Shorea affinis (E)	0.015	0.460	0.826	1.490	**2.301**	0.892
Subcanopy Species						
Palaquium canaliculatum (E)	0.490	0.273	0.172	0.329	0.517	0.291
Semecarpus walkeri (E)	**0.740**	0.849	0.615	0.220	0.103	0.613
Cullenia ceylanica (E)	**1.208**	**3.241**	**3.865**	**3.384**	**4.228**	**3.465**
Myristica dactyloides (E)	**0.675**	**1.403**	**1.763**	**1.677**	0.740	**1.423**
Vitex altissima (NE)	0.655	**1.190**	0.445	0.048	0.000	**0.684**
Chaetocarpus castanocarpus (E)	0.158	0.749	0.829	0.259	0.010	0.596
Shorea cordifolia (E)	0.107	0.317	**0.954**	**0.764**	**1.218**	0.643
Cullenia rosayroana (E)	0.179	0.356	0.483	0.700	**1.195**	0.528
Understory Tree Species						
Mallotus fucescens (E)	**0.841**	0.098	0.000	0.000	0.000	0.077
Garcenia hermonii (E)	**0.670**	**2.450**	**1.805**	**1.198**	**1.870**	**1.982**
Podadenia thwaitesii (E)	**0.315**	0.180	0.175	0.092	0.062	0.159
Xylopia championii (E)	0.221	**0.675**	**0.724**	**0.554**	**0.866**	**0.679**
Dillenia retusa (NE)	0.072	**0.266**	0.029	0.000	0.000	0.130
Humboldtia laurifolia (E)	0.009	0.174	**0.414**	0.460	**0.653**	0.325
Strombosia nana (E)	0.006	0.038	0.177	**0.493**	0.390	0.173
Cinnamomum capurucorundum (E)	0.025	0.029	0.018	0.054	**0.162**	0.045
Shrub and Treelet Species						
Agrostistachys hookeri (E)	**1.497**	**0.665**	0.124	0.005	0.000	**0.391**
Urophyllum ellipticum (E)	**0.149**	**0.356**	0.111	0.013	0.009	0.198
Schmacheria castanaefolia (E)	**0.554**	**0.428**	**0.173**	0.005	0.004	**0.261**
Psychotria nigra (E)	0.054	0.237	0.162	0.007	0.001	0.153
Glochidion acuminatus (E)	0.028	0.038	**0.244**	0.002	0.000	0.083
Agrostistachys intramarginalis (E)	0.003	0.127	**0.561**	**0.885**	**1.149**	**0.456**
Memecylon rosayroana (E)	0.092	0.130	0.112	**0.073**	0.074	0.110
Memecylon sp. D (?)	0.000	0.001	0.017	**0.054**	**0.113**	0.026
Memecylon arnottianum (E)	0.000	0.028	0.028	0.018	**0.147**	0.040
Lianas						
Dalbergia pseudosisoo (NE)	**0.035**	**0.047**	**0.032**	0.007	0.012	**0.033**
Uncaria thwaitesii (E)	**0.028**	**0.042**	**0.034**	**0.028**	0.014	**0.034**
Ventilago maderaspatana (NE)	**0.010**	0.002	0.001	**0.002**	0.005	**0.002**
Salacia reticulata (NE)	0.000	0.000	**0.004**	0.001	0.000	0.001

Note: E and NE represent endemic and nonendemic species, respectively. Numbers in bold indicate the three dominant species in each category. Note the preponderance of endemic species among these dominants.

152 trees were the most common and *C. ceylanica* had the highest basal area. The species dominating the various life form groups (as measured by basal area) differed according to elevation class (table 10.2).

Species–Area and Species–Individual Curves

The species–area curve rises gently over the first half-hectare and thereafter, steeply (fig. 10.6). The species–individual curve on the other hand, rises steeply at first and gently after about 50,000 trees.

Species Diversity Within Different Life Forms

Considering the different life form groups in the plot, the subcanopy trees had the highest Shannon–Wiener diversity, with an H' of 3.13, while $H' = 2.81$ for treelets and shrubs, 2.33 for canopy trees, and 2.05 for understory trees (table 10.3). Evenness (J) was also greatest for subcanopy tree species and lowest for understory tree species. On the other hand, Fisher's α was highest for treelets and shrubs and lowest for canopy trees followed by that for lianas.

The Shannon–Wiener diversity for each of the life form groups was highest for trees between 430 and 459 m and between 460 and 489 m (table 10.3) with two exceptions: in the understory tree group and in the liana group. On the other hand, Fisher's α did not show a consistent trend with elevation within each life form group.

Rank Abundance Curves

When trees were segregated by both life form and elevation range, the distribution of species over trees, with three exceptions, generally followed the log series when the models were tested using χ^2 (Magurran 1988). The exceptions were the curves representing subcanopy species at elevations 430–459 m and 460–489 m and treelets and shrubs at <430 m, all of which fitted the log normal model (fig. 10.7). The most common understory tree species included *Humboldtia laurifolia* (Leguminosae) with 49% of the understory trees; the most common treelet/shrub species included *Agrostistachys intramarginalis* (Euphorbiaceae) with 29% and *A. hookeri* (Euphorbiaceae) with 11.8% of the trees in its category. There was codominance among the canopy and subcanopy species in most elevation classes, whereas single species dominance was mostly evident among the understory tree species and the treelet/shrub species.

In the different elevation ranges, trees in the lowest range showed a log series distribution; in each of the remaining elevations log normal distributions were observed when curves were fit minimizing χ^2 (fig. 10.8).

High and Low Density Taxa

Forty-six percent of the canopy trees on the plot belonged to the genus *Shorea* (Dipterocarpaceae) and another 39% belonged to *Mesua* (Guttiferae). Forty-nine

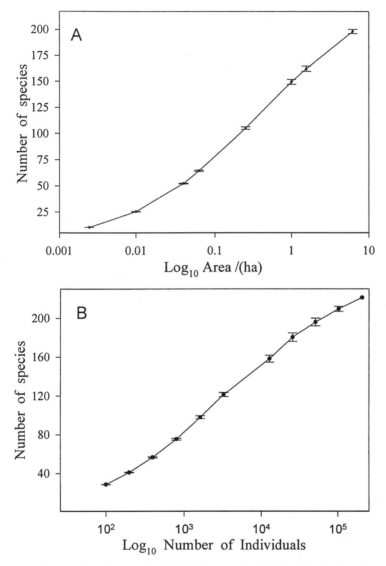

Fig. 10.6. Species–area (A) and species–individual (B) curves for the Sinharaja 25-ha Forest Dynamics Plot. Curve A shows the mean and standard error, obtained using 10,000, 2500, 625, 400, 100, 25, 16, 4, and 1 plots of size 25, 100, 625, 2500, 10,000, 15,625, 62,500 and 250,000 m^2, respectively. A total of 215 species were identified in the whole plot. Curve B shows the mean and standard error, obtained using 415, 210, 104, 52, 26, 13, 7, 4, 2, and 1 samples comprising 500, 1000, 2000, 4000, 8000, 16,000, 30,000, 64,000, 128,000 and 207,540 individuals, respectively.

Table 10.3. Density of Individuals, Species Richness, and Diversity Indices of Species in Different Life Form Groups as Well as All Elevation Ranges in the 25-ha Sinharaja Forest Dynamics Plot

Life Form Groups and Elevation Ranges	Number of Individuals in Areas Sampled	Number of Species in Areas Sampled	Fisher's α	Shannon–Wiener Index	Evenness Index
Canopy Species					
<430 m	1,021	22	3.96	1.46	0.47
430–459 m	12,359	32	3.97	2.56	0.74
460–489 m	13,386	29	3.50	2.28	0.68
490–519 m	12,430	30	3.67	1.95	0.57
>520 m	13,301	29	3.47	1.89	0.56
Subcanopy Species					
<430 m	1,010	38	7.58	2.65	0.73
430–459 m	16,137	49	6.19	3.09	0.80
460–489 m	8,873	48	6.70	3.10	0.80
490–519 m	4,597	45	7.06	2.72	0.72
>520 m	5,681	47	7.33	2.58	0.67
Understory Tree					
<430 m	621	31	6.82	2.84	0.83
430–459 m	16,810	42	5.16	2.40	0.64
460–489 m	13,545	39	4.91	1.80	0.49
490–519 m	7,498	34	4.62	1.44	0.41
>520 m	6,980	30	4.06	1.50	0.44
Shrubs/Treelet					
<430 m	1,941	47	8.83	2.40	0.62
430–459 m	32,172	72	8.85	2.95	0.69
460–489 m	15,098	69	9.39	2.61	0.62
490–519 m	5,823	61	9.47	1.27	0.31
>520 m	6,967	60	8.97	1.50	0.37
Lianas					
<430 m	34	4	1.87	1.29	0.93
430–459 m	655	10	2.17	1.53	0.66
460–489 m	286	8	2.08	1.52	0.73
490–519 m	96	6	2.48	1.54	0.86
>520 m	57	4	1.78	1.25	0.90
All Elevations (424–575 m)					
Canopy	52,497	33	3.40	2.33	0.67
Subcanopy	36,298	51	5.86	3.13	0.80
Understory Tree	45,454	44	4.80	2.05	0.54
Shrubs/Treelet	62,001	77	8.72	2.81	0.65
Lianas	1,128	10	1.90	1.55	0.67

Note: Shannon–Wiener Index calculated using $\log_e$. In part 7 of this volume, Shannon–Wiener is calculated using $\log_{10}$.

percent of the understory trees on the plot belonged to the species *Humboldtia laurifolia* (Leguminosae). Twenty-nine percent of the trees belonging to shrub or treelet species were *Agrostistachys intramarginalis* (Euphorbiaceae) and another 12% were *A. hookeri* (Euphorbiaceae).

The rarest 11 canopy species made up only 1% of canopy tree species stems on the plot and another 17 species comprised the remaining 99%. The rarest 11

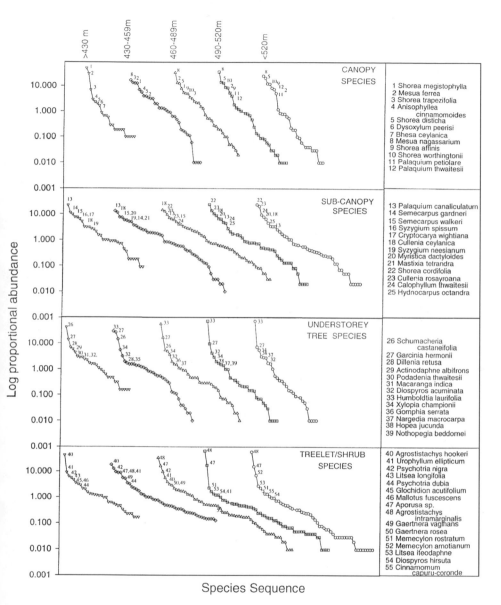

Fig. 10.7. Dominance–diversity curves for different life form groups in different elevation classes in the Sinharaja 25-ha Forest Dynamics Plot and the leading species in each group.

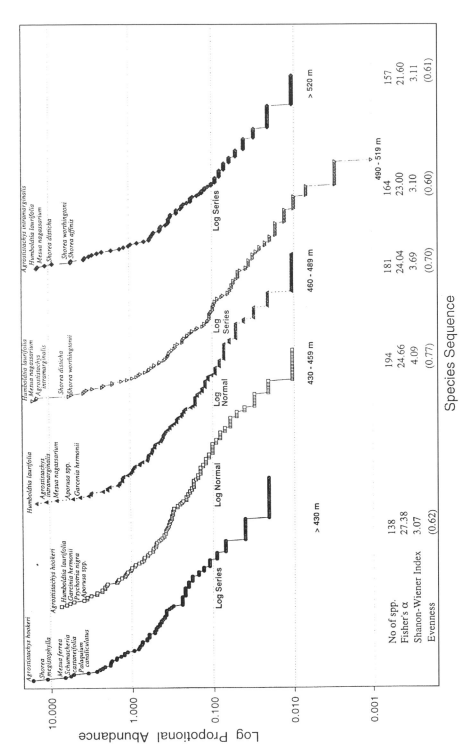

Fig. 10.8. Dominance–diversity curves for the vegetation at different elevation ranges of the Sinharaja 25-ha Forest Dynamics Plot. Species richness, diversity indices, and species evenness index of the vegetation in each elevation range (shown at the lower end of each curve) are shown below each curve.

subcanopy species made up only 1% of subcanopy species trees and another 36 species accounted for the remaining 99%. The rarest 29 species of shrubs and treelets represented 1% of the individuals of this growth form and another 50 comprised the remaining 99%.

The distribution of abundances of nonliana species with four or more stems apiece was roughly log normal (fig. 10.9), although the mode was higher and the tails fatter than the log normal would predict. Eight free-standing species and one liana species were each represented on the plot by a single individual ≥ 1 cm dbh; two free-standing species were each represented by two stems apiece; six free-standing species and two liana were represented by three individuals apiece. Fifty free-standing species and seven liana species were represented on the plot by less than one individual per hectare.

Cluster Analysis

The cluster analysis of 20 × 20 m quadrats revealed two major clusters that accounted for most of the variation among quadrats (table 10.4). Each cluster contained three subclusters. The spatial distribution of plots in each cluster suggested that elevation, aspect, and drainage were the primary factors distinguishing the two major clusters. Plots at lower elevations, 424–475 m, were classified together and separated from those on the higher NE and SW facing slopes and on the ridges of the NE facing slope. Subclusters appear to reflect the microtopography and elevation very closely (fig. 10.10).

Different species dominated the two clusters (table 10.4). In the first cluster, one subcluster was dominated by the canopy tree *Shorea megistophylla* (Dipterocarpaceae) and the other two by the canopy tree *S. trapezifolia* (Dipterocarpaceae). These two canopy species were markedly less common in the second cluster. In the second cluster, *Mesua nagassarium* (Guttifeae) was common among canopy trees, *Humboldtia laurifolia* (Leguminosae) among understory trees, and *Agrostistachys intramarginalis* (Euphorbiaceae) among treelets and shrubs. None of these species was among the dominants in the first cluster. Only among understory trees was a species, *Garcinia hermonii* (Guttiferae) common in all six subclusters. The spatial distribution of many plant species fit these subclusters rather closely (figs. 10.11 and 10.12).

Endemic Species: Density, Basal Area, and Floristic Richness

Eighty-three percent of trees and 88% of the basal area of the Sinharaja Forest Dynamics Plot are contributed by endemic species. Among all size classes, endemic species provide the majority of abundance and basal area values.

Among the plot's 215 species, including lianas, 69% are endemic to Sri Lanka. Forty percent of the liana species, 67% of the treelet and shrub species, 63% of the understory tree species, 77% of subcanopy, and 79% of the canopy species

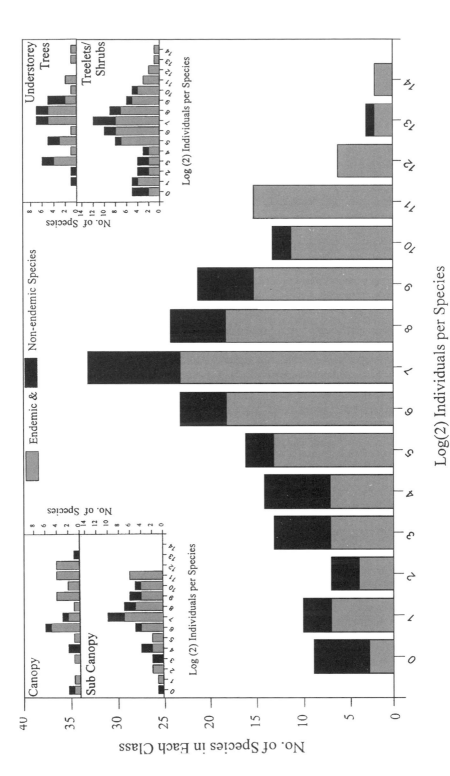

Fig. 10.9. Log normal distribution of the relative abundances of woody species, plotted per octave of abundance, in the Sinharaja 25-ha Forest Dynamics Plot. Insets show the respective distributions for each life form group. Note the preponderance of rare species among the treelet and shrub species.

Table 10.4. Cluster Analysis of 20 × 20 m Plots and the Major Features Representing Each Cluster

Cluster No.	Terrain	Elevation range (m)	Dominant Species in Life Form Groups in Each Cluster			
			Canopy Species	Subcanopy Species	Understory Tree Species	Treelet and Shrub Species
1	Moist valley	424–460	SHORME (2) MESUFE (3)	PALACA (4)	GARCHE (5)	AGROHO (1)
5	NE facing drier, lower slopes	424–475	SHORTR (10)	CULLCE (5) MYRIDA (6) PALACA (7)	GARCHE (3) XYLOCH (8)	APOR (1) PSYCNI (2) UROPEL (4) GAERRO (9)
6	NE facing seepage ways	424–475	SHORTR (6)	SEMEGA (5)	GARCHE (7)	SCHUCA (1) UROPEL (2) PSYCNI (3) PSYCDU (4)
3	SW, NE facing spurs	435–500	MESUNA (2) SHORAF (6)	MYRIDA (8) CULLCE (9)	HUMBLA (1) GARCHE (3) XYLOCH (7)	AGROIN (4) APOR (5)
2	SW facing upper slopes & ridges	465–575	MESUNA (3) SHORDI (4) SHORAF (5) SHORWO (7) PALAPE (11)	PALATH (9) CULLRO (10) SHORCR(8) CULLCE (13)	HUMBLA (1) GARCHE (12)	AGROIN (2) APOR (6)
4	SW facing midslope seepage ways	445–540	MESUFE (2) MESUNA (4) SHORDI (5) SHORME (9)	CULLCE (6) SHORCR (7)	HUMBLA (3) GARCHE (8)	AGROIN (1)

Notes: Numbers in parentheses show the rank of the species when all the life forms within each cluster are considered together. See table 10.2 for full names of species except for *Palaquium petiolare* (PALAPE) *Palaquium hwaitesii* (PALATH) Sapotaceae, *P. thwaitesii* (PALATH) Sapotaceae, *Psychotria dubia* (PSYCDU) Rubiaceae, and *Semecarpus gardneri* (SEMEGA) Anacardiaceae.

Between Cluster Sum of Squares

0 1000 2000 3000

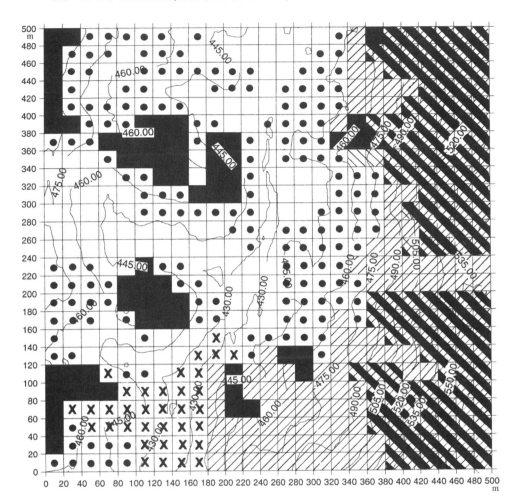

Fig. 10.10. Spatial distribution of subclusters resulting from the multivariate analysis of the Sinharaja 25-ha Forest Dynamics Plot showing their relationship to elevation and topography of the plot. Subclusters 2, 3, and 4 (broad right sloping lines, black fills, and thin left sloping lines respectively) comprise one major cluster; and subclusters 1, 5, and 6 (multiplication sign, dots, and clear areas respectively), the second major cluster.

on the plot are endemic (fig. 10.4). The more common a species, the more likely it is to be endemic (fig. 10.9). Twenty-four of the most common 25 species on the plot are endemic. The only exception is *Mesua nagassarium* (Guttiferae) with 14,881 of the 149,456 trees of the 25 most common species. However, in Sinharaja *M. nagassarium* is represented by the var. *pulchella,* which is endemic to Sri Lanka.

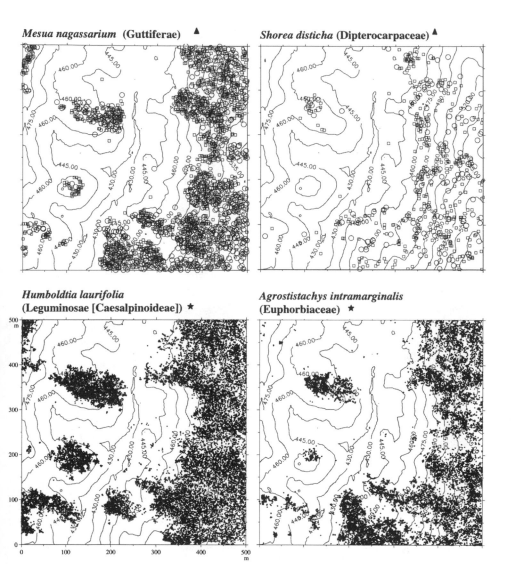

Fig. 10.11. Spatial distribution of individuals of selected canopy (triangles) and understorey (stars) species in the Sinharaja 25-ha Forest Dynamics Plot mostly confined to areas represented by one of the major clusters (subclusters 2, 3, and 4) arising from the cluster analysis.

Discussion

Increase of Tree Density with Elevation

The increase in tree density with increasing elevation over a rather narrow range of 151 m in the Sinharaja Forest Dynamics Plot may be explained by differences

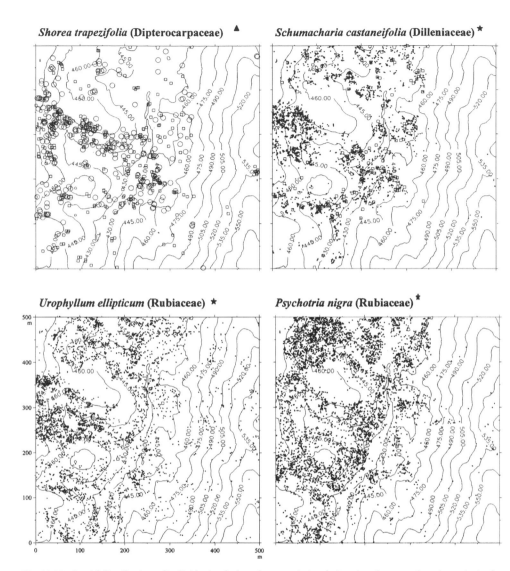

Fig. 10.12. Spatial distribution of individuals of selected canopy (triangles) and understorey (stars) species in the Sinharaja 25-ha Forest Dynamics Plot mostly confined to areas represented by the second major cluster (subclusters 1, 5, and 6) arising from the cluster analysis.

in vegetative structure and gap dynamics in the lower and upper elevations. Trees in the valley are taller and have larger crowns than those on the ridges and upper slopes, where nutrients are more scarce (Ashton 1992; Ashton et al. 1993; Gunatilleke et al. 1996). Trees on ridgetops and the upper slopes have more diffuse crowns, with smaller leaves. Nutrient shortage is likewise associated with

Table 10.5. Tree Diversity on the Deccan Plate and Elsewhere

Site	Plot Size (ha)	Climate and Altitude	N	S	α
Yanomamo, Peru	1.0	Everwet, lowland	580	283	218
Choco, Colombia	1.0	Everwet, lowland	664	252	148
Lambir, Sarawak	1.0	Everwet, lowland	632	245	147
Sinharaja, Sri Lanka	1.0	Everwet, lowland	682	70	20
Western Ghats, India	4.0	Evergreen, seasonal, 450 m	2607	98	20
Eastern Madagascar	0.5	Everwet, 1000 m	539	125	51
Mudumalai, India	1.0	Dry, deciduous, 900 m	278	21	5
Western Madagascar	0.9	Dry, deciduous, lowland	786	45	10

Note: N = number of trees >10 cm dbh on plot; S = number of species included among them; α = Fisher's α, where $S = \alpha \ln (1 + N/\alpha)$. Data for Yanomamo, Peru, from Gentry (1988), for Choco, Colombia, from Faber-Langendoen and Gentry (1991), for Western Ghats from Sinha and Davidar (1988), for Madagascar from Abraham et al. (1996), and for other sites from CTFS.

shorter forest, higher tree density, and smaller leaves in Sarawak and Venezuela (Brunig 1993).

Trees on the ridges root more deeply and widely relative to their size than trees in the valley. At San Carlos de Rio Negro, Venezuela, forest on poorer soil allocates a higher proportion of its biomass to roots (table 6.6 in Leigh 1999). The taller, less deeply rooted trees of the valley blow down more often. This creates more frequent, larger gaps in the valley than those that occur on ridgetops and on upper slopes (Ashton et al. 1995). The increase in large-diameter tree sizes with elevation within the plot also points to greater turnover in the valley.

Low Species Richness

The Sinharaja Forest Dynamics Plot has relatively few tree species compared to plots on other everwet sites. A single hectare of everwet forest such as Yanomamo in Amazonian Peru (Gentry 1988), Bajo Calima in Colombia (Faber-Langendoen and Gentry 1991), Lambir in Sarawak, or Yasuní in Ecuador has more species of trees ≥ 10 cm dbh than the entire 25-ha plot at Sinharaja, including all free-standing trees ≥ 1 cm dbh (table 10.5). Low tree diversity is typical of Sri Lanka's rainforests. Is Sri Lanka's tree diversity so low because Sri Lanka is an island? This seems unlikely because the forests of south India, even those extending over wide areas, also have low tree diversity. Despite the Deccan plate's much greater area, its forests are less diverse than Malagasy forests of corresponding type (table 10.5). Presumably, tree diversity on the Deccan plate is so low because that plate encountered harsh climates as it drifted from Madagascar to Asia (Ashton and Gunatilleke 1987).

Phytosociological studies of trees ≥ 10 cm dbh on 100 quarter-hectare plots distributed over five areas representing the elevation range of the Sinharaja reserve documented 211 species (Gunatilleke and Gunatilleke 1985). The Sinharaja plot, with the same total area, had 167 species ≥10 cm dbh, 141 of which were shared with at least one of the quarter-hectare plots. Within the wet zone of Sri Lanka, the highly dissected terrain, with several parallel ranges of hills, has favored the evolution of species with very restricted distributions. Nonetheless, the 25-ha Forest Dynamics Plot seems to have captured a remarkable fraction of Sinharaja's tree diversity.

Management Challenges

Lowland rainforests in Sri Lanka are fragmented, degraded, and isolated. Judging from interpretations of 1992 Landsat imagery, only about 140,000 ha of rainforest remain in Sri Lanka (Legg and Jewell 1995). Nonetheless, this rainforest is still a repository of much of the country's biological wealth. The abundance of local endemics in Sri Lanka requires the sustainable management of forests set aside for conservation and those destined for exploitation. The abundance of information about Sinharaja accumulated since 1977, including that from the 25-ha Forest Dynamics Plot, and what is known about rainforests elsewhere, makes it opportune to formulate scientific guidelines for the management of both protected and production forests (Ashton et al. 2001).

A logging moratorium was enforced on all rainforests in the wet zone of Sri Lanka, pending the assessment of their biological wealth. Based on this assessment, 30 forest tracts have been set aside for conservation; the remainder are earmarked as production forests. There is also growing realization of the importance of conservation and habitat restoration of the remaining rainforest fragments of Sri Lanka for their biological wealth (Myers et al. 2000; Meegaskumbura et al. 2002).

De Zoysa et al. (1991) and Gunatilleke and Ashton (1987) showed that at least some of the endemic canopy and subcanopy tree species are reappearing in the succession after logging. Much less is known, however, of the response to disturbance associated with logging of endemic species in the understory; and because Sinharaja's tree diversity is highest among understory species, this lack of information must be remedied. De Zoysa et al. (1991) reported that 31 endemic species—9 endangered, 9 vulnerable, 12 rare, and one of unknown status—are distinctly rarer in disturbed sites even after 7 years of recovery. We must learn how to manage the forest so as to preserve these species.

General Implications

(1) At Sinharaja, diversity is greatest among treelets and shrubs, which grow under the shelter of the overstory. If logging is not designed to avoid damaging the understory, this diversity will be threatened.

(2) At Sinharaja, there is great heterogeneity among microsites, not only of tree species composition, but also of forest structure and dynamics. Consequently, one cannot apply a uniform silvicultural method to the whole forest; silvicultural methods must be designed to suit specific habitat types.
(3) The overwhelming proportion of Sri Lankan endemics at Sinharaja, especially among the more common species of the Forest Dynamics Plot and the 100 quarter-hectare plots of Gunatilleke and Gunatilleke (1985), confirms the predominant role endemic trees play in the ecology of this forest.
(4) Ashton et al. (1993) proposed a shelterwood regeneration method for sustained timber production to replace the more detrimental polycyclic system practiced in Sri Lanka today. A field trial has also been initiated to transform monoculture *Pinus* (Pinaceae) plantations into mixed species stands with primary forest canopy species and plants used for purposes other than timber (Ashton et al. 1997; Ashton 1998). Information from the Sinharaja Forest Dynamics Plot on gap regeneration, growth rates, and species habitats should assist these efforts, contributing to the scientific understanding of these forests and facilitating their sustainable management.

Acknowledgments

The authors extend their appreciation to all who have contributed to make the Sinharaja Forest Dynamics Plot a reality. In particular, we thank T. M. N. Jayatissa, M. A. Gunadasa, A. H. M. A. K. Tennakoon, S. Harischandran, T. M. Ratnayaka, W. A. Tennakoon, members of the Youth Exploration Society, Sri Lanka, and the 1994–95 undergraduates of the Botany Special Degree Programme for their assistance in the field; the research managers M. Henson, N. Rajakaruna, and A. Jasentuliyana who initiated the field work; the civil engineers G. Gurusingha, J. Edirisinghe, B. Muthuthambipillai, and their many 1993–96 engineering undergraduates for the physical survey of the site; N. A. B. Naranpanawa, S. Ilangantilleke, A. V. R. A. Manosha, V. Sudharshani, R. Weerasingha, and Nirosha Gunasekara for assistance in data entry, word processing, and graphics; the analysis programmers S. Sumanasekara, M. Almeda, N. Thalangama, and C. Liyanage; M. Nipson and J. Ramjlik for support and coordination from the Harvard Institute for International Development, U.S.; Environment and Forest Conservation Division, Mahaweli Authority, Sri Lanka, for assisting in the preparation of GIS maps; and last but not least, E. Losos of the Center for Tropical Forest Science and J. LaFrankie of the CTFS Asia Program for their continued support; Egbert Leigh for editorial help; the Forest Department, Sri Lanka, for facilities provided and the permission to work in Sinharaja; the University of Peradeniya, Sri Lanka for financial administration of the project; The John D. and Catherine T. MacArthur

Foundation and the Smithsonian Tropical Research Institute, for their generous financial assistance; and the villagers in Kudawa who extended their warm friendship, worked in the field consistently and conscientiously, and helped us in numerous ways to achieve our many goals.

References

Abraham, J. P., R. Benja, M. Randrianasolo, J. U. Ganzhorn, V. Jeannoda, and E. G. Leigh, Jr. 1996. Tree diversity on small plots in Madagascar: A preliminary review. *Revue d'Écologie (La Terre et la Vie)* 51:93–116.

Ashton, P. M. S. 1992. Some measurements of the microclimate within a Sri Lankan tropical rain forest. *Agricultural and Forest Meteorology* 59:217–35.

Ashton, P. S. 1998. A global network of plots for understanding tree species diversity in tropical forests. Pages 47–62 in F. Dallmeier and J. A. Comiskey, editors. *Forest Biodiversity Research, Monitoring and Modeling: Conceptual Background and Old World Case Studies.* UNESCO, Paris; Parthenon Publishing, Pearl River, NY.

Ashton, P. S., and C. V. S. Gunatilleke. 1987. New light on the plant geography of Ceylon I. Historical plant geography. *Journal of Biogeography* 14:249–85.

Ashton, P. M. S., C. V. S. Gunatilleke, and I. A. U. N. Gunatilleke. 1993. A shelterwood method of regeneration for self-sustained timber production in *Mesua-Shorea* forest of south-west Sri Lanka. Pages 255–74 in W. Erdelan, C. Preu, N. Ishwaran, and C. M. Maddumabandara, editors. *Ecology and Landscape Management in Sri Lanka. Proceedings of the International and Interdisciplinary Symposium on Sri Lanka.* Margraf Scientific Books, Weikersheim, Germany.

Ashton, P. M. S., I. A. U. N Gunatilleke, and C. V. S. Gunatilleke. 1995. Seedling survival and growth of four *Shorea* species in a Sri Lankan rain forest. *Journal of Tropical Ecology* 11:263–79.

Ashton, P. M. S., S. Gamage, I. A. U. N. Gunatilleke, and C. V. S. Gunatilleke. 1997. Restoration of a Sri Lankan rainforest: Using Caribbean pine (*Pinus caribaea*) as a nurse for establishing late-successional tree species. *Journal of Applied Ecology* 34:915–25.

Ashton, M. S., C. V. S. Gunatilleke, B. M. P. Singhakumara, and I. A. U. N. Gunatilleke. 2001. Restoration pathways for rain forests in southwest Sri Lanka: A review of concepts and models. *Forest Ecology and Management.* 154:409–30.

Brunig, E. F. 1993. Vegetation structure and growth. Pages 49–75 in F. B. Golley, editor. Ecosystems of the World 14A. *Tropical Rain forest Ecosystems: Structure and Function.* Elsevier, Amsterdam.

Dassanayake, M. D., and F. R. Fosberg. 1980–96. *A Revised Handbook to the Flora of Ceylon.* Vols. I–X. Amerind, New Delhi, India.

De Zoysa, N. D., C. V. S. Gunatilleke, and I. A. U. N. Gunatilleke. 1991. Comparative phytosociology of natural and modified rainforest sites in the Sinharaja MAB reserve in Sri Lanka. Pages 215–24 in A. Gómez-Pompa, T. C. Whitmore, and M. Hadley, editors. *Rain forest Regeneration and Management.* UNESCO, Paris; Parthenon Publishing, Pearl River, NY.

Faber-Langendoen, D., and A. H. Gentry. 1991. The structure and diversity of rain forests at Bajo Calima, Chocó region, western Colombia. *Biotropica* 23:2–11.

Gentry, A. H. 1988. Tree species richness of upper Amazonian forests. *Proceedings of the National Academy of Sciences* 85:156–59.

Gunatilleke, C. V. S., and P. S. Ashton. 1987. New light on the plant geography of Ceylon II. The ecological biogeography of the lowland endemic tree flora. *Journal of Biogeography* 14:295–327.

Gunatilleke, C. V. S., and I. A. U. N. Gunatilleke. 1985. Phytosociology of Sinharaja: A contribution to rain forest conservation in Sri Lanka. *Biological Conservation* 31:21–40.

Gunatilleke, I.A.U.N., and C. V. S. Gunatilleke. 1991. Threatened woody endemics of the wet lowlands of Sri Lanka and their conservation. *Biological Conservation* 55:17–36.

Gunatilleke, C. V. S., I. A. U. N. Gunatilleke, and P. M. S. Ashton. 1995. Rainforest research and conservation: The Sinharaja experience in Sri Lanka. *Sri Lanka Forester* 22:49–60.

Gunatilleke, C. V. S., G. A. D. Perera, P. M. S. Ashton, P. S. Ashton, and I. A. U. N. Gunatilleke. 1996. Seedling growth of *Shorea* section *Doona* (Dipterocarpaceae) in soils from topographically different sites of Sinharaja rain forest in Sri Lanka. Pages 245–65 in M. D. Swaine, editor. *The Ecology of Tropical Forest Tree Seedlings*. UNESCO, Paris; Parthenon Publishing, Pearl River, NY.

Hadley, M., and N. Ishwaran. 1997. Conservation, research and capacity building in the forest of the Lion King, Sri Lanka. Pages 89–102 in *UNESCO, Science and Technology in Asia and the Pacific: Cooperation for Development*. UNESCO Services Stories, Bangkok.

Hubbell, S. P., and R. B. Foster. 1983. Diversity of canopy trees in a neotropical forest and implications for conservation. Pages 25–41 in S. J. Sutton, T. C. Whitmore, and A. C. Chadwick, editors. *Tropical Rain forest: Ecology and Management*. Blackwell Science, Oxford, U.K.

IUCN. 1997. *Designing an Optimum Protected Areas System for Sri Lanka's Natural Forests*. IUCN and World Conservation Monitoring Center, U.K.

IUCN Sri Lanka. 2000. *The 1999 List of Threatened Fauna and Flora of Sri Lanka*. Colombo:IUCN, Sri Lanka.

Jayasuriya, A. H. M. 1999. In situ conservation of biodiversity and a network of conservation areas for Sri Lanka. Pages 27–48 in *Regional Seminar on Forests of the Humid Tropics of South and South-East Asia*. National Science Foundation, Sri Lanka.

Kent, M., and P. Coker. 1992. *Vegetation Description and Analysis*, 1st edition. John Wiley and Sons, London.

Legg, C., and N. Jewell. 1995. A 1:500,000-Scale Forest Map of Sri Lanka: The Basis for a National Forest Geographic Information System. *Sri Lanka Forester Special Issue (Remote Sensing)* 3–21.

Leigh, E. G., Jr. 1999. *Tropical Forest Ecology*. Oxford University Press, New York.

Magurran, A. E. 1988. *Ecological Diversity and Its Measurement*. Chapman and Hall, London.

Manokaran, N., J. V. LaFrankie, K. M. Kochummen, E. S. Quah, J. E. Klahn, P. S. Ashton, and S. P. Hubbell. 1990. *Methodology for the Fifty Hectare Research Plot at Pasoh Forest Reserve*. Research Pamphlet no. 104. Forest Research Institute Malaysia.

Meegaskumbura, M., F. Bossuyt, R. Pethiyagoda, K. Manamendra-Arachchi, M. Bahir, M. C. Milinkovitch, and C. J. Schneider. 2002. Sri Lanka: An Amphibian Hot Spot. *Science* 298:379.

Myers, N., R. A. Mittermeier, C. G. Mittermeier, A. B. Gustav, G. A. B. Da Fonseca, and J. Kent. 2000. Biodiversity hotspots for conservation priorities. *Nature* 403: 853–58.

Pielou, E. C. 1984. *The Interpretation of Ecological Data*. John Wiley and Sons, New York.

SAS Institute Inc. 1989. SAS/GRAPH software: Examples, Version 6. First Edition. SAS Institute Inc., Cary, NC.

———. 1993. SAS/STAT User's Guide, Version 6. Vols 1 and 2. 4th Edition. SAS Institute Inc., Cary, NC.

Seneviratne, G. I., K. Abeynayake, and M. R. K. Lenagala. 1999. Floral diversity of Kalatuwawa—Labugama forest reserve. Pages 27–48 in *Regional Seminar on Forests of the Humid Tropics of South and South-East Asia.* National Science Foundation, Sri Lanka.

Sinha, A., and P. Davidar. 1988. Seed dispersal ecology of a wind dispersed rain forest tree in the Western Ghats, India. *Biotropica* 24:519–25.

Trimen, H. 1891–1900. *Handbook to the Flora of Ceylon.* Vols. 1–5. Dulau, London.

Ward, J. H. 1963. Hierarchical grouping to optimize an objective function. *Journal of American Statistical Association* 58:236–44.

11

Structure, History, and Rarity in a Seasonal Evergreen Forest in Western Thailand

Sarayudh Bunyavejchewin, Patrick J. Baker, James V. LaFrankie, and Peter S. Ashton

Introduction

In 1991 a 50-ha Forest Dynamics Plot (FDP) was established in an area of seasonal dry evergreen forest at the Huai Kha Khaeng Wildlife Sanctuary (HKK) in western Thailand (chap. 27, fig. 27.1). The Western Forest Complex of Thailand, which includes HKK and 16 other wildlife sanctuaries, national parks, and protected areas totaling >10,000 km², is a critical landscape for the conservation of the flora and fauna of southeast Asia (see chap. 27). Nonetheless, surprisingly little is known of the ecology and dynamics of the forests of the Western Forest Complex. Three forest types—seasonal dry evergreen, mixed deciduous, and deciduous dipterocarp—dominate the area and are interspersed across the landscape in mosaic fashion. (A fourth forest type, lower montane evergreen forest, is limited to the highest peaks within the sanctuary and accounts for less than 0.5% of the total area of the sanctuary.)

The mechanisms that differentiate the forest types and maintain the forest mosaic across the landscape at HKK are poorly known. Baker (1997) showed that there were few significant site-related differences in growth and mortality of tree seedlings planted in deciduous forest and evergreen forest gaps. In addition, comparative analyses of soil structure, moisture availability, and nutrient content revealed only minor differences between evergreen and deciduous forest sites (P. J. Baker, unpublished data). Climatic differences, such as total rainfall or length of the dry season, occur at too broad a spatial scale to account for the mosaic of forest types at HKK, which are interspersed at relatively fine scales (0.5–10 km). The influence of physiographic factors, such as topography and landform, and the role of disturbances, such as forest fires and windstorms, on the distribution of forest types is unknown.

To better understand the mechanisms that might account for the forest mosaic, it is critical to establish baseline information on the structure and dynamics of each forest type. In this chapter we report on the results of a detailed census of the most complex of the forest types at HKK—seasonal dry evergreen forest. Smaller study plots are presently being established in the other forest types at HKK and will permit comparison of structure and dynamics in the future. Here we describe the stand structure and floristic composition of the seasonal dry evergreen forest in the FDP, focusing particular attention on the high proportion of rare species within the forest, and compare these patterns with other forests in a global network of large-scale FDPs.

Study Area

The HKK FDP is one of several large-scale plots in southeast Asia, part of a larger network of plots established under the guidance of the Center for Tropical Forest Science (CTFS) of the Smithsonian Tropical Research Institute. The HKK FDP was chosen to represent the climatic limit of evergreen forest in southeast Asia. Rainfall is strongly seasonal with a 6-month dry season extending from November to April. Mean annual rainfall is just under 1500 mm (1983–93 average). The extent and severity of the dry season is variable; some years have sporadic rainfall, others have little or no rain during the entire dry season. Low-intensity surface fires occur at HKK most years, although a given area is probably burnt only once every 3–10 years. Fires swept through the FDP in 1991 and 1998. The methodology for the plot census followed the standard CTFS protocols (Condit 1997). A summary of the basic data obtained from the plot is included in chapter 27.

Results

Basal Area and Tree Density

Mean tree densities from the first census in the HKK FDP for the different size classes were 1613 trees per hectare (≥ 1 cm dbh), 439/ha (≥ 10 cm), and 3.7/ha (≥ 100 cm). Mean basal area was 30.5 m^2/ha (≥ 1 cm dbh), 28.7 m^2/ha (≥ 10 cm dbh), 5.6 m^2/ha (≥ 100 cm dbh). A detailed analysis of the stand structure data is presented in Bunyavejchewin et al. (2001). Both the basal area and tree density of the plot were comparable to other seasonal evergreen forests in Thailand. In northeastern Thailand, seasonal evergreen forest dominated by *Hopea ferrea* (Dipterocarpaceae) had 1168 trees per hectare (>4.5 cm dbh) and 29.1 m^2/ha of basal area, while seasonal evergreen forest dominated by *Shorea henryana* (Dipterocarpaceae) had 1356 trees per hectare (>4.5 cm dbh) and 29.8 m^2/ha of basal area (Bunyavejchewin, 1999). However, the higher density of trees in the HKK FDP is the result of a lower minimum tree size for sampling. When trees <4.5 cm were excluded from the HKK FDP dataset, tree density dropped to 969/ha and basal area decreased slightly to 30.0 m^2/ha.

Floristic Composition

The second census of the HKK FDP included 251 species, 161 genera, and 58 families of trees and shrubs (Bunyavejchewin et al. 2002). The floristic composition of this forest is comparable to published reports for other seasonal evergreen forests in Thailand in both overall diversity and patterns of dominance. The family Dipterocarpaceae dominated the upper canopy of the forest, while species from the Euphorbiaceae, Annonaceae, Lauraceae, and Meliaceae were common in the midstory and understory. Shade-intolerant species indicative of disturbance, such as *Ailanthus triphysa* (Simaroubaceae), *Anthocephalus chinensis* (Rubiaceae), *Duabanga grandiflora* (Sonneratiaceae), *Pterocymbium javanicum* (Sterculiaceae), *Tetrameles nudiflora* (Datiscaceae), and *Toona ciliata* (Meliaceae), were relatively common in the HKK FDP. The HKK plot was less species-rich in absolute terms than other FDPs (e.g., Pasoh, peninsular Malaysia, 814 species; Lambir Hills, Sarawak, 1182; and Barro Colorado Island (BCI), Panama, 300). However, when species diversity was compared on a per tree basis, the HKK FDP had levels of diversity comparable to the aseasonal lowland forests in Malaysia. The floristic structure of the FDP was notable for the high number of species occurring at low densities. Approximately 60% of the species had mean densities of <1 individual per hectare.

Rarity

In one of the early analyses of the BCI FDP in Panama, Hubbell and Foster (1986) investigated patterns of commonness and rarity among tree species. They defined a species as rare if it had <50 individuals within the 50-ha plot—that is, if the mean density was <1/ha. One of the interesting features of the seasonal dry evergreen forest to emerge from the first census of the HKK FDP was the high proportion of rare species. Of the 259 species in the plot, 165 (63.7%) were rare, based on Hubbell and Foster's criterion.

To better assess the nature of the species' abundance patterns and to establish a quantitative basis for comparisons with other tropical forests, we fitted a log normal distribution to the HKK data. The log normal distribution, which has been used to describe species abundance patterns in temperate and tropical ecological communities, often underestimates the number of rare species in tropical forests (Hubbell and Foster 1986; Hubbell 2001). However, fitting the log normal curve to the HKK data was problematic. Although the modal abundance class was the smallest octave (1–2 individuals/species), the distribution of species' abundance had two maxima. We fitted the log normal distribution to both the true mode (i.e., 1–2 individuals/species) and the local maximum at the upper end of the abundance scale (128–256 individuals/species) (fig. 11.1). The true mode of the HKK data was much farther left than the modes of the other plots (data not shown). The curve fitted to the local maximum in the larger abundance classes suggests that the HKK species abundance distribution departs strongly from a log normal distribution due to the considerable overrepresentation of species in the lower abundance classes.

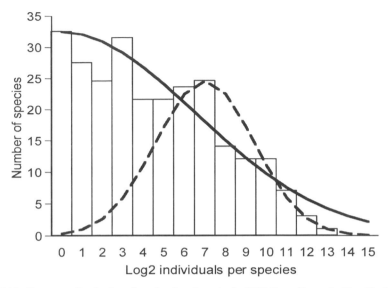

Fig. 11.1. Frequency distribution of species abundance in the HKK Forest Dynamics Plot. The horizontal axis is the abundance in logarithm base 2. The two curves are best-fit log normal distributions. The solid line is fitted to the true mode (1–2 individuals/species); the curve is $S(R) = 32.5 \exp(-0.0121\,R^2)$. The dashed line is fitted to the two abundance classes to the left of the eighth octave and all of the those to the right; the curve is $S(R) = 24.5 \exp(-0.089\,R^2)$.

Comparable data on species abundance were available for the 12 other CTFS FDPs: Mudumalai Forest Reserve (India), Palanan Wilderness Area (Philippines), Ituri Forest (Democratic Republic of Congo), Barro Colorado Island (Panama), Korup National Park (Cameroon), La Planada Nature Reserve (Colombia), Pasoh Forest Reserve (Peninsular Malaysia), Bukit Timah Nature Reserve (Singapore), Lambir Hills National Park (Sarawak), Luquillo Experimental Forest (Puerto Rico), Sinharaja World Heritage Site (Sri Lanka), and Yasuni National Park (Ecuador) (fig. 11.2). Because the FDPs vary in size, a uniform threshold for designating a rare species (e.g., <50 individuals) was inappropriate. Instead we used a generalized approach in which the rarity threshold was equivalent to the number of hectares in the FDP (e.g., in the 16-ha Luquillo plot, any species with <16 individuals on the plot was considered rare). Despite a 13-fold difference in species number and a full order of magnitude difference in tree density, the proportion of rare species was remarkably constant among the FDPs, with the exception of the two most seasonal forests: HKK and Mudumalai. In both of the seasonal forests from Asia, the proportion of rare species in the plots exceeded 60%. The FDP at the Luquillo Experimental Forest also had a relatively high proportion of rare species (~55%). In most of the other FDPs, the proportion

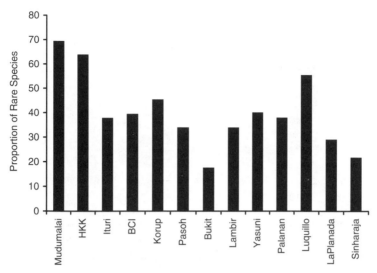

Fig. 11.2. Relative abundance of rare species in 13 of the CTFS Forest Dynamics Plots. A rare species is defined as a species with mean density of $<N$, where N is the number of hectares in the plot. The plots are arranged in order of decreasing seasonality and increasing annual rainfall: Mudumalai, 4 dry months, 1200 mm rainfall; HKK, 6, 1500; Ituri, 3–4, 1700; BCI, 3, 2600; Korup, 3, 5300; Pasoh, 1, 1800; Bukit Timah, 0, 2500; Lambir, 0, 2700; Yasuni, 0, 3100; Palanan, 0, 3500; Luquillo, 0, 3500; La Planada, 0, 4100; and Sinharaja, 0, 5000. The value for Ituri is the average of four 10-ha plots.

of rare species ranged from 30 to 45%, although the FDPs at Bukit Timah and Sinharaja had substantially lower proportions of rare species than the other FDPs. Regressions of proportion of rare species on total number of species and total number of individuals in each FDP showed no relationship between rarity and species richness (fig. 11.3) or tree density (fig. 11.4). What accounts for the HKK, Mudumalai, and Luquillo FDPs having nearly twice the proportion of rare species as the other FDPs?

In their study of commonness and rarity at BCI, Hubbell and Foster (1986) proposed several potential causes of rarity:

1. The required habitat for a species is spatially rare within the plot;
2. Individuals of a species occur in the plot as recent immigrants from external populations—i.e., the FDP population is a sink population;
3. The conditions necessary for recruitment of a species are temporally rare.

In the following sections we address the first and second of these possibilities for the HKK FDP. At present, data are unavailable to consider regeneration niche

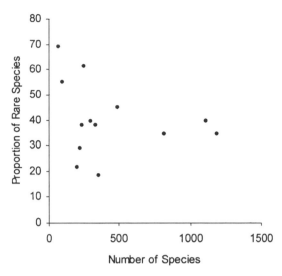

Fig. 11.3. Plot of the proportion of rare species versus the total number of species in each of the 13 forest dynamics plots. The value for Ituri is the average of four 10-ha plots. The relationship between rarity and number of species was nonsignificant (linear regression; $F = 0.77$, $p = 0.39$, $R^2 = 0.070$).

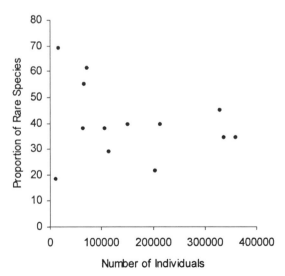

Fig. 11.4. Plot of the proportion of rare species versus the total number of individuals in each of the 13 forest dynamics plots. The value for Ituri is the average of four 10-ha plots. The relationship between rarity and tree density was nonsignificant (linear regression; $F = 0.83$, $p = 0.38$, $R^2 = 0.066$).

Table 11.1. Abundance and Habitat Specialization for Species with at Least 10 Individuals within the Forest Dynamics Plot

	Generalist Species	Specialist Species	Total
Rare (10–99)	**61 (53.2)**	**16 (23.8)**	77
Medium (100–999)	**29 (36.6)**	**24 (16.4)**	53
Abundant (1000+)	15 (15.2)	7 (6.8)	22
Total	105	47	152

Note: Species were defined as specialist or generalist based on a torus-translation analysis of spatial association with four habitats within the plot (see text for details). Observed (expected); values in bold are those with the greatest relative deviation from expected.

specialization (the third cause), although detailed maps of canopy heights to do so will soon be available (e.g., Welden et al. 1991).

Habitat Specialization

Following Hubbell and Foster (1986), we define habitat specialization as specialization for microsite conditions, such as topographic and edaphic conditions, that are unlikely to change significantly over the course of an individual tree's lifetime. Spatial associations between the distribution pattern of a species and the distribution of a particular habitat type provide evidence for habitat specialization. To identify patterns of association between species and habitat types, we classified each 20 × 20 m quadrat within the HKK FDP as ridge and hilltop, slope, stream, or flat. Species were considered habitat specialists if they had a significant positive association with a given habitat type ($p < 0.05$) based on the torus-translation method described by Harms et al. (2001). Analysis was limited to species with > 10 individuals within the plot. In those cases in which a species had a positive association with more than one habitat type, the greater of the associations was taken as the habitat of specialization. Results are presented for order-of-magnitude species abundance classes (10–99, 100–999, and 1000+ individuals).

The results showed that the majority (69%) of species at the HKK FDP were habitat generalists (i.e., no significant positive association for one or more habitat types). When species were compared by abundance class, there were significant differences in the degree of habitat specialization with indifferent species in the smallest abundance class (10–99 individuals) and specialized species in the middle abundance class (100–999) being overrepresented (log-likelihood ratio test; $G = 8.61$, $p = 0.01$) (table 11.1).

The results from the HKK FDP provide an interesting comparison to results obtained from the BCI FDP. First, the proportion of species that were habitat generalists (i.e., indifferent to habitat type) was considerably higher at HKK (69.1%) than at BCI (49.8%). Second, the HKK FDP had an overabundance of habitat-indifferent rare species. In their study of the tree species at BCI, Hubbell and Foster (1986) found the opposite pattern: The most abundant species were more likely to be generalists (i.e., indifferent to habitat type) than specialists. In addition, there were specific differences in the patterns of specialization between the HKK and BCI FDPs. For example, at HKK most specialists were found on the ridges and hilltops (48.9%) or near the stream (23.4%); few were found on the slopes (6.4%). In contrast, most specialists at BCI were found on the flat terrain (58.4%) or on the slopes (33.3%); only a small fraction of species were specialized for hydric habitats (8.3%).

Senescent Populations

If a species exists as a relict population from an earlier forest community or a sink population (i.e., not self-maintaining) from another forest type, its diameter distribution should show a predominance of large trees. Such senescent populations would artificially raise the level of tree species diversity within the study area. Identifying such populations not only would provide a more thorough understanding of local species diversity patterns but also might provide insights into historical forest dynamics.

To identify a senescent population requires two pieces of information: (1) species-specific minimum size at reproductive maturity and (2) the threshold adult/juvenile ratio of a self-maintaining population. For the tree species of the HKK FDP, we classified overstory and midstory trees (>10 m maximum height) as adults if their diameters were >20 cm. While the diameter limit criterion is somewhat arbitrary, it generally correlates with average minimum diameters observed for reproductive individuals of midstory and overstory trees within the FDP (J. V. LaFrankie, personal observation). In addition the diameter limit criterion is similar to that applied to the BCI data (R. Condit, personal communication). We did not include trees that had maximum diameters <20 cm (primarily understory trees and treelets) because of the lack of information on size at onset of reproductive maturity. Little work has been done on the comparative adult/juvenile ratios of senescing vs. regenerating populations. Hubbell and Foster (1986) suggested that if a population had >20% adults it could be considered senescent. For this analysis we have adopted the 20% criterion because neither region-specific nor site-specific data exist regarding threshold levels and because it allows comparisons between the BCI and HKK FDPs.

The results showed that the relative proportion of senescent species increases with decreasing abundance—that is, rare species are more likely to have senescent populations. Of the most common species (1000+ individuals), 34.4% had senescent populations, increasing to 46.4% for occasional species (100–999 individuals), and 66.1% for infrequent species (10–99 individuals). These results are almost identical to those from BCI (Hubbell and Foster 1986). The interpretation of the BCI data indicated that the senescent populations were relatively recent immigrants from nearby secondary forest, but that self-maintaining populations

Table 11.2. Species Abundance and Forest Type Association for All Tree Species with Known Forest Type Associations

	Deciduous Forest Species	Evergreen Forest Species	
Rare (1–50)	**45 (35.7)**	68 (77.3)	113
Common (50+)	**22 (31.3)**	77 (67.7)	99
Total	67	145	212

Note: Observed (expected); values in bold are those with the greatest relative deviation from expected. $P = 0.0006$, $\chi^2 = 11.71$.

had not been able to establish within the plot. At HKK, however, secondary forest occurs only in the sanctuary's buffer zone—more than 15 km from the FDP—and few of the plot species are true secondary forest species.

Nonetheless, given the mosaic distribution of forest types across the HKK landscape, it is possible that species not typically associated with seasonal evergreen forest might exist within the FDP. Indeed, during the preliminary analysis of the 50-ha plot data, it became clear that species associated with deciduous dipterocarp or mixed deciduous forest occurred sporadically throughout much of the plot. If species commonly found in other forest types occurred as recent immigrants or as relict populations that had survived past disturbances, the number of rare species in the seasonal evergreen forest plot would be artificially high. To investigate this possibility we classified each species in the FDP as typical of seasonal evergreen forest, mixed deciduous forest, deciduous dipterocarp forest, or unknown based on Santisuk (1988), the *Flora of Thailand* (Smitinand and Larsen, 1970–97), and a review of the Thai forestry and botany literature (S. Bunyavejchewin, unpublished data). For analysis, we divided species into two groups: evergreen or deciduous forest. A species was considered to be associated with the evergreen forest if it was known to occur in the seasonal evergreen forest, and with deciduous forest if it occurred in one or both of the deciduous forests but not in the seasonal evergreen forest. In this analysis, species with less than 10 individuals were included and species with unknown forest type association were not.

A 2 × 2 contingency analysis showed significant associations between abundance and forest type association (table 11.2). The greatest relative deviations were an observed deficit of common deciduous forest species (22 vs. 31.3) and an excess of rare deciduous species (45 vs. 35.7). Among the evergreen forest species, there was an excess of common species and a deficit of rare species. These patterns could have emerged in one of two ways: (1) the present site had historically supported seasonal dry evergreen forest and the deciduous forest species are recent immigrants or (2) the present site was deciduous forest in the past and has been subsequently replaced with seasonal dry evergreen forest.

The probability that the populations of deciduous forest species within the HKK FDP are sink populations that emigrated from source areas exterior to the plot is relatively small. The deciduous forest species found in the plot exhibit the full range of seed sizes and dispersal mechanisms. For example, *Shorea siamensis* (Dipterocarpaceae), *Terminalia bellerica* (Combretaceae), *Cananga latifolia* (Annonaceae) have relatively large, gravity-dispersed seeds, while *Pterocarpus macrocarpus* (Leguminosae), *Engelhardtia spicata* (Juglandaceae), and *Cratoxylum pruniflorum* (Guttiferae) have light, wind-dispersed seeds. The nearest source population of deciduous dipterocarp forest is located at least 3 km to the east beyond a high ridge, while the nearest mixed deciduous forest lies approximately 1 km south and downhill from the plot. For so many deciduous forest species of such varying dispersal ability to have traveled such great distances from different directions and established into the canopy of seasonal evergreen forest seems unlikely.

A more parsimonious explanation is that prior to the establishment of the current seasonal evergreen forest, a deciduous forest association existed. The shift in composition to seasonal evergreen forest left relict populations of deciduous forest species. Whether the shift occurred suddenly, the consequence of a catastrophic disturbance, or gradually is not clear. Such switches between forest types have been well documented in the temperate zone (e.g., Henry and Swan 1974; Abrams and Downs 1990; Orwig and Abrams 1994), but due to the technical difficulties of reconstructing historical stand dynamics in the absence of annual growth rings, they have never been documented in tropical forests. Certainly, many deciduous forest species in the FDP are represented by senescent populations of extremely large canopy trees. It is possible that deciduous forest species are represented by both emigrant populations and relictual populations from a previous deciduous forest community at the FDP site; however, the species composition and size structure of the plot species suggest that the majority of senescent populations are relictual.

Clearly, species abundance within the HKK FDP is related to forest type affiliation, but are deciduous forest species more likely to have senescent populations than evergreen forest species? When forest type is compared with senescence there are no significant differences ($p = 0.92$, $\chi^2 = 0.011$). In both evergreen and deciduous forest types, the species are evenly divided among senescent and regenerating populations. Nor are there significant differences when rare and common species are analyzed separately (rare species: $p = 0.33$, $\chi^2 = 0.946$; common species: $p = 0.88$, $\chi^2 = 0.023$). The lack of significant differences is interesting in that it highlights the surprisingly high number of senescent populations among evergreen species. However, examination of the evergreen forest species with senescent populations shows that many are long-lived, shade-intolerant species such as *Anthocephalus chinensis* (Rubiaceae), *Duabanga grandiflora* (Sonneratiaceae), *Melia azederach* (Meliaceae), and *Toona ciliata* (Meliaceae). In addition, many of the important structural elements of the seasonal evergreen forest have senescent populations. Examples include *Hopea odorata* (Dipterocarpaceae), *Dipterocarpus alatus* (Dipterocarpaceae), *Litsea cambodiana* (Lauraceae), and *Neolitsea obtusifolia* (Lauraceae).

Senescent populations of many evergreen forest species imply that the conditions that enabled the establishment of many seasonal evergreen forest species were unique and have not existed for a long time. A large disturbance would be necessary to provide the conditions for establishment and canopy recruitment of the long-lived pioneers. Subsequent forest development in the absence of further disturbances would preclude regeneration of such species, leading to senescent populations. While it may be argued that long-lived pioneers established and grew to the canopy as a consequence of gap phase dynamics, a considerable number of species do not appear to be regenerating at sufficient levels in the present-day gaps to ensure future canopy recruitment (P. J. Baker, personal observation). Thus, the lack of significant differences in population structure between evergreen and deciduous forest species, when considered in the light of the species' autoecologies, strengthens the argument for a past disturbance initiating a sudden shift in the community composition at the HKK FDP. A recent study that used a combination of tree-ring records, age estimation, and architectural analyses of the dominant canopy tree species suggested that a catastrophic disturbance in the vicinity of the FDP occurred in the mid-1800s affecting an area of forest >200 ha (Baker 2001), lending support to the conclusions presented here.

While it is uncertain what type of disturbance could have had such a profound affect on the stand of seasonal dry evergreen forest at the HKK FDP, several possibilities exist. Forest fires occur regularly within the sanctuary; however, the fires are typically low intensity and unlikely to create major canopy gaps within the seasonal dry evergreen forest. Humans are another potential disturbance agent. Scant anthropological evidence for historical settlement in the region exists, so we know little of the role and impact of humans on these forests. However, it is worth noting that wars between the Thai and Burmese kingdoms were relatively common in the 1500–1800s and that war parties were known to pass through the regions. Windstorms are known to occur in the region. Several old reports from the British–Burmese forestry literature of the early 20th century describe windstorms and line squalls that flattened several square kilometers of forest in an area relatively close to HKK. More recently, a 30-ha patch of forest was knocked flat by windstorms at HKK in 1987 (T. Prayurasiddhi, personal communication).

The two other forests with high proportions of rare species, Mudumalai and Luquillo, have also been subject to significant disturbances in the past. At Mudumalai, forest fires occur frequently and the high densities of large herbivores, such as elephants, leads to widespread physical damage to many trees. In addition, Mudumalai and the surrounding forests have been subject to logging in the past. These disturbances have had a profound influence on the structure and functioning of the mixed deciduous forest at Mudumalai (see chap. 33). The FDP at Luquillo was established in 1989 immediately after Hurricane Hugo struck Puerto Rico to study the effects of an intense disturbance on a tropical forest ecosystem (see chap. 32). The forest was struck again by Hurricane Georges in 1998. In addition to hurricane impacts, the Luquillo forest has been subject to intensive land use (primarily logging and farming) during the past century.

General Implications

Tropical forests are among the most diverse biotic communities in the world. The network of FDPs established by CTFS has begun to document the nature and scale of tree species rarity throughout much of the tropics. The high proportion of rare tree species in the seasonal dry evergreen forest at HKK provides such an example. The results from the first census underline the importance of both the spatial and temporal aspects of diversity. Specialization for spatially discrete habitats occurred in only 30% of the species for which sufficiently large samples were available; although of the rare species, nearly 80% were habitat generalists. In contrast, Hubbell and Foster (1986) found that the majority of rare species at BCI were specialists; however, habitat specialization *and* regeneration specialization were included in their final determination of species specialization. Our analysis is limited to specialization for habitat. Identifying patterns of regeneration niche specialization based on the results of a canopy height survey will provide a clearer picture of the nature of specialization among rare species at HKK and allow more thorough comparisons with data from BCI and other FDPs.

Much has been written of the highly specialized niche differentiation of tree species in species-rich tropical forests (e.g., Ashton 1969). In this view, patterns of rarity and commonness are a function of available niche space; rare species have little available niche space, common species have relatively more. However, in recent years, some authors have suggested that a majority of tropical tree species may be broad generalists (e.g., Hubbell 1979; Welden et al. 1991; Lieberman et al. 1995). Our analysis, which is necessarily coarse due to the broad habitat definitions, does not provide strong evidence in support of the hypothesis that the success of a species, as measured by its abundance, is positively correlated with habitat specialization. Our data do, however, raise the interesting possibility that rare species may avoid local extinction by being habitat generalists.

Previous studies from CTFS FDPs have highlighted the spatial patterns of rarity in tropical forests and stressed the importance of large reserve size in maintaining populations of rare plants (e.g., Hubbell and Foster 1986; He et al. 1997). Our results suggest that the temporal dynamics of forest development may be an equally important component of rarity at the local scale. The effects of disturbances that occurred long in the past may still be manifested in the current stand structure and species composition. A well-developed understanding of the temporal patterns of stand development is of paramount importance to forest managers and conservationists when attempting to maintain or establish specific structural or compositional features in a stand or across a landscape, particularly if those structures are necessary to maintain current and future populations of threatened or endangered wildlife species.

Acknowledgments

The Huai Kha Khaeng FDP is an ongoing collaborative project of the Thai Royal Forest Department, Harvard University, and the Center for Tropical Forest Science, Smithsonian Tropical Research Institute. We thank the many people who have contributed to the establishment, enumeration, and maintenance of the FDP over the past decade. In particular we remember Dr. Tem Smitinand and Mr. Sueh, both of whom passed away before the completion of the first census. We gratefully acknowledge the support of the staff and workers of the Huai Kha Khaeng Wildlife Sanctuary. For helpful discussions we thank Rick Condit and Bert Leigh. Financial support for the establishment of the HKK FDP was provided by a grant from USAID and the John Merck Fund to Harvard University.

References

Abrams, M. D., and J. A. Downs. 1990. Successional replacement of old-growth white oak by mixed mesophytic hardwoods in southwestern Pennsylvania. *Canadian Journal of Forest Research* 20: 1864–70.

Ashton, P. S. 1969. Speciation among tropical forest trees: Some deductions in light of recent evidence. *Biological Journal of the Linnean Society* 1:155–96.

Baker, P. J. 1997. Seedling establishment and growth across forest types in an evergreen/deciduous forest mosaic in western Thailand. *Natural History Bulletin of the Siam Society* 45:17–41.

———. 2001. *Age Structure and Stand Dynamics of a Seasonal Tropical Forest in Western Thailand.* Ph.D. dissertation, University of Washington, Seattle, WA.

Bunyavejchewin, S. 1999. Structure and dynamics in seasonal dry evergreen forest in northeastern Thailand. *Journal of Vegetation Science* 10: 787–92.

Bunyavejchewin, S., P. J. Baker, J. V. LaFrankie, and P. S. Ashton. 2001. Stand structure of a seasonal evergreen forest at the Huai Kha Khaeng Wildlife Sanctuary, western Thailand. *Natural History Bulletin of the Siam Society* 49: 89–106.

Bunyavejchewin, S., P. J. Baker, J. V. LaFrankie, and P. S. Ashton. 2002. Floristic composition of a seasonal evergreen forest, Huai Kha Khaeng Wildlife Sanctuary, western Thailand. *Natural History Bulletin of the Siam Society* 50: 125–34.

Condit, R. 1997. *Tropical forest census plots: Methods and results from Barro Colorado Island, Panama and a comparison with other plots.* Springer-Verlag. Berlin, Germany.

Harms, K. E., R. Condit, S. P. Hubbell, and R. B. Foster. 2001. Habitat associations of trees and shrubs in a 50-ha neotropical forest plot. *Journal of Ecology* 89:947–59.

He, F., P. Legendre, and J. V. LaFrankie. 1997. Distribution patterns of tree species in a Malaysian tropical rain forest. *Journal of Vegetation Science* 8:105–14.

Henry, J. D., and J. M. A. Swan. 1974. Reconstructing forest history from live and dead plant material: An approach to the study of forest succession in southwest New Hampshire. *Ecology* 55:772–83.

Hubbell, S. P. 1979. Tree dispersion, abundance, and diversity in a tropical dry forest. *Science* 203:1299–1309.

Hubbell, S. P. 2001. *The Unified Neutral Theory of Biodiversity and Biogeography.* Monographs in Population Biology No. 32. Princeton University Press, Princeton, NJ.

Hubbell, S. P., and R. B. Foster. 1986. Commonness and rarity in a neotropical forest: Implications for tropical tree conservation. Pages 205–31 in M. E. Soule, editor. *Conservation Biology: The Science of Scarcity and Diversity.* Sinauer Associates, Sunderland, MA.

Lieberman, M., D. Lieberman, R. Peralta, and G. S. Hartshorn. 1995. Canopy closure and the distribution of tropical forest tree species at La Selva, Costa Rica. *Journal of Tropical Ecology* 11:161–78.

Orwig, D. A., and M. D. Abrams. 1994. Land-use history (1720–1992), composition, and dynamics of oak-pine forests within the Piedmont and Coastal Plain of northern Virginia. *Canadian Journal of Forest Research* 24: 1216–25.

Santisuk, T. 1988. *An Account of the Vegetation of Northern Thailand.* Steiner-Verlag. Stuttgart, Germany.

Smitinand, T., and K. Larsen. 1970–97. *Flora of Thailand.* Vols. 2–6. Forest Herbarium (BKF), Royal Forest Department, Bangkok.

Welden, C. W., S. W. Hewett, S. P. Hubbell, and R. B. Foster. 1991. Sapling survival, growth, and recruitment: Relationship to canopy height in a neotropical forest. *Ecology* 72: 35–50.

12

Stand Structure and Species Diversity in the Ituri Forest Dynamics Plots: A Comparison of Monodominant and Mixed Forest Stands

Jean-Remy Makana, Terese B. Hart, David E. Hibbs, and Richard Condit

Introduction

In tropical forests, does dominance by a single species of canopy tree necessarily give rise to lower diversity among other trees coexisting with it (Hart et al. 1989; Connell and Lowman 1989; Johnston and Gillman 1995)? Does the dominant species of canopy tree cast shade deep enough to inhibit the recruitment and growth of other, more light-demanding canopy species? Or is the dominant species an exceptional intruder in an otherwise diverse forest, a species that has somehow circumvented the factors that limit the abundance of other tree species?

To date, the relationship between single-species canopy dominance and species diversity has only been examined using large-diameter trees in relatively small plots, where sample sizes have not been sufficient to draw robust conclusions. In the Ituri forest of the Democratic Republic of Congo, forests dominated by a single canopy tree species (*Gilbertiodendron dewevrei* [Leguminosae (Caesalpinoideae)]) co-occur with mixed canopy species under the same climatic and soil conditions. Forest Dynamics Plots in each of these forest types allow for the rigorous comparison of tree species diversity across both forest types (hereafter referred to as monodominant and mixed forest). This study compares size class distributions and begins to investigate other phenomena that might be associated with canopy dominance such as stand structure, dominance in the understory, and the dominance and diversity of lianas.

In monodominant forest, *G. dewevrei* may account for up to 90% of large trees (Gérard 1960; Hart et al. 1989; Makana et al. 1998). The strong dominance of this species is thought to partly depend on its ability to cast and tolerate deep shade, making it very difficult for other tree species to establish under its canopy (Richards 1996; Torti et al. 2001). We hypothesize that species that require higher illumination to recruit to mature stage, such as the majority of canopy tree species of the Ituri forest, will be less represented in monodominant stands, whereas small-statured tree species that commonly complete their life cycles in the forest understory will not be affected by the dominance of *G. dewevrei* in the forest canopy.

Methods

In 1993, the Centre de Formation et de Recherche en Conservation Forestière (CEFRECOF) established two 10-ha Forest Dynamics Plots in each of the two main vegetation types of the Ituri forest: the mixed canopy semievergreen forest and the evergreen forest dominated by *G. dewevrei*. The forests surrounding the village of Epulu and the Forest Dynamics Plot are described in greater detail in chapter 28.

For this study, we averaged the results of the two 10-ha plots from each forest type. Two parameters (stem density and species richness) and three dbh classes (≥ 1, ≥ 10, and ≥ 30 cm dbh) were used for most analyses. For the statistical comparison of stem density and basal area, we used nonoverlapping size classes: 1.0–9.9, 10.0–29.9, and ≥ 30 cm dbh. Unless otherwise specified, the results presented in this paper are based on a 1-ha scale from averaged 1-ha subsamples. Species–area curves were constructed by plotting the mean number of species against the area for nonoverlapping square quadrats of varying sizes (Condit et al. 1996b). Quadrats of the following dimensions were used: 5, 10, 20, 40, 80, 100, and 200 m on a side. The last point on each curve corresponded to the mean number of trees and species in the whole 10-ha plots in each forest type. We used the same approach to generate species–individual curves, but instead of area, the mean number of species was plotted against the mean number of individuals for each quadrat size (Condit et al. 1996b). Seven different quadrat sizes were used: 20, 40, 60, 100, 160, 200 m on a side, and 200 × 500 m (whole plot).

The number of species and the density of large *G. dewevrei* individuals (≥ 30 cm dbh) were calculated for each hectare. A regression was calculated for the number of species on the density of *G. dewevrei* to evaluate the impact of single-species dominance, by this canopy species, on species richness in the Ituri forest. The effects of *G. dewevrei* dominance were assessed through regression analyses at three levels. First, we evaluated the effects on overall richness; second, we excluded all trees less than 10 cm dbh; and third, we considered only trees above 30 cm dbh.

Results and Discussion

Stand Structure

The Ituri Forest Dynamics Plot census enumerated a total of 299,139 individuals of shrubs and trees ≥ 1 cm dbh and 19,268 individual lianas ≥ 2 cm dbh. The two forest types differed in size distribution, with mixed forest having more small trees (< 30 cm dbh) and monodominant having more large trees (≥ 30 cm dbh). The difference was significant (Chi-squared $= 10.78$, d.f. $= 2$, $p = 0.005$), particularly in the 10–30 cm dbh size class, where the mixed forest plots had much higher

Table 12.1. Tree Density and Basal Area of Monodominant and Mixed Forests of the Ituri Region

Size Class (cm)	Monodominant Forest		Mixed Forest	
	Mean ± SD	Range	Mean ± SD	Range
	Mean Number of Trees/ha			
1 < 5	5902.8 ±	4913–6941	6810.5 ± 1059.1	5341–8949
5 < 10	582.7 ± 45.2	488–644	862.9 ± 101.2	714–1178
10 < 15	147.9 ± 35.6	88–224	236.0 ± 27.3	183–277
15 < 20	58.6 ± 16.3	36–88	75.0 ± 12.4	56–110
20 < 30	53.3 ± 11.9	31–73	50.0 ± 11.1	33–74
30 < 60	64.4 ± 18.5	41–110	55.2 ± 7.4	38–65
>=60	33.5 ± 7.7	15–48	22.1 ± 5.5	14–35
	Mean Basal Area (m²/ha)			
1 < 5	2.66 ± 0.30	2.19–3.18	3.33 ± 0.70	2.51–4.82
5 < 10	2.23 ± 0.23	1.83–2.64	3.58 ± 0.37	3.04–4.83
10 < 15	1.74 ± 0.46	0.99–2.69	2.79 ± 0.30	2.24–3.27
15 < 20	1.36 ± 0.40	0.80–2.13	1.75 ± 0.30	1.25–2.63
20 < 30	2.54 ± 0.64	1.43–3.74	2.37 ± 0.56	1.60–3.55
30 < 60	9.49 ± 2.64	6.04–15.75	8.02 ± 1.17	5.66–10.86
>=60	17.46 ± 4.68	8.59–23.91	11.36 ± 2.96	6.60–16.86

Note: Each forest type is represented by two 10-ha Forest Dynamics Plots.

density than the monodominant forest plots. The two forest types also showed significant differences in basal area for all three size classes, with mixed forest having higher basal area values than monodominant forest for size classes below 30 cm dbh and monodominant exhibiting a higher value of large trees (table 12.1). Trees <10 cm dbh accounted for 21% of the total basal area in mixed forest, but only 13% of the basal area in monodominant forest.

These differences in tree density and basal area support the observation that the understory is sparser and more open in monodominant forest compared to mixed stands (fig. 12.1), an observation made earlier by Lebrun and Gilbert (1954). There are significantly more gaps in mixed forest than in monodominant forest (Hart et al. 1989), and tree crowns of mixed forest are often not contiguous, allowing more light to reach the understory (Makana et al. 1998). In contrast, the monodominant forest canopy is homogeneous, formed by contiguous crowns of G. dewevrei trees. This dramatically reduces light penetration under the monodominant forest canopy. Only 0.58% of full sunlight reaches the understory of monodominant forest (Torti et al. 2001). Light levels in mixed forest are 1–4% of full sunlight.

Stand structure varied across each plot (fig. 12.1). For trees <10 cm dbh, mixed forest had a greater variation of tree density among individual hectares than monodominant forest, but monodominant stands showed larger variation than mixed for trees both 10–30 cm dbh and ≥30 cm dbh (table 12.1). The basal area for small trees was more variable in mixed forest, but in monodominant

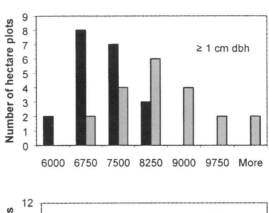

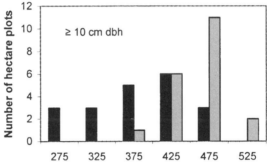

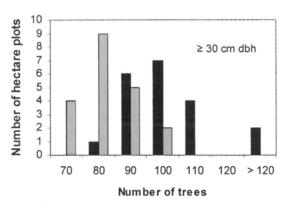

Fig. 12.1. Distribution of stem density in square hectares (100 × 100 m) in monodominant and mixed forests in the Ituri Forest Dynamics Plots ($n = 20$ for each forest type). Mixed (light bars); monodominant stands (dark bars). The x-axis values represent the upper boundaries of each density class.

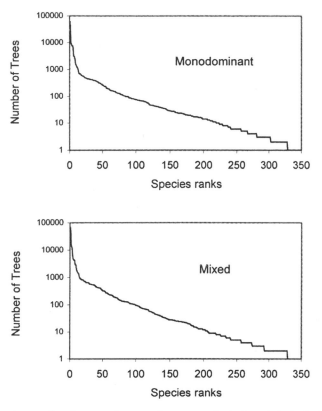

Fig. 12.2. Dominance–diversity curves for monodominant and mixed stands in the Ituri forest. The *y*-axis is the abundance of species in 20 ha (two 10-ha plots) in each forest type. The *x*-axis ranks species according to their abundances (from the most abundant to the least abundant). All individuals ≥1 cm dbh are included.

forest the larger size classes demonstrated greater variation (table 12.1). Studies are only now beginning to examine the association between topography and forest structure in tropical forests. In the Ituri forest, swamp areas appear to have low tree density whereas slopes seem to be particularly densely populated.

Species Richness

Our results refute the assumption that monodominant forests are less diverse than forests with lower dominance. At scales of 1 hectare and above, mixed and monodominant forests had comparable levels of overall species richness among trees ≥1 cm dbh. Within the two 10-ha Forest Dynamics Plots of monodominant forest, a total of 403 species were recorded above the 1 cm dbh limit for

Table 12.2. Number of Species in Subsamples of Forest Dynamics Plots in Monodominant and Mixed Stands in the Ituri Forest, and Comparison of Species Richness between Forest Types as Measured by ANOVA Analysis

Size Class (cm)	Monodominant		Mixed		ANOVA (p-value)
	Mean ± SD	Range	Mean ± SD	Range	
	Number of Species per 1 ha				
≥ 1	178 (19)	142–221	170 (12)	152–191	0.26
≥ 10	56 (24)	18–101	68 (8)	56–85	0.14
≥ 30	14 (8)	1–25	27 (5)	17–38	0.03
	Number of Species per 0.16 ha				
≥ 1	87 (14)	62–141	89 (9)	70–122	0.573
≥ 10	17 (10)	4–42	26 (5)	13–38	0.046
≥ 30	4 (3)	1–11	7 (2)	1–14	0.036
	Number of Species per 0.04 ha				
≥ 1	43 (10)	22–77	47 (7)	26–69	0.248
≥ 10	7 (4)	1–20	11 (3)	1–21	0.045
≥ 30	2 (1)	0–6	2.5 (1)	0–8	0.089

free-standing woody trees. Overall richness for the two 10-ha plots in mixed forest was 410 species. Mean richness per hectare was actually higher for monodominant forest: 178 species versus 170 in mixed forest. While mean number of species per hectare for trees ≥10 cm dbh and ≥30 cm dbh was higher in mixed forest than in monodominant forest, the difference was statistically significant only for trees above 30 cm dbh (table 12.2). At smaller scales, 0.16 ha (40 × 40 m) and 0.04 ha (20 × 20 m), mixed forest was richer than monodominant forest in tree species ≥10 cm dbh, but not in the smaller size classes (table 12.2).

Compared to tropical forests in general, the Ituri forest has reasonably high diversity. The 275 species of trees ≥1 cm dbh averaged across the two 10-ha subplots of monodominant forest is higher than the average of the 10-ha subplots of the mixed forest at Barro Colorado Island (BCI) in Panama (243 per 10 ha), though it is exceeded by the very diverse forests of Amazonian Ecuador (Yasuni, 1006 species per 10 ha) and Southeast Asia (Lambir, 1026; Pasoh, 724). The one other African Forest Dynamics Plot in Korup National Park, Cameroon, also exhibits a relatively high species diversity (378 per 10 ha). The misconception that monodominant forests in Africa are low diversity seems to have originated from smaller plots (<1 ha), where only larger trees were considered. Indeed, the Ituri forests do contain fewer species of the larger trees. The mixed forest averaged 69 species per ha for trees ≥10 cm dbh, while monodominant forest averaged 56 species per ha; this is substantially lower than the 91 species per ha found at BCI. The difference is even more striking when smaller plots are utilized.

Thus, forests in Ituri appear species poor at smaller scales and in the larger size classes, but quite diverse at larger scales and when all trees ≥1 cm dbh are considered, and this holds true for both the monodominant *Gilbertiodendron* stands and the mixed forest. This finding suggests that most of the richness in monodominant forest is accounted for either by rare species or by species with highly clumped distributions, occurring in only a few quadrats (perhaps rare species are limited to specific habitats such as gaps or swamps). We examine these assertions further below, first by considering dominance and rarity, then by examining how dominance relates to diversity.

Relative Abundance of Species

The Janzen–Connell hypothesis for the maintenance of high species diversity in tropical forests suggests low relative abundance for all the species in a given forest community (Janzen 1970; Connell 1971; Hubbell 1979). Other CTFS Forest Dynamics Plots support this assumption. For example, the most abundant species in the two Malaysian Forest Dynamics Plots (Pasoh and Lambir) only accounted for about 3% of the total number of trees in their respective plots (Kochummen et al. 1990; chaps. 31 and 36), and in the 50-ha plot on BCI, the most abundant species represented 16% of the trees (Hubbell and Foster 1986; Condit et al. 1996a; chap. 24).

But the Ituri forest is markedly different in this respect, exhibiting a high degree of dominance. More than half of all the trees ≥1 cm dbh were accounted for by only two species in both forest types (chap. 28). The shrub, *Scaphopetalum dewevrei* (Sterculiaceae), was extremely abundant in both monodominant and mixed stands of the Ituri forest, representing over 40% of all free-standing trees ≥1 cm dbh in both forests. This species only rarely reached a dbh of 10 cm. In the understory, the Ituri forest is dominated not by *Gilbertiodendron* but by *Scaphopetalum*, for both sites, whether or not *Gilbertiodendron* is in the canopy.

At 1 cm dbh, the 8 most common species (2% of the total number of species) represented 75% of the trees, and 72 species (19%) accounted for 95% of the trees in monodominant forest. In mixed stands, 10 species (2.6%) and 70 species (18.5%) made up 75% and 95% of the trees, respectively. In larger size classes, dominance was more concentrated in monodominant than in mixed forest. In monodominant forest, *Gilbertiodendron* accounted for 51% and 71% of trees ≥10 cm dbh and ≥30 cm dbh, respectively. The most common canopy species of mixed forest, *Cynometra alexandri* [Leguminosae (Caesalpinioideae)], represented 16% of the trees at 10 cm dbh and 32% at 30 cm dbh.

While a few species were very abundant, the majority of species were rare (fig. 12.2). Hubbell and Foster (1986) used the cutoff of less than 1 individual per ha to identify "rare" species. In 1995, for the BCI Forest Dynamics Plot, researchers enumerated 33% of its species as rare, while at Ituri, 55% species qualified as rare in each of the 20 ha of mixed forest and the monodominant forest. For trees ≥10 cm dbh, Ituri also had more rare species, with 81% in monodominant forest and

76% in mixed forest, compared to 61% at BCI. (These comparisons are based on a 1000 × 200 m block within the 50-ha plot at BCI to ensure that the contrasts are not due to sample size; the comparison also holds if 10-ha plots at Ituri are compared with 500 × 200 m of the BCI plot). Clearly, species richness in the Ituri forest is enhanced by a high occurrence of rare species.

Gilbertiodendron *Dominance and Species Richness*

Though mean species richness per hectare was nearly equal between mixed and monodominant forests for all diameter limits, variation in the number of species per hectare was much greater in monodominant forest. High canopy dominance is associated with a local reduction in the number of species. To illustrate this relationship, we plotted the number of species against the density of *Gilbertiodendron* ≥30 cm dbh in individual hectares for the four 10-ha plots (fig. 12.3). The negative relationship between species richness and the density of large *Gilbertiodendron* grew stronger as progressively larger-sized trees were considered. When all trees above 1 cm dbh were considered, no significant relationship was observed between species richness and the density of large *Gilbertiodendron* ($r^2 = 0.004$, $p = 0.96$). The r^2 increased to 0.46 and 0.68 for trees ≥10 cm dbh and trees ≥30 cm dbh respectively, and the relationship was significant for both size classes ($p < 0.0001$).

Species–Area and Species–Individual Curves

As in other Forest Dynamics Plots (Condit et al. 1996b), the number of species at Ituri increased rapidly with increasing area from 0.0025-ha quadrats to 4-ha subplots in both forest types at 1 cm dbh (fig. 12.4). The rate of increase declined with area but did not plateau for the largest area, suggesting that more new species would be added if a larger area were surveyed. Species–area curves clearly demonstrate that in small quadrats, monodominant forests have low diversity relative to mixed forests, but monodominant forests "catch up" in larger quadrats. This is most notable for trees ≥10 cm dbh, where curves from the two forests cross (fig. 12.4). It is also evident for trees ≥30 cm dbh, since monodominant forests demonstrate 51% of the richness of mixed forest in 1-ha quadrats, compared to 71% at the 10-ha scale. The curve for a monodominant forest is nearly a straight line whereas that of the mixed forest has a downward concave shape, indicating that the accumulation rate of species decreases with increasing quadrat area in the latter.

Why does the monodominant forest catch up? A simple explanation is that most species are rarer there due to the abundance of *Gilbertiodendron*, and thus more sampling is required in these forests. Comparing a species–individual curve from mixed forest with a curve from monodominant forest when *Gilbertiodendron* is excluded can test this assertion. If this explanation is correct, these species–individual curves will be similar for both forest types, and indeed they

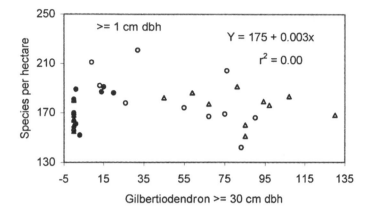

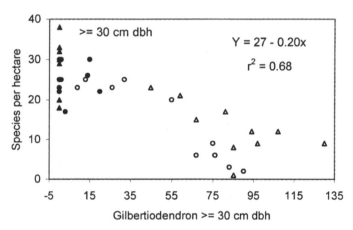

Fig. 12.3. Relationship between the number of species and the density of large *G. dewevrei* in 1-ha subplots in the Ituri forest plots. Open circles and triangles represent monodominant forest 1 and 2, respectively; closed circles and triangles represent mixed forest 1 and 2, respectively.

are (fig. 12.5). When comparing the same number of trees, excluding *G. dewevrei*, monodominant forest was almost as species rich as mixed forest. This is similar to the effects of random deletion of individuals in species-rich plots as shown by Cannon et al. (1998). Thus the difference in the shapes of species–area curves of monodominant and mixed forests at larger diameter limits (i.e., ≥10 cm dbh) is essentially caused by the extremely uneven distribution of species abundance in monodominant forest due to the high dominance of *Gilbertiodendron*. (See chap. 7 for detailed explanations on the effects of species abundance on species accumulation curves.)

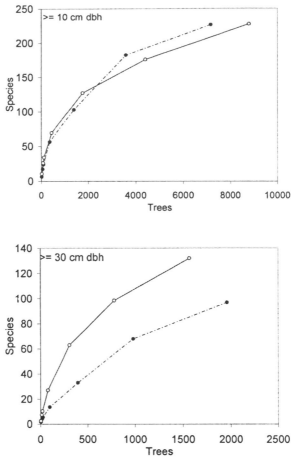

Fig. 12.4. Species–area curves of monodominant forest stands (dotted lines) and of mixed forest stands (solid lines) in the Ituri Forest Dynamics Plots.

Most canopy species in the Ituri forest require some level of canopy opening to establish and grow in the early stages of development. It may be that the deep shade cast in the understory of *G. dewevrei* reduces the likelihood of establishment for other canopy species in monodominant forest as suggested by Torti et al. (2001). But figure 12.5 indicates that most of the canopy tree species present in mixed stands are also present, at very low density, in monodominant stands. Thus the low light levels do not appear to eliminate selected species but rather to displace all species indiscriminately.

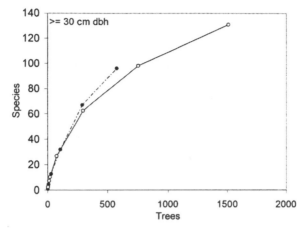

Fig. 12.5. Species–individual curves of monodominant forest stands (dotted lines) and of mixed Forest stands (solid lines) in the Ituri Forest Dynamics Pots. *Gilbertiodendron dewevrei* was excluded from the count.

Liana Density and Diversity

Though an important part of tropical forests, lianas are often neglected in tropical forest studies (Gentry 1991). The enumeration of lianas is challenging because liana individuals are hard to distinguish–they can be connected underground and they split frequently aboveground. Among all the CTFS Forest Dynamics Plots, only two sites, Sinharaja and Ituri, have included complete inventories of climbing plant species. Unfortunately, differences in methodology may make it difficult to compare results from these two studies (Condit 1998). Lianas are now being added to the Korup Forest Dynamics Plot using the identical methodology to that of Ituri. In the methods developed for Ituri, we counted all shoots at breast height, and although this may overestimate the actual number of individuals (Parren et al. submitted for publication), it is the simplest way to assess the abundance and species diversity of climbers.

The density of lianas differed in each of these forest types (table 12.3). Mixed forest, with 642 trees/ha, had nearly twice as many lianas as monodominant forest (322 trees/ha). Lianas are especially abundant in open areas at the forest edge (roadsides and stream sides) and in disturbed environments such as treefall gaps or gaps created by forest exploitation (Parren et al. submitted for publication). We suggest that the higher light levels observed in mixed forest (Torti et al. 2001) explains the higher density of lianas. More detailed studies are required, however, to determine the relationship between forest structure and liana density (e.g., light requirements for regeneration of the most abundant liana species), as Makana et al. (1998) suggested earlier.

Table 12.3. Mean Density and Richness of Lianas per ha in the Two Forest Types of the Ituri Forest Dyanmics Plots

Size Class (cm)	Monodominant Forest	Mixed Forest
	Density (stems/ha)	
≥2 cm	322	642
≥5 cm	53	75
≥10 cm	5	7
	Richness (species/ha)	
≥2 cm	57	58
≥5 cm	21	23
≥10 cm	4	4

Despite the higher number of lianas in mixed forest compared to monodominant forest, the latter had slightly more species in the entire 20 ha: 187 liana species in monodominant forest versus 161 in mixed of lianas ≥2 cm dbh. In small quadrats, though, liana richness was lower in monodominant forest and negatively correlated with *Gilbertiodendron* density (fig. 12.6). This trend echoed the pattern demonstrated by trees; liana species richness at small scales was lower in monodominant than in mixed stands, but at larger scales it caught up. One liana, *Manniophyton fulvum* (Euphorbiaceae), dominated the liana flora in mixed forest, accounting for 23.5% of the stems with 152 individuals per hectare. No species achieved such dominance in monodominant forest. Lianas were a significant part of the overall diversity in the forests encompassed by the plots, representing about one third of the total species in both forest types.

Although African forests may have a higher liana density than Asian or the neotropical forests (Reitsma 1988; Gentry 1991), lianas are an important part of tropical forests in general. Campbell and Newbery (1993) found that a lowland forest of Sabah, Malaysia, had 882 liana stems per hectare, a figure comparable to ours given the smaller size limit used in the Sabah study (0.64 cm dbh). We recommend that liana inventories be included in most of the CTFS Forest Dynamics Plots so that exhaustive comparisons can be made.

Conclusions

The assumption that mixed stands are richer in trees species than monodominant forest was not supported by our results. In areas of 1 hectare or greater, species richness was comparable in these two forest types, although mixed forest was consistently richer when only smaller areas (0.16 ha or less) were considered. A negative correlation was observed between the density of large *G. dewevrei* and the number of species at the hectare scale. Further examinations are required to fully understand the different patterns of species richness between the two forest types.

The comparison of forest structure between mixed and monodominant stands corroborates results obtained from previous studies (Hart et al. 1989; Makana et al. 1998). Mixed forest has a greater density of small trees, whereas larger trees (≥30 cm dbh) are denser in monodominant stands. The understory of both forest types is dominated by the shrub species *S. dewevrei*, whereas *G. dewevrei* forest has a slightly higher richness of lianas. While lianas are much more abundant

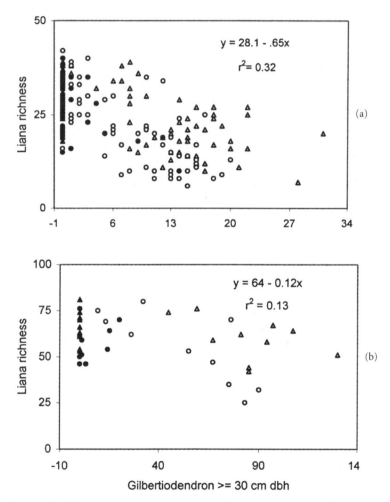

Fig. 12.6. Relationship between the number of species of lianas and the density of large *G. dewevrei* in the Ituri Forest Dynamics Plots. Open circles and triangles represent monodominant forest plots; closed symbols represent mixed forest plots. a. 40 × 40 m quadrats; b. 100 × 100 m quadrats.

in mixed forest than in monodominant forest, the former's liana diversity was dominated by a single species, *Manniophyton fulvum*.

General Implications

A major goal of tropical forest conservation is the maintenance of their high species diversity. In light of this, we consider it a fundamental observation that the monodominant forest of the Ituri region is just as species rich as the nearby mixed forest and it also compares favorably to tropical forests in other parts of the world. However, the strong dominance of *G. dewevrei* in monodominant forest is associated with a reduction in the local abundances of other overstory tree species, demonstrated by the higher quantity of more rare species in Ituri forests than in other tropical sites. Consequently, conservation in monodominant forests might include more emphasis on very large areas, to include viable populations of the non-*Gilbertiodendron* canopy species.

In terms of forest management for the production of timber and other forest products, it is important to recognize the fundamental difference in structure between monodominant and mixed forests. The monodominance of *Gilbertiodendron* appears to be closely tied to low light levels in the understory. We hypothesize that selective harvesting, which tends to open up the canopy and allow more light in the understory, might negatively affect this species. Because *Gilbertiodendron* is recognized as a valuable timber species with extensive local use, studies should be carried out to investigate how *Gilbertiodendron* would react to increased light levels in the understory following selective logging.

Acknowledgments

The welcome that the Conservation and Parks Institution (ICCN) of the Democratic Republic of Congo gave to this large plot project is typical of their willingness to collaborate with outside research and conservation organizations. Through the ICCN we received access to the Réserve de Faune à Okapi and logistical support from its research center, Centre de Formation et de Recherche en Conservation Forestière (CEFRECOF). Support for the field work came from a Smithsonian Institution Scholarly Studies grant to R. Condit and T. B. Hart and the continued strong backing of the Wildlife Conservation Society. The analytical work being undertaken by the first author is possible through a scholarship from the Beinecke Memorial Scholarship Program. We particularly thank C. Ewango and I. Liengola for their rigorous systematic work and we thank all the many field technicians who make this long-term project possible. Two reviewers provided constructive comments on the drafts of this manuscript. This is paper 3329 of the Forest Research Laboratory, Oregon State University.

References

Campbell, E. J. F., and D. M. Newbery. 1993. Ecological relationships between lianas and trees in lowland rain forest in Sabah, East Malaysia. *Journal of Tropical Ecology* 9:469–90.

Cannon, C. H., D. R. Peart, and M. Leighton. 1998. Tree species diversity in commercially logged Bornean rainforest. *Science* 281:1366–68.

Condit, R. 1995. Research in large, long-term tropical forest plots. *Trends in Ecology and Evolution* 10:18–22.

———. 1998. *Tropical Forest Census Plots: Methods and Results from Barro Colorado Island, Panama, and Comparison with Other Plots.* Springer-Verlag, Berlin, Germany.

Condit, R., S. P. Hubbell, and R. B. Foster. 1996a. Changes in a tropical forest with a shifting climate: Results from a 50 ha permanent census plot in Panama. *Journal of Tropical Ecology* 12:231–56.

Condit, R, S. P. Hubbell, J. V. LaFrankie, R. Sukumar, N. Manokaran, R. B. Foster, and P. S. Ashton. 1996b. Species–area and species–individual relationships for tropical trees: A comparison of three 50-ha plots. *Journal of Ecology* 84:549–62.

Connell, J. H. 1971. On the role of natural enemies in preventing competitive exclusion in some marine animals and in rain forest trees. Pages 298–312 in P. J. de Boer and G. R. Gradwell, editors. *Dynamics of Populations. Proceedings of the Advanced Study Institute on Dynamics of numbers in populations, Oosterbeek, Netherlands, 7–18 September 1970.* Pudoc, Wageningen, Netherlands.

Connell, J. H., and M. D. Lowman. 1989. Low-diversity tropical rain forests: Some possible mechanisms for their existence. *American Naturalist* 134: 88–119.

Conway, D. J. 1992. *A Comparison of Soil Parameters in Monodominant and Mixed Forest in Ituri Forest Reserve, Zaire.* Unpublished honors project. University of Aberdeen, Scotland.

Gentry, A. H. 1991. The distribution of climbing plants. Pages 3–50 in F. E. Putz and H. A. Mooney, editors. *The Biology of Vines.* Cambridge University Press, Cambridge, U.K.

Hart, J. A., and P. Carrick. 1996. *Climate of the Réserve de Faune à Okapis: Rainfall and Temperature in the Epulu Sector, 1986–1995.* CEFRECOF Working Paper No. 2. Centre de Formation et de Recherche en Conservation Forestière, Epulu, Democratic Republic of Congo.

Hart, T. B., J. A. Hart, and P. G. Murphy. 1989. Monodominant and species rich forests of the humid tropics: Causes of their co-occurrence. *American Naturalist* 133:613–33.

Hart, T. B., J. A. Hart, R. Dechamps, M. Fournier, and M. Ataholo. 1996. Changes in forest composition over the last 4000 years in the Ituri basin, Zaire. Pages 545–63 in L. J. G. van der Maesen et al., editors. *The Biodiversity of African plants.* Kluwer Academic Publishers, Dordrecht, Netherlands.

Hubbell, S. P. 1979. Tree dispersion, abundance, and diversity in a tropical dry forest. *Science* 203:1299–1309.

Hubbell, S. P., and R. B. Foster. 1986. Commonness and rarity in a neotropical forest: Implications for tropical tree conservation. Pages 205–31 in M. E. Soule, editor. *Conservation Biology: The Science of Scarcity and Diversity.* Sinauer Associates, Sunderland, MA.

Janzen, D. H. 1970. Herbivores and the number of tree species in tropical forests. *American Naturalist* 104:501–28.

Johnston, M., and M. Gillman. 1995. Tree population studies in low-diversity forests, Guyana. I. Floristic composition and stand structure. *Biodiversity and Conservation* 4:339–62.

Kochummen, K. M., J. V. LaFrankie, and N. Manokaran. 1990. Floristic composition of Pasoh Forest Reserve, A lowland rain forest in Peninsular Malaysia. *Journal of Tropical Forest Science* 3:1–13.

Lebrun, J., and G. Gilbert. 1954. *Une Classification Ecologique des Forêts du Congo.* INEAC, Série Scientifique No. 63. Institut National pour l'Etude Agronomique du Congo Belge, Brussels.

Makana, J., T. B. Hart, and J. A. Hart. 1998. Forest structure and the diversity of lianas and understory treelets in monodominant and mixed stands in the Ituri forest, Zaire. Pages 429–46 in F. Dallmeier and J. A. Comiskey, editors. *Forest biodiversity Research, Monitoring and Modeling: Conceptual Background and Old World Case Studies.* Man and the Biosphere Series 20. The Parthenon Publishing Group, Pearl River, NY.

Manokaran, N., and J. V. LaFrankie. 1990. Stand structure of Pasoh Forest Reserve, A lowland rain forest in peninsular Malaysia. *Journal of Tropical Forest Science* 3:14–24.

Parren, M., F. Bongers, G. Caballé, and J. Nabe-Nielsen. Submitted for publication. Lianas: How to study them?

Reitsma, J. M. 1988. *Végétation Forestière du Gabon/Forest Vegetation of Gabon.* Tropenbos Technical Series 1. The Tropenbos Foundation, Ede, Netherlands.

Richards, P.W. 1996. *The Tropical Rain Forests: An Ecological Study.* Cambridge University Press, London, UK.

Torti, S. D., P. D. Coley, and T. A. Kursar. 2001. Causes and consequences of monodominance in tropical lowland forests. *American Naturalist* 157:141–53.

PART 4: Local Variation in Canopy Disturbance and Soil Structure

Introduction

Richard Condit

One fundamental theme in tropical forest ecology is how local variation in environmental conditions—habitat or microhabitat differences, depending on the scale—affects the forest. Treefalls and other disturbances that influence light regime are one type of local variation, and soil structure and moisture regime another. At least some plant species respond strongly to such variation, such as gap-loving pioneer trees or palms that grow only in swamps. Other plant species may respond more subtly to microhabitat differences, creating a shift in species composition between gap and shade or across soil boundaries. Microhabitat variation is also crucial to some animals, such as butterflies that fly mostly in the bright light of gaps, or birds that forage in the dense tangles of treefalls.

The size of Forest Dynamics Plots is ideal for studying this kind of local variation in abiotic conditions. Light gaps typically are much smaller than 25 or 50 ha, so a plot includes a wide variety of gaps. Soil conditions often vary most clearly with topography, and most of the large plots have a complete ridge and valley system. The three chapters in this section (Brokaw et al., Palmiotto et al., and Okuda et al.) address quantitatively the variation in canopy structure and canopy openings across two Central American and two southeast Asian Forest Dynamics Plots. In addition, the latter two articles analyze the variation in canopy structure and canopy openings in relation to soil variation along a ridge–valley catena. Together, these three articles produce a coherent, comparative understanding of microhabitat variation, particularly canopy structure and gaps, in different tropical forests. Attempting one additional step, Okuda et al. and Palmiotto et al. also test whether total tree species diversity varies with canopy and soil conditions in southeast Asia.

Brokaw et al. demonstrate that gaps are larger in the Barro Colorado Island (BCI) plot in Panama than in the Luquillo plot in Puerto Rico. It is now possible to extend the comparison to include the Lambir and Pasoh plots in southeast Asia. For instance, 1–3% of the BCI canopy is <10 m tall, compared to 0.4% at Luquillo; at Lambir, Palmiotto et al. report 2.9%, and at Pasoh Okuda et al. find 2%. Hurricanes evidently restrict canopy height and crown size at Luquillo, so treefall gaps are small. Like BCI, Lambir has many gaps created by large trees falling, a contrast with other sites in Sarawak, which have smaller gaps created when trees die standing (Ashton and Hall 1992). These contrasts are crucial in determining the proportions of trees in different forests that are pioneers. For example, the pioneer guild is much more conspicuous in the BCI plot than at Pasoh (Condit et al. 1999).

The papers by Okuda et al. and Palmiotto et al. provide several other parallel analyses on canopy structure relative to topography and soil, and readers should examine the papers thinking about similarities and differences. Some of the main conclusions are diametrically opposed. For instance, the Pasoh forest is taller on slopes, but Lambir is taller in valleys. Are these real differences between the two forests, or are they due to differing methods of analysis?

Applying similar methods would be valuable. The datasets on canopy structure at Lambir, Pasoh, Luquillo, and BCI are quite comparable, and some very complete analyses on the size and distribution of light gaps and how gaps are distributed relative to topography are now feasible. Perhaps these research groups could work jointly now to produce a comparative analysis. This sort of comparison is the *raison d'etre* of the Center for Tropical Forest Science, and it forms the mechanistic basis for theories about canopy structure. Baseline information on gap structure and topographic variability is crucial when we try to test whether the plant communities are differentiated across microhabitats, and hypotheses about canopy disturbance and dynamism should be fundamental elements of theories about tree diversity and community structure in tropical forests.

References

Ashton, P. S., and P. Hall. 1992. Comparisons of structure among mixed dipterocarp forests of northwestern Borneo. *Journal of Ecology* 80:459–81.

Condit, R., P. S. Ashton, N. Manokaran, J. V. LaFrankie, S. P. Hubbell, and R. B. Foster. 1999. Dynamics of the forest communities at Pasoh and Barro Colorado: Comparing two 50-ha plots. *Philosophical Transactions of the Royal Society of London* 354:1739–48.

13

Disturbance and Canopy Structure in Two Tropical Forests

Nicholas Brokaw, Shawn Fraver, Jason S. Grear, Jill Thompson,
Jess K. Zimmerman, Robert B. Waide, Edwin M. Everham III,
Stephen P. Hubbell, and Robin B. Foster

Introduction

Profile diagrams of tropical forest usually depict a complex interior structure, emergent trees, and an uneven canopy (Whitmore 1975; Richards 1996). But some tropical forests have a simple interior structure, a narrow range of tree heights, and a relatively smooth canopy (Richards 1996). This contrast may reflect environmental influences such as soil type or climate (Bruenig 1970), historical influences such as land use or natural disturbance (Oldeman 1983; Peart et al. 1992), or other factors. In this chapter we explore connections among environment, history, and forest structure by comparing forests with different structures at El Verde, Puerto Rico, and Barro Colorado Island (BCI), Panama.

The forest at El Verde is said to have a particularly smooth canopy due to chronic wind stress and repeated hurricane damage (Odum 1970; Brown et al. 1983; Lugo and Scatena 1995). Severe hurricanes have struck this forest every 50–60 years, on average, since 1700 (Scatena and Larsen 1991), and Puerto Rico has been an island exposed to oceanic storms for at least 19 million years (Schuchert 1935). Odum (1970) argued that this repeated disturbance has resulted in the selection of canopy tree species at El Verde that have small crowns and form smooth stand canopies for mutual protection from wind (cf. Whitmore 1974; Walsh 1996a). The forest on BCI, which is not known to have ever been struck by a hurricane, seems to have an irregular canopy. We document these contrasting canopy structures and examine several likely causes, especially histories of land use and hurricane events.

The short-term effect of hurricanes on forests in the Caribbean are well described (Brokaw and Walker 1991; Walker 1991; Bellingham et al. 1992; Zimmerman et al. 1994). Generally, defoliation is fairly complete, while loss of branches and downed trees are common; however, refoliation and growth of new branches are rapid, and tree mortality is fairly low. But are there long-term effects as suggested by Odum (1970)? Can chronic hurricanes produce a certain inherent forest structure different from that of forests such as BCI? Our paper on tropical forest structure joins recent papers that seek to distinguish the effects of large- versus small-scale disturbances on tropical forest tree dynamics and composition (Lugo and Scatena 1996, Whitmore and Burslem 1998).

Study Areas

The Puerto Rican study area is at El Verde Research Area in the Luquillo Experimental Forest in the Luquillo Mountains of northeastern Puerto Rico (18°20′N, 65°49′W). El Verde is classified as *subtropical wet* in the Holdridge life zone system (Ewel and Whitmore 1973) and *tropical montane* in Walsh's (1996a) tropical climate system (see chapter 32 for more details). Locally this forest is called "tabonuco" after the common name of a dominant tree, *Dacryodes excelsa* Vahl (Burseraceae). Annual wind speed above the canopy averages 4.1 km/hr (Waide and Reagan 1996). Monthly averages at Catalina, the most comparable site in Puerto Rico (5.6 km from El Verde, same exposure) with appropriate data, are 2–6 km/hr (Briscoe 1966; Brown et al. 1983). Soils are largely upland Ultisols, mainly well-drained clays and silty clay loams (Edmisten 1970). Forest in the Luquillo Mountains was damaged by hurricanes in 1928 and 1931, severely in 1932, lightly in 1956, severely by Hurricane Hugo in 1989, lightly by Hortense in 1996, and severely by Georges in 1998 (Crow 1980; Weaver 1989; Scatena and Larsen 1991).

The Panamanian study area is Barro Colorado Island (BCI) in Barro Colorado Nature Monument, Panama (9°09′N, 79°51′W). BCI is classified as "tropical moist" in the Holdridge life zone system (Knight 1975) and "tropical wet" in Walsh's (1996a) climate system (see chap. 24 for more details). Above the canopy, average monthly wind speed varies from 2.4 to 7.8 km/hr and the average annual wind speed is 4.3 km/hr (Environmental Sciences Program, Smithsonian Tropical Research Institute, Panama, www.stri.org/tesp/data.htm). Soils are Oxisols, classified as frijoles clay and characterized as homogeneous, deep red, well-drained, and well-aerated (Knight 1975). Based on a 6-year sample, treefall gaps open up about 0.80% of the forest canopy each year on BCI (Brokaw 1990), a rate similar to other neotropical forests not apparently affected by large-scale disturbance (Hartshorn 1990). BCI is not known to have been damaged by hurricanes, but large (1–5 ha) blowdowns have occurred, and the creation of the lake surrounding the island may have led to stronger winds and more treefalls since 1914 (Foster and Brokaw 1982).

At El Verde we measured canopy height before and after Hurricane Hugo in a 1.08-ha (120 × 90 m) plot. The plot was selected to take advantage of a previously marked 9.0-ha plot, to be accessible, and to accurately represent topography and forest structure in the area. This plot has a 350-m elevation, and its terrain includes ravines, ridges, and somewhat more level areas characteristic of El Verde. The plot was established in 1988 and is mainly within the 16-ha Luquillo Forest Dynamics Plot (chap. 32) established in 1991. The plot area may have been subjected to some light selective logging before 1953, but otherwise it has not been disturbed by humans (García-Montiel 2002; Thompson et al. 2002).

To complement prehurricane data from the 1.08-ha plot at El Verde we recorded the number and size of treefall gaps before Hugo in about 35 ha of forest that included the 1.08-ha plot. This 35-ha plot accurately represented forest that we felt typified the area and was accessible from the research station. As in the 1.08-ha plot, most of this area was lightly and selectively logged before 1953; the remainder (about one third) was harvested for timber (some areas clearcut) or charcoal, or used for coffee and other crops, before 1930, but since then has been covered by forest (García-Montiel 2002; Thompson et al. 2002).

At BCI we measured canopy height in the 50-ha Forest Dynamics Plot. This plot is mostly level and elevations range from 120 to 160 m (chap. 24). Except for 2 ha of secondary growth, the Forest Dynamics Plot has suffered no large-scale clearing in at least 500 years (Piperno 1990). In addition to height measurements in the 50-ha plot, we recorded abundance and size of treefall gaps in 13.4 ha of similar old-growth forest on BCI, selected for its accessibility to trails and spread broadly over the island (Brokaw 1982).

Methods

Canopy Surface

Using a rangefinder and a long pole for sighting, we determined the height interval (fig. 13.1) of the uppermost canopy surface above each point in 5 × 5 m grid systems arrayed in the 1.08-ha study plot at El Verde (475 points) and the 50-ha Forest Dynamics Plot at BCI (20,301 points). We did this in the 1.08-ha plot in May 1989 (before Hurricane Hugo in September 1989), November 1989 (5 weeks after Hurricane Hugo), May 1991, January 1993, October 1994, January 1997 (10 weeks after Hurricane Hortense), and June 1998. On the BCI plot these measurements were taken each year from 1983 through 1995. We calculated means, standard deviations, and variances of canopy surface height from values determined by setting the height for all records within a height interval as the value of the midpoint of the interval. We set records greater than 30 m at 35 m.

Treefall Gaps

We defined a treefall gap at El Verde as a hole in the forest canopy extending down to an average height of about 3 m or less aboveground. The edge of a gap was delineated by the vertical projection of the edge of the canopy foliage. At BCI we used the same definition, except that we counted as gaps holes with

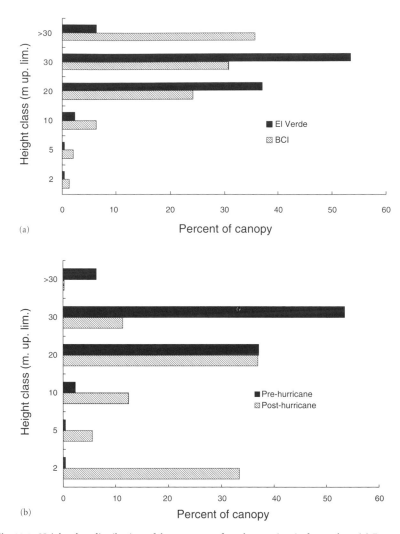

Fig. 13.1. Height-class distribution of the canopy surface above points in forest plots. (a) Forest that is chronically affected by hurricanes: El Verde, Puerto Rico, 16 weeks before Hurricane Hugo in 1989 (1.08-ha plot); and forest not known to be affected by hurricanes: Barro Colorado Island, Panama, in 1989 (50-ha plot). (b) El Verde 16 weeks before Hurricane Hugo and 5 weeks after the hurricane (adapted from Brokaw and Grear 1991). The range of height classes is more peaked (less variable) at El Verde before the hurricane than at BCI, but the El Verde range is greatly variable afterward. The relatively narrow distribution of canopy heights in this small study site at El Verde is verified, in a sense, for the forest as a whole by the data on gaps from large sample sites at El Verde and BCI (fig. 13.4).

vegetation to an average of about 2 m or less in height. The 3-m limit for El Verde permitted inclusion of gaps that were created recently and during a few previous years, whereas the BCI definition included only newer gaps. This difference was intended to compensate for the different sample periods at the two sites (see the discussion that follows).

At El Verde we measured the dimensions of all gaps present in the 35-ha study site in August 1989, 1 month before Hurricane Hugo. On BCI we measured the area of all newly created treefall gaps in the 13.4-ha study site during frequent visits between August 1975 and July 1981 (Brokaw 1982, 1990). For each gap we measured L, the longest axis of the gap (distance between edges), and W, the longest axis perpendicular to L, and then approximated gap area as that of an ellipse: Area $= \pi LW/4$ (Runkle 1992). Where possible at El Verde, we determined if the gapmaking tree had died standing. We did not do this on BCI. In both sites only openings ≥ 20 m^2 were counted as gaps.

Results

Canopy Surface

Before Hurricane Hugo, canopy surface height in the 1.08-ha plot at El Verde was less variable, i.e., smoother, than canopy height in the 50-ha BCI plot (fig. 13.1a, table 13.1). Our measure of variability is the coefficient of variation

Table 13.1. Canopy Surface Heights in Nine Forest Stands

Site	Canopy height		
	Mean (m)	S.D. (m)	CV (%)
Puerto Rico:			
El Verde	20.4	5.52	27.1
Panama:			
Barro Colorado Island	23.0–24.4	9.2–9.8	39.0–40.1
Belize:			
Grano d'Oro	18.2	7.7	42.0
Las Cuevas	18.1	6.4	35.3
New Maria	20.3	6.0	29.5
San Pastor	18.2	7.1	39.2
Thailand:			
Khao Yai Valley	25.5	12.5	48.8
Khao Yai Hilltop	22.0	7.8	35.5
Khao Yai Montane	13.9	5.4	38.9

Notes: El Verde data are from before Hurricane Hugo (this study). Barro Colorado Island data are from 1983, 1989, and 1995 (this study). Belize data are from 496 points in an 18-ha plot at each site (Mallory and Brokaw 1996). Thailand data are from 100 points in a 1-ha plot at each site (Brockelman 1998). The coefficient of variation (CV) of canopy height at El Verde was low compared to the other forests.

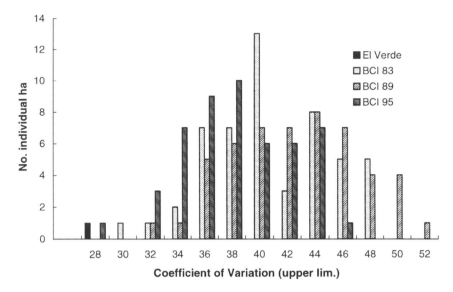

Fig. 13.2. Frequency distributions of coefficients of variation (CV) of canopy height in forest plots at El Verde, Puerto Rico, and at Barro Colorado Island (BCI), Panama. The plot at El Verde was 1.08 ha and was measured once. Plots at BCI were individual hectares in the 50-ha Forest Dynamics Plot, measured several times; the 3 years illustrated were arbitrarily selected (see text). The CV at El Verde was 27.1%; at BCI the mean CVs for 1983, 1989, and 1995 were 40.1, 42.1, and 39.0%, respectively. The relatively low CV of canopy heights in this small study site at El Verde is verified, in a sense, for the forest as a whole by the data on gaps from large sample sites at El Verde and BCI (fig. 13.4).

(CV) of canopy height. (CV is the standard deviation of data as a percent of its mean. This statistic factors out apparent differences in canopy variability due to the greater and lesser potential for variability between forests differing in mean height.) The CV was 27.1% in the plot at El Verde before Hurricane Hugo, lower than the CV in any single hectare in the BCI Forest Dynamics Plot in the years 1983, 1989, and 1995 (fig. 13.2). (These years of BCI data were arbitrarily selected to span values from first to most recent dataset and to include the same year of data as at El Verde; BCI data for other years differed little.)

We statistically tested the differences in variability of canopy surface by comparing the ratio from the formula $S^2_{\log x}/S^2_{\log y}$ (where S^2 = variance of the logs of values in two datasets) with values of the F distribution (Lewontin 1966). The ratios of variability in log canopy height at El Verde before Hurricane Hugo in August 1989 (x) to that at BCI in 1983, 1989, and 1995 (y_i) were 2.92, 3.63, and 2.30 respectively, all $p < 0.0001$; $x\,df = 474$, $y\,df = 20,300$.

After Hurricane Hugo, the canopy surface in the 1.08-ha plot at El Verde was extremely rough. The hurricane sharply reduced mean canopy height while greatly

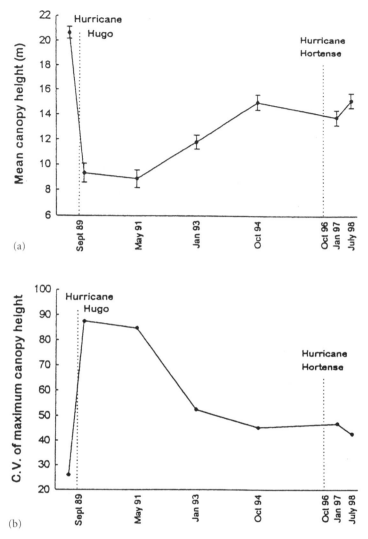

Fig. 13.3. (a) Mean and (b) coefficient of variation (CV) of canopy surface height in a 1.08-ha forest plot at El Verde, Puerto Rico, in relation to the timing of hurricanes. Midpoints of measurement periods are shown. Hurricanes lower the mean canopy height and raise the CV of canopy height, while recovery does the opposite.

increasing its CV (fig. 13.3). Height-class distribution of the canopy surface varied considerably (fig. 13.1b). Note, however, that 37% of the canopy was still 10–20 m in height and 11% between 20–30 m. With regrowth of surviving trees and new regeneration, mean canopy height increased and CV declined until Hurricane Hortense temporarily reversed these trends (fig. 13.3).

Table 13.2. Gap Parameters in Forests at El Verde, Puerto Rico (before Hurricane Hugo), and Barro Colorado Island (BCI), Panama

Gaps	El Verde	BCI
Study Site Size (ha)	~35.0	13.4
Number of Gaps	44	83
Gaps/ha	1.3	6.2
Gap Mean m^2 (S.D.)	49.9 (24.4)	93.2 (78.2)
Gap Median m^2	39.3***	70.4***
Gap m^2/ha	61.8	577.1

***$p < 0.0001$, Wilcoxon Rank Sum Test.

Notes: The data suggest that gaps are both more frequent and larger at BCI. We did not perform statistical tests on some gap data because of biases in methods (see Results) or, in the case of mean gap area, because the distributions were not normal.

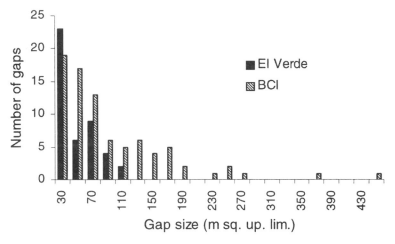

Fig. 13.4. Size-class distribution of treefall gaps in a forest chronically affected by hurricanes and disturbed by humans—El Verde, Puerto Rico, just before Hurricane Hugo in 1989—and in a forest not known to be affected by hurricanes and minimally disturbed by humans: Barro Colorado Island (BCI), Panama. There were more large gaps on BCI.

Treefall Gaps

Before Hurricane Hugo, the El Verde forest had fewer gaps per unit area (35 ha censused), median gap size was significantly less, and there was less area of gap per hectare than on BCI (13.4 ha censused) (table 13.2). The largest gap encountered at El Verde was 117 m^2; at BCI it was 452 m^2. The size-class distributions of gaps differed significantly between El Verde and BCI (fig. 13.4; Kolmogorov–Smirnov test, $D = 0.294$, $p < 0.05$). Of the gapmaking trees at El Verde, 40% appeared to have died standing.

Differences in methods biased our gap totals somewhat. At El Verde we censused gaps just once, whereas on BCI we periodically censused gaps over a period of 6 years. Although we used a looser definition of gap at El Verde in an attempt to include gaps formed within several years before the census, we undoubtedly did not detect some that had filled in, especially small gaps, resulting in underestimates for El Verde. This bias inflates the differences in numbers and total area of gaps (table 13.2). However, we feel that the differences are great enough that the biases alone do not explain them, and these biases would have had little effect on the large difference in gap size between sites. The gap results are also consistent with the independent canopy surface results. In the canopy-height plots, the percentage of low canopy (≤ 2m high), equivalent to gap area, was significantly less at El Verde in August 1989, before Hurricane Hugo (0.4%), than it has been at BCI (e.g., 1.1% in 1983, 3.0% in 1989; $p < 0.05$; test of the equality of two percentages; Sokal and Rohlf 1969).

Discussion

The results showed that a Puerto Rican tabonuco forest had a smoother canopy surface and fewer and smaller treefall gaps (before a recent hurricane) than did a forest on Barro Colorado Island, Panama. The tabonuco forest had a smoother canopy than some other tropical forests as well (table 13.1). Immediately after a hurricane the canopy of the tabonuco stand was rough, but regrowth, largely by sprouting from upright stems (Zimmerman et al. 1994; Walker 1991), appeared to be returning the canopy to a relatively smooth condition (fig. 13.3). These results confirm the several qualitative observations cited above of smooth canopies in Puerto Rican tabonuco forest. Below we discuss several possible, not mutually exclusive, explanations for this phenomenon.

Canopy Surface

One explanation for the smooth canopy at El Verde is that this forest, before Hurricane Hugo in 1989, was a second-growth, relatively even-aged stand of trees that had grown up after the reputedly severe hurricane in 1932 or after human disturbance. Such a stand may not have reached the point when trees develop large crowns and old, big trees begin dying, thereby creating a rough canopy (Oldeman 1983; Riswan et al. 1985; Tyrrell and Crow 1994; Dahir and Lorimer 1996). However, the aerial photographs of El Verde taken in 1936, 4 years posthurricane, and including our El Verde study sites, reveal little evidence of hurricane damage (cf. Foster et al. 1999) and show many large trees, indicating that hurricane damage in 1932 was not severe enough to initiate a uniform secondary succession. Similarly, after Hurricane Hugo, most canopy trees survived (Walker 1995), and the tabonuco forest did not follow classic patterns of second-growth development. On the other hand, since part of the 35 ha at El Verde in which gaps were measured had been intensively disturbed by humans, including clearcutting in the 1920s, land use could account for a relatively smooth canopy 60 years later in part of the study area.

A second explanation for El Verde's smooth, prehurricane canopy is that persistent winds could induce forms of twig and branch development that produce smoothness (Lawton 1982; Telewski 1995). We know that average wind speeds at the canopy surface are no higher at El Verde than at BCI (see above), but we lack perhaps more appropriate data on the persistence of wind or frequency of strong winds. A third explanation is that selective logging before 1953 at El Verde removed some large trees and reduced variability in canopy height (cf. Kammesheidt 1998), but we are uncertain if these putative logging effects would have lasted the 36+ years until our study; such logging effects were obscured after a few decades in a New England (USA) stand (Merrens and Peart 1992). A fourth explanation is that environmental harshness due to drought, poor soil (Bruenig 1973), or some features of a "montane" setting (Walsh 1996b, p. 226; Brockelman 1998) can produce a smooth canopy. These possibilities are hard to evaluate. Extreme drought is uncommon at El Verde (cf. Brown et al. 1983); its soil is not especially infertile (Edmisten 1970); and authors disagree on whether El Verde is a montane site (Ewel and Whitmore 1973; Walsh 1996a). A fifth explanation is that El Verde's smooth canopy may reflect dominance by a few tree species and consequent limited variety of tree architecture (Beard 1946; Richards 1996). But the tabonuco forest canopy seems too species-rich for that phenomenon. For instance, the five most abundant canopy species (normally >21 m height) in the 16-ha Luquillo Forest Dynamics Plot at El Verde comprise 24.1% of all trees ≥10 cm dbh in that plot, while 84 additional species make up the remaining 75.9% (data from Thompson et al. 2002). By comparison, the five most abundant canopy species in the BCI plot comprise 25.7% of all trees ≥10 cm dbh (Loo de Lao, unpublished data).

Lastly, Odum (1970) argued that repeated hurricanes in Puerto Rican tabonuco forest have selected against the emergent habit and for the reduced wind resistance of small crowns, resulting in a smooth forest canopy (smooth canopies do tend to suffer less wind disturbance, Everham and Brokaw 1996; but see Imbert et al. 1996). It is logical that frequent hurricanes would eliminate species with growth forms poorly adapted to violent winds, and it is possible that hurricanes could select for a genetically determined form of a species that best survives storms. However, a more parsimonious explanation of El Verde's smooth canopy is simply that hurricanes, and perhaps more frequent lesser storms, repeatedly prune top and lateral branches of trees that would otherwise grow tall and spread their canopies. (Pérez 1970 described the relationship between stem and crown diameters of trees in Luquillo and some other forests, but did not make comparisons adequate to test these ideas.)

Treefall Gaps

Canopy structure and gap characteristics are two sides of the same coin; thus the relatively smooth canopy in the prehurricane tabonuco forest at El Verde was little broken by gaps. Gaps that did occur were fewer and smaller than those on BCI. Similarly, at Bisley, another tabonuco forest stand in the Luquillo Mountains, gap frequency and size (except in wet valleys) appeared to be less than on BCI (Scatena and Lugo 1995). There are several possible reasons for these differences between tabonuco forest and BCI.

First, treefall gap size can be correlated with size of fallen tree (Brokaw 1982), and there are fewer large trees at El Verde (Brown et al. 1983; Knight 1975), which is probably due to both natural causes and logging. Second, hurricanes may knock down weak trees in a brief period, leaving the more stable ones upright and a forest less prone to treefalls for a long time afterward (Lorimer 1989; Whigham et al. 1999). Third, selective logging may likewise have reduced later treefall gap size and number, but logging ended 36 years before our study. Fourth, topographic differences may produce different treefall rates, but evidence from some forests suggests that slopes usually have more treefalls (Oldeman 1978), whereas El Verde had fewer gaps but has more slopes than the BCI site.

A fifth reason for small gaps at El Verde is that 40% of gapmaking trees in this forest died standing, creating smaller gaps than would similar-sized trees that fall when alive (P. Murphy, personal communication in Odum 1970; Jans et al. 1993). By contrast, only 14% of tree mortality on BCI was standing dead in another study (Putz and Milton 1982), and 11% was standing dead in a Malaysian forest (Palmiotto et al. chap. 14). Exceptional stability of El Verde trees might be explained by their having especially dense, strong wood (cf. Putz et al. 1983). Data are available on wood density (specific gravity) for some of the tree species at El Verde and BCI and for a large sample of neotropical tree species. These data are unsuited to rigorous comparison, as many species are not included, but we present it for heuristic purposes. At El Verde mean wood density (unweighted for abundance) was 0.60 gm/cm^3(s.d. = 0.152) for 62 of 90 tree species that reach 10 cm dbh in the Forest Dynamics Plot (Little and Wadsworth 1964; Little et al. 1974; Reyes et al. 1992; Thompson et al. 2002). On BCI mean wood density was 0.52 gm/cm^3 (s.d. = 0.191) for 41 of 227 species in the Forest Dynamics Plot (FAO 1971; Condit et al. 1996). Lastly, for 470 species in the neotropics, mean wood density was 0.60 gm/cm^3 (s.d. = 0.008) (Reyes et al. 1992). These comparisons suggest that El Verde trees might have overall higher wood densities than trees of BCI, but the comparisons are inconclusive. (Mean wood densities weighted by species abundances in the Forest Dynamics Plots were similar to the above figures but are not reliable, due to lack of data on wood densities of many BCI species.)

Additional reasons for a higher rate of standing death at El Verde could be the strong rooting of the dominant tabonuco tree (Basnet et al. 1992) and the presumed small tree crowns that could reduce both wind drag and crown asymmetry that lead to tree fall (Young and Hubbell 1991).

Conclusion and Comparison with Other Forests

We cannot demonstrate one exclusive cause for the relatively smooth canopy and the low number and small size of gaps at El Verde, before Hurricane Hugo, in comparison with BCI and some other forests. There is no evidence of an environmental difference such as soil or topography (cf. chap. 14) that seems likely to explain these contrasts. But the clear differences in natural and human disturbance histories between El Verde and BCI might have caused the contrasts in forest structure.

El Verde has been struck by hurricanes for millennia; BCI has not, at least in historical time. Arguably, this extreme difference in disturbance should produce contrasting canopy structures. A regime of hurricanes and lesser storms may have selected for species ecotypes that are not tall or emergent, reducing canopy disturbance and gap size when these trees fall. Alternatively, trees extending above the main canopy may simply be periodically pruned to height conformity, and are not evolved ecotypes. It could also be that the smaller size of El Verde tree species reflects taxonomic constraints on the features of those species that, due to biogeographic chance, occur at El Verde, features that are unrelated to wind disturbance but nonetheless produce a smooth canopy and little gap area (Whigham et al. 1999). However, we think it unlikely that the resistance to wind conferred by a smooth canopy (Everham and Brokaw 1996) is merely fortuitous, given Puerto Rico's long exposure to severe storms. Lastly, various human disturbances have surely contributed to, but do not fully explain, the distinctive canopy structure at El Verde.

Elsewhere in the tropics, large-scale wind disturbance seems to have a variable effect on canopy structure, depending on frequency and intensity of these storms (Odom 1970; Walsh 1996a). Philippine forests in Palanan, on the eastern side of Luzon, are qualitatively reported to have a relatively smooth canopy, attributed to the chronic influence of typhoons (Ashton 1993), as at El Verde. In Belize, Central America, forests severely damaged by Hurricane Hattie in 1961 (Friesner 1993) now range in structure from relatively uniform-canopy stands of secondary species to areas where small patches of tall forest are islands in low, liana-covered thicket (Mallory and Brokaw 1996; N. Brokaw, personal observation; table 13.1). These differences are possibly due to variations in local storm intensity and previous disturbance. On St. Vincent, in the Caribbean, hurricanes strike less frequently than in Puerto Rico and apparently produce a "hurricane forest" consisting of low thickets with occasional vine-covered emergent trees (Beard 1945). Storms in Queensland, Australia, produce a similar "cyclone scrub" (Webb 1958).

In forests for which large-scale disturbance is not mentioned, canopy structure also varies and is attributed to several factors. On BCI, we think the relatively rough canopy is explained by large tree size and a treefall regime not associated with large-scale disturbance. In Ecuador, a forest with smooth canopy grows on less fertile soil, possibly limiting tree size, and thus treefall gap size, while a rough-canopy forest is on richer soil (Kapos et al. 1990). Likewise, in Borneo smooth canopies are associated with infertile soil, drought, and relatively low species diversity (Bruenig 1970, 1973). In Thailand smooth canopies are found at high elevation (no mechanism specified), while the presence of typically emergent Dipterocarpaceae produces rough canopies in lowland valleys (Brockelman 1998; table 13.1).

These examples show that there is no simple explanation for canopy structure. We could better explain what controls this feature of tropical forests by (1) studying canopy structure in many forests at different points along environmental and disturbance gradients, (2) using standard methods to measure structure, such as the height measurements used in this paper and remote sensing (Bruenig 1970; Gerard et al. 1998), and (3) measuring canopy structure repeatedly over long periods. We predict the results would show that chronic hurricanes have a lasting and distinctive impact on canopy structure, variations in structure owing partly to site-specific frequency and intensity of hurricanes. This suggests that the canopy structure of forests in hurricane areas will change significantly (O'Brien et al. 1992; Overpeck 1990) if hurricanes increase in frequency and intensity, as predicted by some climate modelers (Emmanuel 1987).

General Implications

Numerous factors, such as site features (topography, soil, climate), species characteristics, and, especially, disturbance history combine to determine the structure of tropical forest canopies. The canopy structure of sites that suffer large-scale disturbance, such as hurricanes (tropical cyclonic storms) and forest clearance, will differ from those that suffer only small-scale disturbances. Hurricanes, predicted to increase in frequency and intensity, will probably change the structure of affected forests.

Work in the Forest Dynamics Plots of the Center for Tropical Forest Science can provide the standardized methods and long-term observations over multiple sites that are needed to understand the determinants of canopy structure and monitor possible shifts in structure associated with climate change.

Acknowledgments

We are grateful to the many field workers who obtained the canopy data at El Verde and Barro Colorado Island, especially Andrés Hernandes. We thank Jack Putz for sending data on wood density. Elizabeth Mallory helped with data analysis. Elizabeth Losos, Douglas Boucher, Fred Scatena, Peter Weaver, Bob Lawton, Sean Thomas, Egbert Leigh, and an anonymous reviewer helped with the manuscript. Suzanne Loo de Lao, Richard Condit, and Peter Weaver provided data and references. Research at El Verde was performed under grant BSR-8811902 from the National Science Foundation to the Institute for Tropical Ecosystem Studies, University of Puerto Rico, and the International Institute of Tropical Forestry, as part of the Long-Term Ecological Research Program in the Luquillo Experimental Forest. The Forest Service (U.S. Dept. of Agriculture) and the University of Puerto Rico provided additional support. Support for Brokaw's work on BCI came from Coulter Fellowships, Hutchinson Fellowships, and the Hinds Fund of the University of Chicago, the Environmental Sciences Program of the Smithsonian Tropical Research Institute, a Faculty Development Grant from Kenyon College, and the National Geographic Society. The canopy survey in the BCI 50-ha plot has been supported logistically and financially by the Smithsonian Tropical Research Institute. This paper is a scientific contribution from the Center for Tropical Forest Science, which is supported by The John D. and Catherine T. MacArthur Foundation.

References

Ashton, P. S. 1993. The community ecology of Asian rain forests, in relation to catastrophic events. *Journal of Biosciences* 4:501–14.

Basnet, K., G. E. Likens, F. N. Scatena, and A. E. Lugo. 1992. Hurricane Hugo: Damage to a tropical rain forest in Puerto Rico. *Journal of Tropical Ecology* 8:47–55.

Beard, J. S. 1945. The progress of plant succession on the Soufriere of St Vincent. *Journal of Ecology* 33:1–9.

Beard, J. S. 1946. The Mora forests of Trinidad, British West Indies. *Journal of Ecology* 33:173–92.

Bellingham, P. J., V. Kapos, N. Varty, J. R. Healey, E. V. J. Tanner, D. L. Kelly, J. W. Dalling, L. S. Burns, D. Lee, and G. Sidrak. 1992. Hurricanes need not cause high mortality: The effects of Hurricane Gilbert on forests in Jamaica. *Journal of Tropical Ecology* 8:217–23.

Briscoe, C. B. 1966. *Weather in the Luquillo Mountains of Puerto Rico.* USDA Forest Service Research Paper ITF-3. Institute of Tropical Forestry, Río Piedras, Puerto Rico.

Brockelman, W. Y. 1998. Study of tropical forest canopy height and cover using a point-intercept method. Pages 521–31 in F. Dallmeier and J. A. Comiskey, editors. *Forest Biodiversity Research, Monitoring and Modeling: Conceptual Background and Old World Case Studies.* Man and the Biosphere Series, Volume 20. Parthenon Publishing Group, Pearl River, NY.

Brokaw, N. V. L. 1982. Treefalls: frequency, timing, and consequences. Pages 101–08 in E. G. Leigh, Jr., A. S. Rand, and D. M. Windsor, editors. *The Ecology of a Tropical Forest: Seasonal Rhythms and Long-Term Changes.* Smithsonian Institution Press, Washington, DC.

Brokaw, N. V. L. 1990. Caída de árboles: frecuencia, cronología y consecuencias. Pages 163–72 in E. G. Leigh, Jr., A. S. Rand, and D. M. Windsor, editors. *Ecología de un Bosque Tropical: Ciclos Estacionales y Cambios a Largo Plazo.* Smithsonian Tropical Research Institute, Balboa, Panama.

Brokaw, N. V. L., and J. S. Grear. 1991. Forest structure before and after Hurricane Hugo at three elevations in the Luquillo Mountains, Puerto Rico. *Biotropica* 23:386–92.

Brokaw, N. V. L., and L. R. Walker. 1991. Summary of the effects of Caribbean hurricanes on vegetation. *Biotropica* 23:442–47.

Brown, S., A. E. Lugo, S. Silander, and L. Liegel. 1983. *Research History and Opportunities in the Luquillo Experimental Forest.* U.S. For. Serv. Gen. Tech. Rep. SO-44. Southern Forest Experiment Station, New Orleans.

Bruenig, E. F. 1970. Stand structure, physiognomy and environmental factors in some lowland forests in Sarawak. *Tropical Ecology* 11:26–43.

Bruenig, E. F. 1973. Species richness and stand diversity in relation to site and succession of forests in Sarawak and Brunei (Borneo). *Amazoniana* 4:293–20.

Condit, R., S. P. Hubbell, and R. B. Foster. 1996. Changes in tree species abundance in a Neotropical forest: impact of climate change. *Journal of Tropical Ecology* 12:231–56.

Crow, T. R. 1980. A rainforest chronicle: A 30 year record of change in structure and composition at El Verde, Puerto Rico. *Biotropica* 12:42–55.

Dahir, S., and C. G. Lorimer. 1996. Variation in canopy gap formation among developmental stages of northern hardwood forests. *Canadian Journal of Forest Research* 26:1875–92.

Edmisten, J. 1970. Soil studies in El Verde rain forest. Pages H79–87 in H. T. Odum and R. F. Pigeon, editors. *A Tropical Rain Forest: A Study of Irradiation and Ecology at El Verde, Puerto Rico.* National Technical Information Service, Springfield, VA.

Emmanuel, K. A. 1987. The dependence of hurricane intensity on climate. *Nature* 362: 483–85.

Everham, E. M., III, and N. V. L. Brokaw. 1996. Forest damage and recovery from catastrophic wind. *Botanical Review* 62:113–85.

Ewel, J. J., and J. L. Whitmore. 1973. *The Ecological Life Zones of Puerto Rico and the U. S. Virgin Islands.* Forest Service Research Paper ITF-18. Institute of Tropical Forestry, Río Piedras, Puerto Rico.

Food and Agriculture Organization of the United Nations (FAO). 1971. *Inventariación y Demostraciones Forestales Panamá: Propiedades y Usos de Ciento Trece Especies Maderables de Panamá. Parte 1: Recopilación de los Resultados de los Ensayos.* FO:SF/PAN 6, Informe Técnico 3.

Foster, D. R., M. Fluet, and E. R. Boose. 1999. Human or natural disturbance: Landscape-scale dynamics of the tropical forests of Puerto Rico. *Ecological Applications* 9:555–72.

Foster, R. B., and N. V. L. Brokaw. 1982. Structure and history of the vegetation of Barro Colorado Island. Pages 67–82 in E. G. Leigh, Jr., A. S. Rand, and D. M. Windsor, editors. *The Ecology of a Tropical Forest: Seasonal Rhythms and Long-Term Changes.* Smithsonian Institution Press, Washington, DC.

Friesner, J. 1993. *Hurricanes and the Forests of Belize*. Forest Planning and Management Project Occasional Series No. 1. Ministry of Natural Resources, Belmopan, Belize.

García-Montiel, D. C. 2002. El legado de la actividad humana en los bosques neotropicales contemporáneos. Pages 97–116 in M. Guariguata and G. Kattan, editors. *Ecología y Conservación de Bosques Neotropicales*. Libro Universitario Regional, Cartago, Costa Rica.

Gerard, F., B. Wyatt, A. Millington, and J. Wellens. 1998. The role of data from intensive plots in the development of a new method for mapping tropical forest types using satellite imagery. Pages 141–158 in F. Dallmeier and J. A. Comiskey, editors. *Forest Biodiversity Research, Monitoring and Modeling: Conceptual Background and Old World Case Studies*. Man and the Biosphere Series, Volume 20. Parthenon Publishing Group, Pearl River, NY.

Hartshorn, G. S. 1990. An overview of neotropical forest dynamics. Pages 585–99 in A. H. Gentry, editor. *Four Neotropical Rain Forests*. Yale University Press, New Haven, CT.

Hubbell, S. P., and R. B. Foster. 1986. Canopy gaps and the dynamics of a neotropical forest. Pages 77–96 in M. J. Crawley, editor. *Plant Ecology*. Blackwell Scientific Publications, Oxford, U.K.

Imbert, D., P. Labbé, and A. Rousteau. 1996. Hurricane damage and forest structure in Guadeloupe, French West Indies. *Journal of Tropical Ecology* 12:663–80.

Jans, L., L. Poorter, R. S. A. R. van Rompaey, and F. Bongers. 1993. Gaps and forest zones in tropical moist forest in Ivory Coast. *Biotropica* 25:258–69.

Kammesheidt, L. 1998. Stand structure and spatial pattern of commercial species in logged and unlogged Venezuelan forest. *Forest Ecology and Management* 109:163–74.

Kapos, V., E. Pallant, A. Bien, and S. Freskos. 1990. Gap frequencies in lowland rainforest sites on contrasting soils in Amazonian Ecuador. *Biotropica* 22:218–25.

Knight, D. H. 1975. A phytosociological analysis of species-rich tropical forest on Barro Colorado Island. *Ecological Monographs* 45:259–84.

Lawton, R. O. 1982. Wind stress and elfin stature in a montane rain forest tree: An adaptive explanation. *American Journal of Botany* 69:1224–30.

Lewontin, R. C. 1966. On the measurement of relative variability. *Systematic Zoology* 15:141–42.

Little, E. L., Jr., and F. H. Wadsworth. 1964. *Common Trees of Puerto Rico and the Virgin Islands*. Agriculture Handbook No. 249. U.S. Department of Agriculture, Forest Service, Washington, DC.

Little, E. L., Jr., R. O. Woodbury, and F. H. Wadsworth. 1974. *Trees of Puerto Rico and the Virgin Islands*. Vol. 2. Agriculture Handbook No. 449, U.S. Department of Agriculture, Forest Service, Washington, DC.

Lorimer, C. G. 1989. Relative effects of small and large disturbances on temperate hardwood forest structure. *Ecology* 70:565–67.

Lugo, A. E., and F. N. Scatena. 1995. Ecosystem-level properties of the Luquillo Experimental Forest with emphasis on the tabonuco forest. Pages 59–108 in A. E. Lugo and C. Lowe, editors. *Tropical Forests: Management and Ecology*. Springer-Verlag, New York.

Lugo, A. E., and F. N. Scatena. 1996. Background and catastrophic tree mortality in tropical moist, wet, and rain forests. *Biotropica* 28:585–99.

Mallory, E. P., and N. V. L. Brokaw. 1996. *Impacts of Silvicultural Trials on Birds and Tree Regeneration in the Chiquibul Forest Reserve*. Consultancy Report No. 20. Forest Planning and Management Project, Forest Department, Belmopan, Belize.

Merrens, E. J., and D. R. Peart. 1992. Effects of hurricane damage on individual growth and stand structure in a hardwood forest in New Hampshire, USA. *Journal of Ecology* 80:787–95.

O'Brien, S. T., B. P. Hayden, and H. H. Shugart. 1992. Global climatic change, hurricanes, and a tropical forest. *Climatic Change* 22:175–90.

Odum, H. T. 1970. Rain forest structure and mineral cycling homeostasis. Pages H3–H52 in H. T. Odum and R. F. Pigeon, editors. *A Tropical Rain Forest: A Study of Irradiation and Ecology at El Verde, Puerto Rico.* National Technical Information Service, Springfield, VA.

Oldeman, R. A. A. 1978. Architecture and energy exchange of dicotyledonous trees in the forest. Pages 535–60 in P. B. Tomlinson and M. H. Zimmerman, editors. *Tropical Trees as Living Systems.* Cambridge University Press, Cambridge, U.K.

Oldeman, R. A. A. 1983. Tropical rain forest, architecture, silvigenesis, and diversity. Pages 139–50 in S. L. Sutton, T. C. Whitmore, and A. C. Chadwick, editors. *Tropical Rain Forest: Ecology and Management.* Blackwell, Oxford, U.K.

Overpeck, J. T., D. Rind, and R. Goldberg. 1990. Climate-induced changes in forest disturbance and vegetation. *Nature* 343:51–53.

Peart, D. R., C. V. Cogbill, and P. A. Palmiotto. 1992. Effects of logging history and hurricane damage on canopy structure in a northern hardwoods forest. *Bulletin of the Torrey Botanical Club* 119:29–38.

Pérez, J. W. 1970. Relation of crown diameter to stem diameter in forests of Puerto Rico, Dominica, and Thailand. Pages B105–22 in H. T. Odum and R. F. Pigeon, editors. *A Tropical Rain Forest: A Study of Irradiation and Ecology at El Verde, Puerto Rico.* National Technical Information Service, Springfield, VA.

Piperno, D. R. 1990. Fitolitos, arqueología y cambios prehistóricos de la vegetación en un lote de cincuenta hectáreas de la isla de Barro Colorado. Pages 153–156 in E. G. Leigh, Jr., A. S. Rand, and D. M. Windsor, editors. *Ecología de un Bosque Tropical: Ciclos Estacionales y Cambios a Largo Plazo.* Smithsonian Tropical Research Institute, Balboa, Panama.

Putz, F. E., and K. Milton. 1982. Tree mortality rates on Barro Colorado Island. Pages 95–100 in E. G. Leigh, Jr., A. S. Rand, and D. M. Windsor, editors. *The Ecology of a Tropical Forest: Seasonal Rhythms and Long-Term Changes.* Smithsonian Institution Press, Washington, DC.

Putz, F. E., P. D. Coley, K. Lu, A. Montalvo, and A. Aiello. 1983. Uprooting and snapping of trees: Structural determinants and ecological consequences. *Canadian Journal of Forest Research* 13:1011–1020.

Reyes, G., S. Brown, J. Chapman, and A. E. Lugo. 1992. *Wood Densities of Tropical Tree Species.* General Technical Report SO-88. U.S. Department of Agriculture Forest Service, Southern Forest Experiment Station, New Orleans.

Richards, P. W. 1996. *The Tropical Rain Forest: An Ecological Study.* Second Edition. Cambridge University Press, Cambridge, U.K.

Riswan, S., J. B. Kenworthy, and K. Kartawinata. 1985. The estimation of temporal processes in tropical rain forest: A study of primary mixed dipterocarp forest in Indonesia. *Journal of Tropical Ecology* 1:171–82.

Runkle, J. R. 1992. *Guidelines and Sample Protocol for Sampling Forest Gaps.* General Technical Report PNW-GTR-283. U. Department of Agriculture, Forest Service, Pacific Northwest Research Station, Portland, OR.

Scatena, F. N., and M. C. Larsen. 1991. Physical aspects of Hurricane Hugo in Puerto Rico. *Biotropica* 23:317–23.

Scatena, F. N., and A. E. Lugo. 1995. Geomorphology, disturbance, and the soil and vegetation of two subtropical wet steepland watersheds of Puerto Rico. *Geomorphology* 13:199–213.

Schuchert, C. 1935. *Historical Geology of the Antillean–Caribbean Region.* Wiley, New York.

Sokal, R. R., and F. J. Rohlf. 1969. *Biometry: The Principles and Practice of Statistics in Biological Research.* W.H. Freeman, San Francisco.

Telewski, F. W. 1995. Wind-induced physiological and developmental responses in trees. Pages 237–63 in M. P. Coutts and J. Grace, editors. *Wind and Trees.* Cambridge University Press, Cambridge, U.K.

Thompson, J., N. Brokaw, J. K. Zimmerman, R. B. Waide, E. M. Everham, III, D. J. Lodge, C. M. Taylor, D. García-Montiel, and M. Fluet. 2002. Land use history, environment, and tree composition in a tropical forest. *Ecological Applications* 12:1344–63.

Tyrrell, L. E., and T. R. Crow. 1994. Structural characteristics of old-growth hemlock-hardwood forests in relation to age. *Ecology* 75:370–86.

Waide, R. B., and D. P. Reagan. 1996. The rain forest setting. Pages 1–16 in D. P. Reagan and R. B. Waide, editors. *The Food Web of a Tropical Rain Forest.* University of Chicago Press, Chicago.

Walker, L. R. 1991. Tree damage and recovery from Hurricane Hugo in Luquillo Experimental Forest, Puerto Rico. *Biotropica* 23:379–85.

Walker, L. R. 1995. Timing of post-hurricane tree mortality in Puerto Rico. *Journal of Tropical Ecology* 11:315–20.

Walsh, R. P. D. 1996a. Climate. Pages 159–205 in P. W. Richards. *The Tropical Rain Forest: An Ecological Study.* Second Edition. Cambridge University Press, Cambridge, U.K.

Walsh, R. P. D. 1996b. Microclimate and hydrology. Pages 206–36 in P. W. Richards. *The Tropical Rain Forest: An Ecological Study.* Second Edition. Cambridge University Press, Cambridge, U.K.

Weaver, P. L. 1989. Forest changes after hurricanes in Puerto Rico's Luquillo Mountains. *Interciencia* 14:181–92.

Webb, L. J. 1958. Cyclones as an ecological factor in tropical lowland rain forest, north Queensland. *Australian Journal of Botany* 6:220–28.

Whigham, D. F., M. B. Dickinson, and N. V. L. Brokaw. 1999. Background canopy gap and catastrophic wind disturbances in tropical forests. Pages 223–52 in L. R. Walker, editor. *Ecosystems of the World 16: Ecosystems of Disturbed Ground.* Elsevier, Amsterdam.

Whitmore, T. C. 1974. *Change with Time and the Role of Cyclones in Tropical Rain Forest on Kolombangara, Solomon Islands.* Paper 46. Commonwealth Forestry Institute, Oxford, U.K.

Whitmore, T. C. 1975. *Tropical Rain Forests of the Far East.* Clarendon Press, Oxford, U.K.

Whitmore, T. C., and D. F. R. P. Burslem. 1998. Major disturbances in tropical rainforests. Pages 549–65 in D. M. Newbery, H. H. T. Prins, and N. D Brown, editors. *Dynamics of Tropical Communities.* Blackwell, Oxford, U.K.

Young, T. P., and S. P. Hubbell. 1991. Crown asymmetry, treefalls, and repeat disturbance of broad-leaved forest gaps. *Ecology* 72:1464–71.

Zimmerman, J. K., E. M. Everham, III, R. B. Waide, D. J. Lodge, C. M. Taylor, and N. V. L. Brokaw. 1994. Responses of tree species to hurricane winds in subtropical wet forest in Puerto Rico: Implications for tropical tree life histories. *Journal of Ecology* 82:911–22.

14

Linking Canopy Gaps, Topographic Position, and Edaphic Variation in a Tropical Rainforest: Implications for Species Diversity

Peter A. Palmiotto, Kristina A. Vogt, P. Mark S. Ashton, Peter S. Ashton, Daniel J. Vogt, James V. LaFrankie, Hardy Semui, and Hua Seng Lee

Introduction

For over a century, scientists such as Alfred Russell Wallace, Charles Darwin, H. N. Ridley, and Daniel Janzen have grappled with questions of species diversity and examined many factors that could explain high diversity, particularly in species-rich tropical forests. In the last few decades, extensive research has been conducted on gap phase dynamics. As a result, the importance of treefall gaps in creating habitat heterogeneity and affecting overall forest dynamics is generally accepted, especially for tropical ecosystems (Ricklefs 1977; Hartshorn 1978; Denslow 1980; Pickett 1983; Whitmore 1984; Pickett and White 1985; Platt and Strong 1989; Putz and Appanah 1987; Mabberley 1992; but see Hubbell et al. 1999). Less attention, however, has been paid to other factors such as soil type, topographic variation, and the efficiency of plants utilizing available resources (Kalliola 1992; Poulsen 1996; Clark et al. 1998; Palmiotto 1998). This study is intended to document the relationship between soil type, topographic position, canopy disturbance, and tree species diversity on 20 × 20 m subplots in 51 ha of mixed dipterocarp forest in the Lambir Hills National Park, Sarawak, Malaysia. The results reported here help elucidate the relationships between microsite heterogeneity and species diversity. This information increases the understanding of the factors that should be monitored in order to sustain and manage these species-rich forests. Before discussing the results from this study, however, it is important to describe how small-scale disturbance (e.g., treefall gaps and landslides) affects microsite heterogeneity and the importance of canopy gaps to the dynamics of southeast Asian tropical forests. A brief summary follows.

Background

Many researchers have discussed the idea that a treefall gap contains a variety of microsites favoring the regeneration of different species (Orians 1982; Brokaw 1985; Brandani et al. 1988; Uhl et al. 1988; Spies and Franklin 1989; Brown and

Whitmore 1992; Vogt et al. 1995). Different microsite conditions within a gap enhance survival or growth of seedlings and saplings of different tree species (Raich and Gong 1990; Ashton 1995; Ashton and Larson 1996). And different tree species can coexist by segregating along microenvironmental gradients in treefall gaps (Denslow 1980; Ashton et al. 1995; Barker et al. 1997; Davies et al. 1998).

The frequency of treefall gaps of various sizes is related to local climate, topography, parent material, and the composition and structure of the vegetation. How the microclimate (i.e., light, humidity, and temperature) varies from a gap's center to its edge depends on the gap's size, orientation, shape, and slope position (Denslow 1980; Hartshorn 1980; Kennedy and Swaine 1992; Putz 1983; Whitmore 1984; Chazdon and Fetcher 1984a, 1984b; P. M. S. Ashton 1992). In addition, within-gap heterogeneity is caused by upturned roots, the fallen bole, and the location of the fallen crown (Orians 1982). In the root zone or tip-up microsite, soil in wet tropical forests is disturbed with relatively little input of organic matter, but light levels increase due to the absence of the tree canopy. The bole zone experiences minimal canopy disturbance, but receives a tremendous input of slowly released organic matter from the tree bole. The crown zone, where the treetop breaks existing vegetation and advanced regeneration, is the zone of highest input of quickly released nutrients in the form of leaves and small branches. These inputs generally decay rapidly and contribute a pulse of nutrients into the system (the small branches having intermediate rates of decay and nutrient release) (Bloomfield et al. 1993; Vogt et al. 1996). Light intensities in the crown zone increase but do not immediately penetrate to the forest floor due to the thick debris (Orians 1982; Brokaw 1985).

In addition to this classical description of the zones within a treefall gap, it is important to recognize the variability associated with treefall gaps in time and space. The light environment in a treefall gap is highly variable (Spies and Franklin 1989), and the gap's influence on microsite conditions can extend well beyond the physical opening into the closed canopy forest (Chen et al. 1992; Laurance and Bierregaard 1997). Additionally, in smaller gaps, root competition by trees bordering the gap can strongly affect regeneration in the gap (Silver and Vogt 1993; Vogt et al. 1995). Gaps also influence the distribution of soil organisms, both large and small, further enhancing habitat heterogeneity. Symbiotic mycorrhizal relationships critical to the uptake of nutrients could be disturbed in gap environments (Janos 1980; Smits 1983). Finally, wind often enlarges gaps aided in part by the asymmetrical nature of bordering tree crowns (Brünig 1964; Putz and Milton 1982; Young and Hubbell 1991).

More dramatically than treefall gaps, landslides create extreme microsite conditions. A growing body of literature suggests that landslides influence ecosystem development, plant succession, and the maintenance of species diversity in tropical forests (Garwood et al. 1979; Guariguata 1990; Walker et al. 1996). Although microsite conditions can be quite heterogeneous within a landslide, relatively discrete zones can be distinguished. In the upper zone, high light levels, exposed weathered bedrock, and mineral soil are left behind after topsoil and plant material slide down the slope (Fernández and Myster 1995; Guariguata 1990). The lower zone or zone of deposition is a heterogeneous mix of broken plants and trees and organic and mineral soil (Guariguata 1990; Walker et al. 1996). Soil nutrients are generally more abundant in the lower zone than the upper zone (Lundgren 1978; Adams and Sidle 1987; Guariguata 1990) and initial colonization and succession proceeds more rapidly in the lower zone (Guariguata 1990; Walker et al. 1996; Myster and Walker 1997).

Clearly, the role of soil heterogeneity in the maintenance of local species richness in primary tropical forests can be extensive. In addition, microsite heterogeneity can be created in soils in situ due to variation in soil texture. Soil texture influences soil moisture and nutrient levels (Brady 1990). These factors in turn cause variation in the magnitude and type of surface organic layers that are produced and the associated microsoil and macrosoil animals that are able to live there. Soil texture also has the potential to indirectly influence the mode of tree death and thereby the microsite conditions that are created when canopy openings occur. Coarser textured, more nutrient poor soils on drought-prone ridges may cause tree species to develop deeper, more extensive root systems, which enables trees to remain standing after they die (Whitmore 1984; Leighton and Wirawan 1986; Ashton and Hall 1992; P. S. Ashton 1992; Gale 1998). Trees growing on finer, more nutrient rich soils, which have a higher moisture retention capacity appear to develop shallow surface coarse root systems and therefore are more prone to windthrow. These windthrown trees create larger canopy openings than those caused by trees that die standing. These factors provide strong circumstantial evidence that soil heterogeneity could enhance tree species diversity. Edaphic conditions, therefore, along with climate and topography are important to consider when examining the patterns of disturbance and their effects on tree species diversity.

Canopy Openings in Southeast Asian Tropical Forests

In tropical forests, numerous pioneer species are dependent on gaps throughout all stages of their life cycle and many shade tolerant species require increased light levels to reach maturity (Ashton 1982; Davies 1996). Species of the family Dipterocarpaceae, which dominate the wet tropical rainforests of southeast Asia, are generally considered shade tolerant, a physiological characteristic of canopy trees that requires light to release growth beyond the seedling stage (Nicholson 1960, 1965; Liew and Wong 1973; Whitmore 1984). Seedling establishment is not thought to be a limiting factor for dipterocarp regeneration because mast fruiting of dipterocarps periodically covers the forest floor with a blanket of seedlings

(Fox 1967, 1973; Poore 1968; Liew and Wong 1973). Seedlings that establish in the forest understory, however, suffer high mortality (Nicholson 1965; Fox 1973), which is generally ascribed to intense competition for abiotic resources, especially light. Shade tolerant dipterocarp seedlings can survive in the understory for many years by allocating resources to root growth and leaf maintenance (Wong and Whitmore 1970; Brown and Whitmore 1992; Palmiotto 1998; Delissio et al. 2002). When a canopy gap opening occurs, seedlings shift their allocation to height growth in order to reach the next stratum of the forest. If a canopy opening does not occur, the seedlings slowly die off, to be replaced by the next mast's cohort of seedlings (Chan 1977; Chan and Appanah 1980; Ashton 1982; Ashton et al. 1988). This scenario assumes that each masting event produces viable seeds that successfully become established. However, successful establishment is not guaranteed due to pre- and post-emergence predation (Chan 1977; Ashton 1982; Curran and Leighton 2000). Consequently, the establishment of large numbers of seedlings is less frequent than the actual interval between mast years (Whitmore 1996). The long interval between successive seedling cohorts has practical implications for the development of timber-harvesting schedules. We are also faced with the theoretical question of how the long interval between successive seedling establishments influences the long-term dynamics of species-rich tropical forests that experience periodic mast fruiting. In conjunction with the timing of successful seedling cohorts, the frequency and size of canopy openings across the landscape must exert a great influence on forest regeneration and tree species composition.

In the mixed dipterocarp forests of the Lambir Hills, the critical factors that influence which species regenerate appear to be gap size, the origin of gap formation, and the timing and frequency of landslides. Landslides in particular expose mineral soil below the humus layer and eliminate regenerating seedling and sapling sources. Previous research on four 0.6-ha plots in Lambir showed that over half of the canopy trees that died during a 25-year period remained standing, decomposing branch by branch (Hall 1991). Proportionately more trees were uprooted and died in clusters on finer textured udult soils than on the coarser textured humult soils (Hall 1991; Gale 1998). These findings suggest a link between the mode of death and soil texture, which could affect our understanding of the dynamics of tropical forests.

Study Site

The study was conducted at the Lambir Hills National Park (4°12′N, 114°E) in the northwest part of the island of Borneo in southeast Asia (chap. 31). The Lambir Hills are a remnant of an uplifted and eroded coastal delta that formed during the early to mid Miocene (10–25 million years ago) (Mulock Houwer 1967; Watson 1985). The sedimentary rock that remained after this uplift is composed of sandstone and shale or clay. The precise lithology of each layer depends on the conditions under which it was laid down (e.g., estuary channel, swamp, delta front) (Watson 1985). The resulting alternation of sandstones and shales characteristic of the area formed a heterogeneous parent material from which the soils of the Lambir Hills have developed. The complex topography has led to landslides that have exposed and mixed these alternating layers (Ohkubo et al. 1995).

Lambir's varied lithology has given rise to soils that vary in depth, sand content, and nutrient concentrations (Ashton 1973; Hirai et al. 1997; Palmiotto 1998). On the younger, sandstone-dominated hills toward the north, shallow Spodosols predominate on the ridges. Lower sandstone cuestas and dip slopes are dominated by soils more than 1 m deep. These are nutrient-poor, yellow-leached Ultisols with a distinct surface horizon of densely rooted raw humus, which this paper refers to as humults (P. S. Ashton 1992; Ashton and Hall 1992). In contrast, in the south and southwest, yellow-red Ultisols are greater than 1 m deep, with relatively higher nutrient content, which this paper refers to as udults. The surfaces of these udult soils have variable accumulations of litter, immediately beneath which exists a leached mineral soil horizon with only shallow and slight humic discoloration (P. S. Ashton 1992; Ashton and Hall 1992).

Forest structure and species composition reflect the large-scale edaphic variability of the park (Ashton 1973; Whitmore 1984). Heath forests of variable stature are found on sandstone ridges to the north while mixed dipterocarp forests grow elsewhere. The forests in the Lambir Hills are exceptionally diverse and have among the highest tree species richness of any surviving forest in the Old World (Davies and Becker 1996; chap. 31). This study took place in the southern 51 ha of the 52-ha Lambir Forest Dynamics Plot.

Methods

Soil Mapping

In the 51-ha study area, soil texture was determined in the field using the hand-method criteria described by Kimmins (1987). Soils were grouped into four textural classes: clay to clay loam, loam, sandy clay loam, and sandy loam. To assess soil texture, an Oakfield soil corer was used to take two or three composite samples from a depth of 5–15 cm at the center of each 20 × 20 m subplot in the 51-ha area, giving a total of 1275 samples. The thickness of the leaf litter and root mat layers was also measured to the nearest centimeter in a circle of 1 m radius about the center of each of these subplots. For data analysis purposes where leaves covered 75% or more of the soil surface but the leaf litter was less than 1 cm thick, a thickness of 0.5 cm was assigned. Where leaf litter covered less than 75% of the soil surface and was less than 1 cm thick, a thickness of 0.1 cm was assigned. Where root mats were present but did not have a measurable thickness greater than 1 cm, a thickness of 0.5 cm was assigned.

Measurements of soil texture and root mat thickness were used to construct a soil map of the 51-ha study area. Soil texture was used as the primary factor for classifying soil type because fine-textured soils tend to be more fertile, coarse-textured soils less fertile (Ashton 1964; Baillie and Ahmed 1984; Hirai et al. 1997; Palmiotto 1998). Root mat thickness was used as a secondary factor as past research has shown that the presence of a measurable root mat is an indicator of higher organic carbon content. For example, humult soils contain 12 kg or more of soil organic carbon per square meter of forest floor in the top 100 cm of mineral soil (Soil Survey Staff 1997). Subplots with soils whose soil texture was classified as loam, sandy clay loam, or sandy loam and were covered by root mats a centimeter or more thick were designated humult soils. Subplots with clay or loam soils lacking a measurable root mat were designated as udult soils. Soils that did not meet the criteria of the udult or humult soil type, as defined above, were classified as intermediate soils. These definitions of soil types simplify the soil heterogeneity present in the Lambir plot, but provide a basic soil type map to which distributions of plant species can be compared.

Mapping Canopy Openings

The pattern and extent of openings in the canopy of the study area were mapped from August 1994 through June 1995. Canopy openings were defined as having a minimum size of 25 m^2 and an average regeneration height of less than 10 m. Two categories of openings were defined based on the height of the regeneration: (1) openings with regeneration <2 m tall (Brokaw 1982) and (2) openings with regeneration 2–10 m tall. The cause of each canopy opening was determined (Clark 1990; Gale 1998) as was the number of trees involved in the gap creation and the number of old treefalls associated with the current gap. Slope, aspect, and stage of forest development (e.g., disturbed, building, mature) were also measured at each 20 × 20 m subplot. A subplot was classified as building forest if it had lower estimated mean canopy height and higher estimated number of smaller diameter trees than mature forest. A subplot was categorized as disturbed only if the majority of the subplot contained a gap or was affected by a landslide. The topographic position (e.g., valley bottom, side slope, and ridgetop) of a particular subplot was classified as a valley bottom or a ridgetop if the subplot's center point was within 30 m of a perennial or intermittent stream or a ridgeline, respectively. Subplots that were not located within 30 m of either a ridge or a stream were classified as side slopes.

Data Analysis

Graphical representations of soil texture, soil type, and canopy disturbance were created using IDRISI geographic information software (Windows Version 2.0). Field drawings of canopy gaps were digitized and their area determined using the

Area command in IDRISI (Eastman 1997). The Thiessen command of IDRISI was used to create soil texture and soil type maps from the field point data collected at the center of each 20 × 20 m subplot. The areas of the resulting square polygons were determined using the Area command in IDRISI (Eastman 1997).

Contingency table analyses were completed on plot variables, mode of disturbance, and edaphic variables using separate chi-square tests. Fisher's α was calculated for each 20 × 20 m subplot from the number N of trees ≥ 1 cm dbh and the number S of species included among them according to the 1997 tree census data, using the equation $S = \alpha \ln(1 + N/\alpha)$ (Fisher et al. 1943). One-way and two-way analyses of variance were performed to test the significance of the differences of Fisher's α among soil textural classes, soil types, and topographic positions using the general linear model (GLM) procedure of the Statistical Analysis System (SAS Institute 1990). When the results from the analysis of variance were significant ($p < 0.05$), a Tukey's studentized range test was used to separate means. Simple linear regression analysis was used to determine whether the Fisher's α was related to the steepness of slope and thickness of organic matter covering the soil, using the REG procedure of SAS (1990).

Results

Soil Heterogeneity

Coarse-textured sandy loam soils dominated the plot and accounted for 66% (33.5 ha) of the plot area (fig. 14.1). In the plot area, 14% (7.3 ha) was classified as intermediate textured loamy soils, 9% (4.7 ha) was sandy clay loam soils and 11% (5.5 ha) was fine textured clay to clay loam soils. Soil texture was not correlated with steepness of slope but was correlated with topographic position ($\chi^2 = 44.8$, $p = 0.001$). Coarser textured soils were more common on ridges than in valley bottoms and finer textured clay loam soils were more common in the valley bottoms than on ridges. Surface organic matter depth averaged 7.4 ± 4.4 cm over the entire plot area and increased in depth as the soil particle sizes became larger (fig. 14.2). Root mats 1 cm in thickness or greater were present in 52% of the subplots (26.5 ha). Since root mat thickness was correlated with soil texture, thicker root mats were found on the ridges where soil texture tended to be coarser compared to the valley bottoms.

Udult soils with fine to medium textures and no measurable root mats covered 9.5 ha out of the 51 ha of the plot (fig. 14.3). These soils were mainly located in the south-central section of the plot where a broad ridge of fine-textured soils with a thin organic surface horizon was bounded by two of the plot's major streams. The second largest patch of udult soil was located at the base of the main east-west ridge just south of the center of the plot following the course of another drainage. Other significant patches of udult soil were located in drainages along the eastern

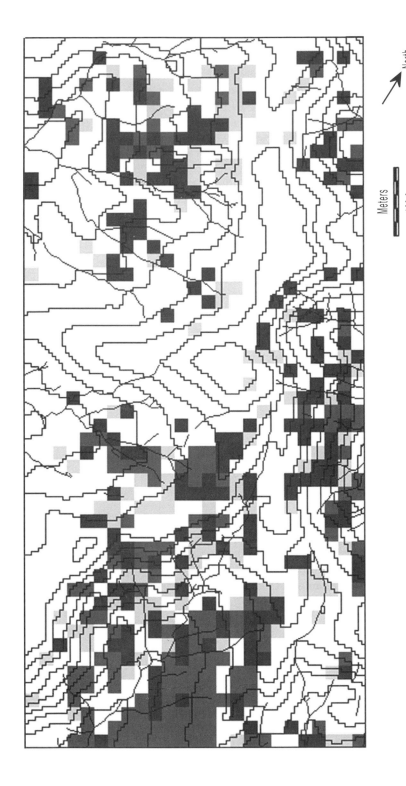

Fig. 14.1. Distribution of soil textural classes in 51 ha of Lambir Hills National Park. Elevation ranges from 120 to 230 m; contour interval = 10 m. Clay–clay loam = gray, loam = dark gray, sandy clay loam = light gray, sandy loam = white.

North

Meters

100.0

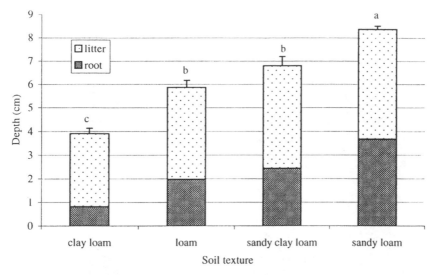

Fig. 14.2. Root and litter depth (mean ± 1 std err) on each soil textural class mapped in 51 ha of Lambir Hills National Park. Different letters denote significant differences between total organic matter depths ($p < 0.05$).

border and in the northwest corner of the plot area. Soils intermediate in their classification between a humult and udult type covered 8.9 ha of the 51-ha study area and were generally found associated with the udult soil type. Humult soils covered 32.6 ha of the 51 ha. These soils dominated the highest main ridges, a gentle yet undulating dip slope in the northwest, and a steep scarp to the east.

Canopy Heterogeneity

A total of 133 canopy openings with regeneration less than 10 m tall were mapped in the 51-ha study area (table 14.1, fig. 14.4). In 14 of these openings, the average height of regenerating vegetation was less than 2 m. Gaps with regeneration less than 2 m tall accounted for 0.3% (0.17 ha) of the study plot and had an average opening of 124 m² (range 28–270 m²). In 119 canopy openings, regeneration was 2–10 m tall. These openings covered 2.9% (1.48 ha) of the study plot and had a mean opening of 125 m² (range 25–2640 m²). The majority of the canopy openings (76.6%) were less than 100 m² in size (fig. 14.5). Canopy openings created by the death of a single tree accounted for 63% of the openings. However, evidence of previous treefalls was associated with 46% of these single-tree canopy openings.

Snap-off and tip-up (uprooted) trees opened the largest number of light gaps and affected the highest percentage of the forest. Snap-off or tip-up trees, alone

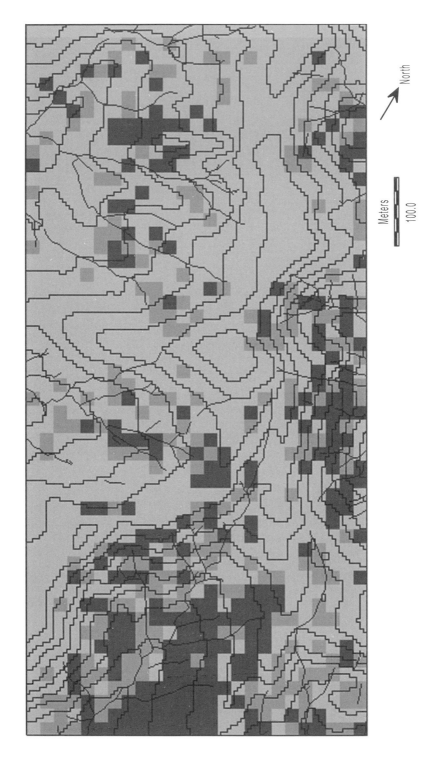

Fig. 14.3. Distribution of soil types in 51 ha of Lambir Hills National Park. Elevation ranges from 120 to 230 m; contour interval = 10 m. Udult = dark gray; intermediate = gray, humult = light gray.

Table 14.1. Frequency and Area of Canopy Openings (Mean ± S.D.), Median and Total Gap Area by Regeneration Height Class and Cause of Canopy Opening in 51 ha of Lambir Hills National Park

Regeneration Height	0–2 m			2–10 m			Total				
		Area (m²)			Area (m²)					Area (m²)	% of
Cause of Opening	N	Mean ± S.D.	Median	N	Mean ± S.D.	Median	N	Mean ± S.D.	Median	(m²)	51 ha
Unknown	1	46		1	325		2	186 ± 197	186	371	0.07
Stream	0			5	87 ± 48	82	5	87 ± 48	82	436	0.09
Branch Fall	0			6	44.8 ± 21	40	6	44.8 ± 21	40	269	0.05
Combination (died standing, tipup, snap)	4	175 ± 90	207	7	62 ± 28	50	11	103 ± 78	70	1,134	0.22
Snapped and Tipups	2	67 ± 33	67	10	113 ± 61	120	12	106 ± 58	105	1,267	0.25
Slide	2	154 ± 165	154	11	480 ± 806	90	13	430 ± 747	90	5,589	1.10
Died Standing	1	28		14	82 ± 103	32	15	77 ± 100	30	1,180	0.23
Tipped Over	0			28	93 ± 81	70	28	93 ± 81	70	2,616	0.51
Snapped	4	132 ± 101	128	37	86 ± 79	61	41	91 ± 81	61	3,726	0.73
Total	14			119			133			16,588	3.25

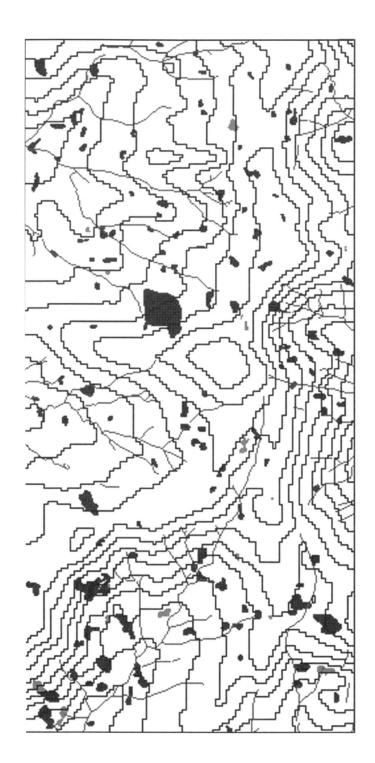

Fig. 14.4. Map of canopy openings in 51 ha of the Lambir Hills National Park. Elevation ranges from 120 to 230 m; contour interval = 10 m. Regeneration height classes: light gray = 0–2 m, dark gray = 2–10 m.

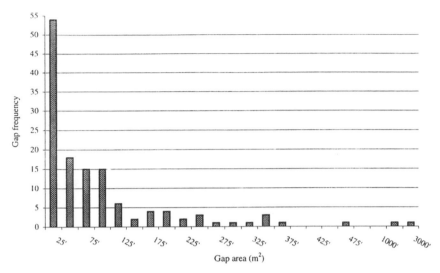

Fig. 14.5. Size distribution of canopy openings grouped in 25-m^2 size classes in 51 ha of the Lambir Hills National Park.

or in combination, caused 61% of the 133 mapped canopy openings and 46% of the total area (table 14.1). Trees that died standing caused 11% of the canopy openings and accounted for only 7% of the open area. Gaps opened by trees that died standing had a smaller median size than gaps opened by any other cause (table 14.1). A combination of trees that died standing and that tipped up or snapped off caused another 8% of the canopy openings. Landslides caused 10% of the canopy openings but occupied 33.7% of the total gap area. The largest of these slides occurred in January 1963 after extremely heavy rainfall that flooded much of northern Sarawak (Watson 1985; Ohkubo et al. 1995). Two of the largest slides, caused by the heavy 1963 rains, created gaps that were still dominated by ferns with only a few clumps of trees less than 10 m tall at the time of the study. Branch falls caused 5% of the canopy openings and streams caused 4%. The causes of 2% of the canopy openings could not be identified.

Subplots classified as mature forest dominated the study plot and covered 64% of the 51-ha area. Building and disturbed subplots covered 32% and 4% of the plot, respectively. The proportion of subplots in mature forest diminished significantly as one progressed up the soil catena from the valley bottoms to the ridgetops ($\chi^2 = 59.6$, $p = 0.001$) (fig. 14.6). No significant relationship was found between the cause of canopy openings and soil texture or soil type. There was, however, a relationship between topographic position and causes of canopy openings. Trees that snapped, the most frequent single cause of the 133 mapped

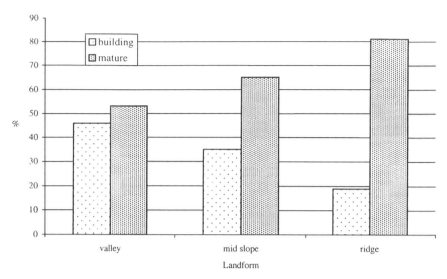

Fig. 14.6. The percentage of forest in mature and building stages of stand development across each landform in 51 ha of the Lambir Hills National Park.

Fig. 14.7. Density of canopy openings per ha caused by trees that died standing, snapped off or tipped up across each landform in 51 ha of the Lambir Hills National Park.

Table 14.2. Result of Two-Way Analysis of Variance for Fisher's α Value for Species Diversity in 20 × 20 m Subplots in Relation to Topographic Position and Soil Type, Texture Class, and Stage of Development at the Lambir Hills National Park

	DF	F	P-value
Soil type	2	8.96	0.0001
Topographic position (TP)	2	42.05	0.0001
Soil type × TP	4	1.86	0.1150
Texture class	3	3.56	0.0138
Topographic position (TP)	2	24.89	0.0001
Texture × TP	6	2.13	0.0477
Stage of Development (SD)	2	7.35	0.0007
Topographic position (TP)	2	10.52	0.0001
SD × TP	4	3.46	0.0080

canopy openings, caused the highest proportion of canopy openings in valley bottoms followed by ridgetops and then the slopes (fig. 14.7). Openings caused by tipped-up trees were the next most common type among the 133 mapped openings. The proportion of openings caused by tipped-up trees was highest in valley bottoms and lowest on ridgetops. In contrast, the proportion of openings caused by trees that died standing was lowest in valley bottoms and highest on ridgetops.

Species Diversity

Topographic position had a significant affect on species diversity as measured by Fisher's α. Valley bottoms had the highest diversity followed by side slopes and ridgetops based on a one-way analysis of variance ($F = 101.11$, $p < 0.0001$). Soil type showed no significant effect on Fisher's α ($F = 1.87$, $p = 0.1544$). However, Fisher's α was significantly higher on subplots with coarse-textured soils than on subplots with finer textured clay loam soils ($F = 4.399$, $p = 0.0044$). Separate two-way analyses of variance confirmed these results showing that valley bottom subplots with coarser textured humult soil had the highest diversity (tables 14.2 and 14.3). On ridgetops and side slopes, soil type had no significant influence on Fisher's α (table 14.3). The average of Fisher's α on disturbed 20 × 20 m subplots was 85.4, which was significantly lower than average for subplots classified as building (95.6) or mature forests (99.2) (table 14.2). Neither slope angle nor organic matter thickness had any relationship with Fisher's α ($r^2 < 0.01$).

Discussion

Canopy Gaps

A clear pattern of gap formation appears from the survey of canopy opening in the Lambir plot. As previous researchers found, distinct differences exist in

Table 14.3. Mean Fisher's α Values for Species Diversity in 20 × 20 m Subplots for Each Soil Type Within Each Topographic Position at the Lambir Hills National Park

Soil Type	Valley Bottom	Side Slope	Ridgetop
Humult	116.0 (1.73, 287)a	98.5 (1.71, 306)a	78.6 (1.55, 223)a
Intermediate	100.7 (2.92, 101)b	92.8 (4.52, 44)a	78.3 (6.20, 14)a
Udult	101.0 (2.17, 182)b	90.2 (3.27, 84)a	74.0 (4.38, 28)a

Notes: Values are least-square means with standard errors and sample size in parentheses. Significant differences between soil type with each topographic position are denoted by different lower case letters.

the size and cause of gaps on ridgetops and valley bottoms (Lawton and Putz 1988; Walker et al. 1991; P. S. Ashton 1992; Poorter et al. 1994; Gale 1998). No overall relationship was detected, however, between the cause of canopy openings and soil texture or soil type, perhaps because the covariation of soil type with topography obscured the relationship at Lambir. An analysis of different research sites that overlie different substrates and subsequently produce different soils may show a relationship between the cause of canopy opening and soil type. Such an examination is justified as an increasing body of research indicates that edaphic and topographic variation may have a strong and predictable influence on forest dynamics and species diversity (Clark et al. 1998, 1999; Gale 1998; Oliveira-Filho et al. 1998; Tuomisto 1998). For example, research focusing explicitly on the mode in which a tree dies, rather than on canopy openings and their related causes, has shown that the proportion of trees that die standing increases, while the proportion of gaps opened by uprooted trees decreases, across the soil catena from valley to ridgetop (Gale 1998). Gale's research examined the underlying mechanisms causing the pattern of canopy openings this study found at Lambir and other studies have found elsewhere. Investigating these mechanisms is critical to developing a broader understanding of how growing space is vacated and which species will reoccupy that growing space.

Canopy Gaps and Topographic Position

When they are well anchored in deep soils, trees tend to die standing rather than being tipped up (Gale 1998). Well-anchored trees tend to occur more often on well-drained ridges such as the coarse-textured humult soils at Lambir. The median gap size of trees that died standing at Lambir was half the median size of gaps from other causes reflecting the tendency of trees that die standing to create smaller treefall gaps (table 14.1). This mode of death results in minimal damage to the surrounding vegetation (Arriaga 1988; Kasenene and Murphy 1991; Krasny and Whitmore 1992; Jans et al. 1993; Midgley et al. 1994; Gale 1998). When trees die standing there is little input of rapidly decomposing organic matter and minimal disturbance to the rhizosphere that would help mix soil nutrients, which would occur when a tree is uprooted. In regions where most trees die standing, the resulting canopy openings favor later successional shade tolerant tree species with a "pessimistic" growth strategy (Kohyama 1987; P. S. Ashton 1992). Species with a pessimistic growth strategy allocate more resources to branch, leaf, and root growth than to height growth, a strategy that enhances survival where light is scarce (Kohyama and Hotta 1990). Moreover, species with higher nutrient-use efficiencies may have a competitive advantage over species with lower nutrient-use efficiencies in low light conditions on nutrient-poor soils (Palmiotto 1998). Finally, the need to reach the canopy favors species with sufficient phenotypic plasticity to increase their height growth efficiency (increase of height/dry matter production per square meter of leaf) when a light gap opens overhead (King 1994).

At Lambir more trees died by uprooting in valley bottoms and on poorly drained midslopes where soils had finer textured soils than on ridgetops. In finer textured soils, roots need not penetrate the soils as deeply to acquire the needed nutrients and moisture, or cannot do so because the deeper soil is episodically poor in oxygen. Therefore, trees are poorly anchored and susceptible to windthrow (Fölster et al. 1976; Gale 1998). Where roots are deep, strong winds tend to cause trees to snap above the ground rather than tip over. Snapped trees were the dominant cause of canopy openings at Lambir where strong convectional storms are prevalent. Soils at Lambir are relatively deep and the steep slopes provide relatively good drainage. Soil conditions favoring deep roots and the dominance of the dipterocarp species with their strong wood account for the high proportion of canopy openings caused by snapped off trees (Hall 1991; Gale 1998). Circumstances that cause most trees to die by uprooting or snapping off of their trunk favor light-demanding species that can more readily use the pulse of nutrients released by these fallen trees. Unlike smaller treefall gaps, which favor a pessimistic growth strategy, the prevalence of large gaps caused by uprooted trees or multiple snapoffs would favor an "optimistic" growth strategy (Kohyama 1987). Optimists apportion more of their resources to height growth, investing in the prospect of more light in the future (Kohyama and Hotta 1990).

Canopy Gaps and Soils

As we have seen, soil and topography can strongly influence a forest's spectrum of sizes and types of treefall gaps. This study found no significant relationship between soil texture or soil type and the causes of canopy openings. On the other hand, 69% of the dead standing trees found when the plot was surveyed for canopy openings were on sandy loam soils, while only 5% were on clay loam soils. Gale's 1998 study, which focused specifically on how trees died, showed that 69% of the trees dying at a site in Brunei on humult soils died standing, compared to only 43% at a different site underlain by udult soils. In Bako National Park on the coast of southern Sarawak where the soils are described as shallow, highly leached and sandy, 84% of the trees died standing (Ashton 1989; Hall 1991; Ashton and Hall

1992; Gale 1998). These soils were described as more similar to the soils found on the ridges than the valley bottoms at Lambir (Hall 1991). These studies support the hypothesis that trees are more likely to die standing on well-drained sandy soils than on less well-drained finer textured soils. This was also the case during the severe droughts of 1982–83, when trees on ridges in East Kalimantan, Indonesia, tended to die standing (Leighton and Wirawan 1986) and during the 1998 drought when canopy trees at Lambir died standing (A. Itoh, personal communication).

Topographic Position, Soils, and Diversity

In the 51-ha study area, valley bottoms were richer in tree species, as measured by Fisher's α, than slopes or ridgetops. As defined in this study, soil type did not significantly influence a subplot's tree diversity when the plot was considered as a whole. However, when the effect of soil type was examined within a particular topographic setting, the highest diversity was found in the subplots on humult soil type in valley bottoms. There was no significant effect of soil type on tree diversity on side slopes or ridgetops, perhaps because the rarity of subplots with udult soils in these settings reduced the power of the statistical tests used. Even though this study found no significant effect of soil type on tree diversity in the study area as a whole, tree diversity was significantly higher on the course-textured soils than on the fine-textured clay loam soils. Therefore, it appears that edaphic conditions do influence tree diversity. Based on the different results obtained for the effects on tree diversity of soil type and soil texture, it seems unwise to analyze species diversity patterns based on soil type alone. The area categorized as a particular soil type may change depending on the criteria used. It is perhaps more appropriate to analyze species distributions based on the spatial variation of soil texture and depth of surface organic layers in association with the topographic variation of the plot (Davies et al. 1998). A combination of descriptive variables is more effective at characterizing the spatial heterogeneity in soil, topography, and the effects of canopy openings of different sizes on the distribution of the different tree species at Lambir.

Moreover, Fisher's α, which is calculated from a plot's number of trees and the number of species included among them, is only one of many ways to measure diversity. At Lambir, where many tree species specialize to different habitats, simple measures of diversity, such as species counts or Fisher's α, do not capture the restriction of different species to different parts of the plot. For example, the well-studied congeneric species *Dryobalanops aromatica* and *D. lanceolata* (Dipterocarpaceae) are each restricted to different soil types in the Lambir plot (Hirai et al. 1997; Itoh et al. 1997). Other species including several Euphorbiaceae (Davies et al. 1998; Debski et al. 2002) also specialize to particular soils or topographic position. If these species are restricted to one soil type, their presence will make measures of species diversity such as species counts and Fisher's α more uniform across the plot without reflecting the forest's true diversity. Such diversity indices will not increase with area predictably because specialists will tend to cancel each other out with increased area (Condit et al. 1996). Clark et al. (1998, 1999) found that in Costa Rica, a majority of species showed highly significant associations with soil type. Analyses of the association of individual species and species groups (e.g., by growth strategy) with edaphic or other variables believed to influence tree distributions must be conducted to make sense of patterns of tree diversity or the factors influencing these patterns. This is especially true in forests hypothesized to have a high proportion of habitat specialists among their tree species, as in the Lambir Hills.

Canopy Gap Synthesis

Trees that died standing had less impact on the forest canopy than the other modes of tree death. Fewer than 30% of the standing dead trees we counted created canopy openings. This proportion would be even lower if all trees that had died standing without creating canopy openings could be identified. On ridges, growing space vacated by dying trees that remain standing is likely to be filled by neighboring trees growing into the gap from the sides rather than high light demanding species. This would explain why a higher proportion of ridge subplots were classified as mature phase rather than building (fig. 14.6). Therefore, due primarily to the lack of open gaps, pioneer tree species may play a less significant role in succession on stable humult soils on ridges and dip slopes at Lambir. In contrast, in valley bottoms on well-structured clay soils with relatively high nutrient concentrations, short-lived pioneers and a guild of fast-growing light hardwoods (including both pioneer and climax species) with single leaders and plagiotropic branching, grow tall enough to reach the main or emergent canopy and play a major part in gap succession. Many of the emergent trees of the mature-phase forest in Lambir's valley bottoms have these characteristics when they are young, but species with these traits are relatively rare on other udult soils and are rare in primary forest on humult soils (Ashton and Hall 1992).

These findings suggest that there is a degree of predictability in the type of species that will be present in different parts of the landscape. This predictability is based on the variation in median gap size created on different topographic positions and soil textural classes in a forest. Although research in the neotropics has suggested that the spatial pattern of canopy opening appears random and gap disturbance does not control species diversity (Hubbell and Foster 1986; Hubbell et al. 1999), it is likely that mechanisms controlling species diversity in Old World tropical forests vary. Direct comparisons across the current system of large-scale Forest Dynamics Plots have the potential to elucidate these differences. At Lambir multiple modes of tree death, edaphic conditions, and land forms greatly increase microsite heterogeneity. These processes appear to contribute to the heterogeneity of regeneration sites and could facilitate the maintenance of site specialization, enhancing Lambir's tree diversity.

General Implications

1. Topographic and edaphic variation in a landscape can influence the mode of tree death, creating a variety of gap microsites. This variety of microsites may support specialist tree species with different growth strategies, thus enhancing tree diversity.
2. If, as this paper suggests, canopy disturbance varies predictably with topography and soil texture, then we may consider how tree diversity and the spectrum of plant growth strategies depend on the modes of canopy disturbance.
3. Examination of individual species distributions by edaphic factors, topographic features, and plant growth strategies may show patterns of habitat specialization that will aid in our understanding of diversity more than analysis of simple indexes of diversity (i.e., Fisher's α or species counts).
4. In addition to surveys on canopy gaps, the CTFS network should develop protocols to measure the mode of tree death so that patterns of tree mortality can be analyzed to develop a more definitive statement on the role of small-scale disturbance and microsite heterogeneity on species diversity in complex forests.

Acknowledgments

We thank the Sarawak Forest Department for permission and support to conduct this research in the Lambir Hills National Park. Additional thanks to Layang Unam, Anthony Ngau Sigan, and Yeeyie Lieskovsky for assistance in the field, Paul Boon for his assistance with GIS analysis, and Jennifer OHara Palmiotto, Akira Itoh, and the book editors for helpful comments on the manuscript. We also thank Suzanne Lao for providing the Fisher's α values and Dr. T. Yamakura for providing topographic data, which are displayed in the figures. This research was supported in part by the Conservation, Food and Health Foundation, Inc., the G. Evelyn Hutchinson Prize, the Tropical Resources Institute at Yale University, the Sigma Xi Scientific Research Society, and the Center for Tropical Forest Science of the Smithsonian Tropical Research Institute.

References

Adams, P. W., and Sidle, R. C. 1987. Soil conditions in three recent landslides in southeast Alaska. *Forest Ecology and Management* 18:93–102.

Arriaga, L. 1988. Natural disturbance and treefalls in a pine–oak forest on the Peninsula of Baja California, Mexico. *Vegetatio* 78:73–79.

Ashton, P. M. S. 1992. Establishment and early growth of advanced regeneration of canopy trees in moist mixed-species forest. Pages 101–22 in M. J. Kelty, B. C. Larson, and C. D. Oliver, editors. *The Ecology and Silviculture of Mixed-Species Forests.* Kluwer Academic Publishers, Dordrecht, Netherlands.

———. 1995. Seedling growth of co-occurring *Shorea* species in the simulated light environments of a rain forest. *Forest Ecology and Management* 72:1–12.

Ashton, P. M. S., C. V. S. Gunatilleke, and I. A. U. N. Gunatilleke. 1995. Seedling survival and growth of four *Shorea* species in a Sri Lankan rainforest. *Journal of Tropical Ecology* 12:1–16.

Ashton, P. M. S., and B. C. Larson. 1996. Germination and seedling growth of *Quercus* (section Erythrobalanus) across openings in a mixed-deciduous forest of southern New England, USA. *Forest Ecology and Management* 80:81–94.

Ashton, P. S. 1964. *Ecological Studies in the Mixed Dipterocarp Forests of Brunei State.* Oxford Forestry Memoirs. Clarendon Press, Oxford, U.K.

———. 1973. *Report on Research Undertaken during the Years 1963–72 on the Ecology of Mixed Dipterocarp Forest in Sarawak.* Mimeo. Sarawak Forest Department. Kuching Sarawak, Malaysia.

———. 1982. Dipterocarpaceae. *Flora Malesiana Series I – Spermatophyta Flowering Plants* 9:237–552.

———. 1989. Species richness in tropical forests. Pages 239–51 in L. B. Holom-Nielsen, I. C. Nielsen, and H. Balslev, editors. *Tropical Forests: Botanical Dynamics, Speciation and Diversity.* Academic Press, London.

———. 1992. The structure and dynamics of tropical rain forest in relation to tree species richness. Pages 53–64 in M. J. Kelty, B. C. Larson, and C. D. Oliver, editors. *The Ecology and Silviculture of Mixed-Species Forests.* Kluwer Academic Publishers, Dordrecht, Netherlands.

Ashton P. S., and P. Hall. 1992. Comparisons of structure among mixed dipterocarp forests of north-west Borneo. *Journal of Ecology* 80:459–81.

Ashton, P. S., T. J. Givnish, and S. Appanah. 1988. Staggered flowering in the Dipterocarpaceae: New insights into floral induction and the evolution of mast fruiting in the aseasonal tropics. *American Naturalist* 132:44–66.

Baillie, I. C., and M. I. Ahmed. 1984. The variability of red yellow podzolic soils under mixed dipterocarp forest in Sarawak, Malaysia. *Malaysian Journal of Tropical Geography* 9:1–13.

Baillie, I. C., P. S. Ashton, M. N. Court, J. A. R. Anderson, E. A. Fitzpatrick, and J. Tinsley. 1987. Site characteristics and the distribution of tree species in mixed dipterocarp forest on tertiary sediments in Central Sarawak, Malaysia. *Journal of Tropical Ecology* 3:201–20.

Barker, M. G., M. C. Press, and N. D. Brown. 1997. Photosynthetic characteristics of dipterocarp seedlings in three tropical rain forest light environments: A basis for niche partitioning? *Oecologia* 112:453–63.

Bloomfield, J., K. A. Vogt, and D. A. Vogt. 1993. Decay rate and substrate quality of fine roots and foliage of two tropical tree species in the Luquillo Experimental Forest, Puerto Rico. *Plant and Soil* 150:233–245.

Brady, N. C. 1990. *The Nature and Property of Soils.* Macmillan, New York.

Brandani, A., G. S. Hartshorn, and G. H. Orians. 1988. Internal heterogeneity of gaps and species richness in Costa Rican tropical wet forest. *Journal of Tropical Ecology* 4:99–119.

Brokaw, N. V. L. 1982. The definition of treefall gap and its effect on measures of forest dynamics. Biotropica 14:158–160.

———. 1985. Treefalls, regrowth, and community structure in tropical forests. Pages 53–69 in S. T. A. Pickett and P. S. White, editors. *The Ecology of Natural Disturbance and Patch Dynamics.* Academic Press, Orlando, FL.

Brown, N. D., and T. C. Whitmore. 1992. Do dipterocarp seedlings really partition tropical rain forest gaps? *Philosophical Transactions of the Royal Society of London. Series B* 335:369–78.

Brünig, E. F. 1964. A study of damage attributed to lightning in two areas of *Shorea albida* forest in Sarawak. *Community Forest Review* 43:134–44.

Chan, H. T. 1977. *Reproductive Biology of Some Malaysian Dipterocarps.* Ph.D. dissertation. University of Aberdeen. Aberdeen, Scotland.

Chan, H. T., and A. Appanah. 1980. Reproductive biology of some Malaysian Dipterocarps. I. Flowering biology. *Malaysian Forester* 43:132–43.

Chazdon, R. L., and N. Fetcher. 1984a. Photosynthetic light environments in a lowland tropical rainforest in Costa Rica. *Journal of Ecology* 72:553–64.

———. 1984b. Light environments in tropical forests. Pages 27–36 in E. Medina, H. A. Mooney, and C. Vazquez-Yanes, editors. *Physiological Ecology of Plants of the Wet Tropics.* Dr. W. Junk, The Hague.

Chen, J., J. F. Franklin, and T. A. Spies. 1992. Vegetation responses to edge environments in old-growth Douglas-fir forests. *Ecological Applications* 2:387–96.

Clark, D. B. 1990. The role of disturbance in the regeneration of neotropical moist forests. Pages 291–315 in K. S. Bawa and M. Hadley, editors. *Reproductive Ecology of Tropical Forest Plants.* UNESCO and Parthenon Publishing Group, Paris.

Clark, D. B., Clark, D. A., and J. M. Read. 1998. Edaphic variation and the mesoscale distribution of tree species in a neotropical rain forest. *Journal of Ecology* 86:101–12.

Clark, D. B., Palmer, M. W., and D. A. Clark. 1999. Edaphic factors and the landscape-scale distributions of tropical rain forest trees. *Ecology* 88:2662–75.

Curran, L. M., and M. Leighton. 2000. Vertebrate responses to spatiotemporal variation in seed production of mast-fruiting Dipterocarpaceae. *Ecological Monographs* 70:101–28.

Condit, R., S. P. Hubbell, and R. B. Foster. 1996. Changes in tree species abundance in a neotropical forest: Impact of climate change. *Journal of Tropical Ecology* 12: 231–56.

Condit, R., S. P. Hubbell, J. V. LaFrankie, R. Sukumar, N. Manokaran, R. B. Foster and P. S. Ashton. 1996. Species–area and species–individual relationships for tropical trees: A comparison of three 50-ha plots. *Journal of Ecology* 84: 549–62.

Davies, S. J. 1996. *The Comparative Ecology of Macaranga (Euphorbiaceae).* Ph.D. dissertation. Harvard University, Cambridge, MA.

Davies, S. J., and P. Becker. 1996. Floristic composition and stand structure of mixed dipterocarp and heath forests in Brunei Darussalam. *Journal of Tropical Forest Science* 8:542–69.

Davies, S. J., P. A. Palmiotto, P. S. Ashton, H. S. Lee, and J. LaFrankie. 1998. Comparative ecology of 11 sympatric species of *Macaranga* in Borneo: Tree distribution in relation to horizontal and vertical resource heterogeneity. *Journal of Ecology* 86:662–73.

Debski, I., D. F. R. P. Burslem, P. A. Palmiotto, J. V. LaFrankie, H. S. Lee, N. Manokaran. 2002. Habitat preferences of Aporosa in two Malaysian forests: Implications for abundance and coexistence. *Ecology* 83:2005–18.

Delissio, L. J., Primack. R. B., P. Hall, and H. S. Lee. 2002. A decade of canopy-tree seedling survival and growth in two Bornean rain forests: Persistence and recovery from suppression. *Journal of Tropical Ecology* 18: 645–658.

Denslow, J. S. 1980. Gap partitioning among tropical rainforest trees. *Biotropica* (*Suppl.*) 12:47–95.

Eastman, J. R. 1997. *IDRISI Users Guide*. Clark Labs for Cartographic Technology and Geographic Analysis, Worcester, MA.

Fernández, D. S., and R. W. Myster. 1995. Temporal variation and frequency disturbance of photosynthetic flux densities on landslides in Puerto Rico. *Journal of Tropical Ecology* 36:73–87.

Fisher, R. A., Corbet, A. S., and C. B. Williams. 1943. The relationship between the number of species and the number of individuals in a random sample of an animal population. *Journal of Animal Ecology* 12:42–58.

Fölster, H., G. de las Salas, and P. Khanna. 1976. A tropical evergreen forest site with perched water table, Magdalena Valley, Columbia. Biomass and bioelement inventory of primary and secondary vegetation. *Oecologia Plantarum* 11:297–320.

Fox, J. E. D. 1967. An enumeration of lowland dipterocarp forest in Sabah. *Malaysian Forester* 30:263–79.

Fox, R. L. 1973. Dipterocarp seedling behavior in Sabah. *Malaysian Forester* 36:205–14.

Gale, N. 1998. *Modes of Tree Death in Four Tropical Forests*. Ph.D. dissertation. University of Aarhus, Aarhus, Denmark.

Garwood, N., D. P. Janos, and N. Brokaw. 1979. Earthquake-caused landslides: A major disturbance to tropical forests. *Science* 205:997–99.

Guariguata, M. R. 1990. Landslide disturbance and forest regeneration in the upper Luquillo mountains of Puerto Rico. *Journal of Ecology* 78:814–32.

Hall, P. 1991. *Structure, Stand Dynamics, and Species Compositional Change in the Mixed Dipterocarp Forests of Northwest Borneo*. Ph.D. dissertation. Boston University, Boston, MA.

Hartshorn, G. S. 1978. Treefalls and tropical forest dynamics. Pages 617–83 in P. B. Tomlinson and M. H. Zimmerman, editors. *Tropical Trees as Living Systems*. Cambridge University Press, Cambridge.

Hartshorn, G. S. 1980. Neotropical forest dynamics. *Biotropica* (*Suppl.*)12:23–30.

————. 1989. Gap-phase dynamics and tropical tree species richness. Pages 65–71 in L. B. Holom-Nielsen, I. C. Nielsen, and H. Balslev, editors. *Tropical Forests: Botanical Dynamics, Speciation and Diversity*. Academic Press, London.

Hirai, H., H. Matsumura, H. Hirotani, K. Sakurai, K. Ogino, and H. S. Lee. 1997. Soils and the distribution of *Dryobalanops aromatica* and *D. lanceolata* in mixed dipterocarp forest. A case study at Lambir Hills National Park, Sarawak, Malaysia. *Tropics* 7:21–33.

Hubbell, S. P., and R. B. Foster. 1986. Canopy gaps and the dynamics of a neotropical forest. Pages 77–96 in M. J. Crawley, editor. *Plant Ecology*. Blackwell Scientific, Oxford, U.K.

Hubbell, S. P., R. B. Foster, S. T. O'Brien, K. E. Harms, R. Condit, B. Wechsler, J. Wright, and S. Loo de Lao. 1999. Light-gap disturbance, recruitment limitation, and tree diversity in a neotropical forest. *Science* 283:540–44.

Itoh, A., T. Yamakura, K. Ogino, H. S. Lee, and P. S. Ashton. 1997. Spatial distribution patterns of two predominant emergent trees in a tropical rainforest in Sarawak, Malaysia. *Plant Ecology* 132:121–36.

Janos, D. P. 1980. Mycorrhizae influence tropical succession. *Tropical Succession* 12:56–64.

Jans, L., L. Poorter, S. A. R. van Rompaey, and F. Bongers. 1993. Gaps and forest zones tropical moist forest in Ivory Coast. *Biotropica* 25:258–69.

Kalliola, R. 1992. *Abiotic Control of the Vegetation in Peruvian Amazon Floodplains: Environment Change and Pioneer Species*. Ph.D. dissertation. University of Turku, Turku, Finland.

Kasenene, J. M., and P. G. Murphy. 1991. Post-logging tree mortality and major branch losses in Kibale Forest, Uganda. *Forest Ecology and Management* 46:295–307.

Kennedy, D. N., and M. D. Swaine. 1992. Germination and growth of colonizing species in artificial gaps of different sizes in dipterocarp rain forest. *Philosophical Transactions of the Royal Society of London. Series B* 335:357–66.

Kimmins, J. P. 1987. *Forest Ecology.* Macmillan, New York.

King, D. A. 1994. Influence of light level on the growth and morphology of saplings in a Panamanian forest. *American Journal of Botany* 81:948–57.

Kohyama T. 1987. Significance of architecture and allometry in saplings. *Functional Ecology* 1:399–404.

Kohyama, T., and M. Hotta. 1990. Significance of allometry in tropical saplings. *Functional Ecology* 4:515–21.

Krasny, M. E., and M. C. Whitmore. 1992. Gradual and sudden forest canopy gaps in Allegheny northern hardwood forests. *Canadian Journal of Forest Research* 22:139–43.

Laurance, W. F., and R. O. Bierregaard. 1997. *Tropical Forest Remnants.* University of Chicago Press, Chicago.

Lawton, R. O., and R. E. Putz. 1988. Natural disturbance and gap-phase regeneration in a wind-exposed tropical cloud forest. *Ecology* 69:764–77.

Leighton, M., and N. Wirawan. 1986. Catastrophic drought and fire in Borneo tropical rain forest associated with the 1982–1983 El Nino southern oscillation event. Pages 75–102 in G. T. Prance, editor. *Tropical Rain Forests and the World Atmosphere.* Westview Press, Boulder, CO.

Liew, T. C., and F. O. Wong. 1973. Density, recruitment, mortality and growth of dipterocarp seedlings in virgin and logged forests in Sabah. *Malaysian Forester* 36:3–15.

Lundgren, L. 1978. Studies of soil and vegetation development on fresh landslide scars in Mgeta Valley, Western Uluguru Mountains, Tanzania. *Geografiska Annaler* 60A:91–127.

Mabberley, D. J. 1992. *Tropical Rain Forest Ecology.* Chapman and Hall, New York.

Midgley, J. J., M. C. Cameron, and W. J. Bond. 1995. Gap characteristics and replacement patterns in the Knysna forest, South Africa. *Journal of Vegetation Science* 6:29–36.

Mulock Houwer, J. A. 1967. Raim Road section, Lambir Hills, Sarawak. Borneo Region, Malaysia. *Geological Survey Bulletin* 9:38–42.

Myster, R. W., and D. S. Fernández. 1995. Spatial heterogeneity of seed rain, seed pool, and vegetative cover on two Monteverde landslides, Costa Rica. *Brenesia* 39–40:137–45.

Myster, R. W., and L. R. Walker. 1997. Plant successional pathways on Puerto Rican landslides. *Biotropica* 13:165–73.

Nicholson, D. I. 1960. Light requirements of seedlings of five species of Dipterocarpaceae. *Malaysian Forester* 23:344–56.

———. 1965. A review of natural regeneration in the dipterocarp forests of Sabah. *Malaysian Forester* 28:4–28.

Ohkubo, T., T. Maeda, T. Kato, M. Tani, T. Tamakura, H. S. Lee, P. S. Ashton, and K. Ogino. 1995. Landslide scars in canopy mosaic structure as large scale disturbance to a mixed-Dipterocarp forest, at Lambir Hills N.P. Sarawak. Pages 172–84 in H. S. Lee, K. Ogino, and P. S. Ashton, editors. *Long-Term Ecological Research in Relation to Forest Ecosystem Management.* Government Printer, Kuching, Malaysia.

Oliveira-Filho, A. T., N. Curi, E. A. Vilela, and D. A. Carvalho. 1998. Effects of canopy gaps, topography and soils on the distribution of woody species in a central Brazilian deciduous dry forest. *Biotropica* 30: 362–75.

Orians, G. H. 1982. The influence of tree-falls in tropical forests in tree species richness. *Tropical Ecology* 23:255–79.

Palmiotto, P. A. 1998. *The Role of Specialization in Nutrient-Use Efficiency as a Mechanism Driving Species Diversity in a Tropical Rain Forest.* D.F. dissertation. Yale University, New Haven, CT.

Pickett, S. T. A. 1983. Differential adaptation of tropical species to canopy gaps and its role in community dynamics. *Journal of Tropical Ecology* 24:68–84.

Pickett, S. T. A., and P. S. White. 1985. *The Ecology of Natural Disturbance and Patch Dynamics.* Academic Press, Orlando, FL.

Platt, W. J. and D. R. Strong. 1989. Special feature: Gaps in forest ecology. *Ecology* 70:535–76.

Poore, M. E. D. 1968. Studies in Malaysian rain forest. I. The forest of Triassic sediments in Jengka Forest Reserve. *Journal of Ecology* 56:143–96.

Poorter, L., L. Jans, F. Bongers, and R. S. A. R. van Rompaey. 1994. Spatial distribution of gaps along three catenas in the moist forest of Tai National Park, Ivory Coast. *Journal of Tropical Ecology* 10:385–98.

Poulsen, A. D. 1996. Species richness and density of ground herbs within a plot of lowland rainforest in north-west Borneo. *Journal of Tropical Ecology* 12:177–90.

Putz, F. E. 1983. Treefall pits and mounds, buried seeds, and the importance of soil disturbance to pioneer trees on Barro Colorado Island, Panama. *Ecology* 64:1069–74.

Putz F. E. and S. Appanah. 1987. Buried seeds, newly dispersed seeds, and the dynamics of a lowland forest in Malaysia. *Biotropica* 19:326–33.

Putz, F. E., and K. Milton. 1982. Tree mortality rates on Barro Colorado Island. Pages 95–100 in E. G. Leigh, Jr., A. S. Rand, and D. M. Windsor, editors. *The Ecology of a Tropical Forest: Seasonal Rhythms and Long-Term Changes.* Smithsonian Institution Press, Washington, DC.

Raich, J. W., and Gong. W. K. 1990. Effects of canopy opening on tree seed germination in a Malaysian dipterocarp forest. *Journal of Tropical Ecology* 6:203–17.

Ricklefs, R. E. 1977. Environmental heterogeneity and plant species diversity: A hypothesis. *American Naturalist* 111:377–81.

Sarawak Water Authority. 1995. *Rainfall Data from Lambir Hills National Park.* Miri, Sarawak, Malaysia.

SAS Institute. 1990. *SAS/STAT User's Guide.* SAS Institute, Cary, NC.

Silver, W. L., and K. A. Vogt. 1993. Fine root dynamics following single and multiple disturbances in a subtropical wet forest ecosystem. *Journal of Ecology* 81:729–38.

Smits, W. T. M. 1983. Dipterocarps and mycorrhiza, an ecological adaptation and a factor in forest regeneration. *Flora Malesiana Bulletin* 36:3926–37.

Soil Survey Staff. 1997. *Keys to Soil Taxonomy.* Pocahontas Press, Blacksburg, VA.

Spies, T. A., and J. F. Franklin. 1989. Gap characteristics and vegetation response in coniferous forests of the Pacific Northwest. *Ecology* 70:543–45.

Tuomisto, H., A. D. Poulsen, and R. C. Moran. 1998. Edaphic distribution of some species of the fern genus *Adiantum* in western Amazonia. *Biotropica* 30:392–99.

Uhl, C., K. Clark, N. Dezzeo, and P. Maquirino. 1988. Vegetation dynamics in Amazonian treefall gaps. *Ecology* 69:751–63.

Vogt, K. A., D. Vogt, H. Asbjornsen, and R. A. Dahlgren. 1995. Roots, nutrients and their relationship to spatial patterns. *Plant and Soil* 168–69:113–23.

Vogt, K. A., D. J. Vogt, P. Boon, A. Covich, F. N. Scatena, H. Asbjornsen, J. L. O'Hara, J. Perez, T. G. Siccama, J. Bloomfield, and J. F. Ranciato. 1996. Litter dynamics along stream, riparian, and upslope areas following Hurricane Hugo, Luquillo experiment forest, Puerto Rico. *Biotropica* 28:458–70.

Walker, L. R., N. V. L. Brokaw, D. J. Lodge, and R. B. Waide. 1991. Ecosystem, plant and animal responses to hurricanes in the Caribbean. *Biotropica* 23:313–513.

Walker, L. R., D. J. Zarin, N. Fetcher, R. W. Myster, and A. H. Johnson. 1996. Ecosystem development and plant succession on landslides in the Caribbean. *Biotropica* 28:566–76.

Watson, H. 1985. *Lambir Hills National Park Resource Inventory and Management Recommendations.* National Parks and Wildlife Office Forest Department Sarawak, Kuching, Sarawak, Malaysia.

Whitmore, T. C. 1984. *Tropical Rain Forest of the Far East.* Clarendon Press, Oxford, U.K.

———. 1996. A review of some aspects of tropical rain forest seedling ecology with suggestions for further inquiry. Pages 3–39 in M. D. Swaine, editor. *The Ecology of Tropical Forest Tree Seedlings.* Parthenon Publishing Group, Paris.

Wong, Y. K., and T. C. Whitmore. 1970. On the influence of soil properties on species distribution in a Malayan lowland dipterocarp forest. *Malaysian Forester* 33:42–54.

Young, T. P., and S. P. Hubbell. 1991. Crown asymmetry, treefalls, and repeat disturbance of broad-leaved forest gaps. *Ecology* 72: 1464–71.

15

Local Variation of Canopy Structure in Relation to Soils and Topography and the Implications for Species Diversity in a Rainforest of Peninsular Malaysia

Toshinori Okuda, Naoki Adachi, Mariko Suzuki,
Nor Azman Hussein, N. Manokaran, Leng Guan Saw,
Amir Husni Mohd Shariff, and Peter S. Ashton

Introduction

Many factors influence the complex structure of rainforests. Two obvious factors are topography and drainage regime, which interact and have additional effects on soil development (Gartlan et al. 1986; Newbery 1991; Ashton and Hall 1992). Most trees do not thrive in areas of impeded drainage. For example, fewer emergent tree species are found in some alluvial regions (Mabberley 1992). Under mesic conditions, roots are shallow so trees tend to be toppled by windstorms, whereas trees on xeric sites usually die standing or are snapped off because they root more deeply in dry soils (Ashton and Hall 1992). For these reasons, differences in soil conditions related to topography may influence the spectrum of gap size, which in turn, may influence tree species richness and diversity in tree community patches. The ways in which gaps form may affect the complexity of canopy surface structure of a tree community, creating a heterogeneity of light environments in the forest understory.

Moreover, gaps formed by falling trees influence the subsequent species composition and diversity of the forest because increased levels of sunlight in the gaps favor the regeneration of many tree species. Relative to the closed canopy, a gap site often has an increased number of species due to the initial increase in the variety of available microenvironments, thereby providing conditions that meet the different physiological demands of various gap species (Ricklefs 1977; Denslow 1980; Denslow et al. 1990; Picket 1983; Brown and Whitmore 1992; Barker et al. 1997).

In this chapter, we used digital elevation models of the canopy surface and topography, combined with data from a tree demographic census in a lowland dipterocarp forest in peninsular Malaysia, to examine how the canopy height changes on a local scale in response to variations in the forest topography and soil type. The study also examines the following two hypotheses (1) tree species diversity increases with decreased canopy height as a result of increased light availability that favors more species and (2) species diversity increases with the increased complexity of canopy structure within local patches that may provide higher variation of microenvironment in the forest floor.

Study Area

The study area is located within the Pasoh Forest Reserve (2°58′N, 102°18′E) in the state of Negeri Sembilan, about 70 km southeast of Kuala Lumpur. The overall vegetation type in the reserve is lowland dipterocarp forest, which is characterized by a high proportion of Dipterocarpaceae (Symington 1943; Wyatt-Smith 1961, 1964). Based on floristic evidence, the core area of the forest in the study area is generally homogeneous, with no evidence of major disturbance. In 1985, a 50-ha Forest Dynamics Plot was established by the Forest Research Institute of Malaysia (FRIM) and the Smithsonian Tropical Research Institute in the primary forest of the Pasoh Forest Reserve. Detailed description of Pasoh Forest Reserve and the tree census of the 50-ha Forest Dynamics Plot are given in chapter 35 and in Manokaran et al. 1990. The initial census totaled 335,352 trees in 814 species, 295 genera, and 78 families (Manokaran et al. 1992).

Methods

Soil and Topography Mapping

Ground elevation was recorded at 20-m intervals in the 50-ha Pasoh Forest Dynamics Plot prior to the initial census. Ground elevation values were interpolated into 10-m intervals using the cubic spline method. Six topography types (ridgetop, higher slope, midslope, lower slope, flatland, and valley/riverine) were assigned to the resulting 10×10 m subplots based on the slope degree ratio and the index of convexity within each subplot (table 15.1). The index of convexity was the difference in elevation between the mean elevation of a focal quadrat and that of an outer quadrat (Yamakura et al. 1995).

The slope degree was the incline rate of a focal quadrat plane $\alpha(P_1, P_2, P_x, P_4)$ against the level plane $\beta(P_1, P_5, P_6, P_7)$ (fig. 15.1). In order to estimate the slope degree, we established a normal vector C to the slope plane α, which is outer

Table 15.1. The Six Topography Types Assigned to 10×10 m Subplots

Topography Type	Slope Degree (SD)	Index of Convexity (IC)
Flat Alluvial Area (Flatland)	<0.05	$-0.3 \leq IC < 0.7$
Ridgetop	<0.05	$IC \geq 0.7$
Valley and Riverine Area	<0.05	$IC < -0.3$
Lower Slope	≥ 0.05	$IC < -0.1$
Midslope	≥ 0.05	$-0.1 \leq IC < 0.7$
Higher Slope	≥ 0.05	$IC \geq 0.7$

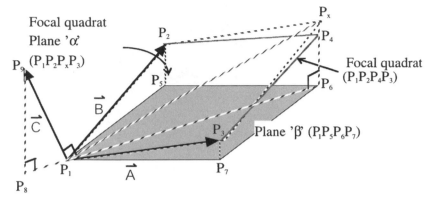

Fig. 15.1. Diagram for determining slope degree of individual subplot. See text for explanation.

product of vectors A and B. The slope degree of plane α can be expressed by the triangle P_x P_1 P_6, which is similar to the triangle P_1 P_9 P_8 formed by vector C. Vectors A and B are formed along two neighboring side lines that extend from corner P_1 of a focal quadrat plane α. The elements of vectors A and B are all determined by the elevation measured at three corners of the focal quadrat (P_1, P_2, P_3), thus the slope degree of plane α against plane β can be obtained by the three-dimensional elements at these three corners. However precisely, another focal quadrat plane can be formed by two other vectors that extend from P_4 for a given focal quadrat, since the hypothetical plane α does not necessarily overlap completely with the focal quadrat (fig. 15.1). Likewise, two other hypothetical planes can exist that extend either from P_2 or P_3. Therefore, the slope degree between the focal quadrat formed by four elevation points (P_1, P_2, P_3, P_4) and its superimposed level plane β is the average value of these four degrees.

The valley/riverine topography type represented depressed or curved areas by streams or valleys between two opposite-facing slopes. The ridgetop topography type was found in areas at the top of two peaks within the plot. About 57% of the Forest Dynamics Plot was covered by the flatland or valley/riverine topography type, whereas the slope topography types (higher slope, midslope, and lower slope) and the ridgetop topography type covered the remaining 43% of the plot. The western and southern half of the study area was mostly covered by the flatland topography type (fig. 15.2).

A soil map of the Pasoh Forest Dynamics Plot was digitized to match the array of 10 × 10 m subplots used to analyze the relationships between soil type and both canopy height and species diversity. Soils in the study area were categorized into 11 series under four major groups based on parent materials and drainage (table 15.2). The hilly parts were mostly covered by soils in Group 1 (BGR series) and Group 2 (TRP and GMI series), whereas flat, alluvial, or riverine areas were

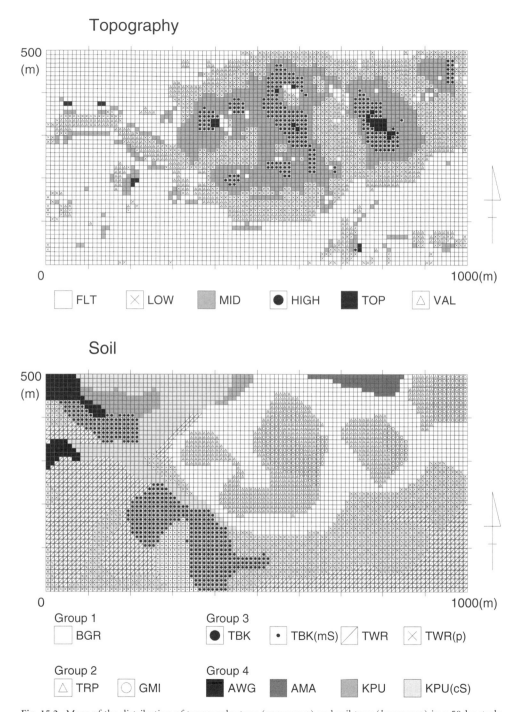

Fig. 15.2. Maps of the distribution of topography type (*upper map*) and soil type (*lower map*) in a 50-ha study area in the Pasoh Forest Reserve. For each 10 × 10 m subplot, one of the following topography types was assigned: ridgetop (TOP), higher slope (HIGH), midslope (MID), lower slope (LOW), flatland (FLT), and valley/riverine (VAL). For the soil types, the hilly parts of the study area were mostly covered by soils in Groups 1 and 2, whereas flat, alluvial or riverine areas were mostly covered by soils in Groups 3 and 4.

Table 15.2. Brief Description of Soils Series in the 50-ha Forest Dynamics Plot in Pasoh Forest Reserve

Major Group	Symbol (Soil Series)	Description	Parent Material	Topographic Specificity in the Plot	Proportional Area in the Plot (%)
Group 1	BGR (Bungor)	Fine sandy clay, weak to moderate medium to fine subangular blocky structures, friable, well-drained haplic Acrisol	Shale	Lower and mid slope	20.8
Group 2	TRP (Terap)	Fine sandy clay to clay with gravel layer of 50–75% mixed subangular and rounded laterites with quartzite, friable, well-drained, perched water table after rain, haplic Acrisol	Reworked	Higher and mid slope and ridetop	10.8
	GMI (Gajah Mati)	Shallow, fine sandy clay to clay with gravel layers of 50–75% mixed subangular and rounded laterites with quartzite, friable, well-drained, perch water table after rain, ferric Acrisol	Reworked	Higher and mid slope	5.7
Group 3	TBK (Tebok)	Fine sandy clay-loam, moderate medium to coarse subangular blocky structures, well-drained, ferric Acrisol	Alluvial	Flatland	8.6
	TBK(mS) (medium sand variant of TBK)	Similar to TBK except the soil texture is medium sandy clay with little fine and coarse sand	Alluvial	Flatland	2.3
	TWR (Tawar)	Coarse sandy clay, moderate medium to coarse subangular blocky structures, moderately to well drained depending on the terrain and location of occurrence	Alluvial	Flatland	18.9
	TWR(p) (pale variant of TWR)	Similar to TWR except perch water table, haplic Acrisol	Alluvial	Flatland	19.4
Group 4	AWG (Awang)	Coarse sandy clay loam to coarse sandy loam with many gray mottles within 50 cm, weak to moderate fine and medium subangular blocky structure, friable, imperfectly to somewhat imperfectly drained soil, ferric Alisol	Alluvial	Flatland and valley	1.9
	AMA (Alma)	Coarse sandy clay loam to coarse sandy loam, poorly to somewhat poorly drained soil, ferric Alisol	Alluvial	Flatland	1.3
	KPU (Kanpon Pusu)	Strong to moderate coarse angular blocky structure, friable, drainage serious (poorly to somewhat poorly drained), fine sandy to clay loam, gleyic-plinthic acrisol	Alluvial	Flatland and valley	2.2
	KPU(cS) (coarse sand variant of KPU)	Similar to KPU except the texture is coarse sandy clay	Alluvial	Flatland and valley	8.1

mostly covered by soils in Group 3 (TWR and TBK series) and Group 4 (AMA, AWG, and KPU series) (fig. 15.2).

Aerial Photographs

Covering the entire 50-ha Forest Dynamics Plot, aerial photographs of the core area of the Pasoh Forest Reserve were taken at a 1/6000 scale in February 1997. To produce a photogrammetric map with submeter accuracy, four 1 × 1 m markers used as ground control points were positioned prior to beginning the photography. Two markers were hung between canopy trees inside the forest while the other two markers were set on the ground outside the forest. The positions of these markers were surveyed using GPS receivers, and traverses were performed using EDM (electronic distance measurement) instruments, which enabled us to measure distances using electromagnetic waves. In addition to the four ground control points, four reference points were placed inside the entire study area to calibrate elevations and coordinates. The final coordinates and elevations were linked to the Malayan Rectified Skew Orthomorphic (MRSO) system and height data.

Based on these aerial photographs and the ground control points, we carried out aerial triangulation to establish the necessary minor photographic control points for stereo digitizing. A digital elevation model of the canopy surface was acquired with an analytical stereo plotter. Only the center position of each stereo model was used in order to achieve relatively reliable stereo interpretation and measurement. Because the center portion of each stereo model is the area of "near nadir"—the location where objects are most vertically captured and where interpretation and measurement are carried out—it is the most accurate as the object being viewed is directly/vertically under the observer's eye. Additionally, viewing at the center of the lens results in the least distortion.

The digitization used a grid pattern with 2.5-m intervals. The precision of the height measurement was less than 0.5 m for clear, well-defined surface objects (e.g., the canopy tower near the study area or 1 × 1 m markers placed in the canopy). The grid data for ground elevation were again interpolated into 2.5-m intervals in order to match the array of the digital elevation model of the canopy surface subgrid system. Canopy height was then obtained by subtracting the ground elevation from the digital elevation model of the canopy surface height for every 2.5-m interval.

Tree Data Analysis

The 1995–97 Pasoh Forest Dynamics Plot census data were used to compare canopy height against species diversity. The data were subdivided into 10 × 10, 20 × 20, or 50 × 50 m subplots, and the number of species and trees were counted within each subplot. The highest point in the canopy surface within each subplot was then obtained from the canopy height values described above.

Hereafter, canopy height refers to the highest canopy point in each subplot. The reason for employing the highest point in each subplot instead of the average value is that the former represents the complexity of canopy layer better than the latter; i.e., the presence of a big, tall tree within the subplot distinctly influences the light and microenvironment of understory, whereas the averaged values tend to mask the variation of canopy height structure.

Fisher's α (Fisher et al. 1943) was calculated for each subplot, and a two-way analysis of variance (ANOVA) was performed to test the significance of the differences between the mean values of Fisher's α in relation to topography and soil. Similarly, the significance of the differences in canopy height was also examined by two-way ANOVA in relation to the topography and soil. Pair-wise differences either in canopy height or Fisher's α between different topography and soil type were examined by the Fisher's Protected LSD model. Simple regression analysis was conducted to analyze the relationship between Fisher's α and canopy height for each topography and soil type. In addition, to analyze the effects of local heterogeneity (complexity) of canopy surface structure on species diversity, variance of canopy height was determined for each subplot size, and the simple linear regression analysis was conducted between the variance of canopy height and Fisher's α. The subplots with higher variance of canopy height values are expected to include physiognomically different patches, while subplots with lower variance of canopy height values are more uniform in height, contained within a large gap, or comprise dense canopy trees of mostly equal height. All statistical analyses were undertaken using Stat View (version 5.0, SAS Institute Inc., Cary, NC).

Parameters like those we employed can be spatially autocorrelated among subplots (Thomson et al. 1996; Nicotra et al. 1999), and that the smaller the subplot, the greater the expected degree of autocorrelation among the subplots. Clark et al. (1996) discussed how to determine the optimal size of subplots and found that sizes greater than 50×50 m minimized the amount of autocorrelation. They proposed that this finding could be applied to old-growth tropical forests. Therefore, we conducted the same analysis for larger subplots using 20×20 and 50×50 m subplots, in addition to the 10×10 m subplots. The canopy height in these enlarged subplots was measured as the height of the highest canopy within the subplot, as in the case of a 10×10 m subplot. Because average crown size in the canopy or emergent layer in the study area was 94.5 m^2, with a range of 3.7 to 886.8 m^2 (Okuda et al. 2003), we conducted the analyses above only to the size of 50×50 m subplots, particularly when examining the variation of canopy height in relation to the spices diversity, topography, and soil types. The soil and topography of a larger subplot were represented by the most abundant type among the 10×10 m subplots. If the number for the top two types (either in soil or topography) in the 10×10 m subplots were the same, then the abundant type was randomly chosen between the two. These random selections were done in 3–12% of the subplots.

Results

Spatial Patterns in Canopy Height

The average canopy height in the study area was 34.7 m and ranged from 11.6 to 60.9 m. The spatial pattern of canopy height varied with location in the study area (fig. 15.3). The northwestern part of the study area was covered by a lower canopy ranging in height from 20 to 40 m, whereas the southwestern and northeastern parts were dominated by a higher canopy. The highest canopy was observed in the southwestern quarter of the area, approximately 260 m east and 80 m north of the southwestern corner of the Pasoh Forest Dynamics Plot. Some patches of depressed canopy, less than 20 m in height and ranging in size from 30 m in radius and larger, occurred along the eastern edge of the study area. A ground survey revealed that these areas represented canopy gaps created by fallen trees. Many trees inside and outside of the study area had been snapped off or toppled by a strong windstorm in June 1995. A comparison of aerial photographs taken before May 1995 and after the storm in February 1997 indicated that the larger gaps had increased in size (N. Adachi, T. Okuda, and N. Manokaran, unpublished).

Variation of Canopy Height with Respect to Topography and Soil Type

Canopy height varied slightly with topography (fig. 15.4, table 15.3). Average canopy height among the six types of topography differed by only 2–2.5 m, but nonetheless exceeded the estimation error. Canopy height was slightly but significantly higher in the sloping parts of the study area (the higher slope, midslope, and lower slope topography types) than in the ridgetop areas or the flat alluvial or riverine areas (the flatland and valley/riverine topography types). Canopy height was more strongly influenced by soil type. The canopy height for soil type AWG averaged 27 m, whereas for the TBK soil type canopy height averaged almost 40 m. Heights were generally lower on the soils with poor drainage in alluvial or riverine areas (Group 4: AWG, AMA, KPU) than on well-drained soils derived from shale (Group 1: BGR) or lateritic parent materials (Group 2: GMI, TRP), which covered the slope topography types (fig. 15.4). The canopy over moderately or well-drained soil types in Group 3 (TBK, TBK(ms), TWR, TWR(p)) was significantly ($p < 0.001$) higher than in the other soil types with poor drainage that developed in alluvial areas (Group 4), suggesting that drainage is a key determinant of canopy height even for soils in flat alluvial areas. It is notable that TBK, whose canopy was highest, is the best-drained soil among those distributed in the alluvial flat area.

Canopy height (m)

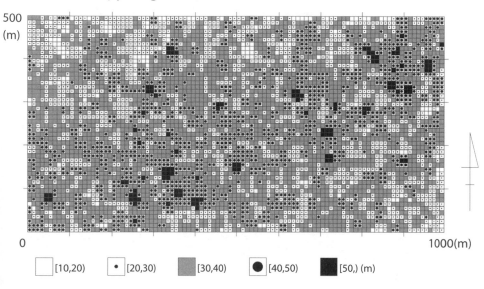

500
(m)

0   1000(m)

☐ [10,20)   ▪ [20,30)   ▨ [30,40)   ● [40,50)   ■ [50,) (m)

Fisher's α

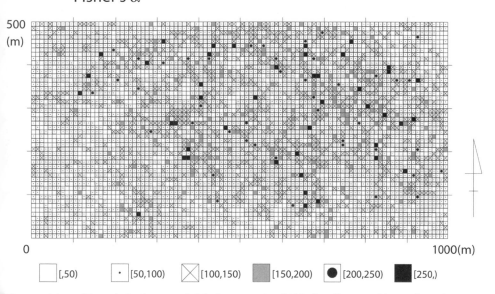

500
(m)

0   1000(m)

☐ [,50)   ▪ [50,100)   ☒ [100,150)   ▨ [150,200)   ● [200,250)   ■ [250,)

Fig. 15.3. Maps of the variation in canopy height (*upper map*) and Fisher's α (*lower map*) in a 50-ha study area in the Pasoh Forest Reserve.

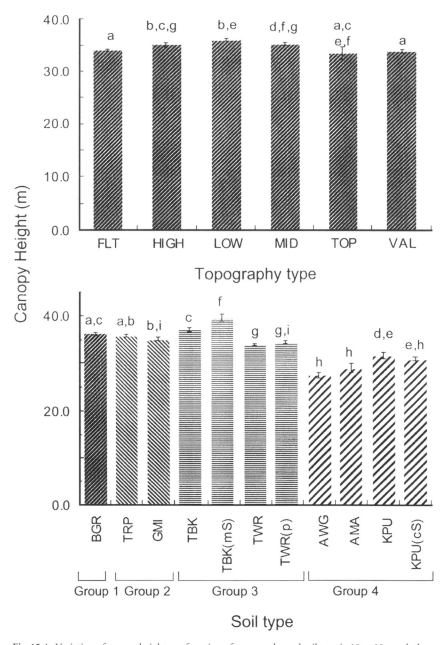

Fig. 15.4. Variation of canopy height as a function of topography and soil type in 10 × 10 m subplots. Bars labeled with different characters are significantly different ($p < 0.05$). Vertical bars represent ±1 standard error. Topography types: ridgetop (TOP), higher slope (HIGH), midslope (MID), lower slope (LOW), flatland (FLT), and valley/riverine (VAL). For the soil types, the hilly parts of the study area were mostly covered by soils in Groups 1 and 2, whereas flat, alluvial or riverine areas were mostly covered by soils in Groups 3 and 4.

Table 15.3. Results of a Two-Way ANOVA for Canopy Height and Fisher's α in Relation to Topography and Soil Types

Parameter	Topography Type (T)	Soil Type (S)	T × S
	10 × 10 m Scale		
Canopy Height			
DF	5	10	33
F-value	1.615	10.904	2.111
P-value	0.152	0.0001	0.0002
Fisher's α			
DF	5	10	33
F-value	5.479	3.306	2.668
P-value	0.0001	0.0003	0.0001
	20 × 20 m Scale		
Canopy Height			
DF	5	10	28
F-value	1.671	3.989	0.914
P-value	0.139	0.0001	0.595
Fisher's α			
DF	5	10	28
F-value	1.031	3.510	1.537
P-value	0.398	0.0001	0.037
	50 × 50 m Scale		
Canopy Height			
DF	4	10	12
F-value	1.382	4.041	0.889
P-value	0.242	0.0001	0.560
Fisher's α			
DF	4	10	12
F-value	5.916	2.802	1.574
P-value	0.0002	0.003	0.103

Note: Degree of freedom (DF) was reduced due to the insufficient number of types for the ANOVA as the scale became larger.

The results of the two-way ANOVA of the interactions between soil type, topography, and canopy height indicated that topography was less important than soil type in determining canopy height (table 15.3). Soil type had consistently significant effects on canopy height for all subplot sizes ($p < 0.0005$), but the effect of topography was not significant in all scaling levels ($p > 0.1$). There was also a highly significant interaction between the two variables ($p = 0.0002$) on the 10 × 10 m scale, but not on a larger scale, indicating that the differences in canopy height cannot be explained by soil type alone in the smaller subplot.

Variation of Species Diversity with Respect to Canopy Height, Topography, and Soil Type

Species diversity (as expressed by Fisher's α was somewhat higher in the northwestern, southwestern, and southeastern parts of the study area where the canopy height was rather low and topography was flat (fig. 15.3). However, Fisher's α for the whole study area showed no significant relationship with canopy height

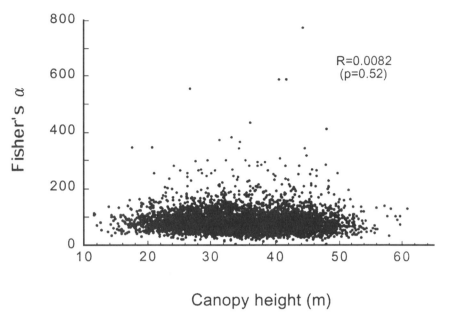

Fig. 15.5. Relationship between canopy height and Fisher's α in 10×10 m subplots ($n = 5000$, $R = 0.0082$, $p = 0.52$).

($R = 0.0082$, $p = 0.52$, fig. 15.5). In addition, regression analyses conducted within each topography and soil type were generally not significant on the three different scales (data are not shown). Similarly, Fisher's α was not significantly correlated with variance of canopy height in all but two cases: the higher slope topography type and the AMA soil type (data are not shown). Both significant cases were observed in the scale of 20×20 m and the relationships were positive (i.e., with higher variance of canopy height, diversity was higher). When all data were pooled, there was no significant relationship between variance of canopy height and Fisher's α at any scale (data are not shown).

In contrast to its relationship with canopy height, the value of Fisher's α was influenced significantly ($p < 0.001$) by differences in topography type except in the 20×20 m subplots (table 15.3). Fisher's α was higher in the upland areas (hill slopes and ridgetops) than in the flatland areas: The species diversity in the ridgetop topography type was significantly higher than in all other topography types (fig. 15.6). The diversity values in flatland topography type were significantly lower than for any other topography types. The differences among the various slope topography types were not significant ($p > 0.05$).

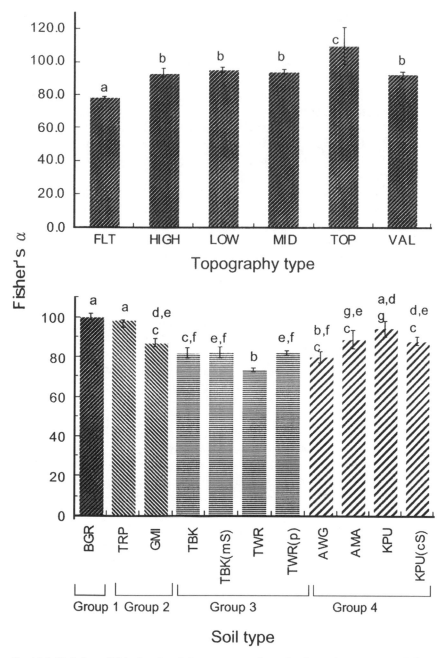

Fig. 15.6. Variation of Fisher's α in relation to topography and soil type in 10×10 m subplots. Bars labeled with different characters are significantly different ($p < 0.05$) Vertical bars represent ± 1 standard error. Topography types: ridgetop (TOP), higher slope (HIGH), midslope (MID), lower slope (LOW), flatland (FLT), and valley/riverine (VAL). For the soil types, the hilly parts of the study area were mostly covered by soils in Groups 1 and 2, whereas flat, alluvial or riverine areas were mostly covered by soils in Groups 3 and 4.

The results of ANOVA showed that the values of Fisher's α were significantly influenced by the differences in soil type regardless of subplot size (table 15.3), while the significance of the effect of topography type on Fisher's α depended on subplot size. Fisher's α was higher for the TRP and BGR soil types, which were distributed primarily on slopes, than for any other type of soils, except KPU. Fisher's α for the GMI soil type, which includes the ridgetop topography type, was not significantly different from that of any other soil types, except TWR. In contrast, Fisher's α was lowest in the TWR soil type (fig. 15.6), which develops in flat alluvial areas.

The results of the two-way ANOVA revealed that both soil type and topography in the scale of 10×10 m subplot size significantly influence species diversity (table 15.3). However, it should be noted that there is an interaction between these two factors, which suggests that neither factor alone can explain the observed variation in species diversity.

Discussion

The results of the present study rejected the hypothesis that local species diversity, as expressed by Fisher's α, decreased with increased canopy height. Our results also rejected another hypothesis that species diversity increased with the complexity of canopy surface structure. The variance of canopy height within subplots patches showed no significant relationship with Fisher's α. In addition, larger subplots (e.g., 50×50 m) with higher variance of canopy height might include various types of forest structure, ranging from gaps to tall canopy, but Fisher's α was not significantly correlated to variation of canopy height. In this respect, these results also suggested that spatial heterogeneity in canopy height did not influence species diversity. Our previous study (Okuda et al. 1997) on the habitat preference of tree species as a function of canopy height showed that in 76 species (about 9% of the total species), sapling (>1 cm in diameter, between the first and third censuses) recruitment occurred preferentially where the canopy height was significantly lower ($p < 0.05$) than the average for the entire study area. Only four species—*Ardisia crassa* (Myrsinaceae), *Cleistanthus myrisanthus* (Euphorbiaceae), *Rinorea anguifera* (Violaceae), and *Shorea maxwelliana* (Dipterocarpaceae)—recruited preferentially under a taller than average canopy. The remainder of the species (90% of those in the plot) were distributed independently of canopy height. This larger set of species swamped any effect of gaps on species diversity (Okuda et al. 1997).

The present study demonstrated that species diversity is influenced strongly by variations in topography and soil type. Soil water availability, which potentially controls gap size and canopy height, may also influence a subplot's species composition. Interactions among environmental factors (e.g., light, soil, water, and nutrients) may cause the observed differences in local-scale diversity. Similar phenomena have been shown in previous studies of floristic changes in relation to topography (Ashton 1964; Austin et al. 1973; Ashton and Hall 1992). Researchers found that forest stature was related to topography, soil depth, and soil water through comparative studies on forest structure among mixed dipterocarp forests in northwestern Borneo. There, potassium and phosphorous concentrations were negatively correlated with understory density and positively correlated with emergent crown diameter; tree allometry was positively correlated with phosphorous and magnesium. In contrast, Newbery et al. (1996) reported that the compositional alternation of tree species did not reflect soil chemistry in a primary lowland dipterocarp forest in Sabah, Malaysia. However, these studies mostly focused on the regional edaphic factors that determine the floristic composition among different forest communities on a larger scale. Can any changes in floristic composition be expected on a fine local scale?

The strong relationship between species diversity and edaphic factors (soil and topography) demonstrated in the present study implies that species composition may change along these two environmental gradients at a local scale. In our previous study, we analyzed the habitat preference of individual tree species in relation to the soil and topography types in the study area, employing chi-square goodness-of-fit tests to examine the significance of differences between the observed and expected frequencies of a species across soil or topography types. In those tests, trees were scored for presence/absence in 50 × 50 m subplots and summed across categories of soil and topography types. According to the analyses, 451 species (55% of the total) showed a habitat preference for one or more of the four soil types used in this paper, while 308 species (38%) showed a preference for one or more of the five topography types. These results suggest strong association between plant distribution and edaphic factors (Okuda et al., unpublished). Such trends of a strong association of species with a topography type have been reported in other tropical forests such the 50-ha Forest Dynamics Plot at Barro Colorado Island, Panama (Hubbell and Foster 1986), the 25-ha Sinharaja Forest Dynamics Plot (chap. 10), and the 50-ha Huai Kha Khaeng Forest Dynamics Plot (chap. 11). In the BCI Forest Dynamics Plot, some plants have been found to be restricted to slopes, a location where the soil contains more water at the end of the dry season (Becker et al. 1988; Condit et al. 1995). Similarly, nonrandom distributions in soil or topography types have been found for mature or juvenile trees in a high proportion of studied or abundant species in other neotropical or subtropical forests (Basnet 1992; Johnston 1992; Clark et al. 1998). In addition, seedling and sapling growth appears to depend on topographical position or soil type (Itoh 1995; Gunatilleke et al. 1996). In peninsular Malaysia, the distribution of rarer species was found to be dependent on soil type (Poore 1968). Therefore, edaphic factors have been shown to have an effect on tree distributions throughout upland tropical forests. If edaphic gradients create substantial differences in the occurrence of tree species, regional heterogeneity in tree species diversity may be expected (Clark et al. 1998), as the present study demonstrated.

Nevertheless, more precise analyses are needed for the habitat preference of each species within the study plot. In addition, the results of this study suggest that further field study is needed to analyze recruitment, mortality, and regeneration of trees as a function of soil type, topography, and light levels (or canopy layering) in the subplots.

General Implications

1. Canopy height of a lowland dipterocarp forest in the Pasoh Forest Reserve in Peninsular Malaysia was not strongly affected by differences in topography, but was significantly influenced by the difference in soil type. Canopy height was lower on poorly drained soils in alluvial or riverine areas than on well-drained soils derived from shale or lateritic parent materials on slopes.
2. Species diversity in the Pasoh Forest Reserve was not significantly correlated with canopy height or variance of canopy height but was significantly affected by differences in topography and soil type.
3. The lack of a clear trend between species diversity and canopy height implies either that the diversity within the two major species groups—gap-adapted species and shade-tolerant species—balance each other as canopy height changes (that is, as the forest ages), or that most of the species are distributed without respect to canopy height.
4. Significant variation of species diversity in relation to soil or topography type implies that niche separation may exist among tree species with respect to differences in soil and topography.

Acknowledgments

We thank Dr. Akio Takenaka (National Institute for Environmental Studies, Japan) and Mr. Hiroyuki Aoki (Tsukuba University) for developing the computer program used for our data analysis. This study is a part of a joint research project between the Forest Research Institute of Malaysia, Universiti Putra Malaysia, and the National Institute for Environmental Studies of Japan (Grant No. E-2(3) from the Global Environment Research Program, Ministry of the Environment of Japan). We are also grateful to Drs. Elizabeth Losos and Egbert G. Leigh (Smithsonian Tropical Research Institute) and to anonymous reviewers for the revision and suggestions that greatly improved the manuscript. The Forest Dynamics Plot at the Pasoh Forest Reserve is an ongoing project of the Malaysian government, initiated by the Forest Research Institute of Malaysia through its former

Director-General, Dato' Dr. Salleh Mohd. Nor, and under the leadership of N. Manokaran, Peter S. Ashton, and Stephen P. Hubbell. Supplementary funding was provided by the National Science Foundation (USA); the Conservation, Food, and Health Foundation, Inc. (USA); the United Nations, through its Man and the Biosphere (MAB) program; UNESCO-MAB grants; UNESCO-ROSTSEA; the continuing support of the Smithsonian Tropical Research Institute, Panama; and the Center of Global Research Center (GGER) of National Institute for Environmental Studies (NIES), Japan.

References

Ashton, P. S. 1964. Ecological studies in the mixed dipterocarp forest of Brunei State. *Oxford Forestry Memoirs* 25:1–75.

Ashton, P. S., and P. Hall. 1992. Comparisons of structure among mixed diperocarp forests of north-western Borneo. *Journal of Ecology* 80:459–81.

Austin, M. P., P. S. Ashton, and P. Greig-Smith. 1973. The application of quantitative methods to vegetation survey III. A re-examination of rain forest data from Brunei. *Journal of Ecology* 60:305–24.

Barker, M. G., M. C. Press, and N. D. Brown. 1997. Photosynthetic characteristics of dipterocarp seedlings in three tropical rain forest light environments: a basis for niche partitioning. *Oecologia* 112:453–63.

Basnet, K. 1992. Effects of topography on the pattern of trees in tabonuco (*Dacryodes excelsa*) dominated rain forest in Puerto Rico. *Biotropica* 24:173–84.

Becker, P., P. E. Rabenold, J. R. Idol, and A. P. Smith, 1988. Water potential gradients for gaps and slopes in Panamanian tropical moist forest's dry season. *Journal of Tropical Ecology* 4:73–184.

Brown, N. D., and T. C. Whitmore, 1992. Do dipterocarp seedlings really partition tropical rain forest gaps? *Philosophical Transactions of the Royal Society of London (Series B)* 335:369–78.

Clark, D. B., D. A. Clark, P. M. Rich, S. Weiss, and S. F. Oberbauer. 1996. Landscape scale evaluation of understory light and canopy structure: Methods and application in a neotropical lowland rain forest. *Canadian Journal Forestry Research* 26, 747–57.

Clark, D. B., D. A. Clark, and J. M. Read. 1998. Edaphic variation and the mesoscale distribution of tree species in a neotropical rain forest. *Journal of Ecology* 86:101–12.

Condit, R., S. P. Hubbell, and R. B. Foster. 1995. Mortality rates of 205 neotropical tree and shrub species and the impact of a severe drought. *Ecological Monographs* 65419–39.

Denslow, J. S. 1980. Gap partitioning among tropical rain forest trees. *Biotropica (Suppl.)* 12:47–55.

Denslow, J. S., J. C. Schultz, P. M. Vitousek, and Strain B. 1990. Growth responses of tropical shrubs to tree fall gap environments. *Ecology* 71:165–79.

Fisher, R. A., A. S. Corbet, and C. B. Williams. 1943. The relation between the number of species and the number of individuals in a random sample of an animal population. *Journal of Animal Ecology* 12:42–58.

Gartlan, J. S., D. M. Newbery, D. W. Thomas, P. G. Waterman. 1986. The influence of topography and soil phosphorus on the vegetation of Korup Forest Reserve, Cameroon. *Vegetatio* 65:131–48.

Gunatilleke, C. V. S., G. A. D. Perera, P. M. S. Ashton, P. S. Ashton, and I. A. U. N. Gunatilleke. 1996. Seedling growth of *Shorea* section *Doona* (Dipterocarpaceae) in soils from topographically different sites of Sinharaja rain forest in Sri Lanka. Pages 245–65 in M. D. Swaine, editor. *The Ecology of Tropical Forest Tree Seedlings.* Parthenon Publishing, New York.

Hubbell, S. P., and R. B. Foster. 1986. Commonness and rarity in a neotropical forest: Implications for tropical tree conservation. Pages 205–31 in M. E. Soule, editor. *Conservation Biology: The Science of Scarcity and Diversity.* Sinauer Associates, Sunderland, MA.

Itoh, A. 1995. Effect of forest environment on germination and seedling establishment of two Bornean rainforest emergent species. *Journal of Tropical Ecology* 11:517–27.

Johnston, M. H. 1992. Soil-vegetation relationships in a tabonuco forest community in the Luquillo Mountains of Puerto Rico. *Journal of Tropical Ecology* 8:253–63.

Mabberley, D. J. 1992. *Tropical Rain Forest Ecology.* Blakie, New York.

Manokaran, N., and J. V. LaFrankie. 1990. Stand structure of Pasoh Forest Reserve, a lowland rain forest in peninsular Malaysia. *Journal of Tropical Forest Science* 3:14–24.

Manokaran, N., J. V. LaFrankie, K. M., Kochummen, E. S. Quah, J. E. Klahn, P. S. Ashton, and S. P. Hubbell. 1990. *Methodology for the Fifty Hectare Research Plot at Pasoh Forest Reserve.* Research Pamphlet No. 104. Forest Research Institute of Malaysia, Kepong, Malaysia.

———. 1992. *Stand Table and Distribution of Species in the Fifty Hectare Research Plot at Pasoh Forest Reserve.* FRIM Research Data. Forest Research Institute of Malaysia, Kepong, Malaysia.

Newbery, D. M. 1991. Floristic variation within kerangas (heath) forest: Re-evaluation from Sarawak and Brunei. *Vegetatio* 96:43–86.

Newbery, D. M., E. J. F. Campbell, J. Proctor, and M. J. Still. 1996. Primary lowland dipterocarp forest at Danum Valley, Sabah, Malaysia. Species composition and patterns in the understory. *Vegetatio* 122:193–220.

Nicotra, A. B., R. L.Chazdon, and V. B. Iriarte 1999. Spatial heterogeneity of light and woody seedling regeneration in tropical wet forests. *Ecology* 80:1908–1926.

Okuda, T., N. Adachi, N. Manokaran, and H. Nor Azman. 1997. Canopy height structure and species composition in a lowland dipterocarp forest in peninsular Malaysia. Pages 107–18 in T. Okuda, editor. *Research Report of the NIES/FRIM/UPM Joint Research Project 1997.* National Institute for Environmental Studies, Tsukuba, Japan.

Okuda, T., M. Suzuki, N. Adachi, E. S. Quah, H. Nor Azman, and N. Manokaran. 2003. Effect of selective logging on canopy and stand structure and tree species composition in a lowland dipterocarp forest in Peninsular Malaysia. *Ecology and Forest Management* 175: 297–320.

Pickett, S. T. A. 1983. Differential adaptation of tropical tree species to canopy gaps and its role in community dynamics. *Journal of Tropical Ecology* 24:68–84.

Poore, M. E. D. 1968. Studies in Malaysian rain forest. I. The forest on Triassic sediments in Jenka Forest Reserve. *Journal of Ecology* 56:143–96.

Ricklefs, R. 1977. Environmental heterogeneity and plant species diversity: A hypothesis. *American Naturalist* 111:376–381.

Symington, C. F. 1943. *Foresters' Manual of Dipterocarps.* Malaysian Forest Record No. 16. Penerbit Universiti Malaya, Kuala Lumpur.

Thomson, J. D., G. Weinblen, B. A. Thomson, S. Alfaro, and P. Legendre. 1996. Untangling multiple factors in spatial distributions: Lilies, gophers, and rocks. *Ecology* 77:1698–715.

Wyatt-Smith, J. 1961. A note on the fresh-water swamp, lowland and hill forest types of Malaya. *Malaysian Forester* 24:110–21.

———. 1964. A preliminary vegetation map of Malaya with description of the vegetation types. *Journal of Tropical Geography* 18:200–13.

Yamakura, T., M. Kanzaki, A. Itoh, T. Ohkubo, K. Ogino, E. Chai, H. S. Lee, and P. S. Ashton. 1995. Topography of large-scale research plot established within a tropical rain forest at Lambir, Sarawak. *Tropics* 5:41–56.

PART 5: The Diversity of Tropical Trees: Background

Introduction

Egbert G. Leigh, Jr.

How can so many kinds of tropical trees coexist in a uniform tract of tropical forest? To find out, we start with three questions. First, the null hypothesis: What happens if all mature trees, regardless of their species, are competitive equivalents and diversity represents a balance between speciation and random extinction? Like null hypotheses in statistics, this hypothesis helps us discern factors that help maintain tree diversity. Second, what are the prerequisites for maintaining a forest's diversity and species composition? Must trees, even of rare species, outcross? How do they do so? Third, how does reduction into isolated fragments affect a forest's diversity and species composition? Can the effects of human-mediated fragmentation reveal anything about the processes organizing the ecology of intact forest?

In the first chapter, Leigh, Condit, and Lao outline Hubbell's (1979, 1997, 2001) neutral theory of forest ecology. Hubbell assumes that trees die at random, regardless of their species, and that seed-parents of their replacements are likewise chosen without reference to their species. Here, diversity represents a balance between speciation and the random extinction of those species for which deaths happen to outnumber births long enough to leave no survivors. The simplicity of the neutral theory's assumptions allows one to elaborate precise null models predicting how fragmenting a forest affects tree diversity, how the number of tree species increases with plot size, the relative abundance of tree species on a plot, and the relationship between the degree of divergence in species composition of two plots and the distance separating them. The neutral theory implies relationships between phenomena previously considered unrelated, such as local speciation, limited seed dispersal, and beta-diversity (species turnover). The neutral theory also provides null hypotheses for elucidating causes of phenomena as diverse as the rate of spread of new species, and changes in tree species composition on fragments newly isolated from intact forest. Testing these null hypotheses reveals that different tree species respond very differently to fragmentation, and that new species can become very common only if they have some advantage over their predecessors.

Many tropical forests are very diverse, and most of their tree species are quite rare. Rare species must maintain enough genetic variability to be able to adapt to environmental change and cope with natural enemies: a circumstance that places a premium on outcrossing. Using animals as pollinators appears to be a prerequisite for maintaining high plant diversity. Plant diversity exploded when, in the Cretaceous, flowering plants developed the ability to attract pollinators that sought out other plants of their species (Crepet 1984).

In this section's second chapter, Stacy and Hamrick assess how far successful pollen travels, and what proportion of a tree's seeds are pollinated by other trees, in 12 tree species on Barro Colorado Island. Many of these species are rare, with less than one reproductive adult per two hectares. Yet, in 10 of their 12 species, over 85% of the seeds were pollinated by other trees. These trees use a variety of animals to convey their pollen: some species employ bees; others, birds; others, bats or hawkmoths; yet others, small insects. Stacy and Hamrick found that as a rule, genes move farther in pollen than in seeds. In most of these species, a substantial proportion of a tree's successful pollen fertilizes recipients over 200 m away. This is not unusual. Thanks to their pollinating wasps, strangler figs, *Ficus obtusifolia*, with one reproductive adult per hectare, fertilize conspecifics several kilometers away (Nason et al. 1996). Such pollinators support ample genetic variation. The strangler fig *Ficus obtusifolia* is heterozygous at 25% of its loci, compared to 21% for the average species of conifer in the north temperate zone (Hamrick and Loveless 1989), even though a conifer species is much more common where it occurs.

In the third chapter, Sean Thomas discusses the effects of reducing a forest to fragments. Fragmenting tropical forest can reveal "ecosystem services," on which its structure, diversity, or productivity depend, that hinge on a continuous expanse of forest. Such services may include pollination, seed dispersal, and protection from wind or herbivores. In the neotropics, forest fragmentation has often had catastrophic impact. In central Amazonia, exposure to drying winds can double or triple tree mortality on newly isolated 1-ha fragments of forest, leading to biomass collapse (Laurance et al. 1997). In Panama's Gatun Lake, fragmentation greatly reduces tree diversity, in part because mammals that bury large seeds, protecting them from insects, cannot survive on small islets (Leigh et al. 1993). In small islets created by Venezuela's Guri Reservoir, leaf-cutter ants have multiplied because their predators cannot travel to or survive on small islets. Furthermore, howler monkeys were trapped on some small islets, and their young cannot disperse. Here herbivore pressure has intensified, at least temporarily, to a level that severely reduces seedling abundance and diversity (Terborgh et al. 2001).

In Thomas's study, near the Pasoh Reserve in Malaysia, the effects of fragmentation appear far less drastic. Here, there is little wind, and fragmentation seems not to increase tree mortality. Even so, fragmenting this forest near Pasoh induces subtle changes in tree species composition. Dipterocarps are much rarer in the fragments than in intact forest. Ballistically dispersed species are more common in fragments, perhaps because these species do not need animals to disperse their seeds and because their seeds are less likely to disperse far enough to land outside the fragment. Ballistically dispersed species tend to occur in clumps. Must they invest more in antiherbivore defense than plants of more scattered species, whose specialist pests find it more difficult to spread from one plant to another? Finally, in five diverse genera, trees whose leaves have lower photosynthetic capacity per unit area are better represented in fragments than in intact forest. Thus fragmentation of the forest near Pasoh has initiated reduction in plant productivity and diversity on the fragments, because fragmentation has disrupted ecosystem services needed for their maintenance.

References

Crepet, W. L. 1984. Advanced (constant) insect pollination mechanisms: patterns of evolution and implications vis-à-vis angiosperm diversity. *Annals of the Missouri Botanical Garden* 71: 607–30.

Hamrick, J. L., and M. D. Loveless. 1989. The genetical structure of tropical tree populations: associations with reproductive biology. Pages 129–46 in J. Bock and Y. B. Linhart, editors. *The Evolutionary Ecology of Plants.* Westview Press, Boulder, CO.

Hubbell, S. P. 1979. Tree dispersion, abundance and diversity in a tropical dry forest. *Science* 203:1299–1309.

———. 1997. A unified theory of biogeography and relative species abundance and its application to tropical rain forests and coral reefs. *Coral Reefs (Suppl.)* 16:S9–21.

———. 2001. *The Unified Neutral Theory of Biodiversity and Biogeography.* Princeton University Press, Princeton, NJ.

Laurance, W. F., S. G. Laurance, L. V. Ferreira, J. M. Rankin-de-Merona, C. Gascon, and T. E. Lovejoy. 1997. Biomass collapse in Amazonian forest fragments. *Science* 278:1117–18.

Leigh, E. G. Jr., S. J. Wright, F. E. Putz, and E. A. Herre. 1993. The decline of tree diversity on newly isolated tropical islands: a test of a null hypothesis and some implications. *Evolutionary Ecology* 7:76–102.

Nason, J. D., E. A. Herre, and J. L. Hamrick. 1996. Paternity analysis of the breeding structure of strangler fig populations: evidence for substantial long-distance wasp dispersal. *Journal of Biogeography* 23:501–12.

Terborgh, J., L. Lopez, P. Nuñez V., M. Rao, G. Shahabuddin, G. Orihuela, M. Riveros, R. Ascanio, G. H. Adler, T. D. Lambert, and L. Balbas. 2001. Ecological melt-down in predator-free forest fragments. *Science* 294:1923–26.

16

The Neutral Theory of Forest Ecology

Egbert G. Leigh, Jr., Richard Condit, and Suzanne Loo de Lao

Introduction

Hubbell (1979, 1997, 2001) developed a neutral theory of tree diversity in connection with his studies of large-scale forest plots in Costa Rica and Panama. By assuming that all trees, regardless of their species, are competitive equivalents, he constructed a precise, mechanistic theory that linked alpha to beta diversity and bound in a common tether the relative abundance of species, the mode of speciation, the effectiveness of seed dispersal, the impacts of fragmentation, and perhaps even the grand patterns of phytogeography (Leigh et al. 1993; Bramson et al. 1998; Hubbell 2001; Condit et al. 2002; Chave and Leigh 2002). Hubbell's theory rivals Corner's (1949, 1958, 1964) durian theory of flowering-plant evolution in scope and ambition, if not in wealth of picturesque detail. Can Hubbell's theory help us make sense of the ecology of tropical forest? If so, how?

Hubbell's fundamental assumption that all trees are competitively equivalent, regardless of their species—that a tree's prospects of death or reproduction do not depend on its species or those of its neighbors—bears little relation to reality. On Barro Colorado Island's 50-ha Forest Dynamics Plot, Hubbell (1998) himself documented the well-known trade-off that tree species face between fast growth in high light and survival in shade (Kitajima 1994; King 1994). On this same plot, Condit et al. (1995, 1996a) showed that different tree species respond very differently to climate change; Harms et al. (2001) showed that some tree species are associated with particular types of habitat; and Ahumada et al. (this volume, chap. 23) showed that in most species, a sapling's chance of surviving its next 12 years is lower when there are more conspecifics among its 20 nearest neighbors.

In this paper, we outline Hubbell's neutral theory in order to answer three questions:

1. Does this theory provide a null hypothesis against whose predictions one can assess the impact of biological processes that the theory ignores?
2. To what extent do this theory's predictions transcend its simplistic assumptions?
3. To what extent does this theory help us frame new questions for further research?

The Theory

Hubbell's neutral theory of forest ecology closely parallels the neutral theory in population genetics of a multi-allelic locus in a haploid population (Chave and Leigh 2002): Hubbell's species correspond to the geneticists' allelic types and Hubbell's speciation corresponds to mutation in the geneticists' "infinite-allele model." In forest ecology, however, trees usually die one by one, whereas most population geneticists assume that separate generations are distinct. We shall introduce successive aspects of the neutral theory in the context of several different problems.

The Impact of Fragmentation

We first consider the neutral theory of how tree diversity decays on completely isolated fragments that are too small for speciation, and how much the species compositions of different fragments change with time. Consider a forest fragment j that contains N_j adult trees. At every time-step, let a randomly chosen tree die, and let another be chosen randomly from the trees alive just before this death to be seed-parent of the dead tree's immediately maturing replacement. Here, the random play of which species the trees dying and recruiting happen to belong to causes tree species to die out, one by one, until there is only one species left to which all N_j trees belong. If a proportion x_{ij} of this fragment's trees belong to species i at time t, then at this time t the probability is $x_{ij}(t)$ that species i is the one that will take over the fragment. Indeed, because there is no bias to the change in numbers of species i, the expected value of $x_{ij}(t+s)$ at all times later than t is $x_{ij}(t)$.

How fast does fragment j's diversity decay? Let time be measured in tree generations. Then each time-step represents $1/N_j$ of a tree generation. Let $F_j(t)$ be the probability that two trees sampled randomly (with replacement) at time t from fragment j are the same species. Then, since x_{ij} is the proportion of trees on fragment j at time t that belong to species i,

$$F_j(t) = \sum_{i=1}^{S} x_{ij}^2(t) \tag{16.1}$$

where S is the total number of species in this system of fragments. We call $F_j(t)$ the relative dominance on fragment j at time t. $H_j(t) = 1 - F_j(t)$, the probability that two trees sampled randomly at time t from fragment j are of different species, is a measure of tree diversity.

How does F_j change from one time-step to the next? $F_j(t + 1/N_j)$ may be expressed (Leigh et al. 1993) as

$$F_j(t)\left(1 - \frac{2}{N_j}\right) + \frac{2}{N_j}\left[\frac{1}{N_j} + \left(1 - \frac{1}{N_j}\right)F_j(t)\right]: \tag{16.2}$$

In a pair of trees sampled at time t, the chance is $1 - 2/N_j$ that one of them will die by time $t + 1/N_j$. If one of these trees does die, the probability is $1/N_j$ that the other will be the replacement's parent, in which case $F = 1$, while the probability is $1 - 1/N_j$ that this replacement's parent is one of the $N_j - 1$ other possibilities, in which case F is unchanged. Therefore,

$$F_j\left(t + \frac{1}{N_j}\right) = F_j(t) + \frac{2}{N_j}\left\{\frac{1}{N_j}[1 - F_j(t)]\right\}; \tag{16.3}$$

$$1 - F_j\left(t + \frac{1}{N_j}\right) = H_j\left(t + \frac{1}{N_j}\right) = H_j(t)\left[1 - \frac{2}{N_j^2}\right]. \tag{16.4}$$

Since m tree generations is mN_j time-steps,

$$H_j(t + m) = H_j\left(t + \frac{mN_j}{N_j}\right) = H_j(t)\left[1 - \frac{2}{N_j^2}\right]^{mN_j}, \tag{16.5}$$

which is nearly $H_j(t)\exp{-2m/N_j}$. Therefore H_j declines toward 0 by a factor of $\exp{-2/N_j}$ per tree generation (cf. Kimura and Crow 1964, eq. 1; Gillespie 1998, pp. 22–26). Because we consider haploids with overlapping generations while geneticists consider diploid populations where successive generations are distinct, we replace the geneticists' $4N$ by N_j.

The average number of generations until extinction of a species now represented on fragment j by $N_j x_{ij}$ trees—provided that this species does not take over the island—is $[N_j x_{ij}/(1 - x_{ij})]\ln(1/x_{ij})$ (Kimura and Ohta 1969, eq. 16).

The codominance $F_{jk}(t)$ of tree species compositions on two fragments j and k at time t is the probability that two trees sampled randomly at time t, one from each fragment, are the same species:

$$F_{jk}(t) = \sum_{i=1}^{S} x_{ij}(t)x_{ik}(t). \tag{16.6}$$

Here, $x_{ik}(t)$ is the proportion of the N_k trees on fragment k at time t which belong to species i. If we know $F_{jk}(t)$, then the expected value of $F_{jk}(t + s)$ is $F_{jk}(t)$ for all future times $t + s$, because the populations of each species change independently on the two fragments, while their expected numbers remain unchanged (Leigh et al. 1993). When s is large enough, only a single species remains on each fragment, and $F_{jk}(t)$ is the probability that the same species takes over both fragments.

*The Balance Between Speciation and Random Extinction
in a Panmictic Population*

Now consider a forest with N mature trees whose dynamics resemble those of the previous section's fragments, only this forest is large enough for speciation to occur there. Like the fragments, this forest is panmictic: that is to say, each tree disperses seeds uniformly through the forest. Let the probability be v that a young tree is a mutant of an entirely new species (the results are little altered if species begin as peripheral isolates containing relatively few trees; see Hubbell 2001). Then

$$F\left(t + \frac{1}{N}\right) = F(t)\left(1 - \frac{2}{N}\right) + \frac{2(1-v)}{N}\left[\frac{1}{N} + \left(1 - \frac{1}{N}F(t)\right)\right]: \quad (16.7)$$

The left-hand side is multiplied by $1 - v$ because the replacement can be of the surviving tree's species only if it is not a mutant representing a new species. At equilibrium, when speciation balances random extinction, $F(t + 1/N) = F(t) = F$, so that

$$\frac{2F}{N} = \frac{2(1-v)}{N}\left[\frac{1}{N} + \left(1 - \frac{1}{N}\right)F\right]; \quad (16.8)$$

$$vF = (1-v)\left[\frac{1}{N}(1-F)\right]. \quad (16.9)$$

Therefore $(Nv + 1 - v)F = 1 - v$; $F \approx 1/(1 + Nv)$ (Kimura and Crow 1964; Gillespie 1998, pp. 27–28). Moreover, the number $S(m)$ of species with m trees apiece in this forest is

$$\frac{Nv}{m}\left(1 - \frac{m}{N}\right)^{Nv-1} \approx \frac{Nv}{m}\exp{-m\left(v - \frac{1}{N}\right)} \quad (16.10)$$

(Appendix 16.1; also see Kimura and Crow 1964, p. 731).

If $Nv > 1$, $S(m) \approx Nvx^m/m$, where $x = 1 - v + 1/N$. This is a log series with $\alpha = Nv$ (Fisher et al. 1943; Watterson 1974; Leigh 1999, app. 8.2). Where this distribution of trees over species applies, the average number S of species in a sample of r trees is $S = Nv \ln(1 + r/Nv)$. Thus Nv satisfies the equation for Fisher's α given in chapter 7 for random samples of any size r from this forest. In this forest, Fisher's α thus serves as a measure of diversity that is independent of sample size.

In a forest that always contains N mature trees, where new species arise as small peripheral isolates from their predecessors, and where speciation is in balance with random extinction, a species now represented by $m \ll n$ trees that began as a peripheral isolate containing $p \ll m$ trees originated an average of $m \ln(N/m)$ generations ago. Furthermore, it takes an average of m generations for such a species to attain m trees for the first time, if it ever does so (Kimura and Ohta 1973, eqs. 11, 17).

Species Turnover

In a forest of indefinite extent with uniform tree density, two trees sampled at random are less likely to be conspecific the further apart they are. It seems likely that most tree species arise from small, local, peripheral subpopulations of their ancestral species (Willis 1922; Mayr 1958). Because seeds disperse only a limited distance from their parents, most species have limited ranges. Even in a uniform habitat without impassible barriers, local speciation, limited seed dispersal, and random extinction will ensure that species composition changes from place to place, changing more the farther one travels (Condit et al. 2002; Chave and Leigh 2002).

   Here, we measure species turnover by the rapidity of decline with increased r of $F(r)$, the probability that two trees randomly chosen from sites a distance r apart are the same species. $F(r)$ is a crude index of species turnover because it depends mainly on the abundance of the commoner trees at these sites. The continued decline of $F(r)$ with increased r must, however, reflect turnover. Otherwise $F(r)$ would approach a value F_∞ once r exceeded the scale of patchiness of the most patchily distributed species.

   To be specific, consider a forest of infinite extent with a uniform density of ρ trees per hectare. Suppose that

1. A tree alive at time t has probability dt of dying by time $t + dt$,
2. The probability that the seed-parent of the dead tree's replacement is between r and $r + dr$ meters distant from the dead tree is $p(r)dr$, where all directions from the dead tree to the replacement's seed-parent are equally likely,
3. The probability is ν that the dead tree's replacement is a mutant of an entirely new species,
4. In this forest, speciation, dispersal, and random extinction are in balance: the probability $F(r)$ that two trees a distance r apart are the same species does not change with time.

   The last assumption, which has nothing to do with neutrality, is the least realistic of the lot. The equilibrium it assumes takes $2/\nu$ generations to attain (Chave and Leigh 2002), a time that can easily exceed the age of the universe. Unfortunately, one must assume equilibrium to solve the equation for $F(r)$ developed in appendix 16.2.

   If the "dispersal kernel" $p(r)dr$ is a radially symmetric bivariate Gaussian $(2r\,dr/\sigma^2)\exp{-r^2/\sigma^2}$, where σ^2 is the mean square dispersal distance of seeds from their parents, then

$$F(r) = \frac{2K_0(\sqrt{4vr^2/\sigma^2})}{\ln(1/v) + \pi r \sigma^2} \tag{16.11}$$

if $r > \sigma/\sqrt{2}$ (Nagylaki 1974; Chave and Leigh 2002), where $K_0(x)$ is the modified Bessel function of order 0 (Olver 1965). Our mean square dispersal distance σ^2 is twice that of Chave and Leigh (2002) because ours is a mean square dispersal distance in a plane, whereas theirs is mean square dispersal along one axis. If $r < \sigma/\sqrt{2}$,

$$F(r) = \frac{\ln(1/v) - r^2\pi^2/6\sigma^2}{\pi\rho\sigma^2 + \ln(1/v)} \tag{16.12}$$

(Chave and Leigh 2002).

If their seeds had been dispersed according to a Gaussian density, British oak trees could never have spread northward so fast when the last glaciers retreated at the end of the last ice age (Skellam 1951; Clark 1998). On the other hand, if their seeds had been dispersed according to a Cauchy density, $p(r)dr = crdr/(c^2 + r^2)^{3/2}$, they could have kept up with the glacial retreat, as they in fact did (Clark et al. 1999). Here, $c\sqrt{3}$ is the median distance of seeds from their parents. The Cauchy density also agrees better than the Gaussian with data on the distribution of seed dispersal distances for a majority of the tree species examined by Clark and colleagues (1999). If seeds are dispersed according to a Cauchy density, then $F(0) = \pi/[12\rho c^2 + \pi]$. If v is so minute that $\sqrt{(rv/c)} \ll 1$ and if r is several times c, then

$$F(r) \approx \frac{6c^2\rho}{r(\pi^2 + 12c^2\rho\pi)} \tag{16.13}$$

(Chave and Leigh 2002).

The Neutral Theory as Null Hypothesis

Fragmentation

To learn how the predictions of the neutral theory concerning the rate of increase of F_j and the constancy of the expected value of F_{jk} match data from real forest fragments, Leigh et al. (1993) censused trees on six islands in Panama's Gatun Lake, each less than 1 ha in area. These six islands have been covered by forest since their isolation from the mainland about 1913 by the newly dammed Gatun Lake. How do diversity and species composition on these islands compare with those on nearby mainland "control plots" of comparable size? Suppose that, in 1910, diversity H_j on each of these islets was equal to the average H in 1980 for mainland control plots with similar numbers of trees ≥ 10 cm dbh. Let a tree generation be

Table 16.1. Decline of Diversity of Trees ≥10 cm dbh on Small Islands in Gatun Lake, 1910–1980

Islands	A 1980	N 1980	S 1980	α 1980	H 1910 Inferred	H 1980 Predicted	H 1980 Observed
Vulture	0.10	59	10	3.45	0.9062	0.8642	0.7573
Camper	0.13	125	19	6.24	0.9212	0.9008	0.7544
Aojeta	0.16	128	26	9.86	0.9212	0.9013	0.8486
Ormosia	0.16	135	32	13.25	0.9212	0.9023	0.8824
NWJG	0.41	340	25	6.22	0.9299	0.9223	0.5852
Annie	0.63	399	37	9.96	0.9299	0.9234	0.5803

Mainland Control Plots (averages)

	N	S	α	H 1910 Inferred	H 1980 Averaged Observations
Mainland Whole	250	47	17.10	(0.9299)	0.9299 ± 0.0267
Mainland Half	125	35	16.14	(0.9212)	0.9212 ± 0.0266
Mainland Quarter	62	23	13.24	(0.9062)	0.9062 ± 0.0343

Notes: Area (A, ha), number of trees ≥10 cm dbh (N), number of tree species (S), Fisher's α (see text), diversity (H) assumed for 1910, predicted for 1980 by the formula $H(1980) = H(1910) \exp -(2.8/N)$, and observed in 1980, for selected small islands and nearby mainland control plots. Data from Leigh et al. (1993).

Table 16.2. Relative Codominance F_{jk} among Small Islands in Gatun Lake, Panama, and Nearby Mainland Control Plots

	M1	M2	M3	M4	Annie	NWJG	Ormosia	Aojeta	Camper
Mainland 2 (M2)	0.0762								
Mainland 3 (M3)	0.0371	0.0441							
Mainland 4 (M4)	0.0387	0.0457	0.0331						
Annie	0.0265	0.0261	0.0362	0.0674					
NWJG	0.0143	0.0098	0.0288	0.0565	0.3901				
Ormosia	0.0068	0.0073	0.0080	0.0122	0.0513	0.1126			
Aojeta	0.0227	0.0222	0.0229	0.0411	0.2146	0.2069	0.0389		
Camper	0.0226	0.0194	0.0265	0.0502	0.2790	0.2645	0.0462	0.1813	
Vulture	0.0409	0.0448	0.0178	0.0327	0.0319	0.0338	0.0284	0.0924	0.1341

Note: Data from Leigh et al. 1993.

50 years, as it is on Barro Colorado. Then $t = 1.4$ tree generations elapsed between 1910 and 1980. Under these assumptions, $H_j(1980)$ is significantly lower than $H_j(1910) \exp[-2t/N_j(1980)] = H_j(1910) \exp[-2.8/N_j(1980)]$, the value of H_j predicted by the null model, for all six small islands studied (table 16.1). Therefore, diversity decline on these islands reflects a *cause* (which may involve, but is not exclusively accounted for by, increased death rate). It did not just happen by chance.

Moreover, F_{jk} is much higher among these islands than among similar-sized pairs of mainland control plots, even though the mainland plots are much closer together (table 16.2). In particular, a similar set of species is taking over these islands. This set includes *Protium panamensis* (Burseraceae), *Swartzia simplex*

(Leguminosae), *Oenocarpus mapora*, a.k.a. *Oenocarpus panamanus* (Palmae), and *Attalea butyracea*, a.k.a. *Scheelea zonensis* (Palmae). The. Learning why these trees are spreading on our small islands will, by reflection, reveal some of the factors that maintain the balance of nature in intact forest. Seeds of these four species are large, yet in three of the four species, insects rarely attack the seeds. Late-falling seeds of the fourth, *A. butyracea*, are also safe from insect attack (Forget et al. 1994). Insects of various other large-seeded tree species, such as *Astrocaryum standleyanum* (Palmae), *Virola surinamensis* (Myristicaceae), and *Dipteryx panamensis* (Leguminosae) must be buried by agoutis to escape destruction (Smythe 1989; Forget and Milleron 1991; Forget 1993). Such tree species are not recruiting on these islets, which do not support agoutis or any other resident mammals (Adler and Seamon 1991). The neutral theory has thus led us to ask whether agoutis are keystone animals for preserving tree diversity in the neotropics.

Causes of Tree Diversity

According to the neutral theory, a species with a large number m of mature trees in a forest with $N \gg m$ trees must have originated an average of m tree generations ago. Many species of old-growth rainforest appear to be represented by over 10 million trees apiece ≥ 10 cm dbh. In the neotropics, a few such species are *Tapirira guianensis* (Anacardiaceae), *Jacaranda copaia* (Bignoniaceae), *Laetia procera* (Flacourtiaceae), *Hymenaea courbaril* (Leguminosae), *Pourouma guianensis* (Cecropiaceae), *Brosimum lactescens* (Moraceae), and *Minquartia guianensis* (Olacaceae). According to the neutral theory, such common species are hundreds of millions of years old. This is almost certainly not true. Falsifying this null hypothesis suggests that to understand tree diversity, we must discover what factors allow at least some new species to spread with nonrandom speed.

Predictions Transcending the Neutral Theory's Validity

In many regions, Fisher's α—defined as the solution of the equation $S = \alpha \ln(1 + N/\alpha)$ where N is the number of trees on a plot and S is the number of species among them—is remarkably independent of plot size, so long as the plot includes several hundred trees (Condit et al. 1996b; Condit, Leigh, and Lao, chap. 7). On Barro Colorado's 50-ha Forest Dynamics Plot in 1995, Fisher's α averaged 34.4 for the 4581 trees ≥ 1 cm dbh on a hectare, 34.2 for the 28,634 such trees on a 250 × 250 m square, and 33.6 for the 114,535 trees on a 500 × 500 m square. It was 33.9 for the 229,069 trees ≥ 1 cm dbh on the whole plot. Even more often, Fisher's α depends little on the lower diameter limit of trees sampled (Condit et al. 1996b). In the first census on the 50-ha Forest Dynamics Plot at Huai Kha Khaeng, Fisher's α was 31.1 for the 79,340 trees ≥ 1 cm dbh, 31.6 for the 37,979

Table 16.3. Number of Trees ≥ 10 cm dbh, N, Number of Species S among Them, Fisher's α, Relative Dominance F, as Observed, and as Estimated by the Formula $1/(\alpha + 1)$ for Selected Hectares on Barro Colorado's 50-ha Forest Dynamics Plot

	N	S	α	F	$1/(\alpha + 1)$
Plateau Hectares					
Hectare 0,2	463	90	33.3	0.0354	0.0292
Hectare 0,3	508	95	34.5	0.0283	0.0282
Hectare 2,1	366	84	34.1	0.0449	0.0285
Hectare 3,1	381	93	39.2	0.0455	0.0249
Hectare 6,2	436	86	32.1	0.0413	0.0302
Hectare 7,2	447	83	30.1	0.0634	0.0322
Hectare 7,3	424	84	31.4	0.064	0.0309
Hectares Overlapping Swamp					
Hectare 2,2	409	92	36.9	0.0309	0.0264
Hectare 3,2	347	89	38.7	0.0323	0.0252
Hectare 4,2	340	99	46.9	0.0275	0.0209
Hectares on Sloping Ground					
Hectare 4,0	408	98	40.9	0.0313	0.0239
Hectare 8,0	402	103	44.8	0.0268	0.0218
Hectare 8,1	414	87	33.6	0.0267	0.0289
Hectare 8,2	407	87	33.9	0.043	0.0287
Hectare 8,3	410	80	29.7	0.042	0.0326

trees ≥ 5 cm dbh, 31.5 for the 21,917 trees ≥ 10 cm dbh, and 32.2 for the 8558 trees ≥ 20 cm dbh (but only 28.9 for the 4172 trees ≥ 30 cm dbh).

These properties of α are predicted by the neutral theory (where size, at least size at maturity, like species, is irrelevant to a tree's competitive ability). As we have seen, the neutral theory also predicts that $F = 1/(\alpha + 1)$, where F is the probability that two trees sampled randomly from a plot are the same species. Nonetheless, F varies much more among hectares than does α, F usually exceeds $1/(\alpha + 1)$ (tables 16.3, 16.4), and F varies far more than α with plot size (Condit et al. chap. 7). Thus the neutral theory's predictions concerning α hold even though other predictions of the neutral theory, to which these are logically linked, are false. Therefore the neutral theory's predictions concerning Fisher's α may be said to transcend the validity of the neutral theory itself.

Hubbell (1997) used two parameters to measure the diversity and reconstruct the distribution over species of trees ≥ 10 cm dbh on a 50-ha Forest Dynamics Plot. He assumed that the plot exchanges migrants with a large but finite, panmictic source pool where speciation balances random extinction, so that the pool's trees are distributed over species according to a log series with $\alpha = N\nu$. Each time that a tree on the plot dies, let the probability be q that its replacement's seed-parent is chosen at random from the source pool, and let $1 - q$ be the probability that it is chosen at random from the plot. From the distribution of abundances of the plot's commoner trees, one can infer Fisher's $\alpha = N\nu$ for the source pool.

Table 16.4. Number of Trees ≥10 cm dbh N, Number of Species S among Them, Fisher's α, Relative Dominance F, as Observed, and as Estimated by the Formula $1/(\alpha + 1)$, and Annual Rainfall P for Sample Hectares in Central Panamá

Plot	N	S	α	F	$1/(\alpha + 1)$	P(mm)
m31 25 km ENE of Colón	498	154	76.3	0.0179	0.0129	3660
L01 N of Ft Sherman	400	63	21	0.0901	0.0455	3280
L03 N of Ft Sherman	366	74	28	0.0303	0.0345	3280
s01 SW of Gatun	530	81	26.7	0.0361	0.0361	3027
m18 BCNM, SW Pena Blanca	431	86	32.3	0.0307	0.03	2580
m10 BCI, 100m W of AVA 20.5	403	78	28.8	0.0465	0.0336	2580
bc1 BCI ha 7.3, plateau	424	84	31.4	0.0638	0.0309	2580
bc3 BCI ha 4.0, slope	408	98	40.9	0.0313	0.0239	2580
m14 BCI, 170 m E of PCS 10.5	381	92	38.5	0.0317	0.0253	2580
m12 BCNM, Buena Vista	521	74	23.6	0.0678	0.0407	2580
m19 Pipeline Rd, Agua Salud	520	89	30.9	0.0467	0.0313	2484
m15 Pipeline Rd, Limbo HC	457	92	34.7	0.0352	0.028	2380
m08 Pipeline Rd, Limbo plot	560	94	32.3	0.0635	0.03	2380
m09 Pipeline Rd, Limbo plot	503	107	41.6	0.0386	0.0235	2380
m17 Pipeline Rd near begin	464	63	19.7	0.0967	0.0483	2280
m16 Pipeline Rd near begin	467	90	33.2	0.043	0.0292	2280
m21 just NE of Gamboa	405	78	28.7	0.0342	0.0337	2180
m26 10 km SE of BCI	490	76	25.2	0.0675	0.0382	2237
m23 Plantation Rd	590	60	16.7	0.1519	0.0565	2180
m27 8 km SE of Gamboa	395	61	20.2	0.052	0.0472	2152
m29 2 km W of Cocoli	357	64	22.7	0.0658	0.0422	1985
c04 Cocoli	298	58	21.5	0.0608	0.0444	1985

Setting the pool's $\alpha = 50$ and $q = 0.10$ for trees ≥10 cm dbh fits the distribution of trees over species on Barro Colorado's 50-ha Forest Dynamics Plot; setting $\alpha = 180$, $q = 0.15$ does the same for Pasoh (Hubbell 1997). If this is true for other rainforest plots, this manner of describing diversity and the distribution of species abundances with two parameters would be an achievement that far transcends the assumptions of his model.

Framing New Questions

In an extensive forest, species composition differs from place to place. As one traverses the Panama Canal from the drier Pacific to the wetter Caribbean side, some tree species drop out, and new ones appear (Pyke et al. 2001). Even in the more uniform expanses of western Amazonia, tree species compositions of 1-ha plots tend to differ more the greater the distance between them (Condit et al. 2002). Thanks to its simple assumptions, the neutral theory is the only theory capable of yielding quantitative predictions about species turnover from a precisely defined, mechanistic model. Usable data on species turnover in the tropics are only now becoming available (Pitman et al. 1999, 2001; Pyke et al. 2001). Can the neutral theory help us order and interpret these data?

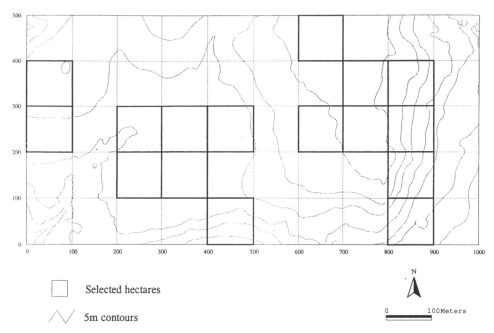

Fig. 16.1. Map of the 50-ha Barro Colorado Island plot, showing hectares used in the similarity analysis. Hectare m, n is the hectare whose western edge is m hectometers (a hectometer is 100 m) from the western edge of the plot and whose southern border is n hectometers from the southern edge of the plot. (Figure prepared by Suzanne Loo de Lao.)

     To find out, let us try to fit the neutral theory's prediction of $F(r)$, the probability that two trees randomly chosen from sites separated by a distance r are the same species, to data for trees ≥ 10 cm dbh in a network of plots in central Panama. First, we consider a selection of 16 old-growth hectares on Barro Colorado's 50-ha plot, which span the variety of habitats on that plot (fig. 16.1, table 16.3). Then we turn to a selection of 22 1-ha plots, including 2 from table 16.3 and 20 others scattered through the Panama Canal area (fig. 16.2, table 16.4).

     How well can the neutral theory account for divergence in vegetation composition among these plots? Because codominance F_{jk} depends not only on the similarity of species composition between plots j and k, but also on their relative dominance, it is better to measure similarity in species composition between plots j and k by $I_{jk} = F_{jk}/(F_j F_k)^{1/2}$ (Nei 1987). $I_{jk} = 1$ when the same species occur on both plots in the same relative abundance, and $I_{jk} = F_{jk} = 0$ when the two plots share no species in common. We measure divergence in species composition between plots j and k by $D_{jk} = -\ln I_{jk}$. $D_{jk} = 0$ when $I_{jk} = 1$ and is higher the less similar the plots. Here, we use $\ln[F(0)/F(r)]$ to predict the average D_{jk} for pairs of plots whose members are separated by a distance of roughly r.

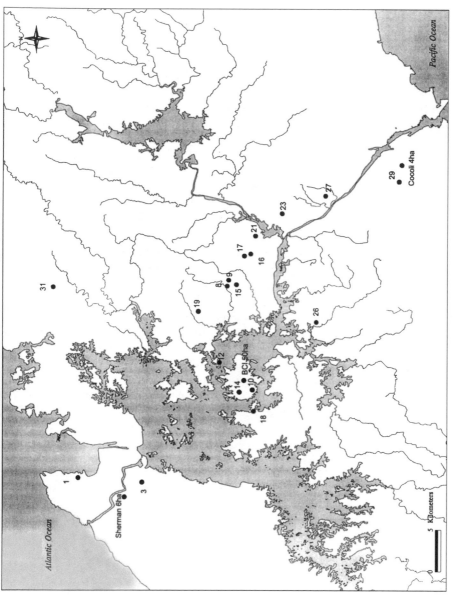

Fig. 16.2. Location of sample hectares in central Panama used in the similarity analysis. (Figure prepared by Suzanne Loo de Lao.)

First, we set $F(0) = \ln(1/\nu)/[\ln(1/\nu) + \pi\rho\sigma^2]$, using equation 16.12. Let $F(0)$ be the relative dominance per hectare on the 48 old-growth hectares of Barro Colorado's 50-ha plot. This is 0.0367 ± 0.0089 (mean $\pm$ SD). Let ρ be the density per square meter of trees ≥ 10 cm dbh on these same hectares. This is 0.0420 ± 0.0035. Since $27.25 = 1/0.0367 = 1/F(0) = 1 + \pi\rho\sigma^2/\ln(1/\nu)$, then $26.25 = 0.1319\sigma^2/\ln(1/\nu)$, $\sigma^2/\ln(1/\nu) = 199$.

To calculate σ, we assume with Hubbell (1997, 2001) that a tenth of the trees ≥ 10 cm dbh on Barro Colorado's 50-ha plot have seed-parents outside the plot. Then, if all trees disperse seeds according to a radially symmetric Gaussian with the same σ, σ must be 59.16 m (app. 16.3). Because $\sigma^2/\ln(1/\nu) = 199$, $\ln(1/\nu) = 17.6$.

If $\sigma/2\sqrt{\nu} = 33{,}800$ m, then $D(r) = \ln[F(0)/F(r)]$ roughly matches the average D_{jk} for pairs of plots separated by a distance close to r (table 16.5). If $\sigma/2\sqrt{\nu} = 33800$ m and $\sigma = 59.16$ m, then $\sqrt{\nu} = 59.16/67,600 = 0.000875$, $\nu = 7.66 \times 10^{-7}$, and $\ln(1/\nu) = 14.1$. On the other hand, if $\sigma/2\sqrt{\nu} = 33,800$ m and $\sigma^2/\ln(1/\nu) = 199$, then $\sigma = 53.3$ m and $\ln(1/\nu) = 14.3$. Fitting Gaussian distributions to observed distributions of seeds about their parents for 65 tree species on Barro Colorado's 50-ha plot, Muller-Landau found that different tree species had very different values of σ, but the mean of these values was 55.15 m, not far from our best-fit value of 53.3 m (Condit et al. 2002; their paper reports σ measured along one dimension, which we multiply by $\sqrt{2}$ to give dispersal in space). Thus the neutral theory's $F(r)$ gives a serviceable account of the trend in the data for divergence in species composition among our selection of plots in central Panama. On the other hand, species composition is less divergent for plots on similar soil than for plots the same distance apart on contrasting soil (Pyke et al. 2001; Condit et al. 2002), falsifying the neutral theory.

How well does $F(r)$ work in other regions? Condit et al. (2002) fitted $F(r)$ to data on networks of plots in central Panama, the environs of the Yasuni Forest

Table 16.5. Comparison of Average of Observed Values of $D(r)$, D_0, with Predicted Values D_{pr}

Sample Hectares on Barro Colorado's 50-ha Plot										
$z \times 100$	0.36	0.71	1.01	1.30	1.54	1.83	2.46			
$K_0(z)$	5.76	5.06	4.72	4.46	4.29	4.11	3.82			
$r = z\sigma/2\sqrt{\nu}$, km	0.12	0.24	0.34	0.44	0.52	0.62	0.83			
$D_{pr}(r)$	0.25	0.38	0.45	0.51	0.55	0.59	0.66			
$D_0(r)$	0.27	0.41	0.38	0.34	0.44	0.48	0.40			
Sample Plots in the Panama Canal Area										
$z = 2r\sqrt{(\nu/\sigma^2)}$	0.02	0.05	0.10	0.20	0.30	0.40	0.50	0.60	0.80	1.00	1.20
$K_0(z)$	4.03	3.11	2.43	1.75	1.37	1.11	0.92	0.78	0.57	0.42	0.32
$r = z\sigma/2\sqrt{\nu}$, km	0.70	1.70	3.40	6.70	10.10	13.40	16.80	20.20	26.90	33.60	40.30
$D_{pr}(r)$	0.61	0.87	1.11	1.44	1.69	1.90	2.08	2.25	2.57	2.87	3.14
$D_0(r)$	0.62	0.87	1.28	1.60	2.35	2.15	1.76	2.20	2.50	2.56	2.99

Note: Predicted values are calculated from $D(r) = \ln[F(0)/F(r)]$, assuming $\sigma/2\nu^{1/2} = 33.8$ km and $z = 2r(\nu/\sigma)^{1/2}$.

Dynamics Plot in Amazonian Ecuador, and the environs of the Parque Nacional Manu in Amazonian Peru. Fitting equation 16.11 to these data, they found that the best-fit values of ν and σ were 4.8×10^{-8} and 57 m in Panama, 3.6×10^{-11} and 77 m near Yasuni, and 1.7×10^{-14} and 103 m near Manu.

Here, too, the neutral theory, or at least its equilibrium assumption, proves false. Why should speciation rates differ by three orders of magnitude between Yasuni and Manu? Moreover, habitat differences do matter. There was much less scatter about the predicted $F(r)$ in the more uniform expanse of western Amazonia than in Panama with its variety of soil types and its steep gradient in climate. Nonetheless, appropriate choice of values for ν and σ enabled $F(r)$ to fit the trend of the data in all three sites for r between 150 m and 50 to 100 km. Here, the neutral theory has proved useful in ordering data and suggesting new questions.

General Implications

1. Hubbell's model serves as a "null standard" for assessing whether observed changes in tree diversity or species composition require invocation of specific causes.
2. Hubbell's model suggests that Fisher's α is a measure of tree species diversity that is independent of sample size. This is true for tropical forests whose diversity and species composition cannot be ascribed to chance.
3. In central Panamá, the two free parameters of Hubbell's model can be adjusted to fit the *average* divergence between the tree species composition of two hectares r kilometers apart with remarkable accuracy. Indeed, Hubbell's model has proved a useful tool for estimating tree species turnover from limited data in western Amazonia as well.

Acknowledgments

We are deeply indebted to Helene Muller-Landau for reading this paper, drawing our attention to misprints, infelicities, and errors, and providing unpublished data; to Jérôme Chave for finding an error in our reading of Nagylaki's (1974) fundamental formula; and to a succession of data analysts for providing plot data in useful form. Finally, we thank the plants and animals of Barro Colorado Island for reminding us of the beauty it is the business of any biologist to understand and communicate.

Appendix 16.1. The Neutral Theory and the Log Series

We will show that, when a speciation rate ν is in balance with random extinction in a forest with a constant number N of mature trees, the mean number of species with m trees apiece is

$$S(m) = \frac{N\nu}{m} \left(1 - \frac{m}{N}\right)^{N\nu-1} = P(p)\frac{1}{N}, \tag{16A.1}$$

where $p = m/N$ and $P(p)dp = N\nu(1 - p)^{N\nu-1}dp/p$ is the expected number of species with between Np and $N(p + dp)$ trees apiece. Notice that

$$\sum_{m=1}^{1} mS(m) = \int_0^1 pP(p)dp = N \tag{16A.2}$$

Moreover, F, the probability that two trees chosen randomly from the forest are conspecific, is

$$\int_0^1 p^2 P(p)dp = \int_0^1 pN\nu(1 - p)^{N\nu-1}dp = \frac{1}{N\nu + 1} \tag{16A.3}$$

as we found in the text by a more direct method.

We will verify the formula for $S(m)$ by showing that it yields the correct value for the probability F_r that r trees sampled randomly from the forest are all conspecific.

Let the probability be F_3 that three trees sampled randomly from the forest are conspecific. When speciation rate ν is in balance with random extinction,

$$F_3 = F_3\left(1 - \frac{3}{N}\right) + \frac{3(1 - \nu)}{N}\left[\frac{2F}{N} + \left(1 - \frac{2}{N}\right)F_3\right] : \tag{16A.4}$$

The chance is $3/N$ that one of our three trees dies in the next time-step. If this happens, the probability is $2/N$ that the replacement descends from one of the two survivors, in which case, if the replacement is not a new species, the probability the three are conspecific is now F. With complementary probability $1 - 2/N$, the replacement descends from another tree; if it is not a new species, the probability that the three trees are conspecific remains F_3. Subtracting $F_3(1 - 3/N)$ from both sides of equation 16A.4 and multiplying the remainder by $N/3$ gives

$$F_3 = (1 - \nu)\left[\frac{2F}{N} + \left(1 - \frac{2}{N}\right)F_3\right]. \tag{16A.5}$$

Subtracting $F_3(1 - \nu)$ from both sides of equation 16A.5 and multiplying the remainder by N, we find $N\nu F_3 = 2(1 - \nu)(F - F_3)$, $F_3 \approx 2F/(N\nu + 2) = 2/(N\nu + 1)(N\nu + 2)$. In analogy with equation 16A.5,

$$F_r = (1 - \nu)\left[\frac{(r - 1)F_{r-1}}{N} + \left(1 - \frac{r - 1}{N}\right)F_r\right]. \tag{16A.6}$$

Thus $F_r \approx (r-1)F_{r-1}/(N\nu + r - 1)$. If $F_{r-1} = (r-2)!\Gamma(N\nu + 1)/\Gamma(N\nu + r - 1)$, where $\Gamma(x)$ is the gamma function, with the property that $\Gamma(x+1) = x\Gamma(x)$ for all positive x, then

$$F_r \approx \frac{(r-1)!\Gamma(N\nu + 1)}{\Gamma(N\nu + r)}. \tag{16A.7}$$

Equation 16A.7 holds for $r = 3$; moreover, if it holds for F_{r-1} it holds for F_r. Thus, by mathematical induction, equation 16A.7 holds for all r. Equation 16A.1 suggests that we may express F_r as

$$F_r = \int_0^1 p^r P(p)dp = \int_0^1 N\nu p^{r-1}(1-p)^{N\nu - 1}dp \tag{16A.8}$$

A look at an integral table shows that equation 16A.8 gives equation 16A.7 for all r, showing that equation 16A.1 is correct.

Appendix 16.2. Deriving an Equation for Finding F(r)

Let the probability F that two trees randomly chosen at time t from an expansive forest are the same species depend only on the distance $r = (x^2 + y^2)^{1/2}$ between them, where one tree is x meters east and y meters north of the other. Thus $F = F(x, y, t) = F(x, -y, t) = F(-x, y, t) = F(-x, -y, t)$. Let time be measured in tree generations, let each new tree have probability ν of being a new species, and let trees be distributed uniformly, with ρ trees per square meter. Then

1. $F(x, y, t + dt) = F(x, y, t)(1 - 2dt) + 2dt(1 - \nu)Q(x, y, t)$. The probability that one of these trees, alive at time t, dies by time $t + dt$, is $1 - 2dt$; while $Q(x, y, t)$ denotes the probability that, if speciation does not occur, the dead tree's replacement is conspecific with the survivor, and $1 - \nu$ is the factor by which the possibility of speciation reduces this probability.
2. At steady state, $F(x, y, t + dt) = F(x, y, t) = F(x, y)$, and $F(x, y) = (1 - \nu)Q(x, y)$.
3. If the seed-parent of the dead tree's replacement is w meters east and z meters north of the tree it replaces, and $x - w$ meters east and $y - z$ meters north of the surviving tree, the probability that the replacement and the survivor are conspecific is $F(x - w, y - z)$, unless $w \approx x$ and $z \approx y$.
4. If $w \approx x$ and $z \approx y$, the probability that the replacement is conspecific with the survivor is

$$F(0, 0) + \frac{1}{\rho dw dz}[1 - F(0, 0)]. \tag{16A.9}$$

This reflects the circumstance that if $-x + dw/2 > w > -x - dw/2$, the probability that the replacement's parent is the survivor among our two trees is $1/\rho dw dz$.

5. The probability that the replacement's parent lies between w and $w + dw$ meters east, and between z and $z + dz$ meters north, of the dead tree is $p(w, z)dwdz$.

6. Therefore, $Q(x, y)$ is

$$\iint_{-\infty}^{\infty} p(w, z) F(x - w, y - z) dw dz + \frac{[1 - F(0, 0)] p(x, y)}{\rho}. \quad (16A.10)$$

7. The desired equation is thus $F(x, y) = (1 - v) Q(x, y)$, where $Q(x, y)$ is replaced by formula 16A.10. Nagylaki (1974) and Chave and Leigh (2002) used Fourier transforms to solve this equation. In testing this theory, we set $p(w, z)dwdz = [dwdz/(\pi\sigma^2)] \exp{-(w^2 + z^2)/\sigma^2}$.

Appendix 16.3. Calculating the Mean Square Dispersal Distance of Seed Dispersal

How can we calculate the mean square dispersal distance σ^2 from the proportion m of seeds falling into a 50-ha plot from parents outside it? First, consider a strip 500 m wide extending north-south through the forest. If mean square dispersal distance is σ^2, what proportion of the seeds falling in the strip come from outside? If almost no seeds from outside reach the middle of the strip, that is to say, if $250 m > 2\sigma$, this question amounts to asking what proportion of the seeds falling into the west half of the strip come from beyond its west boundary.

Choose the strip's west boundary as the y axis, and set $p(x, y)dxdy = \exp{-[(x^2 + y^2)/\sigma^2]dxdy/(\pi\sigma^2)}$. Then the density along the x axis of seeds belonging to a tree on the plot's west boundary is $\exp{-(x^2/\sigma^2)dx/(\pi\sigma^2)^{1/2}}$. Thus half the seeds of trees on the boundary fall into the strip. Similarly, trees a distance z west of the boundary drop a proportion

$$R(z) = \frac{1}{\sqrt{\pi\sigma^2}} \int_z^{\infty} \exp{-(x^2/\sigma^2)}dx \quad (16A.11)$$

of their seeds into the strip. Almost no seeds disperse into the strip from trees $>2\sigma$ west of the boundary: $R(2\sigma) < 10^{-4}$. Therefore, if $2\sigma < 250$ m, the average fraction of seeds falling into the west half of the strip from outside it is

$$\frac{1}{250} \int_0^{\infty} dz \int_z^{\infty} R(x)dx = \frac{1}{250} \left(\frac{\sigma}{2\sqrt{\pi}} \right) \quad (16A.12)$$

(Gradshteyn and Ryzhik 1980, formula 6.281). If a tenth of the seeds falling into the strip's west half come from outside it, then $(1/250)\sigma/(2\sqrt{\pi}) = 1/10$, $\sigma = 50\sqrt{\pi} = 88.6$ m. In fact, this estimate is too large because our strip is only 1000 m long; seeds disperse into it from north and south, as well as from east and west. The fraction of seeds falling into the northwest 500×250 m quarter of the plot from the west and north is $1/10 = (1/500 + 1/250)\sigma/(2\sqrt{\pi}); \sigma = 59.16$ m, $\sigma^2 = 0.35$ ha.

References

Adler, G. H., and J. O. Seamon. 1991. Distribution and abundance of a tropical rodent, the spiny rat, on islands in Panama. *Journal of Tropical Ecology* 7:349–60.

Bramson, M., J. T. Cox, and R. Durrett. 1998. A spatial model for the abundance of species. *Annals of Probability* 26:658–709.

Chave, J., and E. G. Leigh, Jr. 2002. A spatially explicit neutral model of β-diversity in tropical forests. *Theoretical Population Biology* 62:153–68.

Clark, J. S. 1998. Why trees migrate so fast: confronting theory with dispersal biology and the paleorecord. *American Naturalist* 152:204–24.

Clark, J. S., M. Silman, R. Kern, E. Macklin, and J. HilleRisLambers. 1999. Seed dispersal near and far: patterns across temperate and tropical forests. *Ecology* 80:1475–94.

Condit, R., S. P. Hubbell, and R. B. Foster. 1995. Mortality rates of 205 neotropical tree and shrub species and the impact of a severe drought. *Ecological Monographs* 65:419–39.

Condit, R., S. P. Hubbell, and R. B. Foster. 1996a. Changes in tree species abundance in a Neotropical forest: impact of climate change. *Journal of Tropical Ecology* 12:231–56.

Condit, R., S. P. Hubbell, J. V. LaFrankie, R. Sukumar, N. Manokaran, R. B. Foster, and P. S. Ashton. 1996b. Species-area and species-individual relationships for tropical trees: a comparison of three 50-ha plots. *Journal of Ecology* 84:549–62.

Condit, R., N. Pitman, E. G. Leigh Jr., J. Chave, J. Terborgh, R. B. Foster, P. Nuñez V., S. Aguilar, R. Valencia, G. Villa, H. C. Muller-Landau, E. Losos, and S. P. Hubbell. 2002. Beta-diversity in tropical forest trees. *Science* 295:666–69.

Corner, E. J. H. 1949. The durian theory or the origin of the modern tree. *Annals of Botany, N.S.* 13:367–414.

Corner, E. J. H. 1958. The evolution of tropical forest. Pages 34–46 in J. Huxley, A. C. Hardy, and E. B. Ford, editors. *Evolution as a Process*. George Allen and Unwin, London.

Corner, E. J. H. 1964. *The Life of Plants*. World Press, Cleveland.

De Steven, D. 1986. Comparative demography of a clonal palm (*Oenocarpus mapora* subsp. *mapora*) in Panama. *Principes* 30:100–04.

Fisher, R. A., A. S. Corbet, and C. B. Williams. 1943. The relation between the number of species and the number of individuals in a random sample of an animal population. *Journal of Animal Ecology* 12:42–57.

Forget, P.-M. 1993. Post-dispersal predation and scatterhoarding of *Dipteryx panamensis* (Papilionaceae) seeds by rodents in Panama. *Oecologia* 94:255–61.

Forget, P.-M. and T. Milleron. 1991. Evidence for secondary seed dispersal by rodents in Panama. *Oecologia* 87:596–599.

Forget, P.-M., E. Munoz, and E. G. Leigh, Jr. 1994. Predation by rodents and bruchid beetles on seeds of *Scheelea* palms on Barro Colorado Island, Panama. *Biotropica* 26:420–26.

Gillespie, J. H. 1998. *Population Genetics: A Concise Guide.* Johns Hopkins University Press, Baltimore, MD.

Gradshteyn, I. S., and I. M. Ryzhik. 1980. *Tables of Integrals, Series, and Products.* Academic Press, New York.

Harms, K. E., R. Condit, S. P. Hubbell, and R. B. Foster. 2001. Habitat associations of trees and shrubs in a 50-ha neotropical forest plot. *Journal of Ecology* 89:947–59.

Hubbell, S. P. 1979. Tree dispersion, abundance, and diversity in a tropical dry forest. *Science* 203:1299–1309.

———. 1997. A unified theory of biogeography and relative species abundance and its application to tropical rain forests and coral reefs. *Coral Reefs* 16 (Suppl.):S9–S21.

———. 1998. The maintenance of diversity in a neotropical tree community: conceptual issues, current evidence, and challenges ahead. Pages 17–44 in F. Dallmeier and J. A. Comiskey, editors. *Forest Biodiversity Research, Monitoring and Modeling. Conceptual Background and Old World Case Studies.* UNESCO, Paris, and Parthenon Publishing, Pearl River, NY.

———. 2001. *The Unified Neutral Theory of Biodiversity and Biogeography.* Princeton University Press, Princeton NJ.

Kimura, M., and J. F. Crow. 1964. The number of alleles that can be maintained in a finite population. *Genetics* 49:725–38.

Kimura, M., and T. Ohta. 1969. The average number of generations until fixation of a mutant gene in a finite population. *Genetics* 61:763–71.

———. 1973. The age of a neutral mutant persisting in a finite population. *Genetics* 75:199–212.

King, D. A. 1994. Influence of light level on the growth and morphology of saplings in a Panamanian forest. *American Journal of Botany* 81:948–57.

Kitajima, K. 1994. Relative importance of photosynthetic traits and allocation patterns as correlates of seedling shade tolerance of 13 tropical trees. *Oecologia* 98:419–28.

Leigh, E. G., Jr. 1999. *Tropical Forest Ecology: A View from Barro Colorado.* Oxford University Press, New York.

Leigh, E. G., Jr., S. J. Wright, F. E. Putz, and E. A. Herre. 1993. The decline of tree diversity on newly isolated tropical islands: a test of a null hypothesis and some implications. *Evolutionary Ecology* 7:76–102.

Mayr, E. 1958. Change of genetic environment and evolution. Pages 157–180 in J. Huxley, A. C. Hardy, and E. B. Ford, editors. *Evolution as a Process.* George Allen and Unwin, London.

Nagylaki, T. 1974. The decay of genetic variability in geographically structured populations. *Proceedings of the National Academy of Sciences, USA* 71:2932–36.

———. 1976. The decay of genetic variability in geographically structured populations. II. *Theoretical Population Biology* 10:70–82.

Nei, M. 1987. *Molecular Evolutionary Genetics.* Columbia University Press, New York.

Olver, F. W. J. 1965. Bessel functions of integer order. Pages 355–433 in M. Abramowitz and I. A. Stegun, editors. *Handbook of Mathematical Functions.* Dover, New York.

Pitman, N. C. A., J. Terborgh, M. R. Silman, and P. Nuñez V. 1999. Tree species distributions in an upper Amazonian forest. *Ecology* 80:2651–61.

Pitman, N. C. A., J. W. Terborgh, M. R. Silman, P. Nuñez V., D. A. Neill, C. A. Ceron, W. A. Palacios, and M. Aulestia. 2001. Dominance and distribution of tree species in upper Amazonian terra firme forests. *Ecology* 82:2101–17.

Pyke, C. R., R. Condit, S. Aguilar, and S. Lao. 2001. Floristic composition across a climatic gradient in a neotropical lowland forest. *Journal of Vegetation Science* 12:553–66.

Skellam, J. G. 1951. Random dispersal in theoretical populations. *Biometrika* 38:196–218.

Smythe, N. 1989. Seed survival in the palm *Astrocaryum standleyanum*: evidence for dependence upon its seed dispersers. *Biotropica* 21:50–56.

Watterson, G. A. 1974. Models for the logarithmic species abundance distributions. *Theoretical Population Biology* 6:217–50.

Willis, J. C. 1922. *Age and Area*. Cambridge University Press, Cambridge, U.K.

17

Using Forest Dynamics Plots for Studies of Tree Breeding Structure: Examples from Barro Colorado Island

Elizabeth A. Stacy and James L. Hamrick

Introduction

One of the most striking differences between temperate and tropical forests is plant species richness. Visitors to tropical forests have long questioned how trees reproduce in these species-rich sites, where for some species, the nearest conspecific neighbors may be separated by several hundred meters. Prior to the widespread application of hand-pollination experiments and breeding system studies with tropical trees, many thought that tropical trees reproduced primarily or exclusively through self-fertilization (Corner 1954; Baker 1959; Federov 1966).

Ashton (1969) questioned this view. Consistent with his view, studies conducted throughout the tropics during the past 30 years have revealed a strong tendency toward outcrossing in forest trees (e.g., Ashton 1969; O'Malley and Bawa 1987; O'Malley et al. 1988; Dayanandan et al. 1990; Hall et al. 1994; and studies described in this chapter). In many tropical tree species, outcrossing is ensured through apparent self-incompatibility (e.g., Bawa 1974; Opler and Bawa 1978; Chan 1981; Koptur 1984; Perez-Nasser et al. 1993; Sakai et al. 1999) or less commonly, through dioecy (Lewis 1942; Bawa 1974; Bawa et al. 1985). Exceptions exist, however, in species that possess mixed mating systems, where reproduction involves a combination of self-fertilization and outcrossing (Murawski et al. 1990; Murawski and Hamrick 1992b; Murawski et al. 1994a). For both outcrossed species and those with mixed mating systems, a number of important questions remain regarding the nature of breeding in tropical forest trees. Over what distance does pollen move between trees? How frequent are long-distance pollinations? How are mating patterns (including outcrossing rate) affected by the density and spatial distribution of reproductive trees? How do outcrossing rates and patterns of pollen flow vary from one reproductive season to the next? Lastly, what are the consequences of observed mating patterns for population fitness?

Patterns of mating within a population, including individual- and population-level outcrossing rates, comprise the breeding structure. Together, the outcrossing rate of a population and the pattern of pollen movement between plants determine the size of a natural breeding unit and are primary determinants of genetic structure at the population and species levels (Loveless and Hamrick 1984; Hamrick and Godt 1989). Accordingly, the characterization of these mating parameters in tropical tree populations permits insight into observed distributions of genetic diversity and improves prospects for the effective conservation of this diversity. Of particular importance is the characterization of mating parameters in disturbed forest landscapes, where tree population densities as well as flowering phenology and pollinator populations may have been altered. Understanding how tree breeding structure responds to forest disturbance is a first step toward devising strategies for the conservation of tree diversity in disturbed landscapes.

Estimating mating parameters in natural populations typically requires the preliminary mapping, or at least extensive surveying, of populations in order to assess adult density and distribution. Such surveys can be time consuming in tropical forests, where species identification is often difficult. The Center for Tropical Forest Science (CTFS) network of Forest Dynamics Plots affords unprecedented opportunities for population studies of tropical woody species. Since 1982, the first Forest Dynamics Plot—on Barro Colorado Island (BCI), Panama—has hosted a number of studies on the breeding structure and population genetics of its tree species, conducted by researchers at the University of Georgia (e.g., Hamrick and Murawski 1990; Hamrick et al. 1993). In this chapter, we briefly review these studies, detailing selected research to illustrate specifically how Forest Dynamics Plots have been used to estimate patterns of pollen flow in tropical trees. Subsequently, we discuss the potential limitations of, and suggest future directions for, plot-based studies related to tree breeding structure. We close with a consideration of the general implications of these studies for tropical forest ecology and conservation.

Mating System

The studies reviewed here examined the mating systems of 12 tree species occurring within the 50-ha BCI Forest Dynamics Plots (table 17.1). Beyond estimating the proportion of offspring resulting from selfed and outcrossed events in a given reproductive episode (i.e., the classic mating system analysis), these works investigated how breeding structure varied over time, and as a function of the density and distribution of reproductive trees.

The majority of the species studied were outcrossed (table 17.1). Outcrossing was not random; for a given maternal tree, pollen sampling was skewed in favor of one or a few other trees. This deviation from random mating was more pronounced for trees occurring at a low population density, which presumably had fewer mates than those occurring at a higher density. For maternal trees with fewer potential mates, the probability that pollen is sampled evenly among maternal trees is reduced relative to that for trees occurring at a high-density.

Table 17.1. Characteristics and Mean Population Outcrossing Rates of 12 Tree Species Studied on the 50-ha Forest Dynamics Plot on Barro Colorado Island

Species[a]	Family	Stature	Reproductive Adult Density (stems/ha)	Pollinator(s)	t_m(+/− SE)[b]
Beilschmedia pendula	Lauraceae	Canopy	2.1	Bees?	0.918 +/− 0.058[c]
Brosimum alicastrum	Moraceae	Canopy	0.3	Bees, wind	0.875 +/− 0.035[c]
Calophyllum longifolium	Guttiferae	Canopy	0.11[d]	Small insects	1.030 +/− 0.085[e,f]
			0.17[d]		
Cavanillesia platanifolia[g]	Bombacaeae	Canopy	0.18	Hawk moths, bees	0.569 +/− 0.024[h]
			0.08		0.347 +/− 0.025[h]
Ceiba pentandra[g]	Bombacaceae	Canopy	0.24	Bats, birds, bees, beetles, wasps, mammals	0.689 +/− 0.032[I]
Platypodium elegans	Bombacaceae	Canopy	0.5	Bees	0.898 +/− 0.043[c]
			0.34		
Quararibea asterolepis	Bombacaceae	Canopy	6.92	Bats, hawk moths?	1.008 +/− 0.010[h]
Sorocea affinis	Moraceae	Understory	6.51	Small bees, wind	0.969 +/− 0.020[c]
			3.56		1.089 +/− 0.045[c]
Spondias mombin	Anacardiaceae	Canopy	0.13	Small insects	0.989 +/− 0.163[e,f]
			0.14		
Tachigali versicolor	Leguminosae	Canopy	0.1	Bees	0.937 +/− 0.044[j]
Trichilia tuberculata	Meliaceae	Canopy	6	Bees?	1.077 +/− 0.028[c]
Turpinia occidentalis	Staphyleaceae	Subcanopy	0.32[d]	Small insects	1.006 +/− 0.090[f]

[a]All species were functionally hermaphroditic, except for the two species of Moraceae whose study populations appeared to be dioeceous (see Murawski and Hamrick 1990).

[b]Multilocus outcrossing rates were estimated using the multilocus mixed-mating model of Ritland and Jain (1981; Ritland 1990).

[c]Murawski and Hamrick (1991).

[d]Estimate includes minimally reproductive trees.

[e]Outcrossing rates are presented for two consecutive reproductive seasons.

[f]Stacy et al. (1996).

[g]Species with a mixed mating system reveal the relationship between flowering tree density and outcrossing rate.

[h]Murawski et al. (1990).

[i]Murawski and Hamrick (1992a).

[j]Loveless et al. (1998).

Nonrandom mating has also been documented in low-density tree populations in other tropical forests, including Guiana (*Dicorynia guianensis* [Leguminosae {Caesalpinoideae}]; Caron et al. 1998) and Costa Rica (*Carapa guianensis* [Meliaceae]; Hall et al. 1994).

The consequence of low population density was perhaps most severe for species with mixed mating systems. Mixed mating systems, in which a substantial fraction of seeds result from self-fertilization, are atypical for tropical trees. Of the 12 species studied, two large-gap specialists in the Bombacaceae—*Cavanillesia platanifolia* and *Ceiba pentandra*—had mixed mating systems, with population-level estimates of selfing rate ranging from 30% to 65%, where selfing rate $= (1 − t_m) \times 100\%$, and $t_m =$ the multilocus outcrossing rate (Murawski et al.

1990; table 17.1). For *C. platanifolia*, both population- and individual-level estimates of selfing rates were negatively associated with the density of flowering neighbors. The more spatially isolated a tree, the more self-fertilized seeds it produced. For some trees, up to 95% of the seed crop resulted from self-fertilization (Murawski and Hamrick 1992b).

Mating Patterns within Tree Populations

Three studies evaluated effective pollen flow (i.e., pollen movement that results in fertilization) within tree populations occurring on the BCI plot (Hamrick and Murawski 1990; Stacy et al. 1996; Nason and Hamrick 1997). All focused on low-density populations, but increased the size of the target populations by expanding the plot 100 m in all directions, creating an 84-ha study area. In the first two studies (Hamrick and Murawski 1990; Stacy et al. 1996), allozyme electrophoresis and paternity exclusion analysis (Devlin et al. 1988) were used to characterize the distribution of within-population pollen movement and to determine levels of pollen flow derived from outside the 84-ha area. Using similar methods to compare gene flow into continuous- and fragmented-forest populations, the third study (Nason and Hamrick 1997) estimated rates of pollen flow into a cluster of adults of a target species occurring near the center of the 84-ha plot.

Methods

The following describes a general procedure for estimating patterns of pollen movement within tree populations on a permanent plot. The procedure was followed directly in Stacy et al. (1996), but a modified form was used in Hamrick and Murawski (1990) and Nason and Hamrick (1997). To estimate mating patterns, the distribution of pollen movement is assessed individually for each reproductive tree. For each adult, the straight-line distance to the nearest plot edge is determined and treated as the radius of a hypothetical circle encompassing the "known neighborhood" for that tree (fig. 17.1). The known neighborhood is therefore the maximum circular area surrounding each maternal tree within which all potential pollen donors are mapped and genotyped. Trees occurring near the plot center have the largest known neighborhoods. A series of paternity exclusion analyses based on allozyme genotype is then done on the full set of progeny sampled from the maternal tree. In the first run, only the maternal tree is included in the pool of potential fathers. Subsequent runs include additional trees in the paternal pool, one at a time, from the nearest neighbor outward until the edge of the known neighborhood is reached. At the end of each run, the fraction of progeny that cannot be assigned to any tree in the paternal pool is noted. These progeny represent pollen flow events from donors located outside the set of trees included in the

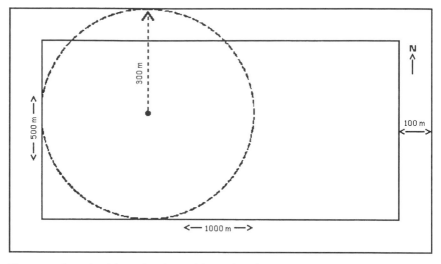

Fig. 17.1. Schematic of the 84-ha plot used for studies of tree population breeding structure. The study plot comprises the 50-ha Forest Dynamics Plot and a 100-m-wide border. Circle depicts an example of a known neighborhood used in the single-tree analysis of mating patterns. Reprinted from Stacy et al. (1996), © by the University of Chicago Press. Reprinted with permission.

analysis. Because allozyme loci typically afford only modest power to discriminate non-true fathers in a paternity analysis, some number of progeny sired by more distant mates will be assigned in error to more proximal trees. Thus, this approach provides minimum estimates of pollen movement distances to each reproductive adult over the spatial scale of its known neighborhood and tends to overestimate the frequency of near-neighbor mating.

Results and Discussion

In the first study, Hamrick and Murawski (1990) investigated mating patterns within a low-density population of the bee-pollinated canopy species, *Platypodium elegans* (Leguminosae [Papilionoideae], in the 84-ha plot over 3 consecutive years. The population of reproductive individuals varied from 8 to 12 individuals per year. All on-plot adults and seeds from all reproductive adults (mean of 18–22 seeds/maternal tree/year) were sampled and genotyped for 15 allozyme loci. Paternity exclusion analysis revealed that although near-neighbor mating was frequent in this low-density population, average pollen movement distances between trees exceeded 300 m in each year of the study. During these 3 years, a minimum of 18-40% of on-plot progeny were sired by pollen donors occurring outside the 84-ha area.

The second study involved two canopy species, *Spondias mombin* (Anacardiaceae) and *Calophyllum longifolium* (Guttiferae), and the subcanopy species, *Turpinia occidentalis* (Staphyleaceae) (Stacy et al. 1996). Though unrelated, these three hermaphroditic species are similar in that all have small, unspecialized flowers presumably pollinated by a diverse array of small insects. Within the extended 84-ha plot, the density of reproductive adults ranged across species from one tree every 7 to 10 ha. All adults and subadults on the plot, as well as progeny from reproductive trees (mean of 28–50 progeny/reproductive tree/year) were sampled and genotyped at several allozyme loci (five to nine loci, depending on species).

A primary goal of this study was to assess the relationship between adult distribution and the pattern of effective pollen movement. The spatial distribution of reproductive trees varied among species. Two of the three populations, however, comprised both clumped and regularly dispersed reproductive trees. The mean distance between nearest reproductive neighbors ranged from 67 to 100 m across species.

This study found a strong association between the local density of reproductive trees and the spatial distribution of cross-pollinations among trees (fig. 17.2). Where flowering adults were evenly spaced relative to the mean population

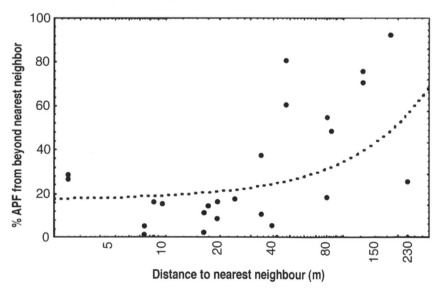

Fig. 17.2. Relationship between adult neighborhood density and the spatial pattern of cross-pollinations for individual maternal trees of three hermaphroditic, small-insect-pollinated tree species. Distance to nearest neighbor depicts an inverse of adult neighborhood density. Values represent two reproductive seasons and show the percentage of incoming apparent pollen flow (%APF) to each maternal tree that is derived from adults located beyond the nearest reproductive neighbor.

distribution, a large fraction of effective pollen moved at least a few hundred meters and well beyond the nearest reproductive neighbors. Conversely, where flowering trees were clumped relative to the mean distribution, the majority of successful crosses were among neighboring trees. For trees occurring within clumps, the percentage of seeds sired by within-clump neighbors ranged from 72% to 100% across species. For *S. mombin*, the distribution of rare and unique alleles near their pollen-donor parents provided a second means for tracking pollen movement. Three adults, each occurring within a clump of reproductive trees, were each heterozygous for a rare or unique allele. With only a single exception seen over two reproductive seasons, these marker alleles were detected exclusively in the progeny of adults occurring within the clump that included the donor tree (within an outcrossing distance of 75, 87, and 138 m of the three donor trees). At a larger spatial scale, both the known-neighborhood method and the tracking of rare alleles from off-plot donors revealed low but substantial levels of pollen flow (5–10%) over at least 300 m for both canopy species in both years. The patterns of effective pollen flow observed for these species were consistent over the 2 years of the study.

Characterization of the spatial distribution of pollen flow within populations of these three species permitted estimation of the area required to encompass a natural breeding unit for low-density canopy species. This was defined as the circular area surrounding a centrally located maternal tree within which 95% of all incoming pollen originates. For low-density species, researchers concluded that a natural breeding unit would extend ≥ 40 ha and ≥ 60 ha for populations characterized by clumped and even distributions of adults, respectively.

In the third study, Nason and Hamrick (1997) estimated levels of apparent and total immigrant pollen flow into a loose cluster of 10 adults of *Spondias mombin* located near the plot center and isolated from other conspecifics by about 200 m. Using an average of 75 progeny per maternal tree genotyped at eight allozyme loci, paternity exclusion analysis revealed a minimum pollen flow of 15% to this cluster. When adjusted for cryptic gene flow events—that is, pollen from off-plot donors that cannot be distinguished genotypically from that of on-plot trees (Devlin and Ellstrand 1990)—they estimated that 44% of the progeny of these trees were sired by pollen donors located ≥ 200 m away.

Together, these studies suggest that while near-neighbor mating is common for insect-pollinated tropical trees occurring in a clumped distribution, as much as 50% of successful pollen may travel >200 m between trees. Moreover, these studies demonstrate that tropical trees with insect pollinators have a substantial potential for pollen flow over a few to several hundreds of meters, or farther. Rates of longer-distance pollen flow (i.e., >300 m) werc greater for the bee-pollinated species (*Platypodium elegans*) than for those presumably pollinated by a variety of small insects (*Calophyllum longifolium*, *Spondias mombin*, and

Turpinia occidentalis). This finding, however, is confounded by the higher genetic exclusion probability, and thus greater power to discriminate paternity, obtained for *P. elegans*, relative to the other species.

If, for insect-pollinated tropical trees, a relationship exists between cross-pollination distance and the size of the insect pollinator, it is not a simple one. Long-distance trap-lining behavior involving interplant movements of thousands of meters has been demonstrated for some neotropical (Janzen 1971) and pale-otropical (Kato 1996) bee species. The long-distance record for pollen transfer from insect-pollinated trees, however, belongs to tropical figs and their tiny wasp pollinators in and around BCI (Nason et al. 1996). By estimating the number of pollen donors contributing to individual maternal families (through paternity exclusion analysis), and the density of trees in male-flowering phase at any given time, Nason et al. (1996) concluded that fig wasps routinely move 4-10 km between flowering trees. They suggest that movement over such great distances by the weak-flying wasps is accomplished with the aid of wind. In striking contrast, however, is the observation in the southeast Asian thrip-pollinated species, *Popowia pisocarpa* (Annonaceae), of pollen-limited seed set in adults separated from flowering conspecifics by only 5 m (Momose et al. 1998). Similarly, in a comparative study of pollinator movements within undisturbed and disturbed habitats, Ghazoul et al. (1998) found that seed set was significantly lower than expected in trees of the self-infertile *Shorea siamensis* (Dipterocarpaceaae) in a disturbed habitat where population density was reduced. The authors suggested that the reduced seed set was due to the failure of the principal pollinators, *Trigona* bees, to move regularly between the more widely spaced trees in the disturbed habitat.

The potential for long-distance pollen transfer has been demonstrated for a variety of tropical pollinators beyond bees and small insects, including bats (Heithaus et al. 1974), butterflies (Murawski 1987), hawkmoths (Haber and Frankie 1989), and hummingbirds (Linhart 1973; Webb and Bawa 1983). Considerably more study will be required, however, before the relationship between pollinator type (and size) and the distribution of pollen flow can be characterized for tropical trees.

Limitations of Plot-Based Studies of Tree Breeding Structure

Genetic Limitations

The studies just reviewed begin to illuminate patterns of paternity in low-density tropical tree populations and reveal ample incidence of cross-pollinations over distances of hundreds of meters. The resolution of studies of pollen movement, however, is a function of the variation in the genetic markers used. Not all study species yield ample allozyme polymorphism, even with substantial effort spent in the lab. Given the methods used to assign paternity, the somewhat restricted polymorphism of allozymes leads to two caveats. One problem is the overestimation of the frequency of near-neighbor mating and the consequent underestimation of long-distance pollen flow. Among neighboring trees, effective mating patterns—including patterns of cross-compatibility—may be considerably more complex than is revealed through the use of genetic data yielding modest exclusion probabilities, where the paternity of some progeny may be misassigned.

A second concern is that limited polymorphism of genetic markers restricts the number of potential fathers that can be evaluated in a paternity analysis, which in turn limits the size of populations available for study. The more individuals that are included in the paternal pool, the lower is the probability of assigning unique gamete genotypes to each. Hence, as the size of the paternal pool increases, the probability of complete exclusion of all non-true fathers for a given progeny decreases. Because of this limitation, studies of the breeding structure of tropical trees have been restricted necessarily to low-density populations (but see below).

Plot Limitations

Even for Forest Dynamics Plots—the largest plots now mapped in the tropics— plot size is insufficient for examining pollen movement for many tree species. For low-density species, 50 ha is not sufficient to capture an effectively interbreeding deme or neighborhood. Even for common species, 25 or 50 ha is too small an area for studies of long-distance gene flow. The studies reviewed above all suggest that pollen flow from beyond the boundaries of a 50-ha plot is not uncommon.

Suggestions for Future Studies of Tree Breeding Structure in Forest Dynamics Plots

1. To date, studies of tree breeding structure have involved only a handful of species and have been carried out in only one of the Forest Dynamics Plots—BCI, Panama. In fact, the bulk of our understanding of breeding systems and gene flow in tropical trees derives almost exclusively from Central American forests. The forests of tropical Asia, Africa, and America differ markedly in the composition and diversity of plant taxa and in the major pollination and seed dispersal agents (Whitmore 1990). Thus, there is little reason to expect that breeding structures observed for neotropical tree species will be characteristic of species occurring in other tropical regions. For example, the first estimates of outcrossing rates of Sri Lankan rainforest trees may indicate a greater incidence of mixed mating systems in these species (Dayanandan et al. 1990; Murawski et al. 1994a, b). As the system of Forest Dynamics Plots is designed to capture a wide variety of forest formations, it offers an excellent backdrop for comparative studies of tree breeding structures across a range of tropical forests. Methods used in BCI studies could serve as guidelines for investigations of mating parameters in the other, less-studied sites.

2. Furthermore, while we are beginning to gain some insight into mating patterns in relatively low-density populations of tropical trees, we still have little understanding of the dynamics of mating in the more common tree species. Past studies of mating patterns have been restricted to low-density populations due to the relatively limited polymorphism of allozyme loci, the traditional tool used for plant mating system and paternity exclusion analyses. Recently, however, hypervariable molecular markers have been developed for population studies of several tree species from throughout the tropics (e.g., Chase et al. 1996a, b; White and Powell 1997; Ujino et al. 1998; Collevatti et al. 1999; Stacy et al. 2001; Jones et al. 2002; Lemes et al. 2002). With these markers, we can study populations with much higher numbers of potential pollen donors. The predominant outcrossing and density-dependent mating patterns observed for low-density tree species may not be characteristic of tropical trees occurring at higher population densities. Pollinator foraging patterns are closely linked to the population density of the host plant (Linhart 1973). Where a tree species is common, local abundance of the food resource may fail to induce long-distance movements between flowering trees. Effective cross-pollinations, therefore, may occur over smaller distances in high-density populations than where reproductive trees are more widely spaced. Alternatively, the relationship between adult population density and patterns of cross-pollination may be confounded by influences of other features such as flowering phenology or the sharing of incompatibility alleles. Additional studies of breeding structure—specifically of the more common tree species—are needed to address these issues.

3. Lastly, Forest Dynamics Plots could be used as controls for studies of the effects of forest disturbance on tree populations. Concern over the impact of anthropogenic disturbance on biodiversity in tropical regions has increased markedly in recent years. Preliminary to an understanding of the effects of forest disturbance on tree species should be the characterization of demographic, genetic, and reproductive parameters in continuous, undisturbed forest. Because Forest Dynamics Plots are situated in undisturbed stands, and because these stands are typically embedded in highly modified or disturbed landscapes, these plots should facilitate the simultaneous study of tree populations in both undisturbed and disturbed forest (e.g., Nason and Hamrick 1997). As Forest Dynamics Plots are both widespread and permanent, they should provide ideal settings for such an approach.

General Implications

1. For tree species with mixed mating systems, the inverse relationship observed between adult neighborhood density and self-fertilization rate indicates that for such species artificial thinning of adult populations (through selective logging,

for example) would result in elevated selfing as long as adult population density remains low. The theoretical consequences of self-fertilization for population fitness are severe. There is a need, however, for empirical studies on the fitness effects of selfing in tropical trees.

2. For outcrossed species, a strong tendency toward near-neighbor mating was observed where reproductive trees were clumped. When coupled with the observation that conspecific tropical trees are typically aggregated in natural stands (Condit et al. 2000) and the assumption that—due to restricted seed dispersal—neighboring adults are related (but see Hamrick et al. 1993), this finding suggests that ample opportunity exists for biparental inbreeding in tropical trees. While the theoretical consequences of biparental inbreeding on population fitness are well known (Uyenoyama 1986, 1988a, b; Lynch 1991), we have little empirical data on the fitness consequences of near-neighbor crosses in tropical forest trees. Three known studies that examined distance-dependent cross-fertility in tropical trees (Koptur 1984; Crome and Irvine 1986; Stacy 2001) all revealed relatively lower fitness in short-distance crosses. Additional, explicit studies of the consequences of near-neighbor mating are required, however, if we are to predict the long-term genetic effects of any forest disturbance expected to restrict longer-distance pollen flow among individuals (e.g., forest fragmentation, timber harvesting).

3. Observations of predominant outcrossing and substantial pollen-mediated gene flow over hundreds of meters indicate that canopy species on BCI occur in large genetic neighborhoods. This conclusion is consistent with the low levels of population genetic structure detected through allozyme studies in 14 tree species on BCI in a 16-km^2 area (Hamrick and Loveless 1989). Together, these observations indicate that tropical tree populations are linked by gene flow to other conspecific populations separated by distances of a few to several kilometers (see 4 below).

4. The natural breeding unit defines the minimum area of continuous forest required to preserve the majority of natural cross-pollinations for a single reproductive adult. The findings of Stacy et al. (1996) indicate that forest fragments smaller than 40–60 ha would fail to permit the bulk of outcrossing events for even a single individual of a low-density canopy species. Extrapolating conservatively to the population level then, forest patches at least 60 ha in size (but probably larger) would be required to permit natural patterns of pollen flow for groups of interbreeding trees occurring at low density.

This figure must be treated as a minimum estimate for the following reason. The natural breeding unit derives loosely from Wright's (1946) genetic neighborhood, but is based on only 95% of effective pollen dispersal, and takes no account of gene flow by seed. The tail of the pollen flow distribution may serve to maintain genetic variation within species by counteracting the effects of random genetic drift within populations. Although estimating the tail of the pollen flow distribution was not plausible in the Forest Dynamics Plot-based studies above, other studies of tropical trees have revealed effective pollen flow exceeding 1 to several kilometers (Nason et al. 1996, 1998; Nason and Hamrick 1997; Apsit 1998; White et al. 1998). It follows, therefore, that isolated forest fragments considerably larger than the projected 60 ha would be required to permit the bulk of natural pollen movement, and hence to preserve genetic diversity of tree species, in fragmented tropical forest landscapes.

Acknowledgments

We thank R. Condit and two anonymous reviewers for helpful comments on earlier versions of this chapter.

References

Apsit, V. J. 1998. *Fragmentation and Pollen Movement in a Costa Rican Dry Forest Tree Species.* Ph.D. dissertation, University of Georgia, Athens, GA.

Baker, H. G. 1959. Reproductive methods as factors in speciation in flowering plants. *Cold Spring Harbor Symposia on Quantitative Biology* 24:177–99.

Bawa, K. B. 1974. Breeding systems of tree species of a lowland tropical community. *Evolution* 28:85–92.

Bawa, K. S., D. R. Perry, and J. H. Beach. 1985. Reproductive biology of tropical lowland rain forest trees. I. Sexual systems and incompatibility mechanisms. *American Journal of Botany* 72:331–45.

Caron, H., C. Dutech, and E. Bandou. 1998. Mating systems of a Guiana tropical forest tree, *Dicorynia guianensis* Amshoff (Caesalpiniaceae). *Genetics Selection Evolution* 30:S153–66.

Chan, H. T. 1981. Reproductive biology of some Malaysian Dipterocarps. III. Breeding systems. *Malaysian Forester* 44:28–36.

Chase, M., R. Kesseli, and K. Bawa. 1996a. Microsatellite markers for population and conservation genetics of tropical trees. *American Journal of Botany* 83:51–57.

Chase, M. R., C. Moller, R. Kesseli, and K. S. Bawa. 1996b. Distant gene flow in tropical trees. *Nature* 383:398–99.

Collevatti, R. G., R. V. Brondani, and D. Grattapaglia. 1999. Development and characterization of microsatellite markers for genetic analysis of a Brazilian endangered tree species *Caryocar brasiliense. Heredity* 83:748–56.

Condit, R., P. P. S. Ashton, P. Baker, S. Bunyavejchewin, S. Gunatilleke, N. Gunatilleke, S. P. Hubbell, R. B. Foster, L. Hua Seng, A. Itoh, J. V. LaFrankie, E. Losos, N. Manokaran, R. Sukumar, and T. Yamakura. 2000. Spatial patterns in the distribution of common and rare tropical tree species: a test from large plots in six different forests. *Science* 288:1414–18.

Corner, E. H. G. 1954. The evolution of tropical forest. Pages 34–46 in J. S. Huxley, A. C. Hardy, and E. B. Ford, editors. *Evolution as a Process.* Allen and Unwin, London, U.K.

Crome, F. H. J., and A. K. Irvine. 1986. "Two bob each way": the pollination and breeding system of the Australian rain forest tree *Syzygium cormiflorum* (Myrtaceae). *Biotropica* 18:115–25.

Dayanandan, S., D. N. C. Attygalla, A. W. W. L. Abeygunasekera, I. A. U. N. Gunatilleke, and C. V. S. Gunatilleke. 1990. Phenology and floral morphology in relation to pollination of some Sri Lankan dipterocarps. Pages 103–33 in K. S. Bawa and M. Hadley, editors. *Reproductive Ecology of Tropical Forest Plants*. UNESCO, Paris and Parthenon, Carnforth, U.K.

Devlin, B., and N. C. Ellstrand. 1990. The development and application of a refined method for estimating gene flow from angiosperm paternity analysis. *Evolution* 44:248–59.

Devlin, B., K. Roeder, and N. C. Ellstrand. 1988. Fractional paternity assignment: Theoretical development and comparison to other methods. *Theoretical and Applied Genetics* 76:369–80.

Federov, A. A. 1966. The structure of the tropical rain forest and speciation in the humid tropics. *Journal of Ecology* 54:1–11.

Frankie, G. W., P. A. Opler, and K. S. Bawa. 1976. Foraging behavior of solitary bees: implications for outcrossing of a neotropical forest tree species. *Journal of Ecology* 64:1049–57.

Ghazoul, J., K. A. Liston, and T. J. B. Doyle. 1998. Disturbance-induced density-dependent seed set in *Shorea siamensis* (Dipterocarpaceae), a tropical forest tree. *Journal of Ecology* 86:462–73.

Haber, W. A., and G. W. Frankie. 1989. A tropical hawkmoth community. *Biotropica* 21:155–72.

Hall, P., L. C. Orrell, and K. S. Bawa. 1994. Genetic diversity and mating system in a tropical tree, *Carapa guianensis* (Meliaceae). *American Journal of Botany* 81:1104–11.

Hamrick, J. L., and M. J. Godt. 1989. Allozyme diversity in plant species. Pages 43–64 in A. H. D. Brown, M. T. Clegg, A. L. Kahler, and B. S. Weir, editors. *Plant Population Genetic, Breeding, and Genetic Resources*. Sinauer, Sunderland, MA.

Hamrick, J. L., and M. D. Loveless. 1989. The genetic structure of tropical tree populations: Associations with reproductive biology. Pages 129–49 in J. H. Bock and Y. B. Linhart, editors. *The Evolutionary Ecology of Plants*. Westview Press, Boulder, CO.

Hamrick, J. L., and D.A. Murawski. 1990. The breeding structure of tropical tree populations. *Plant Species Biology* 5:157–65.

Hamrick, J. L., D. A. Murawski, and J. D. Nason. 1993. The influence of seed dispersal mechanisms on the genetic structure of tropical tree populations. *Vegetatio* 107/108:281–97.

Heithhaus, E. R., P. A. Opler, and H. G. Baker. 1974. Bat activity and pollination of *Bauhinia pauletia*: plant-pollinator coevolution. *Ecology* 55:412–19.

Hubbell, S. P. 1979. Tree dispersion, abundance, and diversity in a tropical dry forest. *Science* 213:1299–1309.

Janzen, D. H. 1971. Euglossine bees as long-distance pollinators of tropical plants. *Science* 171:203–05.

Jones, R. C., J. McNally, and M. Rossetto. 2002. Isolation of microsatellite loci from a rainforest tree, *Elaeocarpus grandis* (Elaeocarpaceae), and amplification across closely related taxa. *Molecular Ecology Notes* 2:179–81.

Kato, M. 1996. Plant-pollinator interactions in the understory of a lowland mixed dipterocarp forest in Sarawak. *American Journal of Botany* 83:732–43.

Koptur, S. 1984. Outcrossing and pollinator limitation of fruit set: Breeding systems of neotropical *Inga* trees (Fabaceae: Mimosoideae). *Evolution* 38:1130–43.

Lemes, M. R., R. P. V. Brondani, and D. Grattapaglia. 2002. Multiplexed systems of microsatellite markers for genetic analysis of mahogany, *Swietenia macrophylla* King (Meliaceae), a threatened neotropical timber species. *Journal of Heredity* 93:287–91.

Lewis, D. 1942. The evolution of sex in flowering plants. *Biological Review* 17:46–67.

Linhart, Y. B. 1973. Ecological and behavioral determinants of pollen dispersal in hummingbird-pollinated *Heliconia. American Naturalist* 107:511–23.

Loveless, M. D., and J. L. Hamrick. 1984. Ecological determinants of genetic structure in plant populations. *Annual Review of Ecology and Systematics* 15:65–95.

Loveless, M. D., J. L. Hamrick, and R. B. Foster. 1998. Population structure and mating system in *Tachigali versicolor*, a monocarpic neotropical tree. *Heredity* 81:134–43.

Lynch, M. 1991. The genetic interpretation of inbreeding depression and outbreeding depression. *Evolution* 45:622–29.

Momose, K., T. Nagamitsu, and T. Inoue. 1998. Thrips cross-pollination of *Popowia pisocarpa* (Annonaceae) in a lowland dipterocarp forest in Sarawak. *Biotropica* 30:444–48.

Murawski, D. A. 1987. Floral resource variation, pollinator response, and potential pollen flow in *Psiguria warscewiczii. Ecology* 68:1273–82.

Murawski, D. A., and J. L. Hamrick. 1991. The effect of the density of flowering individuals on the mating systems of nine tropical tree species. *Heredity* 67:167–74.

———. 1992a. Mating system and phenology of *Ceiba pentandra* (Bombacaceae) in Central Panama. *Journal of Heredity* 83:401–04.

———. 1992b. The mating system of *Cavanillesia platanifolia* under extremes of flowering-tree density: A test of predictions. *Biotropica* 24:99–101.

Murawski, D. A., J. L. Hamrick, S. P. Hubbell, and R. B. Foster. 1990. Mating systems of two Bombacaceous trees of a neotropical moist forest. *Oecologia* 82:501–06.

Murawski, D. A., B. Dayanandan, and K. S. Bawa. 1994a. Outcrossing rates of two endemic *Shorea* species from Sri Lankan tropical rain forests. *Biotropica* 26:23–29.

Murawski, D. A., I. A. U. N. Gunatilleke, and K. S. Bawa. 1994b. The effects of selective logging on inbreeding in *Shorea megistophylla* (Dipterocarpaceae) from Sri Lanka. *Conservation Biology* 8:997–1002.

Nason, J. D., and J. L. Hamrick. 1997. Reproductive and genetic consequences of forest fragmentation: Two case studies of neotropical canopy trees. *Journal of Heredity* 88:264–76.

Nason, J. D., E. A. Herre, and J. L. Hamrick. 1996. Paternity analysis of the breeding structure of strangler fig populations: Evidence for substantial long-distance wasp dispersal. *Journal of Biogeography* 23:501–12.

———. 1998. The breeding structure of a tropical keystone plant resource. *Nature* 391:685–87.

O'Malley, D. M., and K. S. Bawa. 1987. Mating system of a tropical rain forest tree species. *American Journal of Botany* 74:1143–49.

O'Malley, D. M., D. P. Buckley, G. T. Prance, and K. S. Bawa. 1988. Genetics of Brazil nut (*Bertholletia excelsa* Humb. & Bonpl.: Lecythidaceae). II. Mating system. *Theoretical and Applied Genetics* 76:929–32.

Opler, P. A., and K. S. Bawa. 1978. Sex ratios in tropical forest trees. *Evolution* 32:812–21.

Perez-Nasser, N., L. E. Eguiarte, and D. Pinero. 1993. Mating system and genetic structure of the distylous tropical tree *Psychotria faxlucens* (Rubiaceae). *American Journal of Botany* 80:45–52.

Ritland, K. 1990. A series of Fortran computer programs for estimating plant mating systems. *Journal of Heredity* 81:235–37.

Ritland, K., and S. K. Jain. 1981. A model for the estimation of outcrossing rate and gene frequencies using n independent loci. *Heredity* 47:35–52.

Sakai, S., K. Momose, T. Yumoto, M. Kato, and T. Inoue. 1999. Beetle pollination of *Shorea parvifolia* (section *Mutica*, Dipterocarpaceae) in a general flowering period in Sarawak, Malaysia. *American Journal of Botany* 86:62–69.

Stacy, E. A. 2001. Cross-fertility in two tropical tree species: Evidence of inbreeding depression within populations and genetic divergence among populations. *American Journal of Botany* 88:1041–1051.

Stacy, E. A., S. Dayanandan, B. P. Dancik, and P. D. Khasa. 2001. Microsatellite DNA markers for the Sri Lankan rainforest tree species, *Shorea cordifolia* (Dipterocarpaceae), and cross-species amplification in *S. megistophylla. Molecular Ecology Notes* 1:A20.

Stacy, E. A., J. L. Hamrick, J. D. Nason, S. P. Hubbell, R. B. Foster, and R. Condit. 1996. Pollen dispersal in low-density populations of three neotropical tree species. *American Naturalist* 148:275–98.

Ujino, T., T. Kawahara, Y. Tsumura, T. Nagamitsu, H. Yoshimaru, and W. Ratnam. 1998. Development and polymorphism of simple sequence repeat DNA markers for *Shorea curtisii* and other Dipterocarpaceae species. *Heredity* 81:422–28.

Uyenoyama, M. K. 1986. Inbreeding and the cost of meiosis: The evolution of selfing in populations practicing partial biparental inbreeding. *Evolution* 40:388–404.

———. 1988a. On the evolution of genetic incompatibility systems: Incompatibility as a mechanism for the regulation of outcrossing distance. Pages 212–32 in R. E. Michod and B. R. Levin, editors. *The Evolution of Sex: An Examination of Current Ideas.* Sinauer, Sunderland, MA.

———. 1988b. On the evolution of genetic incompatibility systems II. Initial increase of strong gametophytic self-incompatibility under partial selfing and half-sib mating. *American Naturalist* 131:700–22.

Webb, C. J., and K. S. Bawa. 1983. Pollen dispersal by hummingbirds and butterflies: A comparative study of two lowland tropical plants. *Evolution* 37:1258–70.

White, G., and W. Powell. 1997. Isolation and characterization of microsatellite loci in *Swietenia humilis* (Meliaceae): an endangered tropical hardwood species. *Molecular Ecology* 6:851–60.

White, G. M., W. Powell, and D. Boshier. 1998. The dynamics of pollen flow detected in a fragmented population of *Swietenia humilis* (Zucc.) using ISSRs as a marker system. *Biotropica* (Suppl.) 30:S13.

Whitmore, T. C. 1990. *An Introduction to Tropical Rain Forests.* Clarendon Press, Oxford, U.K.

Wright, S. 1946. Isolation by distance under diverse mating systems. *Genetics* 31:39–59.

18

Ecological Correlates of Tree Species Persistence in Tropical Forest Fragments

Sean C. Thomas

Introduction

This volume is concerned primarily with structure and diversity of intact primary forest. Another goal of the Center for Tropical Forest Science (CTFS), however, is to use the understanding gained from large-scale Forest Dynamics Plots to improve techniques of forest conservation and management. These plots also serve as standards or controls against which to measure the impact on forest ecosystems of a wide variety of human activities.

One major impact of human activity is fragmentation of the forest. Agriculture and other human occupations are transforming much of the world's remaining tropical forest into a patchwork of discontinuous fragments (Skole and Tucker 1993). Comparing large Forest Dynamics Plots with nearby fragments allows us to assess the impacts of fragmentation and provides a means for elucidating some of the factors that organize the structure or function of intact forest (Leigh et al. 1993; Terborgh et al. 2001).

So far, most studies of the effects of fragmentation have focused on animal populations (e.g., Lovejoy et al. 1986; Andren 1994; Turner 1996; Laurance 1997a; Zuidema et al. 1996). What research has been devoted to trees has avoided questions that require identifying all of a fragment's trees, focusing instead on how fragmentation affects the genetics of particular tree populations (Nason and Hamrick 1997) or forest microclimate, structure, and dynamics (Kapos 1989; Williams-Linera 1990a, b; Kapos et al. 1993; Turton and Freiburger 1997; Laurance et al. 1997b, 1998a, b; Benitez-Malvido 1998; Sizer and Tanner 1999). Yet trees provide a forest's structure, fuel the function of its ecosystem, and shape its diversity. I focus here on the comparative biology of the response of different tree species to fragmentation (cf. Leigh et al. 1993; Tabarelli et al. 1999). What ecological and physiological attributes enable tree species to persist in small fragments? Conversely, what attributes may predispose tree species to rapid local extinction when their forest is fragmented?

Answers to these questions require not only data on forest fragments themselves but also detailed comparative information on intact forest communities. Studies of fragmentation effects have often been frustrated by a lack of alpha-taxonomic descriptions and regional floras, and in nearly all cases by a lack of species-specific data on life history, growth, and physiological attributes of trees present in the system. In many respects, the large-scale CTFS Forest Dynamics Plots offer an unparalleled opportunity to study fragmentation effects on tropical forest communities, particularly where forest fragments occur near plot sites. In fact, one of the few detailed studies to date examining fragmentation effects on tropical tree species composition was in the vicinity of the first such plot at Barro Colorado Island, Panama (Leigh et al. 1993). Advantages offered by the large-scale plots are many: First, they provide a "control" (large, though unreplicated) for examination of fragmentation effects on tree populations. Second, the plot data can directly contribute comparative information on growth, survivorship, population density, spatial pattern, and habitat specialization for a large number of tree species in the community. Finally, the existence of large-scale plots greatly facilitates studies of many other important aspects of tropical tree biology, from photosynthetic physiology to pollination biology.

This paper addresses the general issue of ecological correlates of species persistence in forest fragments and presents new data from a study conducted in and near Pasoh Forest Reserve, peninsular Malaysia. Since the comparative ecology of tree species persistence in forest fragments has received little previous attention, I first present a series of arguments and hypotheses concerning patterns that might be expected. The topics are introduced in qualitative terms, though many of the issues raised lend themselves to formal mathematical development. The emphasis here is to qualitatively evaluate the potential set of factors involved and to provide a preliminary test of some of these predictions on the basis of data collected in and around Pasoh.

Two Perspectives on Tropical Forest Fragments

In recent years, two quite different pictures of the ecology and natural history of tropical forest fragments have emerged. Studies associated with the Biological Dynamics of Forest Fragments Project, which was established in 1979 near Manaus, Brazil, have elaborated what might be termed a "things fall apart" model of tropical forest fragments (table 18.1). This perspective emphasizes rapid changes in forest physiognomy, the importance of repeated disturbance, erosion of fragment edges, and loss of tree biomass as dominant processes (Lovejoy et al. 1983, 1984, 1986; Ferreira and Laurance 1997; Laurance et al. 1997, 1998a, b). Studies conducted in fragments isolated by flooding of impounded reservoirs have emphasized a similar set of processes, but have also pointed out the potential importance of locally enhanced seed predator and herbivore communities following fragmentation (Terborgh et al. 1997a, b, 2001; Lynam 1997).

Table 18.1. Summary of Predictions for Life History, Morphological, and Physiological Traits Expected to Be Favored in Forest Fragments under the "Things Fall Apart" Model Versus the "More of the Same" Model

Traits	"Things Fall Apart" Model	"More of the Same" Model
Life History		
[†]Population Density in Primary Forest	0	+
[†]Agamospermous Reproduction	+	+
[†]Clonality	0	+
Self-Fertilization	0	+
[†]Dioecy	0	−
Clumped Spatial Distributions	0	+
[†]Habitat Specialization	−	+
High Dispersal Distance	+	−
[†]Early Reproduction	+	+
[†]Demographic Turnover Rate	+	−
Long Lifespan	−	+
Dependence on Biotic Dispersal Agents	−	−
Dependence on Generalist Dispersal Agents	0	+
Dependence on Generalist Pollinators	0	+
Morphology and Growth		
[†]Maximum Tree Height	0	−
Crown Breadth	+	−
[†]Wood Density and Biomechanical Strength	−	+
[†]Growth Rate (under High Resources)	+	0
Epicormic Branching	+	+
Liana Shedding Capacity	+	0
[†]Leaf Size	−	0
Leaf Inclination	+	0
Physiology		
[†]Photosynthetic Capacity (A_{max})	+	0
Light Compensation Point	−	0
Drought Tolerance (Water Use Efficiency)	+	0
Leaf N Content	+	0
[†]Specific Leaf Area (Area:Mass Ratio)	−	0

Note: Directions of predicted effects are indicated as positive (+) negative (−), or no effect or no basis for prediction (0).

[†]Indicates traits examined in the present study.

In contrast, other authors have suggested that isolated forest fragments may persist for decades if not centuries in a relatively undisturbed form, and that tree species loss from such fragments may be quite slow, and determined mainly by random demographic fluctuations. This perspective can thus be labeled a "more of the same" model of tropical forest fragments (table 18.1), in the sense that community change is driven by essentially the same demographic processes found in continuous forest, but with different consequences due to fragment isolation. Such a picture has particularly been emphasized in southeast Asia (Turner et al.

1994, 1996; Turner and Corlett 1996; Corlett and Turner 1997), but similar descriptions have also emerged from studies in Central America (Kellman 1996) and Australia (Harrington et al. 1997). In the case of gallery forests that have been naturally fragmented over long time scales, researchers have suggested that edge effects may help maintain tree species diversity by contributing to environmental heterogeneity (Kellman et al. 1996, 1998).

Leigh et al. (1993) examined the consequences of a species-neutral null model to changes in tree diversity in isolated forest fragments. In this null model, all trees have an equal chance of dying or reproducing regardless of species. Such a model will in general predict that all species have an equal probability of persistence in fragments, at least under identical initial conditions for each species. However, while species neutrality is a useful point of reference, there is ample evidence that tropical trees, even closely related species, differ substantially in terms of vital rates, generation times, and other characteristics (e.g., Rogstad 1990; Moad 1993; Thomas 1996b, c 2003; Davies et al. 1998). For this reason, the impacts of fragmentation, even those due purely to isolation and statistical sampling effects, are also certain to differ among species.

If the natural history of forest fragments followed a "things fall apart" scenario, then one would expect essentially the same traits that are favored by disturbance such as logging or storm damage to be favored by fragmentation (Benitez-Malvido 1998; Sizer and Tanner 1999). Under this view, fragmentation should primarily act to favor the suite of traits encountered in pioneer species, or traits that enable species to endure edge-related biotic or abiotic stress. However, if fragments maintain structural integrity for long periods of time, as emphasized by a "more of the same" model, then quite a different set of characters may be favored. For example, high local population density in primary forest and low rates of demographic turnover will act to buffer species against local extinction due to demographic stochasticity.

Most likely, the effect of tropical forest fragmentation will fall somewhere on a continuum between these two extremes. Thus, fragments will to some extent be affected by increased disturbance, especially soon after isolation, and will exhibit pronounced microenvironmental edge effects and loss of animal populations. However, fragments will also generally recover from this initial phase and remain recognizable as forest that is distinct from areas that have been completely cleared in the recent past. The suite of traits that allow tree species to persist in fragments will thus reflect influences of both disturbance effects and the exacerbated effects of random birth and death events on small populations. The following sections take into account both kinds of processes to develop specific predictions for characteristics that may be favored by forest fragmentation, including aspects of tree life history, growth, morphology, and physiology (table 18.1).

Life History Traits

Population Density and Spatial Pattern

One central life history characteristic that will likely be favored soon after forest fragmentation is that of high population density in primary forest. More common species will, by definition, be at a numerical advantage, even if large increases in tree mortality follow fragment isolation. Over longer time scales, demographic stochasticity also dictates that common species should be favored over rare species. In fact, species with high initial population density will also show longer persistence times even in strictly species-neutral models. For example, if one considers a "null community" in which species abundance follows zero-sum random drift, the expected time to extinction of a given species increases approximately in proportion to population size (Hubbell 2001). Thus, rare species have a higher probability of local extinction, while commoner species will differentially benefit as rarer species are lost from the system. Conversely, a "things fall apart" scenario should eventually result in no correlation (or a negative correlation) between population density in primary forest versus fragments, as gap-dependent or pioneer species that are uncommon in primary forest increase in abundance.

A somewhat less obvious prediction is that tree species exhibiting spatially aggregated distributions may also be favored in fragments. Although the expected time to extinction is linearly related to population size in the null community case, small populations are also subject to exacerbated effects of environmental variation, inbreeding depression, and skewed sex ratios, among other effects (e.g., Gilpin and Soulé 1986). This should generally result in very low probabilities of population survival at low population densities, but a steep increase in survival probability at higher population densities; in other words, an upwardly convex relationship between expected time to extinction and population size. If one considers an ensemble of fragments sampled from an intact community, an evenly distributed species will tend to have uniformly low number of individuals occurring in fragments, while a patchily distributed species will be represented by many individuals in at least some fragments (fig. 18.1). Jensen's inequality (Van Tiel 1984) states that for convex functions the average of the function is larger than the function of the average. Put in other terms, a patchily distributed species will be represented by a viable population in a few fragments, while an evenly distributed species will show uniformly small populations. Thus, a patchily distributed species is expected to show higher average tree densities compared to an evenly distributed species in a set of fragments sampled after some time.

Dispersal and Pollination

Trees that rely on specialized pollinators or seed dispersers may be at a strong disadvantage in forest fragments. Rapid changes in population density have

A. *Croton argyratus*

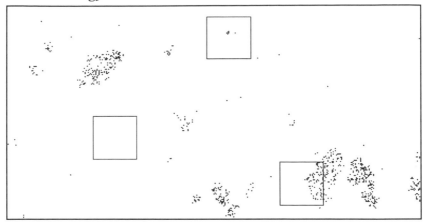

B. *Aporusa nigricans*

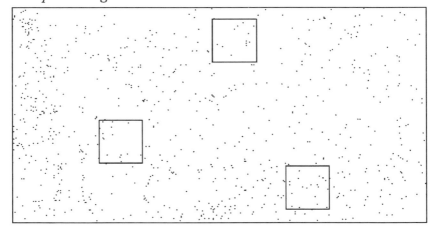

Fig. 18.1. Spatial distributions of trees ≥1 cm dbh for two species at Pasoh Forest Reserve, Malaysia, with three randomly placed 1-ha fragments. (A) *Croton argyratus* (Euphorbiaceae) is a ballistically dispersed species with explosively dehiscing capsules that shows a patchy distribution pattern characteristic of species with this dispersal syndrome. (B) *Aporusa nigricans* (Euphorbiaceae) is a bird-dispersed species showing a more typical spatial distribution. The clumped species is likely to occur at relatively high frequencies in some fragments. If mean time to extinction is a convex function of population size, clumped species such as *C. argyratus* are expected to show a higher average persistence in an ensemble of fragments.

been observed following forest fragmentation in many vertebrate taxa, including important seed dispersal agents such as primates (Lovejoy et al. 1986; Estrada and Coates-Estrada 1996) and birds (Bierregaard and Lovejoy 1988, 1989; Christiansen and Pitter 1997; Warburton 1997). Most often a transient increase in population densities is observed, followed by declines and local extinction (or relocation) of many species. However, many animal species may persist in matrix habitats, and this may strongly influence abundance in fragments as well (Gascon et al. 1999). For example, higher than expected population densities of frugivorous primates have been found in ecotones between primary and secondary forest in some systems (Thomas 1991). Thus, particularly in cases where secondary forest constitutes part of the matrix area, limitations on seed dispersal may be short lived. Less information is available for insect groups that constitute important pollinators. In Amazonian forest fragments, declines in abundance have been found in beetles (Didham et al. 1998) and euglossine bees (Becker et al. 1991), but increased abundance and diversity has been documented in lepidopterans (Brown and Hutchings 1997). It is also important to consider the capacity for animal movements between fragments (e.g., Bierregaard and Dale 1996). Fig wasps (Agaonidae), for example, may be able to travel several kilometers across open habitat (Nason and Hamrick 1997). Certain avian taxa important as seed dispersers, notably hornbills (Buceritodae), are also capable of covering large distances. Thus, while most animal-pollinated and -dispersed trees may be at a disadvantage in forest fragments, specialization for the services of certain pollinators and seed dispersers might be favored. The specifics are certain to vary idiosyncratically among forest types and biogeographic regions, making general predictions difficult.

Seed dispersal patterns may also influence species persistence in fragments through a "seed wastage" effect. In many ecological settings, such as fragments surrounded by agricultural land use, seeds that are dispersed into matrix habitat have no chance of successful establishment. If this is the case, species with long-tailed seed dispersal functions are expected to be at a disadvantage. In contrast, species with short dispersal distances will retain a higher proportion of propagules within the fragment. Thus, species with short dispersal distances, such as ballistically dispersed understory trees, may be favored by several processes: an initial sampling effect, a lack of reliance on specialized seed dispersers, and relatively low seed loss into the matrix habitat.

Apomixis and Selfing

Two ineluctable facets of fragmentation are that remnant populations are small in spatial extent and reproductively isolated (to a greater or lesser degree) from other nearby populations. Life history characteristics that allow isolated populations to persist and/or increase following isolation logically include any traits that facilitate reproduction and recruitment in very small populations. In highly species-rich forest communities, newly isolated fragments will commonly include only a few, or even a single individual, of a given species. For example, within 1-ha (100 × 100 m) subsamples of the Pasoh 50-ha Forest Dynamics Plot, 49% of the species present are represented by five or fewer individuals >1 cm dbh, and 20% are represented by only 1 individual. For such populations, the capacity for clonal or apomictic reproduction and for self-fertilization should be highly advantageous. For isolated populations with only one extant individual, this ability is a necessary condition for any probability of long-term persistence.

Facultative apomixis, usually by means of adventive embryony (in which off-spring are genetically identical to the parent) is found in some tropical trees (Kaur et al. 1978, 1986; Murawski 1996). An exceptional case of obligate asex-ual reproduction via seed has also been inferred for at least one southeast Asian species (Thomas 1997). Clonal growth, while relatively infrequent in tropical canopy trees, is more frequently reported in understory trees, shrubs, and lianas (Peñalosa 1984; Gartner 1989; Sagers 1993) and may be especially prevalent in montane forest types (Kinsman 1990). However, few systematic efforts have been undertaken to assess its frequency in woody tropical species. Of the 820+ tree species represented at Pasoh, only a handful of understory species (e.g., *Garcinia griffithii* [Guttiferae] and *Diospyros* "brown-barked" sp nov. [Ebenaceae], per-sonal observations) are known to spread vegetatively.

Studies of allozyme diversity indicate that tropical trees in primary forest gen-erally have high outcrossing rates (Murawski 1996). Many tropical trees exhibit genetic self-incompatibility (Bawa et al. 1985; Kress and Beach 1993); however, other species can produce viable offspring through selfing. This proportion of self-compatible species may be relatively high in understory species. Kress and Beach (1993) found that 66% of understory flowering plants studied at La Selva were self-compatible, as compared to 25% of canopy trees. Higher altitude forests may also show higher proportions of self-compatible species (Sobrevilla and Arroyo 1982; Tanner 1982; Hernandez and Abud 1987). Current evidence thus suggests that agamospermy, clonality, and capacity for selfing are relatively uncommon characteristics of canopy trees in lowland tropical forests, but may be more pro-nounced among smaller-statured understory trees and shrubs. Other things being equal, such species would be expected to be favored in forest fragments in which populations are represented by one or a few individuals, and where pollinators may also be scarce or absent.

Reproductive Onset

Another reproductive characteristic that may influence persistence in forest fragments is size dependence of reproductive onset, which shows wide varia-tion among species. In a comparative study of Malaysian species, understory to midcanopy trees ranged in size at reproductive onset from <1.0 to 26 cm dbh

(Thomas 1996b), and emergent Dipterocarps reproduced at ~40 cm dbh in primary forest (Thomas and Appanah 1995). Much of this variation is due to absolute differences in tree size, with smaller understory trees and shrubs reproducing at smaller sizes than larger species. However, relative size at onset of maturity (RSOM)—quantified as the ratio of tree height at reproductive onset to maximum tree height for a given species—also shows very wide variation, from ~0.20 to 0.75 (Thomas 1996b). Thus, some canopy trees reaching 30 m or more in height reproduce at smaller sizes than subcanopy species attaining heights of only 15–20 m. Reproduction at small sizes may be an important advantage in small forest fragments, acting to effectively increase local population densities and so lessen the negative impacts of demographic stochasticity.

Mortality and Population Turnover

The "things fall apart" and the "more of the same" models yield diametrically opposed predictions regarding population turnover. If early- successional trees are favored by the initial disturbance event(s), then fragmentation should result in a community that is biased toward species with short lifespans and high rates of mortality and recruitment. However, high demographic turnover should act to increase the probability of local extinction via demographic stochasticity. If fragments remain relatively intact and are not subject to repeated disturbance, early-successional species with high demographic turnover rates may thus actually be at a disadvantage. These alternative predictions may also be examined in terms of gap frequency within fragments. Using a "swiss-cheese" gap definition, gap densities ranging from ~3 to 18 per hectare have been recorded in tropical forests (Brokaw 1985); however, gaps sufficiently large to support recruitment of pioneer species are generally present at much lower densities (e.g., ~0.7/ha, Brokaw 1982). Fragment edges and matrix areas may initially provide good recruitment opportunities for gap-dependent species (e.g., in the first 1–5 years following fragment isolation). However, in many cases, this edge-associated recruitment is likely to rapidly decline, particularly where fragments are surrounded by tree crops, fragment boundaries are intentionally maintained, or edges quickly "seal" due to growth and recruitment at fragment boundaries (e.g., Kapos 1989; Camargo and Kapos 1995). Under this scenario, pioneer species may decline because fragments do not generate sufficient gaps internally to support regeneration requirements of these species.

Morphology and Growth

Size

Tree species size is expected to be negatively related to persistence in small fragments on several grounds. First, a "crown-packing" argument suggests that fragment size may directly place a limit on population size that is simply too low for some species to persist. In the case of dipterocarp forests in southeast Asia, emergent species commonly have crown diameters of 15–20 m (Ashton and Hall 1992), while smaller-statured canopy trees have crown diameters of ~4–10 m. A 1-ha fragment thus has space for crowns of no more than 32 of the largest emergent trees, but for as many as 800 smaller canopy trees. The same fragment would include sufficient physical space for thousands of individuals of small treelets or shrubs. Smaller-statured species also tend to reproduce at smaller sizes than larger species, on either a relative or absolute basis (Thomas 1996b). As noted above, small stature in primary forest tree species is associated with other traits that may facilitate persistence in forest fragments, including low rates of demographic turnover, clonal growth, and capacity for selfing.

Correlates of Resistance to Wind Disturbance

Wind disturbance following fragment isolation has been emphasized in studies of fragments in Amazonia (Lovejoy et al. 1983; Laurance et al. 1997) and Central America (Leigh et al. 1993). Large-statured trees may be more susceptible to windthrow, and this factor may thus favor smaller species in fragments. Other potential morphological correlates of resistance to wind damage include high wood density and biomechanical strength, formation of thickened stems and branches, high stem and branch flexibility, formation of buttresses and/or taproots, and small leaf size (Smith 1972; Zimmerman 1994; Niklas 1998). Both initial exposure and subsequent wind disturbance are also expected to favor species with morphological characteristics allowing recovery from crown breakage, such as capacity for epicormic branching. Narrow crowns may reduce drag, and narrow-crowned species may also be favored under a physical packing argument.

Morphology in Relation to Biotic Interactions

Profuse liana growth in forest fragments is a commonly observed phenomenon (e.g., Turner et al. 1996; Viana and Tabanez 1996; Laurance et al. 2001). Negative effects of lianas on tree growth and survivorship have been documented in a variety of tropical forest systems (e.g., Clark and Clark 1990), and it follows that traits allowing trees to better shed or tolerate lianas might be favored in forest fragments. Such traits may include easily shed bark, flexible trunks, and smooth leaves from which tendrils easily slide off (Putz 1984), and also certain myrmecophytic associations (Fiala et al. 1989).

There has been considerable debate concerning "knock-on" effects on tropical forests in response to changes in animal communities. Intermediate-sized forest remnants may suffer a loss of large predators and thus show increased herbivore densities (e.g., Terborgh 1992; Terborgh et al. 2001). However, studies of tree seedlings in Amazonian forest fragments found reduced herbivory in fragments compared to intact forest soon after fragment isolation (Benitez-Malvido 1995), and no evidence for consistent differences after 6 years (Benitez-Malvido et al. 1999). Thus, while it is certain that changes in biotic interactions will occur in response to fragmentation, the direction and importance of such changes remain largely matters of conjecture. With regard to vertebrates, very small fragments are likely to lose herbivores. This effect could presumably act to favor species with relatively little investment in herbivore defense mechanisms. On the other hand, forest fragments may in some cases serve as refuges for terrestrial vertebrates that forage in matrix habitat. For example, wild pigs (*Sus scrofa*) were frequently observed in forest fragments near Pasoh during the present study. Evidence from both observational and exclosure studies suggests that nest-building and rooting activities of these pigs may have large effects on overall sapling survival, and may favor species that are better able to resprout or re-establish following destructive pig encounters (Ickes et al. 2003; Ickes and Thomas 2003).

Physiology

Sun versus Shade Characteristics

Increases in light and temperature, particularly near fragment edges, result in de-creased soil moisture levels and increased vapor pressure deficit in forest fragments (Kapos 1989; Kapos et al. 1993; Williams-Linera 1990b; Sizer and Tanner 1999). However, these effects also decrease dramatically during the first few years following fragment isolation as trees grow and recruit along fragment edges (Kapos et al. 1997). Recent evidence indicates that in small fragments, litter decomposition rates drop after 12–14 years of isolation (Didham 1997, 1998), and this may in turn reduce nutrient availability. As in other forms of disturbance, fragment isolation may commonly involve an initial pulse of resources, followed by a return to resource levels at or below those found prior to the disturbance event. Physiological characteristics that favor persistence in fragments may thus include traits that allow trees to take advantage of the initial resource flux, as well as traits that allow trees to endure "stressful" conditions associated with fragment edges, such as high temperatures, wind exposure, and low soil moisture. Over longer time periods, traits that enable survival and reproduction under low resource levels may be favored.

Initial impacts of fragment isolation are thus expected to favor tree species with the physiological capacity to take advantage of high light, including high photosynthetic capacity, high leaf N content, low leaf lifespan, and the suite of other traits associated with "sun plants" (Givnish 1988). However, if initial tree responses then result in reduced light (and other resources) for plants within fragment interiors, long-term effects of fragmentation may act to favor shade-tolerant species—at least in a "more of the same" scenario. Repeated disturbance, however, is expected to consistently favor light-adapted physiology. These patterns obviously have an important spatial component as well, with sun-adapted species favored near fragment edges (Kellman et al. 1998). Thus, fragmentation could in theory favor a variety of physiological traits. Another reason to suspect that fragmentation will not act simply to favor species with high-light-adapted traits is that of correlations between physiological and life history traits. For example, small-statured primary forest species, which may be favored in fragments due to sampling and demographic stochasticity effects, tend to exhibit shade-adapted photosynthetic physiology (Thomas and Bazzaz 1999).

Other Microenvironmental Effects

One microenvironmental effect that may be especially persistent in fragments is reduced relative humidity (Kapos 1989; Kapos et al. 1993; Sizer and Tanner 1999). It is likely that interior areas of older fragments are commonly exposed to an environment that is relatively dark and has a lower vapor pressure deficit (VPD) than found in primary forest. Such an environment might substantially alter optimal patterns of stomatal behavior, for example, favoring species that show little stomatal response to VPD (Kapos et al. 1997). Finally, another microenvironmental pattern that is likely to be persistently altered in fragments is that of surface water flow characteristics. Studies in temperate forest systems suggest that upstream disturbance can result in large changes in the flood hydrograph, favoring longer dry periods punctuated by more extreme flooding events (Swank et al. 1988). Such altered patterns of stream hydrology may substantially affect riparian vegetation even within large forest fragments (Kimble 2000).

Methods

Study Area

Pasoh Forest Reserve is located in the south-central portion of peninsular Malaysia (2°58'N, 102°17'E), and includes about 600 ha of primary forest and a 1400-ha buffer zone of forest selectively logged in 1956–59 (Kochummen et al. 1990; Okuda et al. 2003; Chapter 36). In 1970–71 the area immediately surrounding the forest to the north, west, and south sides of the reserve was cleared for oil palm (*Elaeis guinensis* [Palmae]) production. At this time, a large number of forest fragments were formed in areas of similar soil and topography to the reserve, within the watershed of the Pertang River. Surveyed fragments were located on shale-derived soils of the Durian and Batu Anam series, and included some areas of low-lying alluvial soils (Allbrook 1973). These forest fragments now occur within a matrix of land use primarily consisting of tree crops. Oil palm, rubber, and fruit tree plantations are the most common land-use types adjacent to fragments; however, some fragments abut open areas on at least one side, including roads, rice paddy fields, and graveyards.

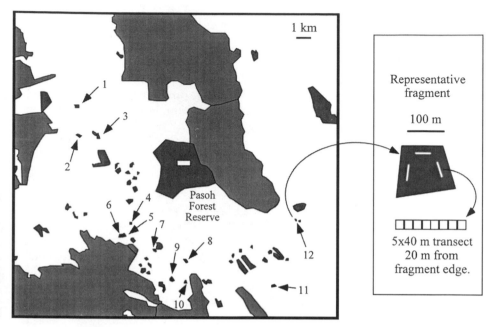

Fig. 18.2. Map of study area indicating location of Pasoh Forest Reserve, and of surveyed ~1-ha fragments isolated from the reserve around 1971. The rectangle within the reserve indicates the location of the CTFS 50-ha Forest Dynamics Plot. Forested areas, shown in gray, are derived from a 1991 Landsat TM image. White areas are primarily oil palm and rubber plantations, but they also include secondary forest, fruit tree plantations, rice paddies, villages, and roads.

From March to July 1995, transects were established within a set of 12 forest fragments within 8 km of Pasoh Forest Reserve (fig. 18.2). An extensive search for extant forest fragments was initially made in the study area. Fragments chosen for sampling were similar in size and shape, each measuring ~1 ha in extent and approximately square. The dimension of the fragments was chosen to match that of the smallest fragments in the Biological Dynamics of Forest Fragments Project outside of Manaus, Brazil, and to provide a maximum probability of fragmentation effects on the forest community. The fragments consisted of primary forest showing either no evidence of prior logging or of limited tree removal. Fragment perimeters were mapped and measured with the aid of a global positioning system (Magellan Nav 5000), compass, and tape measure. A systematic sampling scheme was used to sample fragments, placing transects at 20–25 m from fragment edges. Transects measured 5 × 40 m (enumerated within eight 5 × 5 m subplots), running parallel to the fragment edge (fig. 18.2). In 8 of the 12 fragments, all free-standing trees ≥1 cm diameter at breast height (dbh) were measured and identified in one transect per fragment, following measurement protocols used in the Forest Dynamics Plot survey (Manokaran et al. 1990). In 2–3 additional transects per fragment (for a total of 37 transects), species of five "focal" genera were censused (*Aporusa* [Euphorbiaceae], *Baccaurea* [Euphorbiaceae], *Diospyros* [Ebenaceae], *Garcinia* [Guttiferae], and *Ixora* [Rubiaceae]). Fragments were compared to the Pasoh 50-ha mapped forest plot on the basis of the 1985–87 initial survey data. Demographic turnover rates were calculated as the average of per capita mortality and recruitment over the interval 1987–96 (cf. Phillips et al. 1994).

Analysis

To test for differences in species abundance between fragments and primary forest, predicted values were generated on the basis of tree frequencies observed in primary forest, adjusted for differences in overall tree density (thus, $f = p$ $(A_f/A_p) (D_f/D_p)$, where f is predicted frequency in fragments, p is observed frequency in the 50-ha Forest Dynamics Plot, A_f is the area of the fragment sample, A_p is the area of the primary forest sample, D_f is the total tree density in fragments, and D_p is total tree density in primary forest). G-statistics were used to test for differences between observed and expected tree frequencies (Sokal and Rohlf 1981). For the purposes of these analyses, observed values were pooled over all fragment transects. Expected values were calculated using the 50-ha plot as a whole, and for the east and west halves of the plot. The latter tests, while they do not fundamentally correct for the lack of landscape-scale replication of primary forest samples (which is not possible), provide a qualitative assessment of the importance of coarse-scale spatial variation in species abundance on the results.

Comparative analyses aimed at elucidating correlates of species persistence in forest fragments employed simple general linear model analyses in which species-specific tree frequencies in fragments were the dependent variable, and independent variables included a variety of quantitative and qualitative descriptions for species-specific traits. For the whole community surveys, these variables are restricted to tree density in primary forest, dispersal mode (animal, wind, or ballistic seed dispersal), mating system (dioecious, monoeious, or hermaphrodite), successional status (pioneer versus primary forest, on the basis of previously published descriptions: Corner 1952; Whitmore 1972, 1973; Ng 1978, 1989), and asymptotic height. The latter variable was estimated on the basis of tree diameter distributions in the 50-ha plot, using an empirical relationship between the 97th percentile of tree diameter, and asymptotic tree height as derived from height–diameter relationships (Thomas 1993, 1996a). Analyses for both whole community and focal-species datasets excluded species that were completely absent from either fragment or primary forest samples.

More detailed information on species-specific traits was available for the five focal taxa, which included *Aporusa*, *Baccaurea*, *Diospyros*, *Garcinia*, and *Ixora*. Previous comparative studies of these taxa comprised analyses of growth rates, tree allometry, and wood density (Thomas 1996a), leaf size and shape (Thomas and Ickes 1995), size-dependent reproductive onset (Thomas 1996b), reproductive allometry (Thomas 1996c), sex expression (Thomas and LaFrankie 1993; Thomas 1997), reproductive phenology (Thomas 1993), and photosynthetic physiology (Thomas and Bazzaz 1999). These prior publications give complete methodological and sampling details related to the traits examined as well as the taxonomic data and voucher specimen citations.

Results

Whole Community Surveys

Forest fragments displayed structural characteristics similar to primary forest. On the basis of the whole community survey data, average tree densities of trees ≥ 1 cm dbh were $0.61 \pm 0.08/\text{m}^2$ in fragments, versus $0.64 \pm 0.02/\text{m}^2$ in primary forest. Total basal area was also similar, being 29.6 ± 7.2 m^2/ha in fragments, versus 29.0 ± 2.2 m^2/ha in primary forest. Detailed analyses of fragment versus primary forest structure and species diversity will be presented elsewhere (Thomas, in preparation).

Tree densities in fragments showed a positive correlation ($r = 0.375$; $P < 0.001$) with densities in primary forest (fig. 18.3). Three groups of species stand out from this overall pattern: early-successional pioneer species, ballistically dispersed understory trees, and dipterocarps. Pioneer species are quite uncommon within the Pasoh Forest Dynamics Plot (Kochummen et al. 1990), and even under a very broad definition make up only 3.6% of stems in the 50 ha plot (Davies et al. 2003). Pioneers (under a relatively narrow definition: Thomas 2003) were encountered more frequently than expected in fragments (5.6% of trees, versus 1.4% in primary forest). However, the most common of these species (*Pternandra echinata* [Melastomataceae], *Endospermum malaccense* [Euphorbiaceae], *Mussaendopsis beccariana* [Rubiaceae], and *Porterandia anisophylla* [Rubiaceae]) each made up less than 0.6% of the trees in the overall sample. This set also does not include those species most frequently encountered in roadside secondary forest in the area, such as *Arthrophyllum diversifolium* (Araliaceae), *Fagraea racemosa* (Gentianaceae), *Ficus fistulosa* (Moraceae), *F. glossularioides*, *Macaranga gigantea* (Euphorbiaceae), *M. hypoleuca*, *M. triloba*, *Mallotus paniculatus* (Euphorbiaceae), *Trema tomentosa* (Ulmaceae), and *Vitex pinnata* (Labiatae). Rather, the frequently encountered pioneers in fragments are species characteristically occurring in primary forest and might be better classified as intermediate in successional status.

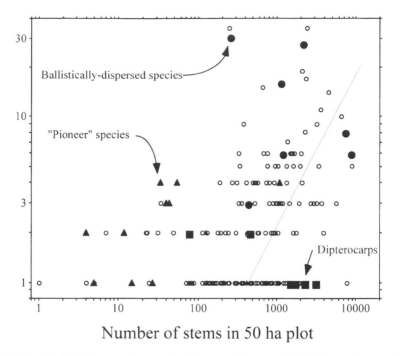

Number of stems in 50 ha plot

Fig. 18.3. Plot of total tree count per species (for complete survey data) in fragments versus tree count in the 50-ha Forest Dynamics Plot at Pasoh Forest Reserve (1987 census data). The line indicates expected frequencies, if tree densities were identical between the two samples. Late-successional animal-dispersed species (open circles) include the vast majority of individuals in both samples, and show a relatively strong correlation between tree densities in fragments versus tree densities in primary forest. Early-successional pioneer species (closed diamonds) show higher than expected frequencies in fragments, but remain relatively uncommon in comparison to other species. Ballistically dispersed species (closed circles) include some of the more common species in primary forest, but are also disproportionately represented in forest fragments. Dipterocarps (closed squares), which represent the only wind-dispersed late-successional species, are greatly underrepresented in fragments.

Ballistically dispersed species were also disproportionately represented in fragments and were numerically much better represented than the pioneer species (16.1% of total trees, versus 12.3% in primary forest). The two most abundant of these species, *Rinorea schlerocarpa* (Violaceae) and *Macaranga lowii*, made up 3.3% and 3.0% of the pooled fragment sample, respectively. (Note that *M. lowii*, although a congener of the secondary forest *Macaranga* species, is morphologically quite distinct and characteristic of primary forest: Whitmore 1972.) Dipterocarp species, which share the uncommon characteristics of wind-dispersal and large adult size, were present at exceedingly low frequencies in fragments (fig. 18.3). Results of the general linear model analysis for the whole community data indicated

Table 18.2. Results of General Linear Model Analysis Examining Correlates of Tree Density in Forest Fragments for the Whole Community Survey Data

Variable	df	SS	MS	F ratio	P
Log Density	1	3.147	3.147	27.558	≤0.0001
Asymptotic Height	1	0.250	0.250	2.192	0.1407
Successional Status	1	0.389	0.389	3.402	0.0669
Dispersal Mode	2	0.949	0.475	4.157	0.0173
Mating System	2	0.156	0.052	0.456	0.7132
Error	165	18.843	0.114		
Total	173	24.582			

Notes: Log species density in forest fragments (pooled tree count) is the dependent variable. Independent variables are log density in primary forest, asymptotic height, successional status (pioneer vs. primary forest species), dispersal mode (animal, wind, ballistic), and mating system (dioecious, monoecious, hermaphroditic).

significant effects of tree density in primary forest and of dispersal mode on tree density in forest fragments (table 18.2). A posteriori contrasts among dispersal mode classes indicated that species characterized by ballistic dispersal were present at higher than expected densities in fragments ($P = 0.005$, $t = 2.885$), whereas wind-dispersed species (mostly dipterocarps) were present at lower than expected densities ($P = 0.021$, $t = 2.332$). Estimated asymptotic height had a marginally significant effect in the analysis ($P = 0.141$), with the trend favoring smaller-statured species in fragments as compared to primary forest.

If average densities in fragments were simply proportional to primary forest densities, the log-transformed relationship should show a slope of 1.0, while a slope >1.0 would indicate that common species are disproportionately represented in fragments. The observed slope of this relationship is significantly <1.0 (reduced major axis slope $= 0.266$; 95% C.I. $= [0.169, 0.368]$; least-squares regression slope $= 0.203 \pm 0.038$). Omitting pioneer species from the sample results in a slope that is slightly steeper but significantly <1.0.

Focal Taxa Surveys

A large proportion of species in the five focal taxa found in the Pasoh Forest Dynamics Plot were also recorded in small forest fragments (table 18.3). Of the 14 species of *Aporusa* identified from the primary forest plot, 12 occurred in fragments, as did one additional species not found in the plot (*A. benthamiana*). Comparable figures for the other genera were 8 of 11 with 1 new species for *Baccaurea* (*B. brevipes*), 19 of 22, with 3 new species for *Diospyros* (*D.* cf. *euphlebia, D. rigida,* and *D. subrhomboidea*), 12 of 19 with 1 new species for *Garcinia* (*G. dumosa*), and 7 of 8 with 1 new species for *Ixora* (*I. javanica*). The total count for these five genera was thus 76 species for the 50-ha primary forest sample, versus 65 species within the set of isolated fragments near the reserve (57 of those species being encountered in systematic surveys within a total sample area of 0.74 ha).

Table 18.3. Frequency of Focal Taxa Species in Small Forest Fragments Compared to Expected Frequencies Based on 50-ha Forest Dynamics Plot Data from Pasoh Forest Reserve

Species	Observed	Expected	G	P	Dir.
Aporusa (Euphorbiaceae)					
A. aurea	17	30.93	20.34	<0.0001[‡]	−
A. benthamiana	1	0.00	0.00		+
A. confusa*	0	0.08	5.01		−
A. falcifera	2	14.05	7.80	0.0251[‡]	−
A. globifera	12	16.94	8.28	0.0052	−
A. lucida (*)	0	9.40	4.48	0.0040[‡]	−
A. lunata (*)	0	0.83	0.38		−
A. microstachya	41	100.74	73.72	<0.0001[‡]	−
A. nervosa*	0	4.23	2.88	0.0896	−
A. nigricans	3	13.56	9.05	0.0026	−
A. prainiana	15	25.23	15.60	0.0001[‡]	−
A."sessile-flowered" sp nov	11	4.97	17.46	0.0000[‡]	+
A. subcaudata	2	36.73	11.64	0.0006[‡]	−
A."swamp" sp nov*	0	[†]			
A. symplocoides	1	4.92	3.19	0.0743	
A. spp	3				
Baccaurea (Euphorbiaceae)					
B. brevipes**	25	0.00	160.94	<0.0001[‡]	+
B. griffithii	1	1.51	0.82		−
B. kunstleri*	0	0.11	4.44		−
B. maingayi*	0	0.45	1.60		−
B. minor	7	1.01	27.17	<0.0001[‡]	+
B. parviflora	30	51.21	32.09	<0.0001[‡]	−
B. pyriformis	0	1.07	0.14		−
B. racemosa	8	18.63	13.52	0.0002[‡]	−
B. ramiflora	2	0.19	9.41		+
B. reticulata	1	5.91	3.55	0.0594	−
B. sp 1*	0	0.04	6.40		−
B. sumatrana*	0	1.98	1.37		−
Diospyros (Ebenaceae)					
D. adenophora	4	2.26	4.58		+
D. andamanica	18	5.91	40.09	<0.0001[‡]	+
D. apiculata	10	22.77	16.46	<0.0001[‡]	−
D. areolata	1	1.97	1.36		−
D. argentea	1	6.86	3.85	0.0497	−
D."brown-barked" sp nov*	0	9.91	4.59	0.0322	−
D. buxifolia	2	8.48	5.78	0.0162	−
D. cauliflora	14	4.66	30.80	<0.0001[‡]	+
D. daemona	2	0.03	17.19		+
D. diepenhorstii	2	0.46	5.86		+
D. cf euphlebia	1	0.00	0.00		+
D. latisepala	4	10.25	7.52	0.0061[‡]	−
D. maingayi (*)	0	1.39	0.65		−
D. nutans	27	34.64	13.45	0.0002[‡]	−
D. penangiana	3	0.16	17.47		+
D. pyrrhocarpa*	0	0.77	0.51		−
D. rigida**	1	0.00	0.00		+
D. rufa*	0	1.89	1.27		−
D. scortechinii	3	48.45	16.69	<0.0001[†]	−

(Continued)

Table 18.3. (Continued)

Species	Observed	Expected	G	P	Dir.
D. singaporensis	1	9.99	4.60	0.0319	−
D. sp 1(*)	0	3.19	2.32		−
D. subrhomboidea **	11	0.00	52.75	<0.0001‡	+
D. sumatrana	22	11.73	27.69	<0.0001‡	+
D. venosa	11	21.52	14.77	0.0001‡	−
D. wallichii	6	1.30	18.31	<0.0001‡	+
D. sp	1				
Garcinia (Guttiferae)					
G. atroviridis	1	0.19	3.32		+
G. bancana	1	1.89	1.27		−
G. "dark" sp nov*	0	1.32	0.55		−
G. dumosa (**)	0	†			
G. forbesii	3	0.92	7.07		+
G. griffithii	9	0.56	50.08	<0.0001‡	+
G. malaccensis	1	13.95	5.27	0.0217‡	−
G. nervosa*	0	8.14	4.19	0.0406	−
G. nigrolineata	2	7.04	5.03	0.0249	−
G. opaca (*)	0	†			
G. parvifolia	6	8.11	3.62	0.0571	−
G. prainiana*	0	0.01	8.60		−
G. pyrifera*	0	0.01	8.60		−
G. rostrata	1	1.56	0.89		−
G. scortechinii	2	5.27	3.88	0.0489	−
G. "small" sp nov (*)	0	11.58	4.90	0.0269‡	−
G. sp 2*	0	1.35	0.59		−
G. sp 4*	0	0.05	5.82		−
G. sp 5*	0	0.01	8.60		−
G. spp	10				
Ixora (Rubiaceae)					
I. concinna	1	14.47	5.34	0.0208‡	−
I. congesta	4	14.39	10.24	0.0014‡	−
I. grandifolia	6	3.42	6.73	0.0095‡	+
I. javanica (**)	0	††			
I. kingstonii*	0	1.36	0.61		−
I. "lanceolate-leaved" sp nov. (*)	0	4.16	2.85	0.0914	−
I. lobbii	2	14.91	8.03	0.0046‡	−
I. nigricans	1	0.38	1.93		+
I. pendula	1	2.50	1.83		−
Total	378	670.09	432.83	<0.0001‡	−

Notes: Counts are for trees ≥1 cm diameter, pooled across surveys of 375 × 40 m transects (totaling 0.74 ha) located in 12 ∼ 1-ha forest fragments. Probability levels for G-test results are given where either observed or expected frequencies are >5. Directions of predicted effects are indicated as positive (+), negative (−), or no effect or no basis for prediction (0).

*Species found in primary forest at Pasoh Forest Reserve, but not observed in forest fragment surveys.

(*) Species found in primary forest at Pasoh Forest Reserve and, while not observed in systematic fragment surveys, were encountered in incidental observations.

**Species found in fragment surveys, but not observed within the 50-ha Pasoh Forest Dynamics Plot.

(**) Species observed in incidental fragment observations (but not in surveys), and not observed within the 50-ha Forest Dynamics Plot.

† Species not distinguished in 50-ha Forest Dynamics Plot survey; expected values for primary forest are not available.

†† Small shrub species not generally attaining 1 cm dbh.

‡ Taxa for which G-tests were significant (P < 0.05) using expected values based on both east and west halves of the 50-ha plot.

These species numbers do not include individuals for which identifications could not be made. In most genera >97% of specimens were identified to species; however, in *Garcinia* 10 of 38 recorded individuals (26%), comprising five or six morphospecies, did not provide good matches to voucher specimens from the 50-ha plot or to other named specimens housed at the herbarium at the Forest Research Institute of Malaysia (KEP). Two other unmatched morphospecies were noted in *Diospyros* and *Aporusa*. Inclusion of these taxa would bring the total species number for fragments to essentially the same value as that found for the 50-ha Forest Dynamics Plot as a whole.

The majority of those species not recorded from fragments were also relatively rare within the 50-ha Forest Dynamics Plot, such as the five unnamed morphospecies of *Garcinia* enumerated during the initial 50-ha Forest Dynamics Plot census (Manokaran et al. 1990). However, some species common within the primary forest plot were absent from the forest fragments. Two notable examples were *Garcinia nervosa* and *Diospyros* "brown-barked" sp nov, (called *D. pendula* in some previous publications: Manokaran et al. 1990). Both are highly distinctive species common in periodically inundated areas of the 50-ha Forest Dynamics Plot. Conversely, a number of species that were completely absent from the 50-ha Forest Dynamics Plot were abundant in forest fragments. Dramatic examples included *Baccaurea brevipes* and *Diospyros subrhomboidea*, both of which were locally abundant in fragments. The latter species has not previously been recorded from the state of Negeri Sembilan (Ng 1978).

Quantitatively, the majority of species in the focal taxa showed lower tree densities in fragments than in primary forest (table 18.3). Of 76 species examined, 55 showed a trend of lower tree densities in fragments, and this pattern was significant in 25 cases. Such quantitative differences were often very large; e.g., *Diospyros scortechinii*, the fifteenth most abundant species in the 50-ha plot, occurred at only 6% of its expected tree density in fragments. However, tree densities were significantly higher than expected in fragments for 10 species, including representatives of each of the five genera. The analysis presented adjusts for the (small) difference in overall tree density between fragments and primary forest, so observed patterns cannot be accounted for by a difference in overall tree mortality or recruitment. The results of the species-specific analyses are consistent with the observed fragment versus primary forest density relationship. The most common primary forest species showed reduced population density in fragments, while a subset of relatively rare species showed substantial increases in population density.

Comparative analyses for the focal taxa indicated significant correlations between tree density in fragments and tree density in primary forest, tree height at onset of maturity, and photosynthetic capacity (table 18.4). Asymptotic height, relative size at onset of maturity, and specific leaf area were also marginally significant in these analyses ($P < 0.10$). The strongest observed pattern was the

Table 18.4. Observed Correlations between Various Aspects of Tree Life History, Morphology, and Physiology versus Density in Forest Fragments

Variable	N	Fragment Stem Density r	Residuals of Fragment Density–Plot Density Relationship r
Primary forest tree density[1]	47	0.372*	
Agamospermy (+/−)[2]	52	−0.109	−0.166
Clonality (+/−)[1]	52	0.107	0.192
Dioecy (+/−)[1]	52	0.131	0.178
Habitat specialist (+/−)[1]	52	0.153	0.219
Asymptotic tree height[3]	27	−0.307	−0.288
Relative size at onset of maturity[4]	27	−0.338	−0.270
Height at onset of maturity[4]	27	−0.409*	−0.332
Demographic turnover rate[1]	47	−0.085	−0.012
Wood density[3]	26	−0.099	−0.108
Average growth rate (1–2 cm size class)[1]	46	0.045	0.087
Leaf size[5]	26	0.066	0.186
Specific leaf area[6]	24	0.375	0.370
Photosynthetic capacity (area basis)[6]	24	−0.599**	−0.550**
Photosynthetic capacity (mass basis)[6]	23	−0.274	−0.256

Notes: Data are for the tree densities in fragments given in Table 18.1. Values for Pearson Product-moment correlations are given for continuous variables; equivalent values for qualitative (+/−) variables are calculated as the square root of the coefficient of determination for a one-way ANOVA for that variable, with the sign indicating direction of the effect (e.g., a positive main effect of agamospermy on fragment tree density would be positive). Sources of data for species-specific determinations of other variables are as indicated. Significance based on regression or ANOVA F-tests: $^*P < 0.05$; $^{**}P < 0.01$.

Sources of data:

[1] Calculated from Pasoh Forest 50-ha Forest Dynamics Plot dataset (1996 recensus data), and/or unpublished field observations.

[2] Species with known agamospermous reproduction: Kaur et al. 1978, Thomas 1997.

[3] Thomas 1996a.

[4] Thomas 1996b.

[5] Average of adult and sapling leaf size: Thomas and Ickes 1995.

[6] Thomas and Bazzaz 1999.

negative correlation between photosynthetic capacity and tree density in fragments (fig. 18.4). Such a relationship could potentially be driven by differences in overall tree densities, if photosynthetic capacity was negatively correlated with density in primary forest. However, there was also a significant negative correlation between the residuals of the fragment density–primary forest density relationship and photosynthetic capacity. This pattern strongly suggests that, at least among these primary forest taxa, shade-tolerant species are differentially favored in fragments.

Discussion

The results presented support several general conclusions. First, a large proportion of species recorded in the Pasoh Forest Dynamics Plot occurred in fragments.

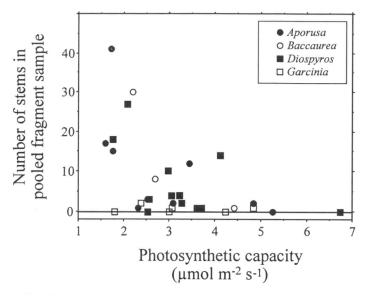

Fig. 18.4. Plot of total tree count per species in fragments versus photosynthetic capacity (leaf area basis) for focal taxa survey data.

While the majority of primary forest species showed reduced abundance in fragments, some species that were rare or absent from the 50-ha plot were common in fragments. Second, it was not simply the case that species associated with high light levels or forest disturbance were favored in fragments. Rather, a range of traits, including relative density in primary forest, dispersal mode, and low photosynthetic capacity, were of importance as correlates of persistence in fragments. Third, many, though not all, of the observed patterns were consistent with the hypothesis that demographic stochasticity is of primary importance as a mechanism that favors certain tree species over others in fragments in this system.

Study Context and Limitations

Before discussing these results at length, it is important to recognize limitations on inferences that can be drawn from this study. Of particular concern are the issues of anthropogenic impacts on forest fragments, and inherent limitations of a sampling design that includes only one replicate of primary forest.

As in any study of anthropogenic disturbance effects, it is critical to place this work in the local human context. Unlike the Biological Dynamics of Forest Fragments Project, the fragments in this study were not isolated and maintained experimentally. Rather, they represent "naturally formed" forest patches subject to human impacts before, during, and after isolation. Historically, most of the fragments in this study were located in areas between habitations and forest cleared for oil palm production by the Malaysian Federal Land Development Agency (FELDA) in 1970–71. For the most part, these fragments were then (and are now) owned collectively by local villages under Malaysian traditional law (for anthropological descriptions see Peletz 1988; Landeen 1992). As noted above (see Methods), the fragments were situated on soils and topography similar to the primary forest reserve at Pasoh; fragment locations appear to be unbiased relative to local topography, occurring on level terrain, parts of small hills, and in low-lying areas on alluvial soils. In terms of anthropogenic disturbance, sampled fragments showed minimal signs of prior felling of large trees, although there was evidence that small trees had been cut for poles in several fragments. Roots of a number of species are in local demand for use in traditional medicines, and saplings are generally uprooted when collected for this purpose (Landeen 1992). Other nontimber forest products harvested in the area include rattan vines (mainly *Calamus* spp. [Palmae]) and a variety of forest fruits (Saw et al. 1988). However, both direct observations and interviews with local residents suggested that the fragments were not heavily utilized at the time of this study, larger areas of forest being preferred for collection of forest products. Although it is likely that continuing human impacts have had some effect on species composition in the fragments, this fact is not necessarily a disadvantage. For many purposes, especially extrapolation of these results to larger areas, it is preferable to obtain a sample representative of the broader landscape in terms of both biological processes and human impacts.

Another important issue is sampling design. Tropical tree species show distributions that are almost always spatially patchy (e.g., He et al. 1997; Condit 2000). Thus, to provide strong inference that differences in species composition between fragments and primary forest are a result of fragmentation per se (either via isolation or edge effects), it would be necessary to sample spatially interspersed sets of both primary forest areas and fragments (Hurlbert 1984). In the present study only one large block of primary forest was sampled simply because only one large block of lowland primary forest was available (fig. 18.2). Other large continuous areas of forest within a ~40-km radius of Pasoh are on hillslopes or heavily impacted by logging; beyond this distance biogeographic differences in forest community types would complicate interpretation of results. For this reason, the statistical tests presented, particularly results for individual species (i.e., table 18.3), should be viewed as a guide to interpreting patterns, keeping in mind that both fragmentation effects and spatial patchiness may influence results.

Traits Favored in Forest Fragments
Of the traits examined, the strongest correlates of tree species persistence in fragments were dispersal mode, tree size at reproductive onset, population density

in primary forest, and photosynthetic capacity; there was also some indication that both early successional and small-statured species were favored. How do these patterns correspond to predictions from the "things fall apart" versus the "more of the same" model? Both models predict that species with ballistic seed dispersal should be favored, either as a result of being unaffected by losses of animal seed dispersers, through sampling effects that favor spatially clumped species (fig. 18.1), or by minimizing "seed wastage." Both models also predict that species showing early reproduction on a size or age basis should be favored. It is therefore not surprising that these two traits should emerge as strong predictors of species persistence in fragments. There was also a relatively strong correlation between species abundance in fragments versus primary forest, an observation consistent with the "more of the same" view. However, there was no evidence that very common primary forest species were disproportionately favored in fragments.

Some trees considered pioneer species showed substantial increases of abundance in fragments. However, considering all species together, successional status was only marginally significant as a correlate of species abundance in fragments (table 18.2). Moreover, the pioneer species favored were not those species characteristically found in roadside secondary forest (exemplified by *Macaranga gigantea*), but rather species of intermediate shade tolerance, such as *Pternandra echinata*, *Endospermum malaccense*, and *Porterandia anisophylla*. These observations thus seem to be a marked contrast to results from Amazonian fragments, where the most characteristic secondary forest species in the area, *Cecropia sciadophylla*, showed a 33-fold increase in fragments (unpublished data cited in Laurance et al. 2001). In addition, among the intensively studied focal taxa, there was a strong negative correlation between photosynthetic capacity and tree density in fragments (fig. 18.4). This pattern is contrary to predictions of the "things fall apart" model, which would lead one to expect an increase in abundance of species physiologically adapted to high-light environments. One potential explanation is that tree growth and canopy closure at the fragment edges may have acted to reduce light levels within fragments over the relevant time scale ($\sim$25 years). Although direct light measurements were not made, censuses of gap frequencies suggest that gaps were slightly less frequent in fragment transects than in primary forest at Pasoh (Thomas, unpublished data). A second possible explanation for higher fragment tree densities in shade-adapted species may be a response to altered patterns of herbivory. It has been speculated that insect herbivores, such as leaf-cutter ants, commonly have increased effects on plant communities in isolated forest fragments (Leigh et al. 1993; Terborgh et al. 2001), and tree species with low photosynthetic rates generally have higher investment in anti-herbivore defenses (Coley and Barone 1996). However, other studies have not detected increased levels of herbivore damage in fragments compared to primary forest (Benitez-Malvido 1998; Benitez-Malvido et al. 1999).

The disproportionate occurrence of ballistically dispersed species in small forest fragments has not been previously reported in the literature. This set of species is a very small proportion of the overall community at Pasoh, which comprises roughly 30 species, mostly in the Euphorbiaceae (*Cleistanthus, Croton, Epiprinus, Koilodepas, Mallotus, Trigonostemon*), but also including members of the Annonaceae (*Anaxagorea*) and Violaceae (*Rinorea*) (see Davies et al. 2003). As noted earlier, ballistically dispersed species have a number of characteristics that may be favored in forest fragments: namely, lack of dependence on animals for dispersal, high local population density, clumped spatial distribution in primary forest, small size, and short dispersal distance. Anecdotally, the predominance of ballistically dispersed Euphorbs has also been noted in 50–100 year-old fragments in the Atlantic coastal forest area of Brazil (Efriam Rodrigues, Departamento de Agronomia Universidade Estadual de Londrina, Brazil, personal communication). Although habitat specialization was not quantified, many of the species showing higher than expected tree densities in fragments were associated with riparian areas (e.g., *Diospyros andamanica, D. cauliflora*, and *Ixora grandifolia*), or hilltops (*Aporusa* "sessile-flowered" sp nov).

Spatially aggregated species may also have been favored in small fragments by deterministic processes. Species that display highly aggregated populations in primary forest may have evolved means of resisting "natural enemies" (Janzen 1970). Moreover, persistence over long time intervals in forest fragments is likely to involve the maintenance of higher local population densities than those found in primary forest. Species that are highly susceptible to species-specific pathogens or herbivores may not be able to maintain sufficiently high population densities to persist in small forest fragments and thus may be eliminated deterministically. There is some evidence for pronounced "distance-dependent" mortality effects in southeast Asian tree species (Chan 1980; Becker and Wong 1985; Okuda et al. 1995; Peters 2003). It is thus tempting to speculate that the virtual absence of dipterocarps from the censused fragments results from this process. This possibility is consistent with observations on a forest fragment in Singapore, where, for example, *Shorea macroptera* (Dipterocarpaceae) was common as a canopy tree but showed no recruitment in small size classes (Turner et al. 1996).

General Implications

In many areas of the tropics, native forests are now represented mainly or entirely by relatively small fragments. It is axiomatic that the conservation value of a large reserve will be higher than that of an otherwise similar small reserve. Environmental organizations, taking their cue from research focusing on birds and mammals, have commonly assumed that small fragments have little or no conservation value to any organism. A few "fragment advocates" have recently challenged this assumption (Turner and Corlett 1996; Renjifo 1999), but data on tree diversity in isolated fragments have been limited to a few case studies, such as the 4-ha Singapore Botanic Gardens' Jungle Reserve (Turner et al. 1996). There is thus a pressing need for more empirical information, particularly on fragments that are more representative of the broader landscape and patterns of human impact. A better understanding of the comparative biology of species represented in fragments is also critical. A fragment dominated by common secondary forest species is clearly of much less conservation interest than a fragment that includes uncommon primary forest species not found in areas where forest has been completely cleared. Moreover, if simple demographic stochasticity is the main mechanism for species loss in fragments, relatively simple models may be constructed to predict fragmentation effects across spatial and temporal scales.

The results of the present study provide a surprisingly hopeful picture for forest fragments in peninsular Malaysia. That the vast majority of primary forest species in a relatively large reserve could also be found in a set of ~1-ha fragments 25 years after isolation clearly indicates a high conservation potential for these fragments. One may even point to examples of the kinds of species that may be protected in a series of widely spaced, small fragments but missed in a single large reserve—species such as *Diospyros subrhomboidea*, found to be locally abundant in a pair of surveyed fragments but not recorded in the Pasoh reserve. This species, a warty-leaved understory treelet in the persimmon genus, may not qualify as an example of "charismatic megaflora"; however, it is certainly representative of the megadiverse plant genera that are characteristic of the Sunda Shelf region (Ashton 1969; Van Steenis 1969; Turner 1997).

Why have studies in the neotropics and southeast Asia yielded such different pictures of fragment biology? Two factors that are almost certainly important are differences in land management and regional disturbance regimes. Windbreaks in matrix habitat have been recommended as a management technique to minimize fragment disturbance (Laurance 1997). In peninsular Malaysia, the single most important land use in areas previously supporting lowland rainforest is fast-growing tree crops; thus wind breaks are effectively created as a matter of course. From a climatological perspective, wind disturbance is predicted to be most important in the 7–20° latitude band, within areas affected by cyclonic, monsoonal, and hurricane patterns. For example, Gatun Lake islands in Panama are obviously and dramatically affected by trade winds, which often show sustained velocities of 5–15 km/hr at canopy height (Leigh et al. 1993). Central Amazonia, while lying near the equator, is nevertheless also strongly affected by prevailing winds (Lovejoy et al. 1986; Laurance et al. 1997). Recent climatological research suggests that the large expanse of the Amazon River, particularly near the confluence of the Rio Negro and the Rio Solomões at Manaus, may be critical in generating temperature gradients that result in the "Amazon River breeze," which commonly exceeds 15 km/hr (Greco et al. 1992; Oliveira and Fitzjarrald 1993). In contrast, the maximum recorded wind speed in a clearing at Pasoh Forest Reserve over a 2-year period was <1 km/hr (Sulaiman et al. 1994), and maximum top-of-canopy values are <5 km/hr (Aoki et al. 1978; Tani et al. 2003). More generally, the interior portion of the Sunda Shelf region of southeast Asia lies within the earth's largest expanse of oceanic "doldrums," and is subject neither to trade winds nor typhoons, although infrequent wind storms in the form of local line squalls do occur (Watts 1954; Whitmore 1984).

Thus, there is reason to believe that differences in climate and land management patterns may result in different patterns of fragment ecology among biogeographic regions. If so, then conservation strategies involving fragments should be tailored to local circumstances. Results of the present study point to the importance of further work on fragment ecology across a range of forest types and biogeographic regions. It is also essential to develop an understanding of the ecological characteristics of species that persist in fragments; therefore, comparative studies of tree biology should be an integral part of future research on fragment dynamics. The CTFS network of Forest Dynamics Plots could provide an invaluable starting point for undertaking this task in representative areas of forest throughout the tropics.

Acknowledgments

The 50-ha Forest Dynamics Plot project at Pasoh Forest Reserve is an ongoing effort of the Malaysian Government, administered and conducted by the Forest Research Institute Malaysia. I thank K. Ickes for expert field and botanical assistance; Dato' Dr. Hj. Salleh Mohd. Nor, S. Appanah, and P. S. Ashton for administrative facilitation; K. M. Kochummen, and J. V. LaFrankie for input on plant identifications; and the inhabitants of Simpang Pertang, Bayai, Ulu Serting, and vicinity for providing access to local forest fragments. The final manuscript was improved by helpful reviews from P. Baker, C. Halpern, E. Leigh, J. Makana, and J. Malcolm. This project was supported by an NSF postdoctoral fellowship grant (BSR-91-01118).

References

Allbrook, R. F. 1973. The soils of Pasoh Forest Reserve, Negeri Sembilan. *Malaysian Forester* 36:22–33.
Andren, H. 1994. Effects of habitat fragmentation on birds and mammals in landscapes with different proportions of suitable habitat. *Oikos* 71:355–79.
Aoki, M. K. Yabuki, and H. Koyama. 1978. Micrometeorology of Pasoh forest. *Malaysian Nature Journal* 30:149–60.

Ashton, P. S. 1969. Speciation among tropical forest trees: Some deductions in the light of recent evidence. *Biological Journal of the Linnaean Society* 1:155–96.

Ashton, P. S., and P. Hall. 1992. Comparisons of structure among mixed dipterocarp forests of north-western Borneo. *Journal of Ecology* 80:459–81.

Bawa, K. S., D. R. Perry, and J. H. Beach. 1985. Reproductive biology of tropical lowland rain forest trees. I. Sexual systems and incompatibility mechanisms. *American Journal of Botany* 72:331–45.

Bazzaz, F. A., and S. T. A. Pickett. 1980. Physiological ecology of tropical succession: A comparative review. *Annual Review of Ecology and Systematics* 11:287–310.

Becker, P., and M. Wong. 1985. Seed dispersal, seed predation, and juvenile mortality of *Aglaia sp.* (Meliaceae) in lowland dipterocarp rain forest. *Biotropica* 17:230–37.

Becker, P., J. S. Moure, and F. J. A. Peralta. 1991. More about euglossine bees in Amazonian forest fragments. *Biotropica* 23:586–91.

Benitez-Malvido, J. 1998. Impact of forest fragmentation on seedling abundance in a tropical rain forest. *Conservation Biology* 12:380–89.

Benitez-Malvido, J., G. Garcia-Guzmán, and I. D. Dossmann-Ferraz. 1999. Leaf-fungal incidence and herbivory on tree seedlings in tropical rainforest fragments: An experimental study. *Biological Conservation* 91:143–50.

Bierregaard, R. O., and V. H. Dale. 1996. Islands in an ever-changing sea: The ecological and socioeconomic dynamics of Amazonian rainforest fragments. Pages 187–204 in J. Schelhas, and G. Greenbery, editors. *Forest Patches in Tropical Landscapes.* Island Press, Washington, DC.

Bierregaard, R., and T. E. Lovejoy. 1988. Birds in Amazonian forest fragments: Effects of insularization. *Acta XIX Congress of International Ornithology* 2:1564–79.

———. 1989. Effects of forest fragmentation on Amazonian understory bird communties. *Acta Amazonica* 19:215–41.

Bierregaard, R. O., T. E. Lovejoy, V. Kapos, A. A. dos Santos, and R. W. Hutchings. 1992. The biological dynamics of tropical rainforest fragments. *Bioscience* 42:859–66.

Brokaw, N. V. L. 1982. The definition of treefall gap and its effect on measures of forest dynamics. *Biotropica* 11:158–60.

———. 1985. Treefalls, regrowth, and community structure in tropical forests. Pages 53–69 in S. T. A. Pickett and P. S. White, editors. *The Ecology of Natural Disturbance and Patch Dynamics.* Academic Press, New York.

Brown, K. S., and R. W. Hutchings. 1997. Disturbance, fragmentation, and the dynamics of diversity in Amazonian forest butterflies. Pages 91–110 in W. F. Laurance and R. O. Bierregaard, editors. *Tropical Forest Remnants: Ecology, Management, and Conservation of Fragmented Communities.* University of Chicago Press, Chicago.

Burkey, T. V. 1989. Extinction in nature reserves: the effect of fragmentation and the importance of migration between reserve fragments. *Oikos* 55:75–81.

Camargo, J. L., and V. Kapos. 1995. Complex edge effects on soil moisture and microclimate in central Amazonian forest. *Journal of Tropical Ecology* 11:205–21.

Chan, H. T. 1980. Dipterocarps II. Fruiting biology and seedling studies. *Malaysian Forester* 43:438–51.

Christiansen, M. B., and E. Pitter. 1997. Species loss in a forest bird community near Lagoa Santa in southeastern Brazil. *Biological Conservation* 80:23–32.

Clark, D. B., and D. A. Clark. 1990. Distribution and effects on tree growth of lianes and woody hemiepiphytes in a Costa Rican tropical wet forest. *Journal of Tropical Ecology* 6:321–31.

Coley, P. D., and J. A. Barone. 1996. Herbivory and plant defenses in tropical forests. *Annual Review of Ecology and Systematics* 27:305–35.

Condit, R., P. S. Ashton, P. Baker, S. Bunyavejchewin, S. Gunatilleke, N. Gunatilleke, S. P. Hubbell, R. B. Foster, L. Hua Seng, A. Itoh, J. V. LaFrankie, E. Losos, N. Manokaran, R. Sukumar, and T. Yamakura. 2000. Spatial patterns in the distribution of common and rare tropical tree species: A test from large plots in six different forests. *Science* 288:1414–18.

Corlett, R. T., and I. M. Turner. 1997. Long-term survival in tropical forest remnants in Singapore and Hong Kong. Pages 333–45 in W. F. Laurance and R. O. Bierregaard, editors. *Tropical Forest Remnants: Ecology, Management, and Conservation of Fragmented Communities.* University of Chicago Press, Chicago.

Corner, E. J. H. 1952. *Wayside Trees of Malaya.* Government Printing Office, Singapore.

Davies, S. J., P. A. Palmiotto, P. S. Ashton, H. S. Lee, and J. V. LaFrankie. 1998. Comparative ecology of 11 sympatric species of *Macaranga* in Borneo: Tree distribution in relation to horizontal and vertical resource heterogeneity. *Journal of Ecology* 86:662–73.

Davies, S. J., Nur Supardi, N. M., LaFrankie, J. V., and P. S. Ashton. 2003. The trees of Pasoh Forest: Stand structure and floristic composition of the 50-ha forest research plot. Pages 35–50 in T. Okuda, N. Manokaran, Y. Matsumoto, K. Niiyama, S. C. Thomas and P. S. Ashton, editors. *Pasoh: Ecology of a lowland rain forest in southeast Asia.* Springer-Verlag, Tokyo.

Didham, R. K. 1997. The influence of edge effects and forest fragmentation on leaf litter invertebrates in Central Amazonia. Pages 55–70 in W. F. Laurance and R. O. Bierregaard, editors. *Tropical Forest Remnants: Ecology, Management, and Conservation of Fragmented Communities.* University of Chicago Press, Chicago.

———. 1998. Altered leaf-litter decomposition rates in tropical forest fragments. *Oecologia* 116: 97–406.

Didham, R. K., P. M. Hammond, and N. E. Stock. 1998. Beetle species responses to tropical forest fragmentation. *Ecological Monographs* 68:295–323.

Estrada, A. and R. Coates-Estrada. 1996. Tropical rain forest fragmentation and wild populations of primates at Los Tuxtlas, Mexico. *International Journal of Primatology* 17:759–83.

Ferreira, L. V., and W. F. Laurance. 1997. Effects of forest fragmentation on mortality and damage of selected trees in central Amazonia. *Conservation Biology* 11:797–801.

Fiala, B., U. Maschwitz, Y. P. Tho, and A. J. Helbig. 1989. Studies of a South East Asian ant–plant association: Protection of *Macaranga* trees by *Crematogaster borneensis. Oecologia* 79:463–70.

Gartner, B. L. 1989. Breakage and regrowth of *Piper* species in rain forest understory. *Biotropica* 21:303–07.

Gascon, C., T. E. Lovejoy, R. O. Bierregaard, J. R. Malcolm, P. C. Stouffer, H. L. Vasconcelos, W. F. Laurance, B. Zimmerman, M. Tocher, and S. Borges. 1999. Matrix habitat and species richness in tropical forest remnants. *Biological Conservation* 91:223–29.

Gilpin, M. E., and M. E. Soulé. 1986. Minimum viable populations: processes of species extinction. Pages 19–34 in M. E. Soulé, editor. *Conservation Biology: The Science of Scarcity and Diversity.* Sinauer Associates, Sunderland, MA.

Givnish, T. J. 1988. Adaptation to sun and shade: A whole plant perspective. *Australian Journal of Plant Physiology* 15:63–92.

Greco, S., S. Ulanski, M. Garstang, and S. Houston. 1992. Low-level nocturnal wind maximum over the central Amazon basin. *Boundary-Layer Meteorology* 58:91–115.

Harrington, G. H., A. K. Irvine, F. H. J. Crome, and L. A. More. 1997. Regeneration of large-seeded trees in Australian rainforest fragments: A study of higher-order interactions. Pages 292–303 in W. F. Laurance and R. O. Bierregaard, editors. *Tropical Forest Remnants: Ecology, Management, and Conservation of Fragmented Communities.* University of Chicago Press, Chicago.

He, F., P. Legendre, and J. V. LaFrankie. 1997. Distribution patterns of tree species in a Malaysian tropical rain forest. *Journal of Vegetation Science* 8:105–14.

Hernandez, H. M., and Y. C. Abud. 1987. Notas sobre la ecologia reproductiva de arboles en un bosque mesofilo de montana en Michoacan, Mexico. *Boletin de la Sociedad Botanica de Mexico* 47:5–35.

Hubbell, S. P. 2001. *The Unified Neutral Theory of Biogeography and Biodiversity.* Princeton University Press, Princeton, NJ.

Hurlbert, S. H. 1984. Pseudoreplication and the design of ecological field experiments. *Ecological Monographs* 54:187–211.

Ickes, K., S. J. DeWalt, and S. Appanah. 2001. Effects of native pigs (*Sus scrofa*) on the understory vegetation in a Malaysian lowland rain forest: An exclosure study. *Journal of Tropical Ecology* 17:191–206.

Ickes, K., S. J. DeWalt, and S. C. Thomas. 2003. Resprouting of woody saplings following stem snap by wild pigs in a Malaysian rain forest. *Journal of Ecology* 91:222-233.

Ickes, K., and S. C. Thomas. 2003. Native, wild pigs (Sus scrofa) at Pasoh and their impacts on the plant community. Pages 507–520 in T. Okuda, N. Manokaran, Y. Matsumoto, K. Niiyama, S. C. Thomas and P. S. Ashton, editors. *Pasoh: Ecology of a lowland rain forest in southeast Asia.* Springer-Verlag, Tokyo.

Itoh, A., T. Yamakura, K. Ogino, H. S. Lee, and P. S. Ashton. 1997. Spatial distribution patterns of two predominant emergent trees in a tropical rainforest in Sarawak, Malaysia. *Plant Ecology* 132:121–36.

Janzen, D. H. 1970. Herbivores and the number of tree species in tropical forests. *American Naturalist* 104:501–29.

Kapos, V. 1989. Effects of isolation on the water status of forest patches in the Brazilian Amazon. *Journal of Tropical Ecology* 5:173–85.

Kapos, V., G. Ganade, E. Matsui, and R. L. Victoria. 1993. $\partial^{13}C$ as an indicator of edge effects in tropical rainforest reserves. *Journal of Ecology* 81:425–32.

Kapos, V., E. Wandelli, J. L. Camaro, and G. Ganade. 1997. Edge-related changes in environment and plant responses due to forest fragmentation in central Amazonia. Pages 33–44 in W. F. Laurance and R. O. Bierregaard, editors. *Tropical Forest Remnants: Ecology, Management, and Conservation of Fragmented Communities.* University of Chicago Press, Chicago.

Kaur, A., C. O. Ha, K. Hong, V. E. Sands, H. T. Chan, E. Soepadmo, and P. S. Ashton. 1978. Apomixis may be widespread among trees of the climax rain forest. *Nature* 271: 75–88.

Kaur, A., K. Jong, V. E. Sands, and E. Soepadmo. 1986. Cytoembryology of some Malaysian dipterocarps, with some evidence of apomixis. *Botanical Journal of the Linnean Society* 92:75–88.

Kellman, M. 1996. Redefining roles: Plant community reorganization and species preservation in fragmented systems. *Global Ecology and Biodiversity Letters* 5:111–16.

Kellman, M., R. Tackaberry, and J. Meave. 1996. The consequences of prolonged fragmentation: Lessons from tropical gallery forests. Pages 37–58 in J. Schelhas and R. Greenberg, editors. *Forest Patches in Tropical Landscapes.* Island Press, Covelo, CA.

Kellman, M., R. Tackaberry, and L. Rigg. 1998. Structure and function in two tropical gallery forest communities: Implications for forest conservation in fragmented systems. *Journal of Applied Ecology* 35:195–206.

Kimble, M. 2000. *Variation in Riparian Ecosystem Biodiversity and Community Structure in Response to Watershed Condition.* Ph.D. thesis, University of Washington, Seattle, WA.

Kinsman, S. 1990. Regeneration by fragmentation in tropical montane forest shrubs. *American Journal of Botany* 77:1626–33.

Kochummen, K. M., J. V. LaFrankie, and N. Manokaran. 1990. Floristic composition of Pasoh Forest Reserve, a lowland rain forest in peninsular Malaysia. *Journal of Tropical Forest Science* 3:1–13.

Kress, W. J., and J. H. Beach. 1993. Flowering plant reproductive systems at La Selva Biological Station. Pages 161–82 in L. McDade, K. S. Bawa, G. Hartshorn, and H. A. Hespenheide, editors. *La Selva: Ecology and Natural History of a Neotropical Rain Forest.* University of Chicago Press, Chicago.

Landeen, L. 1992. *Midwifery and Traditional Medicine in Negeri Sembilan: A Study of Women's Knowledge of Medicinal Plants in a West Malaysian Village.* BA thesis. University of Oregon, Eugene, OR.

Laurance, W. F. 1997a. Responses of mammals to rainforest fragmentation in tropical Queensland: A review and synthesis. *Wildlife Research* 24:603–12.

Laurance, W. F. 1997b. Hyper-disturbed parks: Edge effects and the ecology of isolated rainforest reserves in tropical Australia. Pages 71–83 in W. F. Laurance and R. O. Bierregaard, editors. *Tropical Forest Remnants: Ecology, Management, and Conservation of Fragmented Communities.* University of Chicago Press, Chicago.

Laurance, W. F., S. G. Laurance, L. V. Ferreira, J. M. Rankin de Merona, C. Gascon, and T. E. Lovejoy. 1997. Biomass collapse in Amazonian forest fragments. *Science* 278:1117–18.

Laurance, W. F., L. S. Ferreira, and S. G. Laurance. 1998a. Rain forest fragmentation and the dynamics of Amazonian tree communities. *Ecology* 79:2032–40.

Laurance, W. F., L. S. Ferreira, J. M. Rankin de Merona, S. G. Laurance, R. W. Hutchings, and T. E. Lovejoy. 1998b. Effects of forest fragmentation on recruitment patterns in Amazonian tree communities. *Conservation Biology* 12:460–64.

Laurance, W. F., Perez-Salicrup, P. Delamonica, P. M. Fearnside, S. D'Angelo, A. Jerozolinski, L. Pohl, and T. E. Lovejoy. 2001. Rain forest fragmentation and the structure of Amazonian liana communities. *Ecology* 82:105–16.

Leigh, E. G., S. J. Wright, E. A. Herre, and F. E. Putz . 1993. The decline of tree diversity on newly isolated tropical islands—A test of a null hypothesis and some implications. *Evolutionary Ecology* 7:76–102.

Lovejoy, T. E., R. O. Bierregaard, J. M. Rankin, and H. O. R. Shubart. 1983. Ecological dynamics of tropical forest fragments. Pages 377–84 in S. L. Sutton, T. C. Whitmore, and A. C. Chadwick, editors. *Tropical Rain Forest: Ecology and Management.* Blackwell, Oxford, U.K.

Lovejoy, T. E., J. E. Rankin, R. O. Bierregaard, K. S. J. Brown, L. H. Emmons, and M. E. VanderVoort. 1984. Ecosystem decay of Amazon forest fragments. Pages 285–325 in M. H. Nitecki, editor. *Extinctions.* University of Chicago Press, Chicago.

Lovejoy, T. E., R. O. Bierregaard, A. B. Rylands, J. R. Malcolm, C. E. Quintela, L. H. Harper, K. S. Brown, A. H. Powell, G. V. N. Powell, H. O. R. Shubart, and M. B. Hayes. 1986. Edge and other effects of isolation on Amazon forest fragments. Pages 257–85 in M. E. Soulé, editor. *Conservation Biology: The Science of Scarcity and Diversity.* Sinauer Associates, Sunderland, MA.

Lynam, A. J. 1997. Rapid decline of small mammal diversity in monsoon evergreen forest fragmnts in Thailand. Pages 222–40 in W. F. Laurance and R. O. Bierregaard, editors. *Tropical Forest Remnants: Ecology, Management, and Conservation of Fragmented Communities.* University of Chicago Press, Chicago.

Malcolm, J. R. 1994. Edge effects in central Amazonian forest fragments. *Ecology* 75:2438–45.

Manokaran, N., and J. V. LaFrankie. 1991. Stand structure of Pasoh Forest Reserve, a lowland rain forest in peninsular Malaysia. *Journal of Tropical Forest Science* 3: 14–24.

Manokaran, N., J. V. LaFrankie, K. M. Kochummen, E. S. Quah, J. Klahn, P. S. Ashton, and S. P. Hubbell. 1990. *Methodology for the 50-ha Research Plot at Pasoh Forest Reserve.* Research Pamphlet. Forest Research Institute of Malaysia, Kuala Lumpur, Malaysia.

Moad, A. S. 1993. *Dipterocarp Sapling Growth and Understory Light Availability in Tropical Lowland Forest, Malaysia.* Ph.D dissertation, Harvard University, Cambridge, MA.

Murawski, D. A. 1996. Reproductive biology and genetics of tropical trees. Pages 457–93 in M. Lowman, and N. Nadkarni, editors. *Forest Canopies.* Academic Press, New York.

Murcia, C. 1995. Edge effects in fragmented forests: Implications for conservation. *Trends in Ecology and Evolution* 10:58–62.

Nason, J. D., and J. L. Hamrick. 1997. Reproductive and genetic consequences of forest fragmentation: Two case studies of neotropical canopy trees. *Journal of Heredity* 88:264–76.

Newmark, W. D. 1990. Tropical forest fragmentation and the local extinction of understory birds in the Eastern Usambara Mountains, Tanzania. *Conservation Biology* 5:67–78.

Ng, F. S. P., editor. 1978. *Tree Flora of Malaya.* Vol. 3. Longman, Kuala Lumpur, Malaysia.

———. 1989. *Tree Flora of Malaya.* Vol. 4. Longman, Kuala Lumpur, Malaysia.

Niklas, K. J. 1998. The influence of gravity and wind on land plant evolution. *Review of Palaeobotany and Palynology* 102:1–14.

Okuda, T., N. Kachi, S. K. Yap, and N. Manokaran. 1995. Spatial pattern of adult trees and seedling survivorship in *Pentaspadon motleyi* in a lowland rain forest in peninsular Malaysia. *Journal of Tropical Forest Science* 7:475–89.

Okuda, T., M. Suzuki, N. Adachi, K. Yoshida, K. Niiyama, Nur Supardi, M. N., Hussein, N. M., Manokaran, N., and M. Hashim. 2003. Logging history and its impact on forest structure and species composition in the Pasoh Forest Reserve—Implications for the sustainable management of natural resources and landscapes. Pages 15–34 in T. Okuda, N. Manokaran, Y. Matsumoto, K. Niiyama, S. C. Thomas and P. S. Ashton, editors. *Pasoh: Ecology of a lowland rain forest in southeast Asia.* Springer-Verlag, Tokyo.

Oliviera, A. P., and D. R. Fitzjarrald. 1993. The Amazon River breeze and the local boundary layer: I. Observations. *Boundary-Layer Meteorology* 63:41–162.

Peletz, M. 1988. *A Share of the Harvest.* University of California Press, Berkeley, CA.

Peñalosa, J. 1984. Basal branching and vegetative spread in two tropical rain forest lianas. *Biotropica* 16:1–9.

Peters, H. A. 2003. Neighbour-regulated mortality: the influence of positive and negative density dependence on tree populations in species-rich tropical forests. *Ecology Letters* 6:757–765.

Phillips, O. L., P. Hall, A. H. Gentry, S. A. Sawyer, and R. Vasquez. 1994. Dynamics and species richness of tropical rain forests. *Proceedings of the National Academy of Sciences* 91:2805–09.

Putz, F. E. 1984. How trees avoid and shed lianas. *Biotropica* 16:19–23.

Reich, P. B., D. S. Ellsworth, and C. Uhl. 1995. Leaf carbon and nutrient assimilation and conservation in species of differing successional status in an oligotrophic Amazonian forest. *Functional Ecology* 9:65–76.

Renjifo, L. S. 1999. Composition changes in a subandean avifauna after long–term forest fragmentation. *Conservation Biology* 13:1124–39.

Rogstad, S. H. 1990. The biosystematics and evolution of the *Polyalthia hypoleuca* species complex (Annonaceae) of Malesia. II. Comparative distributional ecology. *Journal of Tropical Ecology* 6:387–408.

Sagers, C. L. 1993. Reproduction in neotropical shrubs: The occurrence and some mechanisms of asexuality. *Ecology* 74:615–18.

Saw, L. G., J. V. LaFrankie, K. M. Kochummen, and S. K. Yap. 1991. Fruit trees in a Malaysian rain forest. *Economic Botany* 45:120–36.

Shafer, C. L. 1995. Values and shortcomings of small reserves. *BioScience* 45:80–88.

Sizer, N. and E. V. J. Tanner. 1999. Responses of woody plant seedlings to edge formation in a lowland tropical rainforest, Amazonia. *Biological Conservation* 91:135–42.

Skole, D. L., and C. J. Tucker. 1993. Tropical deforestation and habitat fragmentation in the Amazon: Satellite data from 1978 to 1988. *Science* 260:1905–10.

Smith, A. P. 1972. Buttressing of tropical trees: A descriptive model and new hypothesis. *American Naturalist* 106:32–46.

Sobrevilla, C. and M. T. K. Arroyo. 1982. Breeding systems in a montane tropical cold forest in Venezuela. *Systematics and Evolution* 140:19–38.

Sokal, R. R., and F. J. Rohlf. 1981. *Biometry.* 2nd edition. Freeman, New York.

Strauss-Debenedetti, S., and F. A. Bazzaz. 1996. Photosynthetic characteristics of tropical species of different successional stages: What patterns emerge? Pages 162–86 in S. S. Mulkey, R. L. Chazdon, and A. Smith, editors. *Tropical Forest Plant Ecophysiology.* Chapman and Hall, New York.

Sulaiman, S., Nik, A. R., and J. V. LaFrankie. 1994. *Pasoh Climatic Summary (1991– 1993).* FRIM Research Data, No. 3. Forest Research Institute Malaysia. Kuala Lumpur, Malaysia.

Swank, W. T., L. W. Sieft, and J. E. Douglass. 1988. Streamflow changes associated with forest cutting, species conversions, and natural disturbances. Pages 297–312 in W. T. Swank and D. A. Crossley, editors. *Forest Hydrology and Ecology at Coweeta.* Springer-Verlag, New York.

Tabarelli, M., W. Mantovani, and C. A. Peres. 1999. Effects of habitat fragmentation on plant guild structure in the montane Atlantic forest of southeastern Brazil. *Biological Conservation* 91:119–27.

Tanner, E. V. J. 1982. Species diversity and reproductive mechanisms in Jamaican trees. *Biological Journal of the Linnean Society* 18:263–78.

Tani, M, Rahim Nik, A., Ohtani, Y, Yasuda, Y, Sahat, M. M., Kasran, B., Takanashi, S., Noguchi, S., Yusop, Z., and T. Watanabe. 2003. Characteristics of energy exchange and surface conductance of a tropical rain forest in peninsular Malaysia. Pages 73–88 in T. Okuda, N. Manokaran, Y. Matsumoto, K. Niiyama, S. C. Thomas and P. S. Ashton, editors. *Pasoh: Ecology of a lowland rain forest in southeast Asia.* Springer-Verlag, Tokyo.

Terborgh, J. 1992. Maintenance of diversity in tropical forests. *Biotropica* 24:283–92.

Terborgh, J., L. Lopez, and J. S. Tello. 1997a. Bird communities in transition: The Lago Guri Islands. *Ecology* 78:1494–01.

Terborgh, J., L. Lopez, and J. Tello, D. Yu, and A. R. Bruni. 1997b. Transitory states in relaxing ecosystems of land bridge islands. Pages 256–74 in W. F. Laurance and R. O. Bierregaard, editors. *Tropical Forest Remnants: Ecology, Management, and Conservation of Fragmented Communities.* University of Chicago Press, Chicago.

Terborgh, J., L. Lopez, P. Nunez, M. Rao, G. Shahabuddin, G. Orihuela, M. Riveros, R. Ascanio, G. H. Adler, T. D. Lambert, and L. Balbas. 2001. Ecological meltdown in predator-free forest fragments. *Science* 294:1923–26.

Thomas, S. C. 1991. Population densities and patterns of habitat use among anthropoid primates of the Ituri Forest, Zaire. *Biotropica* 23:68–83.

———. 1993. *Interspecific Allometry in Malaysian Rainforest Trees*. Ph.D dissertation, Harvard University, Cambridge, MA.

———. 1996a. Asymptotic height as a predictor of growth and allometric characteristics in Malaysian rain forest trees. *American Journal of Botany* 83:556–66.

———. 1996b. Relative size at reproductive onset in rain forest trees: A comparative analysis of 37 Malaysian species. *Oikos* 76:145–54.

———. 1996c. Reproductive allometry in Malaysian rain forest trees: Biomechanics vs. optimal allocation. *Evolutionary Ecology* 10:517–30.

———. 1997. Geographic parthenogenesis in a tropical rain forest tree. *American Journal of Botany* 84:1012–15.

———. 2003. Comparative biology of tropical trees: A perspective from Pasoh. Pages 171–194 in T. Okuda, N. Manokaran, Y. Matsumoto, K. Niiyama, S. C. Thomas and P. S. Ashton, editors. *Pasoh: Ecology of a Lowland Rain Forest in Southeast Asia.* Springer-Verlag, Tokyo.

Thomas, S. C., and S. Appanah. 1995. On the statistical analysis of size-dependent reproductive onset in dipterocarp forests. *Journal of Tropical Forest Science* 7:412–18.

Thomas, S. C., and F. A. Bazzaz. 1999. Asymptotic height as a predictor of photosynthetic characteristics in Malaysian rain forest trees. *Ecology* 80:1607–22.

Thomas, S. C., and K. Ickes. 1995. Ontogenetic changes in leaf size in Malaysian rain forest trees. *Biotropica* 27:427–34.

Thomas, S. C., and J. V. LaFrankie. 1993. Sex, size, and interyear variation in flowering among dioecious trees of the Malayan rain forest. *Ecology* 74:1529–37.

Turner, I. M. 1996. Species loss in fragments of tropical rain forest: A review of the evidence. *Journal of Applied Ecology* 33:200–19.

———. 1997. A tropical flora summarized—A statistical analysis of the vascular plant diversity of Malaya. *Flora* 192:157–63.

Turner, I. M., and R. T. Corlett. 1996. The conservation value of small, isolated fragments of lowland tropical rain forest. *Trends in Ecology and Evolution* 11:330–33.

Turner, I. M., H. T. W. Tan, Y. C. Wee, A. B. Ibrahim, P. T. Chew, and R. T. Corlett. 1994. A study of plant species extinction in Singapore: Lessons for the conservation of tropical biodiversity. *Conservation Biology* 8:705–12.

Turner, I. M., K. S. Chua, J. S. Y. Ong, B. C. Soong, and H. T. W. Tan. 1996. A century of plant species loss from an isolated fragment of lowland tropical rain forest. *Conservation Biology* 10:1229–44.

Turton, S. M., and H. J. Freiburger. 1997. Edge and aspect effects on the microclimate of a small tropical forest remnant of the Atherton Tableland, Northeastern Australia. Pages 45–54 in W. F. Laurance and R. O. Bierregaard, editors. *Tropical Forest Remnants: Ecology, Management, and Conservation of Fragmented Communities.* University of Chicago Press, Chicago.

Van Steenis, C. G. G. J. 1969. Plant speciation in Malesia, with special reference to the theory of non-adaptive saltatory evolution. *Biological Journal of the Linnaean Society* 1:97–133.

Van Tiel, J. 1984. *Convex Analysis: An Introductory Text*. Wiley, New York.

Viana, V. M., and Tabanez, A. A. J. 1996. Biology and conservation of forest fragments in the Brazilian Atlantic moist forest. Pages 151–67 in J. Schelhas and G. Greenbery, editors. *Forest Patches in Tropical Landscapes*. Island Press, Washington, DC.

Warburton, N. H. 1997. Structure and conservation of forest avifauna in isolated rainforest remnants in tropical Australia. Pages 190–206 in W. F. Laurance and R. O. Bierregaard, editors. *Tropical Forest Remnants: Ecology, Management, and Conservation of Fragmented Communities*. University of Chicago Press, Chicago.

Watts, I. E. M. 1954. Line squalls of Malaya. *Journal of Tropical Geography* 3:1–14.

Whitmore, T. C., editor. 1972. *Tree Flora of Malaya*. Vol. 1. Longman, Kuala Lumpur, Malaysia.

———. 1973. *Tree Flora of Malaya*. Vol. 2. Longman, Kuala Lumpur, Malaysia.

Whitmore, T. C. 1984. *Tropical Rain Forests of the Far East*. 2nd edition. Clarendon Press, Oxford, U.K.

———. 1989. Canopy gaps and the two major groups of forest trees. *Ecology* 70:536–38.

Williams-Linera, G. 1990a. Origin and early development of forest edge vegetation in Panama. *Biotropica* 22:235–41.

———. 1990b. Vegetation structure and environmental conditions of forest edges in Panama. *Journal of Ecology* 78:356–73.

Williams–Linera, G., V. Dominguez-Gastelu, and M. E. Garcia-Zurita. 1998. Microenvironment and floristics of different edges in a fragmented tropical rainforest. *Conservation Biology* 12:1091–1102.

Willis, E. O. 1974. Populations and local extinctions of birds on Barro Colorado Island, Panama. *Ecological Monographs* 44:153–69.

Zimmerman, J. K., E. E. Everham, R. B. Waide, D. J. Lodge, C. M. Taylor, and N. V. L. Brokaw. 1994. Responses of tree species to hurricane winds in subtropical wet forest in Puerto Rico: Implications for tropical tree life histories. *Journal of Ecology* 82:911–22.

Zuidema, P. A., J. A. Sayer, and W. Dijkman. 1996. Forest fragmentation and biodiversity: The case for intermediate-sized conservation areas. *Environmental Conservation* 23:290–97

PART 6: The Diversity of Tropical Trees: The Role of Pest Pressure

Introduction

Egbert G. Leigh, Jr.

This section's five chapters all examine one explanation for why there are so many kinds of tropical trees: the impact of species-specific pests. We cannot claim a sound understanding of how so many kinds of trees coexist in tropical forests until we understand why trees are so much more diverse in the tropics than in the temperate zone. Therefore, we consider only theories that can explain why tree diversity is so much lower at higher latitudes. Many factors allow trees of different species to coexist. Many of these factors allow more tree species to coexist in the tropics than in the temperate zone. For example, most forests have light gaps and shaded habitats, with different species in each. Light-gap microclimates differ more from those of the shaded understory in the tropics than in the temperate zone, allowing higher tree diversity in the tropics (Ricklefs 1977). Similarly, most forests have both canopy and understory tree species, yet tropical forests have an extra layer or two (Terborgh 1985), again allowing higher tree diversity in the tropics. As was shown earlier in the book, such physical factors contribute materially to the diversity of tropical trees. These factors, however, appear unable to account for the 5-fold increase in the number of tree species on a tropical over a temperate-zone forest hectare.

On the other hand, the devastating effects of pests and pathogens on tropical plants is a familiar story. The famous cry of a Brazilian politician—either Brazil will kill the leaf-cutter ant or this ant will kill Brazil—is just one example of how dismaying pest pressure can be for farmers in the tropics. Is tree diversity so much higher in the tropics because the pressure from specialized pests is so much heavier in tropical climates (Janzen 1970; Connell 1971)? In the tropics, no winter reduces or annihilates the activity of pests: insect herbivory, at least, is more evenly spread throughout the year at tropical latitudes (Wolda 1983). Consequently young tropical leaves are more heavily eaten, despite being much more poisonous, than their temperate-zone counterparts (Coley and Barone 1996; Coley and Kursar 1996).

Does the higher diversity of tropical trees reflect the greater impact of species-specific pests and pathogens (Janzen 1970, Connell 1971)? Most of the damage from pests and disease is inflicted by organisms that specialize on particular genera, if not species, of plants (Barone 1998, Novotny et al. 2002). Like any plague, such pests spread more easily where their hosts are closer together (Ridley 1930, Gilbert 2002). The five chapters in this part are all devoted to one question: Do seeds, seedlings, or saplings of a species suffer higher mortality where they are more common, or closer to their parents, thereby making room for plants of other species? All five chapters find evidence for mutual repulsion among conspecifics.

Two questions remain, however. First, can we predict how much differential herbivory on seedlings closer to their parents or siblings is needed to maintain a given level of tree diversity? To get at this question, one must keep track of the ages, sizes, and spatial distribution of the various individuals of a species—not merely their total numbers—to predict the effect of diminished survival near conspecifics on species diversity. This task is a bugbear of theoreticians. Second, is mutual repulsion among conspecifics always caused by pests or pathogens? None of these five chapters investigates causes of the mutual repulsion they demonstrate. Documenting mutual repulsion among conspecifics is not quite a demonstration of the impact of pest pressure. There is, however, abundant evidence that pests and pathogens do cause mutual repulsion among conspecifics (Ridley 1930, Augspurger 1984, Howe et al. 1985, Gilbert 2002). We know of no other plausible mechanisms of "mutual repulsion at a distance." Moreover, pest pressure cannot enhance tree diversity unless it causes mutual repulsion among conspecifics.

The first two chapters in this part explore ways of assessing whether young plants suffer higher mortality where adult trees or other young of their species are closer or more common. Itoh and his colleagues measured the total seed production of a dipterocarp, *Dryobalanops aromatica*, on the 52-ha Lambir Forest Dynamics Plot, and estimated the distribution of these seeds about their parents. They divided their plot into 1-ha quadrats, calculated the number of seeds falling into each quadrat from the locations and sizes of the adults nearby, and counted the number of saplings of this species in each quadrat. They found that the ratio of saplings to fallen seeds was highest for quadrats with an intermediate basal area of *D. aromatica*, as if in the areas where adults were rare, the habitat was unsuitable for seedlings, whereas in areas with very many adults, mortality of seedlings or saplings was heavier.

Muller-Landau and her colleagues examined the effects of the abundance and nearness to parents of seeds and seedlings on their survival prospects for the most common mature forest tree, *Trichilia tuberculata* (Meliaceae), and a pioneer tree, *Miconia argentea* (Melastomataceae), on Barro Colorado Island's 50-ha Forest Dynamics Plot. They measured seed fall and seedling emergence for *Trichilia* and survival of buried seeds for *Miconia*. The germination rate of *Trichilia* seeds and the survival rate of its seedlings were lower where seed fall was heavier and conspecific adults closer. *Miconia* seeds died faster where they were more common or closer to conspecific adults. Saplings of both species also died faster where basal area of conspecific adults was higher, but nearby adult conspecifics depressed survival of seeds and seedlings far more severely than they depressed sapling survival. The most striking feature of this chapter was the different types of data needed to demonstrate the depressing effect of nearby conspecifics for a gap species as opposed to a mature forest species.

The section's last three chapters seek to assess the role of pest pressure, as manifested by the proportion of species whose saplings recruit less frequently or die faster near conspecific adults, in maintaining tree diversity on different 50-ha Forest Dynamics Plots.

John and Sukumar asked what governs the distribution and abundance of different tree species in the dry forest at Mudumalai in south India. Habitat segregation plays little role in maintaining tree diversity on the 50-ha plot at Mudumalai. Fire, by contrast, strongly influences tree species recruitment. Proximity to conspecifics diminished recruitment and, to a lesser extent, mortality of some species. In three common species with frequent recruits, the proportion of saplings that are conspecific with a given adult is lower near that adult than at a distance. In two of these species, adults depress the recruitment of conspecific saplings in their neighborhood. On the other hand, in some of these same species, mortality is lower in hectares with a higher basal area of conspecifics.

Wills and his colleagues found that repulsion among conspecifics was much more prevalent among the tree species on the 50-ha Forest Dynamics Plots at Barro Colorado Island and Pasoh than seems to be the case at Mudumalai. They divided the 50-ha plots at both Barro Colorado and Pasoh into 10 x 10 m quadrats. They selected 100 pairs of species, each pair including one from Barro Colorado and an equally common one from Pasoh. For each of these species, they asked how the number of saplings recruiting onto a quadrat was influenced by the density and basal area of conspecific adults on that quadrat. In each plot's 100 tested species, there was a significant tendency for the number of saplings recruiting per conspecific adult on a quadrat to be lower on quadrats with a higher density or basal area of conspecific adults as if, in these species, pests diminish recruitment of a species' saplings more severely near conspecific adults. Wills and his colleagues showed that this density dependence was not due to artifacts such as saplings recruiting to quadrats with few adult conspecifics from nearby ones with many. The quadrat analysis of Wills et al. is powerful enough to detect the impact of adult trees on the recruitment of conspecifics nearby, but it is too weak to detect their impact on the mortality of conspecifics because, for a tree near a quadrat's edge, it replaces close neighbors across the border by other, more distant, neighbors in the quadrat (Peters 2003).

Finally, Ahumada and his colleagues asked what features of a sapling's nearest neighbors living on Barro Colorado Island's 50-ha plot influenced its prospects of surviving. On average, each additional conspecific among a sapling's 20 nearest neighbors in 1982 increased its chance of dying before 1995 by roughly 2%.

Moreover, very rare species suffered most from the presence of an additional con-specific among its neighbors. At the other extreme, saplings of the most common species of canopy tree, *Trichilia tuberculata* (Meliaceae), survived equally well re-gardless of how many conspecifics were among its 20 nearest neighbors. *Trichilia tuberculata* was the first tree species on Barro Colorado Island whose population was shown to be limited by its own density (Hubbell et al. 1990), but survival of *Trichilia* saplings appears to reflect the density of conspecifics at larger scales.

The chapter by Ahumada and colleagues provides compelling evidence for the role of mutual repulsion among conspecifics in maintaining tree diversity. Even so, improvements have been made upon the analysis of Ahumada and colleagues. A parallel study has considered the number of conspecifics within 10 m or so of the focal saplings (the *density* of nearby conspecifics) rather than the number of conspecifics among its 20 nearest neighbors, because conspecifics in a low-density neighborhood should exert less influence. In fact, the probability, averaged over all saplings, that a sapling living on Barro Colorado's plot in 1982 survives to 1995 was significantly lower the higher the density of conspecifics nearby (Hubbell et al. 2001). Moreover, in over half the species on Barro Colorado common enough to analyze, survival was significantly lower where the density of conspecific neighbors was higher (Peters 2003). The same was true in over half the species common enough to test on Pasoh's 50-ha plot (Peters 2003). At Pasoh, furthermore, a new factor comes into play. Wills and Green (1995) proposed that, for a fixed density of conspecific neighbors, *increasing* the density of heterospecific neighbors enhances a sapling's survival prospects, because the benefit of being hidden from specialized pests by these heterospecifics outweighs the harm from crowding. In other words, a specialized pest is less likely to find a suitable host when it is lost in a herd of immune, inedible plants, a phenomenon Wills and Green (1995) call "herd immunity." In contrast to Barro Colorado, saplings at Pasoh with a given density of conspecific neighbors benefit, on the average, from having more heterospecific neighbors (Peters 2003). Are the benefits of herd immunity greater in more diverse forests?

In sum, this section shows how the network of 50-ha Forest Dynamics Plots has made possible studies that are setting us firmly on the road to a definitive answer to the question: Why are there so many kinds of tropical trees?

References

Augspurger, C. K. 1984. Seedling survival of tropical tree species: interactions of dispersal distance, light-gaps, and pathogens. *Ecology* 65:1705–12.

Barone, J. A. 1998. Host-specificity of folivorous insects in a moist tropical forest. *Journal of Animal Ecology* 67:400–09.

Coley, P. D. and J. A. Barone. 1996. Herbivory and plant defenses in tropical forests. *Annual Review of Ecology and Systematics* 27:305–35.

Coley, P. D., and T. A. Kursar. 1996. Anti-herbivore defenses of young tropical leaves: Physiological constraints and ecological trade-offs. Pages 305–36 in S. S. Mulkey, R. L. Chazdon, and A. P. Smith, editors. *Tropical Forest Plant Ecophysiology.* Chapman and Hall, New York.

Connell, J. H. 1971. On the role of natural enemies in preventing competitive exclusion in some marine animals and in rain forest trees. Pages 298–312 in P. J. den Boer and G. Gradwell, editors. *Dynamics of Numbers in Populations.* Centre for Agricultural Publication and Documentation, Wageningen, Netherlands.

Gilbert, G. S. 2002. Evolutionary ecology of plant diseases in natural ecosystems. *Annual Review of Phytopathology* 40:13–43.

Howe, H. F., E. W. Schupp, and L. C. Westley. 1985. Early consequences of seed dispersal for a Neotropical tree (*Virola surinamensis*). *Ecology* 66:781–91.

Hubbell, S. P., J. A. Ahumada, R. Condit and R. B. Foster. 2001. Local neighborhood effects on long-term survival of individual trees in a neotropical forest. *Ecological Research* 16:859–75.

Hubbell, S. P., R. Condit, and R. B. Foster. 1990. Presence and absence of density dependence in a neotropical tree community. *Philosophical Transactions of the Royal Society of London, B* 330:269–81.

Janzen, D. H. 1970. Herbivores and the number of tree species in tropical forests. *American Naturalist* 104:501–28.

Novotny, V., S. E. Miller, Y. Basset, L. Cizek, P. Drozd, K. Darrow, and J. Leps. 2002. Predictably simple: assemblages of caterpillars (Lepidoptera) feeding on rainforest trees in Papua New Guinea. *Proceedings of the Royal Society of London* B 269: 2337–44.

Peters, H. A. 2003. Neighbour-regulated mortality: the influence of positive and negative density dependence on tree populations in species-rich tropical forests. *Ecological Letters* 6:757–65.

Ricklefs, R. E. 1977. Environmental heterogeneity and plant species diversity: a hypothesis. *American Naturalist* 111:376–81.

Ridley, H. N. 1930. *The Dispersal of Plants Throughout the World.* L. Reeve, Ashford, Kent, UK.

Terborgh, J. 1985. The vertical component of plant species diversity in temperate and tropical forests. *American Naturalist* 126:760–76.

Wills, C. and D. R. Green. 1995. A genetic herd-immunity model for the maintenance of MHC polymorphism. *Immunological Reviews* 145:263–292.

Wolda, H. 1983. Spatial and temporal variation in abundance in tropical animals. Pages 93–105 in S. L. Sutton, T. C. Whitmore, and A. C. Chadwick, editors. *Tropical Rain Forest: Ecology and Management.* Blackwell Scientific, Oxford, U.K.

19

An Approach for Assessing Species-Specific Density-Dependence and Habitat Effects in Recruitment of a Tropical Rainforest Tree

Akira Itoh, Naoki Rokujo, Mamoru Kanzaki, Takuo Yamakura, James V. LaFrankie, Peter S. Ashton, and Hua Seng Lee

Introduction

Density-dependent effects on population dynamics are important factors in the maintenance of species diversity in tropical rainforests (e.g., Janzen 1970; Connell 1971). Density-dependent effects depress the recruitment of seedlings of a given species in patches where many conspecifics are present, leading to a less aggregated distribution of that species. Habitat preference or niche specialization is another important mechanism that explains species coexistence in some tropical rainforests (e.g., Clark et al. 1998). This theory predicts that species recruitment should be limited in unsuitable sites where the current density of the species is low. Thus habitat specialization, unlike density-dependent effects, promotes the clumping of individuals. These two mechanisms can affect the population simultaneously, though their relative importance may differ depending on the species and spatial scales.

A proper method for measuring recruitment success is needed to evaluate the effects of density dependence and habitat differences (Condit et al. 1992; Hubbell and Foster 1986). The simplest measure is the number of juveniles or recruits. However, the effects of limited seed dispersal and density dependence cannot be separated by an analysis based only on recruit number (Wills et al. 1997). Wills et al. (1997) measured recruitment per capita (the number of recruits per reproductive adult tree) to successfully detect density dependence with a novel statistical method. However, the per capita recruitment is an overestimate at sites with few or no reproductive adults that are surrounded by sites with many adults (Wills et al. 1997). Therefore, analyses using per capita recruitment may mask habitat effects, which are most likely to be apparent at low-density sites.

A better measure of recruitment is the number of recruits per number of dispersed seeds at each site. This measure requires a good estimate of the spatial variation of seed shadows. There have been many field and theoretical studies on seed dispersal in temperate and tropical forests (e.g., Augspurger and Franson 1988; Greene and Johnson 1989; Okubo and Levin 1989; Andersen 1991; Willson

1993; Masaki et al. 1994; Shibata and Nakashizuka 1995; Clark et al. 1998). Most studies, however, have been conducted on single, small plots (Clark et al. 1999). To analyze density dependence and habitat effects, larger-scale seed input data must be included, as well as the densities of various trees and the site conditions.

The objective of this chapter is twofold. First, we demonstrate a method for estimating seed input on a large scale (i.e., 52 ha) that is based on the seed dispersal patterns of individual trees and the spatial distribution of fruiting trees. Second, we conduct a preliminary analysis of density and habitat effects on the recruitment of an emergent rainforest tree, using the seed input estimate.

Methods

Species and Study Site

This study was conducted in a mixed dipterocarp forest in Lambir Hills National Park (4°N 114°E), Sarawak, Malaysia. The tropical climate is aseasonal and humid with a mean annual precipitation of about 2664 mm (chap. 31). All trees ≥1 cm diameter at breast height (dbh) were mapped, identified to species, and measured at dbh in a 52-ha, 500 × 1040 m Forest Dynamics Plot (Yamakura et al. 1995; 1996).

Dryobalanops aromatica Gaertn. f. (Dipterocarpaceae) is an emergent tree that grows up to 60 m tall and 2 m in diameter. It is distributed in the Malay Peninsula, Sumatra, and Borneo (Ashton 1982). This was the most common canopy species in the 52-ha plot, dominating small patches of the upper canopy structure (Itoh et al. 1995). *D. aromatica* produces one-seeded, wind-dispersed fruits (5–7 g fresh weight) with five sepal wings (4–7 cm long) (hereafter the fruit is referred to as 'seed' in this paper). This species flowers and fruits more frequently than most other dipterocarp species. Fruiting trees were observed at Lambir every year between 1990 and 1998, except for 1995 (A. Itoh, personal observation). Heavy fruiting, however, occurred only in 1991, 1996, and 1997, when many other species also fruited heavily after periods of mass flowering.

Seed Dispersal of Isolated Trees

To estimate the size dependency in seed production and seed dispersal of *D. aromatica*, we estimated the seed dispersal curves for trees of different sizes. We used the data of isolated trees to estimate the seed dispersal pattern of each tree. We selected isolated individuals to overcome the issue of distinguishing the mother trees where seed shadows of several trees overlapped (Clark et al. 1999; see also chap. 22). Finding isolated but non-outlier individuals in a population is not as difficult in tropical forests as in temperate forests because population densities of tropical trees are generally low.

We selected three isolated fruiting trees within the 52-ha plot, each over 60 m from other fruiting conspecifics. One of the three sample trees fruited in 1991 and the other two in 1996; both years were general flowering periods in Lambir. We pooled the data of the 2 years in this study because we did not have enough sample trees in each year to estimate size dependence. The three sample trees in our study were not considered to be outliers in terms of seed production and seed shadows because they are under closed canopy on sandy ridges where most individuals of this species occur (Itoh et al. 2003). Diameter at breast height and height of the sample trees were measured using a diameter tape and a laser rangefinder (Bushnell LYTESPEED 400), respectively. A belt transect of 1 m in width (5 m for the 1991 fruiting tree) was established 22–40 m from the base of each mother tree; the transects were terminated where no seeds were found within 5 m further at the time of first seed census. We selected the directions of the transects so as to minimize the slope inclination along the transects. The transects were divided into 1 × 1 m quadrats (5 × 5 m for the 1991 tree), and all mature seeds dispersed in each quadrat were labeled using numbered flags. The seed census began at the peak period of seed fall and continued at 5–10 day intervals until no new seeds were recorded (4–6 times for about 1.5 months). During the census periods, 1.9–3.8% of the labeled seeds for each sample tree disappeared probably due to predation by rodents (N. Rokujo unpublished data; A. Itoh unpublished data). No seed predation by bearded pigs (*Sus barbatus*) was observed, though they are major seed predators of dipterocarp seeds in Borneo (Curran et al. 1999). This was probably due to the low density of bearded pigs in Lambir, which may be caused by the small area of the park (7000 ha) and the high hunting pressure (Watson 1985). Though we did not know the predation rates before the establishment of the transects, seed predation may have been low enough not to affect estimates of seed fall.

The seed dispersal pattern was evaluated using a Weibull distribution model (Rokujo 1998) for each mother tree. The density function of seeds dispersed x meters from a mother tree, $f(x)$, is expressed by the following equation:

$$f(x) = \frac{1}{n}\frac{m}{a^m}x^{m-1}\exp\left[-\left(\frac{x}{a}\right)^m\right], \tag{19.1}$$

where m is a shape parameter, a is a scale parameter, and n is a normalization constant; that is, $f(x)dA$ is the expected proportion of the total seed fall to be found in a small area dA at a distance x from the parent. Assuming that there is no directional bias in the dissemination of seed (isotropic seed dispersal), n is obtained by integrating the Weibull distribution arc-wise and with distance:

$$n = 2\pi \int_0^\infty x\frac{m}{a^m}x^{m-1}\exp\left[-\left(\frac{x}{a}\right)^m\right]dx = 2\pi a\Gamma\left(\frac{1}{m}+1\right), \tag{19.2}$$

where Γ (-) is the gamma function. Then, seed density at a distance of x meters, $\rho(x)$, is the total seed production of the mother tree (N) times the seed density function:

$$\rho(x) = Nf(x) = \frac{N}{2\pi a \Gamma\left(\dfrac{1}{m}+1\right)} \frac{m}{a^m} x^{m-1} \exp\left[-\left(\frac{x}{a}\right)^m\right]. \tag{19.3}$$

The Weibull distribution model is equivalent to the mechanistic seed dispersal model of Greene and Johnson (1989) if the distribution of wind speed along one direction follows the Weibull distribution. Their model, however, assumes log normal distribution of wind speed in all directions. It is known that frequency of wind speed often follows the Weibull distribution. We adopted the Weibull distribution model because it could express the observed seed dispersal curves in this study, which had a mode at some distance from the mother trees. The Weibull distribution model has a mode at $x > 0$ when $m > 1$. When $m = 1$, the model is equivalent to an exponential model. Seed dispersal models that have the highest seed density under the mother tree, e.g., exponential, Gaussian, or 2Dt models (Clark et al. 1999; see also chap. 22), did not fit to our data.

The parameters N, m, and a were estimated for each mother tree using the data from the belt transect, seed density of each 1-m^2 seed quadrat [$\rho(x)$], and the distance from the mother tree to the center of the quadrat [x]. A nonlinear regression (the Gauss–Newton method) was adopted to minimize the sum of squares of the difference between observed and estimated seed densities using SYSTAT. The 95% confidence limits of the parameters were calculated based on asymptotic standard errors or the Wald statistics (Wilkinson 1997). We also presented Pearson's r^2 values between the model predictions of seed densities and actual seed densities for the purpose of illustrating "goodness of fit". To calculate these correlations, densities were first transformed as log (seed number + 1), to reduce deviations from normality (Zar 1974).

Estimating Seed Shadows in the 52-ha Plot

Combining the position and size of fruiting *D. aromatica* trees with the size-dependent parameters of the Weibull distribution model mentioned above, we estimated the seed shadows of *D. aromatica* in the 52-ha plot in a mast seeding year as follows.

First, fruiting was checked for all *D. aromatica* trees ≥30 cm dbh ($N = 393$) using binoculars in February 1997, a mast seeding year (details of the methods are described in Itoh et al. 2003). Then, the seed dispersal curve of each fruiting tree was estimated using the Weibull distribution, equation (19.3). To estimate the parameters of each tree, we used the size dependency of the parameters obtained from the three sample trees through the following equations.

We used an exponential function for the relationship between tree height H(m) and the scale parameter a,

$$a = \beta \exp[\alpha H] \qquad (19.4)$$

where a and β are coefficients. The exponential function was adopted because mean wind speed, which largely determines the value of a, increases exponentially with the height from the ground (Inoue 1963). The values of a and β were estimated by nonlinear regression using SYSTAT. Since tree height data were not available for all the fruiting trees, we estimated the height of each tree H(m) from its dbh D(cm) using the following equation (Yamakura et al. 1986), which was obtained from 35 sample $D.$ $aromatica$ trees in the plot ($D = 7.2$–148 cm, $r^2 = 0.9405$).

$$\frac{1}{H} = \frac{1}{1.614 D} + \frac{1}{59.8}. \qquad (19.5)$$

We assumed a linear relationship between total seed production N and D,

$$N = cD, \qquad (19.6)$$

where c is a coefficient of proportionality. Though the fecundity of an individual tree is often proportional to basal area rather than diameter (Clark et al. 1999; see also chap. 22), our data did not fit a linear equation with basal area, that is a quadratic equation with diameter, as will be shown later in figure 19.2. We used a size-independent value for the shape parameter m by averaging the m values of three sample trees. The values of m may not depend on tree size since the shape of seed dispersal curves of wind dispersed species may be determined mostly by the morphology of seeds, which in turn determines falling speed. Actually, the values of m differed little among the three sample trees with different sizes as illustrated in figure 19.2. Although the sample size was too small ($n = 3$) to evaluate the significance of size-dependent dispersal parameters, we tentatively used them to estimate the parameters of seed dispersal in the following analysis. In order to summarize the seed shadows in the 52-ha plot, we divided the plot into 5×5 m subquadrats ($N = 20,800$), and calculated the total number of seeds dispersed from all the fruiting trees into each subquadrat. We calculated the number of seeds dispersed from a fruiting tree in each focal subquadrat (25 m^2 in area) by using equation (19.3) to estimate seed density (m^{-2}) for the center of each focal subquadrat and multiplying that number by 25; that is, we assumed that seed density at any point in a focal subquadrat was the same as that expected in the center of that quadrat. Although the density $\rho(x)$ is theoretically positive for all distances x, we ignored quadrats farther than x m where $\rho(x) < 0.0001$ seeds/m^2. Utilizing a finer division of the plot, i.e., 1×1 m, and a smaller cutoff for distance limit in estimation, i.e., $\rho(x) < 0.00001$, the seed shadows were also calculated to detect the sensitivities of the results mentioned below. We found only small differences between the two results. We considered only seeds from trees in the plot and not those from trees off the plot. This limitation may result in an underestimation of seed input for some quadrats near the border of the plot (see also chap. 22). To minimize this effect, we excluded quadrats less than 20 m from the plot border in summarizing the results, thus the total area used for this summary was 46 ha. Our models predict that seed densities at $x > 20$ m are <0.4, <1.9, <3.8, and <5.6 per m^2 for trees of 80, 100, 120, and 140 cm dbh, respectively.

Evaluating Density Dependence Effects

We obtained a preliminary estimate of density-dependent recruitment by comparing potential seed input and current sapling density for 5×5 m quadrats with various local conspecific densities. In this study, we defined saplings as trees with dbh ≥ 1cm and <2 cm. The potential seed input is defined as the mean density of seed input from all adults (trees with dbh ≥ 30 cm) in the 52-ha plot to the focal quadrat during a single event of mass-fruiting. The potential seed input PS of a quadrat was calculated by

$$PS = \sum_i g(D_i)\rho_i(x_i), \tag{19.7}$$

where D_i is the dbh of the ith adult, $g(D)$ is a size-dependent weighting function representing the probability of fruiting of an adult of size D, $\rho_i(x)$ is the Weibull dispersal model of the ith adult, and x_i is the distance from the ith adult to the center of the focal quadrat. The weighting function $g(D)$ was used because smaller trees may fruit less frequently than larger ones over the long term. Thus, PS can be considered as a long-term mean density of seed input at the focal quadrat from all current adults in the plot. Since no data on the long-term size-dependent fruiting frequency were available, we assumed that it is the same as the relationship between dbh and fruiting probability observed in 1997. The function $g(D)$ was expressed by a modified logistic regression model (Thomas 1996):

$$g(D) = \frac{k}{1 + \exp[a + b\ln(D)]}, \tag{19.8}$$

where k is the maximum fruiting probability and a and b are species-specific coefficients. The coefficient values were estimated using the data of fruiting in 1997 by a nonlinear regression analysis with SYSTAT.

To evaluate the effects of local conspecific density on recruitment, we compared the ratio of mean density of current saplings to PS among quadrats with different conspecific density. First, we calculated the total basal area of $D.$ $aromatica$ trees (dbh ≥ 2 cm) within a circle of radius of 5, 10, 15, and 20 m for each 5×5 m quadrat.

The quadrats were grouped into six classes according to the local conspecific basal area (BA) for each radius as BA $= 0$, $0–10$, $10–10^2$, $10^2–10^3$, $10^3–10^4$, and $>10^4$ cm^2. Then, we calculated the ratio of the mean sapling density to the mean *PS* for each local BA class. This ratio suggests the probability of sapling recruitment expected from the current seed input. Finally, we examined whether the number of current saplings in each local BA class was significantly different from that expected from the potential seed input in each BA class using the chi-square test. The expected value in the *i*th basal area class (SE_i) was calculated by:

$$SE_i = NS \frac{PS_i}{\sum PS}, \tag{19.9}$$

where *NS* is the total number of saplings in the plot, $\sum PS$ is the sum of *PS* in all quadrats, and PS_i is the sum of *PS* in the quadrats of the *i*th basal area class. The sapling density of all species, including *D. aromatica*, was also calculated for each quadrat to check the effect of *D. aromatica*'s basal area on the total number of saplings. The analysis was done for the 46-ha area excluding the quadrats less than 20 m from the plot border.

In the analysis of density effects, we used the ratio of current sapling density to the potential seed input as an index of recruitment success. This is based on the assumption that current saplings resulted from the same distribution of seeds in the past that are observed today. This assumption, however, has clear limitations because saplings generally represent a much wider temporal scale than do seeds because saplings may sit in the understory and change little in size. Moreover, seed shadows in the past were not necessarily the same as the present ones if some adults have recently died. Thus, results must be interpreted with caution and considered as preliminary.

Results

Seed Dispersal of Individual Trees and Trees Fruiting in 1997

The density of dispersed seeds had a mode at 5–10 m from the base of the mother tree and dispersal distances were relatively short (20–40 m) in all cases (fig. 19.1). The Weibull distribution model provided a satisfactory fit to the data (fig. 19.1, table 19.1). Values of *a* and *N* increased with tree size, while *m* was more or less constant having large overlaps in the 95% confidence limits among the three sample trees (table 19.1, fig. 19.2). Exponential and linear functions appeared to fit the relationship between tree height and *a*, and that between dbh and *N*, respectively (fig. 19.2), although the sample size was too small to evaluate the significance of the fit.

Of 393 *D. aromatica* trees ≥ 30 cm dbh in the 52-ha plot, 143 (36%) fruited in 1997. The proportion of fruiting individuals increased with diameter in

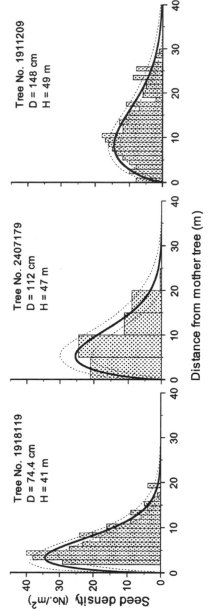

Fig. 19.1. Relationships between the distance from mother trees and densities of dispersed seeds for three *Dryobalanops aromatica* trees. Solid lines are regression lines based on the Weibull-distribution seed dispersal model. Dotted lines present 95% confidence interval. No data for the distance 0–1.8 m for Tree No. 1918119.

Table 19.1. The Estimated Coefficients of the Weibull-Distribution Model of Seed Dispersal for Three *Dryobalanops aromatica* Trees

Tree Number	dbh (cm)	Height (m)	n	a	m	N	r^2
1918119	74.4	41	20	6.75 ± 0.703	1.49 ± 0.256	11,942 ± 2465	0.913
2407179	112	47	8	9.68 ± 1.71	1.59 ± 0.306	17,681 ± 5900	0.960
1911209	148	49	30	14.26 ± 1.63	1.66 ± 0.21	21,329 ± 4662	0.653

Notes: 95% Wald confidence limits are also shown. N: total seed production; a: a scaling parameter; m: a shape parameter; r^2: measures of fit for fitted dispersal models; n: sample size or the number of seed census quadrats.

increments between 30 and 60 cm dbh, then became constant at larger diameters (fig. 19.3). The coefficient b in the logit model was significantly larger than zero, indicating size-dependent fruiting ($p < 0.05$).

We estimate that the *D. aromatica* trees dispersed about 1.17 million seeds in the 46-ha plot in 1997 (table 19.2). Less than half the 46-ha plot was predicted to have seed densities ≥ 0.0001 per m^2. The cumulative area of high seed density (≥ 10 seeds/m^2) was only 3.7 ha (8.1% of the plot); however, 84% of the seeds were dispersed in this area. Dispersed seeds were largely concentrated around mother trees due to short dispersal ranges (figs. 19.4 and 19.5). There were areas with *D. aromatica* adults where no seed input was expected in 1997 because no adults had fruited there. Overlap among seed shadows was rather low. Ninety-eight percent of the 46-ha plot was expected to receive seeds from fewer than five mother trees, and 80% of the seeds were expected to be dispersed in such sites (table 19.3). Fifty-eight percent of the plot's area was expected to receive no seeds at all (table 19.3).

Potential Seed Input and Sapling Density

The mean potential seed input increased with the local basal area of *D. aromatica* for all radii (radius = distance from the center of focal tree) (fig. 19.6). As was expected from the limited seed dispersal of *D. aromatica* mentioned above, the mean potential input increased more rapidly for larger radius in local basal area calculations because quadrats far from adults, which have small basal areas, would receive far fewer seeds. The mean sapling density also increased with the local basal area except for larger local basal area classes of the 5-m radius. However, total sapling numbers in each local basal area class were significantly different from those expected from the potential seed input for all radii ($X^2 = 4432$–5398, $df = 5$, $p < 0.001$). The sapling–seed ratio had a mode at the basal area class 10^2–10^3 cm^2 for all radii except 5 m, which showed a mode at 10–10^2 cm^2. This indicated that there were less saplings at both smaller and larger local basal area classes than expected from the potential seed input. In contrast to sapling density of *D. aromatica*, mean sapling densities for all

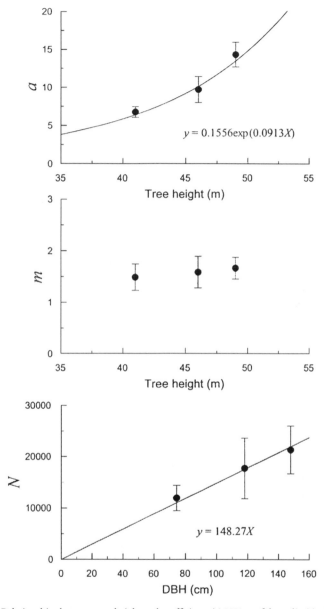

Fig. 19.2. Relationships between tree height and coefficients (±95% confidence limit) of the Weibull-distribution model for three *Dryobalanops aromatica* trees. Solid lines are regression lines of exponential and linear functions for *a* and *N*, respectively.

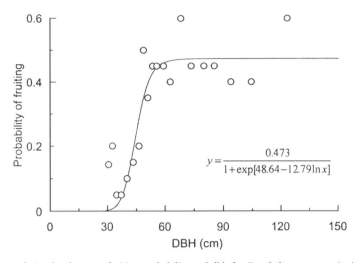

Fig. 19.3. Relationship between fruiting probability and dbh for *Dryobalanops aromatica* in a mast fruiting year, 1997. Circles are means of fruiting probability and dbh bound into sets of $n = 20$ by rank order. The curve is estimated based on a logistic regression function (eq. 19.9).

Table 19.2. Estimated Area of Various Seed Dispersal Densities in the 46-ha Plot

Seed Density (No./m²)	Area		Total Seeds Dispersed	
	ha	%	No.	%
<0.0001	26.5	57.6	—	0
<0.001	2.3	5.0	9	0.00
<0.01	2.6	5.7	102	0.01
<0.1	3.0	6.5	1,224	0.10
<1	3.7	8.0	14,756	1.26
<10	4.3	9.3	171,812	14.6
<100	3.7	8.0	940,416	80.2
≥100	0.04	0.1	44,734	3.81
Total	46	100	1,173,053	100

species together were more or less constant (0.25–0.3 per m²) and independent of *D. aromatica*'s basal area for all radii (data not shown).

Discussion

Estimate of Seed Shadows

Our estimate of seed shadows still presents several possible problems. First, more sample trees are needed for an accurate assessment of the relationship between dispersal parameters (fig. 19.2). Although the estimate of one transect for each

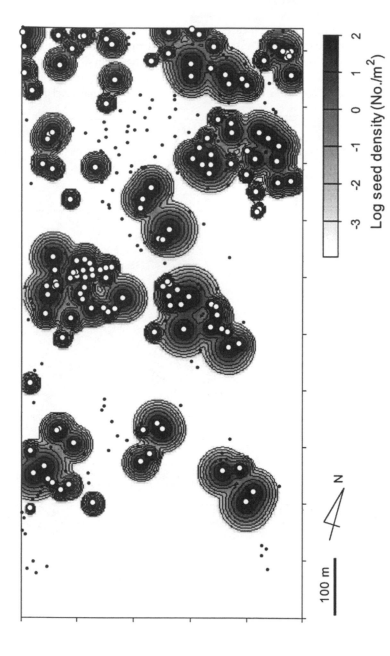

Fig. 19.4. Estimated spatial distributions of seed dispersal of *Dryobalanops aromatica* in a 52-ha plot at Lambir Hills National Park in a mast fruiting year, 1997. Open and closed circles respectively indicate fruiting and nonfruiting trees (≥30 cm dbh).

Log seed density (No./m²)

-3 -2 -1 0 1 2

100 m

N

sample tree was quite good in all cases (fig. 19.1), we need to sample in several directions to confirm our model's assumption of isotropic seed dispersal.

Second, extrapolation of the observed relationship between dbh and N to all fruiting trees may include large errors. Because we selected only heavily fruiting trees—their fruit production being not quite twice the normal level (A. Itoh

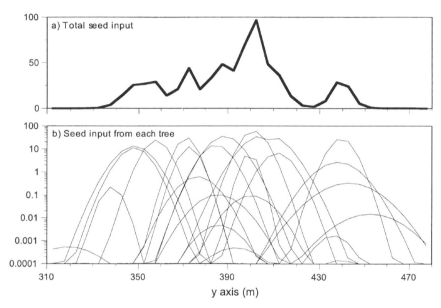

Fig. 19.5. An example of total seed input (a) and estimated individual seed shadows (b) of mother trees of *Dryobalanops aromatica*.

Table 19.3. Distribution of Areas in the 46-ha Plot Based on the Number of Mother Trees from Which Seeds Were Dispersed to Each Area

No. Mother Trees	Area		Total Seeds Dispersed	
	ha	%	No.	%
0	26.5	57.5	—	0
1	8.2	17.7	157,764	13.4
2	5.9	12.7	360,222	30.7
3	3.0	6.6	223,905	19.1
4	1.4	3.1	170,806	14.6
5	0.66	1.4	142,369	12.1
6	0.23	0.49	52,601	4.5
7	0.10	0.21	24,910	2.1
8	0.06	0.12	20,696	1.8
9	0.03	0.05	12,992	1.1
10	0.01	0.02	5,433	0.46
>10	0.01	0.01	1,355	0.12
Total	46	100	1,173,053	100

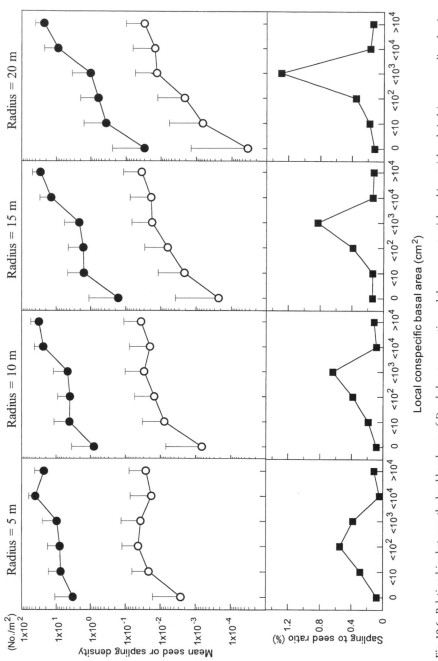

Fig. 19.6. Relationships between the local basal area of *Dryobalanops aromatica* and the mean potential seed input (closed circle), mean sapling density (1 cm $\leq$ dbh < 2 cm) (open circles), and sapling–seed ratio (squares). Each column is based on a different radius in calculation of the local basal area. Vertical bars are the standard deviations.

personal observation)—our estimate of total seed production (1.29 million per 52-ha) could be an overestimate. In contrast to the coefficients a and m, which are mostly determined by characteristics such as seed weight, seed shape, and height of the mother tree, N, may differ even among trees of the same size depending on the degree of fruiting. Thus, for an accurate estimate of the relationship between N and tree size, we may need many more sample trees than the number required for estimates of a and m. However, if we already have good estimates of a and m, we need only a few samples from each tree to estimate N.

The third potential problem of our model is the effect of topography. In this study, we assumed the plot's topography was flat. However, the actual topography of the Lambir plot is quite steep and complex (Yamakura et al. 1995). The dependence of a values on tree height suggests that topography might significantly affect the dispersal pattern. If a tree stands on a steep slope or cliff, the downhill and uphill dispersal patterns may be significantly different. Below, we consider a simple case where a mother tree stands at the edge of a 5-m cliff. In our model, a can be expressed by

$$a = \frac{\alpha H}{F},\tag{19.10}$$

where F is the mean falling speed of a seed (m/s), H is tree height (m), and α is a scaling parameter of the Weibull distribution for wind speed. Supposing that α is constant and independent of topography, a for the downhill direction is slightly larger than that for flat topography and is written as

$$a' = \frac{\alpha(H+5)}{F}.\tag{19.11}$$

Since m and N are independent of topography, the seed dispersal curve in the downhill direction shifts to a slightly greater distance. An example of this change in the dispersal curve is shown in figure 19.7 using the data of a sample tree ($H = 41$ m). A more complicated adjustment is therefore needed for sloping topography.

Density Dependence in Recruitment

The preliminary results of the density-dependence analysis demonstrated the importance of taking into account seed dispersal when performing an analysis of recruitment. The sapling density of *D. aromatica* was positively related to conspecific basal area and showed no sign of site- or density-dependent effects on recruitment. However, the sapling–seed ratio provided a different view of sapling recruitment. The sapling–seed ratio showed a peak at an intermediate basal area class (fig. 19.6), suggesting an increase in sapling recruits at sites with small conspecific basal areas and a decrease at those with large basal areas.

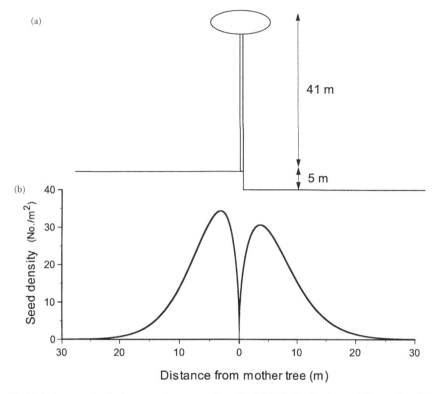

Fig. 19.7. An example of adjustment for topography of the Weibull-distribution seed dispersal model. (a) A 41-m mother tree stands at the ridge of a 5-m cliff. (b) Estimated seed shadows are shown in the uphill direction (*left*) and in the downhill direction (*right*). Note that scales are different for (a) and (b).

However, the changes in the sapling–seed ratio cannot be explained by sapling competition alone since the density of saplings of all species together was not correlated with the basal area of *D. aromatica*. In addition, the density of *D. aromatica* saplings was not so large even on the quadrats with the highest basal area that *D. aromatica* dominates the whole sapling community. Some conspecific density-dependent mortality may occur between the stages of seed dispersal and sapling establishment, and this may regulate the sapling density of *D. aromatica*.

At the local scales of 15–20 m radius, the sapling–seed ratio decreased at quadrats with conspecific basal areas $> 10^3$ cm^2, which is equivalent to an adult of 36 cm dbh. It is notable that this ratio decreased even in the quadrats with 10–10^2 cm^2 conspecific basal areas for a 5-m radius, which is probably because 86%

of these quadrats also had a conspecific basal area $> 10^3$ cm^2 within a 20-m radius. These results suggest that the presence of an adult within at least 20 m reduces the probability of sapling recruitment for *D. aromatica.* The mechanisms of the density effect on recruitment of *D. aromatica* saplings are not clear at present.

There are several possible explanations for the increase in the sapling–seed ratio at quadrats with smaller conspecific basal areas. First, the potential seed input could be overestimated for the quadrats farther from mother trees. This could occur because the Weibull distribution has positive seed densities for all distances up to infinity though we had no data for the sites >40 m from mother trees. To check this, we recalculated the potential seed input using only trees less than 30 m from each quadrat. As was expected, the mean potential seed input decreased for small basal area sites, but positive relationships were still found in these sites for all local scales, where the radius was 5–20 m (data not shown). Therefore, the positive relationships are unlikely to be an artifact due to the overestimation of seed input at sites farther from mother trees.

The second possible explanation is the observed site preference of *D. aromatica* in the Lambir plot. *D. aromatica* is an apparent edaphic specialist, whose distribution is mostly restricted to sandy ridges in the plot (Itoh et al. 2003). The quadrats with a low basal area of *D. aromatica* were mostly in habitat unsuitable for that species; seed and seedling mortality would be high under such subplots. In order to examine the effects of habitat on seeds and seedling mortality, we have started monitoring the mortality and growth of seedlings of all tree species including *D. aromatica* for 1300 2 × 2 m seedling plots established at every 20 × 20 m grid point in the Forest Dynamics Plot in Lambir.

Another possibility is that adults in suitable habitats, i.e., sandy ridges, would produce seeds more frequently than those in unsuitable sites. Itoh et al. (in press) reported that the fruiting probability of *D. aromatica* adults was significantly higher on higher elevations and sandy soil sites in 1997 at Lambir. If trees on sandy ridges would produce seeds more frequently, long-term seed input on these sites should be larger than those on unsuitable habitats. Our calculation of potential seed input did not include habitat dependence in seed production. Further studies are required to examine how the observed habitat dependence in fruiting affects local seed input and relationships between local conspecific basal area and recruitment.

This study showed the significance of seed dispersal data in analyzing population dynamics, though a proper statistical analysis is needed to confirm the density dependence and habitat effects observed in this study (e.g., Wills et al. 1997). Our preliminary analysis suggested that both density and habitat may simultaneously limit recruitment of *D. aromatica* in the Forest Dynamics Plot in Lambir. This is partly because our study plot included highly heterogeneous habitats (Yamakura et al. 1996) and the study species is a clear habitat specialist (Itoh et al. 2003).

Density dependence might be more important than habitat effects in more uniform habitats and for habitat generalist species. It would be interesting to compare the relative importance of habitat and density-dependent effects among different sites and different species using the Forest Dynamics Plots in various regions.

General Implications

1. This study illustrates the importance of seed input data for the assessment of recruitment success. Recruitment success in a local site should be assessed as recruits per seed, because recruitment can only occur where seeds land. Seed input data are especially significant for the analysis of density dependence and habitat effects at low-density sites where few or no adults are present.

2. Large-scale Forest Dynamics Plots provide basic data for seed shadow estimates (i.e., the sizes, location, and species identification of all trees). We showed one possible method, which utilized the seed dispersal patterns of isolated trees and spatial distribution of fruiting adults in the Lambir Forest Dynamics Plot. This method must be refined to develop more reliable estimates of seed input of other tropical rainforest species.

3. Our preliminary analysis of density dependence on the recruitment of a common canopy species suggested that both density-dependent and habitat effects may simultaneously limit recruitment. Relative importance of the two effects may differ between tree species in a forest and between forests with differing habitat heterogeneity. It would be interesting to compare the relative importance among the Forest Dynamics Plots across the tropics.

References

Andersen, M. 1991. Mechanistic models for the seed shadows of wind-dispersed plants. *American Naturalist* 137:476–97.

Appanah, S. 1985. General flowering in the climax rain forests of south-east Asia. *Journal of Tropical Ecology* 1:225–40.

Augspurger, C. K., and Franson, S. E. 1988. Input of wind–dispersed seeds into light–gaps and forest sites in a neotropical forest. *Journal of Tropical Ecology* 4:239–52.

Ashton, P. S. 1982. Dipterocarpaceae. *Flora Malesiana Series 1* 9:237–552.

Clark, D. B., Clark, D. A., and Read, J. M. 1998. Edaphic variation and the mesoscale distribution of tree species in a neotropical rain forest. *Journal of Ecology* 86:101–12.

Clark, J. S., Macklin, E., and Wood, L. 1998. Stages and spatial scales of recruitment limitation in southern Appalachian forests. *Ecological Monographs* 68:213–35.

Clark, J. S., Beckage, B., Camill, P., Cleveland, B., HillerRisLambers, J., Lichter, J., McLachlan, J., Mohan, J., and Wyckoff, P. 1999. Interpreting recruitment limitation in forests. *American Journal of Botany* 86:1–16.

Condit, R., Hubbell, S. P., and Foster, R. B. 1992. Recruitment near conspecific adults and the maintenance of tree and shrub diversity in a neotropical forest. *American Naturalist* 140:261–86.

Connell, J. H. 1971. On the role of natural enemies in preventing competitive exclusion in some marine animals and in rain forest trees. Pages 298–312 in P. J. den Boer and G. R. Gradwell, editors. *Dynamics of Populations. Proceedings of the Advanced Study Institute on Dynamics of Numbers in Populations.* Center for Agricultural Publishing and Documentation, Wageningen, The Netherlands.

Curran, L. M., Caniago, I., Paoli, G. D., Astianti, D., Kusneti, M., Leighton, M., Nirarita, C. E., and Haeruman, H. 1999. Impact of El Niño and logging on canopy tree recruitment in Borneo. *Science* 286:2184–88.

Greene, D. F., and Johnson, E. A. 1989. A model of wind dispersal of winged or plumed seeds. *Ecology* 70:339–47.

Hubbell, S. P., and Foster, R. B. 1990. Structure, dynamics, and equilibrium status of old-growth forest on Barro Colorado Island. Pages 520–41 in A. Gentry, editor. *Four Neotropical Forests.* Yale University Press, New Haven, CT.

Inoue, E. 1963. On the turbulent structure of airflow within crop canopies. *Journal of the Meterological Society of Japan* 41:317–26.

Itoh, A. 1995. *Regeneration Processes and Coexistence Mechanisms of Two Bornean Emergent Dipterocarp Species.* Ph.D. dissertation. Kyoto University, Kyoto, Japan.

Itoh, A., Yamakura, T., Ogino, K., Lee, H. S., and Ashton, P. S. 1995. Population structure and canopy dominance of two emergent dipterocarp species in a tropical rain forest of Sarawak, East Malaysia. *Tropics* 4:133–141.

Itoh, A., Yamakura, T., Ohkubo, T., Kanzaki, M., Palmiotto, P., Tan, S., and Lee, H. S. 2003. Spatially aggregated fruiting in a Bornean emergent tree. *Journal of Tropical Ecology* 19:531–38.

Itoh, A., Yamakura, T., Kanazaki, M., Ohkubo, T., Palmiotto, P. A., LaFrankie, J. V., Ashton, P. S., and Lee, H. S. 2003. Importance of topography and soil texture in spatial distribution of two sympatric emergent dipterocarp species in a Bornean rain forest. *Ecological Research* 18:307–320.

Janzen, D. H. 1970. Herbivores and the number of tree species in tropical forests. *American Naturalist* 104:501–28.

Manokaran, N., and LaFrankie, J. V. 1990. Stand structure of Pasoh Forest Reserve, a lowland rain forest in peninsular Malaysia. *Journal of Tropical Forest Science* 3:14–24.

Masaki, T., Kominami, Y., and Nakashizuka, T. 1994. Spatial and seasonal patterns of seed dissemination of *Cornus controversa* in a temperate forest. *Ecology* 75:1903–10.

Momose, K., Nagamitsu, T., Sakai, S., Inoue, T., and Hamid, A. A. 1994. Climate data in Lambir Hills National Park and Miri Airport, Sarawak. Pages 28–39 in T. Inoue and A. A. Hamid, editors. *Plant Reproductive Systems and Animal Seasonal Dynamics: Long-Term Study of Dipterocarp Forests in Sarawak.* Center for Ecological Research, Kyoto University, Kyoto, Japan.

Okubo, A., and Levin, S. A. 1989. A theoretical framework for data analysis of wind dispersal of seeds and pollen. *Ecology* 70:329–38.

Rokujo, N. 1998. *Seed Dispersal and Seedling Establishment of Tropical Trees.* Masters thesis. Osaka City University, Osaka, Japan.

Shibata, M., and Nakashizuka, T. 1995. Seed and seedling demography of four co-occurring *Carpinus* species in a temperate deciduous forest. *Ecology* 76:1099–08.

Thomas, S. C. 1996. Relative size at onset of maturity in rain forest trees: A comparative analysis of 37 Malaysian species. *Oikos* 76:145–54.

Watson, H. 1985. *Lambir Hills National Park: Resource Inventory with Management Recommendations.* National Parks and Wildlife Office, Forest Department, Kuching, Sarawak, Malaysia.

Wilkinson, L. 1997. *SYSTAT: The System for Statistics.* SYSTAT, Inc., Chicago.

Wills, C., Condit, R., Foster, R. B., and Hubbell, S. P. 1997. Strong density- and diversity-related effects help to maintain tree species diversity in a neotropical forest. *Proceedings of the National Academy of Science* 94:1252–57.

Willson, M. F. 1993. Dispersal mode, seed shadows, and colonization patterns. *Vegetatio* 107/108:261–80.

Yamakura, T., Hagihara, A., Sukardjo, S., and Ogawa, H. 1986. Aboveground biomass of tropical rain forest stands in Indonesian Borneo. *Vegetatio* 68:71–82.

Yamakura, T., Kanzaki, M., Itoh, A., Ohkubo, T., Ogino, K., Chai, E. O. K., Lee, H. S., and Ashton, P. S. 1995. Topography of a large-scale research plot established within the Lambir rain forest in Sarawak. *Tropics* 5:41–56.

Yamakura, T., Kanzaki, M., Itoh, A., Ohkubo, T., Ogino, K., Chai, E. O. K., Lee, H. S., and Ashton, P. S. 1996. Forest structure of the Lambir rain forest in Sarawak with special reference to the dependency of physiognomic dimensions on topography. *Tropics* 6:1–18.

Zar, J. H. 1974. *Biostatistical Analysis.* Prentice-Hall, Englewood Cliffs, NJ.

20

Seed Dispersal and Density-Dependent Seed and Seedling Survival in *Trichilia tuberculata* and *Miconia argentea*

Helene C. Muller-Landau, James W. Dalling, Kyle E. Harms, S. Joseph Wright, Richard Condit, Stephen P. Hubbell, and Robin B. Foster

Introduction

Tropical Forest Dynamics Plots were established to improve our understanding of the structure and dynamics of tropical forests and the population biology of tropical tree species. Such factors include density-dependent mortality due to the action of species-specific pests (Janzen 1970; Connell 1971), niche partitioning with respect to habitat (Ashton 1969; Ricklefs 1977) and recruitment limitation (Hurtt and Pacala 1995; Hubbell et al. 1999). Janzen–Connell effects are posited to maintain species diversity by giving species an advantage when they are relatively rare; habitat specialization can maintain diversity if some species are competitively dominant in different areas (Leigh 1996).

The large-scale patterns evident in the datasets collected in the Barro Colorado Island Forest Dynamics Plots (chap. 24) reveal the presence of both Janzen–Connell effects and habitat specialization (Hubbell and Foster 1983; Condit 1998; chap. 2). Higher sapling mortality in areas of high conspecific density in the common tree species *Trichilia tuberculata* (Meliaceae) and *Alseis blackiana* (Rubiaceae) suggests the operation of Janzen–Connell effects at this stage; however, most species showed no significant effects (Hubbell et al. 1990). Decreased per capita sapling recruitment near adults or in areas of high conspecific density has also been interpreted as reflecting the action of Janzen–Connell effects at earlier life stages (Hubbell and Foster 1986; Condit et al. 1992). Similarly, differential mortality rates in the various habitats of the plot (slope, plateau, swamp) suggest there may be habitat specialization among adults, while differential recruitment might reflect habitat-specific adaptations among seeds and seedlings of different species (Harms 1997). Most species, however, do not show significant differences in mortality or recruitment rates across habitat types (Welden et al. 1991, Harms 1997).

Yet the Forest Dynamics Plot data alone provide limited insight into the seed and seedling stages, which is when Janzen–Connell effects, habitat-specific mortality, and many other processes are thought to act most strongly. Most mortality is concentrated at early stages, when individuals are most highly vulnerable because of their small size and low stored reserves; thus, differences are most likely to arise then and are more likely to matter. The original hypotheses of Janzen and Connell focused on seed predation and seedling herbivory, respectively, and numerous studies of smaller seeds and seedlings have found substantial effects on survival and growth at those stages (Augspurger 1983, 1984; Augspurger and Kelly 1984; Clark and Clark 1984; Schupp 1988b, 1988a; Schupp and Frost 1989; Schupp 1992; Harms et al. 2000). Because Forest Dynamics Plot censuses include only individuals greater than 1 cm in diameter, they provide no direct evidence for processes at smaller size classes. The indirect evidence from patterns within data on larger size classes should be interpreted with caution, since different combinations of processes can give rise to the same patterns.

Spatial and demographic patterns at larger size classes are jointly influenced by dispersal distances, microhabitat preferences, and Janzen–Connell effects (Hamill and Wright 1986). For example, a pattern of recruits spread widely relative to parents may reflect moderate dispersal combined with strong Janzen–Connell effects, moderate dispersal combined with specialization upon a sparsely distributed habitat, or simply long dispersal distances. Without specific knowledge of the dispersal, habitat preferences, or Janzen–Connell effects at smaller size classes, inferences drawn from the pattern alone are tenuous at best. They are made even more problematic by the fact that 1-cm diameter saplings recruiting into the census are actually quite old—the mean age has been estimated at 17 years (Hubbell 1998). Thus, by the time many are recorded as recruits, the adults that produced them may have died or disappeared.

Studies of seeds and seedlings thus provide valuable complementary information necessary for understanding important processes such as Janzen–Connell effects. The location of such studies within Forest Dynamics Plots, where the locations and sizes of adults are known in a large area, has the added advantage of making possible investigation of seed dispersal and Janzen–Connell effects on an unprecedented scale. Indeed, studies of seed fall and seedling recruitment within the 50-ha plot on BCI clearly indicate the presence of negative density dependence at the seedling recruitment stage for all 53 species investigated (Harms et al. 2000). Furthermore, information on these early life stage processes can thereby enable better interpretation of, and even prediction of, patterns in the Forest Dynamics Plot data.

In this chapter, we describe studies of seed and seedling biology for two species of contrasting life history strategies. *Trichilia tuberculata* is relatively large-seeded, shade tolerant, and the commonest canopy tree on the plot; *Miconia argentea*]Melastomataceae] is a small-seeded, light-demanding pioneer. The methods of each study were tailored to the particular biology of the species, and thus were somewhat different from each other. We show how both studies can be integrated with the Forest Dynamics Plot data to provide a coherent picture of the early life stages of these species, including their seed dispersal patterns, regeneration habitat preferences, and density dependence at early life stages

For each species we address these questions:

1. What are the patterns of seed dispersal? How many seeds are produced and how far are they dispersed? In what habitats does regeneration occur?
2. Are Janzen–Connell effects evident at the seed, seedling, and sapling stages? If so, at what spatial scales and with what relative strength?
3. Are spatial and demographic patterns in the Forest Dynamics Plot data consistent with predictions based upon seed and seedling biology?

Materials and Methods

Study Site and Species

Trichilia tuberculata (formerly *Trichilia cipo*) is a medium-sized dioecious tree in the family Meliaceae, common throughout much of the neotropics. Its fruits are capsules 11–18 mm long, orange at maturity, with 3–4 valves folding back to expose a shiny red aril covering 1–2 seeds, which have an average dry weight of 0.15 g (Croat 1978; S. J. Wright, unpublished data). The fruits mature mostly in September and October and are dispersed primarily by mammals and large birds (Croat 1978; Leighton and Leighton 1982; S.J. Wright, unpublished data). *Trichilia* is shade tolerant (Welden et al. 1991) and its seedlings germinate in both shade and sun within 2–3 weeks (Howe 1980; Garwood 1983; De Steven 1994). It is the most common canopy species in the BCI Forest Dynamics Plot, occurring throughout the plot, in shade and gaps, and on the drier plateau and wetter slopes (Hubbell and Foster 1983).

Miconia argentea is a medium-sized monoecious tree in the family Melastomataceae (Croat 1978). Its fruits are berries, 4–8 mm in diameter, blue-purple at maturity, with 1–80 seeds per fruit, each weighing 0.08 mg (Croat 1978; Dalling et al. 1998b). They mature from January to June, and are dispersed by mammals and birds (Poulin et al. 1999), with secondary dispersal by ants (Dalling et al. 1998a, 1998b). Although seed mortality rates are high, some seeds remain dormant for up to 5 years until stimulated to germinate by conditions of high light availability, such as those of a gap; thus, seedlings are found almost exclusively in gaps (Dalling et al. 1998b). It is the second most common pioneer tree within the Forest Dynamics Plot (Dalling et al. 1998b).

This study was conducted in the seasonally moist tropical forest of the 50-ha Forest Dynamics Plot on Barro Colorado Island, Panama. The Forest Dynamics Plot data themselves were used to investigate survival of 1–1.9 cm diameter saplings of both species, to provide information on the locations of adults for the seed dispersal analyses, and in calculations of local conspecific basal area density for the survival analyses.

Seed Traps and Seedling Plots for Trichilia tuberculata

In December 1986 two hundred seed traps were placed along the 2.7 km of trails within the 50-ha Forest Dynamics Plot (fig. 20.1a). Each seed trap consisted of a square, 0.5-m² PVC frame supporting a shallow, open-topped, 1-mm nylon-mesh bag, suspended 0.8 m above the ground on four PVC posts. The average distance between nearest neighbor seed traps was 18.9 ±3.6 m (SD). Beginning in January 1987 and continuing to the present, seed traps were emptied weekly and damaged traps replaced or repaired as needed. All seeds, fruits and seed-bearing fruit fragments >1 mm in diameter falling into the traps were identified to species and recorded. Fruits were categorized as aborted, immature, damaged, fragments, and mature. Only data on seeds and mature fruits falling between January 1, 1987, and January 1, 1998, were used in this analysis (Wright and Calderón 1995; Muller-Landau et al. 2002).

On the three sides of each seed trap away from the nearest trail, and therefore away from the narrow path used to reach the seed trap, seedling census plots were established in January–March 1994. Each seedling-plot was 1 × 1 m, and 2 m distant from its associated seed trap. All seedlings <50 cm tall were identified, measured, and marked. Seedling censuses were repeated in January–March in 1995, 1996, 1997, and 1998. During each seedling census all previously marked seedlings were remeasured, and all new recruits into the seedling plots were marked, measured, and identified.

Soil Seed Samples and Seedling Plots for Miconia argentea

The seeds of Miconia argentea are smaller than the mesh size of the seed traps, so an unknown proportion pass through the traps undetected. We therefore present data from a separate study conducted to examine seed rain and seed survival specifically in this species, as well as in Cecropia insignis (Cecropiaceae) (Dalling et al. 1997, 1998b, 2002). Soil samples were taken at two locations below the crown, and at 5, 10, 20, and 30 m from the crown edge along each of four transects radiating from the crown center of four Miconia trees and from four other points (Cecropia trees), for a total of 192 sampling sites (fig. 20.1b). Samples were taken using a 10.3-cm diameter, 3-cm deep soil corer, yielding a 250-cm³ soil sample at each site. Survival of ungerminated Miconia seeds in the soil was determined by comparing samples taken in May 1993, shortly after the end of the fruiting season, with those from February 1994, shortly before the start of the following fruiting season. Soil samples were placed in a greenhouse and the viable seed density estimated from counts of seedlings that emerged over the following 6 weeks.

(a) *Trichilia tuberculata*

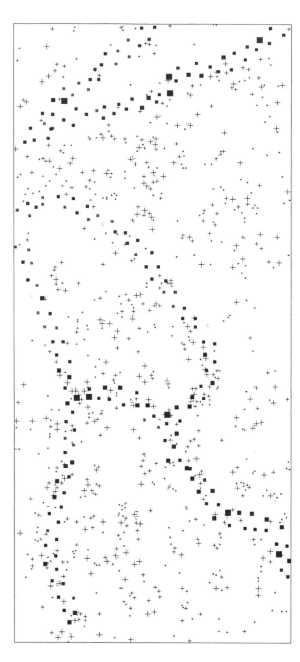

- `.` trees 20-30 cm dbh
- `+` trees 30-40 cm dbh
- `+` trees 40-80 cm dbh
- `□` traps with 0 seeds
- `⊠` traps with 1-10 seeds
- `■` traps with 11-100 seeds
- `■` traps with 101-1000 seeds
- `■` traps with 1001-10000 seeds

Fig. 20.1. Map of the sampling sites and adult trees of the study species. (a) Seed traps and adults of *Trichilia tuberculata*. (b) Soil seed samples and adults of *Miconia argentea*.

(b) *Miconia argentea*

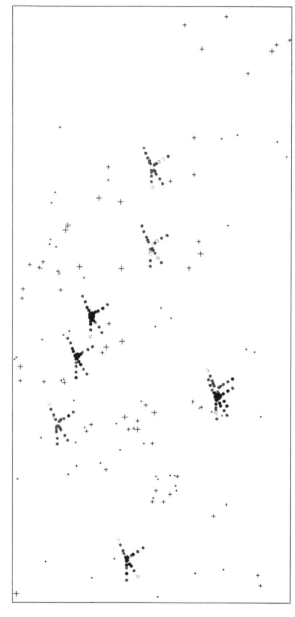

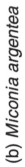

.	trees 6.7-10 cm dbh
+	trees 10-20 cm dbh
+	trees 20-40 cm dbh
○	soil samples with 0 seeds
•	samples with 1-10 seeds
•	samples with 11-100 seeds
●	samples with 101-1000 seeds

Fig. 20.1. (*Continued*)

Since *Miconia* recruits exclusively in gaps, its seedlings are poorly represented in the 600 m² of regularly placed seedling plots described above, most of which are in the understory. Mortality data are, however, available for young *Miconia* seedlings 10–50 cm tall that were censused in March 1996 and again in March 1997 in 53 gaps that formed between 1993 and 1995 (Dalling et al. 1998a). Seedlings were mapped to the nearest meter and marked.

Seed Production and Dispersal Analyses

We used the data on the location of and number of seeds in seed traps or soil seed samples and on sizes and locations of adults within the Forest Dynamics Plot to fit the probability of seed arrival as a function of distance from an adult tree and to fit fecundity as a function of tree size. Starting from a set of parameters specifying these functions, we calculated expected seed rain into a given trap as the sum of contributions from conspecific adult trees on the plot, with those contributions determined by their distances from the trap and their sizes, according to the parameter values. We then searched for parameter values that produced the best fit to the observed seed rain, using maximum likelihood methods (Ribbens et al. 1994; Clark et al. 1999).

Dispersal kernels—functions giving the probability density of seeds at different distances from the parent—were fitted using the 2Dt model introduced by Clark et al. (1999):

$$f(x) = \frac{p}{\pi u \left(1 + \frac{x^2}{u}\right)^{p+1}}$$

where $f(x)$ is the probability density of seeds at a distance x from a parent tree, and p and u are fitted parameters; that is, $f(x)\,dA$ is the expected proportion of the total seed fall to be found in an area dA a distance x from the parent tree. This model provided a better fit than exponential or Gaussian models, which were also tested. Since the parameters of this model are not easily interpretable, we present median dispersal distances, in addition to parameter values, for the best-fit curves (see the appendix for the formula). Contributions to seed rain from trees off the plot were estimated by assuming a uniform density of adult trees there equivalent to that found on the plot, and again weighting by distance (e.g., for *Trichilia*, we assumed 2.05 cm² of reproductive basal area on every square meter of land).

Fecundity was assumed to be proportional to basal area, with a single fitted parameter β for seed production per square centimeter of basal area. Adult size data from the 1985, 1990, and 1995 censuses were used, as appropriate; sizes were interpolated between census years using the assumption of constant absolute growth rates. Since no data were available on exactly which adults were reproductive during the time of the study, or in the case of *Trichilia*, on which were even females, all adults were included as potential parents. On the basis of data collected by S. J. Wright (unpublished), adults were defined as trees having diameters greater than 2/3 of the adult cutoff originally estimated by Robin Foster (unpublished); that is, we included *Trichilia* trees greater than 20 cm, and *Miconia* trees greater than 6.67 cm. We further used our estimates of seed production per unit basal area to estimate total seed production of each species on the plot, simply by multiplying the species-specific estimates of seed production per unit basal area β by the total basal area of adults on the Forest Dynamics Plot.

The distribution of observed values for seed rain into traps around expected values was assumed to follow a negative binomial distribution (Clark et al. 1999). The overdispersion parameter k of the negative binomial was thus the fourth and final fitted parameter. Low values of k correspond to high variances in observed values around the expected values, reflecting clumping of seed rain.

For *Trichilia*, dispersal kernels and seed production functions were fitted to the counts of seed equivalents falling into seed traps; seed equivalents were defined as the number of seeds plus 1.7 times the number of unopened mature fruits, since the mean seed-to-fruit ratio for *Trichilia* is 1.7 (S. J. Wright, unpublished data). (Fractional values for trap contents were rounded to the nearest integer.) Best fits were found both for all years combined and separately for each calendar year when possible (sample sizes proved too small for reliable estimates in 1993 and 1996). Since fruit production and seed fall in *Trichilia* occur in the months August–November, separate calendar years represent distinct fruiting seasons. For *Miconia*, functions were fitted to densities of viable seeds found in the soil samples taken in May 1993, at the end of the fruiting season.

For each analysis, we used a likelihood ratio test to compare the best-fit model against a null model that assumed uniform expected seed rain across the plot (Rich 1988). For illustration of the goodness of fit, we also present Pearson's r^2 values for the fit of model predictions of seed densities to actual seed densities. For calculations of these correlations, densities were first transformed as log (seed number $+$ 1), to reduce deviations from normality (Zar 1974).

Survival Analyses

We examined the dependence of survival at three stages upon several measures of local conspecific density. For *Trichilia*, three dependent variables were examined: the seed-to-seedling survival rate, the first-year seedling survival rate, and the survival of saplings 1–1.9 cm in diameter between 1990 and 1995. The seed-to-seedling survival rate was calculated for each trap by dividing the total number of new seedlings in the three 1-m^2 plots associated with a seed trap in 1995–98 by the number of seeds estimated to have fallen in those plots in 1994–97 given seed rain into the nearby seed trap. The estimated number of seeds was taken to be 6 times the total number of seeds falling into the nearby 0.5-m^2 seed trap in

1994–97, where the multiplier 6 corrects for the different areas of the seedling plots and seed traps. First-year seedling survival probability was calculated for each trap using pooled data for seedlings new in 1995, 1996, 1997, and 1998, weighting all seedlings equally. For *Miconia*, the three dependent variables were survival of seeds from May 1993 to February 1994, the survival of seedlings 10–50 cm tall from March 1996 to March 1997, and the survival of saplings 1–1.9 cm in diameter between 1990 and 1995.

We examined the relationship of seed and seedling survival to several measures of local conspecific density. Total basal area, total reproductive basal area, total number of individuals, and total reproductive individuals of conspecifics within circles of radius 5, 15, and 30 m were used, for a total of 12 independent variables (using the central seed trap as the basis for calculations for *Trichilia* seedlings). Because the results of these analyses were very similar, we present results only for the total basal area. For seed-to-seedling survival and seed survival, we also tested dependence upon local seed density.

The dependence of survival upon local conspecific density was first analyzed using logistic regression. Specifically, we assumed binomial errors and fit the function

$$P(\text{survival}) = \frac{1}{1 + \exp(a + bX)},$$

where X is a measure of local conspecific density. To fit the data on sapling survival over 5 years, this function was raised to the 5th power. Each seed, seedling, and sapling was treated as an independent data point. The effect of conspecific density was considered significant if the presence of the bX term significantly improved the fit for survival according to a log likelihood ratio test (a chi-square test on 2 times the difference in the log likelihoods) (Hilborn and Mangel 1997). In cases where residual deviance exceeded the residual degrees of freedom, residual deviance was rescaled by the residual degrees of freedom and an F-test rather than a chi-square was used to test hypotheses (Crawley 1993). Data were analyzed using JMP statistical software (SAS Institute, Inc., Cary, NC) and the GLIM statistical package (Numerical Algorithms Group, Oxford, U.K.).

Eleven sites at which *Trichilia* seedlings were found but no seeds had been recorded were excluded from the analyses of seed-to-seedling survival for *Trichilia* because they violated the assumptions of the logistic regression. Similarly, we excluded 13 *Miconia* sites where viable seeds were found in the February 1994 samples even though no viable seeds were found in the May 1993 samples. In the three cases where the density of *Trichilia* seeds was less than the density of seedlings but still nonzero, the survival probability was set to 1. These exclusions and changes bias our analyses against finding negative density dependence, since survival probabilities must have been high for seeds or seedlings to be found at sites with initially low seed densities.

Where there are multiple seeds or seedlings per trap or soil sample, the assumption that each seed or seedling is an independent data point is more problematic than usual. In these cases, we also tested the relationship using Spearman rank correlations between the survival rate of seeds and the local density of conspecifics. These analyses treat seed traps, soil seed samples, and seedling plots, rather than individual seeds and seedlings, as independent data points. Since nearest neighbor samples are an average of 19 and 6 m apart, for *Trichilia* and *Miconia*, respectively, we did not further incorporate spatial autocorrelation in the analyses. (For an alternative approach that explicitly incorporates any spatial autocorrelation, see Hubbell et al. 2001.)

Given that the various independent variables are also correlated, we conducted a post-hoc path analysis to separate the influences of seed density and adult basal area within 15 m upon seed-to-seedling survival in *Trichilia* and upon seed survival in *Miconia*. In these analyses, as in the Spearman rank correlations, seed traps and seed samples were the units of analysis. Because path analysis depends upon the data being normally distributed, we first applied an arcsine-square-root transformation to survival probability, and log-transformed seed and basal area densities. The transformed variables were more normally distributed than the untransformed ones; however, they still failed tests of normality (Kolmogorov–Smirnov and Shapiro–Wilks W tests, $p < 0.05$), and so the results of these analyses must be interpreted with caution.

Results

Seed Production and Dispersal

Miconia had more seed rain, higher estimated seed production per unit basal area, and longer estimated dispersal distances than *Trichilia*. A total of 28,276 seed equivalents of *Trichilia* were captured in the total 100-m² area of seed traps during the 11 years of the study; for *Miconia*, 3527 viable seeds were recovered from 3-cm deep soil seed samples covering a total area of 1.6 m² in May 1993, and 544 in February 1994 (fig. 20.1). *Miconia* seed densities must have been even higher, since there is high mortality of seeds, especially below the crown where only 20% of seeds are incorporated into the seed bank (Dalling et al. 1997). Estimated dispersal and fecundity models provided a good fit to the data, explaining 47% of the spatial variation in *Trichilia*, and 64% in *Miconia* (table 20.1). Estimated median dispersal distances were 6.5 m for *Trichilia* and 52 m for *Miconia* (fig. 20.2a). Because the *Miconia* samples were of viable seeds in the soil, the dispersal distance for this species is better interpreted as the median distance of viable seeds from their parents. This is probably considerably higher than the median dispersal distance

Table 20.1. Parameters, Dispersal Distances, and Measures of Fit for Fitted Dispersal Models

Species	Year	No. seeds	β (seeds cm^{-2} yr^{-1})	p	u (m^2)	k	Median Distance (m)	r^2	p
			Fitted Dispersal Parameters						
Miconia argentea	1993	3527	6109	0.16	39.9	1.76	51.5	.64	<0.0001
Trichilia tuberculata	All	28276	18.1	0.94	38.3	0.73	6.5	.47	<0.0001
	1987	2165	14.6	0.84	31.9	0.37	6.4	.22	<0.0001
	1988	3608	23.6	0.74	25.6	0.33	6.2	.26	<0.0001
	1989	1575	12.7	1.72	123.7	0.28	7.9	.20	<0.0001
	1990	1572	12.1	0.58	14.4	0.21	5.8	.17	0.0003
	1991	1844	12.2	0.95	46.0	0.32	7.0	.24	<0.0001
	1992	7067	50.3	0.92	26.9	0.47	5.5	.34	<0.0001
	1993	46	—	—	—	—	—	—	NS
	1994	2093	16.8	0.97	34.3	0.32	5.9	.30	<0.0001
	1995	956	7.7	0.51	18.9	0.31	7.4	.18	0.0003
	1996	421	—	—	—	—	—	—	NS
	1997	6936	52.2	0.94	32.8	0.43	5.9	.28	<0.0001

Notes: No. seeds is the total number of seed equivalents captured in the traps in that year. β is the fitted fecundity parameter, giving seed production in seeds per cm^2 basal area of reproductively sized adults per year; p and u are parameters of the fitted Clark 2Dt dispersal kernel (see methods); and k is the fitted clumping parameter of the negative binomial error distribution (smaller values reflect greater clumping). The median dispersal distance of the fitted dispersal kernels is also presented for ease of comparison. The r^2 values are for Pearson correlations of actual versus fitted log (seed number + 1); p-values are for likelihood ratio test comparisons of the fitted dispersal kernels with the nonspatial null models.

owing to seed dormancy, which allows seeds from previous years' seed rain to persist for several years, and to the higher levels of seed mortality near adults (Dalling et al. 1997).

Estimated seed production per unit basal area was 18 seeds per cm^2 per year in *Trichilia*, while for *Miconia* it was 6100 seeds per cm^2 per year. This corresponds to an average production of 12,000 seeds by a *Trichilia* tree with a diameter of 30 cm, and 4.3 million seeds for a similarly sized *Miconia* tree. Since *Trichilia* is dioecious, and assuming that half the adults are female, this suggests a female of this size produces an average of 24,000 seeds. Again, the numbers for *Miconia* must be interpreted with more caution because they are based on samples that include some viable seeds dispersed in previous years and miss some dead seeds from the current year.

Given a total basal area of *Trichilia* adults in the Forest Dynamics Plot of 75.5 m^2 in 1995, total seed production on the plot is estimated at 13.7 million seeds per year, or 27.3 seeds per m^2 of the plot per year. This corresponds well to the mean 25.7 seeds per m^2 per year captured in the seed traps. For *Miconia*, total basal area was 2.09 m^2, and thus total seed production is estimated at 128 million seeds per year, or an average of 255 seeds per m^2 per year. Because *Miconia* samples were taken disproportionately under and near fruiting trees, the average seed density predicted in the samples is expected to be considerably higher; it is 1350 seeds per m^2, which compares with the mean observed density of 2200 seeds per m^2.

Considerable interannual variation in seed rain is evident; the coefficient of variation (CV) for annual seed rain in *Trichilia* is 93%, with substantially higher seed fall in the El Niño years of 1992 and 1997 (Wright et al. 1999). When models were fitted to *Trichilia* seed rain data for single fruiting seasons, good fits were obtained for all years except the two with the lowest fruit set: 1993 and 1996. In those years, the majority of sites near adult trees had low (often zero) seed densities, probably reflecting almost complete fruit failure of many trees. Despite the considerable interyear variation in seed production by *Trichilia*, the seed shadows were similar in shape across all years for which they were fitted, as reflected in similar median dispersal distances (table 20.1, fig. 20.2b).

Seed, Seedling and Sapling Survival

Overall survival rates of *Trichilia* were higher than those of *Miconia* in both stages where they could be compared. The overall seed-to-seedling transition probability (encompassing seed and early seedling survival) of *Trichilia* was many times higher than that of *Miconia*. In the seedling plots placed along trails, the mean density of new *Trichilia* seedlings observed was 1.01 seedlings per m^2 per year; for *Miconia* it was 0.0033 per m^2 per year (just 8 seedlings total). Given the estimated seed input, this suggests mean transition probabilities of 3.7% for *Trichilia*, and 0.00079% for *Miconia*. Transition probabilities for *Miconia* were better in the gap plots, where seedling densities were 21 times higher than average, at 0.069 seedlings per m^2 (124 seedlings in 1800 m^2), but even so, the transition probability remains much lower than that of *Trichilia*, just 0.017%. Seedling survival rates for the two species in the datasets analyzed here are not comparable because the *Trichilia* seedlings were all in their first year, while the *Miconia* seedlings included older and larger individuals. The annual survival rate of 1–2 cm dbh saplings was 98.0% for *Trichilia*, and 84.4% for *Miconia*.

Survival rates were found to be strongly negatively associated with conspecific density in seeds and saplings, but weakly or not at all among seedlings. Sample sizes, and thus power, were lowest for the seedling analyses. Results were similar for total basal area, adult basal area, and number of adults; total tree number, however, was not significantly negatively associated with survival, and sometimes it was significantly positively associated (results not shown). Of the distance classes examined, conspecific density within 15 m was more consistently and more strongly associated with survival than was density within 5 or 30 m. For seed survival, seed density itself was a more strongly associated factor.

Seed-to-seedling survival of *Trichilia* and seed survival in *Miconia* were negatively associated with conspecific density in most analyses (tables 20.2 and 20.3).

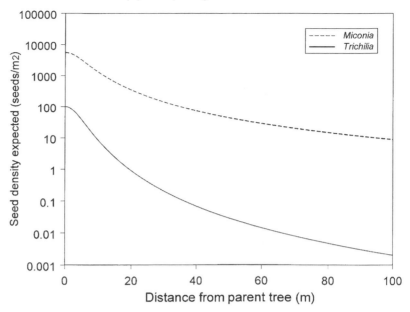

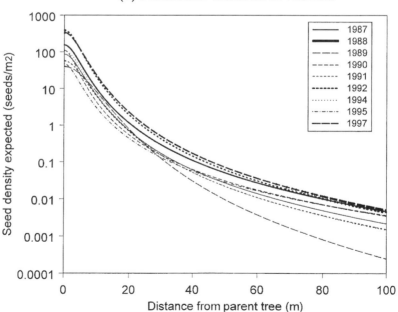

Fig. 20.2. Estimated seed shadows for an adult tree of 30 cm dbh. (a) Comparison of *Miconia* and *Trichilia*. (b) Interannual variation in *Trichilia*; the top two lines are from the El Niño years of 1992 and 1997.

Table 20.2. Parameters of the Logistic Regressions of Survival upon Conspecific Basal Area Density within 5, 15, and 30 m, and with Seed Density

Survival	N Individual	N Sites	BA within 5 $b \pm SE$	BA within 15 $b \pm SE$	BA within 30 $b \pm SE$	Seed Density $b \pm SE$
Trichilia:						
Seed-to-Seedling	60,961	183	.35 ± .02***	.83 ± .04***	.08 ± .07 NS	**1.26 ± .03***
First-Year Seedling	395	97	.02 ± .09 NS	**.43 ± .18***	−.02 ± .38 NS	
Sapling	5,129		.09 ± .01***	.10 ± .02***	**.23 ± .04***	
Micoia:						
Seed	4,250	170	.28 ± .03***	.36 ± .04***	.17 ± .07*	**1.13 ± .08***
Seedling	124		.75 ± .79 NS	.23 ± .28 NS	−.37 ± .19*	
Sapling	403		.11 ± .05*	**.09 ± .02**	.03 ± .03 NS	

Notes: (*) signifies $p < 0.05$, (**) $p < 0.01$, and (***) $p < 0.001$. NS and italics indicate regressions that are not significant at the 0.05 level. Bold indicates that the indicated regressions explain more of the variation in survival for that stage and species than the other ones tested.

Table 20.3. Correlation Coefficients for the Spearman Rank Correlations of Survival upon Conspecific Density

Survival	N Sites	BA within 5 r		BA within 15 r		BA within 30 r		Seed r	
Trichilia:									
Seed-to-seedling	183	−.14	NS	−.45	***	−.32	***	−.67	***
First-year seedling	97	−.01	NS	−.13	NS	−.05	NS	−.09	NS
Miconia: Seed	170	.07		.17	**	.32	***	.12	NS

Notes: (*) signifies $p < 0.05$, (**) $p < 0.01$, (***) $p < 0.001$, NS not significant at the 0.05 level.

For seed-to-seedling survival in *Trichilia*, both the Spearman rank correlations and the logistic regressions identified negative density-dependent effects. Both revealed a strong negative effect of conspecific seed density (tables 20.2 and 20.3) and both suggested a negative relationship between seed-to-seedling survival and local conspecific basal area; but in the latter case, the two methods differed in which distances were significantly associated. Logistic regression results for seed-to-seedling survival need to be interpreted with caution, since the seed-to-seedling transition data were chronically overdispersed; under these conditions, p-values for hypothesis testing are not exact. Moreover, residual deviance was very high; independent variables explained, at most, 14% of total deviance. The post-hoc path analysis indicated that variation in seed density accounts for much more of the variation in survival rate than does variation in local basal area: the standard partial regression coefficient of survival rate upon seed density was −0.51, compared with −0.08 for survival rate upon adult basal area density within 15 m.

For seed survival in *Miconia*, the logistic regressions and Spearman rank correlations gave conflicting results as to the sign of density dependence. Logistic regression showed a strong negative effect of local seed density, and weaker negative effects of conspecific basal area (table 20.2). In contrast, the Spearman rank test indicated that seed survival rates were significantly *positively* correlated with conspecific densities within 30 m (table 20.3). These Spearman rank results were disproportionately influenced by sites having small numbers of seeds; yet at such sites, the estimate of seed mortality was poorer simply because there were fewer data. When sites with few seeds were excluded from the analysis (81 of 170 sites having 5 or fewer seeds initially), the Spearman correlation coefficients became negative (results not shown). With all sites included, the post-hoc path analysis showed a positive standard partial regression coefficient for the relationship between basal area within 15 m and survival, and a negative one between seed density and survival.

Seedling survival showed little significant relationship to local conspecific density in either species (tables 20.2 and 20.3). However, the power of these tests was low due to the small sample sizes, as is clearly evident in the large standard errors on the logistic regression parameter estimates (table 20.2). Slight overdispersion was evident in the data for *Trichilia*, so F-tests on rescaled deviances were used in place of chi-square tests.

Sapling survival was negatively associated with conspecific basal area density in both species (table 20.2). The magnitude of this effect was smaller on a per-year basis than among seeds and seedlings, as reflected in the lower estimates of the slope parameter b (table 20.2). For *Trichilia*, the strongest relation was with total basal area within 30 m; for *Miconia*, total basal area within 15 m was best related.

Discussion

Seed and Seedling Biology of Trichilia *and* Miconia

The abundant, small seeds of *Miconia argentea* travel much farther than the fewer and larger seeds of *Trichilia tuberculata*, more than half of which remain within 11 m of the parent tree. Despite the fact that it is 30 times less common on the plot, *Miconia* is estimated to produce more than 10 times as many seeds as *Trichilia* there. *Miconia*'s higher seed production and longer dispersal distance aid it in reaching its required and relatively rare regeneration habitat–gaps. Because of the longer dispersal distance and lower adult abundance, a much larger proportion of *Miconia*'s seeds end up far from adults, where they may be able to escape Janzen–Connell effects.

Seed and sapling survival were negatively density dependent in both species, suggesting the operation of Janzen–Connell effects at multiple stages. A previous study of somewhat larger seedlings of *Trichilia* found significant density-dependent effects on survival (Shamel 1998); the failure to detect density dependence among seedlings here most likely reflects the relatively low power of the analysis. Previous studies have also shown that density-dependent effects continue to at least 4-cm diameter saplings for *Trichilia* (Hubbell et al. 1990). Again, smaller sample sizes for the less abundant *Miconia* reduce the power to detect similar effects in that species.

Density dependence in seed-to-seedling survival in *Trichilia tuberculata* and in seed survival of *Miconia argentea* appears to be mediated most directly by the initial local density of seeds. Insofar as survival is also negatively correlated with basal area, this is due for the most part to the correlation between basal area density and seed density. This suggests that the agent of density-dependent mortality may be responding to local seed density. Dalling et al. (1998b) present evidence that fungal pathogens are responsible for much seed mortality among *Miconia*.

Seedling survival in both species showed no strong or very significant density-dependent effects. The most significant of these weak relationships was between seedling mortality and local tree density in *Trichilia*. If this result holds for larger sample sizes, it would suggest that the agents responsible for density-dependent seedling mortality also prey upon older individuals, particularly saplings (saplings dominate in the counts of tree numbers, while large adults dominate basal area measures). Saplings may pose an elevated risk of transmission of pests because their foliage and the associated pests and pathogens are in the understory, where seedlings too are located. Some studies suggest that there is strong stratum fidelity among phytophagous insects in neotropical forests and that the same plant species has different insect herbivores in the understory and in the canopy (Basset et al. 1999); however, a study of two tree species on Barro Colorado Island found that nearly the same suite of chewing insects attacked both juvenile and adult conspecifics (Barone 2000). In any case, it is not clear that insects are the agents responsible for Janzen–Connell effects at this stage in these species. Some of the older seedlings of *Trichilia* display a progressive die-back pattern leading to death that has the appearance of being caused by a pathogen (D. DeSteven, personal communication). Yet an intensive effort including field observations, cultures, and greenhouse experiments to find pathogens causing significant problems on *Trichilia* seedlings and saplings, using methods that found diseases in virtually every other species investigated, yielded nothing (G. S. Gilbert, personal communication).

The lower magnitude of density-dependent effects in *Miconia* relative to *Trichilia* (regression coefficients in table 20.1) is consistent with the idea that seedling survival of pioneer species depends mainly upon light availability (Augspurger 1984). Studies by Augspurger (Augspurger 1983, 1984, Augspurger and Kelly 1984) indicate that local seedling density and distance to parent can influence seedling mortality in general and disease-induced mortality in particular in a manner consistent with Janzen–Connell effects. However, Augspurger and Kelly (1984) found that the effects of light and pathogens interact: within the shade, disease mortality was significantly higher in high density areas. Within the sunlight, by contrast, disease was not significantly associated with density. Thus, density-dependent effects should be less important at the seedling stage for light-demanding species whose seedlings survive only in high light conditions. If juveniles are more likely to encounter high light conditions near adults, because of the contagion of gap formation (Young and Hubbell 1991), then this may enhance survival near adults, potentially further countering any Janzen–Connell effects.

Implications for Spatial Patterns

Studies of the spatial patterns of adults and saplings within the plot have concluded that there must be strong density dependence in *Trichilia* at earlier life stages, just as was found in the current study. Hubbell and Foster (1986) and Hubbell et al. (1990) both found the correlations between local densities of juveniles and of adults to be significantly negative. Given that dispersal is primarily local, we would expect local densities of juveniles of nonpioneers to be roughly proportional to those of adults in the absence of density-dependent effects. Condit et al. (1992), found that the density of *Trichilia* recruits is significantly lower than average in sites within 10 m of adults, higher than average between 10 and 45 m from the nearest adult, and somewhat lower than average beyond 50 m (as would be expected based on limited dispersal distance). The results of this study, combined with our results on seed dispersal, suggest that density dependence before the 1-cm sapling stage is very strong in *Trichilia*—strong enough that it more than compensates for the higher density of seeds that fall near adults.

Studies of plot spatial patterns have found no evidence of density dependence in *Miconia*. Because of its long dispersal distances, we would not expect a positive correlation between the local densities of adults and juveniles in the absence of density dependence, and thus, testing for such effects is more complicated. Condit et al. (1992) found significantly higher densities of *Miconia* recruits than the mean at sites within 40 m of the nearest adult, with significantly lower densities at distances greater than 55 m. This latter result is consistent with the dispersal distances observed here, and the evidence that the differences in mortality seen in *Miconia* are smaller than the differences in seed density; thus, the density-dependent survival is masked by differences in seed arrival probabilities even in this very well dispersed species.

Further studies should employ simulations of population dynamics to quantitatively evaluate whether seed dispersal distances, habitat requirements, and Janzen–Connell effects at various stages, as documented here, adequately explain spatial patterns in these species, and to test the relative importance of each. This will require combining models of growth, mortality, reproduction, and seed dispersal, so that the amount of time spent in each stage, and the changing spatial pattern of live adults, can be accounted for. Such an exercise will make it possible to examine the overall importance of effects at each stage. While the per-year impact of density dependence is largest at the seed stage, the many years spent in the sapling stage are likely to make the cumulative impact of density dependence among saplings more important.

Conclusions

Seed shadows, regeneration requirements, and Janzen–Connell effects contribute to spatial patterns in the 50-ha plot for both of the species examined here. In *Trichilia tuberculata*, dispersal is relatively local, regeneration requirements are broad, and density-dependent effects are strong. Thus, there is a dearth of saplings in the immediate vicinity of adults where Janzen–Connell effects are strongest, a surplus of saplings at intermediate distances that are within dispersal distance and subject to lower density-dependent mortality, and a dearth of saplings at large distances to which dispersal does not typically reach. In *Miconia argentea*, dispersal distances are long, regeneration occurs only in gaps (which are rare), and density-dependent effects are weak. Because density-dependent mortality is weak compared with the decline in seeds arriving with distance from parent, there is no deficit of saplings near reproductive adults. Because dispersal distances are long, saplings are common even far from adults and the density of saplings drops only beyond 50 m (Condit et al. 1992).

For the two species studied here, density-dependent effects are strongest at the seed stage. Local density of seeds themselves is the strongest correlate of seed mortality in *Miconia* and *Trichilia*, while local tree density within 15 and 30 m has small and marginally significant negative effects on seedling survival in *Trichilia* but not *Miconia*. It is clear that the strength of Janzen–Connell effects varies among species, and there is a need to examine in more detail how differences in life histories, seed and seedling traits, and phenology might influence the strength and timing of density-dependent effects.

Studies of seed and seedling biology on the one hand and of large-scale patterns in larger size classes on the other provide complementary information on the strength and importance of density-dependent effects. In particular, studies of seed rain and seed and seedling survival within Forest Dynamics Plots make the analysis of seed production, dispersal distances, and density-dependent effects easier and more powerful, since the locations of all nearby conspecific adults (possible parents and pest reservoirs) are known. At the same time, the results illustrate that spatial and demographic patterns reflect multiple influences and that analyses of such patterns for density dependence must be informed by knowledge of dispersal strategies and habitat preferences.

General Implications

· Information on seed and seedling biology of species, and of their natural history more generally, is needed to inform and correctly interpret analyses of Forest Dynamics Plot data.
· Studies of seeds and seedlings within Forest Dynamics Plots make possible powerful analyses of dispersal and density dependence, and provide insight into important aspects of population ecology that cannot be examined with the Forest Dynamics Plot data alone.
· The presence of Janzen–Connell effects and the importance of seed dispersal for escaping them have important implications for conservation and management. Dispersal curves represent the net effects of many dispersal agents. Selective removal of one or a few particularly efficacious dispersers (e.g., toucans, monkeys) might have a disproportionate impact on recruitment if remaining dispersers carry seeds shorter distances, to sites with higher conspecific seed densities, where seeds and seedlings will suffer higher mortality.

Acknowledgments

Osvaldo Calderón collected and identified the *Trichilia tuberculata* seeds and fruits. Eduardo Sierra and Andrés Hernandéz censused the *Trichilia* seedlings. Steve Paton manages the database that includes the *Trichilia* seed and seedling data. Katia Silvera, Felix Matias, and Arturo Morris helped collect the *Miconia* seed and seedling data. Rolando Peréz, Suzanne Loo de Lao, and many other Smithsonian employees and contractors have gathered and processed the 50-ha plot data over the years. Simon Levin, Stephen Pacala, James Clark, and Charles McCullogh gave helpful advice on dispersal model fitting and evaluation. Bert Leigh, two anonymous reviewers, and Philippe Tortell provided helpful comments on the manuscript. We are pleased to acknowledge the support of the National Science Foundation (a graduate fellowship to HCM and grant DEB 9509026 to JWD and SPH), the Smithsonian Institution (a predoctoral fellowship to HCM and an Environmental Sciences Program grant to SJW), and the Andrew W. Mellon Foundation (a postdoctoral fellowship to KEH and a grant to Simon Levin).

Appendix

The formula for the median dispersal distance is obtained by solving for x_m in the following equation:

$$\int_0^{x_m} 2\pi x f(x)dx = \frac{1}{2}$$

where $f(x)$ is the dispersal kernel. For the Clark 2Dt dispersal kernel used in this paper, this is straightforward:

$$\int_0^{x_m} \frac{2pxdx}{u\left(1+\dfrac{x^2}{u}\right)^{p+1}}\, dx = \frac{1}{2}$$

$$\left. \frac{-1}{\left(1+\dfrac{x^2}{u}\right)^{p}} \right|_{x=0}^{x=x_m} = \frac{1}{2}$$

$$1 - \frac{1}{\left(1+\dfrac{x_m^2}{u}\right)^{p}} = \frac{1}{2}$$

$$1 + \frac{x_m^2}{u} = 2^{1/p}$$

$$x_m = \sqrt{u\left(2^{1/p}-1\right)}$$

References

Ashton, P. S. 1969. Speciation among tropical forest trees: Some deductions in the light of recent research. *Biological Journal of the Linnean Society* 1:155–96.

Augspurger, C. K. 1983. Seed dispersal of the tropical tree, *Platypodium elegans*, and the escape of its seedlings from fungal pathogens. *Journal of Ecology* 71:759–71.

—————. 1984. Seedling survival of tropical tree species: Interactions of dispersal distance, light-gaps, and pathogens. *Ecology* 65:1705–12.

Augspurger, C. K., and C. K. Kelly. 1984. Pathogen mortality of tropical tree seedlings: Experimental studies of the effects of dispersal distance, seedling density, and light conditions. *Oecologia* 61:211–217.

Barone, J. A. 2000. Comparison of herbivores and herbivory in the canopy and understory for two tropical tree species. *Biotropica* 32:307–17.

Basset, Y., E. Charles, and V. Novotny. 1999. Insect herbivores on parent trees and conspecific seedlings in a Guyana rain forest. *Selbyana* 20:146–58.

Clark, D. A., and D. B. Clark. 1984. Spacing dynamics of a tropical rain forest tree: Evaluation of the Janzen–Connell model. *American Naturalist* 124:769–88.

Clark, J. S., M. Silman, R. Kern, E. Macklin, and J. HilleRisLambers. 1999. Seed dispersal near and far: Patterns across temperate and tropical forests. *Ecology* 80:1475–94.

Condit, R. 1998. *Tropical Forest Census Plots*. Springer-Verlag, Berlin, and R.G. Landes, Georgetown, TX.

Condit, R., S. P. Hubbell, and R. B. Foster. 1992. Recruitment near conspecific adults and the maintenance of tree and shrub diversity in a neotropical forest. *American Naturalist* 140:261–86.

Connell, J. H. 1971. On the roles of natural enemies in preventing competitive exclusion in some marine animals and in rain forest trees. Pages 298–312 in P. J. den Boer and G. R. Gradwell, editors. *Dynamics of Populations, Proceedings of the Advanced Study Institute on Dynamics of Numbers in Populations, Oosterbeek, 1970.* Centre for Agricultural Publishing and Documentation, Wageningen, Netherlands.

Crawley, M. J. 1993. *GLIM for Ecologists.* Blackwell Scientific Publications, Oxford, U.K.

Croat, T. B. 1978. *Flora of Barro Colorado Island.* Stanford University Press, Stanford, CA.

Dalling, J. W., S. P. Hubbell, and K. Silvera. 1998a. Seed dispersal, seedling establishment and gap partitioning among pioneer trees. *Journal of Ecology* 86:674–89.

Dalling, J. W., H. C. Muller–Landau, S. J. Wright, and S. P. Hubbell. 2002. Role of dispersal in the recruitment limitation of neotropical pioneer species. *Journal of Ecology* 90: 714–27.

Dalling, J. W., M. D. Swaine, and N. C. Garwood. 1997. Soil seed bank community dynamics in seasonally moist lowland tropical forest, Panama. *Journal of Tropical Ecology* 13:659–680.

Dalling, J. W., M. D. Swaine, and N. C. Garwood. 1998b. Dispersal patterns and seed bank dynamics of pioneer trees in moist tropical forest. *Ecology* 79:564–78.

De Steven, D. 1994. Tropical tree seedling dynamics: recruitment patterns and their population consequences for three canopy species in Panama. *Journal of Tropical Ecology* 10:369–83.

Garwood, N. C. 1983. Seed germination in a seasonal tropical forest in Panama: A community study. *Ecological Monographs* 53:159–81.

Hamill, D. N., and S. J. Wright. 1986. Testing the dispersion of juveniles relative to adults: A new analytic model. *Ecology* 67:952–57.

Harms, K. E. 1997. *Habitat-Specialization and Seed Dispersal-Limitation in a Neotropical Forest.* PhD dissertation. Princeton University, Princeton, NJ.

Harms, K. E., S. J. Wright, O. Calderón, A. Hernández, and E. A. Herre. 2000. Pervasive density-dependent recruitment enhances seedling diversity in a tropical forest. *Nature* 404:493–95.

Hilborn, R., and M. Mangel. 1997. *The Ecological Detective: Confronting Models with Data.* Princeton University Press, Princeton, NJ.

Howe, H. F. 1980. Monkey dispersal and waste of a neotropical fruit. *Ecology* 61:944–59.

Hubbell, S. P. 1998. The maintenance of diversity in a neotropical tree community: Conceptual issues, current evidence, and challenges ahead. Pages 17–44 in F. Dallmeier and J. A. Comiskey, editors. *Forest Biodiversity Research, Monitoring and Modeling: Conceptual Background and Old World Case Studies.* Man and the Biosphere Series, Volume 20. Parthenon Publishing, Pearl River, NY.

Hubbell, S. P., and R. B. Foster. 1983. Diversity of canopy trees in a neotropical forest and implications for conservation. Pages 25–41 in S. L. Sutton, T. C. Whitmore, and A. C. Chadwick, editors. *Tropical Rain Forest: Ecology and Management.* Blackwell Scientific Publications, Oxford, U.K.

—————. 1986. Biology, chance, and history and the structure of tropical rain forest tree communities. Pages 314–29 in J. Diamond and T. J. Case, editors. *Community Ecology.* Harper and Row, New York.

—————. 1990. Structure, dynamics, and equilibrium status of old-growth forest on Barro Colorado Island. Pages 522–41 in A. Gentry, editor. *Four Neotropical Forests.* Yale University Press, New Haven, CT.

Hubbell, S. P., R. Condit, and R. B. Foster. 1990. Presence and absence of density dependence in a neotropical tree community. *Philosophical Transactions of the Royal Society of London B* 330:269–81.

Hubbell, S. P., R. B. Foster, S. T. O'Brien, K. E. Harms, R. Condit, B. Wechsler, S. J. Wright, and S. Loo de Lao. 1999. Light-gap disturbances, recruitment limitation, and tree diversity in a neotropical forest. *Science* 283:554–57.

Hubbell, S. P., J. A. Ahumada, R. Condit, and R. B. Foster. 2001. Local neighborhood effects on long-term survival of individual trees in a neotropical forest. *Ecological Research* 16:859–75.

Hurtt, G. C., and S. W. Pacala. 1995. The consequences of recruitment limitation: Reconciling chance, history and competitive differences between plants. *Journal of Theoretical Biology* 176:1–12.

Janzen, D. H. 1970. Herbivores and the number of tree species in tropical forests. *American Naturalist* 104:501–528.

Leigh, E. G., Jr. 1996. Introduction: Why are there so many kinds of tropical trees? Pages 63–66 in E. G. Leigh, Jr., A. S. Rand, and D. M. Windsor, editors. The ecology of a tropical forest: Seasonal rhythms and long-term changes. *The Ecology of a Tropical Forest: Seasonal Rhythms and Long-Term Changes.* Smithsonian Institution Press, Washington, DC.

Leighton, M., and D. R. Leighton. 1982. The relationship of size of feeding aggregate to size of food patch: Howler monkeys (*Alouatta palliata*) feeding in *Trichilia cipo* fruit trees on Barro Colorado Island. *Biotropica* 14:81–90.

Muller-Landau, H. C., S. J. Wright, O. Calderón, S. P. Hubbell, and R. B. Foster. 2002. Assessing recruitment limitation: Concepts, methods and case studies from a tropical forest. Pages 35–53 in D. J. Levey, W. R. Silva, and M. Galetti, editors. *Seed Dispersal and Frugivory: Ecology, Evolution and Conservation.* CAB International, Wallingford, Oxfordshire, U.K.

Poulin, B., S. J. Wright, G. Lefebrve, and O. Calderon. 1999. Interspecific synchrony and asynchrony in the fruiting phenologies of congeneric bird–dispersed plants in Panama. *Journal of Tropical Ecology* 15:213–27.

Ribbens, E., J. A. Silander, Jr., and S. W. Pacala. 1994. Seedling recruitment in forests: Calibrating models to predict patterns of tree seedling dispersion. *Ecology* 75:1794–806.

Rich, J. A. 1988. *Mathematical Statistics and Data Analysis.* Wadsworth & Brooks, Pacific Grove, CA.

Ricklefs, R. E. 1977. Environmental heterogeneity and plant species diversity: A hypothesis. *American Naturalist* 111:376–81.

Schupp, E. W. 1988a. Factors affecting post-dispersal seed survival in a tropical forest. *Oecologia* 76:525–30.

————. 1988b. Seed and early seedling predation in the forest understory and in treefall gaps. *Oikos* 51:71–78.

————. 1992. The Janzen–Connell model for tropical tree diversity: Population implications and the importance of spatial scale. *American Naturalist* 140:526–30.

Schupp, E. W., and E. J. Frost. 1989. Differential predation of *Welfia georgii* seeds in treefall gaps and the forest understory. *Biotropica* 21:200–03.

Shamel, S. I. 1998. *The Effects of Conspecific Adult Density on Seedling Survival: A Test of the Janzen–Connell Hypothesis.* BA thesis, Princeton University, Princeton, NJ.

Welden, C. W., S. W. Hewett, S. P. Hubbell, and R. B. Foster. 1991. Sapling survival, growth, and recruitment: Relationship to canopy height in a neotropical forest. *Ecology* 72:35–50.

Wright, S. J., and O. Calderón. 1995. Phylogenetic patterns among tropical flowering phenologies. *Journal of Ecology* 83:937–48.

Wright, S. J., C. Carrasco, O. Calderón, and S. Paton. 1999. The El Niño Southern Oscillation, variable fruit production and famine in a tropical forest. *Ecology* 80: 1632–47.

Young, T. P., and S. P. Hubbell. 1991. Crown asymmetry, treefalls, and repeat disturbance of broad-leaved forest gaps. *Ecology* 72:1464–71.

Zar, J. H. 1974. *Biostatistical Analysis.* Prentice-Hall, Englewood Cliffs, NJ.

21

Distance- and Density-Related Effects in a Tropical Dry Deciduous Forest Tree Community at Mudumalai, Southern India

Robert John and Raman Sukumar

Introduction

Studying the interdependence of spatial distribution and dynamics in tree communities is crucial to our understanding of the coexistence of tree species in tropical forests (Pacala 1997). Strong spatial patterns in mortality and recruitment have been demonstrated in many studies (Clark and Clark 1984; Condit et al. 1992; Okuda et al. 1997). These studies suggest that seedling survival increases with distance from the parent tree, in accordance with a model proposed by Janzen (1970) and Connell (1971). Other studies have shown that the spatial distribution and density of trees can influence pest and pathogen attacks (Gilbert et al. 1995), and that strong density- and diversity-dependent effects can maintain species diversity (Wills et al. 1997; chap. 23 in this volume; Harms et al. 2000). It is increasingly evident that many species have short dispersal distances and that their spatial scales of resource uptake are small, resulting in competitive effects that are largely local (Hubbell 1998). Thus, plants respond dramatically to local variations in spatial structure such as those caused by treefalls or other small-scale disturbances (Pacala 1997).

While several studies have been carried out in tropical moist forests, few studies have investigated spatial patterns in demographic parameters in highly seasonal dry forests (Hubbell 1979; Murphy and Lugo 1986; Martijena and Bullock 1994). Tropical dry forests harbor fewer species, are less complex in structure, and are perhaps subject to greater environmental stress than moist forests (Murphy and Lugo 1986). However, it is not known if distance, density, and diversity effects on life history parameters of tree species that are believed to be important in influencing the diversity and dynamics of moist forests are also important in dry forests. Intuitively, it would seem that in such forests, large-scale natural disturbances caused by droughts and fires could be more important in structuring the plant community than density dependence (Sukumar et al. 1992; Van Groenendael et al. 1996).

We are carrying out a long-term study on forest dynamics in a 50-ha Forest Dynamics Plot in a tropical dry deciduous forest in Mudumalai Wildlife Sanctuary, southern India (chap. 33). Aspects being studied include forest diversity, structure, and function, the influence of fire and large mammalian herbivores on forest dynamics, life histories of tree species, and maintenance of species diversity (Sukumar et al. 1992, 1998; Joshi et al. 1997; John et al. 2002). The main questions addressed in this chapter are (1) What are the spatial patterns of mortality and recruitment of common species in the community and what is the likely influence of the observed patterns on the spatial distribution of the tree species? (2) Are there distance, density, and diversity effects on mortality, recruitment, and population growth and, if present, can they maintain diversity in the system?

Methods

Study Site

Our study was carried out in a 50-hectare Forest Dynamics Plot in a tropical deciduous forest in southern India (chap. 33). Species diversity, abundance, and basic demographic parameters of the tree species in this plot have been reported (Sukumar et al. 1992, 1998; chap. 33). In this paper, we carried out analyses of spatial patterns in the demography of 15 common species—10 canopy trees and 5 understory trees—from the community of about 70 species. The most common canopy trees include *Lagerstroemia microcarpa* (Lythraceae), *Tectona grandis* (Verbenaceae), *Terminalia crenulata* (Combretaceae), and *Anogeissus latifolia* (Combretaceae). The most common understory trees are *Kydia calycina* (Malvaceae), *Cassia fistula* (Leguminosae [Caesalpinoideae]), and *Xeromphis spinosa* (Rubiaceae). *L. microcarpa* is mainly bee pollinated and wind dispersed; it attains a height of about 20 m and a maximum dbh of 73 cm. *T. grandis* (teak) is an important timber tree with a height ranging between 15 and 20 m and a maximum size of 113 cm dbh. Pollination is mainly by butterflies and seed dispersal by an explosive mechanism. *T. crenulata* is a bee-pollinated, wind-dispersed tree attaining a height of 20 m and a maximum dbh of 113 cm. *A. latifolia* is typical of dry forests and is wind pollinated, with an explosive dispersal mechanism. It reaches a height of about 15 m and a maximum dbh of 72 cm, but is considerably smaller in the very dry forests. *K. calycina* grows in dense clumps and is bee pollinated and dispersed by wind; while *C. fistula* is bee pollinated and animal dispersed. *X. spinosa* also occurs in dense clumps in the understory, is bee pollinated and wind dispersed. These three understory trees attain maximum sizes of about 27, 60, and 59 cm dbh respectively.

Topography and Species Distributions

The topography of the plot is mildly undulating and varies in altitude between 980 and 1120 m above mean sea level. To examine associations of species with topographical features, we correlated species richness with slope and elevation. The slope of the plane made by each quadrat (20 × 20 m) was calculated as the mean of the slopes of the planes of the four triangles obtained by drawing the diagonal bisectors of the quadrat. The elevation of each quadrat was calculated as the mean of the elevation of the four corners. We correlated species richness of all species and abundance of each of the 15 focal species in each quadrat against slope and elevation. We carried out both parametric and nonparametric Spearman rank-order correlations, but results from each test were similar so we report only the former.

Spatial Dependence of Juvenile Mortality Rates

For each species, for each year during 1989–96, juvenile mortality rate was examined as a function of the distance to the nearest conspecific adult, nearest heterospecific adult, and the nearest large tree regardless of species. To do this, we calculated for each juvenile the distance to the nearest conspecific adult and grouped the juveniles into discrete distance classes. We then compared the mortality rates of juveniles by two-way contingency tables, making pair-wise comparisons for each class against every other and against the total sample. A G-test of independence was then conducted on each two-way table. The critical values for G are obtained from the χ^2-distribution with 1 degree of freedom (Sokal and Rohlf 1981). This analysis was repeated for the nearest heterospecific large tree and the nearest large tree irrespective of species. Dry-season grass fires occurred in the plot in 5 of the 8 years during 1989–96. Since we found particularly low densities of tall grasses, the main fuel for fires, underneath dense stands of *L. microcarpa*, we also examined juvenile mortality rates of all species as a function of the distance to the nearest *L. microcarpa* adult.

Finally, we carried out linear regressions with juvenile mortality as the dependent variable and distance to conspecific adults as the independent variable. Distance to heterospecific large trees, to any large tree, and *L. microcarpa* adults are not considered here because the 2 × 2 table analyses did not reveal any significant effects caused by these factors.

Spatial Dependence of Recruitment Rates

Following the procedure outlined by Condit et al. (1992), we examined the spatial variation in recruitment probabilities by calculating recruitment probability at discrete distance intervals from conspecific adults. Recruits were defined as all new individuals ≥ 1 cm dbh recorded in any given census. The recruitment probability of a focal species i was defined as the ratio of the number of recruits of species i to that of all recruits. For each species considered in the study, four distributions were calculated (as in Condit et al. 1992):

1. The distribution of recruits of the focal species i in each distance class d from the nearest conspecific adult, S_{id}. The total number of recruits of species i is therefore ΣS_{id} (the summations over distance).

2. The distribution of all recruits in each distance class d from the nearest adult of the focal species i, A_{id}. The total number of recruits is ΣA_{id} (this was not the same for all i because in each case a different number of recruits was left out of the analysis for being found closer to a boundary of the plot than to the nearest adult).

3. The expected distribution of recruits for the focal species, E_{id}. This is the expected distribution if the recruits of the focal species show no tendency to be attracted or repelled at any distance from the nearest conspecific adult. It is obtained as A_{id} normalized to the total number of recruits of species i, given by $E_{id} = A_{id} \Sigma S_{id} / \Sigma A_{id}$.

4. The distribution of the normalized index of recruitment probability for the focal species. This is expressed as $R_{id} = S_{id}/E_{id}$, the ratio of the observed abundance of recruits of the focal species to the expected abundance. If $R_{id} > 1$ ($S_{id} > E_{id}$), recruits of species i were overrepresented at distance d, and if $R_{id} < 1$ they were underrepresented.

The distribution of recruits of the focal species S_{id} was compared to the distribution of recruits of all species A_{id}, using a χ^2 test based on a 2×2 contingency table. Recruits were tallied for two time intervals, 1988–90 and 1992–94. This was not done for 1990–92 and 1994–96 because there were extensive fires during those intervals, resulting in very few recruits. Two common canopy trees *Terminalia crenulata* and *Anogeissus latifolia* did not recruit during 1988–96 and were therefore not included.

Density and Diversity Effects on Mortality and Recruitment

We employed a quadrat-based analysis to examine density and diversity dependence on mortality, recruitment, and intrinsic rates of population increase. We used two measures of density in each quadrat: basal area and abundance (trees ≥ 1 cm dbh). Basal area weights large trees more than small trees and is an indicator of biomass, whereas abundance represents crowding of trees in the neighborhood. Diversity was expressed using Fisher's α. To calculate the mortality rate for each species we used the discrete-time measure, $m = 1 - (N_t / N_o)^{1/t}$, where N_o is the initial population size and N_t is the number of survivors after time t (Sheil 1995); t was always 1. We expressed recruitment rate on a per adult basis, as the ratio of the number of recruits to the number of adult trees for each species in each quadrat. Quadrats with zero adults were not included. Intrinsic rate of increase was expressed as $(b - d)/N$, where b is the number of recruits, d is the number of deaths, and N is the initial population size of that species (Wills et al. 1997).

The analysis was carried out using 100 × 100 m quadrats; smaller quadrats could not be used due to the low sample size of recruits. In each quadrat, for each species, we obtained the mortality rate, recruitment rate, and intrinsic rate of increase and performed nonparametric Spearman's rank-order correlations with conspecific and heterospecific basal area in that quadrat. Correlations were then repeated with the same demographic parameters but with conspecific and heterospecific abundance, and overall diversity. Recruitment, mortality, and diversity could be calculated for each year, but since girth measurements were taken only once every 4 years, basal areas could be calculated only for 1988, 1992, and 1996. For other years, basal areas were calculated using the most recent measure of dbh for each tree.

Since mortality was strongly size dependent and because size distributions differed considerably between species, we tested mortality in two size classes: small trees (1 to <10 cm dbh) and large trees (≥10 cm dbh).

Results

Topography and Species Distributions

Species richness was marginally higher on steeper slopes and at higher elevations, as indicated by regressions of richness per 20 × 20 m quadrat. Though we have not measured water potential gradients, it is likely that steep slopes retain more soil moisture in the dry season enabling recruitment of seedlings that need higher moisture levels to establish.

The most abundant canopy tree, *Lagerstroemia microcarpa*, was not associated with slopes but was significantly more abundant at lower elevations, though the association was weak. The next three most abundant canopy trees, *Anogeissus latifolia*, *Tectona grandis*, and *Terminalia crenulata*, were negatively associated with slopes, preferring plateaus, and the latter two species were also significantly more abundant at higher elevations. Among the other canopy trees, *Grewia tilifolia* (Tiliaceae) and *Radermachera xylocarpa* (Bignoniaceae) were more abundant on steep slopes and higher elevations, while *Stereospermum personatum* (Bignoniaceae) and *Syzygium cumini* (Myrtaceae) were significantly more abundant at lower elevations. Positive affinity to slope sites was also found in two common understory trees, *Kydia calycina* and *Cassia fistula*. Though topographical features have a significant influence in many species, the associations are weak, judging by the strength of the correlations ($r^2 < 0.05$ in every case). Eight species—five canopy and three understory—showed no affinity for slope, while three showed no association with elevation.

It must be added here that this method for examining affinities with topographical features implicitly assumes that individual trees of a species are independently distributed in space. Spatial autocorrelation in tree distributions violates this assumption and could result in spurious correlations with topographical features. This means that we are likely to have fewer and/or weaker correlations than what have been reported above. In any case, the correlations were weak in most species and we conclude that distributions of tree species are only weakly influenced by topography in this forest.

Spatial Dependence of Juvenile Mortality

Nonrandom juvenile mortality was observed in a few species, but there was no general pattern (fig. 21.1). Juvenile mortality rates in *C. fistula* and *L. microcarpa*

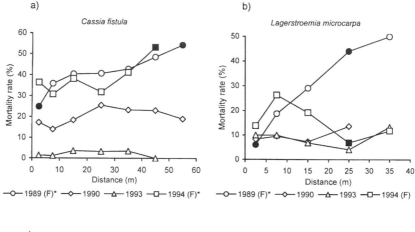

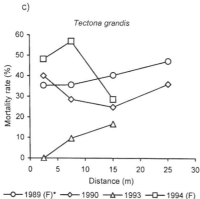

Fig. 21.1. Juvenile mortality rate at different distances from conspecific adults for 4 years. Years in which fires occurred are indicated by (F). Filled symbols indicate that the mortality rate in a given class was significantly different from the total sample. When significant, the shorter distances had lower, and the larger distances higher, mortality rates than the total sample (except in 1994 for *L. microcarpa* when mortality at 25 m was lower). Significant linear regressions are indicated by an asterisk.

were significantly higher at larger distances and lower at very short distances (<10 m) from conspecific adults (G-tests and parametric regressions, $p < 0.05$). A similar but weak effect was also found in *T. grandis* (fig. 21.1c). However, this distance dependence of juvenile mortality was not present for all years tested in these species and was absent in other species.

Distance to the nearest heterospecific large tree or the nearest large tree of any species did not affect juvenile mortality. Juvenile mortality in *C. fistula* and *K. calycina* was lower near adults (<10 m) of *L. microcarpa* only in years when extensive fires occurred in the plot (G-tests, $p < 0.05$). No such spatial dependence was found in years when no fires occurred, and such a "protective" effect due to *L. microcarpa* was absent for juveniles of other species.

Spatial Dependence in Recruitment Rates

In *Cassia fistula*, a common understory tree, recruitment peaked at an intermediate distance from conspecific adults, but there was no inhibition of recruitment near conspecifics (table 21.1, fig. 21.2a). At larger distances, the observed number of recruits was lower than the random expectation (χ^2 tests, $p < 0.05$).

A significant inhibition of recruitment at distances <5 m and 5–10 m from conspecific adults was found in the most common canopy tree, *L. microcarpa* (fig. 21.2b) and another common canopy tree, *T. grandis* (fig. 21.2c). These two species also had significantly higher recruitment probabilities than expected at 15–30 m and 10–25 m from conspecific adults, respectively. In two common understory trees, *K. calycina* and *X. spinosa*, there was a slight excess of recruitment at 10–20 m and 20–30 m respectively, but the sample sizes at other distances were too low to perform tests (table 21.1). All other species had fewer or no recruits at all and were therefore not analyzed. These include *A. latifolia* and *T. crenulata*, each with more than 1500 individuals in the plot, which did not recruit during 1988–96.

Quadrat-Based Tests for Density and Diversity Dependence on Recruitment

Conspecific basal area had a strong and significant negative effect on recruitment in two common canopy trees, *L. microcarpa*, and *T. grandis*, and a common understory tree, *C. fistula*. Recruitment was also strongly negatively correlated with conspecific abundance in these species, but these were weaker than basal area effects (table 21.2). Only six other species had ≥10 recruits. We observed significant negative correlations between recruitment and conspecific density in five of these species (*Cordia obliqua* (Boraginaceae), *Cordia wallichii* (Boraginaceae), *K. calycina*, *R. xylocarpa*, and *X. spinosa*), but sample sizes of recruits were low so we do not report these results.

Significant positive correlations with heterospecific basal area and recruitment were obtained for *L. microcarpa* and *T. grandis*, but these were weaker than

Table 21.1. Results of χ^2-Tests Comparing Recruitment of a Focal Species with Recruitment for All Species Combined at Various Distances from Adults of the Focal Species

Species	Sample Size		Distance Class (m)										Syndrome
	Recruits	Adults	0–5	5–10	10–15	15–20	20–25	25–30	30–35	35–40	40–50	>50	
Cassia fistula	322	364	−	+	++	++	+	−	−−	−	−−	−−	PR
	617	415	+	+	++	−	+	−	−	−	−	−−	PR
Kydia calycina	37	192	+	+	+	++	−	+	−	+	+	−−	NP
Lagerstroemia microcarpa	172	1743	−−	−−	+	++	++	+	−	−	−	−	R
	148	1896	−	−−	+	++	+	++	+	+	×	×	R
Tectona grandis	125	1632	−	−−	++	+	++	−	+	−	×	×	R
	134	1647	−−	−−	++	++	++	++	−	×	×	×	R
Xeromphis spinosa	43	626	+	−	−	−	+	++	−	−	−	−−	PR

Notes: For each distance class, statistically significant overrepresentation is indicated by ++, while nonsignificant overrepresentation is shown as +. Similarly, for underrepresentation at any distance class, significant and nonsignificant tests are indicated by −− and − respectively. The syndromes are PR-partially repelled, NP-no pattern, R-repelled. "×" indicates low sample size, and tests were not done for these cases. For each species, the top row is for 1988–90 and the bottom row for 1992–94, except for *Kydia calycina* and *Xeromphis spinosa* (with single rows) for which the data are for 1988–90.

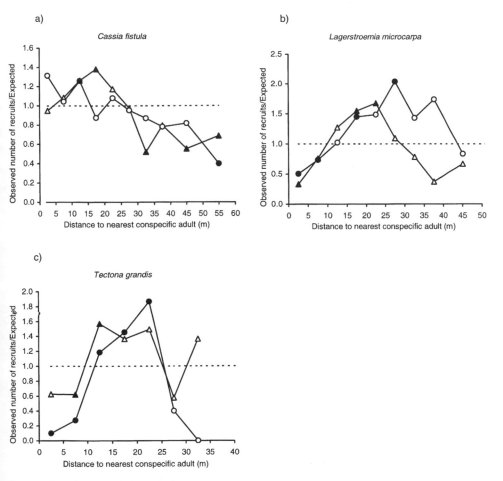

Fig. 21.2. Recruitment probability as a function of distance from conspecific adults. Triangles (Δ) indicate 1988–90 recruitment; circles (O), 1992–94 recruitment. Filled symbols indicate that the distribution of recruits of the focal species was significantly different from that of all species combined, at the 5% level (χ^2-tests).

conspecific effects. In contrast to heterospecific basal area, heterospecific abundance had strong positive effects on recruitment in *L. microcarpa* in 5 of the 8 years, but these correlations were nonsignificant in *C. fistula* and *T. grandis*. There were very few significant correlations between recruitment and diversity, except for a few positive correlations in *T. grandis* (table 21.2) and *X. spinosa*.

We perform several correlations here, and therefore some correlations are expected to appear significant purely by chance with a certain probability. However, since the significant correlations were usually highly significant, most correlations

Table 21.2. Density and Diversity Dependence in Recruitment

Species		1989	1990	1991	1992	1993	1994	1995	1996
Cassia fistula	R	76	285	80	223	570	356	486	216
	CB	+	−	+	+	−	−−−	−−−−	+
	CA	+	−	+	+	−−−	−	−−−−	+
	HB	−−−−	−−−−	+	−	+	+	−	−
	HA	+	−−−−	−	−	−	−	−	−
	DI	+	−	−	−	−	−	+	+
Lagerstroemia microcarpa	R	58	143	56	74	55	118	104	35
	CB	−−−	−−−	−−−	−−−−	−−−	−−−	−−−	−−−
	CA	−−−	−−−	−	−−−−	−−−	−−−	−−−	−−−
	HB	+	+	+++	+++	+++	+++	−	+
	HA	+	+	+++	−−	++++	++++	++++	+++
	DI	+	+	−−	−	−	−	−	+++
Tectona grandis	R	57	82	50	33	78	71	94	11
	CB	−−−−	−−−−	−−−	−−−−	−−−	−−−	−−−	−−−
	CA	−−−	−−−	−−−	−−−−	−	−	−	−−
	HB	++	−	+	+	−	−	+	++
	HA	−	−	+	+	−	+	+	+
	DI	+	++++	++++	+	+	+	+	+

Notes: The number of recruits (R) for each species in each year is listed together with results of Spearman's rank-order correlations of conspecific basal area (CA), conspecific abundance (CB), heterospecific basal area (HB), heterospecific abundance (HA), and diversity (DI) with recruitment rate. Nonsignificant correlations are indicated by a single symbol + or − showing the sign of the correlations. The signs − −, − − −, and − − − − indicate significant negative correlations at α-levels 0.05, 0.01, and 0.001 respectively. Significance is indicated in the same way for positive correlations. Years in which fire occurred are indicated in bold.

remained significant even after adjusting the significance level for the number of correlations performed.

The within-quadrat density- and abundance-dependent effects reported above were checked for "mass" effects (Shmida and Wilson 1985) that can be caused by recruits from parents outside a given quadrat. This effect would be especially important when quadrats having few seed-producing adults are surrounded by quadrats with many such adults. The effect would be stronger at small quadrat sizes and would lead to overestimates of recruitment in low-density quadrats. We therefore conducted partial parametric regressions of recruitment rate against (1) the number of conspecific adult trees within the quadrat, (2) the number of adults in the four neighboring quadrats, and (3) the number of adults in the eight neighboring quadrats (Wills et al. 1997). The first partial regression measures the within-quadrat effect and the latter two the mass effects. The four- and eight-neighbor regressions gave similar results and we report only the eight-neighbor results here.

In *C. fistula*, all except one of the partial regressions were nonsignificant for both the within-quadrat and the mass effects. For 1995, when recruitment was relatively high, a significant negative within-quadrat effect was found but the mass effect was nonsignificant. Similarly, in *L. microcarpa*, recruitment was significantly lower in quadrats with higher basal area of conspecifics while the mass effect was nonsignificant. All other partial regressions were nonsignificant in this species. In *T. grandis*, recruitment was also significantly lower in quadrats with higher basal area of conspecifics in 5 years, when most recruitment occurred. Significant positive mass effects were found in 2 years and a negative mass effect in 1 year in this species, but these mass effects were much weaker than within-quadrat effects judging by the strength of the regressions. All other species had too few recruits to test for mass effects and we conclude that mass effects are relatively weak for 1-ha quadrats in the species tested.

Quadrat-Based Tests for Density-, Abundance-, and Diversity-Dependence of Mortality

Since most tree species had very few individuals in the smaller size classes, we could test only six species for density dependence in mortality among small trees. Most correlations were nonsignificant except for a few negative correlations in *L. microcarpa* and *C. fistula*. Correlations were consistently negative in *Lagerstroemia microcarpa* but variable in other species (data not shown).

We tested 10 species for density dependence in mortality of large trees. The results for six species with the most consistent patterns are shown in table 21.3. Although most correlations were nonsignificant, the sign of the correlations with conspecific density were consistently positive in most species. These include *A. latifolia*, *L. microcarpa*, *E. officinalis*, *X. spinosa*, and *T. crenulata*. In *K. calycina*,

Table 21.3. Density and Diversity Dependence in Large-Tree Mortality

Species		1989	1990	1991	1992	1993	**1994**	1995	1996
Anogeissus latifolia	CB	+	+	+	+	– –	+	+	+
	CA	+	+	+	+	– –	+	+	+
	HB	–	–	–	–	+ +	–	–	–
	HA	–	–	+	+	+	–	+	+ +
	DI	–	–	–	–	–	–	–	–
Kydia calycina	CB	– –	– –	– – –	–	–	–	+	–
	CA	–	–	– – –	– –	–	–	+	– –
	HB	–	–	+	–	+ +	+	+	–
	HA	–	+ + +	+	–	–	– –	–	–
	DI	–	–	–	– –	–	–	– –	–
Lagerstroemia microcarpa	CB	+	+	+	+	+	+	+	+ +
	CA	+	+	+	+	+	+	+	+ +
	HB	–	–	–	– –	–	–	–	– –
	HA	+	–	–	–	–	–	– –	–
	DI	+ + +	+	–	+	–	–	–	–
Emblica officinalis	CB	– –	+	+	+	×	+	+	+ +
	CA	–	+	+	–	×	+	+	+ +
	HB	–	+	–	+	×	–	+	+
	HA	–	+	+	–	×	–	+	–
	DI	–	– –	–	–	×	+	–	–
Xeromphis spinosa	CB	+	+	+	+	+	+	+	–
	CA	+	+	+	+	+	+	+	–
	HB	–	+	–	+	–	–	+	+ + +
	HA	+	–	+	–	+	–	+	–
	DI	+ +	– –	+	– –	+	+ +	–	+
Terminalia crenulata	CB	– –	+	+	+	+	+	+	+
	CA	+	– –	+	–	+	+	+	+
	HB	–	+	+ +	–	+	– –	–	–
	HA	+	+ + +	+	–	+	+	+	–
	DI	– –	+	– – – –	–	–	–	–	–

Notes: Spearman's rank-order correlations of conspecific basal area (CB), conspecific abundance (CA), heterospecific basal area (HB), heterospecific abundance (HA), and diversity (DI) with mortality. Nonsignificant correlations are indicated by a single symbol + or – showing the sign of the correlations. The signs – –, – – –, and – – – – indicate significant negative correlations at α-levels 0.05, 0.01, and 0.001 respectively. Significance is indicated in the same way for positive correlations. Years in which fire occurred are indicated in bold. X indicates that correlations were not obtained due to low sample sizes.

however, these correlations were mostly negative. Correlations of large-tree mortality with heterospecific density and diversity were mostly nonsignificant and variable in sign with no consistent patterns.

Density- and Diversity-Dependence on the Intrinsic Rate of Population Increase

Here we discuss only those species that were included in the recruitment analysis. Intrinsic rates of population increase were largely negatively correlated with conspecific basal area. Positive correlations were obtained only when fire occurred, whereas the correlations were negative in the absence of fire (table 21.4). This pattern was strong in *Tectona grandis* and *Lagerstroemia microcarpa*. In *Cassia fistula*, the intrinsic rate of increase was negatively correlated with conspecific basal area in 1993 and 1995 when fire was absent and positively correlated in 1994 when fire was present. The correlations with conspecific abundance were similar to those with conspecific basal area, but the effects were weaker.

Correlations involving heterospecific density were largely nonsignificant in most species, with no consistent direction in the sign of the correlations; only in *L. microcarpa* were these correlations consistently positive. Correlations with diversity were mostly nonsignificant and variable in sign (table 21.4).

Discussion

Spatial Distribution and Topography

Our preliminary analysis on the influence of slope and elevation on species richness and the distribution of trees indicates that they have only a weak effect. Thus, the clumped distribution of trees in the plot is unlikely to be due to spatial heterogeneity in habitat. Joshi et al. (1997) examined subregions of differing density in the 50-ha plot for self-affinities, as an indicator of clumping in individual species, and for cross-affinities, as an indicator of associations between pairs of species. While positive associations between species are possibly the result of biological interactions, they may also arise as a result of specialization of these species to a similar habitat. Results show that among the common species examined, only three species—*A. latifolia*, *T. grandis*, and *C. fistula*—had positive self-affinities as well as positive cross-affinities. These results plus our observations seem to indicate that niche specialization or niche differentiation is limited in this forest at the 50-ha scale and that most species appear as habitat generalists with respect to topography and elevation.

Spatial Patterns of Mortality in Juvenile Trees

Seasonal grass fires are common in dry deciduous forests throughout tropical Asia (Stott et al. 1990). Dry-season grass fires in Mudumalai have caused

Table 21.4. Density and Diversity Dependence in the Intrinsic Rate of Population Increase

Species		1989	1990	1991	1992	1993	**1994**	1995	**1996**
Cassia fistula	CB	–	–	+	+	– – –	++	– – –	+
	CA	– – –	–	+	–	– – –	++	– – –	–
	HB	+	– –	+	– – –	+	+	–	–
	HA	+++	– – – –	+	–	–	+	– – –	+
	DI	–	+	–	–	–	–	+	+
Lagerstroemia microcarpa	CB	–	– – –	+	–	– – –	–	–	++++
	CA	–	– – – –	+	–	– –	+	–	++++
	HB	++	+	+	+	+	+	+	–
	HA	+	++	+	+	++	+	+	– – –
	DI	– –	+	–	– –	–	– –	+	+
Tectona grandis	CB	+	–	++	+++	– –	+	– –	++++
	CA	– –	–	+	++	– –	+	–	++
	HB	+	–	–	–	–	–	+	–
	HA	–	–	–	+	– –	–	++	–
	DI	–	++	+	– –	+	–	+	–

Notes: Spearman's rank-order correlations of conspecific basal area (CB), conspecific abundance (CA), heterospecific basal area (HB), heterospecific abundance (HA), and diversity (DI) with intrinsic rate of increase. Nonsignificant correlations are indicated by a single symbol + or – showing the sign of the correlations. The signs – –, – – –, and – – – – indicate significant negative correlations at α-levels 0.05, 0.01, and 0.001 respectively. Significance is indicated in the same way for positive correlations. Years in which fire occurred are indicated in bold case.

extensive mortality among juvenile trees and kept recruitment low, resulting in size distributions that appear relatively deficient in small trees (Sukumar et al. 1998). Though some nonrandom patterns of juvenile mortality were found, with mortality rates lower when the trees were close to adults, these were not ubiquitous. It occurred in a few species in a few instances. There is considerable interannual variation in the spatial pattern and intensity of fire (unpublished fire maps), and this probably gave rise to some nonrandom effects. Studies in tropical savannas, where fires are common, indicate that forest trees have higher establishment under canopy cover than in open grassland (Kellman and Miyanishi 1982; Kellman 1985; Bowman and Panton 1993; Hoffmann 1996). Higher establishment of juvenile trees underneath the canopy could result in increased clumping of tree distributions. Though there are traces of a similar effect in this forest, it could be largely obliterated by the very severe fires that generally occur in Mudumalai, killing most juvenile trees.

Density and Diversity Effects on Mortality

The quadrat-based analyses indicate that small-tree mortality was largely unrelated to density and abundance of conspecific or heterospecific trees except in *L. microcarpa* and *C. fistula*. Our field observations and some qualitative data on ground cover do indicate that grass densities were generally lower under dense patches of *L. microcarpa* and *C. fistula*. The low grass densities probably caused a lower incidence and/or intensity of fire in these quadrats, resulting in negative density dependence. This is also consistent with the spatial patterns of juvenile mortality in these species (discussed above).

Mortality rates among large trees were low, despite fires, and although correlations of large-tree mortality with density and diversity were mostly nonsignificant, the sign of the correlations with conspecific density were consistently positive indicating the nature of the effect. Since death rates among large trees are low, the correlations were tested for 4-year-interval mortality rates too. These correlations were significant and strongly positive in *L. microcarpa* and *E. officinalis* and remained nonsignificant but positive in other species.

During the 3-year period 1988–91, the population of *K. calycina* declined by over 66% from an initial population of 5175 individuals, largely due to elephant browsing, especially of mid-sized trees (Sukumar et al. 1998). Correlations indicate negative dependence of large-tree mortality with both conspecific basal area and conspecific abundance. The spatial distribution of *K. calycina* was highly clumped and by 1992, only some dense clusters of trees had survived. Mortality in this species was widespread, but in 1990 and 1991, some quadrats with a very high abundance of trees had lower mortality leading to a reversal in the sign of density dependence.

The patterns of mortality in the smaller sized trees were largely driven by fire, and they probably obscure other density-dependent effects caused by conspecific or heterospecific trees that would have otherwise been present. Fire-related mortality is much lower among large trees (Sukumar et al. 1998), but consistent positive correlations indicate that the observed density dependence among large trees could regulate populations in this forest.

Spatial Patterns of Recruitment

Recruitment rates were generally low except in two canopy trees, *L. microcarpa* and *T. grandis*, and one understory tree, *C. fistula*. In these species, there was inhibition of recruitment close to adults in the canopy trees, and a slight, but significant, excess of recruitment close to adults in *C. fistula*. All three showed a peak in recruitment probability at some distance away from adults. These correspond to the "repelled" pattern (for *L. microcarpa* and *T. grandis*) and the "partially repelled" pattern (for *C. fistula*) described by Condit et al. (1992). In the canopy trees, recruitment appears to have been inhibited up to a distance of about 10–12 m and peaks occurred between 20–35 m. Of the two time intervals examined for recruitment, 1989–90 and 1993–94, extensive fire occurred in 1989, smaller fires in 1990 and 1994, and none in 1993, resulting in a lesser extent of fire in the latter interval. Considering this, the differences in the recruitment probability distributions for the two periods are remarkable. The distribution for 1993–94 appears shifted toward larger distances relative to that for 1989–90. The effect is strong in *L. microcarpa* and weaker in *T. grandis* and *C. fistula*. The increase in juvenile mortality with distance, in the presence of fire, probably explains the shift in recruitment probability toward shorter distances. The inhibitory effect is nevertheless present, though the point at which the recruitment probability cuts the expected line is shifted slightly. Vegetative sprouting does occur in these species and would be expected to lead to "attracted" syndromes of recruitment (Condit et al. 1992; Hoffmann 1996). In particular, *C. fistula* recruits extensively by vegetative coppicing from old rootstocks and this probably explains the "partially repelled" syndrome, with no inhibition or slight excess in recruitment close to adults in this species. However, we have noted that initial establishment in these species is usually from seed, and seedlings that are burned down usually coppice from underground rootstocks.

While several studies examining Janzen–Connell effects have been done in species-rich moist forests (Condit et al. 1992; Connell et al. 1984; Clark and Clark 1984; Burkey 1994; Cintra 1997; Okuda et al. 1997), few studies have come from dry forests. An earlier study in a dry forest in Costa Rica had used static data only (Hubbell 1979). Some of these studies offer support, while others contradict the predictions of the Janzen–Connell model. Recently, research showed that fungal pathogens inhibit recruitment near conspecific adults in a temperate forest tree, *Prunus serotina* (Rosaceae), supporting the Janzen–Connell hypothesis and indicating that the mechanism could be present and important even in relatively species-poor forests (Packer and Clay 2000; see also Lambers et al. 2002).

We examined only a few species due to sample size limitations and found significant reduction in recruitment probability close to adults in two very common canopy trees and a partially repelled syndrome in a common understory tree. These species together accounted for about 42% of all individuals in the 50-ha plot. As we obtain more data, we might be able to carry out a more comprehensive analysis.

Density and Diversity Effects on Recruitment and Intrinsic Rates of Increase

In addition to these distance effects, we also found strong negative density dependence in recruitment in these species. Recruitment was lower in quadrats with a high basal area of conspecific trees. The abundance of conspecific trees also had a negative effect but this was weaker and less common. Recruitment was poor in most species, but in the species that did produce recruits, strong density dependence in recruitment was found. The presence of conspecific density effects and the absence of effects due to basal area of heterospecific trees suggest that abiotic stress factors are not of overriding importance and that pests and pathogens are probably an important influence on recruitment. The higher recruitment rates for *L. microcarpa* in quadrats with a greater number, but not basal area, of heterospecific trees is consistent with this argument.

Furthermore, the presence of strong density dependence indicates that factors bringing about density dependence are time-lagged in their action, arguing strongly for the role of pests and pathogens and a smaller role for seed predators in causing density dependence (Hubbell 1998). The role of fungal pathogens and symbiotic fungi in influencing the diversity of complex communities has been emphasized in several studies (Hubbell 1998; chaps. 22 and 23). The role of pests in dry forests is probably evident in the periodic outbreaks of a stem-borer Cerambycid beetle that has caused death in *Shorea robusta* (Sal) (Dipterocarpaceae) over large areas of deciduous forest in central India (Beeson 1941). These forests are silviculturally managed for Sal timber at the expense of other species, thus decreasing the diversity of the community and increasing the density of Sal. A related beetle killed all adults of a related species, *Shorea roxburghii* (Dipterocarpaceae), in the 50-ha Forest Dynamics Plot in Mudumalai (Sukumar et al. 1992). The presence or role of other insect pests and pathogens including fungi are unknown in this forest and need to be investigated.

Intrinsic rates of increase summarize the combined effects of mortality and recruitment on population growth. Conspecific density and abundance effects on intrinsic rates of increase were generally negative in the absence of fire, indicating that population growth was lower in quadrats with higher conspecific density, an effect that would promote coexistence of species. On the other hand, when extensive fires occurred, intrinsic rates of increase were higher in quadrats of higher conspecific density, an effect that would promote common species.

In conclusion, we found significant distance- and density-related effects on mortality and particularly on recruitment that can promote coexistence in a few common species in the dry deciduous forest at Mudumalai. Grass fires have not only elevated mortality rates and suppressed recruitment but also seemingly reversed the patterns of distance and density dependence such that some common species would have increased in abundance. The combined effect of these patterns could result in an increase in dominance and decrease in diversity in this forest.

Due to small sample sizes and low turnover rates in many species, we could examine only a few species in this study. However, these species account for a substantial proportion of the individuals in the community. Finally, the ultimate factors responsible for these distance and density effects remain to be investigated. Fire plays an important role through its spatial and interannual variation, interacting with biotic factors. Clearly, more detailed analytical and experimental studies are required to identify the forces that influence diversity in this tropical dry deciduous forest.

Influence of Fire on Diversity and Dynamics

We do not know to what extent fire has been important as an ecological stress factor in the evolution of the tropical deciduous forests in this region. The ability and propensity of many species to reproduce vegetatively from underground root suckers following fire point to a history of influence by fire. Human intervention by way of logging and cultivation probably opened the forest to invasion by grasses, thereby altering fire regimes. It is certain that most, if not all, present fires are human-caused and their intensity and frequency are associated with the presence of grasses. The analyses in this paper indicate that frequent fires lower recruitment and act in ways that promote a few common species, increasing dominance, which could potentially decrease diversity in the system. Importantly, in the absence of fire, the patterns of mortality and recruitment are such that they decrease dominance and thus could promote the coexistence of species. In addition, some of the observed nonrandom spatial patterns in mortality and recruitment due to fire in some species could promote clumping in tree distributions.

General Implications

1. Distance- and density-related effects on mortality and recruitment in some common species point to the existence of biotic forces that promote coexistence and maintain diversity in this relatively species-poor dry forest community.

2. Abiotic effects like nutrient depletion or moisture stress may be present but appear to have only a weak influence on diversity at this scale.
3. Frequent ground fires have kept recruitment at low levels and influence mortality and recruitment in ways that promote dominance, which can potentially decrease diversity.
4. Detailed studies including experimental manipulation are needed to understand the various forces that govern tree diversity in dry forests and to help devise plans for conserving diversity in such forests.

Acknowledgments

This study was funded by the Ministry of Environment and Forests, Government of India. We thank the Tamilnadu Forest Department for research permissions and the Tamilnadu Electricity Board for assistance with field accommodations. It is a pleasure to thank H. S. Suresh and H. S. Dattaraja for their long association with the Mudumalai Forest Dynamics Project, which has made this work possible. We are indebted to Stephen Hubbell, Richard Condit, and Elizabeth Losos for their ideas and encouragement, which has inspired much of this work. We also thank C. M. Bharanniah, Shivaji, several field assistants, and the Smithsonian Tropical Research Institute's Center for Tropical Forest Science for their help at various stages.

References

Becker, P. L., W. Lee, E. D. Rotham, and W. D. Hamilton. 1985. Seed predation and the co-existence of tree species: Hubbell's model re-visited. *Oikos* 44:382–90.

Beeson, C. F. 1941. *The Management and Control of Forest Insects in India and Neighboring Countries.* Government of India Publication, New Delhi, India.

Bowman, D. M., and J. S. Panton. 1993. Factors that control monsoon-rainforest seedling establishment and growth in north Australian Eucalyptus savanna. *Journal of Ecology* 81:297–304.

Burkey, T. V. 1994. Tropical tree species diversity: A test of the Janzen–Connell model. *Oecologia* 97:533–40.

Cintra, R. 1997. A test of the Janzen–Connell model with two common tree species in Amazonian forest. *Journal of Tropical Ecology* 13:641–58.

Clark, D. A., and D. B. Clark. 1984. Spacing dynamics of a tropical rain forest tree: Evaluation of the Janzen–Connell model. *American Naturalist* 124:769–88.

Condit, R., S. P. Hubbell, and R. B. Foster. 1992. Recruitment near conspecific adults and the maintenance of tree and shrub diversity in a neotropical forest. *American Naturalist* 140:261–86.

Connell, J. H. 1971. On the role of natural enemies in preventing competitive exclusion in some marine mammals and in rain forest trees. Pages 298–313 in P. J. den Boer and G. R. Gradwell, editors. *Dynamics of Populations.* Centre for Agricultural Publishing and Documentation, Wageningen, Netherlands.

Connell, J. H., J. G. Tracey, and L. J. Webb. 1984. Compensatory recruitment, growth, and mortality as factors maintaining rain forest tree diversity. *Ecological Monographs* 54:141–64.

Gilbert, G. S., S. P. Hubbell, and R. B. Foster. 1995. Density and distance-to-adult effects of a canker disease of trees in a moist tropical forest. *Oecologia* 98:100–8.

Harms, K. E., S. J. Wright, O. Calderón, A. Hernández, and E. A. Herre. 2000. Pervasive density-dependent recruitment enhances seedling diversity in a tropical forest. *Nature* 404:493–95.

Hill-Ris-Lambers, J., J. S. Clark, and B. Beckage. 2002. Density-dependent mortality and the latitudinal gradient in species diversity. *Nature* 417:732–35.

Hoffman, W. 1996. The effects of fire and cover on seedling establishment in a neotropical savanna. *Journal of Ecology* 84:383–93.

Hubbell, S. P. 1979. Tree dispersion, abundance, and diversity in a tropical dry forest. *Science* 203:1299–1309.

———. 1980. Seed predation and the coexistence of tree species in tropical forests. *Oikos* 35:214–29.

———. 1998. The maintenance of diversity in a neotropical tree community: conceptual issues, current evidence and challenges ahead. Pages 17–44 in F. Dallmeier and J. A. Comiskey, editors. *Forest Biodiversity Research, Monitoring and Modeling: Conceptual Background and Old World Case Studies.* Man and the Biosphere Series, Volume 20. Parthenon Publishing Group, Pearl River, NY.

Janzen, D. H. 1970. Herbivores and the number of tree species in tropical forests. *American Naturalist* 104:501–528.

John, R., H. S. Dattaraja, H. S. Suresh, and R. Sukumar. 2002. Density dependence in common tree species in a tropical dry forest in Mudumalai, southern India. *Journal of Vegetation Science* 13:45–56.

Joshi, N. V., H. S. Suresh, H. S. Dattaraja, and R. Sukumar. 1997. The spatial organization of plant communities in a deciduous forest: a computational–geometry–based analysis. *Journal of the Indian Institute of Science* 77:365–73.

Kellman, M. 1985. Forest seedling establishment in Neotropical savannas: transplant experiments with *Xylopia frutescens* and *Calophyllum brasiliense. Journal of Biogeography* 12:373–79.

Kellman, M., and K. Miyanishi. 1982. Forest seedling establishment in Neotropical savannas: observations and experiments in the Mountain Pine Ridge savanna, Belize. *Journal of Biogeography* 9:193–206.

Martijena, N. E., and S. H. Bullock. 1994. Monospecific dominance of a tropical deciduous forest in Mexico. *Journal of Biogeography* 21:63–74.

Murali, K. S., and R. Sukumar. 1993. Leaf flushing phenology and herbivory in a tropical dry deciduous forest, southern India. *Oecologia* 94:114–19.

Murphy, P. G., and A. E. Lugo. 1986. Ecology of a tropical dry forest. *Annual Review of Ecology and Systematics* 17:67–88.

Okuda, T., N. Kachi, S. K. Yap, and N. Manokaran. 1997. Tree distribution pattern and fate of juveniles in a lowland tropical rain forest: implications for regeneration and maintenance of species diversity. *Plant Ecology* 131:155–71.

Pacala, S. W. 1997. Dynamics of plant communities. Pages 532–55 in M. J. Crawley, editor. *Plant Ecology.* Blackwell Scientific, Oxford, U.K.

Packer, A., and K. Clay. 2000. Soil pathogens and spatial patterns of seedling mortality in a temperate tree. *Nature* 404:278–81.

Shmida, A., and M. W. Wilson. 1985. Biological determinants of species diversity. *Journal of Biogeography* 12:1–20.

Sokal, R. R., and F. J. Rohlf. 1981. *Biometry*. W. H. Freeman, New York.

Stott, P. A., J. G. Goldammer, and W. L. Werner. 1990. The role of fire in the tropical lowland deciduous forests of Asia. Pages 32–43 in J. G. Goldammer, editor. *Fire in the Tropical Biota: Ecosystem Processes and Global Changes*. Springer-Verlag, Berlin.

Sukumar, R., H. S. Dattaraja, H. S. Suresh, J. Radhakrishnan, R. Vasudeva, S. Nirmala, and N. V. Joshi. 1992. Long-term monitoring of vegetation in a tropical deciduous forest in Mudumalai, southern India. *Current Science* 62:608–16.

Sukumar, R., H. S. Suresh, H. S. Dattaraja, and N. V. Joshi. 1998. Dynamics of tropical deciduous forest: population changes (1988 through 1993) in a 50–ha plot at Mudumalai, southern India. Pages 495–506 in F. Dallmeier and J. A. Comiskey, editors. *Forest Biodiversity Research, Monitoring and Modeling: Conceptual Background and Old World Case Studies*. Man and the Biosphere Series, Volume 20. Parthenon Publishing Group, Pearl River, NY.

Van Groenendael, J. M., S. H. Bullock, and L. Alfredo Pérez–Jiminéz. 1996. Aspects of the population biology of the gregarious tree *Cordia eleagnoides* in Mexican tropical deciduous forest. *Journal of Tropical Ecology* 12:11–24.

Wills, C. R., R. Condit, R. Foster, and S. P. Hubbell. 1997. Strong density- and diversity-related effects help maintain tree species diversity in a neotropical forest. *Proceedings of the National Academy of Sciences* 94:1252–57.

22

Comparable Nonrandom Forces Act to Maintain Diversity in Both a New World and an Old World Rainforest Plot

Christopher Wills, Richard Condit, Stephen P. Hubbell, Robin B. Foster, and N. Manokaran

Introduction

A central question facing ecologists, particularly those working in the tropics, is to determine the forces responsible for maintaining the extraordinary diversity in ecosystems such as rainforests or tropical reefs. There are, in general, two possibilities. First, each species may be maintained by some equilibrium of forces, such as specialization for particular niches, interactions with other species at the same trophic level, or interactions with predators and pathogens. (It is important to note, as Gillett (1962) has done, that such interactions create their own niches.) Alternatively, such interactions may be present but have appreciable effects on relatively few species. The abundance of species observed in tropical ecosystems would then simply be a reflection of rapid speciation and long periods of time during which the introduction of new species through migration and evolution and loss through local or global extinction may have reached equilibrium.

This "species drift" hypothesis (Hubbell 1995) has been called into question on two fronts. First, when speciation rates in the tropics have been measured, they are not unusually high (Roy et al. 1998). Second, strong nonrandom spatial and temporal distributions of life-history parameters have been observed among many species, particularly in well-studied rainforest plots. These take the form of statistical surveys of patterns of species recruitment and mortality (Clark and Clark 1984; Connell and Lowman 1989; Condit et al. 1992a; Hubbell and Foster 1992; Condit et al. 1994; Wills et al. 1997), and experimental manipulations that are designed to determine the roles of pathogens and distance from parental trees in recruitment. Some of these studies have provided strong support for increased survival of seedlings or saplings with distance from the parental tree (Augspurger 1983; Clark and Clark 1984). These observations tend to support the model of interactions between hosts and pathogens or seed predators suggested by Janzen (1970) and Connell (1971). Others, however, have provided evidence for more complex relationships (Burkey 1994; Forget 1994; Houle 1995; Dalling et al. 1998).

Experimental investigations are perforce confined to a few species. Statistical investigations have also tended to be confined to a few species, those that show the strongest nonrandom effects. Therefore, the majority of species may not be subject to nonrandom forces, and the species that have been investigated so far may be exceptions. In Mendelian populations, only a small fraction of allelic variants have yet been demonstrated to be subject to selective forces (Lewontin 1974; Wills 1981), and a similar situation may hold for forest trees.

Wills et al. (1997) showed that for the well-studied Barro Colorado Island (BCI) Forest Dynamics Plot, nonrandom processes affecting both recruitment and mortality are widespread and affect many different species. It is shown here that very similar but not identical forces are at work on an Old World forest in Malaysia. In both plots, especially at BCI, the pattern of forces tends to maintain diversity. This tendency can be demonstrated by an analysis of the changes in diversity measures over time. The study's results agree with an analysis, based on distances rather than quadrats, that found that about one-eighth of the commonest tree species at Pasoh showed significantly "repelled" distributions—that is, there were more recruits than expected with increased distance from conspecifics (Okuda et al. 1997). The present study, because it is able to detect frequency dependence even when the overall distribution pattern of a species is "clustered," finds more widespread density dependence than was reported in that earlier analysis.

This chapter employs a number of statistical approaches that allow nonrandom patterns to be detected in large datasets in which recruitment and mortality have been followed over time. It represents a departure from conclusions reached by earlier analyses of the BCI forest, which were based on the distribution of species abundances (Hubbell and Foster 1986; Condit et al. 1992a).

At the end of the chapter, some possible explanations for the patterns seen will be offered, along with suggestions for experimental investigation. Statistical methods used in these analyses are explained in detail to allow replication of these analyses in other plots as the data become available.

Methods

The Plots and Their Characteristics

Complete censuses over time have provided information on recruitment and mortality in two 50-ha (1.0 × 0.5 km) Forest Dynamics Plots, one on Barro Colorado Island in Panama (BCI) (Condit et al. 1992b) and the other in the Pasoh Nature Reserve in peninsular Malaysia (Manokaran et al. 1992; Condit et al. 1996). The censuses we used were separated by 8 years in the BCI plot and 10 years in the Pasoh plot (chaps. 24 and 36). All trees above 1 cm diameter at breast height (dbh) were counted, so the data do not include information about small-seedling recruitment and mortality.

Overall Similarities and Differences Between the Two Forests

The two plots have a number of properties in common. The most abundant BCI species are more common than the most abundant species at Pasoh, but aside from this, the shapes of the species abundance curves are similar in both (Condit et al. 1996). The distributions of 100 tree species in each plot that were matched for abundance (see below) were examined by Ripley's K (Ripley 1977; Haase 1995). The species in each forest plot show a wide variety of distributions, ranging from extremely clustered to slightly overdispersed, with a predominance of species showing significant amounts of clustering. However, the distribution of K values is the same in both plots (data not shown). Any differences found between the plots are therefore not the result of different patterns of tree distribution, as might, for example, be the case if most species tend to be clustered in one plot and most tend to be overdispersed in the other. The characteristics of the spatial distributions are also apparently not affected by the greater density of the trees in the Pasoh forest.

Rainfall at the BCI forest, although greater than at Pasoh, is more seasonal (chaps. 24 and 36). The BCI plot is less diverse than the Pasoh plot, with 314 species found during the census period compared with 823 at Pasoh. The density of trees at BCI is also substantially lower; it had 235,424 trees ≥ 1 cm dbh in 1982, at the start of the census period, compared with Pasoh's 335,323 in 1986, at the start of the census period.

Rate of Turnover in the Two Forests

One very important property that distinguishes the two plots, and one which will play a large role in the interpretation of the analyses to be reported here, is the substantial difference between BCI and Pasoh in the rates of recruitment and mortality per unit time. Both of these rates are substantially higher in the BCI forest. Expressed as a fraction of the total trees present at the beginning of the census period, the BCI recruitment rate was 0.038 per year between 1982 and 1990, compared with only 0.013 between 1986 and 1996 at Pasoh, while the BCI mortality rate was 0.028 versus 0.017 at Pasoh during the same periods. These rates can vary widely from one census period to another; but in general, the BCI rates are substantially higher than the Pasoh rates. In part, this reflects the effects of a severe El Niño. In proportionate terms, there is also more species turnover at BCI than at Pasoh—during the census period 10 species were lost and 10 were gained at BCI, while only 5 were lost and 9 gained at Pasoh even though it had more than twice as many species. This appears not to be due to large differences in the numbers of rare species in the two plots: 29 species at BCI had only one or two representatives at the start of the census period, while 42 at Pasoh had only one or two representatives.

Establishment of Two Comparable Datasets

To compare the two plots directly it was necessary to establish comparable sets of data having properties as similar to each other as possible. This was done as follows. The lists of species abundances in the two plots were compared, and 100 species were chosen from the BCI plot, each of which was matched with a species of comparable abundance in the Pasoh plot. These two subsets of species did not include the most abundant species in either plot, since these species could not be matched with species having comparable numbers in the other plot. Indeed, the first pair of species that had comparable numbers consisted of the fifth most abundant species at BCI, *Alseis blackiana* (Rubiaceae), and the fourth most abundant at Pasoh, *Ardisia crassa* (Myrsinaceae). The pairs of matched species included approximately equal numbers of canopy and subcanopy species.

These 100 matched species pairs comprised 116,419 trees in the BCI subset and 115,296 trees in the Pasoh subset. These made up less than a half and about a third of the total number of trees in the BCI and Pasoh plots, respectively. For the two subsets, total numbers of recruits and deaths were determined; those at Pasoh were found to be 49% and 74%, respectively, of those at BCI. To compensate for these differences, 51% of the BCI recruits and 26% of the trees at BCI that died were discarded at random, in order to make the recruitment and mortality numbers in the two subsets comparable. This in turn made the correlations obtained from the two datasets directly comparable because they dealt with equivalent sets of numbers. This process would not have been justified if there were, in either of the subsets, a significant correlation between the numbers of a species and the proportion of recruitment or mortality. When plots were made of numbers in the species against proportion of recruitment or mortality (before adjustment), no significant relationships were detected, indicating that a bias would not be introduced if data were discarded at random. When all the recruitment and mortality data in the BCI subset were used, differences similar to those reported in this paper were also seen (data not shown).

Correlations Between Abundance Measures and Life-History Parameters

Both plots were divided into quadrats and analyzed. The data presented here are from analyses of the plots after division into 10 × 10 m quadrats, yielding 5000 quadrats. Similar analyses carried out on 5 × 5, 20 × 20, and 50 × 50 m quadrats yielded similar results, with frequency-dependent effects "peaking" at different quadrat sizes in different species (Wills et al. 1997). Results from the 10 × 10 m quadrats are shown here because these procedures yielded the most significant correlations overall.

Both parametric and nonparametric statistics were used, and in order to determine their significance, they were compared to statistics obtained from repeated randomizations of the data subsets. These procedures produced randomized datasets that are comparable in numbers and overall patterns of distribution to the original datasets. This allowed a comparison of the distributions of statistics obtained from repeated randomization of the datasets with the statistics obtained from the actual datasets.

The two correlation methods used were the standard parametric correlation and the nonparametric Wilcoxon signed-rank correlation. Weighting was not used. Correlations were obtained between the following datasets:

1. Correlations involving recruitment:
    a. Numbers of recruits of a given species that appeared in a quadrat during the census period (N_R) correlated with total numbers of that same species in the quadrat at the beginning of the census period (N_S).
    b. N_R correlated with the total basal area of that species at the beginning of the census period (BA_S). This latter number was obtained by adding the cross-sectional basal areas at breast height of all the trees of that species in the quadrat, which is correlated with the biomass of that species in the quadrat.
    c. N_R correlated with the total numbers of all trees belonging to other species in the quadrat (N_T). This examined the effect on recruitment of overall crowding in the quadrat. Similar correlations in which only the other species present in the reduced dataset were employed gave statistically indistinguishable results.
    d. N_R correlated with the total basal areas of all the trees of other species in the quadrat (BA_T). This measure gives an approximate idea of the biomass of trees present in the quadrat, except for the species being examined.
2. Correlations involving mortality. Corresponding correlations of mortality with numbers and areas were obtained. If N_M is the number of trees of a given species that died during the census period, correlations were obtained for:
    e. N_M with N_S.
    f. N_M with BA_S.
    g. N_M with N_T.
    h. N_M with BA_T.

Methods of Randomization

Many of these correlations have an autocorrelation component. For example, mortality and recruitment *numbers* will be expected to be larger in quadrats with many trees of that species. It is therefore essential to compare these correlations with control correlations, employing randomized sets of data in which the magnitudes of these autocorrelations will be the same and can be factored out. The distributions of these control correlations can be used, as will be seen below, to provide tests of significance for the actual correlations.

These control correlations were constructed by two "scrambling" methods, in which the data were randomized while preserving many properties of the dataset. Scramblings were carried out 1000 times in each case, and correlations were obtained from each set of scrambled data.

a) Total Scrambling. In this method, the coordinates in the plot of all the trees of a given species were preserved, but their properties were scrambled. The dbh of a tree at the beginning and end of the census period, which indicates whether that tree survived the census period, died, or was recruited, were switched with the properties of another tree of the same species elsewhere in the plot. Thus, a tree that might have survived during the entire census period in the actual dataset might be switched with a tree that died or that was recruited. Because no information was lost during the scrambling process, the forest that resulted had the same number of recruits and the same number of trees that had died, but the numbers of recruits and deaths in a given quadrat were usually different from the numbers actually observed.

A rapid computer method for carrying out this scrambling is to use the C algorithm that scrambles a list by choosing an item from a list at random, adding it to a new list, and shortening the old list by the item that was removed. In other programming languages, a string of symbols can be constructed using different symbols for recruits, mortality, and survivors. As each symbol is chosen from the string at random, the original string is shortened by that symbol.

b) Stratified Scrambling. Allen Herre and Joseph Wright (personal communication) were properly concerned that such total scrambling, which randomizes trees that have very different properties, might introduce spurious correlations because the resulting distribution of tree sizes in each quadrat might be very different from the original distribution. Such an effect would cast doubt on the values of the correlations, particularly on those involving recruitment or mortality with total basal area. To determine whether such an effect was indeed contributing to the correlations, randomizations were carried out in a stratified fashion, by dividing up each species into three size categories: less than 2 cm dbh, between 2 and 10 cm dbh, and greater than 10 cm dbh. Scrambling was then carried out within each of these size categories. Again, the real correlations were compared with the distributions of 1000 scrambled correlations. No significant differences were seen using the two types of scrambling, indicating that the possible biases suggested by Herre and Wright are not present or are too small to be detectable.

To determine whether the distributions of the scrambled correlations were sufficiently normal that they could be used in t-tests, 20 species were picked at random from each of the matched data subsets. Total and stratified parametric and nonparametric correlations were examined from these 20 species, a total of

1280 sets of 1000 correlations each. The Kolmogorov–Smirnov procedure (Zar 1984) was used to test each set for deviations from normality. The abundance significant at the 0.05, 0.01, and 0.001 levels were not significantly different from random expectation.

Diversity Measures

Both forests are of course very diverse, and it can be assumed that if nonrandom processes are operating, an equilibrium level of diversity will be reached. Two questions are asked in this study. First, what is the distribution of diversities across quadrats, and does this distribution differ between the two forests? Second, is diversity increasing or decreasing with time, and is there any indication that regions of the forest with low diversity are increasing in diversity with time to a greater extent than would be expected by chance?

Five measures of diversity were employed: number of species, the Shannon, evenness (Pielou 1975), Simpson's (Simpson 1949), and McIntosh's (McIntosh 1967) indexes. These diversity measures were obtained for all quadrats in the matched sets of data, and the means and distributions compared. Fisher's α could not be employed because in many quadrats the number of species was the same as the number of trees; in such cases Fisher's α goes to infinity.

If a tendency exists for the selection of diversity in either or both of these rainforests, leading to an equilibrium between selection for diversity and its loss by random factors over time, this should be detectable by the following analysis. A correlation of the diversity in the quadrats at the start of the census period with the change in diversity in each quadrat during the time of the census should yield a negative relationship—in other words, the least diverse quadrats should show the largest increases in diversity over time. Obviously, such an analysis again suffers from a statistical problem. A quadrat with low diversity is likely to increase markedly in diversity simply because of regression toward the mean. This is particularly the case if the distribution of diversities in the various quadrats is negatively skewed.

This difficulty was corrected for by the same scrambling procedure used earlier. Again, the forests were scrambled 1000 times, the initial diversities of all the quadrats and their changes over time were determined, and parametric and nonparametric correlations between these values were obtained. Total and stratified scrambling methods were employed.

Results

Plots of the Data

The analyses carried out on the two matched subsets of the forest are presented here as distributions of t-values. This allows the results for all 100 of the species examined in each subset to be summarized in a single distribution. The t-values were obtained by dividing the difference between the actual correlation and the mean of the scrambled correlations for a given species by the standard deviation of the distribution of scrambled correlations.

If the trees in the forest show no relationship between numbers or areas and the life-history parameter being examined—recruitment or mortality—then the actual correlations should fall near the means of the correlations for the scrambled datasets. The resulting t-values should have a mean not significantly different from zero and a standard deviation—and variance—of approximately one.

Figure 22.1a shows an example of such a distribution, for N_M (numbers of deaths per quadrat) versus N_S (numbers of trees of that species in the quadrat). Data from both matched subsets are shown in the figure. The correlations used in this analysis are nonparametric, and the control correlations used to obtain the data in the figure were obtained by the stratified scrambling method.

In both the matched subsets, the mean of the distribution of t-values is approximately zero, and the majority of the t-values lie within ± 3 standard deviations. The variance in each subset, although it is slightly larger than 1, is not substantially larger. Finally, the distribution of t-values in both subsets is symmetrical—the values have no noticeable skew.

Figure 22.1c shows a very different situation. In this figure, numbers of recruits (N_R) have been correlated with total numbers of that species (N_S) in each quadrat. Again, nonparametric correlations and stratified scrambling were used. Here, in contrast to the data presented in figure 22.1a, the means of the t-values are substantially less than zero in both datasets, the variance is substantially greater than one, and there is a pronounced negative skew. In the great majority of species, the correlation between numbers of recruits and total numbers in the quadrat is more negative than the mean of the correlations in 1000 scrambled forests. In some species, the difference between the observed correlation and the distribution of scrambled correlations is very large. As in the previous figure, the shapes of the distributions in the two matched forests are essentially the same. Both forest plots show a strong tendency for high recruitment in quadrats with few pre-existing trees of that species and low recruitment in quadrats with larger numbers of pre-existing trees.

Figure 22.1g shows a striking relationship that was found in both forests and that may give a clue to the kinds of nonrandom forces that are acting on these forests. The figure shows plots from both forests, again using nonparametric correlations and stratified scrambling. The figure examines numbers of recruits (N_R) versus numbers of all trees of other species in the quadrat (N_T). This plot is contrasted in figure 22.1h with numbers of recruits (N_R) versus the basal areas of all other species in the quadrat (BA_T).

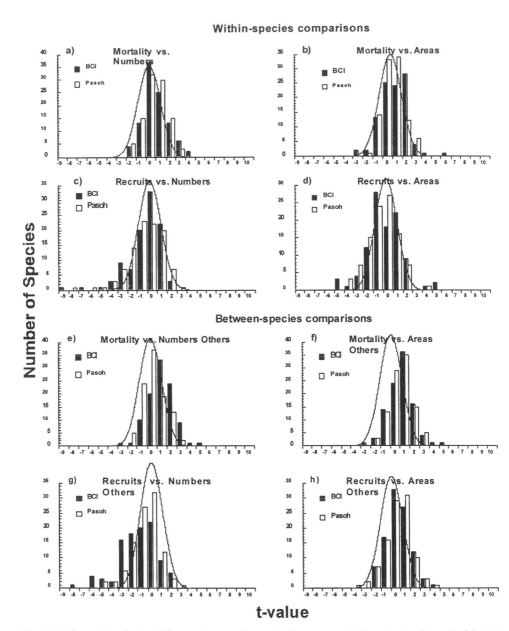

Fig. 22.1. The t-values for the difference between observed and 1000 scrambled correlations for each of the 100 species in the two matched subsets of the Pasoh and BCI data. The data plotted are for comparisons of recruitment and mortality versus numbers and areas within and between species. The bell curves show the distributions of t-values to be expected if the correlations were entirely random.

The two distributions in the figure are very different. In essence, crowding by other species has a strong effect on recruitment, but the effect is much stronger in the BCI than in the Pasoh data. This means that at BCI, sheer numbers of trees of whatever species interfere with recruitment in the great majority of the species examined. But total basal area has no noticeable effect on recruitment (fig. 22.1h); the distribution of t-values has a mean of approximately zero and a variance of approximately one, as would be expected for a situation in which there is no nonrandom effect.

This negative effect of overall crowding on recruitment suggests (but does not prove) that frequency-dependent interactions between hosts and pathogens are important in both forests and are particularly strong at BCI. If the effects on recruitment were due to exhaustion of nutrients, shading, or other abiotic factors, there should also have been a correlation with basal area, which was not seen.

Summary of the Entire Dataset

In general, for all the correlations examined, there are no statistically detectable differences in significance levels with either stratified or total scrambling methods. As would be expected, parametric correlations tend to be slightly larger, in either the positive or negative direction, than nonparametric correlations. There are no large differences between the significances of the correlations that employ numbers of recruits or numbers of deaths and those that employ the fraction of recruits or the mortality fraction. Significant correlations involving the latter fractions had been reported earlier for the BCI forest (Wills et al. 1997), and the current studies show that the correlations found using those comparisons are comparable in significance and direction to those found when the numbers rather than the fractions are used.

Generally, though not always, when the mean of the distribution of t-values for the correlations is significantly different from zero, the distribution is skewed significantly in the same direction. Examples of such skews are seen in figures 22.1c and 22.1g. Furthermore, if there is a large observed deviation from zero of the distribution of t-values, the variance of the t-values tends to be large as well.

These trends are summarized in table 22.1. Recruitment results are presented first. Strong negative correlations occur between recruitment and all quadrat measurements except for the basal areas of trees belonging to other species in the quadrat. Thus, both forests show the striking pattern illustrated in figures 22.1g and 22.1h, in which there is a negative effect of crowding but not of basal area on recruitment.

The correlations involving mortality are weaker. It might be expected, by analogy with the effects seen on recruitment, that there would be a positive correlation between mortality and numbers of the same species, but such a correlation is not detectable. However, some correlations involving mortality are significant, and

Table 22.1. Overall Properties of the Distributions of t-Values

Correlation	BCI			Pasoh		
	Diff of Mean of t-dist	Skew		Diff of Mean of t-dist	Skew	Anomalies
Recruitment:						
N_R vs. N_S	$(-)$***	$(-)$***		$(-)$***	$(-)$***	
N_R vs. BA_S	$(-)$***	$(-)$***		$(-)$***	$(-)$***	Only np signif.
N_R vs. N_T	$(-)$***	$(-)$***		$(-)$***	$(-)$***	
N_R vs. BA_T	n.s.	n.s.		n.s.	n.s.	
Mortality:						
N_M vs. N_S	n.s.	n.s.		n.s.	n.s.	
N_M vs. BA_S	n.s.	n.s.		n.s.	n.s.	
N_M vs. N_T	$(+)$***	n.s.		$(-)$**	n.s.	Sign change betw. forests
N_M vs. BA_T	$(+)$*	n.s.		$(+)$*	n.s.	Only p signif.

Notes: The general tendencies observed are given for each of the 12 types of correlation. N_R = numbers of recruits, N_M = numbers of deaths, N_S = number of trees of a given species in a quadrat at the outset of the census, A_S = basal areas of the trees of a given species in a quadrat at the outset, N_T = number of trees in the quadrat other than the species being examined, A_T = area of trees in the quadrat other than the species being examined. *, **, and *** represent significance levels of 0.05, 0.01, and 0.001, respectively (two-tailed t-tests). p = parametric correlations, np = nonparametric (Spearman rank) correlations. Note on the significance levels used in the table. Because a thousand scramblings were employed to determine the significance of each real correlation, a very accurate estimate of the distribution of expected correlations could be obtained for each species. The t-values reported here are the number of standard deviations away from the mean of the scrambled correlations that the real correlation lies. It is therefore legitimate to use a t-test with 998 degrees of freedom to determine the level of significance of the actual correlation; and it is these levels of significance that are summarized in the table.

unlike the patterns seen with recruitment, these patterns of significance are sometimes different between the two matched subsets of data.

Correlations between the fraction of mortality and the basal area of the same species tend to be significantly positive in the BCI subset, but are not significant in the Pasoh subset. Correlations between mortality numbers or mortality fraction and the total number of other species in the quadrat are significantly positive at BCI but significantly negative at Pasoh. And correlations between mortality numbers or mortality fraction and the basal areas of the other species in the quadrat are positive in both subsets, although the significance is marginal.

Finally, although the differences tend to be small, the variances of the BCI t-value distributions are usually larger than the comparable variances for the Pasoh t-value distributions. Only a minority of the F-values obtained from the ratio of the BCI and Pasoh variances reach significance (data not shown). The greater variances observed at BCI are consistent with the observation, detailed in the next section, that because of the more rapid turnover of trees in the BCI forest, there is a greater disturbance of diversity and a stronger tendency for low-diversity quadrats in the BCI forest to return to a state of high diversity.

Differences in Diversity Between the Two Matched Subsets

One very large difference between the matched subsets can be detected by an examination of diversity measures. Because the subsets have been matched for

Table 22.2. Variance in Diversity Measures across 10-m Quadrats in the Two Matched Subsets

Diversity	BCI			Pasoh			
	Mean	Variance	Skew	Mean	Variance	Skew	$F_{BCI/Pa}$
Shannon	3.61	0.282	−1.22	3.71	0.171	−0.64	1.65***
Evenness	0.945	0.0026	−4.00	0.951	0.0012	−2.21	2.21***
Simpson's	0.946	0.0032	−5.04	0.954	0.00097	−2.98	3.30***
McIntosh's	0.885	0.0070	−2.73	0.896	0.0035	−1.53	1.99***

*** $p = 0.001$.

Table 22.3. Correlations Between the Diversity Measures in Each Quadrat at the Outset of the Census Period and the Change in Diversity Measures over the Time Period of the Census

Correlations	BCI				Pasoh			
	p	t	non-p	t	p	t	non-p	t
Shan vs. Δ Shan	−0.32	**−5.34**	−0.29	**−5.26**	−0.21	−1.11	−0.18	**−5.45**
Even vs. Δ Even	−0.40	**−4.42**	−0.28	**−5.19**	−0.25	−1.87	−0.28	**−2.59**
Simp vs. Δ Simp	−0.40	**−2.34**	−0.24	**−3.47**	−0.20	−0.32	−0.24	−1.57
McIn vs. Δ McIn	−0.36	**−4.16**	−0.26	**−4.62**	−0.25	−1.10	−0.25	**−2.01**

Note: To determine significance of the correlations, they were compared with the distribution of 1000 correlations calculated after total scrambling of the data. Significant t-values are represented by bold numbers.

numbers of trees and numbers of species, the diversities exhibited by an average quadrat in the two subsets should be very similar. But it is possible to ask whether the diversities from quadrat to quadrat vary more in one subset than in the other— that is, whether the variance in diversity from one quadrat to another is different in the two plots.

Table 22.2 shows that there are substantial differences in this variance exhibited by the two matched data subsets. Because all four of the evenness measures used exhibit some degree of correlation, it is not surprising that all should behave in the same way in each of the two matched data subsets. For each of these measures, the variance in diversity from quadrat to quadrat is much greater at BCI. Furthermore, the distributions of diversity statistics across quadrats are strongly negatively skewed. This bias is more pronounced at BCI than at Pasoh and reflects the fact that some quadrats at BCI have much lower diversity than the least diverse quadrats at Pasoh.

Changes in Diversity with Time

We can now go a step further in this analysis and ask what happens to these diversity measures over the course of the censuses. Is the diversity being maintained over time, and if so, can this process be measured?

Table 22.3 shows correlations, both parametric and nonparametric, between diversity measures at the outset of the census and the change in those measures

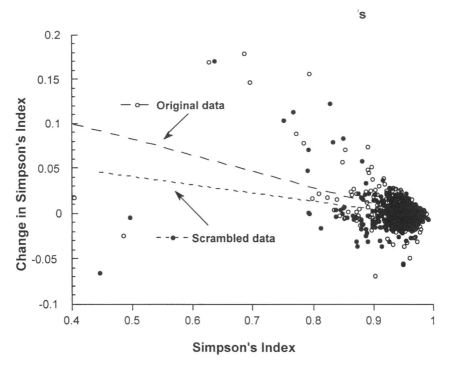

Fig. 22.2. Simpson's index versus change in Simpson's index over time, for the BCI dataset. The plot shows the real data and one set of scrambled data. Regression lines have been fit to the data, to show the difference in slope between the real and scrambled datasets.

during the census period. To determine their significance, these correlations were compared to the distribution of 1000 correlations obtained from scrambled data. The randomized datasets analyzed for this table were scrambled using the total data scrambling method; as with the earlier analyses, stratified scrambling gave similar results.

The correlations are all substantially negative, showing that the means of the scrambled correlations are all less negative than the actual values. To enable the reader to visualize the effects of scrambling on the data, a typical analysis is shown in figure 22.2. Simpson's index for BCI is plotted against the change in Simpson's index over time, and one set of scrambled data is plotted for comparison. In both the original and the scrambled data, low diversity quadrats at the beginning of the census tend to increase substantially in diversity with time. This would be expected in such negatively skewed data, simply because of regression toward the mean. But the actual data show a significantly more pronounced increase in diversity in low-diversity quadrats than the scrambled data, indicating that nonrandom pressures are operating on the forest that increase diversity when diversity is low. Table 22.3 shows that this increase in diversity over time is more significant at BCI than at Pasoh, suggesting that the pressures maintaining diversity at BCI are stronger than those at Pasoh. Even though the Pasoh plot is more diverse, the faster turnover at BCI must contribute to its lower diversity.

Discussion

Sources of Statistical Error: Correlations Between Adjacent Quadrats

If a quadrat with a high density of trees of a given species is surrounded by quadrats with a lower density, it would be expected that recruits from the quadrat in the center would be found in the adjacent quadrats, artificially inflating the recruitment rate of the surrounding quadrats. This could lead to a spurious negative correlation between recruitment and numbers of trees already present (Chesson 1986). A test for this possibility was carried out (Wills and Condit, unpublished). Correlations between rates of recruitment in 1000 pairs of adjacent quadrats, picked at random, in the 10 commonest species at BCI were determined, and none of the correlations were significant. Recruitment into adjacent quadrats therefore does not appear to contribute significantly to the correlations reported here.

Sources of Statistical Error: Differences in Scatter in the Real and Scrambled Data

As pointed out by J. Wright (personal communication), scrambling the data may alter the proportion of the total in a quadrat that is survivors, but does not alter the total. In the real data, the variance of these proportions is likely to be greater than it is in the scrambled data because of nonrandom factors. This will produce a weaker correlation, and the difference between the observed and expected correlation will therefore tend to be negative, but this may entirely be due to stochastic processes.

It is possible to determine whether the differences in correlations are entirely the result of scatter, by examining the parametric correlations and slopes of the regression lines. This has been done elsewhere (Wills and Condit 1999), and the parametric correlations and slopes yield results that are very similar to those obtained for the nonparametric analysis. The effects reported here therefore appear to be real and not simply the result of scatter.

Consequences of Host–Pathogen Models

The best known of the host–pathogen models is that of Janzen (1970) and Connell (1971), who independently proposed a density-dependent model with a spatial component. They suggested that adult trees are likely to be surrounded by pathogens and seed predators that have evolved to become specialists on that particular host species. The result will be that young trees of the same species will be less likely to survive when they are near the adult than when they are distant from it. As a result of such density dependence, no one species will become too numerous in the forest, and this should lead to the maintenance of diversity.

Janzen thought that this model appears to require an overdispersed spacing of trees of a given species in the forest, and indeed (starting with the often-quoted passage by Alfred Russel Wallace in his 1878 *Tropical Nature*), many observers have remarked on the apparently overdispersed distribution of tree species in a mature forest. There are two problems with this. First, overdispersal turns out to be something of an illusion, presumably brought on by the overwhelming complexity of the forest as seen by an observer passing through it. For at least several Forest Dynamics Plots, the majority of tree species tend to be clumped rather than overdispersed (Hubbell et al. 1990; Condit et al. 2000), although this clumping may not be apparent for many species, particularly rare ones, until careful censuses have been carried out.

Second, there is a theoretical problem with overdispersal as a mechanism for the maintenance of diversity. If only a small region surrounds a given mature tree in which the germination of seeds or survival of saplings are reduced, then the Janzen–Connell mechanism might permit the maintenance of a few species but not the hundreds of species commonly seen in a mature forest. Theoretical investigations and observations and computer simulations of overdispersed ant species in a California desert (Ryti and Case 1992) reinforce the conclusion that for a forest to maintain large numbers of species, the basal area over which pathogens exert their inhibitory effects must be very extensive. Although inhibitory effects of proximity to conspecific adults have been detected in a number of studies (Augspurger 1983; Clark and Clark 1984; Condit et al. 2000), the effects extend at the most about 10 m, not enough to produce an extremely overdispersed ecological checkerboard of the type needed to maintain hundreds of different tree species.

However, the Janzen–Connell model need not be an entirely spatial one; it can easily be modified to include temporal and life-cycle factors. One factor of likely importance is the period of susceptibility of hosts to pathogens, which is often far greater in seedlings than in adults. Such differential sensitivity has been exploited in the use of "trap cultivars" which can trap pathogens by using less sensitive stages of the host's life cycle to attract them and stop them from damaging the more sensitive stages (Talekar and Nurdin 1991). A second factor is the length of time taken by the pathogens to build up high enough numbers to do damage. It does, however, require that pathogens not be uniformly distributed throughout the forest.

Were the pathogens of a particular host species uniformly distributed, then no host tree could become sufficiently large without first running a gauntlet of its pathogens. Thus, if the distant seedlings are to survive the sensitive period of their life cycle more often than nearby ones, the distribution of pathogens must be a clumped one. Because of this clumped distribution, if a host seedling becomes established some distance away from its conspecifics, it will take some time for pathogens to "find" this isolated host. By that time the host tree may be established well enough such that it can resist the onslaught of the pathogens. Hubbell et al. (1990) proposed that waves of infection passed from tree to tree by root contact may result in such time-lagging, and Gilbert et al. (1994) found evidence for such time-lagging in the transmission of root canker.

One clear prediction of the Janzen–Connell effect is that most of the density-dependent mortality should affect immature trees. This prediction is borne out by the results presented in this chapter.

Time-lagging of pathogens will result in the continual and simultaneous buildup and breakdown of clusters of host species in the rainforest, and this process should have reached a steady state in a mature forest. Low numbers or densities should lead to a positive local intrinsic rate of natural increase, while high numbers or densities should lead to a negative local intrinsic rate of natural increase. For the majority of the commonest species in the forest, this prediction is also borne out by the results presented here. As has been pointed out elsewhere, such an epidemiological approach to rainforest diversity allows many different species, at a wide variety of abundances, to be maintained simultaneously. It also suggests mechanisms by which the biochemical and morphological diversity of host (tree) species in the forest can actually increase over time (Wills 1996; Wills et al. 1997).

In addition to pathogens, there are many other possible sources for a larger number of ecological niches. One is spatial or temporal fluctuation in niche number (McLaughlin and Roughgarden 1993). For example, the normal extreme light gradient from top to bottom of the canopy in dense tropical forests might be disturbed because of a local treefall, an outbreak of a disease, or a fire, allowing saplings in the understory to take advantage of the changed light conditions. These saplings have often persisted for years or decades in the understory (Welden et al. 1991), perhaps providing a repository of species diversity. If members of this diverse understory were to grow up to fill gaps, they might provide small "islands" of diversity that would permit the maintenance of high species numbers. However, such islands of diversity associated with gaps do not appear to be common in the lowland tropical rainforest examined here (Hubbell et al. unpublished data), though the true importance of the very complex question of gap ecology remains to be determined.

A second source of niches may be specific interactions between tree species, such that one species facilitates the survival of another and vice versa (Callaway 1995). There is vast literature on such interactions for both animals and plants, but no good assessment of how these interactions might contribute to the maintenance of diversity in an ecosystem as a whole (Rosenzweig 1995). As was found earlier (Wills et al. 1997), between-species interactions tend to be much weaker in the Barro Colorado Island rainforest than within-species interactions, and unlike within-species interactions, the few that are significant tend to fall into equal numbers of positive and negative interactions.

Another argument supporting the role of pathogens is that only a minority of species at BCI show a strongly clumped distribution (Condit et al. 1992a), even though the majority should have shown strong clumping if seed dispersal were solely a function of distance.

Forces Acting on Recruitment in the Two Plots and Time-Lagging of Recruitment

The most striking feature of the data presented here is that strong negative relationships exist between recruitment and numbers of conspecifics, and between recruitment and basal area of conspecifics, in both of the matched datasets. These effects are very strong, and as reported earlier for BCI, involve the majority of species (Wills et al. 1997). Even stronger negative effects have been observed in collection-station-based comparisons between numbers of seeds of a given species and numbers of small seedlings of that same species at BCI (Harms et al. 2000).

A second important feature of the data is that both matched subsets display a very strong negative relationship between overall crowding and recruitment. If the total number of trees present in a quadrat, regardless of their species, is large, very few recruits of any species will be found. This result argues strongly that the effect of crowding on recruitment is time-lagged. If any unusual degree of crowding were to have an immediate negative effect on recruitment, then it would be difficult for quadrats with different degrees of crowding to arise. It might be argued that if large numbers of recruits suddenly appear in a quadrat as a result of treefall gaps, for example, then further recruitment might be repressed. However, figure 22.3 shows that the recruits over time across quadrats in the BCI subset are not distributed in a bimodal fashion. There is therefore no indication that recruitment takes place in sudden bursts and involves only a minority of quadrats. The deficit in recruitment, as a quadrat grows more crowded, appears to take place gradually.

The effect of crowding on recruitment would be expected on the assumption that under such conditions one or more physical resources are limiting—light, minerals, other nutrients, even water. However, as was shown in table 22.1, no significant relationship exists between recruitment and the total biomass (as measured by total basal area) of all the other species in the quadrat. If the availability of energy and nutrients is driving this relationship, it might be expected that these nutrients would be in shorter supply in quadrats with high basal area.

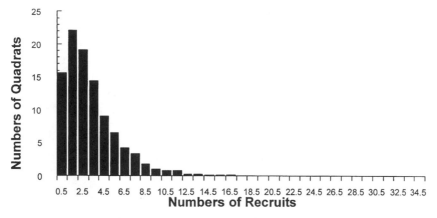

Fig. 22.3. Recruits over time across quadrats in the BCI subset.

Forces Acting on Mortality in the Two Plots

In general, correlations between mortality and numbers or basal area are weaker and less consistent between the two data subsets. One might expect positive relationships between mortality and numbers or basal areas of conspecifics, since recruitment is inhibited by conspecific numbers and basal area. However, aside from a puzzling strong relationship between mortality fraction and conspecific basal area that is seen only in the BCI subset, such relationships are not found. This suggests that once trees have reached 1 cm dbh, mortality is largely random. Nonrandom mortality may be taking place among seedlings of less than 1 cm dbh, however, and this may contribute to the strong density effects that are seen in recruitment.

However, significant relationships occur between mortality and overall numbers or overall basal area. In the BCI subset these results are consistent—both crowding and high basal area appear to increase mortality. The correlation with basal area shows that unlike the situation with recruitment, mortality is greater in quadrats with high total basal area. But in the Pasoh subset, the relationships are different. There tend to be positive correlations between mortality and basal area, but negative ones between mortality and numbers—that is, there is relatively more mortality in less crowded quadrats! Whatever may be causing this phenomenon, the increased recruitment in less crowded quadrats is apparently enough to overcome it.

Differences in Diversity Patterns in the Two Subsets

One substantial difference between the two matched subsets is the much larger variance seen among quadrats for diversity measures in the BCI subset (table 22.2).

This difference is likely the result of a more rapid turnover of trees in the BCI forest. Although differences in turnover have been minimized in the matched subsets for purposes of statistical analysis, the diversities seen at the outset of the census period will reflect the different actual rates of turnover in the two forests. The turnover of trees at BCI is roughly twice that at Pasoh per unit time.

Such rapid turnover at BCI will result in a more substantial sampling effect on quadrat diversity than is seen at Pasoh. This effect, however, is countered by the stronger tendency of the less diverse BCI quadrats to increase in diversity over time (table 22.3 and fig. 22.2). This increase in diversity must primarily be the result of the frequency-dependent effects that are operating with roughly the same strength in the two forests. As a result, Pasoh shows a less variable distribution of diversities than BCI. The effects of these various factors are currently being explored through simulations (Wills, in preparation).

Nature of the Forces Acting on the Two Plots

We have argued elsewhere (Wills et al. 1997) that the time-lagged effect of crowding on recruitment, coupled with the lack of effect of total basal area on recruitment, supports the argument that biotic factors play an important role in the observed frequency dependence. One testable hypothesis that could explain this phenomenon is that disease organisms are likely to be more plentiful in quadrats that are crowded with trees, something that can be thought of as the "nursery school" effect.

Gillett (1962) pointed out that pests have the effect of increasing the underlying complexity of an ecosystem. Their activity can create a wide variety of ecological niches that may be difficult for field workers to detect without a great deal of effort. A tree of a given species that is growing in a region in which it is attacked by a large number of pathogens or predators, or is aided by the presence of a variety of symbionts, is living in a very different ecological niche from one that is growing in a relatively pathogen-free region or one that is surviving without the benefit of symbionts. Further, the mix of pathogens and symbionts may change over quite short distances in the forest, providing even more niches. One example is provided by fungal pathogens, which are only just beginning to be understood in these forests. Although mycorrhizae tend not to be pathogens in tropical New World forests and may be associated with a wide variety of tree species, these fungi have profound ecological effects and are likely to influence the distribution of pathogenic fungi. Mycorrhizal associations of great complexity (Jackson and Mason 1984) have been followed in detail in tropical and temperate forests (Allen et al. 1995) and in commercially grown trees (Jackson et al. 1995). Connell and Lowman (1989) suggest that interactions of trees with particularly beneficial types of mycorrhizae may lead to monodominance of a tree species. The reciprocal possibility that more complex interactions with mycorrhizae may lead to diversity has also been considered (Fitter and Garbaye 1994; Allen et al. 1995). These authors point out that loss of a mycorrhizal species can lead to changes in host diversity.

Mycorrhizae make up only a small subset of the fungi in a rainforest environment. Preliminary results from transplant experiments involving *Tetragastris panamensis* (Burseraceae), *Calophyllum longifolium* (Rubiaceae), and *Brosimum alicastrum* (Moraceae) show that each is parasitized by a different mix of phytophthoran and other fungi (J. Davidson and A. Herre, personal communication). Sometimes, however, pathogens can affect a wide range of host species. In one recent study (Schardl et al. 1997), a large number of fungal species in the genus *Epichloë* that interact with a wide variety of temperate grass species were studied at the molecular level. The fungal species ranged in their effects from pleiotropic to antagonistic symbionts. The situation was complicated, but some of the pathogenic species tended to affect a wide range of hosts, some of which were in different tribes, while the pleiotropic symbionts tended to be confined to, and adapted to, particular host species. The pathogenic fungi destroy seed production in their host species and can only be transmitted horizontally as spores through the agency of a dipteran fly, *Phorbia phrenione*. There would therefore be strong selection for them to be able to affect a variety of hosts. The beneficial fungal species can be transmitted vertically, and there is no such pressure on them.

A similar situation in the neotropics, in which a pathogen has been found on a remarkably wide range of hosts, is seen in the witches' broom pathogen, the agaric fungus *Crinipellis perniciosa*, which is a commercially important disease of cocoa. Strains of this fungus affect may different host species in addition to cocoa—hosts that have been discovered so far are solanaceous plants, the unique family represented by the shrub *Bixa orellana* (Bixaceae), and a large number of liana species. Griffith and Hedger (1994a, b) found that the liana biotype outcrosses readily, while the others largely do not. This outcrossing may be connected to the fact that the fungus can spread rapidly among the many different liana species that tend to grow readily in open parts of the rainforest. The other host species affected by this fungus, in contrast, are scattered thinly throughout the forest. Such interactions, as they are understood in depth, will allow a greater understanding of frequency-dependent effects in these complex ecosystems.

General Implications

1. This study illustrates a novel statistical approach for examining density- and basal area-dependent effects on recruitment and mortality. In order to create comparable datasets for a correlation analysis between the two Forest Dynamics Plots, subsets of the data sets were chosen from each, resulting in 100 matched pairs of species. This statistical approach allows testing of a variety of hypotheses against randomized data, through the application of repeated total and stratified scrambling which was applied to each dataset as a control. Additionally,

this study verifies that the differences in correlations are not due to the result of scatter. This statistical method is thus applicable to other comparable datasets in tropical forests.

2. Correlation analyses of the BCI and Pasoh plots suggest that crowding by all species, not just conspecifics, affect recruitment negatively. This negative effect of overall crowding on recruitment suggests that frequency-dependent interactions between hosts and pathogens are important in both forests, and are particularly strong at BCI.

3. At BCI, there is a tendency for there to be a positive correlation between mortality of a species and density of other species in the same quadrat, while negative correlations tend to be observed at Pasoh. We suggest that this discrepancy in results may be attributable to the size of a sapling, such that once trees reach 1 cm dbh, mortality is largely random.

4. This study shows that, in both forests, diversity is enhanced over time by nonrandom mechanisms (i.e., pathogens), and that the stochastic effects of increased turnover have had a larger effect on the BCI plot. This may be connected with the more rapid turnover of species in each quadrat at BCI. Even though the Pasoh plot is more diverse, some of the difference in diversity from BCI may simply be because sampling plays a smaller role in plots like Pasoh that have slower turnover. It will be important to search for this effect in other plots.

Acknowledgments

I thank Allen Herre, Dawn Field, Kyle Harms, Robert John, Egbert Leigh, David Metzgar, Trevor Price, Joseph Wright, and Neville Yoon for discussions and helpful comments. The Smithsonian Tropical Research Institute in Panama and the Forest Research Institute Malaysia provided overall logistical and financial support for the censuses, with particular thanks to Drs. I. Rubinoff and Salleh Mohm. Nor. The BCI project has been supported by grants from the National Science Foundation, the Smithsonian Scholarly Studies Program, the Smithsonian Tropical Research Institute, The John D. and Catherine T. MacArthur Foundation, the World Wildlife Fund, the Earthwatch Center for Field Studies, the Geraldine R. Dodge Foundation, and the W. Alton Jones Foundation. The Pasoh project was supported by the Forest Research Institute Malaysia, the National Science Foundation, the Rockefeller Foundation, and the John Merck Fund. The data analyses presented here were unexpectedly supported by a grant from the National Institutes of Health (to C.W.).

References

Allen, E. B., M. F. Allen, D. J. Helm, J. M. Trappe, R. Molina, and E. Rincon. 1995. Patterns and regulation of mycorrhizal plant and fungal diversity. *Plant and Soil* 170:47–62.

Augspurger, C. K. 1983. Seed dispersal of the tropical tree *Platypodium elegans*, and the escape of its seedlings from fungal pathogens. *Journal of Ecology* 71:759–71.

Burkey, T. V. 1994. Tropical tree species diversity: A test of the Janzen–Connell model. *Oecologia* 97:533–40.

Callaway, R. 1995. Positive interactions among plants. *Botanical Review* 61:306–49.

Chesson, P. L. 1986. Environmental Variation and the Coexistence of Species. Pages 240–56 in J. Diamond and T. J. Case, editors. *Community Ecology.* Harper and Row, New York.

Clark, D. A., and D. B. Clark. 1984. Spacing dynamics of a tropical rainforest tree: evaluation of the Janzen–Connell model. *American Naturalist* 124:769–88.

Condit, R., S. P. Hubbell, and R. B. Foster. 1992a. Recruitment near conspecific adults and the maintenance of tree and shrub diversity in a neotropical forest. *American Naturalist.* 140:261–286.

———. 1992b. Stability and change of a neotropical moist forest over a decade. *Bioscience* 42:822–828.

———. 1994. Density dependence in two understory tree species in a neotropical forest. *Ecology* 75:671–705.

Condit, R., S. P. Hubbell, J. V. LaFrankie, R. Sukumar, N. Manokaran, R. B. Foster, and P. S. Ashton. 1996. Species–area and species–individual relationships for tropical trees: A comparison of three 50–ha plots. *Journal of Ecology* 84:549–62.

Condit, R., P. Ashton, P. Baker, S. Bunyavejchewin, S. Gunatilleke, N. Gunatilleke, S. Hubbell, R. Foster, A. Itoh, J. LaFrankie, L. Hua Seng, E. Losos, N. Manokaran, R. Sukumar, T. Yamakura. 2000. Spatial patterns in the distribution of common and rare tropical tree species: A test from large plots in six different forests. *Science* 288:1414–8.

Connell, J. H. 1971. On the roles of natural enemies in preventing competitive exclusion in some marine animals and in rain forest trees. Pages 298–310 in P. J. den Boer and G. R. Gradwell, editors. *Dynamics of Populations. Proceedings of the Advanced Study Institue on Dynamics of Numbers in Populations, Oosterbeek, 1970.* Center for Agricultural Publishing and Documentation, Wageningen, Netherlands.

Connell, J. H., and M. D. Lowman. 1989. Low–diversity tropical rain forests: Some possible mechanisms for their existence. *American Naturalist* 134:88–119.

Dalling, J. W., M. D. Swaine, and N. C. Garwood. 1998. Dispersal patterns and seed bank dynamics of pioneer trees in moist tropical forest. *Ecology* 79:564–78.

Fitter, A. H., and J. Garbaye. 1994. Interactions between mycorrhizal fungi and other soil organisms. *Plant and Soil* 159:123–132.

Forget, P. M. 1994. Recruitment pattern of *Vouacapoua americana* (Caesalpiniaceae), a rodent-dispersed tree species in French Guiana. *Biotropica* 26:408–19.

Gilbert, G. S., S. P. Hubbell, and R. B. Foster. 1994. Density and distance-to-adult effects of a canker disease of trees in a moist tropical forest. *Oecologia* 98:100–08.

Gillett, J. B. 1962. Pest pressure, an underestimated factor in evolution. Pages 37–46 in D. Nichols, editor. *Taxonomy and Geography, A Symposium.* London Systematics Association, London.

Griffith, G. W., and J. N. Hedger. 1994a. The breeding biology of biotypes of the witches' broom pathogen of cocoa, *Crinipellis perniciosa. Heredity* 72:278–89.

———. 1994b. Spatial distribution of mycelia of the liana (L–) biotype of the agaric *Crinipellis perniciosa* (Stahel) singer in tropical forest. *New Phytologist* 127:243–59.

Harms, K. E., S. J. Wright, O. Calderón, A. Hernández, and E. A. Herre. 2000. Pervasive density-dependent recruitment enhances seedling diversity in a tropical forest. *Nature* 404:493–95.

Haase, P. 1995. Spatial pattern analysis in Ecology based on Ripley's K-function: Introduction and methods of edge-correction. *Journal of Vegetation Science* 6:575–82.

Houle, G. 1995. Seed dispersal and seedling recruitment: The missing link(s). *EcoScience* 2:238–44.

Hubbell, S. P. 1995. Towards a theory of biodiversity and biogeography on continuous landscapes. Pages 171–99 in G. R. Carmichael, G. E. Folk, and J. L. Schnoor, editors. *Preparing for Global Change: A Midwestern Perspective.* Academic Publishing, Amsterdam.

Hubbell, S. P., R. Condit, and R. B. Foster. 1990. Presence and absence of density dependence in a neotropical tree community. *Philosophical Transactions of the Royal Society of London (Series B)* 330:269–82.

Hubbell, S. P., and R. B. Foster. 1986. Biology, chance and history in the structure of tropical rainforest tree communities. Pages 314–29 in T. J. Case and J. Diamond, editors. *Community Ecology.* Harper and Row, New York.

———. 1992. Short–term dynamics of a neotropical forest: Why ecological research matters to tropical conservation and management. *Oikos* 63:48–61.

Jackson, R. M., and P. A. Mason. 1984. *Mycorrhiza.* E. Arnold, London.

Jackson, R. M., C. Walker, S. Luff, and C. McEvoy. 1995. Inoculation and field testing of Sitka spruce and Douglas fir with ectomycorrhizal fungi in the United Kingdom. *Mycorrhiza* 5:165–73.

Janzen, D. H. 1970. Herbivores and the number of tree species in tropical forests. *American Naturalist* 104:501–28.

Lewontin, R. C. 1974. *The Genetic Basis of Evolutionary Change.* Columbia University Press, New York.

Manokaran, N., J. V. LaFrankie, K. M. Kochummen, E. S. Quah, J. Klahn, P. S. Ashton, and S. P. Hubbell. 1992. *Stand Table and Distribution of Species in the 50–ha Research Plot at Pasoh Forest Reserve.* Forest Research Institute of Malaysia, Kepong, Malaysia.

McIntosh, R. P. 1967. An index of diversity and the relation of certain concepts to diversity. *Ecology* 48:392–404.

McLaughlin, J. F., and J. Roughgarden. 1993. Species interactions in space. Pages 89–98 in R. E. Ricklefs and D. Schluter, editors. *Species Diversity in Ecological Communities.* University of Chicago Press, Chicago.

Okuda, T., N. Kachi, S. K. Yap, and N. Manokaran. 1997. Tree distribution pattern and fate of juveniles in a lowland tropical rain forest: Implications for regeneration and maintenance of species diversity. *Plant Ecology* 131:155–71.

Pielou, E. C. 1975. *Ecological Diversity.* Wiley, New York.

Ripley, B. D. 1977. Modelling spatial patterns. *Journal of the Royal Statistical Society* B39:177–92.

Rosenzweig, M. L. 1995. *Species Diversity in Space and Time.* Cambridge University Press, New York.

Roy, K., D. Jablonski, J. W. Valentine, and G. Rosenberg. 1998. Marine latitudinal diversity gradients: Tests of causal hypotheses. *Proceedings of the National Academy of Sciences* 95:3699–702.

Ryti, R. T., and T. J. Case. 1992. The role of neighborhood competition in the spacing and diversity of ant communities. *American Naturalist* 139:355–74.

Schardl, C. L., A. Leuchtmann, K. R. Chug, D. Penny, and M. R. Siegel. 1997. Coevolution by common descent of fungal symbionts (*Epichloë* spp.) and grass hosts. *Molecular Biology and Evolution* 14:133–43.

Simpson, E. H. 1949. Measurement of diversity. *Nature* 163:688.

Talekar, N. S., and F. Nurdin. 1991. Management of *Anomala cupripes* and *Anomala expansa* in soybean by using a trap cultivar in Taiwan. *Tropical Pest Management* 37:390–92.

Welden, C. W., S. W. Hewett, S. P. Hubbell, and R. B. Foster. 1991. Sapling survival, growth, and recruitment: relationship to canopy height in a neotropical forest. *Ecology* 72:35–50.

Wills, C. 1981. *Genetic Variability.* Oxford University Press, Oxford, U.K.

———. 1996. Safety in diversity. *New Scientist* 149:38–42.

Wills, C., and R. Condit. 1999. Similar non-random processes maintain diversity in two tropical rainforests. *Proceedings of the Royal Society of London (Series B)* 266:1445–52.

Wills, C., R. Condit, R. Foster, and S. P. Hubbell. 1997. Strong density- and diversity-related effects help to maintain tree species diversity in a neotropical forest. *Proceedings of the National Academy of Science* 94:1252–57.

Zar, J.H. 1984. *Bioistatistical Analysis.* Englewood Cliffs NJ, Prentice-Hall.

23

Long-Term Tree Survival in a Neotropical Forest

The Influence of Local Biotic Neighborhood

Jorge A. Ahumada, Stephen P. Hubbell, Richard Condit,
and Robin B. Foster

Introduction

When two of us (Hubbell and Foster) began the Barro Colorado Island (BCI) Forest Dynamics Project two decades ago, our grail was to understand the mechanisms enabling the coexistence of many kinds of trees in tropical moist forests (Hubbell and Foster 1983)—perhaps the "Mount Everest" of questions in community ecology. A fundamental assumption of our quest was that these mechanisms operated in an explicit spatial context and therefore they could not be fully understood without mapping individual trees in their local biotic setting in the forest. First, we expected trees to interact most strongly with their immediate neighbors, so it was reasonable to assume that their long-term fate would depend on their local biotic neighborhood (Hubbell and Foster 1986b). Second, because trees live a long time, the study would have to be long term. And finally, because of the high tree diversity and extreme rarity of many of the tree species, the study would have to be large scale to obtain sample sizes of a majority of the tree species that would be adequate for studying their population dynamics. And thus the 50-ha permanent plot on BCI was born in 1980.

One of our principal goals was to establish a set of protocols for collecting data that would enable us and our successors to test a broad range of hypotheses for the maintenance of high tree species richness. It was and remains of course impossible to predict all data requirements for all possible hypotheses that might be erected in the future. But we decided that all hypotheses would require us to know about not only the larger individuals typically measured (e.g., 10 cm diameter at breast height (dbh) and up), but also small saplings. Accordingly, the protocol involved identifying, measuring, and mapping all trees, saplings, and shrubs greater than 1 cm dbh in a 50-ha Forest Dynamics Plot on BCI. The plot was then recensused every 5 years to record all individual plant survival, growth, and new sapling recruitment. By the end of 2000, the BCI plot had been completely censused five times. In addition to these measurements, an annual census of treefall gaps and canopy height had also been taken since 1983 throughout the BCI plot on a 5-m grid of sampling points, with a 2-year lapse in 1996 and 1997. This information,

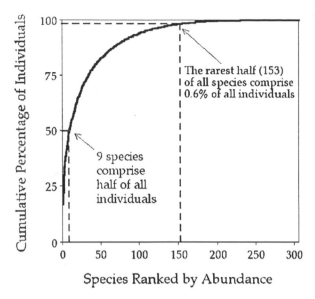

The rarest half (153) of all species comprise 0.6% of all individuals

9 species comprise half of all individuals

Species Ranked by Abundance

Fig. 23.1. Relationship between the cumulative number of individuals sampled and the cumulative number of species in the BCI 50-ha Forest Dynamics Plot, for species ranked in order of abundance from commonest (left) to rarest (right). Nine common species make up half the individuals >1 cm dbh. The rarest halves of all species (153 species) collectively make up only 0.6% of all individuals.

combined with data on tree distribution, survival, and growth rates, has been very useful for categorizing tree species to life history guild (Hubbell and Foster 1986a, b; Welden et al. 1991). These data is also critical for understanding the dynamics of canopy tree replacement processes. Beginning in 1988, a study of seed rain and seedling germination was also added, involving 200 seed traps and 600 seedling germination plots (three at each trap site) along the permanent trails through the plot (Harms et al. 2000; Hubbell et al. 1999). The BCI plot contains a steady-state number of approximately 240,000 trees of free-standing woody plants > 1 cm dbh, and slightly more than 300 species (chap. 24). As expected, many species are quite rare. Half (153) of the species collectively comprise less than 1% of all trees (fig. 23.1). However, some species achieve very high abundances: the nine most abundant species together make up approximately 50% of all trees. Because of the large size of the plot, 124 species achieve population sizes of 200 individuals >1 cm dbh or more, which is adequate for many species- and population-level analyses, including the survival analysis presented here.

Although the BCI data are spatially explicit, we have only just begun to use them to test hypotheses about how the survival of a focal individual tree depends on the structure and composition of the tree's immediate biotic environment. In

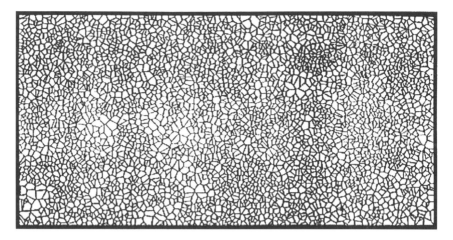

Fig. 23.2. Dirichlet polygonal tesselation of individual tree neighborhoods for trees >30 cm dbh in the BCI 50-ha Forest Dynamics Plot, showing the imperfect hexagonal packing of trees; after Hubbell et al. (1990).

this paper we examine how the probability of a tree surviving for 13 years, from the first census in 1982 to the fourth recent census in 1995, depends on the structure and species composition of its 20 nearest neighbors. The reason for choosing 20 neighbors is as follows. The neighborhood of an individual plant is formally defined by the Dirichlet polygonal tesselation of the plot, shown in figure 23.2 for trees over 30 cm dbh in the BCI plot. First-order neighbors are those that share a polygon side with the focal plant; second-order neighbors are those that share a polygon side with one or more of the first-order neighbors. If the BCI trees were perfectly hexagonally packed, they would have 6 first-order neighbors and 12 second-order neighbors, for a total of 18 first- and second-order neighbors. We chose 20 neighbors as a round number close to the expected number of first- and second-order nearest neighbors under perfect hexagonal packing. Although in the BCI forest, trees are of different sizes and do not exhibit perfect hexagonal packing, they are nevertheless surprisingly close. The mean number ± 1 SD of first-order neighbors is 5.21 ± 0.46, and the number of first- and second-order neighbors combined is 18.77 ± 3.25 (fig. 23.3). The 20-nearest-neighbor size was chosen because it occupies an area approximately equal to the size of the understory patch involved in the replacement of a single canopy tree (8–10 m in crown diameter). We report analyses of variable neighborhood sizes and the spatial extent of neighborhood effects elsewhere (Hubbell et al. 2001).

To test hypotheses about the effects of the local biotic neighborhood on long-term focal plant survival, we employ the formal statistical method of logistic

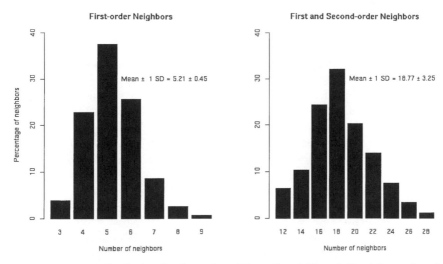

Fig. 23.3. Frequency distributions for the number of first-order neighbors *(left)* and the number of first- and second-order neighbors combined *(right)* in the BCI tree community.

regression (Neter et al. 1996); we compute the odds of a tree's survival, and the odds ratio as a function of four independent variables describing the 20-nearest-neighbor biotic environment of a focal plant. Because some readers may not be familiar with logistic regression, we briefly review the statistical procedures in the methods section.

The importance of spatially explicit data in plant ecology has been recognized for a long time (Harper 1977), but long-term, spatially explicit data on plant communities at the individual plant level are rare. Most theoretical models of forest dynamics work from the spatial arrangement of individuals (Botkin 1993; Shugart 1984; Pacala et al. 1996), where the dynamic unit is the gap made by a single treefall. These models are largely driven by light competition, and therefore relative plant growth rates under different shade intensities is the dominant response variable. This suggests that the initial relative size or height of the focal plant among its immediate neighbors may be an important predictor of the plant's probability of long-term survival.

However, many other potential neighborhood factors can affect the fate of a plant besides competition for light. In tropical forests, for example, one of the major hypotheses for the maintenance of diversity focuses on density-dependent predator and pathogen attack in an explicit spatial context. Janzen (1970) and Connell (1971) independently proposed that host-specific seed and seedling predators might promote tree diversity in tropical forests by preventing any one species from recruiting near parent trees and becoming locally monodominant, thereby opening space for other species. This hypothesis is supported by new evidence for strong and pervasive local intraspecific density dependence in the BCI plot in the seed-to-seedling transition (Harms et al. 2000). There is also evidence for weaker density dependence in the >1 cm dbh saplings of many tree species in the plot (Wills et al. 1997; chap. 22), on spatial scales ranging up to several hectares. However, whether such density dependence can be detected in the immediate biotic neighborhood of a focal plant over long time periods has never been tested on the spatial scale of very local tree neighborhoods. An important question is whether the commonness or rarity of species in the BCI forest can be explained by or at least correlated with the strength of the density dependence measured during the course of the replacement of a single tree in the forest. This is important because the critical question is whether the adult tree population of a species in the forest is regulated by density-dependent forces operating throughout the entire canopy tree replacement process.

Another variable of considerable current interest is the potential stabilizing effect of diversity itself. The potential relationship between diversity and stability has had a long and controversial history in ecology (May 1973; Pimm 1991), but evidence from African savannas and old-field plant communities in the temperate zone suggests that more diverse species assemblages may be more resistant to various stresses (e.g., drought) and less likely to lose species richness for a given level of stress (McNaughton 1985; Tilman et al. 1997). In the case of tropical forests, we expect that higher local species richness might promote higher survival of individual trees for two theoretical reasons. One is frequency-dependent predator or pathogen attack, conferring a survival and/or growth advantage on locally rare species because they are harder to find and/or because their low densities do not sustain an attack or epidemic (Wills et al. 1997). Another is that a greater local diversity of functional groups and resource exploitation strategies may stabilize local assemblages (Tilman et al. 1997). Can the effect of species richness per se on a focal plant's survival be detected on the spatial scale of individual canopy and subcanopy trees?

Another potentially important variable is the variation in the quality of sites for plant survival, irrespective of species. For example, regenerating treefall gaps are places of high thinning mortality. They are good places for the growth of the few survivors, but most plants in treefall gaps die. We can assess variation in microsite quality, for whatever reason, by measuring the number of a focal plant's original neighbors that survive, irrespective of species. By this measure, high-quality sites are those in which most of all of the plant's 20 neighbors in 1982 are still alive in 1995, and low-quality sites are those in which few neighbors survived. The fate of the focal plant can then be assessed as a function of the overall neighborhood survival rate, while not including the fate of the focal plant in the measurement of neighborhood survival.

To our knowledge, this is the first long-term study to analyze the survival of individual trees as a function of their immediate biotic neighborhood, as well as the first study of any duration that includes four neighborhood variables simultaneously. Virtually all individual-based studies of neighborhood competition have been conducted on herbaceous plant communities (reviewed in Weiner 1995), and they have focused mainly on competition for light. Studies on tree communities have focused mainly on temperate trees, and many of the studies have been literature-parameterized models of tree growth under light competition (Botkin 1993; Shugart 1984), although more recent models have been parameterized from field measurements (Pacala et al. 1996). The latter study, however, made the assumption that light competition was the driving mechanism determining the fate of an individual tree. The importance of the other neighborhood variables for long-term survival of individual trees has never before been assessed by direct measurement.

Methods

The BCI Forest Dynamics Project (FDP) was established on Barro Colorado Island, a 15-km² former hilltop located in artificial Gatun Lake in the Panama Canal (see chap. 24). The western half of the island, including a relatively level plateau on the summit, is covered with old-growth forest that may have been selectively logged in the last century. In 1980, a 50-ha permanent plot was surveyed in the old-growth forest on the plateau. Humans lived on the plot in pre-Columbian times, but they never cleared any part of it for agriculture (Piperno 1990). By 1982, the first census of all free-standing woody plants ≥1 cm dbh (excluding woody climbers) was completed. Data collected included tree diameter (dbh), species identity, x and y coordinates (estimated precision to about 0.5 m), and a series of measurements about plant vegetative and reproductive condition. (For details see Condit 1998; part 7 introduction.) Beginning in 1983, the height of the canopy was estimated by range finder over the corners of every 5-m grid point in the plot. Heights were classified into six height intervals that were more precise for low canopies and less so for high because higher canopy heights were harder to measure precisely. For a subset of 50 tree species, tree height–dbh relationships were separately determined (O'Brien et al. 1995). These measurements revealed a very tight interspecific relationship between diameter and height, indicating that relative diameter can be used as a reliable indicator of relative plant size and height.

From the map coordinates of each plant, the 20 nearest neighbors of every focal plant alive in 1982 were found, using a computer program. For each of these 2 individual neighborhoods, we computed the following: (1) the number of the original neighbors still alive in 1995 (Microsite Quality), (2) the number of the 20 neighbors that had smaller tree diameters (dbh) in 1982 (Relative Plant Size), (3) the number of neighbors that were of the same species as the focal plant (Conspecific Density), and (4) the number of different species among the 20 neighboring plants in 1982 (Species Richness). The dependent variable was the binary state variable: In 1995 was the focal plant still alive (1) or dead (0)?

Binary survival data are the standard data form used in statistical analysis of binary data, and logistic regression is the standard method of analysis. Although the data are binary, the real question is how the independent variables affect the probability that a plant will live or die. In logistic regression analysis the odds of survival are formally defined as the ratio of the probability of surviving to the probability of dying. Consider the case of a single independent variable, say Microsite Quality. Let s be the probability of survival (the dependent or Y variate), and let X_i be Microsite Quality. Then the logistic function for the probability of survival as a function of Microsite Quality, and its log odds, are

$$Y = \frac{e^{(\beta_0 + \beta_1 X_i)}}{1 + e^{(\beta_0 + \beta_1 X_i)}} = s_i$$

$$\log\left(\frac{s_i}{1 - s_i}\right) = \beta_0 + \beta_1 X_i = \pi_i$$

Note that by taking the natural logarithm of the odds, one obtains a linear function of the independent variable. Call this function the logit transform, and symbolize this function by the Greek letter pi (π_i). The behavior of the logistic function for Microsite Quality is shown in figure 23.4. Note that there are no intermediate Y values between 0 and 1 because the data observed indicate that the focal plant is either alive (1) or dead (0) in 1995. In figure 23.4 there are many observations of focal plants alive or dead at each value of the X variate, a fact that is obscured in the graph; but the *proportion* of focal plants still alive out of all plants increases as the number of neighbors alive in 1995 (Microsite Quality) increases. The logistic function is therefore fitting how the proportion alive or estimated probability of survival changes with Microsite Quality.

One can now generalize the logistic model to accommodate several to many independent variables, and if desired, their interactions. However, in the present analysis, we found little evidence for strong interactions among the independent variables, so we chose to ignore interactions. Therefore the generalized logistic model used can be written as

$$Y = \frac{e^{(\beta_0 + \beta_1 X_1 + \beta_2 X_2 + \cdots \beta_n X_n)}}{1 + e^{(\beta_0 + \beta_1 X_1 + \beta_2 X_2 + \cdots \beta_n X_n)}}$$

An important tool of logistic regression is the ability to statistically isolate the contribution of each independent variable to a focal plant's survival. This is done using

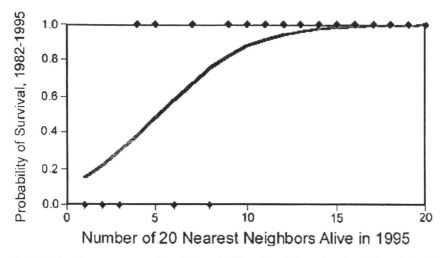

Fig. 23.4. Sample one-way regression of the probability of survival as a function of the number of the original 20 nearest neighbors of a focal plant in the BCI plot that survived until the last census in 1995. The logistic function estimates the fraction surviving for each number of neighbors surviving, even though the observations are binary; either the focal plant was alive in 1995 (value of 1) or it was dead (value of 0). Note that in general there are many superimposed observations for each number of surviving neighbors, which cannot be seen from the graph.

the *odds ratio*, which is not the same as the odds. Recall that the odds are defined as the probability of surviving divided by the probability of dying, for a given value of the independent variable. The odds ratio is defined as $\exp[\beta_i(X_i + 1) - \beta_i(X_i)]$. The most important property of the odds ratio is that it is independent of the partial value of X_i; it measures the partial effect of variable X_i on the odds of survival. The numerical value of the odds ratio for X_i is $\exp(\beta_i)$. One only needs simple algebra to confirm that this is so:

$$\pi_i[X] = \beta_0 + \beta_1 X$$
$$\pi_i[X + 1] = \beta_0 + \beta_1(X + 1)$$
$$\pi_i[X + 1] - \pi_i[X] = \beta_1$$

The dataset was analyzed at two different levels: (1) at the guild level, using all species whose guild is known regardless of their abundance, and (2) at the species level, for the 124 species that had a minimum 1982 population size of 200 individuals. For both analyses, species were divided into (a) four growth-form classes (shrubs: adults <4 m tall; understory trees: adults 4–10 m tall; midstory trees: adults 10–20 m tall; and canopy trees: adults >20 m tall) and (b) three functional classes based on relative shade tolerance: shade-intolerant pioneers (distribution of saplings strongly skewed toward low-height canopy sites), shade-tolerant species (distribution of saplings is strongly skewed toward high-height canopy sites), and intermediate species (distribution of saplings is skewed toward intermediate-height canopy sites [10–20 m]). In addition, for the guild level analysis, all species were separated into three abundance classes (species whose population sizes were on the order of $<10^2$, 10^3, or $>10^4$ individuals). The classification of species into shade-tolerance guilds was based on Welden et al. (1991) and subsequent analyses of growth rate in BCI tree species (Hubbell et al. unpublished). For each guild, we calculated the odds ratios for each species, and calculated their averages and standard deviations to test for significant departures from no effect ($\beta_i = 0$ or $\exp(\beta_i) = 1$). Significant differences in the odds ratios among guilds were tested using standard ANOVA but with unequal sample sizes in each class. The β_i data were fit using maximum likelihood procedures using a simplex method algorithm (Nelder and Mead 1965).

Results

The results of the survival analysis for all four neighborhood factors considered in one combined model are listed in table 23.1 for each of the 124 species for which there were over 200 individuals on the plot. Among these 124 species, the guild structure and number of species were as follows: For the growth form functional guilds, there were 20 shrubs, 28 understory trees, 35 midstory trees, and 41 canopy trees. For the three light-requirement guilds, 10 species were classified as light-demanding gap species; 15 as intermediate species; and 99 as shade-tolerant mature-phase species.

Tests for pairwise interactions among neighborhood variables did not reveal strong interaction effects among any of the variables, compared with the simple effects of these variables. For example, figures 23.5 and 23.6 show the joint probability distribution of survival as a function of Microsite Quality and Relative Plant Size or Conspecific Density, respectively. For this reason we chose to limit the analyses to the four main neighborhood variables, without considering their interaction terms. We are also thereby justified in presenting the univariate results graphically in what follows.

Figure 23.7 shows the odds ratios for all 124 species considered together for the four neighborhood variables. The most important variable was Microsite Quality, with a mean odds ratio of 1.108 ± 0.007. The second most important neighborhood variable was Conspecific Density, with a mean odds ratio of 0.923 ± 0.019. Relative Plant Size was third, with a mean odds ratio of 1.021 ± 0.005. The last variable was Species Richness, with a mean odds ratio of 1.011 ± 0.003. Thus, the overall quantitative importance of these four neighborhood variables ranged over about an order of magnitude.

Table 23.1. Odds Ratios for Four Neighborhood Variables on the Odds of Survival of 124 Tree and Shrub Species in the BCI Plot

Species	Life Form	Light Guild	MQ	RPS	CD	SR
Acalypha diversifolia	S	G	1.009	0.990	0.770	0.963
Adelia triloba	U	I	0.907	1.091	0.820	0.068
Alchornea costaricensis	T	G	1.015	1.252	0.820	1.012
Alibertia edulis	U	S	1.207	1.013	0.829	1.107
Alseis blackiana	T	S	1.083	1.050	0.891	1.049
Anaxagorea panamensis	S	S	1.311	1.002	0.868	0.965
Andira intermis	T	I	1.079	0.988	0.592	1.093
Annona acuminata	S	S	1.012	0.994	0.866	0.992
Apeiba membranacea	T	I	1.063	1.140	0.433	1.013
Aspidosperma cruenta	T	S	1.235	0.977	0.806	0.993
Astrocaryum standleyanum	M	S	1.017	1.184	1.301	1.019
Bactris coloniata	S	S	1.032	1.143	1.135	0.849
Bactris major	U	S	1.144	1.069	1.185	1.154
Beilschmiedia pendula	T	S	1.127	1.052	0.955	1.004
Brosimum alicastrum	T	S	1.102	0.965	1.560	0.899
Calophyllum longifolium	T	S	1.181	0.976	0.921	1.014
Capparis frondosa	S	S	1.158	1.015	0.829	0.972
Casearia aculeata	U	S	1.046	1.005	0.507	1.009
Casearia arborea	T	I	1.076	1.085	0.948	0.993
Casearia sylvestris	N	S	1.161	1.016	0.773	0.953
Cassipourea elliptica	M	S	1.130	0.994	1.616	1.009
Cecropia insignis	T	G	1.169	1.179	1.094	1.027
Cestrum megalophyllum	S	I	1.148	0.951	1.048	1.040
Chrysoclamys eclipes	U	S	1.179	1.046	0.955	1.025
Chrysophyllum argenteum	T	S	1.072	1.071	0.951	1.089
Coccoloba manzanillensis	U	S	1.078	1.030	0.501	0.993
Conostegia cinnimomifolia	S	S	1.034	0.875	0.787	0.958
Cordia bicolor	M	S	1.030	1.128	0.923	1.042
Cordia lasiocalyx	M	S	1.082	1.015	0.818	1.051
Coussarea curvigemmia	U	S	1.141	0.987	0.810	1.058
Croton billbergianus	U	G	1.089	1.022	0.867	1.019
Cupania sylvatica	M	S	1.187	1.031	0.364	0.867
Desmopsis panamensis	U	S	1.099	0.931	1.001	0.992
Drypetes standleyi	T	S	1.156	1.054	0.921	0.996
Erythrina costaricense	U	S	1.069	1.051	1.507	0.995
Erythroxylon ma (ery2ma)	M	S	1.091	1.016	0.974	0.938
Eugenia coloradensis	T	S	1.195	1.005	0.840	1.031
Eugenia galalonensis	U	S	1.129	0.933	0.906	0.962
Eugenia nesiotica	M	S	1.135	0.988	0.928	0.874
Eugenia oerstediana	M	S	1.122	1.020	1.027	1.003
Faramea occidentalis	U	S	1.181	0.971	0.938	0.987
Garcinia intermedia	M	S	1.202	0.957	0.987	0.958
Garcinia madurno	M	S	0.999	9.918	0.821	1.091
Guapira standleyanum	T	S	1.137	1.164	1.060	1.046
Guarea guidonia	M	S	1.116	1.052	0.998	1.005
Guarea sp. nov.	M	S	1.100	1.007	1.058	1.025
Guatteria dumatorum	T	S	1.068	0.999	0.994	1.015
Guettarda foliacea	U	S	1.087	1.045	0.497	1.035
Gustavia superba	M	I	1.017	1.193	0.894	0.877

(Continued)

Table 23.1. (Continued)

Species	Life Form	Light Guild	MQ	RPS	CD	SR
Hasseltia floribunda	M	S	1.057	1.077	1.046	1.018
Heisteria concinna	U	S	1.113	1.026	0.885	0.995
Herrania purpurea	U	S	1.128	1.008	1.449	1.046
Hirtella triandra	M	S	1.158	1.041	0.881	1.046
Hybanthus prunifolius	S	S	1.139	0.942	1.041	1.031
Inga co	M	I	1.008	0.957	0.759	1.017
Inga goldmanii	T	S	1.064	0.966	0.603	1.053
Inga marginata	T	I	1.051	1.060	0.805	1.005
Inga pezezifera	T	I	1.156	0.938	1.141	1.083
Inga quaternata	M	S	1.065	1.011	0.975	1.072
Inga sapindoides	M	S	1.096	1.034	0.683	1.040
Inga sp. nov.	U	S	1.126	1.019	0.651	0.955
Inga umbellifera	T	S	1.075	0.961	0,813	0.958
Jacaranda copaia	T	G	1.026	1.241	0.980	1.058
Lacistema aggregatum	U	S	1.092	1.017	0.997	1.032
Laetia thamnia	U	S	1.095	1.004	1.206	1.010
Licania platypus	T	S	1.300	1.020	0.517	1.081
Lonchocarpus latifolius	T	S	1.067	1.012	0.736	0.985
Malmea sp (malmsp)	S	G	1.235	0.915	0.478	1.241
Maquira costaricana	M	S	1.137	0.947	1.000	0.955
Miconia affinis	U	I	0.973	0.943	0.867	1.006
Miconia argentea	M	G	1.085	1.080	0.925	0.998
Miconia nervosa	S	S	0.972	0.848	0.812	1.056
Mouriri myrtilloides	S	S	1.139	0.980	0.906	1.000
Nectandra cissifolia	T	S	1.118	1.041	0.832	0.955
Ocotea cernua	M	S	1.103	1.057	0.583	1.042
Ocotea punctata (ocotpu)	T	S	1.046	1.027	0.833	1.048
Ocotea obtusifolia	T	I	1.050	1.035	1.128	1.019
Ocotea whiteii	T	S	1.089	1.097	0.955	1.002
Oenocarpus mapoura	M	I	1.053	1.208	0.965	1.092
Olmedia aspera	U	S	1.027	1.090	0.970	1.026
Ouratea lucens	S	S	1.145	0.990	0.849	0.961
Palicourea guianensis	S	G	1.016	0.988	1.058	1.027
Perebia xanthophylla (perexa)	M	S	1.129	1.104	0.667	1.035
Petagonia macrocaarpa	U	S	1.075	1.085	1.021	0.970
Picramnia latifolia	U	S	1.079	1.009	0.804	0.994
Piper aequale	S	S	1.122	0.996	0.931	1.074
Piper coloradensis	S	S	1.071	1.015	0.931	1.019
Platymiscium pinnatum	T	S	1.101	1.038	0.675	0.986
Poulsenia armata	T	S	1.050	1.084	1.150	1.017
Pouteria reticulata	T	S	1.101	1.051	0.771	1.022
Prioria copaifera	T	S	1.201	1.028	0.826	0.992
Protium costaricensis	M	S	1.052	0.999	0.658	1.049
Protiium panamense	M	S	1.202	0.983	0.922	1.015
Protium tenuifolium	M	S	1.165	1.009	0.941	1.011
Psychotria horizontalis	S	S	1.163	0.964	1.035	0.997
Psychotria marginata	S	S	1.198	0.957	1.008	0.950
Pterocarpus rohrii	T	S	1.063	0.976	0.955	0.983
Quararibea asterolepis	T	S	1.166	1.042	0.953	0.985

Table 23.1. (Continued)

Species	Life Form	Light Guild	MQ	RPS	CD	SR
Randia armata	U	S	1.132	1.045	1.055	1.021
Rinoria sylvatica	S	S	1.178	1.015	0.983	0.988
Senna da	U	I	0.975	1.023	1.183	1.217
Simarouba amara	T	I	1.114	1.039	0.989	1.016
Siparuna pauciflora	U	S	0.990	1.073	0.811	1.003
Sloanea terniflora	T	S	1.283	1.027	0.792	0.901
Socratea exorrhiza	M	S	1.073	1.106	1.034	1.075
Sorocea affinis	S	S	1.070	0.971	0.938	1.042
Stylogyne standleyanum	S	S	1.043	1.006	0.958	1.069
Swartzia simplex var. grandiflora	U	S	1.212	0.967	0.783	0.902
Swartzia simplex var. ochnacea	U	S	1.245	0.947	0.686	0.953
Tabebuia rosea	T	S	0.998	1.054	2.336	0.926
Tabernaemontana arborea	M	S	1.078	1.012	0.922	1.060
Tachigali versicolor	T	S	1.141	0.991	0.853	1.009
Talesia nervosa	U	S	1.144	0.957	0.962	1.052
Talesia princeps	M	S	1.037	0.999	0.670	0.928
Tetragastris panamensis	T	S	1.205	1.031	0.841	0.986
Trichilia pallida	M	S	1.155	1.051	0.739	0.968
Trichilia tuberculata	T	S	1.117	1.009	1.005	1.029
Triplaris cumingiana	M	I	1.088	1.118	0.928	1.005
Unonopsis pittieri	M	S	1.110	1.040	0.846	0.971
Virola surinamensis	T	S	1.115	1.036	1.749	1.104
Xylosma macrocarpa	M	S	1.344	0.989	0.602	0.896
Zanthoxylum belizense	T	G	1.105	1.112	0.968	1.088
Zanthoxylum panamense	T	G	1.019	1.077	0.831	1.086
Zanthoylum procerum	M	I	0.998	0.997	1.229	1.159

Notes: For each neighborhood variable, the numbers are the proportional increase (>1) or decrease (<1) in the odds of survival of a focal plant of a given species that occurs by adding one individual in that variable, while holding the other variables constant. The odds ratios were obtained by a simplex minimization algorithm. MQ = microsite quality. RPS = relative plant size. CD = conspecific density. SP = species richness. Life forms: S = shrub, U = understory tree, M = midstory tree, T = canopy tree. Light guild: G = gap pioneer species, S = shade-tolerant species, I = intermediate species.

These results for all species combined mask significant, often large, differences among guilds in their responses to the four variables. In figure 23.8, we present some of the more salient differences. Figure 23.8a shows the effects of Microsite Quality on the probability of survival of shade-tolerant and gap species. Species in both guilds respond strongly to increasing neighborhood survival rates, but the effect is strongest in the shade-tolerant species. Survival in shade-tolerant species increased from about 10% when no 1982 neighbors survived, to about 80% when all 20 neighbors survived to 1995 (odds ratio = 1.19). Figure 23.8b shows the response of these two guilds to Relative Plant Size. When both guilds are presented to the same scale, it is obviously much more critical for gap species to be larger than their neighbors than it is for shade-tolerant species. However, when the scale of variation in survival is expanded (fig. 23.8c), we see that the shade-tolerant species do, in fact, respond to Relative Plant Size, although to a

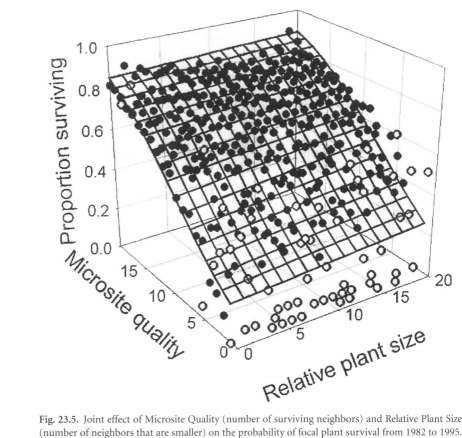

Fig. 23.5. Joint effect of Microsite Quality (number of surviving neighbors) and Relative Plant Size (number of neighbors that are smaller) on the probability of focal plant survival from 1982 to 1995. Closed circles represent points supported by >5 observations. Open circles represent points supported by ≤5 observations. Note the lack of significant interaction of the effects.

smaller extent. The effect of Conspecific Density on gap and shade-tolerant species is shown in figure 23.8d. Species in both guilds show a consistent negative effect, but gap species suffer a much stronger negative effect from conspecific effects than do shade-tolerant species.

Selected additional guild comparisons are shown in figures 23.8e–h. In figure 23.8e, we see that trees have an increasing survival when they are larger than their neighbors, in contrast with the pattern in shrubs, which show reduced survival when they are bigger than their neighbors. Conspecific Density effects are also different in trees and shrubs (fig. 23.8f). Trees consistently show negative

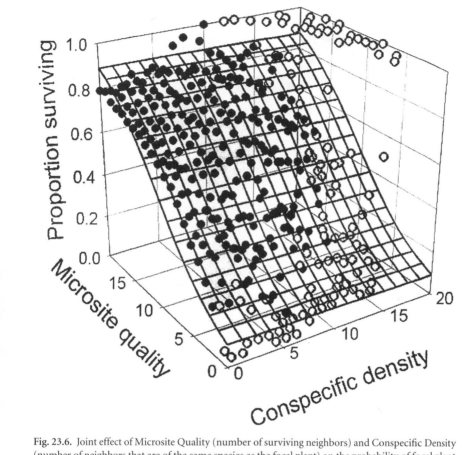

Fig. 23.6. Joint effect of Microsite Quality (number of surviving neighbors) and Conspecific Density (number of neighbors that are of the same species as the focal plant) on the probability of focal plant survival from 1982 to 1995. Closed circles represent points supported by >5 observations. Open circles represent points supported by ≤5 observations. Note the lack of significant interaction of the effects.

effects, whereas shrubs collectively show low conspecific effects. This is largely due, however, to weak conspecific density effects in the most common shrub species; other shrub species do show negative conspecific effects (see below). Negative conspecific effects are also much stronger in rare tree species than in common species (fig. 23.8g), which may be a major cause for their differences in abundance. Consider, for example, a focal plant with five conspecific neighbors. A focal plant of a species with fewer than 100 trees on the plot has a 30% chance of surviving from 1982 to 1985, whereas a focal plant of a species with more than

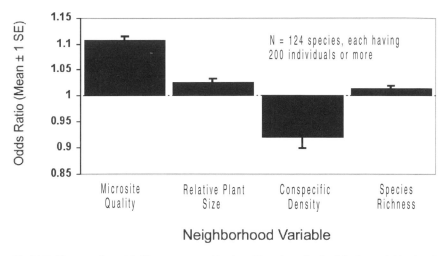

Fig. 23.7. The overall partial effect, as measured by the odds ratios, of each of the four neighborhood variables on focal plant survival in the BCI 50-ha Forest Dynamics Plot. The means are computed across 124 species each of which had a minimum population size of 200 individuals. The error bars are one standard error of the mean. All effects are significant at $p < 0.05$ and all but Species Richness are significant at $p < 0.001$.

10,000 individuals has better than a 70% chance of surviving. Finally, figure 23.8h shows that shade-tolerant species show a weak positive effect of being in more species-rich neighborhoods, and gap species show a weak negative effect.

These univariate relationships, however, do not fully control for interactions among variables, however weak they are, so it is preferable to analyze the odds ratios for each factor, which specifies the effect of adding one individual to a neighborhood factor (e.g., one more neighbor that survives), while holding all other factors constant. The average odds ratios from the subset of 124 species with more than 200 individuals separated according to growth form guilds are shown in figure 23.9; those for the light-requirement guilds, in figure 23.10. For the growth form guilds, Microsite Quality emerges once again as the most important variable (fig. 23.9a). The odds ratios for Microsite Quality range from a low of about 1.098 ± 0.010 in understory treelets to a high of 1.113 ± 0.008 in canopy trees. In the light-requirement guilds, Microsite Quality and Conspecific Density are approximately equal in importance. The odds ratio for Microsite Quality are lowest in the intermediate guild (1.044 ± 0.018) and highest in the shade-tolerant guild (1.120 ± 0.008). However, Conspecific Density is essentially of the same importance (but a negative effect) to all three guilds defined by light requirement. The odds ratio for conspecific density ranges from a low magnitude of about

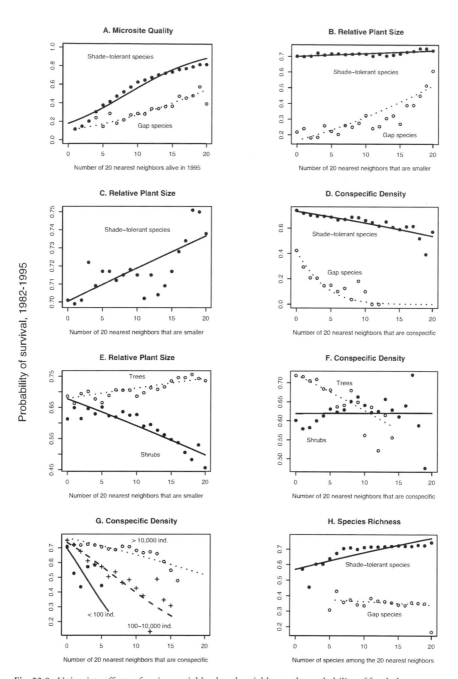

Fig. 23.8. Univariate effects of various neighborhood variables on the probability of focal plant survival in different guilds.

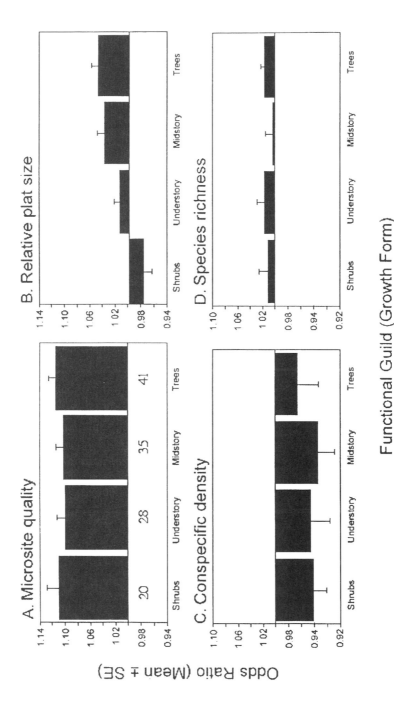

Fig. 23.9. Effect of the four neighborhood variables on the odds ratios of survival for the four growth forms. Odds ratios are means for 124 species; error bars are one standard error of the mean.

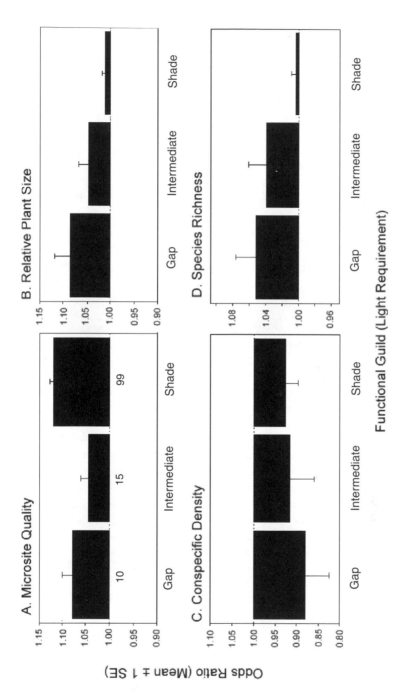

Fig. 23.10. Effect of the four neighborhood variables on the odds ratios of survival for the three light-requirement guilds. Odds ratios are means for 124 species; error bars are one standard error of the mean.

0.927 ± 0.030 in shade-tolerant trees, to a high magnitude of 0.877 ± 0.051 in gap species.

The guilds respond very differently to Relative Plant Size (figs. 23.9b, 23.10b). Shrubs show a negative effect of being larger than their neighbors (odds ratio: 0.975 ± 0.013). The odds ratios shift to progressively larger positive effects with increasing plant stature, reaching 1.047 ± 1.011 in canopy trees. The largest difference in response to Relative Plant Size, however, is seen among the light-requirement guilds (fig. 23.10b). Gap species show the largest positive effect (odds ratio: 1.083 ± 0.035), whereas shade-tolerant species have the smallest positive effect (odds ratio: 1.0113 ± 0.006).

Species Richness has a weak but generally positive effect on focal plant survival (figs. 23.9d, 23.10d). Among growth form guilds, however, these effects were significant only in the canopy tree guild (odds ratio: 1.017 ± 0.005). Among light-requirement guilds, the largest positive effect is on gap species (odds ratio: 1.052 ± 0.024), and the smallest (and a nonsignificant) effect is on shade-tolerant species (odds ratio: 1.003 ± 0.005).

Discussion

These results demonstrate the importance of the local biotic neighborhood in determining the long-term survival of individual trees and shrubs in the BCI forest. In all species and guilds the two dominating variables were Microsite Quality and Conspecific Density, which consistently had positive and negative effects, respectively, on focal plant survival; these effects were often large. For at least some guilds and species, all neighborhood variables (Microsite Quality, Relative Plant Size, Conspecific Density, and Species Richness) were significant.

One of the most intriguing results is the importance of the variable Microsite Quality, which was largely unexpected. A posteriori, it is perhaps not surprising that sites where survival is high irrespective of species are also sites of high survival of the focal plant and vice versa. But this conclusion is unsatisfactory as a mechanism because it leaves unanswered the question of the cause or causes of "good" versus "bad" sites for survival. We do not fully know what causes variation in good and bad survival sites, but we have some clues from the respective spatial distribution of these sites. The location of low-survival sites (three or fewer neighbors surviving from 1982 to 1995) are shown in figure 23.11a, and the location of high-survival sites (all 20 neighbors surviving) are shown in figure 23.11b. It is immediately apparent from comparing the distributions that the bad survival sites are far more clumped than the good survival sites, a fact that can be confirmed formally using Ripley's K statistic (fig. 23.12). Comparing these distributions with the maps of canopy gaps reveals that most (but not all) of the bad sites are in places of chronic gap disturbance, which explains their high

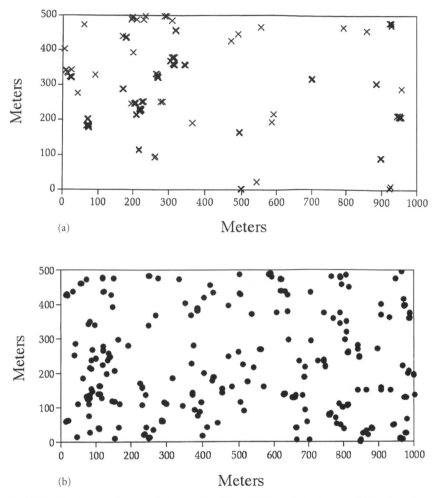

Fig. 23.11. Distribution of low-survival microsites (a) and high-survival microsites (b), in which three or fewer neighbors of focal plants survived from 1982 to 1995 in the BCI 50-ha Forest Dynamics Plot.

aggregation. Conversely, the canopy maps also reveal that most of the good survival sites are in mature high-canopy forest and consist of sapling neighborhoods of shade-tolerant species located in the understory. Growth rate analyses on the same neighborhoods confirm these interpretations (Hubbell et al. unpublished). We will visit each of these sites to measure light levels and other site variables to further characterize good and bad survival sites.

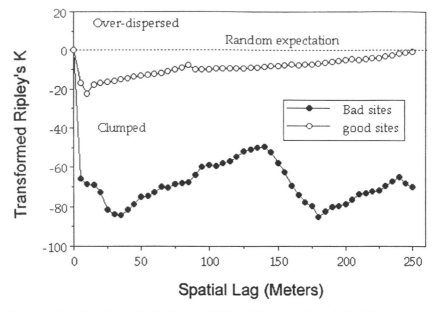

Fig. 23.12. Spatial analysis of the distribution of high- and low-survival sites in the BCI 50-ha Forest Dynamics Plot. Departures from random distribution as a function of radial distance from a given point (the spatial lag) are measured by Ripley's K parameter (transformed). High-survival (good) sites are less clumped than are low-survival (bad) sites.

Another noteworthy result was the strength and pervasiveness of negative density dependence. Previous studies of density dependence in BCI trees have taken a quadrat-based approach (e.g., Wills et al. 1997; chap. 22), but we wished to assess whether the effects of density dependence on individual plants could be detected within those plants' immediate neighborhoods. A few species exhibited positive effects of conspecific density, but most showed negative effects (table 23.1). Whether these conspecific effects arise from intraspecific competition or from density-dependent attack by host-specific enemies, or some combination of the two, is not clear at this point. However, we predict that enemies, particularly fungal pathogens, are largely responsible for these effects. We base this on several inferences and observations. First, if intraspecific competition were the primary cause, one might predict that common species, subject to higher densities of conspecifics, would also suffer the greatest effects of conspecific density, but the opposite was the case. Rare species, not common species, suffered the strongest negative conspecific effects. This finding of a correlation between the intensity of density-dependence and rarity further suggests that differential susceptibility to their predators and pathogens is one of the major causes of differing relative species abundances in the BCI forest. However, this explanation works well for trees, but less well for shrubs, which as a guild suffer much weaker conspecific density effects. Why this should be so is currently unknown.

Another surprise was the relative weakness of the effects of relative plant size on survival—relative to microsite quality and conspecific density. Relative plant size is the central variable in most current forest dynamics models. Relative plant size, i.e., how many of a focal plant's neighbors were smaller, was very important to gap species, which are by definition shade intolerant. However, the effect on survival of shade-tolerant species was more than an order of magnitude less important. However, the importance of this variable is greater when we consider relative growth rates rather than survival. We report elsewhere that there are strong effects of relative plant size on growth rates of shade-tolerant tree species. Focal plants that are larger than all of their 20 nearest neighbors exhibit relative growth rates 3.5 times higher than when they are smaller than all 20 nearest neighbors (Hubbell et al. 2001). One of the more interesting findings was the differing responses of shrubs and trees to being larger than their neighbors. Shrubs survived consistently more poorly when they were the largest plant in their neighborhood, whereas trees survived consistently better. Shrubs are understory specialists and evidently do not do well when they are the canopy layer for other plants. To our knowledge, this is the first quantitative relative fitness evidence of the understory specialization of shrubs. When shrubs are the largest plants, it usually means that they are in a light gap and survived the treefall. Shrubs do not establish in gaps (they germinate in the shade), so such gap-located shrubs are suddenly thrust into a very different physical and biotic environment. Observations on herbivory on the most common shrub in the BCI plot, *Hybanthus prunifolius* (Violaceae), show that gap plants are repeatedly defoliated by insects, but understory plants are rarely defoliated (Foster, unpublished). Thus, the poor performance of shrub species in gaps may have more to do with herbivory than with an inability of shrubs to thrive in high-light environments in the absence of pest pressure.

The fourth and final variable considered was neighborhood species richness. This variable was the weakest of all four neighborhood variables. It was significant in only two guilds—canopy trees and gap species. Nevertheless, the odds ratios showed that a canopy tree had a 1.7% increase and gap species had a 5.2% increase in mean focal plant survival over 13 years for every additional species in their respective neighborhoods. Recall that gap species suffered stronger effects of density dependence as well. Because these are odds ratios, they are the partial effects of each variable, controlling for all the other variables. Thus, we have extracted a signal of both density and frequency dependence from the neighborhood survival analyses, at least for these two guilds. It should be noted, however, that the effects of increased species richness are not uniformly positive in all species. Perusal of table 23.1 reveals many species in which the effect of species richness on focal plant survival is in fact negative. Therefore, the importance of frequency dependence and a rare-species advantage in maintaining tree diversity in tropical forests must still be regarded as very much an open question.

Conclusions and General Implications

These results amply justify our original assumption in 1980 that the fundamental mechanisms controlling tree diversity in the BCI forest, and in tropical forests in general, can only be understood in an explicitly spatial context. The results show that the long-term survival of a tree in the BCI forest is strongly affected by the trees that immediately surround it. What is particularly exciting is that the empirical results provided many effects not anticipated by previous students of forest dynamics. A tandem analysis of neighborhood effects on focal plant growth also yielded interesting and unexpected results (Hubbell et al. unpublished).

These findings on focal plant survival and growth can be directly translated into dynamic models of canopy tree replacement processes. However, the neighborhood effects reported here are on an extremely small spatial scale, appropriate for the replacement of a single canopy tree. For modeling, we also need to know how these effects on plant survival and growth decay with distance from the focal plant. A separate analysis, reported elsewhere, shows that the effects of the four neighborhood variables spatially decay with distance at different rates (Hubbell et al. unpublished). For example, the effects of Relative Plant Size decay very fast with distance, whereas the effects of Conspecific Density decay more slowly.

These analyses of focal plant survival and growth are still insufficient for full model development for the BCI forest. In addition we require information on fecundity of the tree species, seed dispersal, and the processes that affect seed germination and seedling survival and growth. However, the present results and the combined results of many other researchers on the BCI forest suggest that we are now closer than ever to the grail of understanding the mechanisms of tree species coexistence in this renowned tropical forest.

Acknowledgments

We thank Suzanne Loo de Lao, Rolando Peréz, Elizabeth Losos, Ira Rubinoff, and more than 100 field assistants for their long-term help and support of the BCI Forest Dynamics Project. We also acknowledge the support of the Smithsonian Tropical Research Institute, the National Science Foundation, The John D. and Catherine T. MacArthur Foundation, the John Simon Guggenheim Foundation, the Pew Charitable Trusts, the Andrew W. Mellon Foundation, and numerous other private organizations and individuals for their direct and indirect support of this work.

References

Botkin, D. B. 1993. *Forest Dynamics: An Ecological Model.* Oxford University Press, Oxford, U.K.

Condit, R. 1998. *Tropical Forest Census Plots: Methods and Results from Barro Colorado Island, Panama and A Comparison with Other Plots.* Springer-Verlag, Berlin, Germany.

Connell, J. H. 1971. On the role of natural enemies in preventing competitive exclusion in some marine animals and in rain forest trees. Pages 298–312 in P. J. den Boer and G. R. Gradwell, editors. *Dynamics of Populations. Proceedings of the Advanced Study Institute on Dynamics of Numbers in Populations.* Center for Agricultural Publishing and Documentation, Wageningen, The Netherlands.

Harms, K. E., S. J. Wright, O. Calderon, A. Hernandez, and E. A. Herre. 2000. Pervasive density-dependent recruitment enhances seedling diversity in a tropical forest. *Nature* 404:493–95.

Harper, J. 1977. *The Population Biology of Plants.* Academic Press, London, U.K.

Hubbell, S. P. and R. B. Foster. 1983. Diversity of canopy trees in a neotropical forest and implications for conservation. Pages 25–41 in S. L. Sutton, T. C. Whitmore, and A. C. Chadwick, editors. *Tropical Rain Forest: Ecology and Management.* Blackwell, Oxford, U.K.

———. 1986a. Commonness and rarity in a neotropical forest: implications for tropical tree conservation. Pages 205–30 in M. E. Soule, editor. *Conservation Biology: The Science of Scarcity and Diversity.* Sinauer Associates, Sunderland, MA.

———. 1986b. Biology, chance and history and the structure of tropical rain forest tree communities. Pages 314–29 in J. Diamond and T. J. Case, editors. *Community Ecology.* Harper and Row, New York.

Hubbell, S. P., R. B. Foster, S. T. O'Brien, K. E. Harms, R. Condit, B. Wechsler, S. J. Wright, and S. Loo de Lao. 1999. Light-gap disturbances, recruitment limitation, and tree diversity in a neotropical forest. *Science* 283:554–57.

Hubell, S. P., J. A. Ahumada, R. Condit and R. B. Foster. 2001. Local neighborhood effects on long-term survival of individual trees in a neotropical forest. *Ecological Research* 16(5):859–75.

Janzen, D. H. 1970. Herbivores and the number of tree species in tropical forests. *American Naturalist* 104:501–28.

May, R. M. 1973. *Stability and Complexity of Model Ecosystems.* Princeton University Press, Princeton, NJ.

McNaughton, S. J. 1985. Ecology of a grazing ecosystem: The Serengeto. *Ecological Monographs* 55:259–94.

Nelder, J. A., and R. Mead. 1965. A simplex method for function minimization. *Computer Journal* 7:308–13.

Neter, J., M. H. Kutner, C. J. Nachsteim, and W. Wasserman. 1996. *Applied Linear Regression Models.* Irwin, Chicago.

O'Brien, S. T., S. P. Hubbell, P. Spiro, R. Condit, and R. B. Foster. 1995. Diameter, height, crown, and age relationships in eight neotropical tree species. *Ecology* 76:1926–39.

Pacala, S. W., C. D. Canham, J. Saponara, J. A. Silander, Jr., R. K. Kobe, and E. Ribbens. 1996. Forest models defined by field measurements: Estimation, error analysis, and dynamics. *Ecological Monographs* 66:1–44.

Pimm, S. 1991. *Balance of nature? Ecological Issues in the Conservation of Species and Communities.* University of Chicago Press, Chicago.

Piperno, D. 1990. Fitolitos, arquelogia y cambios prehistoricos de la vegetacion e un lote de 50 hectareas de la isla de Barro Colorado. Pages 153–156 in E. G. Leigh, S. A. Rand, and D. M. Windsor, editors. *Ecología de un Bosque Tropical: Ciclos Estacionales y Cambios a Largo Plazo.* Smithsonian Tropical Research Institute, Balboa, Panama.

Shugart, H. H. 1984. *A Theory of Forest Dynamics.* Springer-Verlag, New York.

Tilman, D., J. Knops, D. Wedin, P. Reich, M. Ritchie, and E. Siemann. 1997. The influence of functional diversity and composition on ecosystem processes. *Science* 277:1300–02.

Weiner, J. 1995. Following the growth of individuals in crowded plant populations. *TREE* 10:360.

Welden, C.W., S. W. Hewett, S. P. Hubbell, and R. B. Foster. 1991. Sapling survival, growth and recruitment: Relationship to canopy height in a neotropical forest. *Ecology* 72:35–50.

Wills, C., R. Condit, R. B. Foster, and S. P. Hubbell. 1997. Strong density and diversity-related effects help to maintain tree species diversity in a neotropical forest. *Proceedings of the National Academy of Science* 94:1252–57.

PART 7: Forest Dynamics Plots

Introduction

Elizabeth C. Losos and Suzanne Loo de Lao

Thousands of permanent tree plots, large and small, are scattered across the globe. Comparing them, however, is often impractical or even misleading due to their incongruent methods. Little consistency exists among plots in the minimum cutoff size, measurement intervals, focal taxa, life forms, plot shape, and plot size. The Center for Tropical Forest Science (CTFS) network of Forest Dynamics Plots is unique in maintaining a large, pantropical dataset of comparable tree demographic data. The advantages of this standardized dataset are numerous. Most obviously, data can be compared among sites for regional- and global-scale analyses, as illustrated in part 2 of this volume. From a practical perspective, standardized data also allow statistical and analytical tools to be easily transferred from one study to another, saving researchers the effort of reinventing these tools for each site.

In part 7 we present, for the first time, standardized qualitative and quantitative descriptions of 15 Forest Dynamics Plots. The 25-ha Khao Chong Forest Dynamics Plot in peninsular Thailand is not included because data from its first census are not yet available. As background, in this introduction we first briefly describe the methodology used to establish a Forest Dynamics Plot, including modifications made at different sites.

Getting Started

Establishing a large-scale Forest Dynamics Plot is an enormous undertaking that should not be taken lightly. Before initiating a new project, scientists and managers should consider three questions: First, do the scientific questions of interest require a large plot? A Forest Dynamics Plot can provide a unique type of spatially explicit demographic, dynamic data that can shed light on otherwise insoluble questions. Examples of these have been explored throughout this volume. That is not to say, however, that these plots represent the best approach for all research avenues. If one is interested, for example, in mapping regional species distributions and beta diversity, then setting up a large number of small plots across a range of habitats would be more effective.

Second, does host-country institutional support exist for the project? Are host-country scientists and institutions able and interested in carrying out a large-scale, long-term project? Given the spatial and temporal scale of these large-plot projects, they must be firmly rooted in a host-country institution in order to be sustainable. Plots within the CTFS network are developed by a host-country institution, using mostly host-country labor, under the supervision of host-country scientists. Foreign institutions, scientists, and graduate students frequently contribute to the management, fund-raising, training, and analysis, but they do not provide the backbone of the Forest Dynamics Plot projects.

Finally, are sufficient resources available to establish a Forest Dynamics Plot? The investment of time, labor, and money is enormous, especially for the initial census. Condit (1998) estimated the people-hours and material costs (see tables 7I.1 and 7I.2). In general, the first census of a 50-ha Forest Dynamics

Table 7I.1. Estimates of Labor Needed to Complete a Census and Recensus of a 50-ha Forest Dynamics Plot

Phase of Work	Person-Months/50 ha	Person-Months/ha
Topography	30	0.6
Mapping and identification	375–500	7.5–10
Data entry (double)	28	0.6
Recensus	127	2.5
Recensus data entry	14	0.3

Notes: Data from Condit (1998). The table does not include higher-level positions such as field director or post-doctoral fellow to supervise field activities.

Table 7I.2. Costs of Equipment and Supplies Needed to Complete a Census and Recensus of a 50-ha Forest Dynamics Plot

Item	Cost/Unit ($US)	Units Needed	Total Cost ($US)
Surveying compass	1,500	1	1,500
PVC stakes for 20 × 20 grid*	1.20	1,400	1,680
PVC stakes for 5 × 5 grid*	0.33	21,000	7,000
Tags* (unit = tree)	0.05	400,000	20,000
Tag string (unit = tree)	0.007	400,000	2,800
Tag punching machine	1,375	1	1,375
Computers, printers	2,000–4,000	4	12,000
Waterproof paper (unit = sheet)	0.085	24,000	2,040
Plain paper (unit = sheet)	0.009	24,000	216
Calipers	25	10	250
Dbh tape	30	10	300
Colored flagging (unit = meter)	0.022	20,000	444
Compass	8	10	80
Binoculars	100–300	2	400
Clipboards, nails, hammers, rope, tape measures, ladder			~1,000
Shipping			~5,000
Total			$56,185.00

Notes: Asterisks indicate items for which use of homemade goods can lead to considerable savings, as long as local labor is inexpensive. Shipping costs are obviously highly variable. Note that the table does not include vehicle costs, and many plots require full-time vehicle use. Data adapted from Condit (1998).

Plot costs between \$100,000 and \$450,000. While many plots have been initiated before the entire sum was raised (or the budget increased after the project was underway), it is best to secure nearly full funding before embarking on the first enumeration.

Selecting a Location

Many factors play into the decision of where to locate a plot. Principal investigators may be especially eager to study certain features of the forest, whether it is the hyper-diversity of Yasuní, Ecuador, the high endemism of Sinharaja, Sri Lanka, or the management requirements of Mudumalai, India. These research interests can affect which sites within a country or region are most attractive. On a broader scale, CTFS also encourages the selection of sites that fit into CTFS's global criteria. CTFS aims to replicate a range of climatic, topographic and geologic geologic, and natural disturbance conditions across Asia, Latin America, and Africa (see part 2, introduction). For example, the 25-ha Khao Chong Forest Dynamics Plot currently being established in peninsular Thailand was selected as an intercontinental replication of the climatic, soil, and disturbance characteristics of the BCI Forest Dynamics Plot in Panama. The 16-ha hurricane-prone plot in Luquillo, Puerto Rico, provides a useful comparison to the 16-ha typhoon-prone plot in Palanan, Philippines.

Ease of access is another critical factor influencing site selection. Traveling over the Sierra Madre Mountains to the Palanan plot or through the dense Congo Basin to the Ituri plots in the Democratic Republic of Congo require lengthy and arduous travel. Given the large number of workers and long timeframe of each project, such difficult travel adds time and money to a project. More accessible sites—those near paved highways such as Pasoh or Lambir in Malaysia—allow workers and scientists to come and go with ease. However, a trade-off must often be made between access and site protection. Hard-to-reach sites are typically undisturbed and well protected due to their remoteness. Easily reached sites tend to be more vulnerable to human disturbance. Ultimately, the investment to set up a plot will not pay off if the forest is destroyed or badly degraded before carrying out a recensus.

The next step is to determine the size and the specific location for a Forest Dynamics Plot. In general, CTFS scientists have used the rule of thumb that approximately half of the species within a Forest Dynamics Plot should contain at least 100 individuals, thus allowing statistically rigorous analyses at the population level (see chap. 2). Of course it is impossible to know the composition and abundance of trees in a plot before it is enumerated, so these figures need to be estimated from available information. The more homogeneous the area, the smaller the sample size required. The desired degree of heterogeneity will, in turn, depend upon the questions scientists wish to address. The first two plots, in Panama and peninsular Malaysia, were specifically selected for their homogeneous habitat (see chap. 2). Since then, many researchers have sought more heterogeneous habitats, in part to be able to evaluate how habitat features influence species distributions. The Forest Dynamics Plot in Sinharaja, Sri Lanka, for example, has especially pronounced ridges, slopes, and valleys (chaps. 10 and 37). The plot in Lambir Hills, Sarawak overlays two distinct soil types (chaps. 15 and 31). Some scientists have placed plots across multiple forest types, such as in the Ituri forest in the Democratic Republic of Congo with its mosaic of monodominant and mixed forest stands (chaps. 12 and 28) or Nanjenshan Nature Reserve in Taiwan with its typhoon-prone forests on the leeward and windward side of the mountains (chap. 34). A mosaic of repeated features such as two or more valley/ridge systems or a patchwork of soil types are especially useful for spatial analyses of variation in species distribution and demography across habitats.

Field Protocol

This section briefly describes the field protocol for establishing a large Forest Dynamics Plot. For more details, see Manokaran et al. (1990) and Condit (1998). Further extensive though unpublished documentation of the protocol used at every site can also be obtained by directly contacting the principal investigators at each site (see contact information in chaps. 24 through 38).

For the CTFS network to achieve its global potential, every Forest Dynamics Plot must generate strictly comparable data. This does not require, however, identical field protocols or data entry programs for each site. Each program must adapt the methodology to the local geography, workforce skill level, social norms, and available resources. Consider, for example, the siting of corner stakes within each quadrat. Due to dense understory vegetation, some Forest Dynamics Plots require permanent, highly visible, closely spaced stakes—one every 5 m. For those in more open forests, permanent markers at the corners of the 20 × 20 m quadrats suffice. While this may affect the time and money spent on censusing and recensusing, it should have no impact on the output of data. Moreover, the type of stake used varies dramatically from site to site. The original stakes used in BCI and Pasoh were aluminum canes, painted red on the top for easy detection. More recent plots have utilized the cheaper and lighter polyurethane (PVC) tubes, painted neon pink or yellow to further increase their visibility. PVC tubes are typically 1.5 m in length (with about 0.5 m buried into the ground) and about 2 cm in diameter. In three forests frequented by elephants—herbivores with a penchant for pulling up and hurling foreign objects—PVC poles do not last long. Scientists at the Mudumalai plot in India, with one of the highest density of elephants in the world, rely on carved stones for quadrat corner markers; in Huai Kha Khaeng in Thailand, cement plugs are the preferred material; in Ituri in DR Congo, a variety of options have been tried including wooden stakes and metal poles set in cement plugs.

Field methods for a large CTFS Forest Dynamics Plot include six main phases: topographic survey and map gridding, tree enumeration, botanical identification, field checking, recensusing, and data management.

Topographic Survey and Map Gridding

Once the plot site is selected, the next step is to survey the terrain and place stakes every 20 m. A detailed topographic map of the plot generates information on elevation, slope, and aspect, all crucial elements to understanding the relationship between topography and tree distribution.

For some plots the surveying team is composed of biologists; at others, engineers. While either option can work well, it is important to have at least one biologist that is familiar with the scientific objectives of the project supervising the surveying.

Because of the large size of Forest Dynamics Plots, meter tapes and compasses cannot ensure adequate precision. Where there are slopes, corrections need to be made in order to measure the horizontal distance between markers. Professional surveying practices and tools, such as theodolite compasses, ensure that the changes in elevation are correctly incorporated into calculations of distance between two markers. Several projects have considered using global positioning systems (GPS) to locate all corner stakes, but this is still difficult in closed tropical forests, where reception is poor. Nonetheless, it is extremely valuable to have the exact reading of at least three GPS points within or nearby the plot, in order to locate the plot in regional maps or remote sensing images.

The longest axis of the plot is surveyed first—1000 m in length for a 50-ha plot—and this usually serves as the central axis running through the middle of the plot. Sets of lines separated by 20 m are then surveyed perpendicular to the central axis. For a 50-ha plot, this would entail 51 lines, measuring 250 m on either side of the central axis. Markers are planted in the forest floor every 20 m along these perpendicular lines, thus creating a grid of 20 m × 20 m quadrats. The surveying team measures relative elevation at each of these 20-m posts.

It is important that these lines, and especially the central axis, are placed with great accuracy, as even a negligible error at one end can translate into a large error at the other. Frequent side checks of the perpendicular lines are necessary. While accuracy is of great importance, it must be balanced with the need to minimize the man-made disturbance to the vegetation—the very subject under investigation. The surveying team should *not* cut swaths of vegetation along grid lines for a clean sighting of a line, as this would create unacceptable damage to the plot. When a branch or herbaceous plant blocks the line of view, the vegetation should be pulled back temporarily so as not to damage any plants. To this end, various Forest Dynamics Plot programs have independently developed means of dealing with interfering vegetation. In Ecuador, a trail 5 m outside the plot's perimeter was cut to ensure that lines were sited as straight as possible. The trail was subsequently used to steer traffic outside of the plot, with a few trails entering from the perimeter. The project leaders of the Sinharaja plot in Sri Lanka created a novel solution for this problem: Within the University of Perideniya, the Department of Botany teamed up with the Department of Engineering to survey the 25-ha Forest Dynamics Plot in Sinharaja. The plot survey was a large field exercise for engineering undergraduates, who were challenged to survey lines as accurately as possible while creating minimal damage to the forest. The student surveying teams used heavy ropes to pull vegetation away from the line of sight. Once surveyed, the ropes were released and the vegetation sprang back into place. When immovable obstacles, such as trees, were encountered in the line of view, the surveyors sited a line one or several meters to the side and then, after passing the obstacle, sited back to the original trajectory.

The placement of quadrat markers and the measurement of relative elevation at each 20-m marker provides the data necessary to construct a topographic map of the plot, which can be constructed using one of several commercial software programs.

Once the 20 m × 20 m grid has been established, these quadrats are further subdivided into 5 × 5 m subquadrats. The markers at the 5-m subquadrat corners are typically smaller and sometimes less permanent than the quadrat corner markers. The surveying team can carry out this exercise, or the tree mapping teams can take on this activity just prior to mapping (see the discussion on censusing trees that follows).

Censusing Trees

Tree mapping, also known as plot enumeration, is the core census activity. The aim of plot enumeration is to locate every free-standing woody tree and large shrub whose diameter is greater than 1 cm at the height of 1.3 m (diameter breast height or dbh). Each tree is measured to the nearest millimeter at its girth, mapped within a 5 × 5 m subquadrat, assessed for noteworthy characteristics, and given a numbered tag. The data produced from this stage represent the heart of the plot datasets. Because the initial plot enumeration of a 50-ha plot can include hundreds of thousands of individuals and last for 2 or more years, maintaining a high level of accuracy, consistency, and enthusiasm during plot enumeration is a constant challenge.

The mapping teams are typically comprised of recent university graduates in biology or forestry, forestry technicians, or locals from nearby communities. Often members of the topographic survey team also join in this mapping phase. Most Forest Dynamics Plots have multiple mapping teams working concurrently. Each team has its own 20-m column, which they advance through until it is completed (20 × 500 m for 25- or 50-ha plots). Then the team starts on another column. Within each column, mapping is done at the scale of the 20 × 20 m quadrat, advancing systematically through each of the sixteen 5 × 5 m subquadrats.

A mapping team has roughly six repetitive responsibilities: Recording data, attaching tags, reading off tag numbers, measuring dbh, recording tree locations on a map, and painting a line at the point of measurement (POM). Forest Dynamics Plot programs have different philosophies on the most efficient and effective way of carrying out these chores. In the two African sites—the Ituri and Korup Forest Dynamics Plots—the teams are made up of as many as six people working together, each with a separate task. More typically, as in La Planada Forest Dynamics Plot in Colombia, a pair of workers split all the tasks.

At most sites, the mapping team sweeps through the entire plot, mapping and measuring all trees except those with extremely large girths or tall buttresses that require that the trees' diameter be measured well above 1.3 m. If the diameter cannot be accurately measured using standard techniques or the point of measurement cannot be easily reached from the ground, the mapper will record the individual as a 'big tree' and pass it by in the first sweep. After the entire plot has been enumerated, one mapping team will return with tree climbers, ladders, and specialized tools such as oversized calipers to measure the diameter and height of all of the big trees.

Differences do occur in the details of the enumerations due to variation in forests and institutional settings. Many of these differences do not affect the final data. For example, at most plots, trees are drawn by hand on a map, which is later digitized to obtain the x, y coordinates of all trees. In a few sites, such as Sinharaja and Ituri, the x, y coordinates of every tree are measured directly in the field within each subquadrat. Similarly, at some plots, every single tree is painted at its point of measurement (POM), while at other plots, mappers paint only those trees whose POMs deviate from the standard 1.3 m location.

Some activities, however, do alter the resulting data and hamper comparisons. Palms, for example, are not included in all of the Forest Dynamics Plots. In Sarawak, Malaysia, there are so few free-standing palms that this family has not been included in the Lambir Forest Dynamics Plot (though it is being added in the most recent census). Palms cannot be—and are not—ignored in neotropical sites, due to their high abundance. In Yasuní, palms represent almost 8% of the basal area for trees over 1 cm dbh. Another example includes species that are clonal or multiple stemmed. At Pasoh, only one stem of multiple-stemmed individuals is measured. At Huai Kha Khaeng, every stem larger than 1 cm dbh is measured and tagged individually. One of the aims of CTFS is to eliminate or minimize such differences through standardized protocols.

Scientists at each site also collect additional information that is of particular relevance to that site—such as hurricane damage in Luquillo and elephant damage in Mudumalai—or is of particular scientific interest—such as canopy height at BCI and Pasoh, seedfall in Yasuní, Pasoh, and BCI, ground vegetation cover in Ituri, and crown form and diameter in Lambir. The addition of lianas to the census is a significant modification made in Ituri, Korup, and Sinharaja. The former two sites are using an identical method that maps all climbers whose diameter is greater than 1 cm; at Sinharaja, all lianas with a diameter greater than 2 cm are mapped. Currently, a standardized study of soil nutrients is being carried out at 10 CTFS sites. Recording data also differs among sites. For most plots, mapping teams document all enumeration information on paper data sheets in the field. For example, at BCI, mappers record the subquadrat location, tag number, measurement in millimeters, and special codes on one sheet; on another page, they note trees with multiple stems greater than 1 cm dbh; on a third page, mappers list problems; and on the last page, they draw a map of tree locations. At all sites, data are later entered into computer files, either by the tree mapping teams on their days off or during the evenings, or by data entry specialists. In the Bukit Timah Forest Dynamics Plot, one of the censuses was conducted using an early model of hand-held computers instead of paper data sheets and maps. The advantage of computerization is obvious: Data entry is avoided, thus saving time and money and eliminating potential errors that arise from data transcription. However, several computer glitches and faulty data entry during the Bukit Timah recensus resulted in serious errors in the location of many trees. Because the errors were invisible to the mappers, many were not detected until the subsequent census several years later. At that point, they could not be easily resolved. Consequently, future censuses at Bukit Timah returned to paper. More recently, better hand-held computers with more advanced computer programs, and further experience with these tools have improved the prospects of field computers for Forest Dynamics Plot enumeration. They will likely be a viable option in the near future. Nan Chen, working at a CTFS-associated plot in Khao Yai National Park, Thailand, has developed a computer program for rapid field data entry with easy data verification.

Botanical Identification

Identifying every single tree to the species level is typically the most challenging and most time consuming exercise when setting up a Forest Dynamics Plot, particularly in the poorly described, hyper-diverse forests such as those in Yasuní National Park or Lambir Hills National Park. The botanical identification of the 25-ha Yasuní Forest Dynamics Plot, for example, took a team of botanists more than 5 years to complete. The botanical team typically consists of a senior botanist—an expert taxonomist who oversees the entire identification process and trains other botanists—and one or several less experienced but enthusiastic junior botanists. At many sites, an assistant who can climb trees also accompanies the team.

Training the botanical team typically takes place in the first few hectares of the plot. In Yasuní, for example, the first 2 ha were used to train the botanical team. During this stage, the team works closely together collecting at least a leaf specimen from every tagged individual (except for small individuals with few leaves) and taking notes on the specific characteristics of each plant. During this training stage, botanists repeatedly return to individual trees in order to familiarize themselves with traits and variation within individual species. Specimens are brought back to an herbarium and sorted into morphospecies. These copious collections provide botanists with a good opportunity to assess the range of variation within the more common species. The senior botanist identifies or verifies the scientific name for as many of the morphospecies as possible. The rest are sent out to experts elsewhere in the country or abroad or, less frequently, outside experts are brought to the plot for field identifications.

Many of the specimens collected during this initial stage are used to make a reference collection. Almost every Forest Dynamics Plot has a nearby herbarium where sterile and fertile collections are dried, stored, and organized for easy reference. Many plots have one set in the field and a better set (mostly fertile) at the home institution or national herbarium. Because of the vast numbers of collections made, most sterile specimens are not retained at the herbarium. Fertile collections are typically shared with national herbaria and, when possible, with international ones.

At the completion of the training stage, all the individuals within the first 2 ha or so are sorted into morphospecies and, to the extent possible, given at least tentative species names. For most plots, after this point, the botanical team no longer collects a specimen for every individual tree if they can confidently and consistently recognize the species in the field. As the botanical team becomes more and more familiar with the flora, the intensity of collecting continues to diminish. However it is critical that the botanical teams remain conservative as to when they stop collecting specimens for a species, as overconfidence can lead to errors that are very costly to correct later.

Ideally the botanical team and the tree mapping teams work concurrently, with the botanists visiting a quadrat soon after its trees have been tagged and measured. The greater the time lapse between the visit of the two teams, the higher the probability of trees dying during the intervening period (thus increasing the difficulty of tree identification). This problem is largely avoided in sites such as Pasoh, where the tree mapping teams collect a sterile specimen of every individual tree except the most common ones. Here, the majority of the identification is done by expert taxonomists in the herbarium, except when individual trees need to be field checked for verification.

Over time, many members of the tree mapping teams also become familiar with the common species. In sites such as Sinharaja, the botanical survey is used as a training exercise for students on the mapping teams, who attempt to identify every tree in the field as they collect enumeration data. These mapping teams bring in a specimen of every individual, which are rechecked in the herbarium by the senior botanist. Thus, the mappers are given continual and rapid feedback on their species identifications. At many sites, some mappers become so skilled that they are transferred to botanical teams.

Data Management

A well-designed database can be very helpful for managing the data generated during plot censuses. The three types of data generated by a Forest Dynamics Plot census—elevational data from the topographic mapping, demographic and spatial information from the enumeration, and species identification—must be converted into a functional electronic format. Though each site adopts its own data management system, they all generally follow a three-step process: First, transfer handwritten data from the field to an electronic format; second, correct errors in the database; and, third, reorganize the database so it is more manageable for queries and analysis. During the first stage, information from each of the field datasheets is entered by hand or, for the maps, digitized into an electronic format. At most sites, data are entered twice by two different individuals, and the datasets are then compared with each other for data-entry errors.

The second step entails checking for errors and correcting the original database. Data entry programs have data-screening algorithms that detect inappropriate entries from data collection, data recording, or data transcription. Examples of these errors are described more fully later in this chapter.

The third step of the process is the storage of the information in a more efficient and manageable form. For the BCI Forest Dynamics Plot, the data are reorganized into four types of relational files:

1. *Permanent File* This file includes permanent data—that is, data that do not change throughout the censuses—such as tag number, species code, and location of an individual, i.e., its x, y coordinates within the plot.

2. *Census Files* Each census has a separate file that contains, for each tree, a tag number, girth measurement, a status code indicating whether the individual is alive, dead, or now below the 1 cm dbh cutoff (for example, if the main stem broke since the previous census), a code indicating whether the POM has changed, the number of multiple stems greater than 1 cm dbh, the date of census for an individual tree, and other general codes, such as whether the tree is leaning, has a broken stem, or reason for death.
3. *Multiple Stem Files* Each census has a separate file that contains the tag and dbh measurements of each multiple-stemmed individual.
4. *Species Identification File* This file includes a list of species codes and their complete names (species, genus, and family), plus any other information available on the species, such as growth form, size when species produces flowers and fruits, mode of fruit dispersal, and whether the species is pioneer, intermediate, or shade tolerant.

The files are linked to each other by the individual tree tag number, and to the Species Identification File by the species code. In addition to the above categories, the data are also saved as ascii/text files in the same format but separately for each species.

Over the last few years, the data, saved as text files, have been formatted to be used specifically by the statistical computing and graphics programming language R (a close relative of the commercial program S-Plus), though the data can be used by virtually any database software. The Center for Tropical Forest Science encourages and works with all Forest Dynamics Plot programs to organize their data in this standardized format. The advantage of having consistent data files is that they can be easily compared among sites and software developed for one site can be used with others. This does not mean that the data entry programs have to be identical. Individual Forest Dynamics Plot programs may continue to use their own program for data entry and data management that suits their own needs. What is critical is that the dataset can be reformatted into the standardized CTFS format for comparison with other plot data.

Source of Data Errors

A concern of any permanent plot study is the accuracy of its data. Problems arise from inaccurate data collection, recording, and transcription. The data collected from large Forest Dynamics Plots are different in many ways from the data collected by other types of research. For example, studies in biomedical research usually generate databases with few subjects, but many variables. Problems that frequently arise in this type of study are low sample size and high dropout rates.

Forest Dynamics Plots, with their large number of trees, are vulnerable to different types of errors:

Field Errors

On one hand, Forest Dynamics Plots have the advantage of reducing error because, due to the length of the studies, workers become increasingly experienced and consistent with their measurement and identification abilities. On the other hand, because the initial census takes a long time to complete, worker fatigue can set in and data quality decline. Regular assessments of data precision are critical. Different checking methods have been developed at different sites to assess the accuracy of the enumeration. For many Forest Dynamics Plots, field team leaders regularly revisit a pre-defined subset of the quadrats and remeasure a portion of trees in each. At the BCI Forest Dynamics Plot, the mapping teams are periodically assigned to completely remeasure a certain number of 5×5 m quadrats that were mapped initially by a different mapping team. The data are compared and error estimates calculated. Error checking of species identification is also crucial. Senior botanists check the accuracy of field identifications by revisiting randomly selected quadrats or a random sample of individuals. During the first census of the Pasoh Forest Dynamics Plot, the senior botanist reexamined all trees ≥ 10 cm dbh in some hectares and made a complete check of particular taxa.

Data Entry

One of the main duties of the database manager is to make sure the data are entered into the computer as accurately as possible during the censuses. One key means of ensuring accuracy is to carry out double entry by two different persons. Double entry has been found more effective and efficient than manual checking of single data entry. Database managers at many sites have developed data entry programs that screen for as many errors as possible. A program developed for the BCI Forest Dynamics Plot, for example, checks that records are unique for every tree in all the appropriate files; verifies that dbh measurements fall within a feasible size range, preferably according to the species if this information is available; locates duplicate records and tag numbers; checks date records against the time interval when the quadrat was censused; checks for legal species codes; checks that all multiple-stemmed plants have their measurements entered from the multiple-stem forms; identifies missing records; and verifies that all plants have been mapped. A common error that occurs in the field is writing the wrong tag number in one of the forms. Fortunately, many of these errors are easily fixed in the office. Others may require a field visit, either immediately or at the time of the subsequent recensus. For example, an unnaturally large increase or decrease in a dbh measurement may be due to an error of measurement in either the current census or the previous census. The suspect tree would have to be checked in the field. A plant found dead in a previous census but 'alive' in the current one would also need to be field checked.

Recensus

Dynamic data on tree growth, mortality, and recruitment are the most valuable information produced by Forest Dynamics Plots. Such data are generated during the plot recensuses by remeasuring tagged trees from the previous census, assessing mortality, and adding new trees that reach the 1 cm dbh cutoff. Because scientists are often anxious to obtain dynamic data from the plots as quickly as possible, the first recensus may occur in as little as 3 years from the beginning of the first census. Thereafter, most plots are recensused at 5-year intervals. One exception is the Mudumalai Forest Dynamics Plot, which is recensused every 4 years instead of 5; species mortality and recruitment are recorded annually at Mudumalai. This accelerated schedule is possible, in part, because tree density is comparatively low in this dry-forest plot.

If well planned, the recensus of a Forest Dynamics Plot takes substantially less time and resources than the original census. The savings are due, in part, to the fact that the project leaders and workers—many of whom may have worked in the previous census—are more familiar with the flora and terrain. Many of the logistical problems have already been solved. It is also due, of course, to the fact that much less work is involved.

For the recensus of the BCI Forest Dynamics Plot, the mapping team takes five datasheets with them in the field: a map for each 20 × 20 m quadrat, which has the location and tag number for every tree included in the previous census; a sheet for recording dbh measurements of tagged trees, with pre-printed information on tag numbers, species codes, and previous dbh measurements from the previous census; and three empty datasheets for recording new recruits, multiple-stemmed trees, and problems. At BCI, the mapping team also takes a list of the species codes and names with them, since many of them can identify the most common species. The botanist later rechecks the identification of every recruit.

The mapping team remeasures every tree. The painted lines at the POM from the previous census help ensure that the measurement is made at exactly the same point as in the last census, which is critical for growth calculations. If a tree has a string, wire, or nail, but its tag has disappeared, then the mapper replaces the tag with the original number, as confirmed by the tree's mapping coordinates. If a tree shows no sign of a previous tag, but the mapper suspects that the tree was included in the prior census, then the problem is recorded on the problem page. In these cases, the original number is typically used only if the tree is ≥1 cm.

Determining whether a tree is alive or dead can be tricky. The mapping team carefully searches for sprouts and living leaves on trees that appear to have died since the last census. In addition, during every census, tagged trees disappear. This may be because they have died and decomposed; it may be that the tag has fallen off the tree and the individual is not readily re-identified (especially problematic if the tree occurs in the middle of a light gap); or the tag number might be a mistake from the previous census. In that case it is important that dead trees are still included on maps in the next census, so that trees do not miraculously come back to life. Many special cases arise during the recensus, especially related to resprouts, multiple-stemmed trees, and buttresses. Condit (1998) describes in detail how these issues are handled for the BCI Forest Dynamics Plot.

During a recensus, the botanical team follows the mapping teams again. The botanists can work much faster in a recensus than in the original census because species identifications have largely been worked out and because the number of trees in need of identification is much smaller. Species identification is limited to the new recruits or individuals from the last census with questionable identification.

The chapters that follow describe in more detail the status of a recensus for individual plots.

Plot Census Data

For each of the 15 Forest Dynamics Plots discussed in chapters 24–38, we present basic tree demographic data in seven tables. The tables allow comparison among the plots based on standardized quantitative metrics. In the descriptions that follow below, we explain the general protocol used to calculate each table's data. Exceptions to the general protocols are noted in the individual chapters.

Table 1: Climate Data

These numbers are based on data from the closest reliable, long-term weather station available. Average Daily Temperature Maximum (ADTMx) represents one month averages of the daily maximum temperatures. Average Daily Temperature Minimum (ADTMn) represents one month averages of the daily minimum temperatures. Daily solar radiation, measured in Watts per square meter (Q), is included for those sites where these data were collected near the plots.

Table 2: Plot Census History

Number of Trees includes all woody, free-standing trees or large shrubs greater than 1 cm diameter at breast height (dbh) or greater than 10 cm dbh, including unidentified individuals, in a Forest Dynamics Plot. *Number of Species* represents the number of morphospecies in the plot among trees and large shrubs greater than 1 and 10 cm dbh excluding unidentified individuals. Unidentified individuals are those that were identified only to genus or family, if that, and cannot be distinguished as a separate species by a taxonomist.

The census dates cover the period when the mapping teams initiated their activities until the completion of the full census of the plot. The census dates do not include the period of checking and follow-up visits to the plot.

Table 3: Summary Tally

Density and diversity parameters were calculated for the plot as a whole, and for average hectares. Hectare means for these parameters are averages of parameters calculated for nonoverlapping single hectares. In general, for these tables, basal area and number of trees are based on all individuals, identified or not. Basal area (BA) was calculated using the diameter at breast height for all individuals (identified and unidentified) and all stems greater than 1 cm dbh of multiple-stemmed individuals, except where noted in the text. N is the number of trees and large shrubs, including identified and unidentified individuals. Numbers of species (S), genera (G), and families (F) include only trees identified to species or morphospecies. The Shannon–Wiener or Diversity Index (H') was calculated using individuals identified to species or morphospecies as:

$$H' = -\sum_{i=1}^{S} (p_i)(\log p_i)$$

where S = Number of species and p_i = the proportion of identified trees belonging to species i. Note that $\log_{10}$ was used rather than $\log_e$, as was used in chapter 10.

Fisher's α was calculated as:

$$S = \alpha \ln\left(1 + \frac{N^*}{\alpha}\right),$$

where S = the number of species and N^* = the number of individuals identified to species or morphospecies (the box in chapter 7 describes how to calculate Fisher's α). Fisher's α is infinite when the number of individuals equals the number of species, or when there are no individuals in that category. Consequently, α could not be calculated for some of the single hectares containing individuals of the largest size class (>60 cm dbh). For plots in which Fisher's α could not be calculated for every single hectare, a table footnote indicates how many hectares were used in the calculation.

In calculating diversity indices as hectare means, both α and H' were first calculated for each hectare and then averaged.

Table 4: Rankings by Family

The top 10 families for basal area (BA), number of trees and large shrubs (Trees), and identified species (Species) are presented for each Forest Dynamics Plot. In cases where there is a tie for the tenth place, all families are included. For each of the 10 families with highest basal area, the family's total basal area, the percentage of the plot's basal area belonging to that family, and the percentage of the plot's trees belonging to that family are tabulated. For each of the plot's 10 most abundant families, the family's total number of individuals, and the percentage of the plot's trees belonging to that family are tabulated. For each of the plot's 10 most diverse families, the family's total number of species and the percentage of the plot's species belonging to that family are tabulated. Family nomenclature follows Mabberley (1997). See the notes in the introduction to part 2 for more information on nomenclature.

Table 5: Rankings by Genus

The top 10 genera for total basal area (BA), number of trees and large shrubs (Trees) and number of identified species (Species) are tabulated for each Forest Dynamics Plot. In the case of ties for the tenth place, all genera are included. For each of the 10 genera with highest basal area, their total basal area, the percentage of the plot's total basal area belonging to that genus, and the percentage of the plot's total number of individuals belonging to that genus are tabulated. For each of the 10 most abundant genera, the total number of individuals in the genus and the percentage of the plot's trees belonging to that genus are tabulated. The number of species is recorded for each of the plot's 10 most diverse genera. Family and genus nomenclature follow Mabberley (1997). See the notes in the introduction to part 2 for more information on nomenclature.

Table 6: Rankings by Species

The 10 species with the highest total basal area (BA) and the 10 species with the highest number of individuals (Trees) are tabulated for each Forest Dynamics Plot. For the 10 species with the highest basal area, their total basal area and the proportion of the plot's total basal area belonging to that species are tabulated. For the 10 most abundant species, the total number of individuals in the species and the percentage of the plot's total number of individuals belonging to that species are tabulated. Family nomenclature follows Mabberley (1997). See the notes in the introduction to part 2 for more information on nomenclature.

Table 7: Tree Demographic Dynamics

The average growth rate between a given pair of censuses was calculated as follows. For each tree, the diameter increment between censuses was divided by the interval (in years) between the dates it was measured. For multiple-stemmed trees, this quotient was calculated for the largest surviving stem. The average growth rate for a given size class is the average of the growth rates of every measured tree whose dbh was in the appropriate size range at the first census. Trees whose annual growth rate was greater than 7.5 cm/year or whose relative annual growth rate was less than −5% per year were not included in the growth calculations, under the assumption that these records represent measurement errors.

Mortality rate of trees within a given diameter class between a given pair of censuses is $(1/T)$ times the natural logarithm of the proportion of trees in this size class at the first census that survived to the second. Here, T is the average census interval in years between diameter measurements for those trees in this size class that survived to the second census.

Recruitment rate of trees into a given size class between a given pair of censuses is $(1/T)\ln[(S+R)/S]$ where S is the number of trees in this size class at the first census that survived to the second, and R is the number of new trees appearing in this size class between censuses.

Basal area loss within a given size class includes basal area lost from mortality and from stem breakages to below 1 cm dbh. The rate of basal area loss within a size class between a given pair of censuses is $(1/T)$ times the total basal area of trees in this size class at the first census that died, or declined to <1 cm dbh, by the second. All stems of multiple-stemmed individuals were included, unless noted in the text.

The basal area gain within a size class between a given pair of censuses is $(1/T)$ times the total gain in basal area of trees in that size class at the first census. All stems of multiple-stemmed trees are included in this calculation of basal area gain plus new recruits into that size class, except where noted in the text. Trees with recorded diameter increment exceeding 7.5 cm/yr, or whose diameter shrank by more than 5% a year, were excluded from this calculation under the assumption that these records represent measurement errors.

Plot Figures

Graphical representation of the Forest Dynamics Plots' geographical setting and topography are also presented in a standardized format for each of the 15 sites.

Figure 1: Plot Location

The geographic map shows the approximate location of each Forest Dynamics Plot within its country or province. The country maps have only cities and rivers designated. An inset box depicts the protected area in which the Forest Dynamics Plot is located, with a small black square or rectangle representing the location of the plot. Note that the plot is not usually depicted to scale.

Figure 2: Topographic Map

The second figure illustrates the topography of each Forest Dynamics Plot. Unless otherwise indicated, all plots have 5-m contour intervals (black lines). Plots with relatively homogenous topography also have 2.5-m contour intervals (dashed lines).

Figure 3: Perspective Map

The final figure for each plot is a perspective map, also illustrating the sites's contours. To help visualize topographic variation within the plots, the scale of the z-axis has been doubled for the perspective maps.

Notes

As in part 2, in an effort to provide taxonomic consistency across Forest Dynamics Plots, plant genus and family names for each site have been standardized using the dictionary of vascular plants by D. J. Mabberley (1997) (see part 2, notes).

The principal investigators of the original census of each Forest Dynamics Plot are indicated with asterisks in the author's list on the title page of each chapter.

References

Condit, R. 1998. *Tropical Forest Census Plots.* Springer-Verlag, Berlin; R. G. Landes, George-town, TX.

Mabberley, D. J. 1997. *The Plant-Book: A Portable Dictionary of the Vascular Plants.* Cambridge University Press, Cambridge, U.K.

Manokaran, N., J. V. LaFrankie, K. M. Kochummen, E. S. Quah, J. E. Klahn, P. S. Ashton, and S. P. Hubbell. 1990. *Methodology for the Fifty Hectare Research Plot at Pasoh Forest Reserve.* Research Pamphlet No. 104. Forest Research Institute of Malaysia, Kepong, Malaysia.

24

Barro Colorado Island Forest Dynamics Plot, Panama

Egbert G. Leigh, Jr., Suzanne Loo de Lao, Richard Condit,
Stephen P. Hubbell, Robin B. Foster, and Rolando Pérez

Site Location, Administration, and Scientific Infrastructure

Barro Colorado Island (BCI) is a 1500-ha island in the Panama Canal (fig. 24.1).
The island and surrounding mainland peninsulas comprise the Barro Colorado
Nature Monument (BCNM), a fully protected national biological reserve since
1923. Both the BCNM and the Barro Colorado Island Forest Dynamics Plot are
managed by the Smithsonian Tropical Research Institute. A permanent scientific
station located on the island has well-equipped laboratories, computer facilities,
housing accommodations, dining facilities, a conference room, and a canopy
tower.

A roughly 6-ha Forest Dynamics Plot has been established in wetter but less
diverse forest within the limits of Fort Sherman, near the Caribbean end of the
Panama Canal, and a 4-ha plot has been established in dry forest at Cocoli, near
the Canal's Pacific terminus. In these two plots all individuals $\geq$1cm dbh of
free-standing woody tree species have been marked, measured, and identified.
In addition, trees $\geq$10 cm dbh have been censused on a number of plots to
document the impact of soil type and the effect of the rainfall gradient from the
drier Pacific to the wetter Caribbean shore. These plots include twenty-five 1-ha
plots scattered throughout the Panama Canal area, two 0.25-ha plots about 50 km
west of Cocoli, and seven 1-ha plots in diverse, wet forest between 25 and 75 km
east of Ft. Sherman (Pyke et al. 2001).

Climate

From 1972 to 1989, the annual rainfall recorded in BCI's upper laboratory clearing
averaged 2551 mm/year, with a maximum of 4469 mm in 1981 and a minimum
of 1815 mm in 1976 (Windsor 1990; table 24.1). The dry season is severe, lasting
from sometime in December into April or May. In the open, the average diurnal
temperature maximum (ADTMx) is 31.1°C, the average diurnal temperature
minimum (ADTMn) is 23.2°C, and the average (1984–89) solar radiation (Q) is
181 W/m^2 (Windsor 1990).

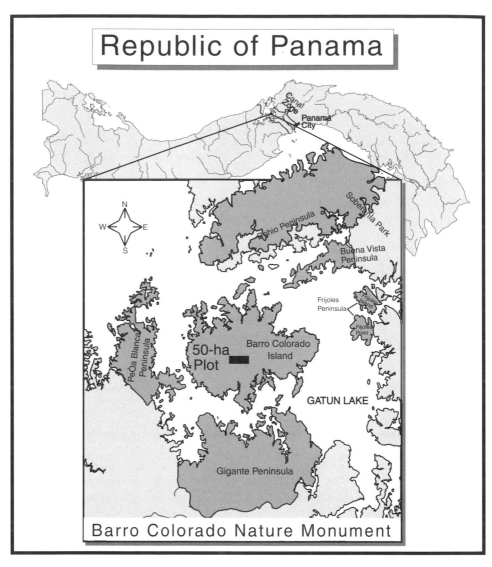

Fig. 24.1. Location of the 50-ha Barro Colorado Island Forest Dynamics Plot.

Topography and Soil

The 50-ha Forest Dynamics Plot is located on Barro Colorado Island, mostly on a level plateau at 140 m above sea level (ranging from 120 to 160 m above sea level). The fringes of the plot, however, include gentle slopes falling away from the plateau, generally 7°–20° in inclination (figs. 24.2 and 24.3.)

Table 24.1. BCI Climate Data

	Jan	Feb	Mar	Apr	May	Jun	Jul	Aug	Sep	Oct	Nov	Dec	Total/ Averages
Rain (mm)	71	37	23	106	245	275	237	322	309	364	360	202	2551
ADTMx (°C)	31.0	31.3	31.9	32.3	31.8	30.9	30.7	30.7	30.9	30.7	30.5	30.7	31.1
ADTMn (°C)	22.8	22.9	23.1	23.5	23.7	23.5	23.4	23.2	23.1	22.8	23	22.9	23.2
Q (W/m²)	209	218	242	220	185	154	157	158	167	137	153	174	181.2

Note: Climate data were measured from 1972–1989 at BCI's upper laboratory clearing, 1.3 km northeast of the 50-ha plot's NE corner.

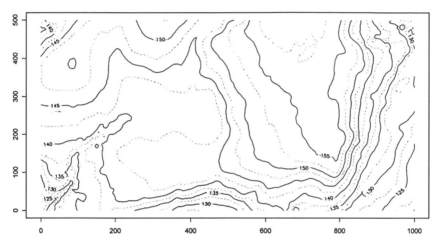

Fig. 24.2. Topographic map of the 50-ha Barro Colorado Island Forest Dynamics Plot with 5-m contour intervals (solid line) and intermediate 2.5 m intervals (dotted lines).

The parent rock beneath the plateau is an andesitic cap. This cap accumulates water and produces springs along the slopes around the plot, where the cap reaches the soil surface. Thus, particularly toward the end of the dry season, soils of slopes contain more moisture available to plants than those of the plateau (Becker et al. 1988). The soils under most of the plot are well-weathered kaolinitic Oxisols with low cation exchange capacity, although the plot also contains a 2-ha seasonal swamp. The 40-ha Conrad catchment, which largely overlaps the 50-ha plot, exports an average of about 110 kg of suspended solids and 110 kg of solutes per hectare per year, representing an average erosion rate of 0.01 mm/year (R. Stallard personal communication). The 50-ha plot supports a vegetation characteristic of fertile soil (Foster and Brokaw 1982).

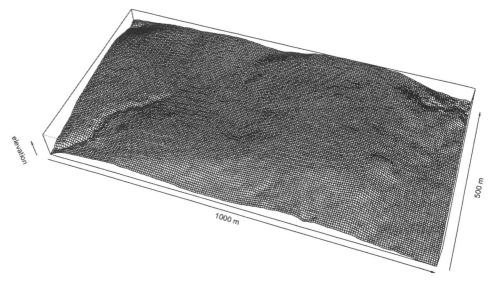

Fig. 24.3. Perspective map of the 50-ha Forest Dynamics Plot.

Forest Type and Characteristics

Barro Colorado Island is covered by semideciduous lowland moist forest, approximately half of which is mature. The canopy is generally 20–40 m tall, with some emergent trees reaching 50 m (Bohlman personal communication). The forest is estimated to hold 281 ± 20 Mg/ha of aboveground biomass, lianas included. A third of this biomass is stored in trees larger than 70 cm dbh (Chave et al. 2003). Leaf-area index is approximately 7 (Leigh 1999). Over one quarter of the canopy tree species are deciduous for part of the dry season. However, normally only 10% of the canopy is deciduous at the peak of leaf loss in March (Condit et al. 2000), although this figure varies according to the severity of the dry season. The understory is evergreen (Croat 1978). From August 1969 through July 1971, the total dry weight of fine litter falling on and near the plot (as measured in over one hundred $1/12$-m^2 tubs) was 1151 g/m^2/year, of which 611 g was leaves; from 1988 through 1991, the total fall of fine litter on Poacher's Peninsula, 1.4 km south of the southeast corner of the plot, as measured in sixty 0.25-m^2 litter traps, averaged 1181 g/m^2/year, of which 743 g was leaves.

Tree species composition is relatively homogeneous across the plot. The most common canopy trees are *Quararibea asterolepis* (Bombacaceae) and *Trichilia tuberculata* (Meliaceae), which together account for only one eighth of the plot's

Table 24.2. BCI Plot Census History

Census	Dates	Number of Trees (≥ 1 cm dbh)	Number of Species (≥ 1 cm dbh)	Number of Trees (≥ 10 cm dbh)	Number of Species (≥ 10 cm dbh)
First	March 1981–July 1983	235,341	305	20,881	238
Second	January 1985–October 1985	242,088	306	20,719	237
Third	February 1990–February 1991	244,059	303	21,233	229
Fourth	January 1995–February 1996	229,049	301	21,455	227
Fifth	January 2000–October 2000	213,802	300	21,205	226

Note: Five censuses have been completed, the next census is expected to begin in January 2005.

Table 24.3. BCI Summary Tally

Size Class (cm dbh)	Average per Hectare							50-ha Plot				
	BA	N	S	G	F	H'	α	S	G	F	H'	α
≥ 1	32.1	4581	169	120	48	1.62	34.6	301	180	58	1.71	34.2
≥ 10	27.8	429	91	74	36	1.66	35.6	227	149	52	1.86	35.4
≥ 30	19.7	82	35	33	23	1.38	23.9	142	101	43	1.72	28.5
≥ 60	10.8	17	11	11	8	0.95	17.8*	67	56	28	1.54	17.2

*Based on 46 hectares.

Notes: BA represents basal area in m^2, N is the number of individual trees, S is number of species, G is number of genera, F is number of families, H' is Shannon–Wiener diversity index using $\log_{10}$, and α is Fisher's α. Basal area includes all multiple stems for each individual. All individuals were identified. Data are from the fourth census.

basal area. *Hybanthus prunifolius* (Violaceae), an understory shrub, and saplings of the understory tree *Faramea occidentalis* (Rubiaceae), together make up nearly one third of the trees on the plot between 2 and 5 m tall. Species diversity tends to be lower on flat areas than on slopes, which are wetter at the end of the dry season (Becker et al. 1988). Some species, such as *Calophyllum longifolium* (Guttiferae), *Virola surinamensis* (Myristicaceae), *Guatteria dumetorum* (Annonaceae), *Drypetes standleyi* (Euphorbiaceae), and *Piper* spp. (Piperaceae) are more common on or are restricted to slopes. Other species, such as the oil palm *Elaeis oleifera* (Palmae), are restricted to the seasonal swamp, and *Gustavia superba* (Lecythidaceae) dominates the 2 ha of secondary forest, an area of unusually low diversity. For census data and rankings, see tables 24.2–24.7.

Flowering peaks just after the onset of the rains in May (Foster and Brokaw 1982). Fruit fall peaks around the beginning of the rainy season, with a lesser peak in September (Foster and Brokaw 1982). Seed germination peaks shortly after the onset of the rainy season (Garwood 1983). Leaf fall peaks at the end of the rainy season and well into the dry season. Leaves flush around the beginning of the rainy season and evergreens flush at the turn of the year (Leigh and Windsor 1982).

Table 24.4. BCI Rankings by Family

Rank	Family	Basal Area (m²)	% BA	% Trees	Family	Trees	% Trees	Family	Species
1	Bombacaceae	183.4	11.4	1.0	Rubiaceae	46,715	20.4	Leguminosae	37
2	Leguminosae	159.0	9.9	7.5	Violaceae	38,489	16.8	Rubiaceae	31
3	Rubiaceae	145.8	9.1	20.4	Leguminosae	17,185	7.5	Moraceae	21
4	Meliaceae	122.7	7.6	7.3	Annonaceae	16,690	7.3	Flacourtiaceae	15
5	Euphorbiaceae	107.4	6.7	1.8	Meliaceae	16,611	7.3	Euphorbiaceae	12
6	Moraceae	97.4	6.1	3.6	Burseraceae	10,974	4.8	Melastomataceae	12
7	Lauraceae	62.9	3.9	1.9	Melastomataceae	8,825	3.9	Lauraceae	10
8	Palmae	62.6	3.9	1.3	Moraceae	8,167	3.6	Palmae	10
9	Burseraceae	59.1	3.7	4.8	Guttiferae	6,340	2.8	Annonaceae	9
10	Bignoniaceae	53.2	3.3	0.3	Chrysobalanaceae	5,529	2.4	Guttiferae	9

Notes: The top 10 families for trees ≥1 cm dbh are ranked in terms of basal area, number of individual trees, and number of species, with the percentage of trees in the plot. Data are from the fourth census.

Table 24.5. BCI Rankings by Genus

Rank	Genus	Basal Area (m²)	% BA	% Trees	Genus	Trees	% Trees	Genus	Species
1	*Trichilia* (Meliaceae)	101.9	6.3	5.8	*Hybanthus* (Violaceae)	36,064	15.7	*Inga* (Leguminosae)	15
2	*Quararibea* (Bombacaceae)	99.0	6.2	1.0	*Faramea* (Rubiaceae)	27,136	11.8	*Ficus* (Moraceae)	12
3	*Alseis* (Rubiaceae)	67.6	4.2	3.6	*Trichilia* (Meliaceae)	13,386	5.8	*Psychotria* (Rubiaceae)	11
4	*Hura* (Euphorbiaceae)	63.7	4.0	0.0	*Desmopsis* (Annonaceae)	11,761	5.1	*Piper* (Piperaceae)	8
5	*Prioria* (Leguminosae)	61.9	3.9	0.6	*Alseis* (Rubiaceae)	8,171	3.6	*Miconia* (Melastomataceae)	7
6	*Faramea* (Rubiaceae)	59.8	3.7	11.8	*Mouriri* (Melastomataceae)	7,130	3.1	*Casearia* (Flacourtiaceae)	5
7	*Ceiba* (Bombacaceae)	53.8	3.3	0.0	*Protium* (Burseraceae)	6,767	3.0	*Bactris* (Palmae)	4
8	*Virola* (Myristicaceae)	48.0	3.0	0.9	*Psychotria* (Rubiaceae)	5,692	2.5	*Cupania* (Sapindaceae)	4
9	*Oenocarpus* (Palmae)	47.9	3.0	0.8	*Swartzia* (Leguminosae)	5,481	2.4	*Eugenia* (Myrtaceae)	4
10	*Jacaranda* (Bignoniaceae)	38.7	2.4	0.1	*Hirtella* (Chrysobalanaceae)	5,084	2.2	*Nectandra* (Lauraceae)	4
								Ocotea (Lauraceae)	4
								Protium (Burseraceae)	4
								Zanthoxylum (Rutaceae)	4

Notes: The top 10 tree genera for trees ≥1 cm dbh are ranked by basal area, number of individual trees, and number of species with the percentage of trees in the plot. Data are from the fourth census.

Table 24.6. BCI Ranking by Species

Rank	Species	Number Trees	% Trees	Species	Basal Area (m^2)	% BA	% Trees
1	*Hybanthus prunifolius* (Violaceae)	36,064	15.8	*Trichilia tuberculata* (Meliaceae)	99.4	6.2	5.6
2	*Faramea occidentalis* (Rubiaceae)	27,136	11.9	*Quararibea asterolepis* (Bombacaceae)	99.0	6.2	1.0
3	*Trichilia tuberculata* (Meliaceae)	12,818	5.6	*Alseis blackiana* (Rubiaceae)	67.6	4.2	3.6
4	*Desmopsis panamensis* (Annonaceae)	11,761	5.1	*Hura crepitans* (Euphorbiaceae)	63.7	4.0	0.1
5	*Alseis blackiana* (Rubiaceae)	8,171	3.6	*Prioria copaifera* (Leguminosae)	61.9	3.9	0.6
6	*Mouriri myrtilloides* (Melastomataceae)	7,130	3.1	*Faramea occidentalis* (Rubiaceae)	59.8	3.7	11.9
7	*Hirtella triandra* (Chrysobalanaceae)	5,044	2.2	*Ceiba pentandra* (Bombacaceae)	53.8	3.4	0.0
8	*Psychotria horizontalis* (Rubiaceae)	4,860	2.1	*Oenocarpus mapora* (Palmae)	47.9	3.0	0.8
9	*Garcinia intermedia* (Clusiaceae)	4,299	1.9	*Jacaranda copaia* (Bignoniaceae)	38.7	2.4	0.1
10	*Tetragastris panamensis* (Burseraceae)	4,139	1.8	*Anacardium excelsum* (Anacardiaceae)	34.1	2.1	0.0

Notes: The top 10 tree species for trees ≥ 1 cm dbh are ranked by number of trees and basal area. The percentage of the total population is also shown. Data are from the fourth census.

Fauna

The 15 km^2 of Barro Colorado Island maintain 40 species and about 5 tons/km^2 of nonflying mammals. The largest carnivores include one resident puma (*Puma concolor*) and more than 20 ocelots (*Leopardus pardalis*). The island's largest herbivores are the tapir (*Tapirus bairdii*), deer (*Mazama americana*), and collared peccary (*Tayassu tajacu*). Barro Colorado Island also has 1300 howler monkeys (*Alouatta palliata*), about 300 white-faced monkeys (*Cebus capucinus*), 1500 agoutis (*Dasyprocta punctata*), and several thousand sloths, mostly three-toed sloths (*Bradypus variegatus*), but with at least 1000 two-toed sloths (*Choloepus hoffmanni*) (Glanz 1982; Leigh 1999). It also has 73 species of bats. Excluding sloths, which comprise some 2.5 tons/km^2 (Glanz 1982), Barro Colorado Island has 2.3 tons/km^2 of nonflying mammals, which annually eat over 40 tons dry weight of food, mostly fruit, per hectare (Leigh 1999). BCI also has about 100 kg/km^2 of birds (about 1300 pairs/km^2) eating about 5.5 tons/km^2/year of food. About half of these pairs are arboreal insectivores, and another quarter are omnivores. Together, they eat 2.4 tons/km^2/year of folivorous insects, an amount that is enough to play a dominant role in limiting insect damage (Leigh 1999).

Table 24.7. BCI Tree Demographic Dynamics

Size Class (cm dbh)	Growth Rate (mm/yr)			Mortality Rate (%/yr)			Recruitment Rate (%/yr)			BA Losses (m^2/ha/yr)			BA Gains (m^2/ha/yr)		
	82–85	85–90	90–95	82–85	85–90	90–95	82–85	85–90	90–95	82–85	85–90	90–95	82–85	85–90	90–95
1–9.9	1.02	0.82	0.58	2.65	2.25	2.56	4.70	3.43	2.14	0.10	0.08	0.10	0.27	0.35	0.18
10–29.9	3.26	2.32	2.14	2.60	1.96	1.88	3.74	3.42	2.93	0.21	0.14	0.14	0.42	0.44	0.32
≥30	6.47	6.07	3.88	3.38	2.07	1.90	3.71	2.37	2.33	0.82	0.35	0.33	0.54	0.50	0.35

Natural Disturbances

Most canopy disturbances are small treefall gaps created when one or a few trees fall (Brokaw 1990; Hubbell et al. 1999). The return time for any particular point in old forest being included in a treefall gap is 126 years (Brokaw 1990). Barro Colorado Island is outside the hurricane belt, although local wind storms sometimes fell a hectare or more of forest. The return time of such wind storms for any particular site is 1000–5000 years (Leigh 1999).

Roughly once every decade or two, an El Niño decreases rainfall during the rainy season, and the following dry season is unusually severe. Mortality was elevated substantially during the 1982–85 census interval due to an unusually severe 1983 dry season associated with the powerful El Niño event that year (Condit et al. 1995). But growth and recruitment were also high during the drought, apparently enhanced by the extra light that entered the forest as a result of the higher mortality (Condit et al. 1992a, 1999). While such events sometimes increase tree mortality, especially among moisture-loving species, to date, El Niño droughts have not had a significant effect on the structure of the canopy (Condit et al. 1992a).

Human Disturbance

About half of Barro Colorado Island, including almost all the 50-ha Forest Dynamics Plot, has been continuously forested for at least 1000 years. In pre-Columbian times, there were two small camps—both over 600 years old—on the site of the 50-ha plot, but there is no evidence of agriculture or forest clearing (Piperno 1990). Most mahogany trees (*Swietenia macrophylla*, Meliaceae) were removed over a century ago. Most of the remaining half of BCI, including 2 ha of the 50-ha plot was cleared by settlers for farms during the 19th century. These camps were abandoned around the turn of the 20th century. No logging or extraction of forest products has occurred since 1923.

Around 1910, the Chagres River was dammed for the Panama Canal. Barro Colorado Island, cut off from the mainland by the rising Lake Gatun, became an island. Afterward, the number of bird and mammal species present on BCI declined. Since the 1920s, several large mammals and birds, and a few small forest birds have gone extinct either due to early hunting or to isolation from the mainland (Enders 1939; Willis 1974; Karr 1982; Leigh 1999; Robinson 1999). White-lip peccaries disappeared from Barro Colorado Island around 1930. Poaching pressure on Barro Colorado Island accelerated greatly in 1932. Pumas were still common there just after World War II, but no puma was sighted on BCI after 1958 until they began to recolonize in 1998. The first jaguar sighting on BCI was in 1983; they are rare visitors. Two harpy eagles were introduced to Barro Colorado Island in 1999 though neither stayed longer than a year. Barro Colorado Island was effectively protected from hunters from about 1980 onward. At present, the faunal community is largely intact.

The closest forest edge, Wetmore's Cove, is approximately 1 km south of the Forest Dynamics Plot's southern boundary. As an island, the forest is completely surrounded by water.

Plot Size and Location

BCI is a 50-ha, 1000 × 500 m plot; the long axis lies east-west. Northwest corner is at 9°9′20.7″ N, 79°51′18.6″ W; southwest corner is at 9°9′4.5″ N, 79°51′19.1″ W.

Funding Sources

The Barro Colorado Island Forest Dynamics Plot has been funded chiefly by the U.S. National Science Foundation, the Smithsonian Tropical Research Institute, and The John D. and Catherine T. MacArthur Foundation.

References

Barone, J. A. 1998. Host-specificity of folivorous insects in a moist tropical forest. *Journal of Animal Ecology* 67:400–09.

Becker, P., P. E. Rabenold, J. R. Idol, and A. P. Smith. 1988. Water potential gradients for gaps and slopes in a Panamanian tropical moist forest's dry season. *Journal of Tropical Ecology* 4:173–84.

Brokaw, N. V. L. 1987. Gap-phase regeneration of three pioneer tree species in a tropical forest. *Journal of Ecology* 75:9–19.

———. 1990. Caída de árboles: frecuencia, cronología y consecuencias. Pages 163–172 in E. G. Leigh, Jr., A. S. Rand, and D. M. Windsor, editors. *Ecología de un Bosque Tropical: Ciclos Estacionales y Cambios a Largo Plazo.* Smithsonian Tropical Research Institute, Balboa, Panama.

Chave, J., R. Condit, S. Lao, J. P. Caspersen, R. B. Foster and S. P. Hubbell. 2003. Spatial and temporal variation of biomass in a tropical forest: results from a large census plot in Panama. *Journal of Ecology* 91:240–52.

Condit, R. 1998. *Tropical Forest Census Plots.* Springer-Verlag, New York.

Condit, R., P. S. Ashton, N. Manokaran, J. V. LaFrankie, S. P. Hubbell, and R. B. Foster. 1999. Dynamics of the forest communities at Pasoh and Barro Colorado: comparing two 50-ha plots. *Philosophical Transactions of the Royal Society of London* 354:1739–48.

Condit, R., S. P. Hubbell, and R. B. Foster. 1992a. Short-term dynamics of a neotropical forest: change within limits. *Bioscience* 42:822–28.

———. 1992b. Recruitment near conspecific adults and the maintenance of tree and shrub diversity in a neotropical forest. *American Naturalist* 140:261–86.

———. 1995. Mortality rates of 205 neotropical trees and shrub species and the impact of a severe drought. *Ecological Monographs* 65:419–39.

———. 1996. Changes in tree species abundance in a neotropical forest: impact of climate change. *Journal of Tropical Ecology* 12:231–56.

————. 1996. Assessing the response of plant functional types in tropical forests to vegetation change. *Journal of Vegetation Science* 7:405–16.

Condit, R., R. Sukumar, S. P. Hubbell, and R. B. Foster. 1998. Predicting population trends from size distributions: a direct test in a tropical tree community. *American Naturalist* 152:495–509.

Condit, R., K. Watts, S. A. Bohlman, R. Pérez, R. B. Foster, and S. P. Hubbell. 2000. Quantifying the deciduousness of tropical forest canopies under varying climates. *Journal of Vegetation Science* 11:649–58.

Croat, T. B. 1978. *Flora of Barro Colorado Island.* Stanford University Press, Stanford, CA.

Enders, R. K. 1935. Mammalian life histories from Barro Colorado Island, Panama. *Bulletin of the Museum of Comparative Zoology* 78:385–502.

————. 1939. Changes observed in the mammal fauna of Barro Colorado Island. 1929–37. *Ecology* 20:104–06.

Foster, R. B., and N. V. L. Brokaw. 1982. Structure and history of vegetation of Barro Colorado Island. Pages 67–82 in E. G. Leigh, Jr., A. S. Rand, and D. M. Windsor, editors. *The Ecology of a Tropical Forest.* Smithsonian Institution Press, Washington, DC.

Foster, R. B., and S. P. Hubbell. 1990. Estructura de la vegetación y composición de especies en un lote de cincuenta hectareas de la Isla de Barro Colorado. Pages: 141–52 in E.G. Leigh, Jr., A. S. Rand, and D. M. Windsor, editors. *Ecología de un Bosque Tropical: Ciclos Estacionales y Cambios a Largo Plazo.* Smithsonian Tropical Research Institute, Balboa, Panama.

Garwood, N. 1983. Seed germination in a seasonal tropical forest in Panama: a community study. *Ecological Monographs* 53:159–81.

Glanz, W. E. 1982. The terrestrial mammal fauna of Barro Colorado island: censuses and long-term changes. Pages 455–68 in E. G. Leigh, Jr., A. S. Rand, and D. M. Windsor, editors. *The Ecology of a Tropical Forest.* Smithsonian Institution Press, Washington, DC.

Harms, K. E., R. Condit, S. P. Hubbell, and R. B. Foster. 2001. Habitat associations of trees and shrubs in a 50-ha neotropical forest plot. *Journal of Ecology* 89:947–59.

Harms, K. E., S. J. Wright, O. Calderon, A. Hernandez, and E. A. Herre. 2000. Pervasive density-dependent recruitment enhances seedling diversity in a tropical forest. *Nature* 404:493–95.

Hubbell, S. P., R. B. Foster, S. T. O'Brien, K. E. Harms, R. Condit, B. Wechsler, S. J. Wright, and S. Loo de Lao. 1999. Light-gap disturbances, recruitment limitation, and tree diversity in a neotropical forest. *Science* 283:554–57.

Janzen, D. H., and P. S. Martin. 1982. Neotropical anachronisms: the fruits the gomphotheres ate. *Science* 215:19–27.

Johnsson, M. J., and R. F. Stallard. 1989. Physiographic controls on the composition of sediments derived from volcanic and sedimentary terrains on Barro Colorado Island, Panama. *Journal of Sedimentary Petrology* 59:768–81.

Karr, J. R. 1982. Population variability and extinction in the avifauna of a tropical land bridge island. *Ecology* 63:1975–78.

Kitajima, K. 1994. Relative importance of photosynthetic traits and allocation patterns as correlates of seedling shade tolerance of 13 tropical trees. *Oecologia* 98:419–28.

King, D. A. 1994. Influence of light level on the growth and morphology of saplings in a Panamanian forest. *American Journal of Botany* 81:948–57.

————. 1998. Relationship between crown architecture and branch orientation in rain forest trees. *Annals of Botany* 82:1–7.

King, D. A., E. G. Leigh, Jr., R. Condit, R. B. Foster, and S. P. Hubbell. 1997. Relationship between branch spacing, growth rate, and light in tropical forest saplings. *Functional Ecology* 11:627–35.

Kursar, T. A. 1989. Evaluation of soil respiration and soil CO_2 concentration in a lowland moist forest in Panama. *Plant and Soil* 113:21–29.

Kursar, T. A., S. J. Wright, and R. Radulovich. 1995. The effects of the rainy season and irrigation on soil water and oxygen in a seasonal forest in Panama. *Journal of Tropical Ecology* 11:497–516.

Leigh, E. G., Jr. 1999. *Tropical Forest Ecology*. Oxford University Press, New York.

Leigh, E. G., Jr., and D. M. Windsor. 1982. Forest production and regulation of primary consumers on Barro Colorado Island. Pages 111–22 in E. G. Leigh, Jr., A. S. Rand, and D. M. Windsor, editors. *The Ecology of a Tropical Forest*. Smithsonian Institution Press, Washington, DC.

Mabberley, D. J. 1997. *The Plant-Book: A Portable Dictionary of the Vascular Plants*. Cambridge University Press, Cambridge, U.K.

Martin, P. S. 1973. The discovery of America. *Science* 179:969–74.

———. 1984. Prehistoric overkill: the global model. Pages 354–403 in P. S. Martin and R. G. Klein, editors. *Quarternary Extinctions, a Prehistoric Revolution*. University of Arizona Press, Tucson, AZ.

Muller-Landau, H. C., S. J. Wright, O. Calderón, S. P. Hubbell, and R. B. Foster. 2002. Assessing recruitment limitation: Concepts, methods and case studies from a tropical forest. Pages 35–53 in D. J. Levey, W. R. Silva, and M. Galetti, editors. *Seed Dispersal and Frugivory: Ecology, Evolution and Conservation*. CAB International, Wallingford, Oxfordshire, U.K.

Nagy, K. A., and G. G. Montgomery. 1980. Field metabolic rate, water flux and food consumption in three-toed sloths, *Bradypus variegatus. Journal of Mammalogy* 61:465–72.

Peters, H. A. 2003. Neighbour-regulated mortality: the influence of positive and negative density dependence on tree populations in species-rich tropical forests. *Ecological Letters* 6:757–65.

Piperno, D. R. 1990. Fitolitos, arqueología y cambios prehistóricos de la vegetación en un lote de cincuenta hectáreas de la isla de Barro Colorado. Pages 153–56 in E. G. Leigh, Jr., A. S. Rand, and D. M. Windsor, editors. *Ecología de un Bosque Tropical: Ciclos Estacionales y Cambios a Largo Plazo*. Smithsonian Tropical Research Institute, Balboa, Panama.

Piperno, D. R., and D. M. Pearsall. 1998. *The Origins of Agriculture in the Lowland Neotropics*. Academic Press, San Diego, CA.

Pyke, C. R., R. Condit, S. Aguilar, and S. Lao. 2001. Floristic composition across a climatic gradient in a neotropical lowland forest. *Journal of Vegetation Science* 12:553–66.

Robinson, W. D. 1999. Long-term changes in the avifauna of a tropical forest isolate, Barro Colorado Island, Panama. *Conservation Biology* 13:85–97.

Robinson, W. D. 2001. Changes in abundance of birds in a neotropical forest fragment over 25 years: a review. *Animal Biodiversity and Conservation* 24.2:51–65.

Smith, A. P., K. P. Hogan, and J. R. Idol. 1992. Spatial and temporal patterns of light and canopy structure in a lowland tropical moist forest. *Biotropica* 24:503–11.

Sternberg, L. S. L., S. S. Mulkey, and S. J. Wright. 1989 Ecological interpretation of leaf carbon isotope ratios: Influence of respired carbon dioxide. *Ecology* 70:1317–24.

Willis, E. O. 1974. Populations and local extinctions of birds on Barro Colorado Island, Panama. *Ecological Monographs* 44:153–69.

Wills, C., and R. Condit. 1999. Similar non-random processes maintain diversity in two tropical rainforests. *Proceedings of the Royal Society of London* 266:1445–52.

Wills, C., R. Condit, R. B. Foster, and S. P. Hubbell. 1997. Strong density- and diversity-related effects help to maintain tree species diversity in a neotropical forest. *Proceedings of the National Academy of Science* 94:1252–57.

Windsor, D. M. 1990. Climate and moisture availability in a tropical forest: long-term records from Barro Colorado Island, Panama. *Smithsonian Contributions to the Earth Sciences* 29:1–145.

Wright, S. J., C. Carrasco, O. Calderon, and S. Paton. 1999. The El Niño Southern Oscillation, variable fruit production, and famine in a tropical forest. *Ecology* 80:1632–47.

Wright, S. J., and F. H. Cornejo. 1990. Seasonal drought and the timing of flowering and leaf fall in a neotropical forest. Pages 49–61 in K. S. Bawa and M. Hadley, editors. *Reproductive Ecology of Tropical Forest Plants.* Parthenon Publishing Group, Park Ridge, NJ.

Yavitt, J. B. 2000. Nutrient dynamics of soil derived from different parent material on Barro Colorado Island, Panama. *Biotropica* 32:198–207.

Yavitt, J. B., and S. J. Wright. 2001. Drought and irrigation effects on fine root dynamics in a tropical moist forest, Panama. *Biotropica* 33:421–34.

25

Bukit Timah Forest Dynamics Plot, Singapore

Shawn K. Y. Lum, Sing Kong Lee, and James V. LaFrankie

Site Location, Administration, and Scientific Infrastructure

The 2-ha Bukit Timah Forest Dynamics Plot, established in 1993, lies within the Bukit Timah Nature Reserve, located in the center of the small island state of Singapore (Lum and Sharp 1996; fig. 25.1). The reserve encompasses approximately 125 ha of forest on the slopes of Bukit Timah hill, rising to 163 m above sea level. The reserve, two thirds of which is primary forest, lies within the center of the island, only 8 km from the city center and immediately adjacent to a highly developed urban area. Heavy construction of new residential complexes continues around the base of the hill. Since 1985, a six-lane expressway has separated the nature reserve from a 2600-ha parcel of 50-year old secondary forest of the Central Water Catchment.

Bukit Timah received protection since the mid-1800s and was gazetted a nature reserve by 1939 (Corlett 1992). It is now a totally protected area governed by the Singapore National Parks Board. The nature reserve includes an extensive network of trails and a visitor center at the entrance. Two quarries on the edge of the reserve have only recently been closed.

The 2-ha plot is located in primary forest, just off of one of the reserve's major hiking trails. A second 2-ha Forest Dynamics Plot is currently being established nearby in secondary forest.

The Bukit Timah Forest Dynamics Plot is a collaboration between the National Institute of Education of Nanyang Technological University and the Center for Tropical Forest Science.

Climate

The climate in Singapore is equatorial. The mean monthly temperature ranges from 23.1 to 30.7°C. The mean annual rainfall of Singapore is 2473 mm (97-year record, central and southern stations combined, Watts 1955; see also Dale 1963 and table 25.1). During the last century, the year-to-year variation was significant: a low of 1605 mm in the year 1889 to a high of 3451 mm in 1914. In approximately half of the years, total rainfall differed from the mean value by more than 10%. Patterns of wind, rain, and temperature follow two monsoonal wind systems:

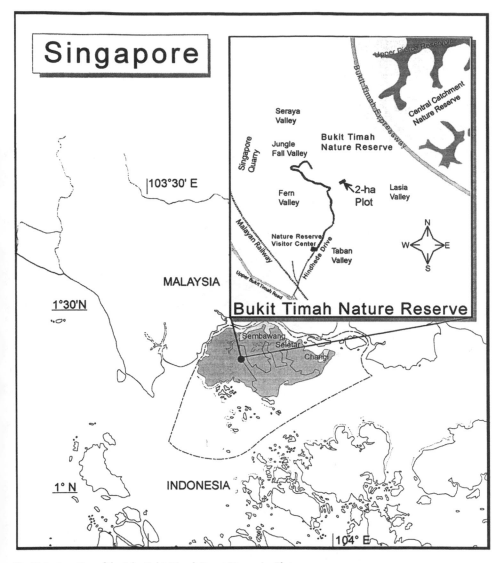

Fig. 25.1. Location of the 2-ha Bukit Timah Forest Dynamics Plot.

December to March weather is dominated by northeasterly monsoonal winds, and June to September weather is dominated by southwesterly monsoonal winds. Despite the relatively everwet, aseasonal conditions, evapotranspiration commonly exceeds rainfall for a 1–3 month period around February (Nieuwolt 1965).

Table 25.1. Bukit Timah Climate Data

	Jan	Feb	Mar	Apr	May	Jun	Jul	Aug	Sep	Oct	Nov	Dec	Total/ Averages
Rain (mm)	256	184	212	200	200	171	158	179	177	216	249	271	2473
ADTMx (°C)	30.7	30.7	30.8	31.1	31.2	30.8	30.5	30.5	30.5	30.5	30.3	30.3	30.7
ADTMn (°C)	22.7	22.7	23.0	23.0	23.1	23.3	23.3	23.3	23.1	23.3	23.1	23.1	23.1

Note: Average daily maximum and minimum (ADTMx, ATDTMn) temperature, data are based on Singapore meteorological records from 1930–1941 and 1947–1960 using the mean of five stations (Dale 1963).

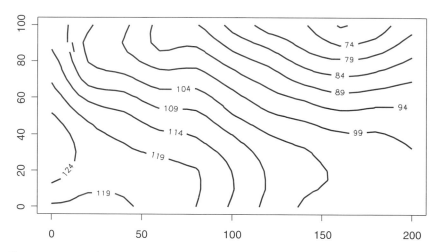

Fig. 25.2. Topographic map of the 2-ha Bukit Timah Forest Dynamics Plot with 5-m contour intervals.

Topography and Soil

At 163 m above sea level, the summit of the Bukit Timah Nature Reserve is the highest point in Singapore. The reserve is underlain by mid-Triassic granite (termed Bukit Timah Granite) which forms old, wet, and highly weathered (pale-udult) soils of the rengam series in the Ultisols order (Ives 1977). Humults, highly weathered Ultisols with a shallow A° organic horizon, dominate convex surfaces on ridges. The soils on the slopes are well-drained sandy loams or sandy-clay loams, which are acidic (pH 3.5–3.8 near the surface, increasing to 4.0–4.2 at 30 cm), relatively nutrient poor, and phosphorus limited (Grubb et al. 1994). See figures 25.2 and 25.3.

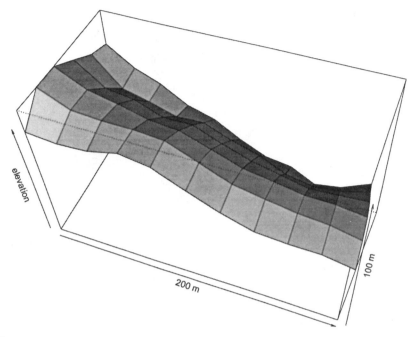

Fig. 25.3. Perspective map of the 2-ha Bukit Timah Forest Dynamics Plot.

Forest Type and Characteristics

The forest of Bukit Timah represents the typical coastal hill forest of the southern Malay Peninsula (Wyatt-Smith 1963), characterized by an upper canopy of mixed species, especially species of the family Dipterocarpaceae, mainly *Shorea curtisii* and *Dipterocarpus caudatus*. Patches of *S. curtisii* are restricted to the steep ridges. Other dominant emergent species include *Ixonanthes reticulata* and *I. icosandra* (Ixonanthaceae), *Artocarpus* spp. (Moraceae), and *Gluta wallichii* (Anacardiaceae). Although today most of the forest in the Bukit Timah Nature Reserve is of broken and irregular stature, the 2-ha Forest Dynamics Plot is comprised of many large-diameter timber trees with a basal area of more than 30 m²/ha. The overall species diversity in the Forest Dynamics Plot is high with 321 species distributed among 60 families. Epiphytes are poorly represented but lianas are numerous, especially thorny climbing palms known as rattans. Thirty species of palms are documented by Corlett (1995), the sixth largest family of plants in the nature reserve. Nineteen of the 30 palm species are rattans. Examples of rattans

Table 25.2. Bukit Timah Plot Census History

Census	Dates	Number of Trees ($\geq$1 cm dbh)	Number of Species ($\geq$1 cm dbh)	Number of Trees ($\geq$10 cm dbh)	Number of Species ($\geq$10 cm dbh)
First	May 1993–June 1993	12,668	321	813	165
Second	November 1995–April 1996	12,892	335	843	170
Third	March 1997–September 1997	11,571	322	761	163
Fourth	March 2003–July 2003	11,918	329	843	160

Note: Four censuses have been completed of the Bukit Timah Forest Dynamics Plot. The next census is expected in 2005.

Table 25.3. Bukit Timah Summary Tally

Size Class (cm dbh)	Average per Hectare							2-ha Plot				
	BA	N	S	G	F	H'	α	S	G	F	H'	α
$\geq$1	34.6	5959	276	152	58	1.90	60.0	329	170	62	1.94	62.6
$\geq$10	30.3	422	113	78	37	1.77	51.2	160	103	46	1.86	58.5
$\geq$30	22.8	102	41	36	21	1.36	25.1	63	51	29	1.46	31.2
$\geq$60	11.5	20	10	8	6	0.85	8.4	15	11	8	0.96	8.9

Notes: BA represents basal area in m^2, N is the number of individual trees, S is number of species, G is number of genera, F is number of families, H' is Shannon–Wiener diversity index using $\log_{10}$, and α is Fisher's α. Basal area was calculated using only the largest stem of multiple-stemmed individuals. 252 individuals were not identified to species or morphospecies. Data are from the fourth census.

Table 25.4. Bukit Timah Rankings by Family

Rank	Family	Basal Area (m^2)	% BA	% Trees	Family	Trees	% Trees	Family	Species
1	Dipterocarpaceae	26.0	38.4	6.0	Burseraceae	1667	14.5	Euphorbiaceae	35
2	Moraceae	6.4	9.5	8.9	Euphorbiaceae	1243	10.8	Annonaceae	24
3	Euphorbiaceae	4.1	6.0	10.8	Moraceae	1027	8.9	Myrtaceae	22
4	Ixonanthaceae	3.1	4.6	0.5	Ebenaceae	803	7.0	Rubiaceae	18
5	Lauraceae	2.8	4.2	2.7	Guttiferae	771	6.7	Guttiferae	15
6	Burseraceae	2.7	4.0	14.5	Dipterocarpaceae	691	6.0	Lauraceae	13
7	Anacardiaceae	2.6	3.8	4.1	Anacardiaceae	471	4.1	Myristicaceae	13
8	Leguminosae	2.4	3.5	0.9	Myrtaceae	435	3.8	Sapotaceae	13
9	Rhizophoraceae	2.3	3.4	2.4	Annonaceae	419	3.6	Burseraceae	11
10	Rubiaceae	1.7	2.6	2.8	Sapotaceae	399	3.5	Leguminosae	11
					Ulmaceae	399	3.5		

Notes: The top 10 families for trees $\geq$1 cm dbh are ranked in terms of basal area, number of individual trees, and number of species. Data are from the fourth census.

found in Bukit Timah include *Plectocomia elongata, Daemonorops* spp., and *Calamus* spp. Other palms include *Oncosperma horridum*, which reaches canopy height, *Licuala* spp., and one that is not found anywhere else in Singapore, the delicate *Rhopaloblaste singaporensis*. For census data and rankings, see tables 25.2–25.7.

Table 25.5. Bukit Timah Rankings by Genus

Rank	Genus	Basal Area (m^2)	% BA	% Trees	Genus	Trees	% Trees	Genus	Species
1	*Shorea* (Dipterocarpaceae)	20.2	29.9	4.7	*Santiria* (Burseraceae)	1057	9.2	*Syzygium* (Myrtaceae)	21
2	*Dipterocarpus* (Dipterocarpaceae)	4.1	6.1	0.9	*Streblus* (Moraceae)	824	7.1	*Diospyros* (Ebenaceae)	9
3	*Streblus* (Moraceae)	3.8	5.7	7.1	*Diospyros* (Ebenaceae)	803	7.0	*Aporusa* (Euphorbiaceae)	8
4	*Ixonanthes* (Ixonanthaceae)	3.1	4.6	0.5	*Calophyllum* (Guttiferae)	589	5.1	*Garcinia* (Guttiferae)	8
5	*Artocarpus* (Moraceae)	2.6	3.9	1.7	*Shorea* (Dipterocarpaceae)	537	4.7	*Polyalthia* (Annonaceae)	8
6	*Koompassia* (Leguminosae)	1.9	2.8	0.2	*Dacryodes* (Burseraceae)	480	4.2	*Artocarpus* (Moraceae)	7
7	*Santiria* (Burseraceae)	1.8	2.7	9.2	*Syzygium* (Myrtaceae)	430	3.7	*Memecylon* (Melastomataceae)	7
8	*Syzygium* (Myrtaceae)	1.5	2.3	3.7	*Gironniera* (Ulmaceae)	399	3.5	*Calophyllum* (Guttiferae)	6
9	*Litsea* (Lauraceae)	1.5	2.3	1.3	*Pimelodendron* (Euphorbiaceae)	361	3.1	*Elaeocarpus* (Elaeocarpaceae)	6
10	*Hopea* (Dipterocarpaceae)	1.4	2.1	0.3	*Gluta* (Anacardiaceae)	342	3.0	*Litsea* (Lauraceae)	6
								Xanthophyllum (Xanthophyllaceae)	6

Notes: The top 10 tree genera for trees ≥1 cm dbh are ranked by basal area, number of individual trees, and number of species. Data are from the fourth census.

Table 25.6. Bukit Timah Ranking by Species

Rank	Species	Number Trees	% Trees	Species	Basal Area (m^2)	% BA	% Stems
1	*Streblus elongatus* (Moraceae)	824	7.1	*Shorea curtisii* (Dipterocarpaceae)	17.1	25.3	4.3
2	*Santiria apiculata* (Burseraceae)	756	6.6	*Dipterocarpus caudatus* (Dipterocarpaceae)	4.1	6.1	0.9
3	*Diospyros lanceifolia* (Ebenaceae)	628	5.4	*Streblus elongatus* (Moraceae)	3.8	5.7	7.1
4	Shorea curtisii (Dipterocarpaceae)	495	4.3	*Ixonanthes reticulata* (Ixonanthaceae)	3.1	4.6	0.4
5	*Dacryodes rostrata* (Burseraceae)	470	4.1	*Shorea ochrophloia* (Dipterocarpaceae)	2.1	3.2	0.3
6	*Gironniera parvifolia* (Ulmaceae)	381	3.3	*Koompassia malaccensis* (Leguminosae)	1.9	2.8	0.2
7	*Pimelodendron griffithianum* (Euphorbiaceae)	361	3.1	*Hopea mengarawan* (Dipterocarpaceae)	1.4	2.1	0.3
8	*Gluta wallichii* (Anacardiaceae)	342	3.0	*Campnosperma auriculata* (Anacardiaceae)	1.4	2.1	0.1
9	*Gynotroches axillaris* (Rhizophoraceae)	232	2.0	*Gynotroches axillaris* (Rhizophoraceae)	1.3	1.9	2.0
10	*Calophyllum ferrugineum* (Guttiferae)	228	2.0	*Pellacalyx saccardianus* (Rhizophoraceae)	1.1	1.6	0.4

Notes: The top 10 tree species for trees ≥1 cm dbh are ranked by number of trees and basal area. Data are from the fourth census.

Table 25.7. Bukit Timah Tree Demographic Dynamics

Size Class (cm dbh)	Growth Rate (mm/yr)		Mortality Rate (%/yr)		Recruitment Rate (%/yr)		BA Losses (m²/ha/yr)		BA Gains (m²/ha/yr)	
	93–95	95–03	93–95	95–03	93–95	95–03	93–95	95–03	93–95	95–03
1–9.9	0.94	0.50	1.44	0.78	0.03	2.21	0.06	0.04	0.26	0.18
10–29.9	2.86	2.35	1.30	1.42	3.27	3.16	0.09	0.10	0.33	0.29
≥30	4.19	3.79	1.59	1.81	3.49	2.34	0.37	0.43	0.60	0.46

Note: Because information on the census date for individual trees was not available for the third census, the same date (June 15, 1997) was used for all records from that census.

Fauna

The annual census of Singapore's native birds conducted by the Singapore Bird Group (Nature Society) found 207 species, of which 127 are resident (unpublished data). Over 300 species of butterflies (excluding skippers) are found on the island of Singapore (Fleming 1975). Until recently, the mammals of Singapore were poorly surveyed. In the late 1990s, a survey by the Nature Society (Singapore) Vertebrate Study Group documented 44 species of mammals, including 17 bats, in Singapore (unpublished data).

Natural Disturbance

A forest fragment is more exposed to winds once it has lost the intervening protection of surrounding vegetation (Saunders et al. 1991). This can result in greater mortality of trees due to windthrows, decreased air moisture and increased temperature, and disturbed soil. The problem is exacerbated at Bukit Timah because it is a hill, with parts of the forest exposed at the edge of quarry cuts. While lightning strikes and local wind storms are infrequent, according to local naturalists, they are increasing in frequency in likely response to the fragmentation and isolation of the nature reserve. Emergent trees are blown down at a high rate and are not being replaced (Corlett 1995). Two large trees in the plot have been hit by lightning since 1993. As more big trees are felled, the survivors become increasingly conspicuous, potentially increasing their risk of becoming future lightning conductors (Lum and Sharp 1996).

Human Disturbance

Local extirpations have been relatively widespread (Corrlett 1992; Turner et al. 1996; Brook et al. 2003). Of the 44 mammal species, 60% are threatened with local extirpation. The number of extirpated mammal species is not known with certainty, but estimates suggest approximately 20 species, including all of the larger animals—elephant (*Elephas maximus*), tiger (*Panthera tigris*), tapir (*Tapirus indicus*), and most primates—have been lost. Of the primates, the banded leaf monkey (*Presbytis femoralis*), which was extirpated from Bukit Timah in 1987, is represented by perhaps only one troop in the adjacent Central Water Catchment. Also in the Central Water Catchment, the long-tailed macaque (*Macaca fascicularis*) is represented by an overabundant population that varies between 500 and 1000 individuals among 30 troops. Lim (1992, 1997) records 70 native forest bird species that have been extirpated since the arrival of Sir Stamford Raffles in 1817, including the large seed dispersers such as hornbills, trogons, broadbills, and large pigeons. Besides species extirpations, significant shifts in composition are hinted at by recent surveys (Brook et al. 2003, Nature Society [Singapore] Vertebrate Study Group, unpublished data). An example of this is that the most abundant species of forest rats in Bukit Timah are not the spiny rats (*Maxomys* spp.) of primary forests, but rather *Rattus annandalei bullatus*, a species found in secondary forest and forest margins.

Direct human disturbance has been varied. Until recently, the hunting of tiger (*Panthera tigris*), deer (*Cervus* spp.), pigs (*Sus* spp.), flying foxes (*Pteropus* spp.), and wild birds was common. Specialized timber extraction was widespread by the mid-19th century and led to the enactment of the first tropical forestry regulations in southeast Asia. A paved road, passable for automobiles, was built to the summit of Bukit Timah in 1924. Near the summit, large granite quarries were built during the early 20th century and the open pits still remain. The Battle of Bukit Timah, at the commencement of the Japanese Occupation (1942–45), was accompanied by extensive short-term encampments near the base and scattered damage within the forest. The reserve is crisscrossed with paths and an asphalt road, affecting the internal environment by changing drainage routes, opening up the canopy, and allowing human disturbance. Various researchers fear a drying out of Bukit Timah (e.g., Corlett 1995). Wee (1995), for example, in his studies on the ferns of Bukit Timah, expressed fear that many moisture-sensitive shade ferns will disappear over time.

The nature reserve has more than a quarter million visitors a year. The increasing number of casual weekend visitors poses a threat to the delicate nature reserve. An announcement in 1995 that there would be 4000 new low-rise housing units by the year 2015 at the foot of Bukit Timah as part of the development plans for Bukit Panjang suggested that human intrusions into the nature reserve could only multiply further (Lum and Sharp 1996). Also of note is the risk from introduced alien plants, such the aggressive weed*Clidemia hirta* (Melastomataceae), which are increasingly common in the nature reserve, though not yet in the 2-ha plot. Exotic bird species such as white-throated laughing thrush (*Garrulax leucolophus*) and lineated barbet (*Megalaima lineata*) are also starting to multiply at the edge of the reserve. These birds, which prey on eggs and nestlings, are a big threat to the dwindling population of native forest birds (L. K. Wang, personal communication.).

Plot Size and Location

The 2-ha Forest Dynamics Plot, 200 × 100 m, its long axis along a ridge in the southeast-northwest direction, is located at approximately 1°15′N, 103°45′E.

Funding Sources

The Bukit Timah Dynamics Plot has been funded chiefly by National Institute of Education of Nanyang Technological University and the Smithsonian Tropical Research Institute.

References

Brook, B. W., N. S. Sodhi, and P. K. L. Ng. 2003. Catastrophic extinctions follow deforestation in Singapore. *Nature* 424:420–423.

Chan, L., and R. T. Corlett. 1999. Biodiversity in the nature reserves of Singapore. *Gardens' Bulletin Singapore* 49:145–47.

Corlett, R. T. 1992. The ecological transformation of Singapore 1819–1990. *Journal of Biogeography* 19:411–20.

———. 1995. The future of Bukit Timah Nature Reserve. *Gardens' Bulletin Singapore. Supplement* No. 3:165–68.

Dale, W. L. 1963. Surface temperature in Malaya. *Journal of Tropical Geography* 17:55–71.

Fleming, W. A. 1975. *Butterflies of West Malaysia and Singapore.* Longmans, Kuala Lumpur, Malaysia.

Grubb, P. J., I. M. Turner, and D. F. Burslem. 1994. Mineral nutrient status of coastal hill dipterocarp forest and *Adinandra belukar* in Singapore: Analysis of soil, leaves and litter. *Journal of Tropical Ecology* 10:559–77.

Ives, D. W. 1977. *Soils of the Republic of Singapore.* New Zealand Soil Survey Report 36. New Zealand Soil Bureau, Wellington, New Zealand.

Keng, H. 1990. *The Concise Flora of Singapore: Gymnosperms and Dicotyledons.* Singapore University Press, Singapore.

Lim, K. S. 1992. *Vanishing Birds of Singapore.* The Nature Society of Singapore, Singapore.

———. 1997. *Birds: An Illustrated Field Guide to the Birds of Singapore.* Sun Tree Publishing, Singapore.

Lum, S., and I. Sharp. 1996. *A View from the Summit: The Story of Bukit Timah Nature Reserve.* Nanyang Technological University, Singapore.

Mabberley, D. J. 1997. *The Plant-Book: A Portable Dictionary of the Vascular Plants.* Cambridge University Press, Cambridge, U.K.

Nieuwolt, S. 1965. Evaporation and water balances in Malaya. *Journal of Tropical Geography* 20:34–53.

Saunders, D. A., R. J. Hobbs, and C. R. Margules. 1991. Biological consequences of ecosystem fragmentation: A review. *Conservation Biology* 5(1):18–32.

Turner, I. M., and R. T. Corlett. 1996. The conservation value of small, isolated fragments of lowland tropical rain forest. *Trends in Ecology and Evolutionary Biology* 11:330–33.

Turner, I. M., K. S. Chua, J. S. Y. Ong, B. C. Soong, and H. T. W. Tan. 1996. A century of plant species loss from an isolated fragment of lowland tropical rain forest. *Conservation Biology* 10:1229–44.

Watts, I. E. M. 1955. The climate of west Malaysia and Singapore. *Journal of Tropical Geography* 7:1–71.

Wyatt-Smith, J. 1963. *A Manual of Malayan Silviculture for Inland Forest.* Yau Seng Press, Kuala Lumpur, Malaysia.

Wee, Y. C. 1995. Pteridophytes. *Gardens' Bulletin Singapore. Supplement* No. 3:61–70.

26

Doi Inthanon Forest Dynamics Plot, Thailand

Mamoru Kanzaki, Masatoshi Hara, Takuo Yamakura, Tatsuhiro
Ohkubo, Minoru N. Tamura, Kriangsak Sri-ngernyuang, Pongsak
Sahunalu, Sakhan Teejuntuk, and Sarayudh Bunyavejchewin

Site Location, Administration, and Scientific Infrastructure

The Doi Inthanon Forest Dynamics Plot is located in the well-protected and well-developed montane forest of Doi Inthanon National Park, Chiang Mai province, northern Thailand (fig. 26.1). The park, established in 1972 and located 50 km to the southwest of the city of Chiang Mai, comprises 48,240 ha and ranges from 400 m up to 2565 m above sea level.

The 15-ha plot was initiated in 1996 and is maintained by a collaborative project involving scientists from Kasetsart University (Thailand), the Royal Forest Department (Thailand), Maejo University (Thailand), Osaka City University (Japan), Kyoto University (Japan), Utsunomiya University (Japan), and Chiba Natural History Museum, and Institute (Japan). The park and the plot are easily reached by car. A park-operated guesthouse and food shop are available in the park headquarters as well as a simple laboratory, supported by Japanese funding. Scientists from Kasetsart University have also established 45 small plots, 40 × 40 m in size, to investigate changes in vegetation along the park's altitudinal gradient (Teejuntuk et al. 2003).

Climate

Doi Inthanon has an average annual rainfall of 1908 mm, ranging from 1229 to 2561 mm over a 7-yr period (measured in 1993–99 at the Royal Project Office of Doi Inthanon Natonal Park, 1300 m above sea level). This montane forest has a typical tropical monsoonal climate with a 5- to 6-month dry season. Rainfall increases with altitude (Santisuk 1988), and clouds frequently cover the mountains above 1500 m, even in the dry season. See table 26.1.

Topography and Soil

The 15-ha Doi Inthanon Forest Dynamics Plot is located on the middle slope (about 1700 m above sea level) of Doi Inthanon mountain. Within the 15-ha plot,

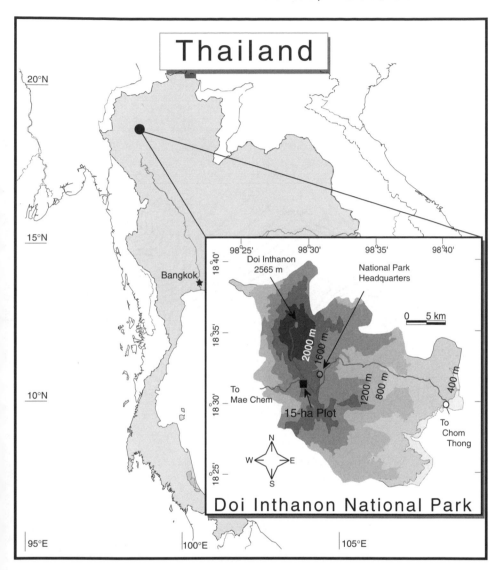

Fig. 26.1. Location of the 15-ha Doi Inthanon Forest Dynamics Plot.

slope inclination varies from 0° to 45.6° and elevation differs by 78 m (figs. 26.2 and 26.3). Moderately inclined slopes cover most of the plot.

   Bedrock within the plot is comprised of post-Silurian and post-Permian granite and pre-Cambrian gneiss. The substratum rocks in Doi Inthanon have produced a coarse sandy loamy soil (Pendleton 1962). Soil in the plot is coarse, sandy, and

Table 26.1. Doi Inthanon Climate Data

	Jan	Feb	Mar	Apr	May	Jun	Jul	Aug	Sep	Oct	Nov	Dec	Total/Averages
Rain (mm)	3	14	31	92	247	228	271	355	371	212	74	9	1908
ADTMx (°C)	23.8	25.9	28.5	28.9	27.7	25.9	25.2	25.2	25.7	25.7	23.4	22.7	25.7
ADTMn (°C)	11.5	13.4	16.9	18.3	18.6	18.2	18.1	18.0	17.8	16.3	14.2	11.8	16.1

Note: Average daily maximum and minimum temperatures (ADTMx and ADTMn) and mean monthly rainfall were determined from 1993–99 climate data obtained from the nearest weather station located at 1300 m above sea level and operated by the Royal Project Office of Doi Inthanon National Park.

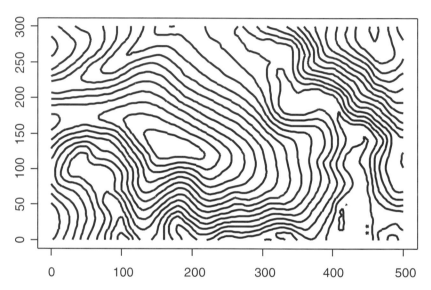

Fig. 26.2. Topographic map of the 15-ha Doi Inthanon Forest Dynamics Plot with 5-m contour intervals.

well drained, except along streams where drainage is poor. The Forest Dynamics Plot contains various geographical features including ridges, permanent streams, rectilinear slopes, and flat terrain along streams.

Forest Type and Characteristics

According to Santisuk (1988), altitudinal zonation of vegetation in the mountains of northern Thailand consists of two zones: the lowland zone (0 to 1000 m),

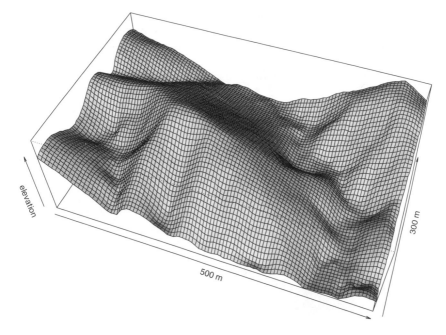

Fig. 26.3. Perspective map of the 15-ha Doi Inthanon Forest Dynamics Plot.

dominated by deciduous forest, and the montane zone (above 1000 m), dominated
by evergreen forest. At approximately 1800 m, the montane zone is further divided
into lower montane forest (LMF) and upper montane forest (UMF). The UMF is
distinguished from the LMF by shorter canopy height and lower species richness,
but tree species tend to be similar between the two forest types. The vegetation of
the park accurately represents the altitudinal zonation. The Doi Inthanon Forest
Dynamics Plot (1700 m altitude) is located near the transition zone from LMF to
UMF. The structure of the forest varies in relation to the topography of the plot.
The forest canopy is dense and reaches from 15 to 30 m in height with the tallest
trees of the plot reaching more than 50 m and the maximum tree size reaching
more than 175 cm dbh. Saplings and pole size trees are abundant. Many small
gaps (<1000 m^2) are scattered throughout the plot. Aboveground biomass in the
plot is estimated to be 570 tons/ha, roughly equivalent to those in lowland tropical
rain forests (Yamakura et al. 1986, 1996). The huge emergent trees contribute to
the high biomass of the site.

In terms of basal area, the Fagaceae and Lauraceae are the dominant families within the plot. The dominant species is *Mastixia euonymoides* (Cornaceae), with individual trees growing to a maximum height of over 50 m. This emergent species overtops the crowns of Fagaceae and Lauraceae trees. The other characteristic species in the forest are *Nyssa javanica* (Nyssaceae) and *Manglietia garretii*

Table 26.2. Doi Inthanon Plot Census History

Census	Dates	Number of Trees (≥ 1 cm dbh)	Number of Species (≥ 1 cm dbh)	Number of Trees (≥ 10 cm dbh)	Number of Species (≥ 10 cm dbh)
First	February 1997–March 2000	73655	162	7785	112

Notes: The first census of the 15-ha Forest Dynamics Plot was completed in March 2000 and identification was completed in March 2001. The next census began in 2003.

Table 26.3. Doi Inthanon Summary Tally

Size Class (cm dbh)	Average per Hectare							15-ha Plot				
	BA	N	S	G	F	H'	α	S	G	F	H'	α
≥ 1	39.8	4910	104.9	73.1	41.9	1.62	19.0	162	105	57	1.75	19.9
≥ 10	36.1	519	66.6	50.4	31.1	1.59	20.5	106	76	43	1.74	18.6
≥ 30	26.3	128	38.2	30.2	22.2	1.40	18.9	90	61	39	1.62	19.7
≥ 60	11.3	20.8	9.9	9.0	7.9	0.88	8.3	37	28	23	1.18	11.0

Notes: BA represents basal area in m^2, N is the number of individual trees, S is number of species, G is number of genera, F is number of families, H' is Shannon–Wiener diversity index using log$_{10}$, and α is Fisher's α. Basal area includes all multiple stems for each individual. Individuals are counted using their largest stems. Data are from the first census. 981 trees were excluded from the calculation of species diversity indices because they were not identified to species or morphospecies.

Table 26.4. Doi Inthanon Rankings by Family

Rank	Family	Basal Area (m^2)	% BA	% Trees	Family	Trees	% Trees	Family	Species
1	Fagaceae	122.7	20.1	13.1	Lauraceae	10,797	14.7	Lauraceae	25
2	Lauraceae	92.2	15.1	14.7	Fagaceae	9,683	13.1	Rubiaceae	13
3	Cornaceae	87.4	14.3	1.3	Euphorbiaceae	8,785	11.9	Myrsinaceae	11
4	Euphorbiaceae	51.3	8.4	11.9	Rubiaceae	7,340	10.0	Fagaceae	8
5	Magnoliaceae	35.8	5.9	0.9	Guttiferae	5,995	8.1	Euphorbiaceae	8
6	Guttiferae	32.4	5.3	8.1	Myrtaceae	3,689	5.0	Rosaceae	6
7	Myrtaceae	25.7	4.2	5.0	Theaceae	3,428	4.7	Theaceae	5
8	Rubiaceae	17.0	2.8	10.0	Meliaceae	3,043	4.1	Moraceae	5
9	Oleaceae	11.5	1.9	1.5	Myrsinaceae	2,729	3.7	Meliaceae	4
10	Aceraceae	11.4	1.9	1.7	Rutaceae	2,681	3.6	Rutaceae	4

Notes: The Top 10 families for trees ≥ 1 cm dbh are ranked in terms of basal area, number of individual trees, and number of species, with the percentage of trees in the plot. Data are from the first census.

Table 26.5. Doi Inthanon Rankings by Genus

Rank	Genus	Basal Area (m²)	% BA	% Trees	Genus	Trees	% Trees	Genus	Species
1	Quercus (Fagaceae)	74.6	12.2	2.8	Calophyllum (Guttiferae)	5995	8.1	Litsea (Lauraceae)	9
2	Mastixia (Cornaceae)	72.8	12.9	0.9	Castanopsis (Fagaceae)	5481	7.4	Prunus (Rosaceae)	5
3	Manglietia (Magnoliaceae)	35.7	5.9	0.9	Mallotus (Euphorbiaceae)	4939	6.7	Lasianthus (Rubiaceae)	4
4	Calophyllum (Guttiferae)	32.4	5.3	8.1	Litsea (Lauraceae)	4688	6.4	Ardisia (Myrsinaceae)	4
5	Litsea (Lauraceae)	28.5	4.7	6.4	Syzygium (Myrtaceae)	3689	5.0	Castanopsis (Fagaceae)	3
6	Cryptocarya (Lauraceae)	27.2	4.5	3.1	Psychotria (Rubiaceae)	3055	4.1	Psychotria (Rubiaceae)	3
7	Syzygium (Myrtaceae)	25.7	4.2	5.0	Melicope (Rutaceae)	2647	3.6	Symplocos (Symplocaceae)	3
8	Drypetes (Euphorbiaceae)	25.7	4.2	2.6	Symplocos (Symplocaceae)	2441	3.3	Lithocarpus (Fagaceae)	3
9	Castanopsis (Fagaceae)	25.4	4.2	7.4	Heynea (Meliaceae)	2407	3.3	Elaeocarpus (Elaeocarpaceae)	3
10	Lithocarpus (Fagaceae)	22.8	3.7	2.9	Cryptocarya (Lauraceae)	2320	3.1	Cinnamomum (Lauraceae)	3

Notes: The top 10 tree genera for trees ≥1 cm dbh are ranked by basal area, number of individual trees, and number of species with the percentage of trees in the plot. Data are from 15 ha of the first census.

(Magnoliaceae), both long-lived, deciduous pioneers. A gymnosperm, *Podocarpus neriifolius* (Podocarpaceae), a small bamboo (*Melocanna* sp. [Gramineae]), a small palm (*Pinanga* sp. [Palmae]), a rattan (*Calamus viminalis* var. *cochinchinensis* [Palmae]), and temperate species such as *Prunus* spp. (Rosaceae) and *Betula alnoides* (Betulaceae) also occur in the plot (Hara et al. 2002). One new species of *Ophiopogon* (Convallariaceae) was found in the plot and named *O. siamensis* (Tamura, 1998). For census data and rankings, see tables 26.2–26.6.

Fauna

Doi Inthanon National Park is home to at least 65 mammal species, including primates, the Indian civet (*Viverra zibetha*), barking deer (*Muntiacus muntijak*), the Asiatic bear (*Ursus thibetanus*), native cats, bats, and squirrels (Gray et al. 1991). The park is also famous for its rich bird fauna, with 383 bird species reported (Gray et al. 1991). In the 15-ha Forest Dynamics Plot, rodents, a wild boar, and many birds have been observed but no scientific census has been conducted.

Table 26.6. Doi Inthanon Rankings by Species

Rank	Species	Number Trees	% Trees	Species	Basal Area (m²)	% BA	% Trees
1	Calophyllum polyanthum (Guttiferae)	5995	8.1	Mastixia euonymoides (Cornaceae)	72.8	11.9	0.9
2	Mallotus khasianus (Euphorbiaceae)	4939	6.7	Quercus eumorpha (Fagaceae)	51.5	8.4	1.6
3	Castanopsis calathiformis (Fagaceae)	3993	5.4	Manglietia garretii (Magnoliaceae)	35.7	5.9	0.9
4	Syzygium angkae ssp. angkae (Myrtaceae)	2647	3.6	Calophyllum polyanthum (Guttiferae)	32.4	5.3	8.1
5	Melicope pteleifolia (Rutaceae)	2407	3.3	Quercus brevicalyx (Fagaceae)	23.0	3.8	1.2
6	Heynea trijuga (Meliaceae)	1968	2.7	Cryptocarya densiflora (Lauraceae)	20.9	3.4	1.8
7	Psychotria symplocifolia (Rubiaceae)	2682	3.6	Syzygium angkae ssp. angkae (Myrtaceae)	15.4	2.5	3.6
8	Eurya nitida var. nitida (Theaceae)	1795	2.4	Drypetes sp. (Euphorbiaceae)	15.1	2.5	1.4
9	Symplocos macrophylla ssp. sulcata var. sulcata (Symplocaceae)	1733	2.4	Nyssa javanica (Cornaceae)	14.5	2.4	0.4
10	Lindera metcalfiana (Lauraceae)	1588	2.2	Mallotus khasianus (Euphorbiaceae)	13.9	2.3	6.7

Notes: The top 10 tree species for trees ≥1 cm dbh are ranked by number of trees and basal area. The percentage of the total population is also shown. Data are from 15 ha of the first census.

Natural Disturbance

There are no large-scale disturbances such as wind storms, volcanoes, fire, or floods in the plot. Canopy trees mainly die standing.

Human Disturbance

The 15-ha plot is well protected from human disturbances, except for moderate animal hunting and plant collecting by local peoples. However, Hmong people who have inhabited the park since the end of the 19th century have practiced shifting cultivation and destroyed a huge area of the park's montane zone, especially below 1500 m. This montane forest area has been replaced by secondary evergreen oak forest and pine plantations (*Pinus kesiya* [Pinaceae]). Because shifting cultivation practices have been halted by the government, the Hmong now participate in flower, fruit, and vegetable cultivation in permanent fields through the aid of projects within the Royal Project Foundation and the Thai Royal Forest Department.

Populations of tiger (*Panthera tigris*), sambar deer (*Cervus unicolor*), elephant (*Elephas maximus*), and other large mammals are either locally extinct or drastically reduced due to hunting and habitat loss.

Plot Size and Location

Doi Inthanon is a 15-ha, 500 × 300 m plot; its long axis lies approximately north-south (S16°W). The northwest corner of the plot is located at 18°31′34″N, 98°29′39″E.

Funding Sources

The Doi Inthanon Forest Dynamics Plot has been funded by the Japanese Ministry of Education, Culture, Sports, Science and Technology and Nippon Life Insurance Foundation.

References

Gray, D., C. Piprell, and M. Graham. 1991. *National Parks of Thailand.* Communications Resources Ltd., Bangkok.

Hara, M., M. Kanzaki, T. Mizuno, H. Noguchi, K. Sri-ngernyuang, S. Teejuntuk, C. Sungpalee, T. Ohkubo, T. Yamakura, P. Sahunalu, P. Dhanmanonda, and S. Bunyavejchewin. 2002. The floristic composition of tropical montane forest in Doi Inthanon National Park, northern Thailand, with special reference to its phytogeographical relation with montane forests in tropical Asia. *Natural History Research* 7:1–17.

Kanzaki, M., and P. Sahunalu. 1998. Doi Inthanon: New forest dynamics plot in a Thai montane forest. *Inside CTFS.* Center for Tropical Forest Science, Washington, DC.

Mabberley, D. J. 1997. *The Plant-Book: A Portable Dictionary of the Vascular Plants.* Cambridge University Press, Cambridge, U.K.

Pendleton, L. 1962. *Thailand, Aspects of Landscape and Life.* Duell, Sloan and Pearce, New York.

Rasmussen, J. N., A. Kaosa-ard, T. E. Boon, M. C. Diaw, K. Edwards, S. Kadyschuk, M. Kaosa-ard, T. Lang, P. Preechapanya, K. Rerkasem, and F. Rune. 2000. *For Whom and for What? Principles, Criteria and Indicators for Sustainable Forest Resources Management in Thailand.* Danish Centre for Forest, Landscape and Planning, Copenhagen.

Santisuk, T. 1988. *An Account of the Vegetation of Northern Thailand.* Franz Steiner Verlag Wiesbaden GMBH, Stuttgart, Germany.

Sri-ngernyuang, K., M. Kanzaki, T. Mizuno, H. Noguchi, S. Teejuntuk, C. Sungpalee, M. Hara, T. Yamakura, P. Sahunalu, P. Dhanmanonda, and S. Bunyavejchewin. 2003. Habitat differentiation of Lauraceae species in a tropical lower montane forest in northern Thailand. *Ecological Research* 18:1–14.

Tamura, M. N. 1998. A new species of the genus *Ophiopogon* (Convallariaceae) from Thailand. *Acta Phytotaxonomica et Geobotanica* 49:27–32.

Teejuntuk, S., P. Sahunalu, K. Sakurai, and W. Sungpalee. 2003. Forest structure and tree species diversity along an altitudinal gradient in Doi Inthanon National Park, Northern Thailand. *Tropics* 12:85–102.

Yamakura, T., A. Hagihara, S. Sukardjo, and H. Ogawa. 1986. Aboveground biomass of tropical rain forest stands in Indonesian Borneo. *Vegetatio* 68:71–82.

Yamakura, T., M. Kanzaki, A. Itoh, T. Ohkubo, K. Ogino, E. Chai. H. S. Lee, and P. S. Ashton. 1996. Forest structure of a tropical rain forest at Lambir, Sarawak with special reference to the dependency of its physiognomic dimensions on topography. *Tropics* 6:1–18.

27

Huai Kha Khaeng Forest Dynamics Plot, Thailand

Sarayudh Bunyavejchewin, Patrick J. Baker, James V. LaFrankie,
and Peter S. Ashton

Site Location, Administration, and Scientific Infrastructure

The Huai Kha Khaeng (HKK) Wildlife Sanctuary is a 278,000-ha reserve in the western part of Thailand. HKK is one of 17 Wildlife Sanctuaries and National Forests that together comprise Thailand's Western Forest Complex (WFC). The WFC covers an area of approximately 1,870,000 ha along the west-central portion of Thailand, making it the largest contiguous area of protected forest in continental southeast Asia. In recognition of their critical role in the conservation of the flora and fauna of continental southeast Asia, HKK and Tung-Yai Naresuan, a neighboring wildlife sanctuary, which together constitute the core of the WFC, were awarded UNESCO World Heritage Site status in 1991. The two sanctuaries form the largest area of protected land in the Indo–Burmese biogeographic region.

The 50-ha HKK Forest Dynamics Plot is located in the center of the northern half of the HKK Wildlife Sanctuary at the Khlong Phuu Research Station (fig. 27.1). The 50-ha HKK plot and associated research station are administered by the Royal Forest Department of Thailand. Facilities at the research station are limited due to the extremely remote location of the site. The research station can accommodate 6–8 visitors. It has no electricity, running water, laboratory, greenhouse, or computer facility. Less than 5 km from the 50-ha Forest Dynamics Plot is another, smaller (16-ha) Forest Dynamics Plot, located in mixed deciduous forest.

Climate

Mean annual rainfall is 1476 mm, based on a 10-year average (1983–93) from a weather station approximately 4 km from the plot. The extent and severity of the dry season (less than 100 mm of precipitation a month) is variable; some years have sporadic rainfall during the dry season, others have little or no rain during the entire dry season. On average, though, there is a 6-month dry season lasting from November to April (table 27.1).

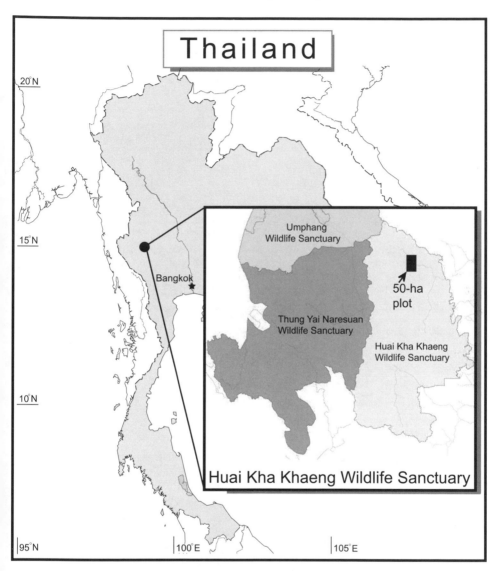

Fig. 27.1. Location of the 50-ha Huai Kha Khaeng Forest Dynamics Plot.

Topography and Soil

The HKK plot is characterized by gently sloping terrain, punctuated by small areas with steep slopes (figs. 27.2 and 27.3). Altitude in the plot ranges between 549 and 638 m above sea level. A low hill bisects the long axis of the plot,

Table 27.1. HKK Climate Data

	Jan	Feb	Mar	Apr	May	Jun	Jul	Aug	Sep	Oct	Nov	Dec	Total/ Averages
Rain (mm)	6	30	39	82	226	120	123	155	278	360	47	10	1476
ADTMx (°C)	31.8	29.1	32.7	34.1	35.2	30.5	30.9	29.7	28.8	28.5	26.6	26.5	30.4
ADTMn (°C)	16.8	15.0	15.0	17.4	20.2	20.4	21.8	21.5	19.5	18.7	14.9	11.6	17.7
Q (W/m^2)	174.8	209.6	213.4	223.7	195.9	162.4	155.7	136.8	160.8	145.0	142.5	146.3	172.2

Notes: Rainfall is an average of 1983–1993 readings at the Kapook Kapiang Ranger Station, which is approximately 4 km north of the HKK Forest Dynamics Plot. Average daily minimum and maximum temperatures (ADTM$_n$ and ADTM$_x$) and solar radiation (Q) data are mean values from automated readings from 1992 to 1994. ADT = average daily temperature.

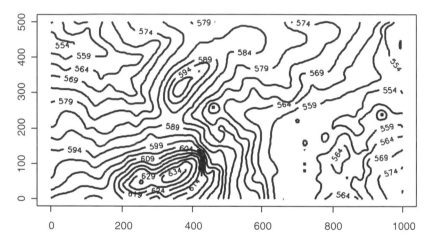

Fig. 27.2. Topographic map of the 50-ha Huai Kha Khaeng Forest Dynamics Plot with 5-m contour intervals.

dividing the plot into a southwest-facing dry side and a north-facing mesic side (Bunyavejchewin et al. 1998). An ephemeral stream runs through the northernmost edge of the plot. The plot soils are pale yellow-grey siliceous, highly weathered Ultisol soils. Soil analyses from the nearby 16-ha plot indicate that the parent material is a residuum of granite porphyry. Soil textures are sandy loam in the surface horizon and sandy-clay loam in the subsurface horizon. Clay accumulates in the lower horizons, from 40 cm downward, due to leaching and weathering of granite in upper horizons. The soil pH is neutral to slightly acidic (5.2–6.8), increasing in acidity with depth (Lauprasert 1988). Litterfall was collected between

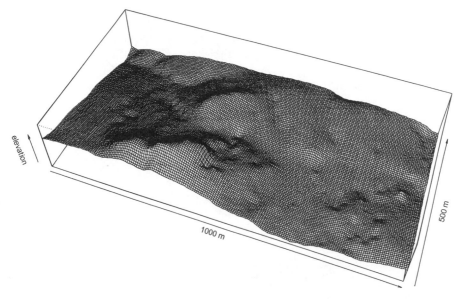

Fig. 27.3. Perspective map of the 50-ha Huai Kha Khaeng Forest Dynamics Plot.

February 1995 and January 1996 on three 1-ha plots. The mean litterfall rate for the plots was 1085 g/m^2/year.

Forest Type and Characteristics

The HKK sanctuary includes a mosaic of several forest types including seasonal dry evergreen, mixed deciduous, and deciduous dipterocarp forests, as well as a small area of variably dry to mesic montane forests. The 50-ha HKK Forest Dynamics Plot is situated in an area of seasonal dry evergreen forest dominated by the emergent *Hopea odorata* (Dipterocarpaceae), various Annonaceae species, and mesic species such as *Acer oblongum* (Aceraceae). The forest has numerous species that are widely distributed throughout, and large isolated figs, *Ficus spp.* (strangling and free standing), which are some of the most conspicuous features of this forest type. The proportion of small trees (1–2 cm dbh) is low in comparison to that of the Malaysian lowland mixed dipterocarp forests. Species richness in the plot is relatively low, with a somewhat elevated proportion of rare species (those with fewer than five individuals in the plot). Canopy height ranges from 40 to 55 m, with occasional trees (mostly dipterocarps) more than 60 m tall. Data are not yet available for leaf area index or total aboveground biomass. For census data and rankings, see tables 27.2–27.7.

Fauna

HKK contains at least 159 species of mammals (60 of which are bats) in 33 families, 379 species of birds in 46 families, 70 species of reptiles in 14 families, 17 species of amphibians in 6 families, and 52 species of fish in 15 families. The wildlife sanctuary has viable populations of elephant (*Elephas maximus*), tiger (*Panthera tigris*),

Table 27.2. HKK Plot Census History

Census	Dates	Number of Trees (≥1 cm dbh)	Number of Species (≥1 cm dbh)	Number of Trees (≥10 cm dbh)	Number of Species (≥10 cm dbh)
First	February 1992–August 1994	79,345	259	21,892	217
Second	January 1999–October 1999	72,509	251	21,875	214

Note: Two censuses have been completed, the next census is expected to begin in January 2004.

Table 27.3. HKK Summary Tally

Size Class (cm dbh)	Average per Hectare							50-ha Plot				
	BA	N	S	G	F	H′	α	S	G	F	H′	α
≥1	31.2	1450	96	78	35	1.50	23.3	251	161	58	1.63	32.6
≥10	29.4	438	65	58	29	1.48	21.3	214	144	54	1.64	32.9
≥30	20.8	82	29	28	18	1.29	17.2	148	106	44	1.61	30.0
≥60	12.1	19	9	9	7	0.83	9.7	77	56	28	1.35	19.9

Notes: BA represents basal area in m^2, N is the number of individual trees, S is number of species, G is number of genera, F is number of families, H' is Shannon–Wiener diversity index using log$_{10}$, and α is Fisher's α. Basal area includes all multiple stems for each individual. 186 individuals were not identified to species or morphospecies. Data are from the second census.

Table 27.4. HKK Rankings by Family

Rank	Family	Basal Area (m^2)	% BA	% Trees	Family	Trees	% Trees	Family	Species
1	Dipterocarpaceae	317.2	21.2	3.8	Euphorbiaceae	15,643	21.6	Euphorbiaceae	32
2	Annonaceae	292.1	19.5	21.5	Annonaceae	15,542	21.5	Leguminosae	17
3	Lauraceae	110.3	7.4	5.3	Sapindaceae	8,552	11.8	Moraceae	17
4	Euphorbiaceae	100.9	6.7	21.6	Lauraceae	3,859	5.3	Sapindaceae	11
5	Sapindaceae	88.5	5.9	11.8	Rubiaceae	3,846	5.3	Lauraceae	10
6	Moraceae	73.8	4.9	0.2	Ebenaceae	3,432	4.7	Meliaceae	10
7	Datiscaceae	64.2	4.3	1.8	Dipterocarpaceae	2,718	3.8	Rubiaceae	10
8	Ebenaceae	59.6	4.0	4.7	Rutaceae	2,565	3.5	Annonaceae	9
9	Lythraceae	56.5	3.8	1.3	Leguminosae	2,240	3.1	Rutaceae	9
10	Meliaceae	49.5	3.3	2.2	Bignoniaceae	1,658	2.3	Lythraceae	8

Notes: The top 10 families for trees ≥1 cm dbh are ranked in terms of basal area, number of individual trees, and number of species, with the percentage of trees in the plot. Basal area is calculated from only the largest stem of multiple-stemmed individuals. Data are from the second census.

Table 27.5. HKK Rankings by Genus

Rank	Genus	Basal Area (m^2)	% BA	% Trees	Genus	Trees	% Trees	Genus	Species
1	Hopea (Dipterocarpaceae)	154.8	10.3	0.4	Croton (Euphorbiaceae)	10,674	14.8	Ficus (Moraceae)	13
2	Miliusa (Annonaceae)	130.7	8.7	2.3	Polyalthia (Annonaceae)	5,583	7.7	Lagerstroemia (Lythraceae)	8
3	Polyalthia (Annonaceae)	82.4	5.5	7.7	Dimocarpus (Sapindaceae)	5,098	7.1	Syzygium (Myrtaceae)	7
4	Ficus (Moraceae)	72.0	4.8	0.1	Orophea (Annonaceae)	4,393	6.1	Diospyros (Ebenaceae)	6
5	Dipterocarpus (Dipterocarpaceae)	65.8	4.4	0.4	Prismatomeris (Rubiaceae)	3,441	4.8	Aporusa (Euphorbiaceae)	5
6	Vatica (Dipterocarpaceae)	65.7	4.4	2.7	Diospyros (Ebenaceae)	3,432	4.7	Cassia (Leguminosae)	4
7	Tetrameles (Datiscaceae)	64.2	4.3	1.8	Phoebe (Lauraceae)	2,453	3.4	Aglaia (Meliaceae)	3
8	Diospyros (Ebenaceae)	59.6	4.0	4.7	Baccaurea (Euphorbiaceae)	2,137	3.0	Albizia (Leguminosae)	3
9	Lagerstroemia (Lythraceae)	56.5	3.8	1.3	Arytera (Sapindaceae)	2,033	2.8	Casearia (Flacourtiaceae)	3
10	Neolitsea (Lauraceae)	45.9	3.1	1.4	Vatica (Dipterocarpaceae)	1,937	2.7	Grewia (Tiliaceae)	3
								Homalium (Flacourtiaceae)	3
								Litsea (Lauraceae)	3
								Mallotus (Euphorbiaceae)	3
								Polyalthia (Annonaceae)	3
								Vitex (Labiatae)	3

Notes: The top 10 tree genera for trees ≥1 cm dbh are ranked by basal area, number of individual trees, and number of species with the percentage of trees in the plot. Basal area is calculated from only the largest stem of multiple-stemmed individuals. Data are from the second census.

leopard (*Panthera pardus*), and wild buffalo (*Bubalus arnee*). Mammal density and biomass have been estimated in a 50–70 km^2 area around the nearby Khao Nang Rum Wildlife Research Station for gaur (*Bos* spp.; 1.8 ind./km^2 and 810 kg/km^2), sambar deer (*Cervus unicolor*; dry season: 1.9 ind./km^2 and 255 kg/km^2; wet season: 4.2 ind./km^2 and 563 kg/km^2), common muntjak (*Muntiacus muntijak*; 3.1 ind./km^2 and 65 kg/km^2), lesser bamboo rat (*Cannomys badius*; 777.4 ind./km^2 and 257 kg/km^2), common tree shrew (*Tupaia glis*; 24.1 ind./km^2 and 4.3 kg/km^2), white-handed gibbon (*Hylobates lar*; 5.4 ind./km^2 and 21.6 kg/km^2), and elephant (0.08 ind./km^2 and 167 kg/km^2). In addition, A. Rabinowitz (personal communication) estimates that tiger density in KNR is 0.01 ind./km^2 and leopard density is 0.04 ind./km^2. There are several endangered animals in the HKK Wildlife

Table 27.6. HKK Rankings by Species

Rank	Species	Number Trees	% Trees	Species	Basal Area (m²)	% BA	% Stems
1	*Croton oblongifolius* (Euphorbiaceae)	10,614	14.7	*Hopea odorata* (Dipterocarpaceae)	154.8	10.3	0.4
2	*Polyalthia viridis* (Annonaceae)	5249	7.3	*Saccopetalum lineatum** (Annonaceae)	130.7	8.7	2.3
3	*Dimocarpus longan* (Sapindaceae)	5098	7.1	*Polyalthia viridis* (Annonaceae)	81.4	5.4	7.3
4	*Orophea polycarpa* (Annonaceae)	4393	6.1	*Vatica cinerea* (Dipterocarpaceae)	65.7	4.4	2.7
5	*Prismatomeris malayana* (Rubiaceae)	3441	4.8	*Tetrameles nudiflora* (Datiscaceae)	64.2	4.3	1.8
6	*Phoebe tavoyana* (Lauraceae)	2453	3.4	*Dipterocarpus alatus* (Dipterocarpaceae)	62.6	4.2	0.4
7	*Baccaurea ramiflora* (Euphorbiaceae)	2137	3.0	*Neolitsea obtusifolia* (Lauraceae)	45.9	3.1	1.4
8	*Arytera litoralis* (Sapindaceae)	2033	2.8	*Lagerstroemia tomentosa* (Lythraceae)	45.3	3.0	1.0
9	*Vatica cinerea* (Dipterocarpaceae)	1937	2.7	*Alphonsea ventricosa* (Annonaceae)	43.4	2.9	1.7
10	*Mitrephora thorelii* (Annonaceae)	1816	2.5	*Arytera litoralis* (Sapindaceae)	36.0	2.4	2.8

*Genus *Saccopetalum* now changed to *Miliusa* according to Mabberley (1997).

Notes: The top 10 tree species for trees ≥1 cm dbh are ranked by number of trees and basal area. The percentage of the total population is also shown. Basal area is calculated from only the largest stem of multiple-stemmed individuals. Data are from the second census.

Table 27.7. HKK Tree Demographic Dynamics

Size Class (cm dbh)	Growth Rate (mm/yr)	Mortality Rate (%/yr)	Recruitment Rate (%/yr)	BA Losses (m²/ha/yr)	BA Gains (m²/ha/yr)
	93–99	93–99	93–99	93–99	93–99
1–9.9	1.84	4.84	5.70	0.06	0.13
10–29.9	2.50	1.89	2.63	0.17	0.31
≥30	3.13	1.92	1.85	0.35	0.31

Note: Basal area is calculated from only the largest stem of multiple-stemmed individuals.

Sanctuary, including the clouded leopard (*Neofelis nebulosa*), white-winged duck (*Cairina scutulata*), and green peafowl (*Pavo muticus*) (Srikosamatara 1993).

Natural Disturbances

Low-intensity surface fires are an important natural disturbance in the HKK area, and affect all size classes of trees. During the dry season, deciduous dipterocarp and mixed deciduous forests are particularly prone to forest fires. They support thick understories dominated by grass and bamboo, respectively, which can be extremely dry and flammable. The effect of fire on forest composition is debatable, but there is concern that the fires are increasing the extent of deciduous dipterocarp forest and decreasing the extent of evergreen forest. Widespread fires occur approximately every 3–10 years. In 1991 and 1998, fire swept through the Forest Dynamics Plot. In 1992, a small portion of the southeast corner of the plot was burned.

Windthrows and elephants are responsible for numerous treefalls and tree damage in the plot. Wild pigs cause small-scale damage to the vegetation by digging into the ground and affecting seedling recruitment.

Human Disturbance

Continental southeast Asia has been populated for at least 10,000 years. Historical population densities in the region of the WFC are unknown. In recent decades there has been considerable population growth, which has led to widespread land conversion for agriculture. Many of the fires that occur in the HKK forest are set by people living in the area. The hunting of animals for food and the extraction of timber and nontimber products (mushrooms, fruits, medicinal plants, honey, and game) are illegal and laws preventing these activities have been enforced with a fair degree of vigilance since 1972, the year HKK was gazetted as a wildlife sanctuary. However, some poaching continues at irregular intervals. Logging within HKK, which was primarily focused on trees over 100-cm girth, was halted upon establishment of the sanctuary; however, logging in the National Forest to the north of HKK, which is within about 5 km of the Forest Dynamics Plot, continued until the late 1980s. The plot itself is located in a block of forest that has never been logged. The distance from the plot to the nearest forest edge varies by direction. The nearest area of nonforest is approximately 20 km to the east and includes the outlying fields of local villages. Forests extend at least 40–50 km to the north, south, and west.

Plot Size and Location

The HKK is a 50-ha, 500 × 1000 m plot; its long axis lies north-south. The northwest corner of the plot is located at 15°37′58.4″N and 99°12′34.1″E. The southeast corner is 15°37′54.7″N and 99°13′28.2″E.

Funding Sources

The HKK Forest Dynamics Plot has been funded in part through grants from the Rockefeller Foundation, the U.S. Agency for International Development (with the assistance of the World Wide Fund for Nature), the John Merck Fund, Conservation Food & Health, and the U.S. National Science Foundation and has been generously supported by the Thai Royal Forest Department.

References

Baker, P. J. 1997. Seedling establishment and growth across forest types in an evergreen/deciduous forest mosaic in western Thailand. *Natural History Bulletin of the Siam Society* 45:17–41.

———. 2001. Age structure and stand dynamics of a seasonal tropical forest in western Thailand. Ph.D. dissertation. University of Washington, Seattle, WA.

Bunyavejchewin, S. 1999. Structure and dynamics in a seasonal dry evergreen forest in northeastern Thailand. *Journal of Vegetation Science* 10:787–92.

———. 2002. *Structure and Composition of a Seasonal Dry Evergreen Forest in Western Thailand.* Ph.D. Dissertation. Osaka City University, Osaka, Japan.

Bunyavejchewin, S., J. V. LaFrankie, P. Pattapong, M. Kanzaki, A. Itoh, T. Yamakura, and P. S. Ashton. 1998. Topographic analysis of a large-scale research plot in seasonal dry evergreen forest at Huai Kha Khaeng Wildlife Sanctuary, Thailand. *Tropics* 8:45–60.

Bunyavejchewin, S., P. J. Baker, J. V. LaFrankie, and P. S. Ashton. 2001. Stand structure of a seasonal evergreen forest at the Huai Kha Khaeng Wildlife Sanctuary, western Thailand. *Natural History Bulletin of the Siam Society* 49:89–106.

Bunyavejchewin, S., P. J. Baker, J. V. LaFrankie, and P. S. Ashton. 2002. Floristic composition of a seasonal evergreen forest, Huai Kha Khaeng Wildlife Sanctuary, western Thailand. *Natural History Bulletin of the Siam Society* 50:125–34.

Bunyavejchewin, S., J. V. LaFrankie, P. J. Baker, M. Kanzaki, P. S. Ashton, and T. Yamakura. 2003. Spatial distribution patterns of the dominant canopy dipterocarp species in a seasonal dry evergreen forest in western Thailand. *Forest Ecology and Management* 175:87–101.

Chamchumroon, V. 1997. *The Morphological Study and Identification of Some Woody Plant Seedlings of Dry Evergreen Forest at Klong Plu, Huai Kha Khaeng Wildlife Sanctuary, Changwat Uthai-thani.* Masters thesis. Kasetsart University, Thailand. (In Thai with English abstract.)

Chormali, P. 2002. *Structure and Litterfall Production in the Tropical Semi-Evergreen Forest at Klong Plu, Huai Kha Khaeng Wildlife Sanctuary, Changwat Uthai Thani.* Masters thesis. Kasetsart University, Thailand. (In Thai with English abstract.)

Condit R., P. S. Ashton, P. Baker, S. Bunyavejchewin, S. Gunatilleke, N. Gunatilleke, S. P. Hubbell, R. B. Foster, A. Itoh, J. V. LaFrankie, H. S. Lee, E. Losos, N. Manokaran, R. Sukumar, and T. Yamakura. 2000. Spatial patterns in the distribution of tropical tree species. *Science* 288:1414–18.

Davies, S. J., S. Bunyavejchewin, and J.V. LaFrankie. 2001. A new giant-leaved *Macaranga* (Euphorbiaceae) from dry seasonal evergreen forest in Thailand. *Thai Forest Bulletin (Botany)* 29:51–57.

Duengkae, P. 1996. *Breeding Biology of Silver-Breasted Broadbill,* Serilophus lunatus *(Gould) in Huai Kha Khaeng Wildlife Sanctuary, Changwat Uthai Thani.* Masters thesis. Kasetsart University, Thailand. (In Thai with English abstract).

Hirsch, P. 1988. Spontaneous land settlement and deforestation in Thailand. Pages 359–76 in J. Dargavel, K. Dixon, and N. Semple, editors. *Changing Tropical Forests.* Centre for Resource and Environmental Studies, Canberra, Australia.

————. 1987. *Participation, Rural Development, and Changing Production Relations in Recently Settled Forest Areas in Thailand.* Ph.D. dissertation. University of London, London, England.

Lauprasert, M. 1988. *The Creation of a Permanent Sample Plot in Dry Evergreen Forest of Thailand and Investigations of a Suitable Plot Size for Permanent Sample Plot Programs.* Masters thesis. International Institute for Aerospace Survey and Earth Sciences, Enschede, Netherlands.

Lekagul, B., and J. A. McNeely. 1977. *Mammals of Thailand.* Sahakarnbhat, Bangkok, Thailand.

Mabberley, D. J. 1997. *The Plant-Book: A Portable Dictionary of the Vascular Plants.* Cambridge University Press, Cambridge, U.K.

Muangkhum, K. 2001. *Home Range and Seasonal Forage of the White-Handed Gibbon* (Hylobates lar (*Linn.*)) *in Huai Kha Khaeng Wildlife Sanctuary.* Masters thesis. Kasetsart University, Thailand. (In Thai with English abstract.)

Nakhasathien, S., and B. Stewart-Cox. 1990. *Nomination of the Thung Yai-Huai Kha Khaeng Wildlife Sanctuary as a UNESCO World Heritage Site.* Royal Thai Forestry Department, Bangkok, Thailand.

Ogawa, H., K. Yoda, and T. Kira. 1961. A preliminary survey on the vegetation of Thailand. *Nature and Life in Southeast Asia* 1:21–158.

Plotkin J. B., M. D. Potts, D. W. Yu, S. Bunyavejchewin, R. Condit, R. Foster, S. Hubbell, J. LaFrankie, N. Manokaran, L. H. Seng, R. Sukumar, M. A. Nowak, and P. S. Ashton. 2000. Predicting species diversity in tropical forests. *Proceedings of the National Academy of Sciences* 97:10850–54.

Rabinowitz, A. 1990. Fire, dry dipterocarp forest, and the carnivore community in Huai Kha Khaeng Wildlife Sanctuary, Thailand. *Natural History Bulletin of the Siam Society* 38:99–115.

————. 1989. The density and behaviour of large cats in a dry tropical forest mosaic in Huai Kha Khaeng Wildlife Sanctuary, Thailand. *Natural History Bulletin of the Siam Society* 37:235–51.

Rabinowitz, A., and S. Walker. 1991. A carnivore community in a dry tropical forest mosaic in Huai Kha Kheng Wildlife Sanctuary, Thailand. *Journal of Tropical Ecology* 7:37–47.

Sponsel, L. E., and P. Natadecha. 1988. Buddhism, ecology and forests in Thailand: Past, present and future. Pages 305–26 in J. Dargavel, K. Dixon, and N. Semple, editors. *Changing Tropical Forests.* Centre for Resource and Environmental Studies, Canberra, Australia.

Srikosamatara, S. 1993. Density and biomass of large herbivores and other mammals in a dry tropical forest, western Thailand. *Journal of Tropical Ecology* 9:33–43.

Stott, P. 1986. The spatial pattern of dry season fires in the savanna forests of Thailand. *Journal of Biogeography* 13:345–58.

28

Ituri Forest Dynamics Plots, Democratic Republic of Congo

Jean-Remy Makana, Terese B. Hart, Innocent Liengola,
Corneille Ewango, John A. Hart, and Richard Condit

Site Location, Administration, and Scientific Infrastructure

The Ituri Forest Dynamics Plots were established in 1994 in the 1,350,000-ha Okapi Faunal Reserve (OFR), Eastern Province, Democratic Republic of Congo, which extends from 1° to almost 3°N latitude and from 28° to almost 30°E longitude (Hart and Carrick 1996; fig. 28.1). Declared a reserve in 1992, the OFR is located in the Ituri River basin, whose forest covers approximately 60,000 km^2.

Among this project's specific goals is to compare the mixed and monodominant forest types and their dynamics on a scale that will facilitate an understanding of their distributions. Despite numerous smaller scale studies, this understanding has remained elusive (Hart et al. 1989; Hart 1995; Torti 1998). To achieve this goal, the Centre de Formation et de Recherche en Conservation Forestière (CEFRECOF) established four 10-ha (200 × 500 m) Forest Dynamics Plots, instead of a single 50-ha plot, in the Okapi Faunal Reserve. Two of the 10-ha plots were placed in mixed Caesalpiniaceous-dominated forest with a base camp on the Edoro Stream, 20 km northwest of OFR's administrative center in the town of Epulu. The other two 10-ha Forest Dynamics Plots were situated in monodominant mbau forest approximately 10 km to the southeast, near a camp located on the Lenda Stream. The four plots are located approximately 150 km from the transition to anthropogenic fire-interface savanna (Makana et al. 1998).

Prior to the establishment of the large plots, twenty-four 625-m^2 small plots were established in both mixed and monodominant forest in 1981–82 to investigate differences in diversity and dominance between the two forest types (Hart et al. 1989).

CEFRECOF's headquarters, located between Lenda and Edoro camps, is constructed of permanent materials and has independent lodging and limited dormitory space. The two camps have simpler accommodations—small tin-roofed rooms of local materials. For the last 15 years, CEFRECOF personnel have received basic support from the Wildlife Conservation Society along with operating funds for the camps, main center, and associated infrastructure.

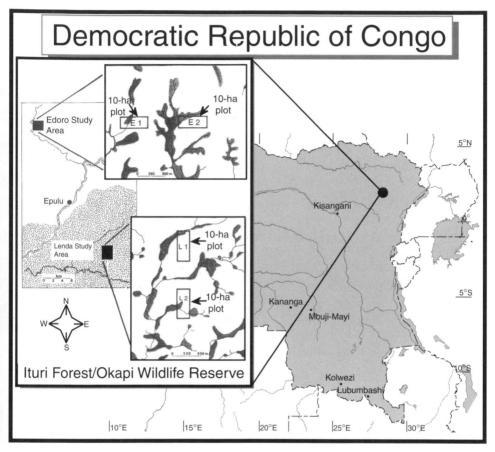

Fig. 28.1. Locations of the four 10-ha Ituri Forest Dynamics Plots. The stippled area in the inset represents monodominant forest. The black areas in the secondary insets represent swamps connected by rivers and streams.

Climate

Over a 9-year period (1987–95), the average annual rainfall recorded at the administrative center of the OFR was 1682 mm, with a maximum of 2085 mm and a minimum of 1304 mm. The dry season lasts 3–4 months. The average annual maximum temperature was 25.5°C (Hart and Carrick 1996; table 28.1).

Topography and Soil

The topography of the area is gentle, with occasional rolling hills containing exposed patches of shallow rocky soils (Hart et al. 1996). The Ituri Forest Dynamics

Table 28.1. Ituri Climate Data

	Jan	Feb	Mar	Apr	May	Jun	Jul	Aug	Sep	Oct	Nov	Dec	Total/ Averages
Edoro:													
Rain (mm)	32	80	103	183	186	152	195	183	192	205	194	80	1785
ADTMx (°C)	25.4	25.2	26.9	27.5	28.3	24.4	24.1	25.8	24.7	24.3	24.4	24.4	25.5
ADTMn (°C)	16.5	17.1	18.1	18.5	18.5	18.3	18.1	18.1	17.9	18.1	18.2	17.8	17.9
Lenda:													
Rain (mm)	51	41	90	195	168	152	165	165	161	226	172	88	1674
ADTMx (°C)	27.9	27.3	28.7	29.0	29.3	29.9	27.7	26.1	27.3	27.0	26.3	27.0	27.8
ADTMn (°C)	17.7	17.3	17.8	18.8	18.9	18.8	18.5	18.4	18.2	18.5	18.6	18.2	18.3

Notes: Mean monthly rainfall and average daily temperature are based on readings taken during 1991–1995 at Edoro and Lenda base camps (Hart and Carrick 1996).

Plots are located approximately 750 m above sea level (asl), with an elevational range from 700 to 850 m asl (figs. 28.2 and 28.3). Both the Lenda (monodominant forest) and Edoro (mixed forest) study areas are relatively flat. Differences in elevation between the lowest and the highest points within the 10-ha plots are 24 m for Lenda-1, 16 m for Lenda-2, 14 m for Edoro-1, and 21 m for Edoro-2 (Makana 1999).

The area's soils are derived from granitic pre-Cambrian shield rock (Laveau 1982) and fall under the order Oxisols, which dominates most of the Congo basin rainforest block in central Africa (Brady 1990). Their texture ranges from loamy sand to sandy clay. The soils are very acidic (mean pH values at 20 cm are 3.96 in mixed forest and 4.17 in monodominant mbau forest) and low in available phosphorus and nitrogen (Hart 1985). Mean soil sand content at Epulu is 64% in mixed forest and 72% in monodominant forest (Hart 1985).

Forest Type and Characteristics

The vegetation in the Ituri forest is composed of two principal types of tropical moist forest: mixed-canopy semievergreen forest (mixed forest) and single-canopy dominant evergreen forest (monodominant forest). There are also small areas of specialized vegetation such as swamp forest that occurs along streams in areas of poor drainage and a xerophyllous flora found in isolated patches such as the dry hilltops in the northern forest (Hart et al. 1996). The canopy in mixed and monodominant forests reaches a height of 30–40m. The dominant species of the monodominant forest is mbau (*Gilbertiodendron dewevrei* (De Wild.) Léonard [Leguminosae (Caesalpinioideae)]), which comprises up to 90% of the large trees in pure stands and forms a homogeneous and continuous canopy. In the mixed forest, the canopy is more heterogeneous and frequently broken by emergent trees more than 40 m in height (Hart et al. 1989). Two species, *Cynometra*

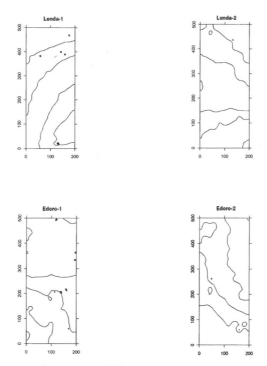

Fig. 28.2. Topographic map of the four 10-ha Ituri Forest Dynamics Plots with 5-m contour intervals (solid line) and intermediate 2.5-m contour interval (dashed line).

alexandri C.H. Wright [Leguminosae (Caesalpinioideae)] and *Julbernardia seretii* (De Wild.) Troupin [Leguminosae (Caesalpinioideae)] occur at high proportions in the canopy of the mixed forest and account for over one third of the basal area of trees above 30 cm dbh (Makana et al. 1998).

The two Lenda plots are dominated by *G. dewevrei* at the canopy level. One of these plots (Lenda-1), however, includes about 2.5 ha of mixed forest. The Edoro plots constitute mixed forest, with the exception of some small monodominant patches at Edoro-1. The most abundant canopy species at Edoro are *C. alexandri* and *J. seretii*. All plots at both study areas contain small patches of temporary or permanent swamp along streams (Makana 1999).

The Lenda monodominant forest and the Edoro mixed forest present different structures. The monodominant forest creates a more even, deep shade than the mixed forest. This results from significant differences in the distribution of free-standing woody stems between different diameter classes. The monodominant forest has more large trees, which result in higher basal area. The deep shade cast by a closed homogeneous canopy in the monodominant forest creates a sparse

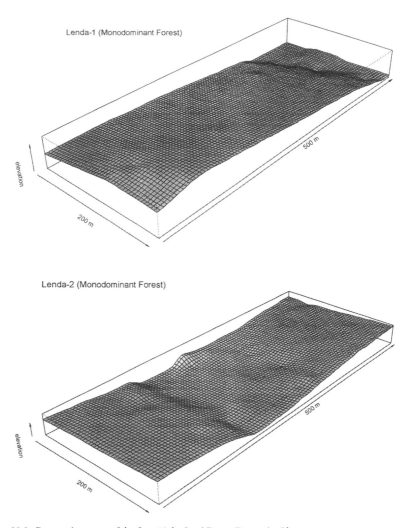

Fig. 28.3. Perspective maps of the four 10-ha Ituri Forest Dynamics Plots.

and open understory, whereas the understory of the mixed forest is more dense. Liana density is twice as high in mixed forest as in monodominant forest (Makana 1999; Makana et al. 1998). In spite of the significant differences in structure and the spatial segregation (Edoro and Lenda sites are more than 20 km apart), the two forest types are remarkably similar with regard to species composition. They share most of very common species in both the canopy and the understory. *C. alexandri* and *J. seretii*, which dominate the canopy of mixed forest, are the

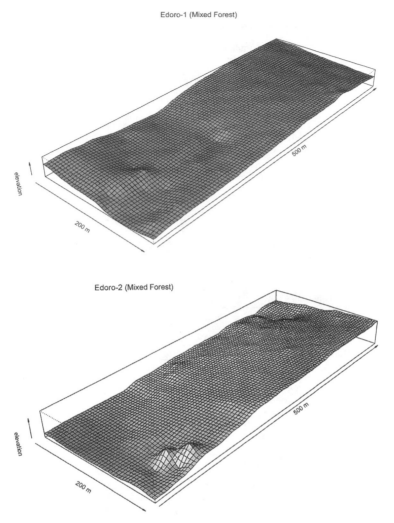

Edoro-1 (Mixed Forest)

Edoro-2 (Mixed Forest)

Fig. 28.3. (Continued)

second and third most common species in monodominant stands. *Scaphopetalum dewevrei* (Sterculiaceae) is a shrub species that represents more than 40% of all stems in each forest (chap. 12). For census data and rankings, see tables 28.2–28.9.

In addition to all tree stems ≥1 cm dbh, plot researchers censused all lianas ≥2 cm dbh. Phenological data, which have been collected for all major dominant tree species in both forest types since 1993, are not yet available.

Table 28.2. Ituri Plot Census History

Census	Plot	Dates	Number of Trees (≥1 cm dbh)	Number of Tree Species (≥1 cm dbh)	Number of Lianas (≥2 cm dbh)	Number of Lianas Species (≥2 cm dbh)	Number of Trees (≥10 cm dbh)	Number of Tree Species (≥10 cm dbh)
First	Combined	Feb 1994–Jun 1996	299,114	420	19,306	243	15,914	278
First	Lenda-1	Feb 1994–Aug 1995*	70,979	327	3,276	139	3,431	188
First	Lenda-2	Oct 1995–Dec 1995	65,893	302	3,162	134	3,721	161
First	Edoro-1	Apr 1994–Jul 1995	76,276	311	6,487	119	4,252	157
First	Edoro-2	Jan 1996–Jun 1996	85,966	312	6,381	146	4,510	146

*169 large trees with dbh >30 cm were measured in July 1996.

Notes: The second census was completed in January 2002.

Table 28.3. Ituri Summary Tally

Size Class (cm dbh)	Average per Hectare							10-ha Plot				
	BA	N	S	G	F	H′	α	S	G	F	H′	α
Lenda-1 (Monodominant Forest)												
≥1	37.0	7097	161	109	37	1.08	29.5	283	179	50	1.13	37.5
≥10	32.0	343	57	46	22	1.07	20.2	168	114	39	1.32	37.0
≥30	26.5	90	13	13	8	0.55	5.1	67	55	22	0.81	16.7
≥60	18.1	34	5	5	3	0.38	2.0	22	21	10	0.56	5.3
Lenda-2 (Monodominant Forest)												
≥1	38.0	6589	157	106	38	1.16	29.0	260	165	49	1.20	34.4
≥10	33.2	372	48	40	20	0.82	15.3	149	102	33	0.96	31.1
≥30	27.5	106	13	12	8	0.39	4.2	62	50	22	0.49	14.4
≥60	16.8	33	4	4	3	0.25	1.5	24	22	15	0.33	6.0
Edoro-1 (Mixed Forest)												
≥1	32.7	7627	148	107	37	1.25	26.0	245	168	46	1.29	31.4
≥10	26.2	425	61	52	26	1.37	19.5	137	102	35	1.53	27.1
≥30	19.3	76	24	23	14	1.04	12.2	80	64	26	1.31	22.5
≥60	11.2	21	9	9	6	0.80	7.1	35	32	16	1.16	12.1
Edoro-2 (Mixed Forest)												
≥1	33.7	8596	150	107	37	1.08	25.9	259	170	43	1.12	32.9
≥10	26.4	451	67	55	26	1.41	21.9	164	116	38	1.56	33.4
≥30	19.4	78	27	25	16	1.13	15.6	102	78	29	1.41	31.3
≥60	11.6	23	11	10	7	0.83	7.6	45	39	19	1.16	16.6

Notes: For each of four 10-ha plots, BA represents basal area in m^2, N is the total number of individual trees, S is the total number of species, G is the number of genera, F is the total number of families, H′ is the Shannon–Wiener diversity index using log$_{10}$, and α is Fisher's α. Basal area includes all multiple stems for each individual. Individuals are counted using their largest stem. All numbers are averages per hectare for 10-ha plots. In the Lenda plots 1 and 2, respectively, 11,380 and 2,507 individuals were not identified to species or morphospecies. In the Edoro plots 1 and 2, respectively, 12,290 and 2,516 individuals were not identified. Data are from the first census.

Table 28.4. Ituri–Lenda (Monodominant Forest) Rankings by Family

Rank	Family	Basal Area (m²)	% BA	% Trees	Family	Trees	% Trees	Family	Species
1	Leguminosae	552.5	74.4	11.3	Sterculiaceae	62,990	46.7	Rubiaceae	50
2	Sterculiaceae	40.6	5.5	46.7	Euphorbiaceae	19,950	14.8	Euphorbiaceae	37
3	Euphorbiaceae	25.8	3.5	14.8	Leguminosae	15,260	11.3	Leguminosae	31
4	Sapindaceae	22.7	3.1	6.6	Sapindaceae	8,942	6.6	Sapotaceae	25
5	Sapotaceae	12.2	1.6	2.3	Ebenaceae	4,398	3.3	Sapindaceae	20
6	Rubiaceae	9.5	1.3	2.5	Guttiferae	3,497	2.6	Annonaceae	15
7	Apocynaceae	9.1	1.2	0.6	Rubiaceae	3,318	2.5	Meliaceae	15
8	Meliaceae	8.8	1.2	0.7	Sapotaceae	3,099	2.3	Ochnaceae	12
9	Irvingiaceae	8.5	1.1	0.2	Annonaceae	2,785	2.1	Sterculiaceae	11
10	Annonaceae	7.3	1.0	2.1	Flacourtiaceae	1,245	0.9	Flacourtiaceae	10

Notes: The top 10 families for trees ≥1 cm dbh are ranked in terms of basal area, number of individual trees, and number of species with the percentage of trees in the combined two plots at Lenda monodominant forest (20 ha). Data are from the first census.

Table 28.5. Ituri–Edoro (Mixed Forest) Rankings by Family

Rank	Family	Basal Area (m²)	% BA	% Trees	Family	Trees	% Trees	Family	Species
1	Leguminosae	278.0	42.4	15.6	Sterculiaceae	70,615	44.3	Euphorbiaceae	57
2	Euphorbiaceae	50.8	7.8	9.7	Leguminosae	24,829	15.6	Rubiaceae	40
3	Sterculiaceae	48.0	7.3	44.3	Euphorbiaceae	15,403	9.7	Leguminosae	29
4	Sapindaceae	35.3	5.4	8.6	Sapindaceae	13,715	8.6	Sapindaceae	22
5	Rubiaceae	31.7	4.8	2.8	Rubiaceae	4,472	2.8	Sapotaceae	21
6	Rutaceae	26.2	4.0	0.1	Annonaceae	4,449	2.8	Annonaceae	17
7	Ebenaceae	20.1	3.1	2.7	Ebenaceae	4,320	2.7	Meliaceae	15
8	Sapotaceae	16.4	2.5	1.3	Tiliaceae	3,442	2.2	Flacourtiaceae	11
9	Tiliaceae	16.3	2.5	2.2	Flacourtiaceae	3,406	2.1	Ochnaceae	11
10	Meliaceae	14.5	2.2	0.7	Sapotaceae	2,004	1.3	Sterculiaceae	10
								Moraceae	9

Notes: The top 10 families for trees ≥1 cm dbh are ranked in terms of basal area, number of individual trees, and number of species with the percentage of trees in the combined two plots at Edoro mixed forest (20 ha). Data are from the first census.

Fauna

Biogeographical evidence suggests that the Ituri forest was an important Pleistocene forest refuge and a center of dispersal for many current east and central African vertebrate taxa (reviews in Thomas 1991; Hart et al. 1996). Presently, the Ituri forest is thought to be Africa's most species-rich area in mammalian fauna (Grubb 1982). In addition to an important number of endemic mammalian species such as the okapi (a forest giraffe, *Okapia johnstoni*), the aquatic genet (*Osbornictis piscivora*), and the owl-faced monkey (*Cercopithecus hamlyni*),

Table 28.6. Ituri–Lenda (Monodominant Forest) Rankings by Genus

Rank	Genus	Basal Area (m²)	% BA	% Trees	Genus	Trees	% Trees	Genus	Species
1	*Gilbertiodendron* (Leguminosae)	469.0	63.2	7.6	*Scaphopetalum* (Sterculiaceae)	61,385	45.5	*Drypetes* (Euphorbiaceae)	10
2	*Julbernardia* (Leguminosae)	44.2	6.0	1.7	*Drypetes* (Euphorbiaceae)	11,088	8.2	*Ouratea* (Ochnaceae)	10
3	*Scaphopetalum* (Sterculiaceae)	32.3	4.3	45.5	*Gilbertiodendron* (Leguminosae)	10,294	7.6	*Psychotria* (Rubiaceae)	7
4	*Cynometra* (Leguminosae)	18.0	2.4	0.3	*Alchornea* (Euphorbiaceae)	7,570	5.6	*Chrysophyllim* (Sapotaceae)	6
5	*Pancovia* (Sapindaceae)	16.6	2.2	5.3	*Pancovia* (Sapindaceae)	7,117	5.3	*Cola* (Sterculiaceae)	6
6	*Drypetes* (Euphorbiaceae)	10.4	1.4	8.2	*Diospyros* (Ebenaceae)	4,398	3.3	*Trichilia* (Meliaceae)	6
7	*Alstonia* (Apocynaceae)	8.4	1.1	0.0	*Garcinia* (Guttiferae)	3,361	2.5	*Rothmannia* (Rubiaceae)	5
8	*Cola* (Sterculiaceae)	7.6	1.0	0.8	*Julbernardia* (Leguminosae)	2,321	1.7	*Chytranthus* (Sapindaceae)	5
9	*Diospyros* (Ebenaceae)	5.8	0.8	3.3	*Aidia* (Rubiaceae)	1,545	1.1	*Albizia* (Leguminosae)	4
10	*Zanthoxylum* (Rutaceae)	5.7	0.8	0.1	*Manilkara* (Sapotaceae)	1,533	1.1	*Beilschmiedia* (Lauraceae)	4
								Dialium (Leguminosae)	4
								Diospyros (Ebenaceae)	4
								Entandrophragma (Meliaceae)	4
								Garcinia (Guttiferae)	4
								Irvingia (Irvingiaceae)	4
								Rinorea (Violaceae)	4
								Rothmannia (Rubiaceae)	4

Notes: The top 10 tree genera for trees ≥1 cm dbh are ranked by basal area, number of individual trees, and number of species with the percentage of trees in the combined two plots in Lenda monodominant forest (20 ha). Data are from the first census.

the forest is home to nine antelope species—including bongo (*Tragelaphus eryceros*), sitatunga (*Tragelaphus spekei*), Bates pygmy antelope (*Neotragus batesi*), and six species of duiker (*Cephalophus* spp.)—elephant (*Loxodonta africana cyclotis*), forest buffalo (*Syncerus caffer nanus*), water chevrotain (*Hyemoschus aquaticus*), African golden cat (*Profelis aurata*), leopard (*Panthera pardus*), giant ground pangolin (*Manis gigantea*), two species of tree pangolin (*Manis tetradactyla* and *M. tricupsis*), giant forest genet (*Genetta victoriae*), bush pig (*Potamochoerus porcus*), giant forest hog (*Hylochoerus meinertzhageni*), tree hyrax (*Dendrohyrax arboreus*),

Table 28.7. Ituri–Edoro (Mixed Forest) Rankings by Genus

Rank	Genus	Basal Area (m²)	% BA	% Trees	Genus	Trees	% Trees	Genus	Species
1	Cynometra (Leguminosaeoideae)	150.2	22.9	2.9	Scaphopetalum (Sterculiaceae)	67,741	42.5	Drypetes (Euphorbiaceae)	10
2	Julbernardia (Leguminosae)	67.2	10.3	9.9	Julbernardia (Leguminosae)	15,847	9.9	Ouratea (Ochnaceae)	10
3	Scaphopetalum (Sterculiaceae)	32.3	4.9	42.5	Pancovia (Sapindaceae)	12,007	7.5	Beilschmiedia (Lauraceae)	6
4	Pancovia (Sapindaceae)	28.8	4.4	7.5	Drypetes (Euphorbiaceae)	7,631	4.8	Trichilia (Meliaceae)	6
5	Zanthoxylum (Rutaceae)	26.2	4.0	0.1	Alchornea (Euphorbiaceae)	6,005	3.8	Chrysophyllim (Sapotaceae)	5
6	Diospyros (Ebenaceae)	20.1	3.1	2.7	Cynometra (Leguminosae)	4,668	2.9	Cola (Sterculiaceae)	5
7	Gilbertiodendron (Leguminosae)	19.9	3.0	0.6	Diospyros (Ebenaceae)	4,320	2.7	Synsepalum (Sapotaceae)	5
8	Erythrophleum (Leguminosae)	17.5	2.7	0.1	Polyalthia (Annonaceae)	2,902	1.8	Celtis (Ulmaceae)	5
9	Hallea (Rubiaceae)	15.2	2.3	0.1	Dasylepis (Flacourtiaceae)	2,855	1.8	Chytranthus (Sapindaceae)	5
10	Cola (Sterculiaceae)	13.3	2.0	0.6	Desplatsia (Tiliaceae)	2,184	1.4	Dialium (Leguminosae)	5

Notes: Top 10 tree genera for trees ≥1 cm dbh are ranked by basal area, number of individual trees, and number of species with the percentage of trees in the combined two plots in Edoro mixed forest (20 ha). Data are from the first census.

at least four tree squirrels (*Paraxerus, Protoxerus, Funisciurus,* and *Heliosciurus*), and two species of flying squirrel (*Anomalurus* and *Idiurus*). There are many primate species in the area, 13 of which are anthropoid primates, including chimpanzee (*Pan troglodytes*), baboons (*Papio anubis*), blue monkeys (*Cercopithecus mitis*), red-tailed monkeys (*Cercopithecus ascanius*), mona monkeys (*Cercopithecus wolfi denti*), gray-checked mangabeys (*Cercocebus albigena*), agile mangabeys (*Cercocebus galeritus agilis*), red colobus (*Colobus badius*), Abyssian black and white colobus (*Colobus guereza*), and Angolan (*Colobus angolensis*). Total primate biomass is 710 kg/km² or 112 individuals/km². Overall primate density is quite low for Africa because of the relative scarcity of folivorous species and the low utilization of mbau forests by primates (Thomas 1991). Using transect survey data, researchers estimated the density of some of the mammalian species in the OFR (at a distance from human activity) as 0.79/km² for elephant, 0.89/km² for chimpanzee, and 0.48/km² for okapi (Hart and Hall 1996). Estimated densities have also been calculated for nine other ungulate species in the OFR (Hart 2001). The Ituri forest also hosts a large number of birds such as hornbills, turacos, flycatchers, bubuls, parrots, drongos, sunbirds, starlings, and warblers. At least 333 bird species are known to occur within the central sector of the OFR.

Table 28.8. Ituri–Lenda (Monodominant Forest) Rankings by Species

Rank	Species	Number Trees	% Trees	Species	Basal Area (m^2)	% BA	% Trees
1	*Scaphopetalum dewevrei* (Sterculiaceae)	61,385	45.5	*Gilbertiodendron dewevrei* (Leguminosae)	469.0	63.2	7.6
2	*Gilbertiodendron dewevrei* (Leguminosae)	10,294	7.6	*Julbernardia seretii* (Leguminosae)	44.2	6.0	1.7
3	*Drypetes bipindensis* (Euphorbiaceae)	8,128	6.0	*Scaphopetalum dewevrei* (Sterculiaceae)	32.3	4.4	45.5
4	*Alchornea floribunda* (Euphorbiaceae)	7,570	5.6	*Cynometra alexandri* (Leguminosae)	18.0	2.4	0.3
5	*Pancovia harmsiana* (Sapindaceae)	7,117	5.3	*Pancovia harmsiana* (Sapindaceae)	16.6	2.2	5.3
6	*Garcinia smeathmannii* (Guttiferae)	3,228	2.4	*Alstonia boonei* (Apocynaceae)	8.4	1.1	0.0
7	*Diospyros bipindensis* (Ebenaceae)	3,134	2.3	*Fagara macrocarpa** (Rutaceae)	5.7	0.8	0.1
8	*Julbernardia seretii* (Leguminosae)	2,321	1.7	*Cola lateritia* (Sterculiaceae)	5.5	0.7	0.3
9	*Aidia micrantha* (Rubiaceae)	1,545	1.1	*Drypetes bipindensis* (Euphorbiaceae)	5.4	0.7	6.0
10	*Polyalthia suaveolens* (Annonaceae)	1,428	1.1	*Cleistanthus michelsonii* (Euphorbiaceae)	5.1	0.7	0.5

*Genus *Fagara* now changed to *Zanthoxylum* according to Mabberley (1997).

Notes: The top 10 tree species for trees ≥1 cm dbh ranked by number and percentage of trees and basal area in the combined two plots of the Lenda monodominant forest (20 ha). Data are from the first census.

Natural Disturbances

Ituri, as a midcontinent forest, is not subject to monsoons. Windfalls, ranging from single treefalls to strings of blowdowns of several hectares, create forest gaps of various sizes in the Ituri forest (Hart 1985; Hart et al. 1989). Frequency of disturbance combined with different tree longevities result in differing stand half-lives for the two forest types, the monodominant forest having a particularly long half-life (Hart 2001). Elephants frequently maintain or enlarge small-scale forest disturbances. There is also a historical record of small-scale fire disturbance; however, the impact of fire is inadequately understood (Hart et al. 1996).

Human Disturbance

Agricultural activities and extraction of nontimber forest products occur to a limited extent in the Okapi Faunal Reserve. Measures are being taken to control poaching and selective logging, though the overall impact of the latter activity has been limited up to the present. Logging is occurring on the forest edge and along the major navigable rivers, but because of lack of transportation and continuing

Table 28.9. Ituri–Edoro (Mixed Forest) Rankings by Species

Rank	Species	Number Trees	% Trees	Species	Basal Area (m²)	% BA	% Trees
1	Scaphopetalum dewevrei (Sterculiaceae)	67,741	42.5	Cynometra alexandri (Leguminosae)	150.2	22.9	2.9
2	Julbernardia seretii (Leguminosae)	15,847	9.9	Julbernardia seretii (Leguminosae)	67.2	10.3	9.9
3	Pancovia harmsiana (Sapindaceae)	12,007	7.5	Scaphopetalum dewevrei (Sterculiaceae)	32.3	4.9	42.5
4	Alchornea floribunda (Euphorbiaceae)	6,005	3.8	Pancovia harmsiana (Sapindaceae)	28.8	4.4	7.5
5	Cynometra alexandri (Leguminosae)	4,668	2.9	Fagara macrocarpa* (Rutaceae)	24.7	3.8	0.1
6	Diospyros bipindensis (Ebenaceae)	4,316	2.7	Diospyros bipindensis (Ebenaceae)	20.1	3.1	2.7
7	Drypetes bipindensis (Euphorbiaceae)	4,282	2.7	Gilbertiodendron dewevrei (Leguminosae)	19.9	3.0	0.6
8	Polyalthia suaveolens (Annonaceae)	2,902	1.8	Erythrophloeum suaveolens (Leguminosae)	17.5	2.7	0.1
9	Dasylepis seretii (Flacourtiaceae)	2,855	1.8	Hallea stipulosa (Rubiaceae)	15.2	2.3	0.1
10	Leptonychia multiflora (Sterculiaceae)	1,832	1.2	Cleistanthus michelsonii (Euphorbiaceae)	12.1	1.9	0.5

*Genus *Fagara* now changed to *Zanthoxylum* according to Mabberley (1997).

Notes: The top 10 tree species for trees ≥1 cm dbh ranked by number and basal area of trees and basal area in the combined two plots of the Edoro mixed forest (20 ha). Data are from the first census.

civil strife, logging has not yet moved into the large blocks of eastern Congolese forest. This can be expected to change with peace and development.

Plot Size and Location

The Ituri study comprises four 10-ha plots, each 500 × 200 m. The two plots in the monodominant forest (Lenda-1 and Lenda-2) lie in a north-south line and are separated by 500 m. This configuration was replicated in the mixed forest (Edoro-1 and Edoro-2) with an east-west orientation (see plot layouts in Makana et al. 1998). Northwest corners of the four plots are as follows. Lenda-1 is north of Lenda-2: 1°19.134′N, 28°38.675′E and 1°18. 609′N, 28°38.670′E. Edoro-1 is west of Edoro-2: 1°33.741′N, 28°30.778′E and 1°33.739′N, 28°31.314′E.

Funding Sources

Funding for the Ituri Forest Dynamics Plots has been provided by the Wildlife Conservation Society and the Smithsonian Institution, with important contributions from the National Geographic Society and Conservation Food & Health Foundation.

References

Brady, N. C. 1990. *The Nature and Properties of Soils*. 10th edition. Macmillan, New York.

Conway, D. J. 1992. *A Comparison of Soil Parameters in Monodominant and Mixed Forest in Ituri Forest Reserve, Zaire*. Honors project. University of Aberdeen, Aberdeen, Scotland.

Grubb, P. 1982. Refuges and dispersal in the speciation of African forest mammals. Pages 537–53 in G. T. Prance, editor. *Biological Diversification in the Tropics*. Columbia University Press, New York.

Hart, J. A. 2001. Diversity and abundance in an African forest ungulate community and implications for conservation. Pages 183–206 in B. Weber, L. White, A. Vedder, and L. Naughton-Treves, editors. *African Rain Forest Ecology and Conservation*. Yale University Press, New Haven, CT.

Hart, J. A., and P. Carrick. 1996. Climate of the Reserve de Faune à Okapi: Rainfall and temperature in the Epulu sector 1986–1995. Unpublished CEFRECOF Working Paper N° 2, Kampala, Uganda.

Hart, J. A., and J. S. Hall. 1996. Status of eastern Zaire's forest parks and reserves. *Conservation Biology* 10:316–24.

Hart, T. B. 1985. *The Ecology of a Single-Species Dominant Forest and a Mixed Forest in Zaire, Equatorial Africa*. Ph.D. dissertation. Michigan State University, East Lansing, MI.

Hart, T. B. 1990. Monospecific dominance in tropical rain forests. *Trends in Ecology and Evolution* 5:6–11.

Hart, T. B. 1995. Seed, seedling and sub-canopy survival in monodominant and mixed forests of the Ituri Forest, Africa. *Journal of Tropical Ecology* 11:443–59.

Hart, T. B. 2001. Forest dynamics in the Ituri Basin (DR Congo). Dominance, diversity and conservation. Pages 154–64 in B. Weber, L. White, A. Vedder, and L. Naughton-Treves, editors. *African Rain Forest Ecology and Conservation*. Yale University Press, New Haven, CT.

Hart, T. B., and J. A. Hart. 1995. The Ituri forest large plot project. *Inside CTFS*. Center for Tropical Forest Science, Washington, DC.

Hart, T. B., and J.-R. Makana. 1999. Dealing with rebels and digitizing maps at the Ituri Forest Dynamics Plots. *Inside CTFS*. Center for Tropical Forest Science, Washington, DC.

Hart, T. B., J. A. Hart, and P. G. Murphy. 1989. Monodominant and species rich forests of the humid tropics: Causes of their co-occurrence. *American Naturalist* 133:613–33.

Hart, T. B., J. A. Hart, R. Dechamps, M. Fournier, and M. Ataholo. 1996. Changes in forest composition over the last 4000 years in the Ituri Basin, Zaire. Pages 545–63 in L. J. G. van der Maesen, X. M. van der Burgt, and J. M. Medenbach de Rooy, editors. *The Biodiversity of African Plants, Proceedings VIVth AETFAT Congress*. Kluwer Academic Publishers, Dordrecht, Netherlands.

Laveau, J. 1982. *Étude Géologique du Haut-Zaïre. Genèse et Évolution d'un Segment Lithosphèrique Archéen*. Musée Royale de l'Afrique Centrale, Annales, série n° 8, Sciences Géologiques 88, Tervuren, Belgium.

Mabberley, D. J. 1997. *The Plant-Book: A Portable Dictionary of the Vascular Plants*. Cambridge University Press, Cambridge, U.K.

Makana, J.-R. 1999. *Forest Structure, Species Diversity and Spatial Patterns of Trees in Monodominant and Mixed Stands in the Ituri Forest, Democratic Republic of Congo.* Masters thesis. Oregon State University, Corvallis, OR.

Makana, J.-R., and S. C. Thomas. 2002. Analysis of commercial timber species in the paleotropics. *Inside CTFS.* Center for Tropical Forest Science, Washington, DC.

Makana, J.-R., T. B. Hart, and J. A. Hart. 1998. Forest structure and diversity of lianas and understory treelets in monodominant and mixed stands in the Ituri Forest, Democratic Republic of the Congo. Pages 429–46 in F. Dallmeier and J. A. Comiskey, editors. *Forest Biodiversity Research, Monitoring and Modeling: Conceptual Background and Old World Case Studies.* UNESCO's Man and the Biosphere Series, Vol. 20. Parthenon Publishing, Pearl River, New York.

Thomas, S. C. 1991. Population densities and patterns of habitat use among anthropoid primates of the Ituri Forest, Zaire. *Biotropica* 23:68–83.

Torti, S. D. 1998. *Causes and Consequences of Monodominance in Tropical Lowland Forests.* Ph.D. dissertation. University of Utah, Salt Lake City, UT.

Torti, S. D., and P. D. Coley. 1999. Tropical monodominance: A preliminary test of ectomycorrhizal hypothesis. *Biotropica* 31:220–28.

Torti, S. D., P. D. Coley, and T. A. Kursar. 2001. Causes and consequences of monodominance in tropical lowland forests. *American Naturalist* 157:141–53.

29

Korup Forest Dynamics Plot, Cameroon

George B. Chuyong, Richard Condit, David Kenfack, Elizabeth C. Losos,
Sainge Nsanyi Moses, Nicholas C. Songwe, and Duncan W. Thomas

Site Location, Administration, and Scientific Infrastructure

Korup National Park was created in 1986 and covers 125,900 hectares, most of
which is evergreen forest. The park includes the former Korup Forest Reserve,
established under British mandate in the 1930s. It is now under the administration
of the Department of Wildlife and Protected Areas in the Ministry of Environment
and Forests (MINEF). The 50-ha Korup Forest Dynamics Plot (KFDP) is located
near the southern end of the park (fig. 29.1). The plot is approximately 60 km
inland from the open Atlantic Ocean in the Bight of Biafra, and about 30 km from
the edge of the mangrove swamps of the Rio Del Rey estuary. It is also 10 km from
the Cameroon–Nigeria border at its closest point.

Both the Korup National Park and the KFDP are administered from the isolated town of Mundemba, which can be reached from Yaoundé (the capital of
Cameroon) or Douala by four-wheel-drive vehicle or by light aircraft. The KFDP
program maintains an office, residence, and limited visitor accommodations in
Mundemba, supervised by a few permanent field staff. The park headquarters
has two small, unequipped laboratories. Travel from Mundemba to the 50-ha
plot requires a 10-km drive on plantation roads along the edge of the park and
a 10-km hike through the park. The research camp, "Chimpanzee Camp," is less
than 1 km from the KFDP and has simple visitor accommodations.

The KFDP program is run by the Center for Tropical Forest Science of the
Smithsonian Tropical Research Institute with the Bioresources Development and
Conservation Programme–Cameroon (BDCPC). The KFDP program also has
a research agreement with the Herbarium at the Limbe Botanical Garden, a
4–5 hour drive from Mundemba in the dry season. Limbe is a coastal town,
about 2 hours drive west of Douala International Airport, and is a good place
to start an expedition to Korup. Travel during the latter part of the wet season
(July–November) is very difficult.

In addition to the KFDP, the Smithsonian Institution/Monitoring and Assessing Biodiversity Program has established a series of 1-ha forest monitoring plots
elsewhere in Cameroon, in the forest reserves of Ejagham and Takamanda and in
the Campo National Park.

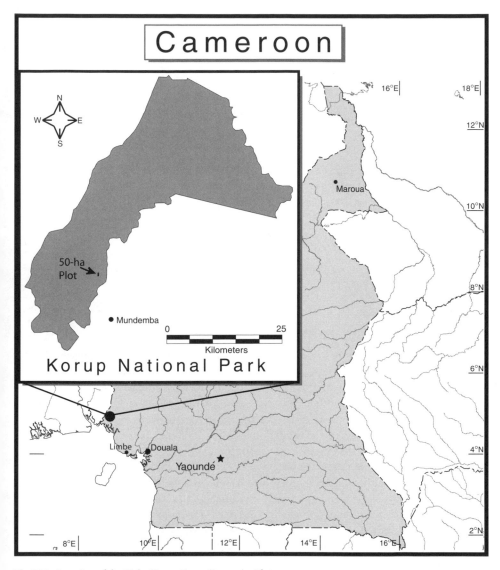

Fig. 29.1. Location of the 50-ha Korup Forest Dynamics Plot.

Climate

Although it is only 5 degrees north of the equator, Korup has a pseudoequatorial climate, with two seasons instead of four. Over a 22-year period, from 1973 to 1994, annual rainfall near the southern (coastal) end of the park, about 15 km

Table 29.1. Korup Climate Data

	Jan	Feb	Mar	Apr	May	Jun	Jul	Aug	Sep	Oct	Nov	Dec	Total/ Averages
Rain (mm)	38	58	221	329	459	564	913	914	691	668	341	76	5272
ADTMx	31.6	32.8	32.6	32.0	31.5	30.4	28.7	27.8	28.8	29.7	30.7	31.1	30.6
ADTMn	21.9	23.0	23.1	23.0	22.8	22.8	22.2	22.4	22.3	22.5	22.8	23.1	22.7

Notes: Mean monthly rainfall and average daily temperature based on data collected about 20 km from the plot from 1973 to 1994.

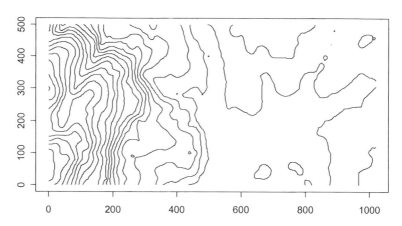

Fig. 29.2. Topographic map of the 50-ha Korup Forest Dynamics Plot with 5-m contour intervals.

south of the plot, ranged from 4027 mm to 6368 mm, and averaged 5272 mm. There is a distinct dry season (average monthly rainfall less than 100 mm) from December to February and a long and intense wet season from approximately May to October, with the heaviest rainfall (exceeding 1000 mm a month in some years) in August. Temperatures vary little throughout the year. The mean monthly maximum temperature in the dry season is 31.8°C and in the wet season 30.2°C. The average annual maximum temperature for the area is 30.6°C (table 29.1). Solar radiation also varies little, ranging from 199 to 248 W/m[2] (Newbery et al. 1998). Many storm fronts move through Korup, especially at the beginning (March, April) and end (September, October) of the rainy season, when the intertropical convergence zone passes. Treefalls in the forest mostly occur during this period.

Topography and Soil

Elevation in Korup National Park ranges from near sea level along the Atlantic coast to 1079 m above sea level (asl) on Mount Yuhan in the hilly center of the park. The KFDP is located on the southern edge of the hills around Yuhan, where

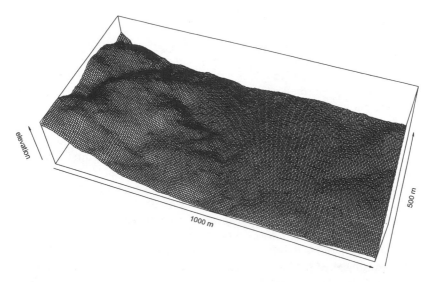

Fig. 29.3. Perspective map of the 50-ha Korup Forest Dynamics Plot.

the terrain starts to slope gently down toward the mangrove swamps of the Rio Del Rey, 30 km to the south. The lowest point in the KFDP is about 150 m asl, with a range of about 90 m between the highest and the lowest points (figs. 29.2 and 29.3).

The plot occupies part of a narrow, flat-bottomed valley surrounded by steep hillsides. It is fairly flat at the low-lying south end and rises steeply within the northern third of the plot. Habitat types include swamp, low-lying gentle slopes, and moderate to steep hillsides with boulders and small rock outcrops. A valley runs east-west through the central portion of the plot with a permanent stream. There are several other permanent and seasonal streams in the plot.

Soils in the southern part of Korup are generally skeletal, sandy (up to 70% sand in some areas), and very nutrient poor through the leaching action of the very high rainfall, which removes nutrients and clay particles. Most of the organic material is in the top few centimeters of the soil profile. As soil depth increases, there are small, but increasing, amounts of clay (Newbery et al. 1997). Soils on the steeper slopes tend to be thin and stony but still very nutrient poor. The soils are mostly derived in situ through the weathering of the metamorphic bedrock. Exposed rocks in the plot have been identified as syenite porphyry, a base-rich form of granite. Along the valley bottom there are swamps on alluvium, with sandy and silty soils, the latter hydric and gleyed.

Forest Type and Characteristics

In the Pleistocene era, the forests of the Korup area formed part of the Cross River–Mayombe Refugium (Maley 1987) that extended, probably discontinuously, along the West African coast from southeastern Nigeria to the Congo Republic. At present, the Korup forests appear very ancient and rich in paleoendemics, though with some puzzling indicators of old disturbance. The forest in the vicinity of the KFDP has been classified as Biafran coastal forest by Letouzey (1968, 1985). Thomas (1996) modified Letouzey's classification and described the coastal forests of south Korup and the southwest part of Mount Cameroon as *Oubanguia alata* (Scytopetalaceae) forests, a vegetation type dominated by this species where the annual rainfall exceeds 4000 mm. This lowland evergreen forest is locally dominated by Leguminosae-Caesalpinioideae that sometimes occur in groves of codominant species: *Didelotia letouzeyi* Pellegr., the rare *Microberlinia bisulcata* A.Chev., *Tetraberlinia bifoliolata* (Harms) Hauman, and the newly described narrow endemic, *T. korupensis* Wieringa.

Floristically, the KFDP is very rich and has many rare species for Cameroon, including some of the most abundant species in the plot. *Oubanguia alata* Bak.f., the species with the highest basal area, is a near-endemic, mostly limited to the wet forests of the Korup–Mount Cameroon area. The plot's most common large tree and the species with second-highest basal area, *Lecomtedoxa klaineana* (Pierre ex Engl.) Dubard (Sapotaceae) is rare in Cameroon and limited to the coastal forest of Cameroon and Gabon. It occurs gregariously in nutrient-poor lowland forest, but lacks ectomycorrhizae. The two most abundant species, *Phyllobotryon spathulatum* Müll. Arg. (Flacourtiaceae) and *Cola semecarpophylla* K. Schum. (Sterculiaceae) are also Lower Guinea endemics with rather limited distribution. *Phyllobotryon* is especially unusual since the flowers are borne on the leaves. By contrast, *Dichostemma glaucescens* Pierre (Euphorbiaceae), third in both abundance and basal area, is widespread and often dominates the wetter forests throughout Lower Guinea. New species are continuously being described from the Korup area (Thomas 1986; Thomas and Gereau 1993; Thomas and Harris 1999; Gereau and Kenfack 2000; Sonké et al. 2002). For census data and rankings, see tables 29.2–29.6.

A more or less continuous canopy layer reaches about 15 to 25 m, with scattered emergents up to 50 m tall. The woody understory is fairly dense with both lianas

Table 29.2. Korup Plot Census History

Year	Dates	Number Trees (≥1 cm dbh)	Number Species (≥1 cm dbh)	Number Trees (≥10 cm dbh)	Number Species (≥10 cm dbh)
First	January 1997–July 1999	329,026	494	24,591	307

Notes: One census has been completed, the second census is scheduled to start in November 2004.

Table 29.3. Korup Summary Tally

Size Class	Average per Hectare						50-ha Plot				
(cm dbh)	BA	N	S	G	F	H'	α	S	G	F	H'	α
≥1	32.0	6581	236	143	45	1.75	48.0	494	235	62	1.93	57.0
≥10	26.1	492	87	66	30	1.48	30.8	307	179	53	1.73	49.4
≥30	16.1	84	35	32	20	1.29	24.0	192	129	43	1.69	41.5
≥60	6.8	11	8	8	7	0.79	21.6*	85	74	32	1.58	28.6

*Fisher's alpha based on 42 hectares, sample size for remaining 8 ha was too low to calculate alpha.

Notes: BA represents basal area in m^2, N is the number of individual trees, S is number of species, G is number of genera, F is number of families, H' is Shannon–Wiener diversity index using log_{10}, and α is Fisher's α. Basal area includes all multiple stems for each individual. 640 individuals were not identified to species or morphospecies. Data are from the first census.

Table 29.4. Korup Rankings by Family

Rank	Family	Basal Area (m^2)	% BA	% Trees	Family	Trees	% Trees	Family	Species
1	Euphorbiaceae	256.6	16.1	13.8	Sterculiaceae	72,262	22.0	Rubiaceae	86
2	Scytopetalaceae	227.1	14.3	4.7	Euphorbiaceae	45,296	13.8	Leguminosae	38
3	Leguminosae	143.6	9.0	5.9	Violaceae	31,386	9.6	Euphorbiaceae	37
4	Sterculiaceae	139.3	8.7	22.0	Flacourtiaceae	31,077	9.5	Sterculiaceae	28
5	Sapotaceae	106.9	6.7	0.4	Rubiaceae	22,945	7.0	Annonaceae	22
6	Olacaceae	104.9	6.6	4.5	Leguminosae	19,378	5.9	Sapindaceae	18
7	Ebenaceae	57.2	3.6	5.8	Ebenaceae	19,139	5.8	Guttiferae	16
8	Flacourtiaceae	53.4	3.4	9.5	Scytopetalaceae	15,551	4.7	Anacardiaceae	15
9	Apocynaceae	45.4	2.9	1.9	Olacaceae	14,657	4.5	Ebenaceae	14
10	Irvingiaceae	41.9	2.6	0.1	Annonaceae	10,705	3.3	Violaceae	14

Notes: The top 10 families for trees ≥1 cm dbh are ranked in terms of basal area, number of individual trees, and number of species, with the percentage of trees in the plot. Data are from the first census.

and small trees. The herbaceous layer, however, is sparse outside of light gaps. One unusual characteristic of the wet forests of Korup is an abundance of small, unbranched trees with terminal rosettes of large leaves (litter-trap treelets), from at least five dicotyledon families.

Small litterfall is estimated at 8.7 Mg/ha/year, of which close to 60% is leaf litter (Chuyong et al. 2000). Leaf litter production is continuous but peaks during the dry season. Litter breakdown and mineralization are very rapid, particularly during the first 3 months of the rainy season.

In terms of phenology, the flowering and fruiting of Korup trees is strongly seasonal. Most species flower from January to July, with a flowering peak at the beginning of the wet season, around March–May, followed by the peak fruiting season. Most species flower and set fruits to varying degrees in most years but there is a marked tendency toward mast fruiting at greater than 1-year intervals. The gregarious Caesalpiniodeae of Korup have been shown to have a 2–3 year mast fruiting pattern (Newbery et al. 1998).

Table 29.5. Korup Rankings by Genus

Rank	Genus	Basal Area (m²)	% BA	% Trees	Genus	Trees	% Trees	Genus	Species
1	*Oubanguia* (Scytopetalaceae)	218.9	13.8	4.6	*Cola* (Sterculiaceae)	70,254	21.7	*Cola* (Sterculiaceae)	23
2	*Cola* (Sterculiaceae)	137.8	8.7	21.7	*Rinorea* (Violaceae)	30,504	9.4	*Diospyros* (Ebenaceae)	14
3	*Lecomtedoxa* (Sapotaceae)	99.3	6.3	0.1	*Phyllobotryon* (Flacourtiaceae)	26,736	8.3	*Rinorea* (Violaceae)	13
4	*Dichostemma* (Euphorbiaceae)	78.8	5.0	5.3	*Diospyros* (Ebenaceae)	19,139	5.9	*Psychotria* (Rubiaceae)	12
5	*Protomegabaria* (Euphorbiaceae)	73.2	4.6	1.0	*Dichostemma* (Euphorbiaceae)	17,251	5.3	*Garcinia* (Guttiferae)	10
6	*Strombosia* (Olacaceae)	64.9	4.1	2.4	*Oubanguia* (Scytopetalaceae)	15,011	4.6	*Trichoscypha* (Anacardiaceae)	10
7	*Diospyros* (Ebenaceae)	57.2	3.6	5.9	*Strombosia* (Olacaceae)	7,876	2.4	*Beilschmiedia* (Lauraceae)	7
8	*Hymenostegia* (Leguminosae)	34.1	2.2	1.3	*Drypetes* (Euphorbiaceae)	7,576	2.3	*Drypetes* (Euphorbiaceae)	6
9	*Vitex* (Labiatae)	31.5	2.0	0.1	*Angylocalyx* (Leguminosae)	5,853	1.8	*Ouratea* (Ochnaceae)	6
10	*Klaineanthus* (Euphorbiaceae)	30.9	2.0	0.6	*Tabernaemontana* (Apocynaceae)	4,256	1.3	*Vitex* (Labiatae)	6

Notes: The top 10 tree genera for trees ≥1 cm dbh are ranked by basal area, number of individual trees, and number of species with the percentage of trees in the plot. Data are from the first census.

Fauna

The forest around the plot is an important habitat for medium- and large-sized mammals. At least 326 species of birds are found in and around the Korup National Park (Thomas 1992, Rodewald 1993), along with over 40 species of terrestrial mammals—including eight diurnal primate species—and more than 200 species of macrofungi. A recent survey conducted by Larsen (1997) indicates that the Korup–Oban Hills area harbors more than 1000 of tropical Africa's 3700 butterfly species. The short rivers, characteristic of the Korup area, support endemic species of fish (Reid 1989; Forbin 1996).

The most common large mammals in the plot are several species of duiker (forest antelope), especially *Cephalophus monticola* and *C. dorsalis*, and various primates, including four species of guenon (*Cercopithecus spp.*), mangabeys (*Cercocebus torquatus*), red colobus (*Colobus badius*), and at least five species of nocturnal primates. Frequent visitors are drills (*Mandrillus leucophaeus*) and bush pigs (*Potamochoerus porcus*). Forest elephants (*Loxodonta africana cyclotis*) and chimpanzees (*Pan troglodytes*) are seen occasionally. Squirrels, flying squirrels, porcupines, hyraxes, and pangolins are all common. The many bat species include large fruit bats, which are important seed dispersers, and small nectar-eating bats.

Table 29.6. Korup Rankings by Species

Rank	Species	Number Trees	% Trees	Species	Basal Area (m²)	% BA	% Trees
1	*Phyllobotryon spathulatum* (Flacourtiaceae)	26,728	8.1	*Oubanguia alata* (Scytopetalaceae)	218.6	13.7	4.5
2	*Cola semecarpophylla* (Sterculiaceae)	24,518	7.5	*Lecomtedoxa klaineana* (Sapotaceae)	99.3	6.2	0.1
3	*Dichostemma glaucescens* (Euphorbiaceae)	17,251	5.3	*Dichostemma glaucescens* (Euphorbiaceae)	78.8	5.0	5.3
4	*Cola praeacuta* (Sterculiaceae)	15,471	4.7	*Protomegabaria stapfiana* (Euphorbiaceae)	73.2	4.6	1.0
5	*Oubanguia alata* (Scytopetalaceae)	14,918	4.5	*Cola praeacuta* (Sterculiaceae)	34.9	2.2	4.7
6	*Cola sp.nov.* (Steculiaceae)	12,366	3.8	*Strombosia pustulata* (Olacaceae)	34.4	2.2	1.3
7	*Cola flavo-velutina* (Sterculiaceae)	8,234	2.5	*Cola semecarpophylla* (Sterculiaceae)	33.3	2.1	7.5
8	*Diospyros preussii* (Ebenaceae)	7,356	2.2	*Klaineanthus gaboniae* (Euphorbiaceae)	30.9	1.9	0.6
9	*Angylocalyx oligophyllus.* (Leguminosae)	5,796	1.8	*Hymenostegia afzelii* (Leguminosae)	27.7	1.7	1.2
10	*Rinorea lepidobotrys* (Violaceae)	5,492	1.7	*Diospyros gabunensis* (Ebenaceae)	27.6	1.7	1.2

Notes: The top 10 tree species for trees ≥1 cm dbh are ranked by number of trees and basal area. The percentage of the total population is also shown. Data are from the first census.

The large boulders and caves in the plot area are important bat roosting habitat, and also serve as nest sites for the rare bird, *Picathartes oreas.* Forest hingeback turtles (*Kinyxis)* are common. Large snakes include Gabon and rhinoceros vipers (*Bitis gabonica* and *B. nasicornis*) and black cobras (*Naja melanoleuca*). The dwarf crocodile (*Ostreolaenus tetraspis*), listed by IUCN as vulnerable, is common in the creeks of Korup and may be present in the plot.

According to Thomas (1992), the most common birds in the understory include species that follow army ant swarms such as the fire-crest alethe (*Alethe diademata*) and the brown-chested alethe (*A. poliocephala*), also sunbirds, greenbuls, forest robins (*Stiphrornis erthrothorax*), and paradise flycatchers (*Terpsiphone rufiventer*). Hornbills of several species are common in the canopy and are important dispersers of canopy fruits. Rare rainforest birds in the plot area include the white-crested tiger bittern (*Tigriornis leucolophus*) found along forest creeks, the crowned hawk-eagle (*Staphanoaetus coronatus*) that hunts monkeys, Sjostedt's barred owlet (*Glaucidium sjoestedti*) that hunts during the day, the white-crested hornbill (*Tockus albocristatus*) that follows monkey troops, the rarely seen black-eared groundthrush (*Turdus cameronensis*), and the IUCN-listed red-headed rockfowl (*Picathartes oreas*). Nine species of kingfisher occur in the forest.

Natural Disturbance

Although the Korup forest is slow-growing with a slow turnover rate of trees, small-scale natural disturbances such as branchfalls and treefalls are frequent. More rarely, windthrows flatten larger areas, up to 1 ha. The origin of these strong winds is something of a mystery since hurricanes do not occur in the area; the events appear to be linked to thunderstorms. There is also evidence of a long-term change in the climate of the area, as demonstrated by the scattered presence of large old trees of species normally found in drier forest. These species have little or no regeneration in the evergreen forest. According to Maley (1987), there is evidence in the pollen record of a drier period about 2000 years ago, so these species may be left over from that event. The forest does not burn naturally, except on rocky areas with very thin soils.

Human Disturbance

Small widely separated villages, with their shifting agricultural fields and small plantations, occur within and around the Korup forest. There has been no commercial logging in the park, only logging for local use and the harvesting of natural plant products. Some plants, notably the Yoruba chewing stick, *Massularia acuminata* (Rubiaceae), and the Hausa stick, *Carpolobia* spp. (Polygalaceae), are overharvested to the north of Korup, but not in the Forest Dynamics Plot where they are less protected and abundant. Use of the plants of the Korup area has been studied by the Korup Project (Thomas et al. 1989). Patches of secondary forest on long-abandoned fields are encountered sporadically, though not in the plot. Apparently, the human population in the area was historically at low density except along the coast, probably because of the very dense vegetation and the very low soil fertility. In the recent past, hunting of large mammals by shotgun was intense. Hunters with shotguns came from the nearby oil palm plantations, and there were a few elephant hunters with heavier caliber weapons. During this period, mammal populations declined drastically, and elephants were more or less extirpated. Since the creation of the national park, hunting has declined to moderate levels as a result of law enforcement, and animal populations are recovering. Elephants have started to visit the Korup Forest Dynamics Plot again in the last few years. However, because of the exceptionally low productivity of the forest, it is likely that mammal biomass is naturally relatively low compared to drier forests.

Plot Dimensions

Korup is a 50-ha, 1000 × 500 m plot; its long axis lies north-south (magnetic north). Post 1800 (plot coordinate: 560, 0), which is located 560 m to the south of the plot's northwest corner (plot coordinate: 0, 0), is at 05°03'86"N and 08°51'17"E.

Funding Sources

The census of the Korup Forest Dynamics Plot has been primarily funded by award number TW327 of the International Cooperative Biodiversity Groups (a consortium of the U.S. National Institutes of Health, U.S. National Science Foundation, and the U.S. Department of Agriculture), with supplemental funding contributed by the Central Africa Regional Program for the Environment (a program of the U.S. Agency for International Development and the Celerity Foundation).

References

Chuyong, G. B., D. M. Newbery, and N. C. Songwe. 2000. Litter nutrients and retranslocation in a central African rain forest dominated by ectomycorrhizal trees. *New Phytologist* 148:493–510.

———. 2002. Litter breakdown and mineralization in a central African rain forest dominated by ectomycorrhizal trees. *Biogeochemistry* 61:73–94.

Courade, G. 1974. *Cameroon Regional Atlas, West 1.* ORSTOM, Paris.

Forbin, I. 1996. Fish biodiversity of lower Korup Basin with emphasis on the development of ornamental fish trade, aquaculture and capture fishery. Consultancy report to the Korup Project, Mundemba, Ndian South West Province, Cameroon.

Gartlan, J. S., D. M. Newbery, D. W. Thomas, and P. G. Waterman. 1986. Studies on the rain forest vegetation of Cameroon. 1: The role of phosphorus in species distribution in Korup Forest Reserve. *Vegetatio* 65:131–48.

Gereau, R. E., and D. Kenfack. 2000. Le genre *Uvariopsis* (Annonaceae) en Afrique tropicale, avec la description d'une espèce nouvelle du Cameroun. *Adansonia* 22(1):39–43

Hawkins, P., and M. Brunt. 1965. Soils and ecology of west Cameroon. A report to the government of West Cameroon. Project Ca/TE/LA, Food and Agriculture Organization, Rome, Italy.

Kingdon, J. 1997. *The Kingdon Field Guide to African Mammals.* Academic Press, London.

Larsen, T. 1997. Butterfly inventory in Korup National Park, Rumpi Hills Forest Reserve and surrounding areas. Interim report to the Korup Project, Mundemba, Cameroon.

Letouzey, R. 1968. *Étude Phytogéographique du Cameroun.* Lechevalier, Paris.

———. 1985. *Carte Phytogéographique du Cameroun, 1:500 000.* Institut de la Recherche Agronomique. Toulouse, France.

Mabberley, D. J. 1997. *The Plant-Book: A Portable Dictionary of the Vascular Plants.* Cambridge University Press, Cambridge, U.K.

Maley, J. 1987. Fragmentation de la forêt dense humide ouest-africaine et extension des biotopes montagnards au quaternaire récent: nouvelles données polliniques et chronologiques: implications paléoclimatiques et biogéographiques. *Palaeoecology of Africa* 18:307–34.

Moyersoen, B., A. H. Fitter, and I. J. Alexander. 1998. Spatial distribution of ectomycorrhizas and arbuscular mycorrhizas in Korup National Park rain forest, Cameroon, in relation to edaphic parameters. *New Phytologist* 139:311–20.

Newbery, D. M., and J. S. Gartlan. 1996. The structural analysis of the rainforest of Korup and Douala-Edea. *Proceedings of the Royal Society of Edinburgh* 1048:177–224.

Newbery, D. M., I. J. Alexander, and J. A. Rother. 1997. Phosphorous dynamics in a lowland African rainforest: the influence of ecto-mycorrhizal trees. *Ecological Monographs* 67:367–409.

Newbery, D. M., N. C. Songwe, and G. B. Chuyong. 1998. Phenology and dynamics of an African rain forest at Korup, Cameroon. Pages 267–308 in D. M. Newbery, H. H. T. Prins, N. D. Brown, editors. *Dynamics of Ecological Communities.* Blackwell Science, Oxford, U.K.

Newbery, D. M., G. B. Chuyong, J. J. Green, N. C. Songwe, F. Tchuenteu, and L. Zimmermann. 2002. Does low phosphorus supply limit seedling establishment and tree growth in groves of ectomycorrhizal trees in a central African rainforest? *New Phytologist* 156:297–311.

Reid, G. M. 1989. *Living Waters of Korup Rainforest: A Hydrobiological Survey Report and Recommendations with Emphasis on Fish and Fisheries.* WWF Report No. 2306/A8:1. Korup Project, Mundemba, Cameroon.

Rodewald, P. G. 1993. *An Annotated Checklist of Birds of Korup National Park and the Project Area.* Korup Research Project No. 3. Wildlife Conservation Society, New York.

Sonké, B., D. Kenfack, and E. Robbrecht. 2002. A new species of the *Tricalysia atherura* group (Rubiaceae) from southwestern Cameroon. *Adansonia* 24:173–77.

Thomas, J. 1992. Birds of the Korup National Park, Cameroon. *Malimbus* 13:11–23.

Thomas, D. W. 1986. Notes on *Deinbollia* from Cameroon. *Annals of the Missouri Botanical Garden* 73:219–21.

Thomas, D. W. 1996. Botanical Survey of the Rumpi Hills and Nta Ali. Report to the GTZ, Germany, and to the Korup Project, Mundemba, Cameroon.

Thomas, D. W., and R. E. Gereau. 1993. *Ancistrocladus korupensis* (Ancistrocladaceae): A new liana form Cameroon. *Novon* 3:494–98.

Thomas, D. W., and D. J. Harris. 1999. New Sapindaceae from Cameroon and Nigeria. *Kew Bulletin* 54:951–57.

Thomas, D. W., D. Kenfack, G. B. Chuyong, S. N. Moses, E. C. Losos, R. S. Condit, and N. C. Songwe. 2003. Tree Species of Southwestern Cameroon: Tree Distribution Maps, Diamter Tables, and Species Documentation of the 50-Hecatre Korup Forest Dynamics Plot. Center for Tropical Forest Science of the Smithsonian Tropical Research Institute and Bioresources Development and Conservation Programme-Cameroon, Washington, D.C.

Thomas, D. W., J. M. Thomas, W. A. Bromley, and Mbenkum Fonki Tobias. 1989. *Korup Ethnobotany Survey.* World Wide Fund for Nature, Godalming, U.K.

30

La Planada Forest Dynamics Plot, Colombia

Martha Isabel Vallejo, Cristián Samper, Humberto Mendoza, and Joel Tupac Otero

Site Location, Administration, and Scientific Infrastructure

La Planada Nature Reserve is located within cloud forests on the western slope of the Andes in the Department of Nariño ($1°17'N$, $78°15'W$), near the Colombian border with Ecuador (fig. 30.1). The 3200-ha reserve was established in 1982 through a collaborative agreement between Fundación para la Educación Superior and the World Wildlife Fund.

La Planada is easily accessible along the road that connects Pasto, the capital of the Department of Nariño, and Tumaco, an important port on the Pacific Ocean. La Planada is located halfway between these towns, just 5 km from the small town of San Isidro. An administrative area inside the reserve contains an administrative center, laboratory for students and researchers, dormitories, tourist center, restaurant, small library, and reference collection of local plant and animals. La Planada also has telephone connections and electricity from a generator.

The 25-ha La Planada Forest Dynamics Plot is located within the reserve, approximately a 1-hour walk from the administrative center. The plot was established by the Instituto de Investigación de Recursos Biológicos Alexander von Humboldt, a recently created national biodiversity research institute set up by public and private organizations in Colombia. While the 25-ha Forest Dynamics Plot is the only tree census plot in the reserve at present, scientists are exploring the possibility of establishing another permanent plot for comparative purposes.

Climate

La Planada Forest Dynamics Plot receives moisture-laden winds from the Pacific coast. A dominant feature of montane climates on the Pacific slopes of the Andes is the occurrence of periods of low cloud cover (Grubb and Whitmore 1966). The dense fog causes a marked decrease in temperature, which greatly reduces the evaporation rates, so that high levels of humidity are maintained. Studies over a 17-year period (1986–2002) indicate that the average annual rainfall is 4087 mm, ranging from 3191 mm to 5315 mm. Rainfall is high throughout the year; however, a marked drier period occurs from the end of June to at the beginning of

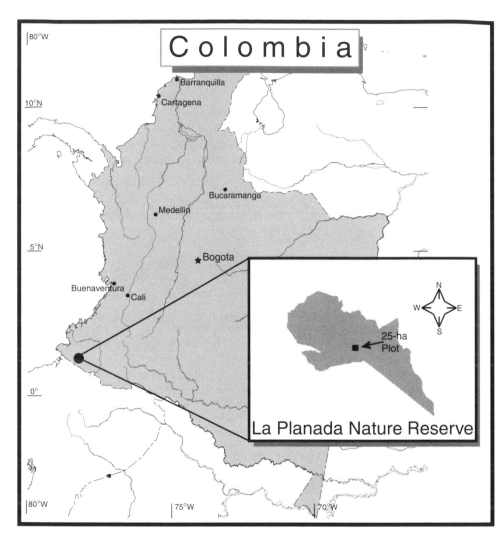

Fig. 30.1. Location of the 25-ha La Planada Forest Dynamics Plot.

September, though never less than 100 mm in a month. At 1800 m elevation, the average temperature is 19°C and the maximum 24.5°C. See table 30.1.

Topography and Soil

The reserve covers an altitudinal range of 1300–2100 m above sea level. La Planada Forest Dynamics Plot was established in a relatively flat area, though the terrain

Table 30.1. La Planada Climate Data

	Jan	Feb	Mar	Apr	May	Jun	Jul	Aug	Sep	Oct	Nov	Dec	Total/ Averages
Rain (mm)	493	388	431	479	411	268	150	160	278	455	447	455	4415
ADTMx (°C)	22.8	23.2	24.1	23.8	24.1	23.8	24.4	25.0	24.2	24.2	23.8	22.8	23.8
ADTMn (°C)	13.6	13.3	13.3	13.4	13.2	12.8	11.5	11.6	12.5	13	13	13.3	12.9

Notes: Mean monthly rainfall and average daily temperature are based on data measured at the La Planada Biological Station during 1986–2002.

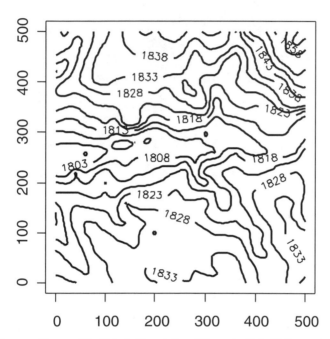

Fig. 30.2. Topographic map of the 25-ha La Planada Forest Dynamics Plot with 5-m contour intervals.

becomes steep near the El Oso and La Calladita streams. The plot's elevation ranges from 1796 to 1891 m above sea level, although the majority lies on a large plateau at 1800 m (figs. 30.2 and 30.3). The plot is crossed from east to west by the El Tejón stream, which receives the waters of four other minor streams (El Oso, La Calladita, Quebrada Juntas, and El Mar). These streams are permanent throughout the year, but there are also several ephemeral underground drainages dependent on factors such as rainfall, erosion, and treefalls.

The predominant soils on the plot are Andisols; these soils were developed from different ranges of volcanic ash in the Pleistocene. The parental material

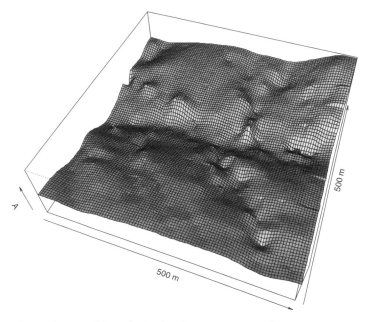

Fig. 30.3. Perspective map of the 25-ha La Planada Forest Dynamics Plot.

of Andisols is piroclastic (geological material of volcanic origin formed by fragments of different rocks), which is variable in composition, grain structure, and
thickness, and strongly influenced by environmental conditions. These soils are
frequently found in the Andean region and are characterized by slow decomposition of organic matter.

Physically, soil texture is mainly sandy, which affects soil aeration, drainage,
dampness retention, fertility, erosion susceptibility, and permeability (Salas and
Ballesteros *unpublished data*, Jaramillo 2002). In the plot, there is a clear relationship between the sand content of the superficial horizon and the suffusion
phenomenon in some areas. Micro- and macroorganisms contribute to the porous
nature of the soil in some well-drained areas.

Forest Type and Characteristics

La Planada Nature Reserve contains more than 890 species of vascular plants,
580 dicots, 192 monocots, and nearly 120 ferns (Mendoza and Ramirez 2001;
Ramirez and Mendoza 2002). Whereas Lauraceae is the most prevalent and

characteristic family in most neotropical montane forests, Rubiaceae and Melastomataceae dominate the La Planada forest between 1300 and 1800 m asl (Gentry
1995).

Located above the cloud line, the La Planada Forest Dynamics Plot is classified
as pluvial premontane forest (Holdridge 1979). Nearly 75% of the woody plant
species of the La Planada Nature Reserve occur in the 25-ha plot (Humberto
Mendoza personal communication). The forest structure in the plot includes
five layers defined by the predominant vegetation type: emergent trees (20–25 m
high), canopy trees (12–15 m), saplings (5–8 m), shrubs and bushes (2–4 m),
and low understory plants (0–2 m). The average height of the canopy is 15 m.
The emergent trees can exceed 25 m in height and 1.5 meters in dbh. Stranglers
and vines colonize almost all of the emergent trees. The very dense understory is
dominated by species of the Rubiaceae family (*Faramea* and *Palicourea*). Hemiepi-
phyte species of the Guttiferae and Moraceae families are also very abundant in
the plot. One of the most important structural characteristics of the La Planada
forest is the dominant presence of epiphytes, especially from the families Orchidaceae, Araceae, and Gesneriaceae (Gentry 1991). Consequently, the frequency
of epiphytes in La Planada is considered to be among the highest in the Andes
(Gentry 1991). For census data and rankings see tables 30.2–30.6.

Table 30.2. La Planada Plot Census History

Census	Dates	Number of Trees (≥ 1 cm dbh)	Number of Species (≥ 1 cm dbh)	Number of Trees (≥ 10 cm dbh)	Number of Species (≥ 10 cm dbh)
First	February 1997–November 1997	115,129	228	14,650	179

Notes: One census has been completed. The second census was initiated in November 2002 and completed in July 2003.

Table 30.3. La Planada Summary Tally

Size Class (cm dbh)	Average per Hectare							50-ha Plot				
	BA	N	S	G	F	H'	α	S	G	F	H'	α
≥ 1	29.8	4605	154	89	44	1.72	30.6	228	121	54	1.78	27.3
≥ 10	23.8	586	88	58	34	1.66	28.6	179	102	49	1.76	28.7
≥ 30	12.9	83	28	24	17	1.23	15.3	92	58	29	1.43	19.7
≥ 60	2.9	7	5	5	5	0.62	8.1*	29	26	19	1.18	9.9

* Fisher's alpha is based on 19 hectares.

Notes: BA represents basal area in m^2, N is the total number of individual trees, S is the total number of species, G is the number of genera, F is the total number of families, H' is the Shannon–Wiener diversity index using $\log_{10}$, and α is Fisher's α. Basal area includes all multiple stems for each individual. 1444 individuals were not identified to species or morphospeices. Data are from the first census.

Table 30.4. La Planada Rankings by Family

Rank	Family	Basal Area (m²)	% BA	% Trees	Family	Trees	% Trees	Family	Species
1	Euphorbiaceae	109.2	14.9	4.4	Rubiaceae	40,161	35.3	Rubiaceae	28
2	Rubiaceae	103.2	14.1	35.3	Melastomataceae	9,789	8.6	Melastomataceae	26
3	Myristicaceae	83.3	11.4	1.8	Palmae	9,707	8.5	Lauraceae	21
4	Leguminosae (sensus lato)	46.1	6.3	5.8	Cyatheaceae	6,861	6.0	Solanaceae	11
5	Lauraceae	41.3	5.6	5.5	Leguminosae	6,540	5.8	Leguminosae	10
6	Melastomataceae	40.9	5.6	8.6	Lauraceae	6,235	5.5	Moraceae	10
7	Palmae	33.3	4.5	8.5	Euphorbiaceae	5,051	4.4	Cyatheaceae	9
8	Bombacaceae	29.8	4.1	1.7	Myrtaceae	3,426	3.0	Euphorbiaceae	9
9	Myrtaceae	29.6	4.0	3.0	Myristicaceae	2,046	1.8	Myrtaceae	9
10	Cyatheaceae	28.7	3.9	6.0	Bombacaceae	1,890	1.7	Araliaceae	6

Notes: The top 10 families for trees ≥ 1 cm dbh. Data are from the first census.

Table 30.5. La Planada Rankings by Genus

Rank	Genus	Basal Area (m²)	% BA	% Trees	Genus	Trees	% Trees	Genus	Species
1	Otoba (Myristicaceae)	83.3	11.5	1.8	Faramea (Rubiaceae)	22,999	20.8	Miconia (Melastomataceae)	9
2	Hyeronima (Euphorbiaceae)	64.6	8.9	2.2	Palicourea (Rubiaceae)	9,707	8.8	Palicourea (Rubiaceae)	9
3	Faramea (Rubiaceae)	51.7	7.2	20.8	Cyathea (Cyatheaceae)	6,673	6.0	Cyathea (Cyatheaceae)	8
4	Alchornea (Euphorbiaceae)	43.1	6.0	2.3	Inga (Leguminosae)	4,439	4.0	Ficus (Moraceae)	8
5	Elaeagia (Rubiaceae)	38.8	5.4	3.9	Prestoea (Palmae)	4,432	4.0	Inga (Leguminosae)	6
6	Inga (Leguminosae)	29.5	4.1	4.0	Elaeagia (Rubiaceae)	4,310	3.9	Piper (Piperaceae)	6
7	Matisia (Bombacaceae)	25.2	3.5	1.6	Miconia (Melastomataceae)	3,899	3.5	Psychotria (Rubiaceae)	6
8	Cyathea (Cyatheaceae)	25.0	3.5	6.0	Aiphanes (Palmae)	3,366	3.0	Faramea (Rubiaceae)	5
9	Prestoea (Palmae)	24.7	3.4	4.0	Ocotea (Lauraceae)	3,126	2.8	Meriania (Melastomataceae)	4
10	Miconia (Melastomataceae)	24.6	3.4	3.5	Alchornea (Euphorbiaceae)	2,514	2.3	Myrcia (Myrtaceae)	4

Notes: Top 10 tree genera for trees ≥ 1 cm dbh. Data are from the first census.

Fauna

La Planada has 243 species of birds, of which 189 are resident, 24 migratory, 24 endemic, and 6 occasional. The mountain toucan (*Andigena laminirostris*), the reserve's flagship bird, is endemic to the Andean region. The toucan barbet (*Semnornis ramphastinus*) is also common in the reserve. Eighty species of mammals—half

Table 30.6. La Planada Rankings by Species

Rank	Species	Number Trees	% Trees	Species	Basal Area (m²)	% BA	% Trees
1	Faramea calyptrata (Rubiaceae)	17,416	15.1	Otoba lehmanii (Myristicaceae)	83.3	11.4	1.8
2	Palicourea pyramidalis (Rubiaceae)	7,131	6.2	Hyeronima oblonga (Euphorbiaceae)	64.6	8.8	2.1
3	Cyathea planadae (Cyatheaceae)	4,995	4.3	Faramea calyptrata (Rubiaceae)	39.1	5.3	15.1
4	Prestoea acuminata (Palmae)	4,432	3.9	Elaeagia utilis (Rubiaceae)	38.8	5.3	3.7
5	Elaeagia utilis (Rubiaceae)	4,310	3.7	Alchornea triplinervia (Euphorbiaceae)	34.2	4.7	1.8
6	Aiphanes erinaceae (Palmae)	3,366	2.9	Matisia boliviarii (Bombacaceae)	25.2	3.4	1.5
7	Faramea sp. (Rubiaceae)	2,488	2.2	Prestoea acuminata (Palmae)	24.7	3.4	3.9
8	Hyeronima oblonga (Euphorbiaceae)	2,421	2.1	Billia colombiana (Hippocastanaceae)	18.8	2.6	1.5
9	Ocotea sp. (Lauraceae)	2,283	2.0	Sloanea aff. gracilis. (Elaeocarpaceae)	16.4	2.2	0.1
10	Otoba lehmanii (Myristicaceae)	2,046	1.8	Inga sp. (Leguminosae)	15.8	2.2	1.7

Notes: Top 10 tree species for trees ≥1 cm dbh. Data are from the first census.

of which are bats—have been identified in the reserve. The mammal most notably associated with the region is the Andean bear (*Tremarctos ornatus*). Since the creation of the reserve, five captive-bred bears have been reintroduced there. Other animals found in the reserve include carnivores such as the ocelot (*Felis pardalis*), tiger cat (*Felis tigrina*), and kinkajou (*Potos flavus*); large herbivores such as collared peccaries (*Tayassu tajacu*); primates such as the mantled howler (*Alouatta palliata*), the Colombian black spider monkey (*Ateles fusciceps*), and the white-faced capuchin (*Cebus capuchinus*); and three species of deer (*Mazama americana, Mazama rufina,* and *Pudu mephistophiles*). Preliminary inventories have recorded approximately 50 species of reptiles and 30 species of amphibians.

Because no good approximations of insect biodiversity exist, the Humboldt Institute in cooperation with the University of Kentucky and Los Angeles County Museum of Natural History has recently launched an inventory of insects in La Planada and 12 other areas inside of Colombian national parks. Findings are not yet available.

Natural Disturbance

Windthrows are the most important natural disturbance within the La Planada Forest Dynamics Plot. During the approximately 8 months of the first plot census, three large forest gaps were created by windthrows. Two windthrows flattened almost 0.16 ha of the 25-ha plot.

Human Disturbance

An estimated 150 ha of the forest in the plateau near the reserve headquarters were cleared for cattle grazing pastures several decades ago. When the reserve was created in 1982, these pastures were abandoned and the process of natural regeneration initiated. At first, the old pastures were thickly covered by *Tibouchina lepidota and T. gleasoniana* (Melastomataceae) shrubs and dense masses of *Disterigma sterophyllum* (Ericaceae), a plant that grows as an epiphyte in the forest. After years of succession, pioneer trees including *Clethra fagifolia* (Clethraceae), *Vismia spp, Clusia cruciata, C. weberbauerii,* and *C. longistyla* (Guttiferae) began to dominate the forest. Several species that grow as epiphytes in the forest are found at ground level in these secondary areas, such as *Stenospermatium robustum* (Araceae), *Schefflera sphaerocoma* (Araliaceae), *Macleania bullata* (Ericaceae), and *M. stricta* (Samper 1992).

The forest surrounding the plot is relatively undisturbed. While there is moderate poaching and agricultural encroachment near the border of the reserve, these activities do not occur within or around the plot. One activity that does affect the La Planada forest is the harvest of palmito (*Prestoea acuminata* [Palmae]), the edible heart of palm extracted by humans in July and December. The exploitation of this palm results in decreased species regeneration due to seedling trampling, physiological changes in the subcanopy, and the creation of poorly drained areas due to the removal of the species. In addition, heart of palm collection also affects the wildlife by altering pollination and dispersal activities and limiting the food source for birds such as cotingas (*Lipaugus criptolophus, Rupicola peruviana*) and mountain toucans (*Andigena laminirostris, Aulacorhynchus haematophygus*), rodents such as *Orizomys albigularis,* and guan (*Chamaepetes goudotii*) (Guzmán 2000). During harvest season, the project supervisor and reserve forester vigilantly patrol the plot and the limits of the reserve to prevent poaching.

The Forest Dynamics Plot can be classified into an area with flat slopes and areas with steep slopes. In the region of flat slopes, selective logging activities occurred about 20 years ago and the forest is currently regenerating. The other region of the plot is characterized by greater slope with minimal human disturbance.

Plot Size and Location

The 25-ha plot is 500×500 m. Its northwest corner (00, 25) is located at $1°09'31.6''$N, $77°59'44.8''$W. The northeast corner (25, 25) of the plot is located at $1°09'31.6''$N, $77°59'28.6''$W. The southwest corner (00, 00) of the plot is located at 1°09′05.3″N, 77°59′44.8″W. The southeast corner (25, 00) of the plot is located at 1°09′15.3″N, 77°59′28.6″W.

Funding Sources

The La Planada Forest Dynamics Plot project is carried out by the Instituto de Investigación de Recursos Biológicos Alexander von Humboldt, through an agreement with the Fundación para la Educación Superior. Financial support was provided by the Ministerio de Medio Ambiente from Colombia and COLCIENCIAS.

Acknowledgments

Technical assistance has been provided through the Center for Tropical Forest Science and the Smithsonian Tropical Research Institute. The establishment of the permanent plot was done in agreement with the Organización para el Desarrollo Campesino de San Isidro in Nariño. We are especially grateful to biologist Constanza Ríos who worked in the beginning of the project, Gilbert Oliva for his assistance in the field, Carol Franco for her assistance with data input into the database, Jorge Humberto Rodríguez and Dolors Armenteras for their collaboration with database design and the use of GIS, Olga Lucia Guzmán and Pilar Amézquita for their cooperation and comments to the document, and all the La Planada reserve staff and workers for their continuing support.

References

Amaya-Marquez, M. 1999. Densidad de Columnea (Gesneriaceae) en relación con la edad de un bosque de neblina (La Planada: Narino, Colombia). *Rev. Acad. Colomb. Cienc. (Suppl.)* 23:123–31.

Escobar, F., and P. Chacón de Ulloa. 2000. Distribución espacial y temporal en un gradiente de sucesión de la fauna de coleópteros coprófagos (Scarabaeidae, Aphodiinae) en un bosque tropical montano. *Revista de Biología Tropical* 48:961–75.

Gentry, A. H. 1991. Vegetación del Bosque de Niebla. Pages 25–51 in C. Uribe, editor. *Bosques de Niebla de Colombia.* Banco de Occidente, Bogotá, Colombia.

———. 1995. Patterns of diversity and floristic composition in neotropical montane forests. Pages 103–26 in S. P. Churchill, H. Balslev, E. Forero, and J. Lutein, editors. *Biodiversity and Conservation of Neotropical Montane Forests.* New York Botanical Garden, New York.

Grubb, P. J., and T. C. Whitmore. 1966. A comparison of montane and lowland forest in Ecuador. II. The climate and its effects in the distribution and physiognomy of the forest. *Journal of Ecology.* 54:303–33.

Guzmán, O. L. 2000. *Propagación Natural y Artificial de P. acuminata (Arecaceae) en la Reserva Natural La Planada, Nariño, Colombia.* Undergraduate thesis. Universidad de Nariño, Nariño, Colombia.

Holdridge, L. R. 1979. *Ecología Basada en Zonas de Vida.* Instituto Interamericano de Ciencias Agrícolas, San José, Costa Rica.

Jaramillo, D. F. 2002. *Introducción a la Ciencia del Suelo.* Universidad Nacional de Colombia, Escuela de Geociencias. Medellín, Colombia.

Knudsen, H. 1995. *Demography, Palm-Heart Extractivism and Reproductive Biology of* Prestoea acuminata *(Arecaeae) in Ecuador.* Masters thesis. University of Aarhus, Aarhus, Denmark.

Mabberley, D. J. 1997. *The Plant-Book: A Portable Dictionary of the Vascular Plants.* Cambridge University Press, Cambridge, U.K.

Mendoza-Cifuentes, H., B. Ramírez-Padilla. 2000. *Flora de La Planada: Guía Ilustrada de Familias y Géneros.* Instituto Alexander von Humboldt, Fundación para la Educación Superior, World Wildlife Fund, Bogota, Colombia.

Mendoza-Cifuentes, H., B. Ramirez-Padilla. 2001. Dicotiledóneas de La Planada, Colombia: Lista de Especies. *Biota Colombiana* 2:59–73.

Murillo, M. M. 1997. *Modelo Sobre la Dinámica de Población de* Prestoea Acuminata *como Herramienta de Manejo Sostenible en un Bosque Montano en la Reserva Natural la Planada, Nariño-Colombia.* Undergraduate thesis. Universidad de los Andes, Mérida, Venezuela.

Otero, J. T. 1996. Biología de *Euglossa nigropilosa* Moure (Apidae: Euglossini) I. Aspectos de nidificación. *Boletín del Museo de Entomología de la Universidad del Valle* 4(1):1–19.

Ramirez-P. B., and H. Mendoza-C. 2002. Monocotiledóneas y Pteridofitos de La Planada, Colombia: Lista de Especies. *Biota Colombiana* 3(2):285–295.

Samper, C. 1992. *Natural Disturbance and Plant Establishment in an Andean Cloud Forest.* PhD thesis. Harvard University, Cambridge, MA.

Stahl, B. 1995. Three new species of *Clavija* (Theophrastaceae). *Novon* 5:370–474.

31

Lambir Forest Dynamics Plot, Sarawak, Malaysia

Hua Seng Lee, Sylvester Tan, Stuart J. Davies, James V. LaFrankie,
Peter S. Ashton, Takuo Yamakura, Akira Itoh, Tatsuhiro Ohkubo,
and Rhett Harrison

Site Location, Administration, and Scientific Infrastructure

Lambir Hills National Park is in the Malaysian state of Sarawak, on the island of
Borneo. Established in 1975, this 6823-ha park lies about 10 km from the South
China Sea and ranges in elevation from ca. 60 m to 465 m above sea level. The
totally protected park is managed by the National Parks and Wildlife Branch of
the Sarawak Forest Department. The research section of the Forest Department
coordinates investigations within the park. The 52-ha Lambir Forest Dynamics
Plot is located along the southern edge of the park, 20 km from the town of
Miri by a good highway (fig. 31.1). In the immediate vicinity of the plot, there are
two permanent residential buildings with rudimentary laboratories and herbaria,
three canopy towers, a canopy walkway, and a construction crane equipped for
biological canopy research.

In 1963, P. S. Ashton, then of the Sarawak Forest Department, established six
0.6-ha plots within Lambir Hills National Park, four of which were made permanent and recensused every 5 years thereafter (Ashton and Hall 1992). Three
and three-quarters of these latter plots are included within the 52-ha Forest Dynamics Plot. The six plots were part of a network of 105 plots (0.6 ha each) laid
out throughout apparently mature lowland mixed dipterocarp forest in Sarawak,
over a distance of 500 × 150 km (Potts et al. 2002).

Climate

The climate of northwestern Borneo is strongly influenced by the Indo-Australian
monsoon. While generally everwet, this region is subject to a shift in monsoonal
winds that often triggers brief droughts. A northerly monsoon occurs from
December to March and a southerly monsoon from May to October (Proctor
et al. 1983). Tropical typhoons do not affect the region.

No long-term weather records in the immediate vicinity of the plot are available. The 30-year mean annual rainfall at the Miri airport, 20 km from the plot,
is 2725 ± 72 mm/year (Malaysian Meteorological Service from 1968–98). The
heaviest rainfalls in the region coincide with the northerly monsoon. Records of

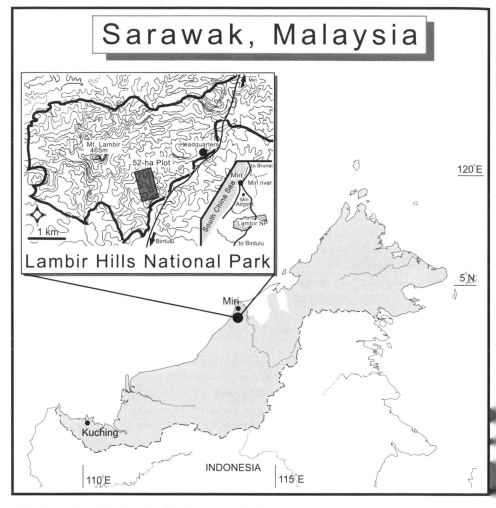

Fig. 31.1. Location of the 52-ha Lambir Forest Dynamics Plot.

monthly rainfall in Miri since 1912 indicate that the recent spate of droughts is unprecedented and the severe drought in 1998 was the worst on record for Miri (Harrison 2001). See table 31.1.

Topography and Soil

The geology of the Lambir Hills area is comprised of sedimentary rocks, bands of clay, and sandstones deposited by rivers and streams in the mid-Miocene era. Steep

Table 31.1. Lambir Climate Data

	Jan	Feb	Mar	Apr	May	Jun	Jul	Aug	Sep	Oct	Nov	Dec	Total/ Averages
Rain (mm)	229	200	182	222	204	153	165	180	198	290	322	319	2664
ADTMx (°C)	28.7	29.4	30.6	31.1	31.4	31.1	30.8	30.6	30.2	29.9	30.1	29.4	30.3
ADTMn (°C)	22.1	22.5	23.4	23.3	23.4	23.1	22.9	22.9	22.9	22.6	22.7	22.6	22.9

Notes: Mean annual rainfall data were taken from the Telecom tower (1985–1997), a relay station of a Malaysian telephone company located inside the park (Telecom, unpublished data). Some of the monthly temperature means were averaged from only 3 years during this period. Temperature data was collected from the Pasoh field station.

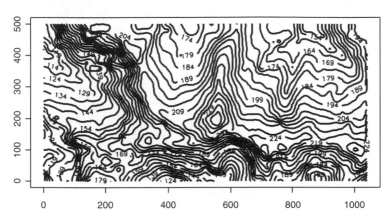

Fig. 31.2. Topographic map of the 52-ha Lambir Forest Dynamics Plot with 5-m contour intervals.

slopes ranging from 23 to 30% cover 85% of the park area (Yamakura et al. 1995; Hazebroek and Abang Kashim 2000). The Lambir Forest Dynamics Plot follows a cuesta ridge, with a gentle dip slope dissected by steep ravines and a precipitous scarp (chap. 19; figs. 31.2 and 31.3). The plot elevation varies from 104 to 244 m above sea level and contains numerous extremely steep slopes and ravines. Most of the plot lies over sandy humult Ultisol soils derived from sandstone. Clay-rich udult Ultisols, derived from shales and bearing relatively high concentrations of phosphorous and magnesium, cover a little less than a third of the plot (Ashton 1973; Baillie et al. 1987; Palmiotto 1998; chap. 14). The plot's udult soils have the following characteristics: Munsell color intensity = 6.0, % clay = 45, % sand = 16, pH = 4.5. HCl extractable concentrations of specific minerals from the udults have the following concentrations (mg/g soil): N = .60, P = 0.12, K = 5.53, Ca = 0.10, Mg = 1.64, % Fe and Al = 11.0. The humult soils have the following characteristics: Munsell color intensity = 6.0, % clay = 21 ± 3, % sand = 69 ± 1, pH = 4.8 ± 0.2. HCl extractable concentrations of specific minerals from the humults have the following concentrations (mg/g soil): N = 0.25 ± 0.03,

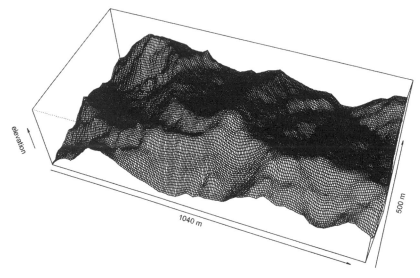

Fig. 31.3. Perspective map of the 52-ha Lambir Forest Dynamics Plot.

P = 0.08 ± 0.04, K = 1.95 ± 1.17, Ca = 0.11 ± 0.06, Mg = 0.61 ± 0.38, % Fe and Al = 8.6 ± 2.5 (Ashton and Hall 1992).

Forest Type and Characteristics

Approximately two-thirds of Lambir Hills National Park, and the entire 52-ha plot, is covered by mature lowland mixed dipterocarp forest. The canopy of this forest type is fairly heterogeneous and tends to be 40–60 m tall, with emergents reaching over 75 m. About 80% of the mixed dipterocarp forest is in a mature phase (Ohkubo et al. 1995). Average aboveground tree biomass in the plot is 520 tons/ha (Yamakura et al. 1996). Mass flowering followed by mass fruiting occurs on a supra-annual cycle, typically beginning in March but also sometimes in August. The park also supports several conspicuously different forest types, such as the lower-stature kerangas forest on white sand near the summit and on the western side of the hills and open secondary forest around the forest margin and on large landslips. The beta diversity of the trees in Lambir Hills is as rich as in any plant community in the Old World. Two of the major factors leading to this diversity are thought to be soil and topographic heterogeneity.

Within the plot, floristic composition and stand structure varies with the change from humult Ultisols in the higher areas of the plot to udult Ultisols in lower portions. The forest community on udult soils, dominated by *Koilodepas*

Table 31.2. Lambir Plot Census History

Census	Dates	Number of Trees (≥ 1 cm dbh)	Number of Species (≥ 1 cm dbh)	Number of Trees (≥ 10 cm dbh)	Number of Species (≥ 10 cm dbh)
First	October 1991–June 1993	346,061	1179	32,662	1008
Second	May 1997–October 1997	359,603	1182	33,175	1003

Notes: Two censuses have been completed, the enumeration of the third census was initiated in August 2003.

Table 31.3. Lambir Summary Tally

Size Class (cm dbh)	Average per Hectare						52-ha Plot				
	BA	N	S	G	F	H'	α	S	G	F	H'	α
≥ 1	43.5	6907	618	201	66	2.40	165.3	1182	287	83	2.65	152.2
≥ 10	37.8	637	247	117	47	2.19	153.6	1003	258	73	2.59	195.1
≥ 30	26.3	119	67	42	25	1.70	73.6	574	186	60	2.28	154.1
≥ 60	13.9	26	16	9	6	1.09	24.7	200	80	36	1.77	64.2

Notes: BA represents basal area in m^2, N is the number of individual trees, S is number of species, G is number of genera, F is number of families, H' is Shannon–Wiener diversity index using $\log_{10}$, and α is Fisher's α. Basal area includes all multiple stems for each individual. Individuals are counted using their largest stem. 12,083 individuals were not identified to species or morphospecies. Data are from the second census.

longifolium (Euphorbiaceae), *Millettia vasta* (Leguminosae), *Dryobalanops lanceolata* (Dipterocarpaceae), and *Hopea dryobalanoides* (Dipterocarpaceae), contained a third fewer trees and basal area yet slightly higher species richness than that on humult soils, dominated by *Dipterocarpus globosus* (Dipterocarpaceae), *Elateriospermum tapos* (Euphorbiaceae), *Dryobalanops aromatica* (Dipterocarpaceae), *Whiteodendron moultonianum* (Myrtaceae), and *Shorea acuta* (Dipterocarpaceae) (Lee et al. 2002). For census history and rankings, see tables 31.2–31.7.

Fauna

Lambir Hills National Park has 366 species of vertebrates (not including fish) (Shanahan and Debski 2002). Sixty-one species of mammals have been recorded, ranging from sun bear (*Helarctos malayanus*) to slow loris (*Nycticebus coucang*), including five fruit bats and five insectivorous bats. The largest herbivores are the sambar deer (*Cervus unicolor*), bearded pig (*Sus barbatus*), and two species of Muntjak or barking deer (*Muntiacus muntjac* and *M. atherodes*). The largest canopy herbivores are the langurs (*Presbytis hosei* and *P. melalophos*). Important frugivores include, on the ground, two species of mouse deer (*Tragulus javanicus* and *T. napu*) and, in the canopy, fruit bats including the large flying fox (*Pteropus*

Table 31.4. Lambir Rankings by Family

Rank	Family	Basal Area (m²)	% BA	% Trees	Family	Trees	% Trees	Family	Species
1	Dipterocarpaceae	908.1	41.0	15.5	Dipterocarpaceae	53,696	15.5	Euphorbiaceae	125
2	Burseraceae	144.5	6.5	6.5	Euphorbiaceae	49,724	14.3	Dipterocarpaceae	87
3	Euphorbiaceae	140.8	6.3	14.3	Burseraceae	22,749	6.5	Lauraceae	77
4	Anacardiaceae	134.5	6.1	5.6	Anacardiaceae	19,322	5.6	Rubiaceae	58
5	Myrtaceae	96.8	4.4	3.4	Rubiaceae	15,669	4.5	Myrtaceae	57
6	Lauraceae	76.3	3.4	3.5	Annonaceae	14,987	4.3	Meliaceae	55
7	Guttiferae	52.8	2.4	2.9	Myristicaceae	12,251	3.5	Annonaceae	54
8	Myristicaceae	51.5	2.3	3.5	Lauraceae	12,053	3.5	Guttiferae	51
9	Leguminosae	46.1	2.1	2.1	Myrtaceae	11,713	3.4	Burseraceae	40
10	Sapotaceae	38.7	1.7	1.6	Guttiferae	9,945	2.9	Myristicaceae	40

Notes: The top 10 families for trees ≥1 cm dbh ranked in terms of basal area, number of individual trees, and number of species, with the percentage of trees in the plot. Data are from the second census.

Table 31.5. Lambir Rankings by Genus

Rank	Genus	Basal Area (m²)	% BA	% Trees	Genus	Trees	% Trees	Genus	Species
1	Shorea (Dipterocarpaceae)	455.4	20.5	6.7	Shorea (Dipterocarpaceae)	23,415	6.7	Shorea (Dipterocarpaceae)	56
2	Dipterocarpus (Dipterocarpaceae)	213.4	9.6	1.5	Dryobalanops (Dipterocarpaceae)	11,497	3.3	Syzygium (Myrtaceae)	53
3	Dryobalanops (Dipterocarpaceae)	166.8	7.5	3.3	Dacryodes (Burseraceae)	11,292	3.2	Diospyros (Ebenaceae)	34
4	Santiria (Burseraceae)	61.5	2.8	2.1	Diospyros (Ebenaceae)	9,419	2.7	Litsea (Lauraceae)	28
5	Gluta (Anacardiaceae)	60.3	2.7	2.5	Vatica (Dipterocarpaceae)	8,890	2.6	Aglaia (Meliaceae)	25
6	Dacryodes (Burseraceae)	60.0	2.7	3.2	Gluta (Anacardiaceae)	8,629	2.5	Xanthophyllum (Xanthophyllaceae)	25
7	Syzygium (Myrtaceae)	52.8	2.4	2.1	Macaranga (Euphorbiaceae)	8,096	2.3	Garciniaa (Guttiferae)	23
8	Allantospermum (Ixonanthaceae)	34.3	1.5	2.2	Allantospermum (Ixonanthaceae)	7,473	2.2	Ficus (Moraceae)	21
9	Whiteodendron (Myrtaceae)	32.0	1.4	1.0	Santiria (Burseraceae)	7,450	2.1	Aporusa (Euphorbiaceae)	18
10	Vatica (Dipterocarpaceae)	31.4	1.4	2.6	Syzygium (Myrtaceae)	7,415	2.1	Knema (Myristicaceae)	17
11								Santiria (Burseraceae)	17

Notes: The top 10 tree genera for trees ≥ 1 cm dbh are ranked by basal area, number of individual trees, and number of species with the percentage of trees in the plot. Data are from the second census.

Table 31.6. Lambir Rankings by Species

Rank	Species	Number Trees	% Trees	Species	Basal Area (m²)	% BA	% Trees
1	*Dryobalanops aromatica* (Dipterocarpaceae)	10,562	3.0	*Dryobalanops aromatica* (Dipterocarpaceae)	155.3	7.0	3.0
2	*Allantospermum borneense* (Ixonanthaceae)	7,473	2.2	*Dipterocarpus globosus* (Dipterocarpaceae)	138.5	6.3	1.0
3	*Vatica micrantha* (Dipterocarpaceae)	6,274	1.8	*Shorea beccariana* (Dipterocarpaceae)	61.6	2.8	1.1
4	*Fordia splendidissima* (Leguminosae)	3,764	1.1	*Shorea laxa* (Dipterocarpaceae)	47.3	2.1	1.0
5	*Shorea beccariana* (Dipterocarpaceae)	3,761	1.1	*Shorea acuta* (Dipterocarpaceae)	41.8	1.9	0.3
6	*Gluta laxiflora* (Anacardiaceae)	3,615	1.0	*Allantospermum borneense* (Ixonanthaceae)	34.3	1.6	2.2
7	*Whiteodendron moultonianum* (Myrtaceae)	3,405	1.0	*Whiteodendron moultonianum* (Myrtaceae)	32.0	1.4	1.0
8	*Shorea laxa* (Dipterocarpaceae)	3,322	1.0	*Shorea curtisii* (Dipterocarpaceae)	28.0	1.3	0.1
9	*Dipterocarpus globosus* (Dipterocarpaceae)	3,318	1.0	*Elateriospermum tapos* (Euphorbiaceae)	22.2	1.0	0.3
10	*Dacryodes expansa* (Burseraceae)	3,302	0.9	*Swintonia schwenkii* (Anacardiaceae)	21.2	1.0	0.3

Notes: The top 10 tree species for trees ≥ 1 cm dbh are ranked by number and percentage of trees and basal area. Data are from the second census.

Table 31.7. Lambir Tree Demographic Dynamics

Size Class (cm dbh)	Growth Rate (mm/yr)	Mortality Rate (%/yr)	Recruitment Rate (%/yr)	BA Losses (m²/ha/yr)	BA Gains (m²/ha/yr)
1–9.9	0.52	1.74	2.72	0.07	0.20
10–29.9	1.63	1.3	1.96	0.14	0.32
≥30	2.7	1.01	1.34	0.23	0.39

Notes: Data recorded between the 1992 and 1997 censuses. Climbers, including free-standing hemi-epiphytic figs, and all palms were not enumerated in the first two censuses.

vampyrus), nine species of squirrel (*Ratufa affinis, Callosciurus prevostii caroli, C. notatus, C. adamsi, Sundasciurus hippurus, S. lowi, Dremomys everetti, Exilisciurus exilis, Rheithrosciurus macrotis*), five species of flying squirrel (*Petaurillus hosei, Petinomys setosus, P. vordermanni, Petaurista petaurista, Aeromys thomasi*), the colugo (*Cynocephalus variegatus*), two macaques (*Macaca fascicularis* and *M. nemstrina*), the Bornean gibbon (*Hylobates muelleri*), and the binturong (*Arctitis bintourong*). The largest mammal predators are civets (*Viverra tangalunga, Paradoxurus hermaphroditus,* and *Hemigalus derbyanus*) and the Oriental small-clawed otter (*Aonyx cinerea*). Some mammals characteristic of Borneo, such as the orangutan, are not found in Lambir, and there is no evidence that they were ever found there. Lambir Hills also hosts 237 species of birds (Shanahan and Debski

2002). Pythons (*Python curtus* and *P. reticulatus*) are important predators and can reach very large sizes (a 6-m skin was collected from the plot).

Invertebrates are also very diverse and include such distinctive species as Rajah Brooke's birdwing (*Trogonoptera brookiana brookiana*) and the giant moth, *Antheraea celebensis* (Saturnidae), but are generally poorly cataloged. With 361 species, ants are one well-studied group that illustrates the great diversity within the park (Yamane unpublished data).

Natural Disturbances

The most important natural disturbances within Lambir Hills National Park are landslips. With continuous heavy rain, the steep soils give way and patches of forest as large as 1-ha can collapse. The very wet winter of 1963 (daily rainfall of 560 mm in May) led to such a large number of slips that today, nearly 40 years later, they are still only in the early stages of recovery (Ohkubo et al. 1995). Unpredictable, severe droughts, usually associated with El Niño Southern Oscillation events, are also very important for their impact on seedling regeneration, tree mortality (especially among larger size classes) (Nakagawa et al. 2000), and phenology with consequences for organisms dependent on plant resources. Harrison (2000) has also demonstrated a breakdown of the fig/fig-wasp pollination system during an El Niño event.

Human Disturbance

Humans have lived in the vicinity of Lambir National Park for a very long time. The Niah Cave archeological site, 50 km down the coast, includes the oldest records of human settlements in southeast Asian rainforest, dated at 40,000 years before present. Within the park boundary, however, no record of cultivation exists. The main products extracted from the forest have been animals—deer, pigs (*Sus barbatus*), porcupine (*Trichys fasciculate*)—timber, rattan, and wild fruits and shoots. While the park is a totally protected area, hunting and extraction of timber and rattan continue from time to time. On the eastern side of the park, illicit entry and encroachment are continuing problems. Sanctioned projects that reduced the park's area include the construction of a radio tower, the main highway (which divided the park in two), and the excavation for a major water pipeline.

There has been a significant increase in the proportion of montane and open- or disturbed-habitat bird species recorded in recent years. Nine mammal and 13 bird species have not been seen since the original park survey in 1985 (Shanahan and Debski 2002). Hornbills, which are richly abundant in the nearby forests of Brunei, are conspicuous in their paucity in Lambir. One species, the vocal helmeted hornbill (*Buceros vigil*) was recorded in 1985 but has not been seen recently, almost certainly as a consequence of illegal hunting. Some of the disappearances and the increase in open- or disturbed-habitat bird species are likely to be a consequence of two factors: the small size of the park and the dominance of nearby secondary vegetation. Forty-five percent of frogs and 37% of reptiles recorded in 1985 (Watson 1985) have not been recorded since, perhaps due to undersampling.

The southeastern side of the plot is about 200 m from the Miri–Kuching truck road, now abandoned, and faces an area of shifting cultivation fields. The rest of the plot is surrounded by relatively undisturbed forest.

Plot Size and Location

The 52-ha, 1040 × 500 m plot lies north-south on its long axis. The northwest corner of the plot is located at 4°11′20″ N, 114°00′56″ E. The northeast corner of the plot is located at 4°11′20.4″ N, 114°00′55.9″ E. The southeast corner of the plot is located at 4°10′59.1″ North, 114°01′26.7″ East. The southwest corner of the plot is located at 4°10′51.0″ N, 114°01′12.6‴″ E (T. Ohkubo and A. Itoh, unpublished data).

Funding Sources

The Lambir Forest Dynamics Plot has been funded by the Sarawak Forest Department, U.S. Agency for International Development, U.S. National Science Foundation, and Monbusho (the Ministry of Education, Science, Culture, and Sports, Japan).

References

Ashton, P. S. 1973. Sarawak mixed dipterocarp forest ecology. Unpublished report to the Government of Sarawak, Malaysia.

Ashton, P. S., and P. Hall. 1992. Comparison of structure among mixed dipterocarp forests of north-west Borneo. *Journal of Ecology* 80:459–81.

Baillie, I. C. 1976. Further studies on drought in Sarawak, East Malaysia. *Journal of Tropical Geography* 43:20–29.

Baillie, I. H, P. S. Ashton, M. N. Court, J. A. R. Anderson, E. A. Fitzpatrick, and J. Tinsley. 1987. Site characteristics and the distribution of tree species in mixed dipterocarp forests on tertiary sediments in central Sarawak. *Journal of Tropical Ecology* 3:201–20.

Davies, S. J. 1998. Photosynthesis of nine pioneer *Macaranga* species from Borneo in relation to life-history. *Ecology* 79:2292–2308.

———. 2001. Tree mortality and growth in 11 sympatric *Macaranga* species in Borneo. *Ecology* 82:920–32.

Davies, S. J., and P. S. Ashton. 1999. Phenology and fecundity in 11 sympatric pioneer species of *Macaranga* (Euphorbiaceae) in Borneo. *American Journal of Botany* 86:1786–95.

Davies, S. J., P. Palmiotto, P. S. Ashton, H. S. Lee, and J. V. LaFrankie. 1998. Comparative ecology of 11 sympatric species of *Macaranga* in Borneo: Tree distribution in relation to horizontal and vertical resource heterogeneity. *Journal of Ecology* 86:662–73.

Harrison, R. D. 1999. *Phenology and Wasp Population Dynamics of Several Species of Dioecious Fig in a Lowland Tropical Rain Forest in Sarawak, Malaysia.* Ph.D. thesis. Kyoto University, Kyoto, Japan.

———. 2000. Repercussions of El Niño: Drought causes extinction and the breakdown of mutualism in Borneo. *Proceedings of the Royal Society (London) Series B* 267:911–15.

———.2001. Drought and the consequences of El Niño in Borneo: A case study of figs. *Research in Population Ecology* 43:63–75.

Harrison, R. D., N. Yamamura, and T. Inoue. 2000. Phenology of a common roadside fig in Sarawak. *Ecological Research* 15:47–61.

Hazebroek, H. P., and Abang Kashim bin Abang Morshidi. 2000. Lambir Hills National Park. Pages 147–76 in *National Parks of Sarawak, Natural History Publications (Borneo).* Kota Kinabalu, Sabah, Malaysia.

Hirai, H., H. Matsumura, H. Hirotani, K. Sakurai, K. Ogino, and H. S. Lee. 1997. Soils and the distribution of *Dryobalanopes aromatica* and *D. lanceloata* in mixed dipterocarp forest: A case study at Lambir Hills National Park, Sarawak, Malaysia. *Tropics* 7:21–33.

Inoue, T., T. Yumoto, A. A. Hamid, H. S. Lee, and K. Ogino. 1995. Construction of a canopy observation system in a tropical rainforest of Sarawak. *Selbyana* 16:24–35.

Inoue, K., M. Kato, and T. Inoue. 1995. Pollination ecology of *Dendrobium setifolium, Neuwiedia borneensis,* and *Lecanorchis multiflora* (Orchidaceae) in Sarawak. *Tropics* 5:95–100.

Ishizuka, S., S. Tanaka, K. Sakurai, H. Hirai, H. Hirotani, K. Ogino, H. S. Lee, and J. J. Kendawang. 1998. Characterization and distribution of soils at Lambir Hills National Park in Sarawak, Malaysia, with special reference to soil hardness and soil texture. *Tropics* 8:31–44.

Itoh, A. 1995a. Effects of forest floor environment on seedling establishment of co-occurring Bornean rainforest emergent species. *Journal of Tropical Ecology* 11:517–27.

———. 1995b. Population structure and canopy dominance of two emergent dipterocarp species in a tropical rain forest of Sarawak, East Malaysia. *Tropics* 4:113–41.

———. 1995c. Regeneration processes and coexistence mechanisms of two Bornean emergent dipterocarp species. Ph.D. thesis. Kyoto University, Kyoto, Japan.

Itoh, A., T. Yamakura, K. Ogino, and H. S. Lee. 1995. Survivorship and growth of seedlings of four dipterocarp species in a tropical rainforest of Sarawak, East Malaysia. *Ecological Research* 10:327–38.

Itoh, A., T. Yamakura, K. Ogino, H. S. Lee, and P. S. Ashton. 1997. Spatial distribution patterns of two predominant emergent trees in a tropical rainforest in Sarawak, Malaysia. *Plant Ecology* 132:121–36.

Itoh, A., T. Yamakura, M. Kanzaki, T. Ohkubo, P. A. Palmiotto, J. V. LaFrankie, J. J. Kendawang, and H. S. Lee. 2002. Rooting ability of cuttings relates to phylogeny, habitat preference and growth characteristics of tropical rain forest trees. *Forest Ecology and Management* 168:275–87.

Itoh A., T. Yamakura, T. Ohkubo, M. Kanzaki, P. A. Palmiotto, J. V. LaFrankie, P. S. Ashton, and H. S. Lee. 2003. Importance of topography and soil texture in spatial distribution of two sympatric dipterocarp trees in a Bornean rain forest. *Ecological Research* 18:307–320.

Itoh A., T. Yamakura , T. Ohkubo, M. Kanzaki, P. A. Palmiotto, S. Tan, and H. S. Lee. 2003. Spatially aggregated fruiting in a Bornean emergent tree. *Journal of Tropical Ecology* 19:531–538.

Itioka, T., M. Nomura, Y. Inui, T. Itino, and T. Inoue. 2000. Difference in intensity of ant defense among three species of *Macaranga* myrmecophytes in a southeast Asian dipterocarp forest. *Biotropica* 32:318–26.

Kato, M. 1996. Plant-pollinator interactions in the understory of a lowland mixed dipterocarp forest in Sarawak. *American Journal of Botany* 83:732–43.

Kato, M., and T. Inoue. 1994. The origin of insect pollination. *Nature (London)* 368:195.

Kato, M., T. Itioka, K. Momose, S. Sakai, S. Yamane, A. A. Hamid, M. B. Merdek, H. Kallang, and T. Inoue. In press. Various population fluctuation patterns of light–attracted beetles in a tropical lowland dipterocarp forest in Sarawak. *Research on Population Ecology.*

LaFrankie, J. V., S. Tan, and P. S. Ashton. 1995. *Species List for the 52-ha Forest Dynamics Research Plot: Lambir Hills National Park, Sarawak, Malaysia.* Miscellaneous Internal Report. Center for Tropical Forest Science, Smithsonian Tropical Research Institute, Singapore.

Lee, H. S., J. V. LaFrankie, S. Tan, T. Yamakura, A. Itoh, and P. S. Ashton. 1999. *The 52-ha Forest Research Plot at Lambir Hills National Park Sarawak, Malaysia. Volume 2: Maps and Diameter Tables.* Sarawak Forest Department Kuching, Sarawak, Malaysia.

Lee, H. S., S. J. Davies, J. V. LaFrankie, S. Tan, T. Yamakura, A. Itoh, T. Ohkubo, and P. S. Ashton. 2002. Floristic and structural diversity of mixed dipterocarp forest in Lambir Hills National Park, Sarawak, Malaysia. *Journal of Tropical Forest Science* 14:379–400.

Lesslar, P., and M. Wannier. 1998. *Destination—Miri: A Geological Tour, Northern Sarawak's National Parks and Giant Caves.* Interactive CD. http://www1.sarawak.com.my/ecomedia_software/

Mabberley, D. J. 1997. *The Plant-Book: A Portable Dictionary of the Vascular Plants.* Cambridge University Press, Cambridge, U.K.

Maschwitz, U., B. Fiala, S. J. Davies, and K. E. Linsenmair. 1996. A south-east Asian myrmecophyte with 2 alternative inhabitants: *Camponotus* or *Crematogaster* as partners of *Macaranga lamellata. Ecotropica* 2:29–40.

Momose, K., T. Nagamitsu, and T. Inoue. 1996. The reproductive biology of an emergent dipterocarp in a lowland rain forest in Sarawak. *Plant Species Biology* 11:189–98.

Momose, K., R. Ishii, S. Sakai, and T. Inoue. 1998. Reproductive intervals and pollinators of tropical plants. *Proceedings of the Royal Society of London* 265:2333–39.

Momose, K., T. Yumoto, T. Nagamitsu, M. Kato, H. Nagamasu, S. Sakai, R. D. Harrison, T. Itioka, A. A. Hamid, and T. Inoue. 1998. Pollination biology in a lowland dipterocarp forest in Sarawak, Malaysia. Characteristics of the plant-pollinator community in a lowland dipterocarp forest. *American Journal of Botany* 85:1477–1501.

Nakagawa, M., K. Tanaka, T. Nakashizuka, T. Ohkubo, T. Kato, T. Maeda, K. Sato, H. Miguchi, H. Nagamasu, K. Ogino, S. Teo, A. A. Hamid, and H. S. Lee. 2000. Impact of a severe drought associated with the 1997–1998 El Niño in a tropical forest in Sarawak. *Journal of Tropical Ecology* 16:355–67.

Nomura, M., T. Itioka, and T. Itino. 2000. Variations in abiotic defense within myrmecophytic and non-myrmecophytic species of *Macaranga* in a Bornean dipterocarp forest. *Ecological Research* 15:1–11.

Ohkubo, T., T. Maeda, T. Kato, M. Tani, T. Yamakura, H. S. Lee, P. S. Ashton, and K. Ogino. 1995. Landslide scars in canopy mosaic structure as a large scale disturbance to a mixed dipterocarp forest at Lambir Hills National Park, Sarawak. Pages 172–84 in H. S. Lee, P. S. Ashton, and K. Ogino, editors. *Long Term Ecological Research of Tropical Rain Forest in Sarawak.* Ehime University, Matsuyama, Japan.

Palmiotto, P. A. 1995. Preliminary characterization of soil texture and organic matter thickness in 52 ha of lowland mixed dipterocarp forest, Lambir Hills National Park, Sarawak Malaysia. Pages 61–67 in H. S. Lee, P. S. Ashton, and K. Ogino, editors. *Long Term Ecological Research of Tropical Rain Forest in Sarawak.* Ehime University, Matsuyama, Japan.

Palmiotto, P. A. 1998. *The Role of Specialization in Nutrient-Use Efficiency as a Mechanism Driving Species Diversity in a Tropical Rain Forest.* Ph.D. thesis, Yale University, New Haven, CT.

Potts, M. D., P. S. Ashton, L. S. Kaufman, and J. B. Plotkin. 2002. The effect of habitat and distance on tropical tree species: A floristic comparison of 105 plots in Northwest Borneo. *Ecology* 83:2782–97.

Primack, R. B., P. S. Ashton, P. Chai, and H. S. Lee. 1985. Growth rates and population structures of Moraceae trees in Sarawak, East Malaysia. *Ecology* 66:577–88.

Proctor, J., J. M. Anderson, P. Chai, and H. W. Vallack. 1983. Ecological studies in four contrasting lowland rain forests in Gunung Mulu national park, Sarawak. *Journal of Ecology* 71:237–60.

Putz, F. E., and P. Chai. 1987. Ecological studies of lianas in Lambir National Park, Sarawak, Malaysia. *Journal of Ecology* 75:523–31.

Sakai, S. 2000. Reproductive phenology of gingers in a lowland dipterocarp forest in Borneo. *Journal of Tropical Ecology* 16:337–54.

Sakai, S., and T. Inoue. 1999. A new pollination system: Dung-beetle pollination discovered in *Orchidantha inouei* (Lowiaceae, Zingiberales) in Sarawak, Malaysia. *American Journal of Botany* 86:56–61.

Sakai, S., M. Kato, and T. Inoue. 1999. Three pollination guilds and variation in floral characteristics of Bornean gingers (Zingiberaceae and Costaceae). *American Journal of Botany* 86:646–58.

Sakai, S., M. Kato, and H. Nagamasu. 2000. *Artocarpus* (Moraceae)—gall midge pollination mutualism mediated by a male-flower-parasitic fungus. *American Journal of Botany* 87:440–45.

Sakai, S., K. Momose, T. Yumoto, M. Kato, and T. Inoue. 1999. Beetle pollination of *Shorea parvifolia* (section Mutica, Dipterocarpaceae) in a general flowering period in Sarawak, Malaysia. *American Journal of Botany* 86:62–69.

Sakai, S., K. Momose, T. Yumoto, T. Nagamitsu, H. Nagamasu, A. A. Hamid, T. Nakashizuka, and T. Inoue. 1999. Plant reproductive phenology over four years including an episode of general flowering in a lowland dipterocarp forest, Sarawak, Malaysia. *American Journal of Botany* 86:1414–36.

Shanahan, M., and S. G. Compton. 2001. Vertical stratification of figs and fig-eaters in a Bornean lowland rainforest: How is the canopy different? *Plant Ecology* 53:121–32.

Shanahan, M., and I. Debski. 2002. Vertebrates of Lambir Hills National Park, Sarawak. *Malayan Nature Journal* 56:103–18.

Watson, H. 1985. *Lambir Hills National Park: Resource Inventory with Management Recommendations.* National Parks and Wildlife Office, Forest Department, Kuching, Sarawak, Malaysia.

Yamada, T., and E. Suzuki. 1999. Comparative morphology and allometry of winged diaspores among the Asian Sterculiaceae. *Journal of Tropical Ecology* 15:619–35.

Yamakura, T., A. Hagihara, S. Sukardjo, and H. Ogawa. 1990. Aboveground biomass of tropical rain forest stands in Indonesian Borneo. *Vegetatio* 68:71–82.

Yamakura, T., M. Kanzaki, A. Itoh, T. Ohkubo, K. Ogino, E. O. K. Chai, H. S. Lee, and P. S. Ashton. 1995. Topography of a large-scale research plot established within a tropical rain forest at Lambir, Sarawak. *Tropics* 5:41–56.

———. 1996. Forest structure of a tropical rain forest at Lambir, Sarawak with special reference to the dependency of its physiognomic dimensions on topography. *Tropics* 6:1–18.

Yumoto, T. 2000. Bird pollination of three *Durio* species (Bombacaceae) in a tropical rainforest in Sarawak, Malaysia. *American Journal of Botany* 87:1181–88.

32

Luquillo Forest Dynamics Plot, Puerto Rico, United States

Jill Thompson, Nicholas Brokaw, Jess K. Zimmerman, Robert B. Waide, Edwin M. Everham III, and Douglas A. Schaefer

Site Location, Administration, and Scientific Infrastructure

The 16-ha Luquillo Forest Dynamics Plot (LFDP) is located near El Verde Field Station in the Luquillo Mountains of northeastern Puerto Rico, approximately 35 km southeast of San Juan (fig. 32.1). The field station and LFDP are within the Luquillo Experimental Forest (established in 1956), which is coterminous with the Caribbean National Forest (CNF). The CNF covers 11,330 ha, of which 4630 ha are designated for noncommercial uses including research (Brown et al. 1983). The CNF was proclaimed a forest reserve in 1903 and has been administered by the U.S. Department of Agriculture Forest Service since 1917 (Brown et al. 1983). El Verde Field Station is administered by the Institute for Tropical Ecosystem Studies, University of Puerto Rico, and has laboratories and housing for visiting researchers.

A large number of smaller tree demographic plots located in and around the Luquillo Mountains are regularly censused by the International Institute of Tropical Forestry (USDA Forest Service). These include eight 0.4-ha plots set up in 1945–46 that are remeasured every 10 years. There are also 83 circular (10-m diameter) plots that were set up in 1987 and are censused every 5 years (a subset of plots are assessed every year for seedlings). In forested areas in other locations in Puerto Rico, there are 50 circular plots of various diameters that are remeasured every 10 years. The Smithsonian Institution's Monitoring and Assessment of Biodiversity Program set up a 1-ha plot in Puerto Rico in 1988.

Climate

The climate is classified as tropical montane in Walsh's (1996) tropical climate system, and as subtropical wet in the Holdridge life zone system (Ewel and Whitmore 1973). Annual rainfall at El Verde averages just over 3500 mm/year (1975–99). On average, no month has less than 200 mm of rain, although a drier season occurs from January through April. Unusually heavy rainfall events occur about

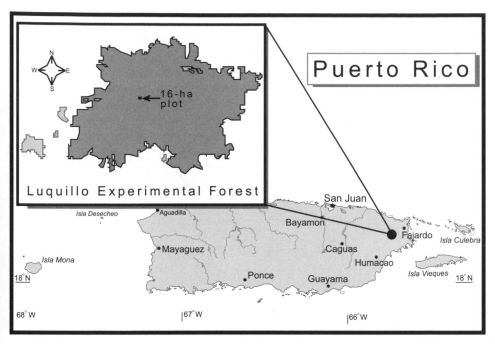

Fig. 32.1. Location of the 16-ha Luquillo Forest Dynamics Plot.

once every 6 years (Fred Scatena personal communication). The highest rainfall recorded at El Verde between 1975 and 2001 (370 mm in less than 12 hours) fell on 17 April 1997. There are also severe droughts, such as in 1994, when less than 70% of the average annual rain fell, and many small streams in the forest stopped flowing entirely for 2 months. Rainfall extremes are not closely linked to the El Niño cycle. Daily average maximum air temperature is 25.2°C, minimum is 20.5°C, and average is 22.8°C (Brown et al. 1983; and Luquillo LTER Internet Site http://luq.lternet.edu). See also table 32.1.

Mean annual wind speed above the canopy is 4.1 km/hr (Waide and Reagan 1996). Monthly averages at Catalina, the most comparable site in Puerto Rico (5.6 km from El Verde, same exposure) with appropriate data, are 2–6 km/hr (Brown et al. 1983).

Average total daily radiation at Bisley (latitude 18.5°18′, longitude 65.5°44′) was 155 +/−59 W/m^2 (mean for 1993 - 02). These data were collected by the USDA Forest Service (Luquillo LTER Internet Site http://luq.lternet.edu). The site is also subject to tropical storms and hurricanes (see below).

Table 32.1. Luquillo Climate Data

	Jan	Feb	Mar	Apr	May	Jun	Jul	Aug	Sep	Oct	Nov	Dec	Total/ Averages
Rain (mm)	233	227	203	232	351	242	307	361	350	288	401	353	3548
ADTMx (°C)	23.0	23.2	24.2	25.3	26.0	27.1	26.7	26.8	26.5	25.8	24.4	23.4	25.2
ADTMn (°C)	18.8	18.7	19.0	19.7	20.7	21.7	21.9	22.0	21.7	21.2	20.6	19.6	20.5

Notes: Raw data are on the Luquillo website for 1975–1999 (http://luq.lternet.edu).

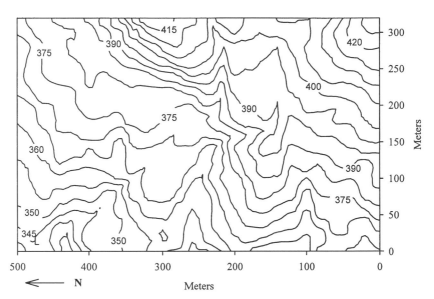

Fig. 32.2. Topographic map of the 16-ha Luquillo Forest Dynamics Plot with 5-m contour intervals.

Topography and Soil

Topography on the LFDP has northwest-running drainages producing steep northeast and southwest-facing slopes, with an elevation across the plot of 333–428 m above sea level (figs. 32.2 and 32.3). The mean slope of the plot is 17% but ranges from 3 to 60%. Soils were formed in residual volcanic ash that fell in the ocean to form volcaniclastic sandstones and siltstones, which were subsequently uplifted (F. Scatena personal communication). Soils are dominated by old, deeply weathered kaolinitic Oxisols (zarzal) and Ultisols (cristal), and young, less-developed Entisols (coloso and fluvaquents) and Inceptisols (prieto) in stream channels. Zarzal, cristal, and prieto are deep clay soils, while coloso and fluvaquents are formed from alluvium in the stream channels (Soil Survey Staff 1995).

Fig. 32.3. Perspective map of the 16-ha Luquillo Forest Dynamics Plot. Modified from Willig et al. (1996), © 1996 by the Association for Tropical Biology. Reprinted with permission.

The plot lies mainly within the 265-ha Quebrada Sonadora Catchment, which exports 410 kg/ha/year of dissolved solids (the sum of inorganic ions excluding silica and bicarbonate, for which data are not available), calculated from data of Schaefer et al. (2000). The closest catchment (within 5 km) for which data on suspended solids are available is the 1782-ha Quebrada Mameyes, which exports 1000 kg/ha/year on average (USGS 1993–98). The flux of suspended solids is sensitive to short-term variations in stream flow. During 1998, the flux was eight times greater than the usual annual rate, mainly as a result of the greater stream flows during Hurricane Georges. Based on a surface soil bulk density of 0.8 g/cm^3 (Soil Survey Staff 1995) and excluding dissolved solid exports of atmospheric origin, the sum of the average exports is equivalent to a soil erosion rate of approximately 0.14 mm/year (the inclusion of silica and bicarbonate fluxes would increase that estimate).

Forest Type and Characteristics

Within the Luquillo Experimental Forest, there are four forest types: tabonuco, colorado, palm-brake, and dwarf forest. These forest types are associated with different soil types and roughly stratified by elevation (Brown et al. 1983). Dwarf (cloud) forest grows on peaks and ridges above 750 m above sea level. Colorado forest grows above the average cloud condensation level (600 m asl), while tabonuco

Table 32.2. Luquillo Plot Census History

Census	Dates	Number of Trees (≥ 1 cm dbh)	Number of Species (≥ 1 cm dbh)	Number of Trees (≥ 10 cm dbh)	Number of Species (≥ 10 cm dbh)
First	August 1990–September 1993	85,607	145	12,917	89
Second	November 1994–October 1996	67,170	138	13,988	86

Notes: Third census was completed in April 2002 (data available for two censuses); the next census is expected to begin in 2005.

We first tagged all trees ≥ 10 cm dbh in the whole plot and then tagged all trees ≥ 1 cm dbh. These data have been combined to comprise the first census. For the second census all stems ≥ 1 cm dbh were measured in the same census.

We measure *Prestoea acuminata* (Palmae) as soon as its youngest leaf arises from the plant at a height of 130 cm from the ground. For the data in this table, however, only those *P. acuminata* that had a "woody bole" at 130 cm from the ground at the time of the census are included.

forest is found below 600 m asl and is best developed on low, protected, well-drained ridges. Palm-brake forest is interspersed within the tabonuco, colorado, and dwarf forests and is limited to areas of steep slopes, poor drainage, and saturated soils (Brown et al. 1983). The 16-ha Luquillo Forest Dynamics Plot lies within the tabonuco forest, a forest named after the dominant tree, *Dacryodes excelsa* Vahl (Burseraceae). Mean canopy height is about 20 m, with the tallest trees nearing 35 m. Canopy height is lower, and forest structure broken, in the years after hurricanes (chap. 13). Few trees in the LFDP exceed 1 m dbh. *Buchenavia tetraphylla* (formerly *B. capitata*) (Aubl.) R. A. Howard (Combretaceae) is one of the few deciduous species in the forest; it drops its leaves all at once and flowers soon afterward (Sastre-de-Jesús 1979). See tables 32.2–32.7 for census data and rankings.

Species composition varies markedly across the plot, depending on land use history (Thompson et al. 2002). *Dacryodes excelsa* characterizes less disturbed areas, while *Casearia arborea* (Rich.) Urb. (Flacourtiaceae) dominates more disturbed areas. The midstory palm, *Prestoea acuminata* (Willd.) H.E. Moore (Palmae), is common in poorly drained areas, including stream channels, and on slopes.

Canopy heights recorded at 6565 grid points (every 5 m) between April 1999 and March 2000 showed 19.1% of points with a canopy height of ≤ 2 m, 33.8% ≤ 10 m, and 47.1% > 10 m. The relatively high percentage of low canopy is a result of the damage caused by Hurricane Georges in September 1998.

Total annual litter fall near the LFDP was 568.25 g/m^2 (mean for 1990–2001), comprised of leaves 419.7 g/m^2, wood 55.3 g/m^2, and miscellaneous 93.19 g/m^2 (Mathew Warren unpublished data).

Table 32.3. Luquillo Summary Tally

Size Class (cm dbh)	Mean per Hectare						16-ha Plot				
	BA	N	S	G	F	H'	α	S	G	F	H'	α
≥ 1	38.3	4171.7	73.3	58	33.6	1.45	13.54	138	102	47	1.41	16.60
≥ 10	34.4	875.7	42.1	37	24.0	1.01	9.27	86	71	39	1.13	12.20
≥ 30	18.3	109.7	22.2	21	15.5	1.08	8.78	57	49	29	1.29	11.30
≥ 60	5.3	11.4	5.6	5.5	5.5	0.64	5.76*	23	22	19	1.01	7.05

*Does not include three 1-ha plots with fewer than nine species, each species with only one individual.

Notes: BA represents basal area in m², N is the number of individual trees, S is number of species, G is number of genera, F is number of families, H' is Shannon–Wiener diversity index using $\log_{10}$, and α is Fisher's α. Basal area includes all multiple stems for each individual. Individuals are counted using their largest stem. Mean values are based on 15 nonoverlapping, 1-ha plots. Data are from the second census. 8 individuals were not identified to species or morphospecies.

Table 32.4. Luquillo Rankings by Family

Rank	Family	Basal Area (m²)	% BA	% Trees	Family	Trees	% Trees	Family	Species
1	Palmae	106.0	17.3	8.8	Rubiaceae	22,874	34.1	Rubiaceae	16
2	Burseraceae	83.6	13.7	3.3	Palmae	5,896	8.8	Melastomataceae	8
3	Flacourtiaceae	50.5	8.3	8.8	Flacourtiaceae	5,889	8.8	Lauraceae	8
4	Combretaceae	44.8	7.3	0.3	Cecropiaceae	4,845	7.2	Flacourtiaceae	7
5	Sapotaceae	40.3	6.6	2.6	Elaeocarpaceae	3,215	4.8	Piperaceae	7
6	Leguminosae	39.8	6.5	2.7	Araliaceae	2,884	4.3	Euphorbiaceae	7
7	Meliaceae	36.2	5.9	2.6	Piperaceae	2,255	3.4	Myrtaceae	7
8	Cecropiaceae	31.6	5.2	7.2	Burseraceae	2,208	3.3	Leguminosae	6
9	Euphorbiaceae	30.3	5.0	2.6	Leguminosae	1,800	2.7	Moraceae	4
10	Elaeocarpaceae	21.9	3.6	4.8	Euphorbiaceae	1,755	2.6	Meliaceae	4
								Sapotaceae	4
								Solanaceae	4

Notes: The top 10 families for trees ≥ 1 cm dbh are ranked in terms of basal area, number of individual trees, and number of species, with the percentage of trees in the plot. Data are from the second census.

Fauna

At El Verde, the diversity of birds (31 resident species, density 33 individuals/ha, Waide 1996) and mammals (13 native species, all of which are bats) is low compared to mainland tropical forests (Reagan and Waide 1996). The Puerto Rican Tody (*Todus mexicanus*), for example, is the only understory insectivorous bird. This low diversity of homeothermic vertebrates is compensated by a high density of frogs (11 species of *Eleutherodactylus*), with the most common, *E. coqui*, at a density of 25,570 individuals/ha (Stewart and Woolbright 1996), and lizards

Table 32.5. Luquillo Rankings by Genus

Rank	Genus	Basal Area (m²)	% BA	% Trees	Genus	Trees	% Trees	Genus	Species
1	Prestoea (Palmae)	100.8	16.5	8.7	Palicourea (Rubiaceae)	13,190	19.6	Piper (Piperaceae)	7
2	Dacryodes (Burseraceae)	78.3	12.8	2.5	Psychotria (Rubiaceae)	8,684	12.9	Ocotea (Lauraceae)	6
3	Buchenavia (Combretaceae)	44.8	7.3	0.28	Prestoea (Palmae)	5,853	8.7	Miconia (Melastomataceae)	6
4	Manilkara (Sapotaceae)	39.8	6.5	2.46	Casearia (Flacourtiaceae)	5,295	7.9	Psychotria (Rubiaceae)	4
5	Guarea (Meliaceae)	34.2	5.6	1.38	Cecropia (Cecropiaceae)	4,848	7.2	Myrcia (Myrtaceae)	3
6	Cecropia (Cecropiaceae)	31.6	5.2	7.2	Sloanea (Elaeocarpaceae)	3,215	4.8	Eugenia (Myrtaceae)	3
7	Inga (Leguminosae)	30.9	5.1	2.19	Schefflera (Araliaceae)	2,665	4.0	Coccoloba (Polygonaceae)	3
8	Casearia (Flacourtiaceae)	29.5	4.8	7.9	Piper (Piperaceae)	2,255	3.4	Casearia (Flacourtiaceae)	3
9	Sloanea (Elaeocarpaceae)	21.9	3.6	4.8	Manilkara (Sapotaceae)	1,650	2.5	Solanum (Solanaceae)	2
10	Homalium (Flacourtiaceae)	17.9	2.9	0.4	Dacryodes (Burseraceae)	1,649	2.5	Palicourea (Rubiaceae)	2
								Micropholis (Sapotaceae)	2
								Inga (Leguminosae)	2
								Guarea (Meliaceae)	2
								Ficus (Moraceae)	2
								Cordia (Boraginaceae)	2
								Citharexylum (Verbenaceae)	2
								Byrsonima (Malpighiaceae)	2

Notes: The top 10 tree genera for trees ≥1 cm dbh are ranked by basal area, number of individual trees, and number of species with the percentage of trees in the plot. Data are from the second census.

(five *Anolis* species), with the most common three species at a density of 25,000 individuals/ha (Reagan 1996). Of consequence for forest dynamics, mammalian herbivores (e.g., deer, tapir, agouti) are absent from the LFDP. The black rat (*Rattus rattus*) and the mongoose (*Herpestes auropunctatus*) are introduced species and the only mammals, other than the native bats, that inhabit the forest.

Table 32.6. Luquillo Rankings by Species

Rank	Species	No. Trees	% Trees	Species	Basal Area (m²)	% BA	% Trees
1	*Palicourea riparia* (Rubiaceae)	13,186	19.63	*Prestoea acuminata* (Palmae)	100.8	16.49	8.71
2	*Psychotria berteriana* (Rubiaceae)	6,416	9.55	*Dacryodes excelsa* (Burseraceae)	78.3	12.81	2.45
3	*Prestoea acuminata* (Palmae)	5,853	8.71	*Buchenavia tetraphylla* (Combretaceae)	44.8	7.33	0.28
4	*Cecropia schreberiana* (Cecropiaceae)	4,848	7.22	*Manilkara bidentata* (Sapotaceae)	39.8	6.50	2.46
5	*Casearia arborea* (Flacourtiaceae)	3,477	5.18	*Guarea guidonia* (Meliaceae)	33.5	5.47	0.86
6	*Sloanea berteriana* (Elaeocarpaceae)	3,215	4.79	*Cecropia schreberiana* (Cecropiaceae)	31.6	5.16	7.22
7	*Schefflera morototoni* (Araliaceae)	2,665	3.97	*Inga laurina* (Leguminosae)	27.9	4.57	1.90
8	*Psychotria brachiata* (Rubiaceae)	2,236	3.33	*Casearia arborea* (Flacourtiaceae)	23.5	3.85	5.18
9	*Piper glabrescens* (Piperaceae)	2,028	3.02	*Sloanea berteriana* (Elaeocarpaceae)	21.9	3.58	4.79
10	*Casearia sylvestris* (Flacourtiaceae)	1,809	2.69	*Homalium racemosum* (Flacourtiaceae)	17.9	2.93	0.39

Notes: The top 10 tree species for trees ≥1 cm dbh are ranked by number and percentage of trees and basal area.

Table 32.7. Luquillo Tree Demographic Dynamics

Size Class (cm dbh)	Growth Rate (mm/yr)	Mortality Rate (%/yr)	Recruitment Rate (%/yr)	BA Losses (m²/ha/yr)	BA Gains (m²/ha/yr)
1–9.9*	1.6	14.05	5.25	0.43	0.031
10–29.9**	3.4	5.0	4.78	0.71	1.85
≥30**	2.7	2.4	2.52	0.23	2.99
P. acuminata * 1–9.9		1.78	11.01	0.003	1.78
P. acuminata ** 10–29.9		1.1	6.14	.006	0.12

Notes: Table shows growth, mortality, and recruitment rates for three size classes of trees and mortality and recruitment rates for *Prestoea acuminata* between the first (1990) and second census (1996). The data include the effects of Hurricane Hugo (September 1989) on recruitment, mortality, and growth rates.

*Calculated between 1993 and 1996, mean time between censuses of each 20 × 20m quadrat was 2.91 years.

**Calculated between 1990 and 1996, mean time between censuses of each 20 × 20m quadrat was 4.46 years.

Growth rate includes all live stems on all plants, including those stems with negative growth.

Basal area losses and gains include all live stems on all plants, including those stems with less than 5% negative growth.

Natural Disturbances

The average return interval of severe hurricanes to the area is 50–60 years (Scatena and Larsen 1991), but the forest was damaged by severe hurricanes in both 1989 (Hugo) and 1998 (Georges). Local topography and local exposure to winds influence the extent of wind damage (Boose et al. 1994). Stands on exposed sites are continually trimmed by the trade winds and as a result they suffer proportionately less damage during major storms than do areas that are usually more sheltered. Individual tree damage varies according to species (Zimmerman et al. 1994). Treefall gaps that are not associated with major storms tend to be fewer and smaller than in some other tropical forests (chap. 13). Landslides also cause natural disturbance. Landslides affect on average 1% of the forest area at any time and are mainly related to heavy rainfall events and the reshaping of stream channels.

Human Disturbance

The LFDP has a varied land use history but has been covered by forest since the 1930s (cf. Foster et al. 1999). Before then, 1.16 ha of the 16-ha plot was farmed; 9.6 ha was variably clearcut, then allowed to regrow but planted in places with coffee or fruit trees; and 5.24 ha has always been in forest but was selectively logged (Thompson et al. 2002). A small area (320 m²) was clearcut in the 1960s as part of an experiment (Odum and Pigeon 1970). In addition, the introduction of the black rat and mongoose to the forest is an indirect form of human disturbance.

Plot Size and Location

LFPD is 16-ha, 500 × 320 m plot, the long axis lies north-south. The southwest corner is located at 18°19′26″ N and 65°49′3″ W.

Funding Sources

The Luquillo Forest Dynamics Plot was established with funds from a U.S. National Science Foundation (NSF) SGER grant (BSR-9015961) to the University of Puerto Rico (UPR) a proposal written by D. Jean Lodge and Charlotte M. Taylor. This work was also supported by LTER grants (BSR-8811902 and BSR-8811764) from NSF to the Institute for Tropical Ecosystem Studies, UPR, and the International Institute for Tropical Forestry, as part of the Long-Term Ecological Research Program in the Luquillo Experimental Forest. The U.S. Forest Service (Department of Agriculture) and UPR gave additional support. Funds were also provided through grants (RII-880291 and HRD-9353549) from NSF to the UPR

Center for Research Excellence in Science and Technology. The recent third census was funded by the Andrew W. Mellon Foundation.

References

Boose, E. R., D. R. Foster, and M. Fluet. 1994. Hurricane impacts to tropical and temperate forest landscapes. *Ecological Monographs* 64:369–400.

Brown, S., A. E. Lugo, S. Silander, and L. Liegel. 1983. Research history and opportunities in the Luquillo Experimental Forest. USDA Forest Service General Technical Report SO–44, Southern Forest Experiment Station, New Orleans, LA.

Edmisten, J. 1970. Soil studies in El Verde rain forest. Pages H79–87 in H. T. Odum and R. F. Pigeon, editors. *A Tropical Rain Forest: A Study of Irradiation and Ecology at El Verde, Puerto Rico.* National Technical Information Service, Springfield, VA.

Ewel, J. J., and J. L. Whitmore. 1973. The ecological life zones of Puerto Rico and the U.S. Virgin Islands. Forest Service Research Paper ITF–18. International Institute of Tropical Forestry, Río Piedras, Puerto Rico.

Foster, D. R., M. Fluet, and E. R. Boose. 1999. Human or natural disturbance: Landscape-scale dynamics of the tropical forests of Puerto Rico. *Ecological Applications* 9:555–72.

Lugo, A. E., and F. N. Scatena. 1995. Ecosystem-level properties of the Luquillo experimental forest with emphasis on the tabonuco forest. Pages 59–108 in A. E. Lugo and C. Lowe, editors. *Tropical Forests: Management and Ecology.* Springer-Verlag, New York.

Mabberley, D. J. 1997. *The Plant-Book: A Portable Dictionary of the Vascular Plants.* Cambridge University Press, Cambridge, U.K.

Odum, H. T., and R. F. Pigeon. 1970. *A Tropical Rain Forest: A Study of Irradiation and Ecology at El Verde, Puerto Rico.* National Technical Information Service, Springfield, VA.

Reagan, D.P. 1996. Anoline lizards. Pages 321–46 in D. P. Reagan and R. B. Waide, editors. *The Food Web of a Tropical Rain Forest.* University of Chicago Press, Chicago.

Reagan, D. P., and R. B. Waide. 1996. *The Food Web of a Tropical Rain Forest.* University of Chicago Press, Chicago.

Sastre-de-Jesús, I. 1979. *Ecological Life Cycle of* Buchenavia Capitata *(Vahl) Eichl., a Late Secondary Successional Species in the Rain Forest of Puerto Rico.* Masters thesis. University of Tennessee, Knoxville, TN.

Scatena, F. N., and M. C. Larsen. 1991. Physical aspects of Hurricane Hugo in Puerto Rico. *Biotropica* 23:317–23.

Schaefer, D. A., W. H. McDowell, F. N. Scatena, and C. E. Asbury. 2000. Effects of hurricane disturbance on stream water concentrations and fluxes in eight tropical forest watersheds of the Luquillo Experimental Forest, Puerto Rico. *Journal of Tropical Ecology* 16:189–207.

Soil Survey Staff. 1995. Order 1 soil survey of the Luquillo Long–Term Ecological Research Grid, Puerto Rico. U.S. Department of Agriculture, Natural Resources Conservation Service, Lincoln, NE.

Stewart, M. M., and L. L. Woolbright. 1996. Amphibians. Pages 273–320 in D. P. Reagan and R. B. Waide, editors. *The Food Web of a Tropical Rain Forest.* University of Chicago Press, Chicago.

Thompson, J., N. Brokaw, J. K. Zimmerman, R. B. Waide, E. M. Everham, III, D. J. Lodge, C. M. Taylor, D. García-Montiel, and M. Fluet. 2002. Land use history, environment, and tree composition in a tropical forest. *Ecological Applications* 12:1344–63.

U.S. Geological Survey. 1993–98. Water resources data for Puerto Rico and the U.S. Virgin Islands, water year 1993 (etc.). U.S. Geological Survey Water Data Report 93–1, 94–1, (etc.), Guaynabo, Puerto Rico.

Waide, R.B. 1996. Birds. Pages 363–98 in D. P. Reagan and R. B. Waide, editors. *The Food Web of a Tropical Rain Forest.* University of Chicago Press, Chicago.

Waide, R. B., and D. P. Reagan. 1996. The rain forest setting. Pages 1–16 in D. P. Reagan and R. B. Waide, editors. *The Food Web of a Tropical Rain Forest.* University of Chicago Press, Chicago.

Walker, L. R., N. V. L. Brokaw, D. J. Lodge, and R. B. Waide. 1991. Special issue: Ecosystem, plant and animal responses to hurricanes in the Caribbean. *Biotropica* (*Suppl.*) 23:313–521.

Walsh, R. P. D. 1996. Climate. Pages 159–205 in P. W. Richards. *The Tropical Rain Forest: An Ecological Study.* Second Edition. Cambridge University Press, Cambridge, U.K.

Willig, M.R., D.L. Moorhead, S. B. Cox, and J. C. Zak, 1996. Functional diversity of soil bacterial communities in the tabonuco forest: Interactions of anthropogenic and natural disturbance. *Biotropica* 28:65–76

Zimmerman, J. K., E. M. Everham, III, R. B. Waide, D. J. Lodge, C. M. Taylor, and N. V. L. Brokaw. 1994. Responses of tree species to hurricane winds in subtropical wet forest in Puerto Rico: Implications for tropical tree life histories. *Journal of Ecology* 82:911–22.

33

Mudumalai Forest Dynamics Plot, India

Raman Sukumar, Hebbalalu Sathyanarayana Suresh,
Handanakere Shivaramiah Dattaraja, Robert John,
and Niranjan V. Joshi

Site Location, Administration, and Scientific Infrastructure

Mudumalai Wildlife Sanctuary (11°36′N, 76°32′E) abuts the northern flank of
the Nilgiri mountain range in the Western Ghats and is contiguous with the
protected areas of Bandipur and Wynaad (Sukumar et al. 1992). Mudumalai was
designated as a wildlife sanctuary in 1940, expanded in 1958, and expanded again
to the current area of 32,100 ha in 1977. The 50-ha Mudumalai Forest Dynamics
Plot is located in Compartment 17 of the Kargudi Range in the Mudumalai Wildlife
Sanctuary, Nilgiris District, Tamilnadu State in southern India (fig. 33.1). This
plot was set up and is managed by the Centre for Ecological Sciences of the Indian
Institute of Science, Bangalore, India.

In addition to the Forest Dynamics Plot, a series of nineteen 1-ha permanent
plots has also been set up along a rainfall gradient in the Mudumalai Wildlife
Sanctuary. Another 1-ha permanent plot has been located outside Mudumalai, in
a montane area at Thaishola, in the upper plateau of the Nilgiri District. A field
station with moderately equipped laboratory, computer, telephone, and accom-
modation facilities is located in Masinagudi village near Mudumalai.

Climate

The western and central parts of Mudumalai receive most of their rainfall from
the southwest monsoon during June-September, while the drier eastern region
also receives significant rainfall from the northeast monsoon during October–
November. Pre-monsoon showers also occur during April and May. The dry
season usually starts in mid-November and lasts until mid-April with 6 consecutive months averaging less than 100 mm precipitation a month. Rainfall ranges
from 700 mm/year at the eastern end of Mudumalai Wildlife Sanctuary to over
1800 mm/year at the western end. The dry season spans 5–6 months in most areas,
but in the low rainfall regions, it extends to about 8 months. Rainfall in the 50-ha
Forest Dynamics Plot averages about 1200 mm/yr, as discerned from 1941–70 and
1990–96 data at Kargudi (5 km south of the plot, von Lengerke 1977) and from

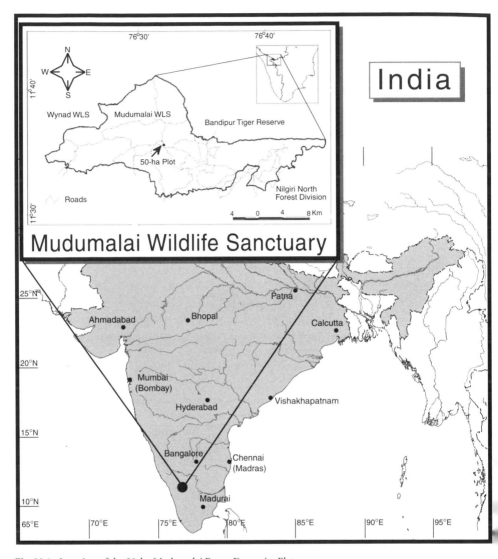

Fig. 33.1. Location of the 50-ha Mudumalai Forest Dynamics Plot.

records of the Indian Institute of Science and the Tamilnadu Electricity Board. See also table 33.1.

Topography and Soil

The 50-ha Mudumalai Forest Dynamics Plot has an undulating terrain with elevation varying between 980 and 1120 m above sea level (figs. 33.2 and 33.3). A

Table 33.1. Mudumalai Climate Data

	Jan	Feb	Mar	Apr	May	Jun	Jul	Aug	Sep	Oct	Nov	Dec	Total/Averages
Rain (mm)	7	9	88	119	129	160	146	130	126	170	115	50	1250
ADTMx (°C)	26.7	28.8	34.0	30.4	29.5	27.0	28.3	29.4	23.1	26.5	27.2	27.8	28.2
ADTMn (°C)	16.1	16.3	19.8	18.0	16.2	18.2	20.1	20.1	14.7	18.4	15.2	14.2	17.3
Solar Radiation	308	367	352	349	345	251	177	179	266	252	312	255	284

Notes: Mean monthly rainfall data from 1990 to 2000, and average daily temperature (ADT) maximums and minimums during 1990–2000. Solar radiation pertains to UPASI Tea Research Station at Gudalur, 15 km (linear distance) to the southwest of the 50-ha plot. Radiation during June–August can be expected to be slightly higher in the plot because of lower cloudiness.

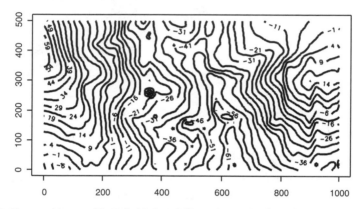

Fig. 33.2. Topographic map of the 50-ha Mudumalai Forest Dynamics Plot with 5-m contour intervals.

small stream, which is joined by a number of smaller streams, drains the 50-ha plot. The main stream crosses the plot in a southeast-northwest direction. All streams are typically dry after the monsoon season. The nineteen 1-ha plots in Mudumalai are located within the same elevation range as the 50-ha plot, while the montane forest plot in Thaishola is located at 1875 m.

The parent rock material of the 50-ha plot is either hornblende biotite gneiss or granite gneiss. The soils of Mudumalai Wildlife Sanctuary have been classified into four orders, Inceptisols, Alfisols, Mollisols, and Entisols (George et al. 1988). The soils of the Mudumalai Forest Dynamics Plot have been classified under order Alfisols, subgroup udic haplustalf and family clayey skeletal. Soils are dark brown to deep black in color, and are slightly acidic with pH values of 6.2–6.6. The total soil organic carbon content is 1.6–2.4%, while the nitrogen, phosphorus, and sulfur contents are 0.16, 0.07, and 0.02%, respectively. The soils are rich in such nutrients as iron, manganese, zinc, and copper.

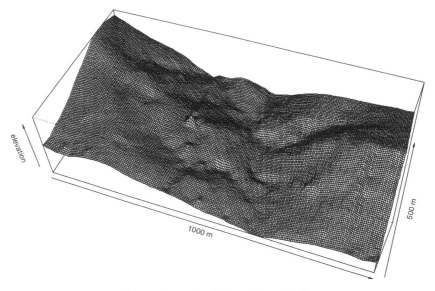

Fig. 33.3. Perspective map of the 50-ha Mudumalai Forest Dynamics Plot.

Forest Type and Characteristics

The strong gradient in rainfall across Mudumalai Wildlife Sanctuary brings about marked structural and floristic differences. Vegetation varies from moist deciduous forest with patches of semievergreen forest in the western region, through dry deciduous forest, to dry thorn forest toward the east. Dry deciduous forest, spread over the central and northern parts, constitutes the largest expanse of the sanctuary. The Mudumalai Forest Dynamics Plot is located in the transition zone between dry and moist deciduous forests. The vertical stratification in the dry deciduous forest is simple, with only a tree canopy layer (~10–18 m) and a ground layer of prominent dense grasses. A layer of midstory trees is also present, but it is sparse and discontinuous. The common canopy trees in the dry deciduous forest are *Tectona grandis* (Labiatae), *Terminalia crenulata* (Combretaceae), and *Anogeissus latifolia* (Combretaceae), though *Lagerstroemia microcarpa* (Lythraceae), characteristic of moist deciduous forest, is the dominant tree in the 50-ha plot. The understory has fewer species with *Kydia calycina* (Malvaceae), *Phyllanthus emblica*, (Euphorbiaceae), and *Catunaregam spinosa* (Rubiaceae) among the common species. A characteristic feature of the dry deciduous forest at Mudumalai is the presence of a dense understory of perennial, tall grasses (Gramineae) such as *Themeda cymbaria* and *Cymbopogon flexuosus*. Grass cover is generally inversely proportional to canopy cover; the grass cover is highest in the dry deciduous forests that have been repeatedly burned. Canopy cover is about 50–75% in the deciduous forests, with higher values in the moister forests.

The canopy in the moist deciduous forest is about 20–25 m high and denser than that of the dry deciduous forest. The midstory is more continuous and grasses may be sparse or absent in the wetter parts. Tree species that are common in moist deciduous forest include *Lagerstroemia microcarpa* (Lythraceae), *Grewia tiliifolia* (Tiliaceae), *Syzygium cumini* (Myrtaceae), *Persea macarantha* (Lauraceae), *Meliosma simplicifolia* (Sabiaceae), and *Olea dioica* (Oleaceae). The low rainfall areas in the eastern parts of Mudumalai promote dry thorn forest or mixed deciduous forest. Here, the forest is defined by a ground layer of thorny evergreen shrubs about 3–5 m tall, with a sparse cover of deciduous trees. The average number of woody plant species per hectare in the dry thorn/deciduous forest is about 35, while it is 25 and 46 in the dry and moist deciduous forests, respectively. The common woody plant species in the dry thorn forest include *Gardenia turgida* (Rubiaceae), *Flueggea lycopyros* (Euphorbiaceae), *Anogeissus latifolia* (Combretaceae), and *Ziziphus xylopyrus* (Rhamnaceae).

The strong 5- to 6-month dry season typically begins in November and lasts until April. Leaves are completely shed in January and February and new leaves are flushed by the end of April (Murali and Sukumar 1993). Presently recovering from selective logging (which ended in 1968), the 50-ha plot comprises secondary forest that is maintained through regular burning of the ground layer. Species diversity varies within the plot and is only weakly associated with topographical features. However, a number of rare species characteristic of moist riparian forest such as *Mangifera indica* (Anacardiaceae), *Mallotus philippensis* (Euphorbiaceae), *Olea dioica* (Oleaceae), and *Bischofia javanica* (Euphorbiaceae) can be found along the stream banks of the plot. In the drier sites, the peak in species flowering extends well into the wet season, while in the medium-to-wetter sites, the peak occurs in the dry season. Bird-pollinated species tend to flower in the dry season, while wind-pollinated species flower in the rainy season when winds are stronger. For the majority of the tree species at Mudumalai, flowering and leaf flush are simultaneous (Murali and Sukumar 1994). Litter production varies along the rainfall gradient from about 750 kg/ha in the dry thorn forest to about 3750 kg/ha in the semievergreen forest patches toward the western part; the peak litter fall of 5030 kg/ha is reached in the deciduous forests of the central part (unpublished data). For census data and rankings, see tables 33.2–33.7.

Fauna

Over 200 species of birds, at least 17 species of amphibians, 42 species of reptiles, and 35 species of nonvolant mammals have been documented in the sanctuary. Mudumalai harbors one of the highest densities of large mammals in Asia

Table 33.2. Mudumalai Plot Census History

Census	Dates	Number Trees (≥ 1 cm dbh)	Number Species (≥ 1 cm dbh)	Number Trees (≥ 10 cm dbh)	Number Species (≥ 10 cm dbh)
First	May 1988–May 1989	25,553	70	15,037	62
	August 1989–January 1990	23,207	69	14,637	62
	June 1990–October 1990	22,146	69	14,295	62
	June 1991–October 1991	19,571	69	13,836	62
Second	June 1992–November 1992	17,628	67	14,047	63
	July 1993–September 1993	17,799	67	13,774	63
	June 1994–August 1994	17,176	69	13,480	63
	June 1995–November 1995	17,266	68	13,256	63
Third	June 1996–September 1996	15,304	65	13,070	63
	June 1997–January 1997	15,791	69	12,876	63
	July 1999–September 1998	16,397	69	12,739	62
	June 1998–October 1999	16,828	70	12,633	62
Fourth	June 2000–December 2000	18,024	71	12,576	63

Notes: Four full censuses have been completed. Unlike most other Forest Dynamics Plots, annual censuses for mortality and recruitment are also undertaken in this 50-ha plot. Tree girths were measured only once every 4 years, in 1988, 1992, 1996, and 2000. One species of bamboo, *Bambusa arundinacea*, was tagged and enumerated but is not included in the calculations reported in this chapter.

Table 33.3. Mudumalai Summary Tally

Size Class (cm dbh)	Average per Hectare							50-ha Plot				
	BA	N	S	G	F	H'	α	S	G	F	H'	α
≥ 1	25.5	366.8	24.7	22.0	16.6	0.944	6.2	71	55	29	1.089	9.4
≥ 10	25.1	244.8	19.8	18.4	14.2	0.870	5.3	63	51	29	1.014	8.7
≥ 30	20.8	105.7	14.0	13.4	11.0	0.830	4.4	50	42	24	0.970	7.6
≥ 60	7.11	15.8	5.4	5.0	4.7	0.561	3.5	28	22	15	0.875	5.7

Notes: BA represents basal area in m^2, N is the total number of individual trees, S is the total number of species, G is the total number of genera, F is the total number of families, H' is the Shannon–Wiener diversity index using $\log_{10}$, and α is Fisher's α. Two individuals were unidentified to species or morphospecies. Data are from the 2000 census.

Table 33.4. Mudumalai Rankings by Family

Rank	Family	Basal Area (m^2)	% BA	% Trees	Family	Trees	% Trees	Family	Species
1	Labiatae	364.8	28.2	10.6	Combretaceae	4761	26.2	Leguminosae	13
2	Combretaceae	355.0	27.5	26.2	Lythraceae	4054	22.3	Moraceae	7
3	Lythraceae	243.9	18.9	22.3	Leguminosae	3504	19.3	Euphorbiaceae	5
4	Tiliaceae	74.3	5.8	2.1	Labiatae	1924	10.6	Malpighiaceae	5
5	Myrtaceae	44.7	3.5	2.2	Sterculiaceae	573	3.2	Combretaceae	4
6	Sapindaceae	32.5	2.5	0.5	Malpighiaceae	566	3.1	Labiatae	4
7	Malpighiaceae	32.5	2.5	3.1	Euphorbiaceae	462	2.5	Oleaceae	3
8	Leguminosae	30.5	2.4	19.3	Bignoniaceae	432	2.4	Sapindaceae	3
9	Bignoniaceae	30.2	2.3	2.4	Myrtaceae	408	2.2	Tiliaceae	3
10	Euphorbiaceae	18.0	1.4	2.5	Tiliaceae	383	2.1	Olacaceae	2

Notes: The top 10 families for trees ≥ 1 cm dbh in 50-ha plot. Data are from 2000 census.

Table 33.5. Mudumalai Ranking by Genus

Rank	Genus	Basal Area	% BA	% Trees	Genus	Trees	% Trees	Genus	Species
1	*Tectona* (Labiatae)	353.3	27.3	10.4	*Lagerstroemia* (Lythraceae)	4054	22.3	*Ficus* (Moraceae)	6
2	*Terminalia* (Combretaceae)	254.1	19.7	14.4	*Cassia* (Leguminosae)	3206	17.7	*Grewia* (Tiliaceae)	3
3	*Lagerstroemia* (Lythraceae)	243.9	18.9	22.3	*Terminalia* (Combretaceae)	2612	14.4	*Terminalia* (Combretaceae)	3
4	*Anogeissus* (Combretaceae)	100.9	7.8	11.8	*Anogeissus* (Combretaceae)	2149	11.8	*Bauhinia* (Leguminosae)	2
5	*Grewia* (Tiliaceae)	74.3	5.8	2.1	*Tectona* (Labiatae)	1883	10.4	*Cassia* (Leguminosae)	2
6	*Syzygium* (Myrtaceae)	44.7	3.5	2.2	*Helicteres* (Sterculiaceae)	547	3.0	*Cordia* (Boraginaceae)	2
7	*Schleichera* (Sapindaceae)	32.2	2.5	0.4	*Catunaregam* (Malphigiaceae)	519	2.9	*Dalbergia* (Leguminosae)	2
8	*Catunaregam* (Malphigiaceae)	27.1	2.1	2.9	*Phyllanthus* (Euphorbiaceae)	422	2.3	*Lagerstroemia* (Lythraceae)	2
9	*Ficus* (Moraceae)	17.7	1.4	0.4	*Syzygium* (Myrtaceae)	408	2.2	*Olea* (Oleaceae)	2
10	*Phyllanthus* (Euphorbiaceae)	16.0	1.2	2.3	*Grewia* (Tiliaceae)	383	2.1	*Ziziphus* (Rhamnaceae)	2

Notes: Top 10 tree genera for trees ≥1cm dbh. Data are from 2000 census.

Table 33.6. Mudumalai Rankings by Species

Rank	Species	Number Trees	% Trees	Species	Basal Area (m^2)	% BA	% Trees
1	*Lagerstroemia microcarpa* (Lythraceae)	3708	21.9	*Tectona grandis* (Labiatae)	353.3	27.3	10.4
2	*Terminalia crenulata* (Combretaceae)	2572	14.0	*Lagerstroemia microcarpa* (Lythraceae)	240.4	18.6	21.9
3	*Cassia fistula* (Leguminosae)	2250	13.7	*Terminalia crenulata* (Combretaceae)	235.8	18.3	14.0
4	*Anogeissus latifolia* (Combretaceae)	2166	11.8	*Anogeissus latifolia* (Combretaceae)	100.9	7.8	11.8
5	*Tectona grandis* (Labiatae)	1885	10.4	*Grewia tilifolia* (Tiliaceae)	74.2	5.7	2.1
6	*Catunaregam spinosa* (Malphigiaceae)	538	2.9	*Syzygium cumini* (Myrtaceae)	44.7	3.5	2.2
7	*Phyllanthus emblica* (Euphorbiaceae)	430	2.3	*Schleichera oleosa* (Sapindaceae)	32.2	2.5	0.4
8	*Syzygium cumini* (Myrtaceae)	390	2.2	*Catunaregam spinosa* (Malpighiaceae)	27.1	2.1	2.9
9	*Grewia tiliifolia* (Tiliaceae)	386	2.4	*Terminalia bellirica* (Combretaceae)	16.4	1.3	0.2
10	*Radermachera xylocarpa* (Bignoniaceae)	317	1.9	*Phyllanthus emblica* (Euphorbiaceae)	16.0	1.2	2.3

Notes: 10 most abundant species in plot for trees ≥1 cm dbh. Data are from 2000 census.

Table 33.7. Mudumalai Tree Demographic Dynamics from 1988 to 2000

Tree size (cm dbh)	Growth (mm/yr)			Mortality (%/yr)			Recruitment (%/yr)			Basal Area Losses (m²/ha/yr)			Basal Area Gains (m²/ha/yr)		
	1988–92	1992–96	1996–2000	1988–92	1992–96	1996–2000	1988–92	1992–96	1996–2000	1988–92	1992–96	1996–2000	1988–92	1992–96	1996–2000
1–9.9	3.691	2.898	3.285	27.966	19.277	6.696	15.756	16.591	15.442	0.09	0.02	0.01	0.01	0.00	0.01
10–29.9	2.766	1.671	2.195	4.044	3.032	1.658	17.876	17.288	16.038	0.14	0.12	0.04	0.07	0.01	0.01
>30	3.744	1.819	3.218	0.732	0.585	0.465	18.631	17.707	16.614	0.14	0.13	0.13	0.12	0.13	0.19

including six species of the larger herbivores: the Asian elephant (*Elephas maximus*), gaur (*Bos gaurus*), sambar deer (*Cervus unicolor*), axis or spotted deer (*Axis axis*), muntjac or barking deer (*Muntiacus muntjak*), and chowsingha or four-horned antelope (*Tetracerus quadricornis*) (Varman and Sukumar 1995). There are three primate species: the common langur (*Semnopithecus entellus*), the bonnet macaque (*Macaca radiata*), and the slender loris (*Loris tardigradus*). Other arboreal animals include the malabar giant squirrel (*Ratufa indica*) and flying squirrel (*Petaurista philippensis*). There are three large predators: the tiger (*Panthera tigris*), leopard (*Panthera pardus*), and dhole or Asiatic wild dog (*Cuon alpinus*) (Venkataraman et al. 1995). One species of bear, the sloth bear (*Melursus uscinus*), also occurs in the sanctuary. Density estimates (mean ± S.E.) are available for the larger herbivores; elephant densities average 2.2 ± 0.6 individuals/km^2, gaur 2.2 ± 0.5/km^2, sambar 4.1 ± 1.1/km^2, and spotted deer 29.1 ± 4.7/km^2 (unpublished data for the year 2001).

Natural Disturbances

Mudumalai has a very high density of browsing large mammals that can cause damage to woody plants through their feeding activity. Yet only one understory tree, *Kydia calycina* (Malvaceae), and one shrub, *Helicteres isora* (Sterculiaceae), have drastically declined in the Mudumalai Forest Dynamics Plot as a result of feeding by elephants (chap. 21). Similarly, deer have damaged, by rubbing their antlers, only a small proportion of juvenile stems of a few species. It is likely that browsing by large mammals such as gaur exerts a greater influence on tree saplings <1 cm dbh, though this has not been monitored. Treefalls during storms are not common but known to occur occasionally during the winter monsoon when cyclones form.

Human Disturbances

The Nilgiri region, including Mudumalai, as with other regions in southern India, has an ancient history of human settlement (Hockings 1989; Prabhakar 1994). Remains of a megalithic culture that prevailed from about 100 A.D. onward can be seen in the midelevation forests. Hunter–gatherer societies such as the Kurubas, Irulas, Paniyas, and Kotas have inhabited this region for several centuries. Until the end of the 18th century, the human populations fluctuated in size in response to disease outbreaks and local strife. From the 19th until the mid-20th century, the presence of malaria certainly discouraged settlers from the plains from moving into this region, and population densities everywhere were lower than current levels. Anthropogenic disturbance to the forests would thus have been inevitable.

While the Mudumalai forests have a history of logging going back to the early part of the 19th century, the systematic extraction of timber began with the organization of the forest department by the British rulers during the mid-19th century (Ranganathan 1941). Elaborate forest working plans were developed and many species including *Tectona grandis* (Labiatae), *Pterocarpus marsupium* (Leguminosae), *Terminalia crenulata* (Combretaceae), *Lagerstroemia microcarpa* (Lythraceae), and *Dalbergia latifolia* (Leguminosae) were extracted for timber. Much of the timber was used for the construction of buildings, bridges, and railway sleepers and in mines. It appears that logging operations opened the forest to invasion by grasses, which in turn introduced anthropogenic fire to the forest. The history of fire before logging began has not been clearly documented (Wilson 1939; Ranganathan 1941). The factors that limited logging were the availability of nearby markets, the availability of elephants to transport the logs, and the prevalence of malaria and dysentery that discouraged people from working in the forests. Elephants were captured and trained for logging operations. Other mammals have been hunted to a limited extent for meat, but no known large mammal or bird has been extirpated from Mudumalai or its adjoining areas in recent times.

Mudumalai Wildlife Sanctuary, including the 50-ha plot, comprises secondary forest that is regenerating after several cycles of selective timber harvest by the Tamilnadu Forest Department for the species listed above. The last round of timber extraction was completed in the sanctuary by the mid-1980s. Records in the forest working plans indicate that fires increased in occurrence and spread due to the invasion and proliferation of grasses under the open canopy of logged forest. Grass densities were observed to decline when the forest canopy became denser under protection from fire (Wilson 1939; Ranganathan 1941).

At present, major disturbances in the sanctuary include frequent grass fires and grazing by livestock. During the extended 6-month dry season, the grasses are desiccated and highly flammable. Human-induced grass fires occur annually to varying extent in the central dry deciduous forest tracts and less frequently in the western (moist deciduous forest) and eastern (dry thorn forest) regions. Cattle grazing and small-scale cultivation are present only around human settlements in western and eastern parts; they are absent in the dry deciduous tracts in the central parts of the sanctuary where the 50-ha plot is located.

Intense and widespread fires usually result in high mortality among juvenile stems and decreased rates of recruitment in that year. Typical of dry, fire-affected forests, vegetative coppicing from burned stems and underground rootsuckers are common among many species. As reflected in the forest structure, the influence of large mammals and fires has kept juvenile tree densities and recruitment at low levels. Since the inception of the Mudumalai Forest Dynamics Plot in 1988, fires have swept through the plot 5 out of 13 years (1989, 1991, 1992, 1994, and 1996).

Better protection by the forest department prevented fires during 1997–2001, but a major fire occurred again during 2002, causing considerable mortality.

Collection of nontimber forest products (NTFPs) is currently banned in the sanctuary, although livestock grazing and collection of fodder is permitted in the eastern part. Until about a decade ago, several products were collected from the sanctuary; these included honey and fruits of *Phyllanthus emblica* (Indian gooseberry, Euphorbiaceae), *Sapindus emarginata* (Soap nut, Sapindaceae), *Acacia sinuata*, and *Tamarindus indica* (tamarind, Leguminosae). More details of the economy of NTFPs are given in Ganesan (1993) and Narendran et al. (2001).

Plot Size and Location

The 50-ha plot is 1000 × 500 m; its long axis lies east-west. The southwest corner of the plot is located at 11°35'48" N and 76°31'45" E.

Funding Sources

The Mudumalai Forest Dynamics Plot has been funded by the Ministry of Environment and Forests, Government of India.

References

Bhaskar, N., M. Balasubramanyam, S. Swaminathan, and A. A. Desai. 1995. Home range of elephants in the Nilgiri Biosphere Reserve, south India. Pages 296–313 in J. C. Daniel and H. S. Datye, editors. *A Week with Elephants.* Bombay Natural History Society, Bombay, and Oxford University Press, New Delhi.

Blasco, F. 1971. Montagnes du sud de L'Inde. *Forêt, Savanes, Ecologie.* Institut Français de Pondicherry, Pondicherry, India.

Ganesan, B. 1993. Extraction of non-timber forest products, including fodder and fuelwood in Mudumalai, India. *Economic Botany* 47:268–74.

George, M., M. Gupta, and J. Singh. 1988. *Forest Soil Vegetation Survey: Report on Mudumalai Forest Division, Tamil Nadu.* Forest Soil-Vegetation Survey, Southern Region, Coimbatore.

Hockings, P., editor. 1989. *Blue Mountains: The Ethnography and Biogeography of a South Indian Region.* Oxford University Press, New Delhi.

John, R. 2000. *Habitat Associations, Density-Dependence, and the Tree Species Diversity in a Tropical Dry Deciduous Forest in Mudumalai, Southern India.* Ph.D. thesis, Indian Institute of Science, Bangalore.

John, R, H. S. Dattaraja, H. S. Suresh, and R. Sukumar. 2002. Density dependence in common tree species in a tropical dry forest in Mudumalai, southern India. *Journal of Vegetation Science* 13:45–56.

Joshi, N. V., H. S. Suresh, H. S. Dattaraja, and R. Sukumar. 1997. The spatial organization of plant communities in a deciduous forest: A computational-geometry-based analysis. *Journal of the Indian Institute of Science* 77:365–75.

Mabberley, D. J. 1997. *The Plant-Book: A Portable Dictionary of the Vascular Plants.* Cambridge University Press, Cambridge, U.K.

Murali, K. S. 1992. *Vegetative and Reproductive Phenology of a Tropical Dry Deciduous Forest, Southern India.* Ph.D. thesis, Indian Institute of Science, Bangalore.

Murali, K. S., and R. Sukumar. 1993. Leaf flushing and herbivory in a tropical deciduous forest, southern India. *Oecologia* 94:114–19.

———. 1994. Reproductive phenology of a tropical dry forest in Mudumalai, southern India. *Journal of Ecology* 82:759–67.

Narendran, K., I. K. Murthy, H. S. Suresh, H.S. Dattaraja, N.H. Ravindranath, and R. Sukumar. 2001. Nontimber forest product extraction: A case study from the Nilgiri Biosphere Reserve, southern India. *Economic Botany* 55:528–38.

Prabhakar, R. 1994. Resource use, culture and ecological change: A case study of the Nilgiri hills of southern India. Ph.D. thesis, Indian Institute of Science, Bangalore.

Ranganathan, C. R. 1941. *Working Plan for the Nilgiris Division.* Government Press, Madras, India.

Silori, C., and B. K. Mishra.2001. Assessment of livestock grazing pressure in and around the elephant corridor in Mudumalai Wildlife Sanctuary, south India. *Biodiversity and Conservation* 10:2181–95.

Sukumar, R. 1986. The elephant populations of India: Strategies for conservation. *Proceedings of the Indian Academy of Sciences (Animal Sciences/Plant Sciences), November 1986 Supplement:* 59–71.

———. 1989. *The Asian Elephant: Ecology and Management.* Cambridge University Press, Cambridge, U.K.

Sukumar, R., H. S. Dattaraja, H. S. Suresh, J. Radhakrishnan, R. Vasudeva, S. Nirmala, and N. V. Joshi. 1992. Long term monitoring of vegetation in a tropical deciduous forest in Mudumalai, southern India. *Current Science* 62:608–16.

Sukumar, R., H. S. Suresh, H. S. Dattaraja, and N. V. Joshi. 1998. Dynamics of a tropical deciduous forest: Population changes (1988 through 1993) in a 50-hectare plot at Mudumalai, southern India. Pp. 495–506 in F. Dallmeier and J. A. Comiskey, editors. *Forest Biodiversity Research, Monitoring and Modeling: Conceptual Background and Old World Case Studies.* Man and the Biosphere Series, Vol. 20, Parthenon Publishing Group, Pearl River, NY.

Suresh, H. S., H. S. Dattaraja, and R. Sukumar. 1996. Tree flora of Mudumalai Sanctuary, southern India. *Indian Forester* 122:507–19.

Suresh, H. S., H. R. Bhat, H. S. Dattaraja, and R. Sukumar. 1999. *Flora of Mudumalai Wildlife Sanctuary.* Centre for Ecological Sciences Technical Report 64. Centre for Ecological Sciences, Indian Institute of Science, Bangalore.

Suresh, H. S., and R. Sukumar. 1999. Phytogeographical affinities of flora of Nilgiri Biosphere Reserve. *Rheedea* 9:1–21.

Varman, K. S., and R. Sukumar. 1993. Ecology of sambar in Mudumalai sanctuary. Pages 273–84 in N. Ohtaish and H. I. Shenoy, editors. *Deer in China: Biology and Management.* Elsevier Science Publications, Amsterdam.

———. 1995. The line transect method for estimating densities of large mammals in a tropical deciduous forest: An evaluation of models and field experiments. *Journal of Biosciences* 20:273–87.

Venkataraman, A. B., R. Arumugam, and R. Sukumar. 1995. The foraging ecology of dhole (*Cuon alpinus*) in Mudumalai Sanctuary, southern India. *Journal of Zoology (London)* 237:543–61.

von Lengerke, J. H. 1977. *The Nilgiris: Weather and Climate of a Mountain Area in South India.* Franz Steiner Verlag, Wiesbaden, Germany.

Watve, M. G. 1992. *Ecology of Host-Parasite Interactions in a Wild Mammalian Host Community in Mudumalai.* Ph.D. thesis, Indian Institute of Science, Bangalore.

Watve, M. G., and R. Sukumar. 1995. Parasite abundance and diversity in mammals: Correlates with host ecology. *Proceedings of the National Academy of Sciences* 92:8945–49.

Wilson, C. C. 1939. *Proceedings of the Chief Conservator of Forests, Nilgiris Division, Tamil Nadu.* Proceedings No. 499. Government Press, Madras, India.

34

Nanjenshan Forest Dynamics Plot, Taiwan

I-Fang Sun and Chang-Fu Hsieh

Site Location, Administration, and Scientific Infrastructure

The Nanjenshan Forest Dynamics Plot is located in the Nanjenshan Nature Reserve of Ken-Ting National Park on the Heng-Chun peninsula at the southern end of Taiwan (fig. 34.1). The reserve is 450 km south of Taipei. The Nanjenshan Nature Reserve comprises more than 2400 ha of forest and is located in the northeastern corner of the park. The reserve was established in 1984 when the Ken-Ting National Park was established and is managed under the authority of Ken-Ting National Park. A 3-ha Forest Dynamics Plot was established in 1989 on the east ridge of Wan-li-te Mountain. In 1998, the plot was expanded to 6 ha. In this chapter, we report only on the first 3 ha.

Four other permanent plots were established in the nature reserve from 1992–99 following the Forest Dynamics Plot methodology: (1) The 2.8-ha Creek Plot located in the foothills of Mt. Nanjenshan; (2) the 1-ha Nanjen Transect (40 × 250 m) extending up an altitudinal gradient from the stream valley to the top of Mt. Nanjenshan; (3) the 2-ha Nanjen Lake Plot located near Nanjen Lake; and (4) the 5-ha Secondary Forest Plot located along the bank of Basajaru Creek. All five permanent plots were designated as one of Taiwan's Long-Term Ecological Research (LTER) sites in 1995.

A field station located at the entrance of the nature reserve is equipped with modern amenities, such as a kitchen and dining facilities, bunk beds, laundry machine, and a small laboratory. The nearest town, Man-Chou, is about 15 km away and is connected to the station by a well-maintained, paved road.

Climate

An automated weather station located inside the plot showed that annual rainfall averages 3582 mm (1996–99), with a maximum of 4240 mm and a minimum of 2536 mm. While there are no distinct dry months (months averaging less than 100 mm rainfall), rainfall is distinctly seasonal. Almost half the annual rainfall occurs during summer (June–August), with more than 70% falling between June and October. The mean annual temperature is 23.5°C, with mean daily temperature ranging from 16.3°C (January) to 30.7°C (July). See table 34.1.

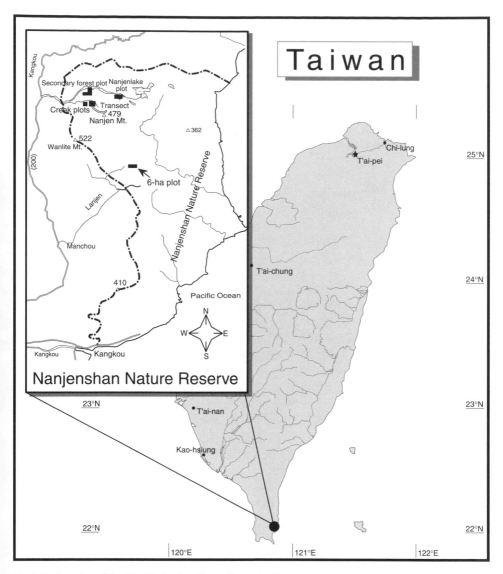

Fig. 34.1. Location of the 6-ha Nanjenshan Forest Dynamics Plot.

Topography and Soil

The terrain of Nanjenshan Nature Reserve is rugged. Several rivers originate in this area, and they dissect the landscape into deep gorges and ravines. The plot has strong relief with slopes typically varying between 30 and 50% grade. Elevation ranges from 300 m above sea level in the protected creek bottom to 340 m atop the most exposed slopes facing the northeastern monsoon winds. A small creek

Table 34.1. Nanjenshan Climate Data

	Jan	Feb	Mar	Apr	May	Jun	Jul	Aug	Sep	Oct	Nov	Dec	Total/ Averages
Rain (mm)	111	107	116	130	206	681	568	463	321	523	206	150	3582
ADTMx (°C)	22.3	22.4	25.9	28.2	28.6	27.7	30.7	30.0	28.3	26.6	24.7	23.2	26.8
ADTMn (°C)	16.3	16.4	18.6	19.3	21.9	23.2	23.8	23.7	22.3	21.7	19.9	18.0	20.3

Notes: Average daily temperatures (maximum and minimum) were recorded from December 1996 to April 1999 (http://lter.npust. edu.tw/).

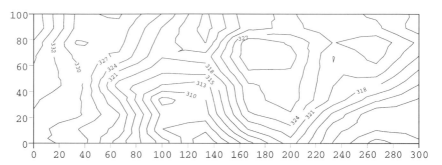

Fig. 34.2. Topographic map of the 3-ha Nanjenshan Forest Dynamics Plot with 3-m contour intervals.

runs north-south through the plot in a steep ravine, dividing the plot into distinct windward and leeward slopes. See figures 34.2 and 34.3.

The exposed bedrock consists of interbedded strata of sandstone and shale of the Miocene. The soils consist of more than 45% clay, and are classified as paleudults or hapludalfs with advanced weathering and pedogenesis (Chen et al. 1997). The soil is strongly acidic with pH values between 4 and 5, and its organic carbon content and cation exchange capacity are low (Chen et al. 1997). Soil depth is relatively shallow, ranging from 40 to 60 cm to bedrock (Sun et al. 1998).

Forest Type and Characteristics

The Nanjenshan forest contains a broad mixture of temperate and tropical species that is attributed to the geography and landscape of the region. The higher altitudinal limit of tropical rainforest on the sides of larger tropical mountains has been reported from all parts of the tropics. Called the Massenerhebung effect, this phenomenon is hypothesized to be the result of the increased heating of larger mountain massifs, which enables lowland plants to extend the range of their altitudinal distribution. In small, isolated mountains and coastal ranges, the opposite tendency has been witnessed; temperate species descend beyond their

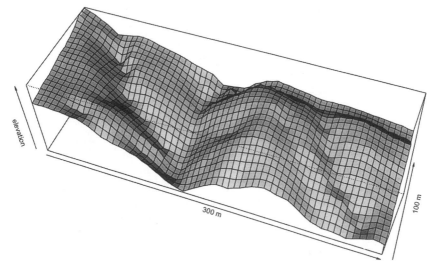

Fig. 34.3. Perspective map of the 3-ha Nanjenshan Forest Dynamics Plot.

typical lower elevation limit. Because the Nanjenshan Forest Dynamics Plot is located on a small, isolated mountain at the southern end of the Central Ridge of Taiwan, which is a large mountain massif, the broad mixture of both temperate and tropical species in the forest may be an indication of the Massenerhebung effect. In addition, the northeasterly monsoon winds are particularly strong and persistent in the Nanjenshan forest area because of a topographic funneling effect. This further decreases air temperatures, thus allowing temperate species to descend to even lower elevations.

Nanjenshan forest is highly differentiated into zones, characterized by abrupt changes in species composition, due to the existence of varying degrees of specialization and generalization to microsite conditions along the stress gradient (Sun 1993). Most species are associated with either the windward or leeward slopes. The 3-ha Nanjenshan Forest Dynamics Plot can be divided into four habitats: windward (0.72 ha), leeward (0.82 ha), intermediate (1.23 ha), and creek (0.23 ha).

The forest on windward slopes tends to be short in stature (3–5 m), unstratified, high in stem density (20,065 trees/ha), and low in per-stem species richness. Leaves are small and thick, and litter fall rates are low (Sun 1993). Dominant species at these sites comprise trees from Fagaceae, Aquifoliaceae, Theaceae, and Winteraceae, such as as *Quercus championii* (Fagaceae), *Ilex cochinchinensis* (Aquifoliaceae), and *Eurya hayatai* (Theaceae).

In the intermediate habitat, trees are not directly exposed to strong winds, though they still suffer moderate wind damage. The average canopy height is between 5 and 8 m with minimal forest stratification, similar to the forest structure of the windward habitat. The intermediate habitat has the lowest species richness per area of the four habitats. Dominant species includes *Ilex cochinchinesis* (Aquifoliaceae), *Castanopsis carlesii* (Fagaceae), and *Illicium arborescens* (Illiciaceae).

In the leeward habitat, the forest grows on protected slopes and is relatively tall (10–15 m), stratified, relatively low in stem density (7505 trees/ha), and much higher in per-stem species richness. Litter fall rates are much higher and leaves are larger and thinner (Sun 1993). Dominant species found in this habitat include *Schefflera octophylla* (Araliaceae), *Psychotria rubra* (Rubiaceae), and *Beilschmiedia erythrophloia* (Lauraceae).

In the creek habitat, trees reach up to 15–20 m. Stem density is the lowest of the four habitats, averaging 4257 trees/ha. Compared to other habitats, the creek habitat has many fewer individuals but each individual grows much larger. This habitat is very similar to the leeward habitat. However, the species composition is quite different. Dominant species at this habitat include *Schefflera octophylla* (Araliaceae), *Astronia ferruginea* (Melastomataceae), and *Wendlandia formosana* (Rubiaceae). For census data and rankings, see tables 34.2–34.7.

Table 34.2. Nanjenshan Plot Census History

Census	Dates	Number of Trees (≥1 cm dbh)	Number of Species (≥1 cm dbh)	Number of Trees (≥10 cm dbh)	Number of Species (≥10 cm dbh)
First	July 1989–February 1991	36,629	118	3162	79
Second	July 1996–February 1997	36,383	125	3272	82

Notes: Two censuses have been completed of 3 ha, the next census is expected to begin in 2004.

Table 34.3. Nanjenshan Summary Tally

Size Class (cm dbh)	Average per Hectare							3-ha Plot				
	BA	N	S	G	F	H'	α	S	G	F	H'	α
≥1	36.3	12,209	104	68	37	1.64	15.6	118	79	41	1.68	15.15
≥10	23.9	1,054	61	44	24	1.49	14.0	79	55	30	1.55	14.69
≥30	5.2	46	13	11	7	0.94	5.5	26	21	14	1.17	5.32
≥60	0.3	1	1	1	1	—	—	2	2	1	0.28	—

Notes: BA represents basal area in m², *N* is the number of individual trees, *S* is number of species, *G* is number of genera, *F* is number of families, *H'* is Shannon–Wiener diversity index using $\log_{10}$, and α is Fisher's α. Basal area includes all multiple stems for each individual. Individuals are counted using their largest stem. All species were identified. Data are from first census.

Table 34.4. Nanjenshan Rankings by Family

					3-ha Forest Dynamics Plot				
Rank	Family	Basal Area (m^2)	% BA	% Trees	Family	Trees	% Trees	Family	Species
1	Fagaceae	11.7	32.3	8.3	Aquifoliaceae	4910	13.4	Lauraceae	11
2	Theaceae	3.4	9.4	8.6	Illiciaceae	4421	12.1	Fagaceae	10
3	Aquifoliaceae	2.7	7.5	13.4	Lauraceae	3254	8.9	Rubiaceae	10
4	Illiciaceae	2.1	5.9	12.1	Theaceae	3145	8.6	Euphorbiaceae	8
5	Lauraceae	1.9	5.3	8.9	Fagaceae	3058	8.3	Theaceae	8
6	Myrtaceae	1.8	5.0	4.7	Rubiaceae	3008	8.2	Moraceae	5
7	Daphniphyllaceae	1.8	5.0	3.3	Euphorbiaceae	2075	5.6	Aquifoliaceae	5
8	Araliaceae	1.7	4.6	1.2	Myrtaceae	1724	4.7	Myrtaceae	4
9	Oleaceae	1.5	4.2	3.0	Celastraceae	1388	3.8	Myrsinaceae	4
10	Euphorbiaceae	1.2	3.4	5.6	Symplocaceae	1328	3.6	Rutaceae	4

Notes: The top 10 families for trees ≥1 cm dbh are ranked in terms of basal area, number of individual trees, and number of species, with the percentage of trees in the plot. Data are from the first census.

Fauna

Nanjenshan has 14 species of mammals, including 6 species of bats, all of which are insectivores. The only large mammal in the nature reserve is the Taiwanese macaque (*Macaca cyclopis*), which mainly feeds on leaves and fruits. There are 45 species of birds, a figure significantly lower than other tropical forests. Other taxa include 19 species of amphibians and 28 species of reptiles (Hou and Huang 1999). Insects are more abundant and diverse, representing at least 464 species and 85 families. Fungi are another diverse group found in this area; over 368 species have been identified.

Natural Disturbance

The north-easterly winter monsoon is the major environmental factor in this region, which starts in the middle of October and lasts until late February. The monsoon winds are particularly strong and persistent in this area because Nanjenshan Nature Reserve is located at the first break in the central mountain massif that runs the length of Taiwan. Average wind speed is greater than 5 m/sec in the windward slope and less than 2.5 m/sec in the leeward slope during monsoon season. During nonmonsoon season (March to September), average wind speed is less than 2 m/sec.

Typhoons are also very common in this region. They can arrive as early as May or as late as November. When a typhoon directly hits the forest, it strips away many trees and creates landslides and huge gaps. Due to the destructive nature of typhoons, they greatly affect the regeneration processes of the forest. In contrast, monsoon winds do not cause massive and direct destructive damage to the forest.

Table 34.5. Nanjenshan Rankings by Genus

					3-ha Forest Dynamics Plot				
Rank	Genus	Basal Area (m²)	% BA	% Trees	Genus	Trees	% Trees	Genus	Species
1	*Castanopsis* (Fagaceae)	15.9	14.6	2.8	*Ilex* (Aquifoliaceae)	4910	13.4	*Ficus* (Moraceae)	6
2	*Quercus* (Fagaceae)	12.8	11.8	3.9	*Illicium* (Illiciaceae)	4421	12.1	*Ilex* (Aquifoliaceae)	5
3	*Ilex* (Aquifoliaceae)	8.2	7.5	13.4	*Psychotria* (Rubiaceae)	2029	5.5	*Lasianthus* (Rubiaceae)	5
4	*Illicium* (Illiciaceae)	6.4	5.9	12.1	*Eurya* (Theaceae)	1461	4.0	*Castanopsis* (Fagaceae)	4
5	*Daphniphyllum* (Daphniphyllaceae)	5.5	5.0	3.3	*Syzygium* (Myrtaceae)	1423	3.9	*Persea* (Lauraceae)	4
6	*Schefflera* (Araliaceae)	5.0	4.6	1.2	*Quercus* (Fagaceae)	1417	3.9	*Syzygium* (Myrtaceae)	4
7	*Osmanthus* (Oleaceae)	4.6	4.2	3.0	*Antidesma* (Euphorbiaceae)	1349	3.7	*Quercus* (Fagaceae)	3
8	*Syzygium* (Myrtaceae)	4.3	3.9	3.9	*Symplocos* (Symplocaceae)	1328	3.6	*Neolitsea* (Lauraceae)	3
9	*Lithocarpus* (Fagaceae)	3.6	3.3	1.1	*Daphniphyllum* (Daphniphyllaceae)	1192	3.3	*Ardisia* (Myrsinaceae)	3
10	*Pasania* (Fagaceae)	2.8	2.5	0.5	*Microtropis* (Celastraceae)	1172	3.2	*Glochidion* (Euphorbiaceae)	3

Notes: Top 10 tree genera for trees ≥1 cm dbh. Data are from the first census.

Table 34.6. Nanjenshan Rankings by Species

Rank	Species	Number Trees	% Trees	Species	Basal Area (m²)	% BA	% Stems
1	*Illicium arborescens* (Illiciaceae)	4421	12.1	*Castanopsis carlesii* (Fagaceae)	14.5	13.3	2.5
2	*Ilex cochinchinensis* (Aquifoliaceae)	2248	6.1	*Illicium arborescens* (Illiciaceae)	6.4	5.9	12.1
3	*Psychotria rubra* (Rubiaceae)	2029	5.5	*Quercus longinux* (Fagaceae)	6.3	5.8	2.5
4	*Eurya hayatai* (Theaceae)	1385	3.8	*Daphniphyllum glaucescens* (Daphniphyllaceae)	5.5	5	3.3
5	*Antidesma hiiranense* (Euphorbiaceae)	1349	3.7	*Schefflera octophylla* (Araliaceae)	5.0	4.6	1.2
6	*Daphniphyllum glaucescens* (Euphorbiaceae)	1192	3.3	*Quercus championii* (Fagaceae)	4.8	4.4	0.7
7	*Microtropis japonica* (Celastraceae)	1172	3.2	*Osmanthus marginatus* (Oleaceae)	4.6	4.2	3.0
8	*Ilex matsudai* (Aquifoliaceae)	1170	3.2	*Lithocarpus amygdlifolius* (Fagaceae)	3.6	3.3	1.1
9	*Osmanthus marginatus* (Oleaceae)	1090	3.0	*Schima superba* (Theaceae)	3.1	2.8	1.1
10	*Ilex uraiensis* (Aquifoliaceae)	1085	2.9	*Elaeocarpus sylvestris* (Elaeocarpaceae)	2.8	2.6	0.5

Notes: The top 10 tree species for trees ≥1 cm dbh are ranked by number of trees and basal area. Data are from the first census.

Table 34.7. Nanjenshan Tree Demographic Dynamics for 1991–97

Size Class (cm dbh)	Growth Rate (mm/yr) 1991–1997	Mortality Rate (%/yr) 1991–1997	Recruitment Rate (%/yr) 1991–1997	BA Losses (m²/ha/yr) 1991–1997	BA gains (m²/ha/yr) 1991–1997
1–9.9	1.14	1.77	2.0	0.20	0.58
10–29.9	1.62	1.38	3.0	0.29	0.48
≥30	2.74	2.03	4.1	0.11	0.15

The monsoon winds blow continuously, 24 hours a day, 7 days a week, and exert a chronic stress on the forest that may have a different and a potentially greater cumulative effect on forest structure and dynamics than typhoons.

Human Disturbance

Human activities in this area date back to 1000–1500 years ago. Several archeological sites have been excavated at the northern part of the reserve. Evidence of human activities that occurred 50 to 60 years ago includes extraction of *Cinnamomum* (Lauraceae), the collection of forest orchids, and the creation of betel palm (*Areca catechu*, Palmae) plantations. However, most human activities were restricted to the northern part of the nature reserve, or below 250 m elevation. At higher elevations, where the permanent plot is located, the area's remoteness ensured little anthropogenic disturbance of the reserve prior to its designation as a national park. After the establishment of the park, all hunting, poaching, and illegal land use were forbidden. However, hunters are occasionally seen inside the reserve. To what extend their activities affect the forest dynamics is unknown.

The permanent plot is situated in the center of the nature reserve, which is composed of more than 2000 ha of continuous forest. The forest edge nearest to the plot is about 2 km away. The reserve is surrounded primarily by secondary forest, betel palm plantations, coconut plantations, and rice fields.

Plot Size and Location

Nanjenshan is a 3-ha, 300 × 100 m plot, whose long axis lies east-west. The northwest corner of the plot is at 22°03′34.1″ N and 120°51′09.1″ E. The southeast corner is at 22°03′30.8″ N and 120°51′19.6″ E.

Funding Sources

The Nanjenshan Forest Dynamics Plot has been funded by the National Science Council of Taiwan and Kenting National Park.

References

Chang, K. M., H. W. Chang and C. C. Hwang. 1996. Anatomy on *Satsuma albida* (H. Adams, 1870) from Taiwan (Pulmonata: Camaenidae). *Bulletin of Malacology, ROC* 20:25–30.

Chen, Yu–Yun. 1998. *Litter Decomposition in a Lowland Rain Forest of Nanjenshan.* Department of Botany, National Taiwan University. (In Chinese with English summary.)

Chen, Z. S., C. F. Hsieh, F. Y. Jiang, T. H. Hsieh, and I. F. Sun. 1997. Relations of soil properties to topography and vegetation in a subtropical rain forest in southern Taiwan. *Plant Ecology* 132:229–41.

Hara, M., K. Hirata, M. Fujihara, K. Oono, and C. F. Hsieh.1997. Floristic composition and stand structure of three evergreen broad-leaved forests in Taiwan, with special reference to the relationship between micro-landform and vegetation pattern. *Natural History Museum Research Special Issue* 4:81–112.

Hou, P. C. L., and S. P. Huang. 1999. Metabolic and ventilatory responses to hypoxia in two altitudinal populations of the toad, *Bufo bankorensis. Comparative Biochemistry and Physiology A* 124:413–21.

Hsieh C. F., I. F. Sun, and C. C. Yang. 2000. Species composition and vegetation pattern of a lowland rain forest at the Nanjenshan LTER site, southern Taiwan. *Taiwania* 45 (1):107–19.

Hsieh C. F., J. C. Wang, and C. N. Wang. 1999. *Staurogyne debilis* (T. Anders) C. B. Clarke (Acanthaceae) in Taiwan. *Taiwania* 44 (2):305–09.

Huang, H. N., I. F. Sun, and C. P. Ho. 1998. Establishment of forest general light index model. *Tunghai Journal* 39:65–77.

Jiang, Fei–Yu. 1991. *Environmental Influences on the Soil Properties and Pedogenesis in Nanjen mountains, Taiwan*. Department of Agricultural Chemistry, National Taiwan University. (In Chinese with English summary.)

Lai, I–Ling. 1996. *Seedling Recruitment and Understory Patterns of Nanjenshan Subtropical Rain Forest*. Department of Botany, National Taiwan University. (In Chinese with English summary.)

Li, S. P., and C. F. Hsieh. 1997. Seedling morphology of some woody species in a subtropical rain forest of Southern Taiwan. *Taiwania* 42:207–38.

Liu, J. C., Z. S. Chen, G. N. White, and J. B. Dixon. 1998. Goethite in an Alfisol and an Ultisol of southern Taiwan. *Journal of the Agricultural Association of China New Series* 182:83–98.

Liu, Shiang-Yao. 1994. *Litterfall Production and Nutrient Content in Nanjenshan Subtropical Rain Forest*. Department of Botany, National Taiwan University. (In Chinese with English summary.)

Mabberley, D. J. 1997. *The Plant-Book: A Portable Dictionary of the Vascular Plants.* Cambridge University Press, Cambridge, U.K.

Shih, H. T., H. K. Mok, H. W. Chang, and S. C. Lee. 1999. Morphology of *Uca formosensis* Rathbun, 1921 (Crustacea: Decapoda: Ocypodidae), an endemic fiddler crab from Taiwan, with notes on its ecology. *Zoological Studies* 38:164–77.

Su, Mong-Whai. 1993. *The Leaf Structure of the Canopy of Nanjenshan Subtropical Rain Forest*. Department of Botany, National Taiwan University. (In Chinese with English summary.)

Sun, I. F. 1993. *The Species Composition and Forest Structure of a Subtropical Rain Forest at Southern Taiwan*. Ph.D. thesis, University of California, Berkeley, CA.

Sun, I. F., C. F. Hsieh, and S. P. Hubbell. 1996. The structure and species composition of a subtropical monsoon forest in southern Taiwan on a steep wind-stress gradient. Pages 147–69 in I. M. Turner, C. H. Diong, S. S. L. Lim, and P. K. L. Ng, editors. *Biodiversity and the Dynamics of Ecosystems*. DIWPA Series Volume 1. Center for Ecological Research, Kyoto University, Kyoto.

Sun, I. F., C. F. Hsieh, and S. P. Hubbell. 1998. The structure and species composition of a subtropical monsoon forest in southern Taiwan on a steep wind-stress gradient. Pages 599–626 in F. Dallmeier and J. A. Comiskey, editors. *Forest Biodiversity Research, Monitoring and Modeling: Conceptual Background and Old World Case Studies*. Man and the Biosphere Series, Vol. 20, Parthenon Publishing Group, Pearl River, NY.

Tsai, C. C., and Z. S. Chen. 2000. Lithologic discontinuity of Ultisols along a toposequence in Taiwan. *Soil Science* 165:587–96.

Tsai, C. C., and Z. S. Chen. 2001. Soil characteristics and variability of an Ultisol toposequence in Nanjenshan subtropical forest ecosystem of Taiwan. *Food Science & Agriculture Chemistry* 3:13–22.

Wang, Kuo-Hsiung. 1993. *The Seedling Survivorship and Growth of Four Species in Nanjenshan Subtropical Rain Forest*. Department of Botany, National Taiwan University. (In Chinese with English summary.)

Wu, Shan-Huah. 1998. *Short-term Dynamics of a Subtropical Rain Forest in Nanjenshan*. Department of Botany, National Taiwan University. (In Chinese with English summary.)

35

Palanan Forest Dynamics Plot, Philippines

Leonardo L. Co, Daniel A. Lagunzad, James V. LaFrankie,
Nestor A. Bartolome, Jeanmaire E. Molina, Sandar L. Yap,
Hubert G. Garcia, John P. Bautista, Edmundo C. Gumpal,
Robert R. Araño, and Stuart J. Davies

Site Location, Administration, and Scientific Infrastructure

The 250,000-ha Palanan Wilderness Area was created in 1979. It lies in the Province of Isabela on the Island of Luzon, and was recently redefined as the Northern Sierra Madre Natural Park (NSMNP). It is a totally protected area under the administration of the Philippine Department of Environment and Natural Resources. Although described as a wilderness area, it includes the historic seaside town of Palanan with several thousand permanent residents. Palanan is one of the last spots in the Philippines where one can traverse pristine habitats from seashore and mangrove, through lowland forests to montane and mossy forests. All terrestrial habitats found in the Philippines are represented here, except peat swamps and semideciduous rainforest.

The 16-ha Forest Dynamics Plot is located about 4 km from Palanan town, near Villa Robles (fig. 35.1). The plot was established in 1994 through a scientific consortium that included Conservation International, the College of Forestry at Isabela State University (based in the nearby town of Cabagan, Isabela), and the Center for Tropical Forest Science in concert with the Arnold Arboretum of Harvard University.

Access to the site is via small commercial propeller airplanes to the town of Palanan, and from there by foot to the plot itself. The consortium has built some modest wooden facilities, powered at night for lighting by a small electric generator, for researchers and staff to sleep and eat adjacent to the plot. A permanent stream runs next to the buildings providing fresh water year round. Food and supplies can be obtained in Palanan town but must be carried to the site.

Climate

No long-term weather data are available for the immediate vicinity of Palanan. The most comparable data have been collected at a station of the Philippine Meterological Service in Casiguran, about 80 km south of the plot. In Casiguran, the 30-year mean annual rainfall is 3379 mm, with a minimum of 1347 mm and a

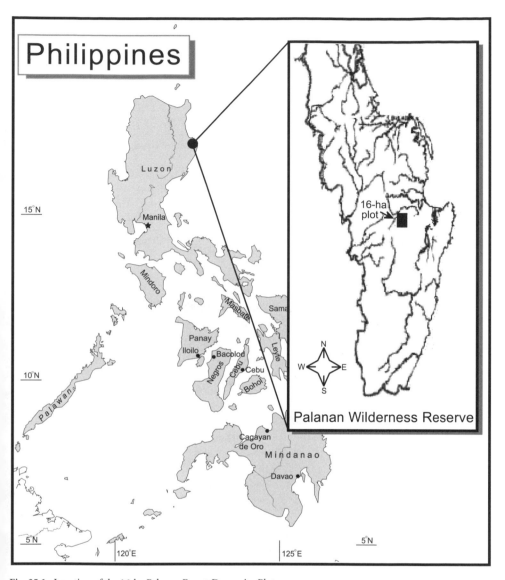

Fig. 35.1. Location of the 16-ha Palanan Forest Dynamics Plot.

maximum of 6841 mm. The rainfall in Palanan is thought to be somewhat higher, perhaps a 30-year mean of 5000 mm. The very wet year of 2000 had a measured rainfall at Palanan town center of more than 8000 mm (exact data are not available). The 4 months from January to April are relatively dry, though no month averages less than 100 mm. The mean annual temperature in Casiguran is

Table 35.1. Palanan Climate Data

	Jan	Feb	Mar	Apr	May	Jun	Jul	Aug	Sep	Oct	Nov	Dec	Total/ Averages
Rain (mm)	220	152	166	129	229	218	274	217	287	510	525	453	3379
ADTMx	26.9	28.5	30.2	32.1	33.4	33.6	33.0	33.0	32.0	31.1	28.6	27.7	30.8
ADTMn	18.7	19.2	20.1	21.5	22.6	23.2	23.0	23.0	22.9	22.0	20.8	19.9	21.4

Notes: Mean monthly rainfall and average daily temperature are based on data measured at Casiguran, 1966 to 1995.

26.1°C, with an average diurnal maximum of 30.8°C and a minimum of 21.4°C (table 35.1). Typhoons hit Palanan both from the southwest and the northeast, generally from August to November, and the prevailing wind direction is southerly from May to September and northerly from October to April.

Topography and Soil

The Northern Sierra Madre Natural Park is located near the northern end of the Sierra Madre mountain range, which runs along most of Luzon's eastern shore, from Cagayan Province at the northern tip of the island to Quezon Province, south of Manila. The range is a highly complex structure of volcanic, limestone, ultra-basic, and uplifted granites, all interdigitated with tongues of limestone appearing amid the ultrabasic outcrops.

In the vicinity of the Palanan Forest Dynamics Plot, clays, limestone, and ultra-basics are patchily distributed and support forests of markedly different composition and stature. Extensive outcrops of ultrabasics support a stunted heath-like lowland forest, especially about 15 km to the north and south of Palanan. Although no soil profile data are currently available for the plot, soils in the area mostly appear to be clay loams ranging from yellowish to reddish-brown, generally overlain with partly decomposed detrital materials 0.3–1 cm thick. Along deep gully bottoms, dark soil rich in humus is occasionally found. Exposed bedrock seen within the plot is mostly shale, although boulders of certain igneous rocks occur, especially along the eastern fork of Ditalad Creek.

The Palanan Forest Dynamics Plot straddles a low broad ridge, with an elevation range of 85–140 m above sea level (figs. 35.2 and 35.3). Trending diagonally along a northwest to southeast direction, the ridge lies roughly perpendicular to the general direction of typhoons originating from the Pacific. As a result, the plot has marked windward and leeward sides, which are reflected in the structure of the plot's forest cover. Flanking the plot are two branches of Ditalad Creek, a tributary of the Palanan River, one of the major river systems within the park. Gullies carved by this creek provide sheltered spots with higher diversity and taller trees than on the more exposed broad ridge summits.

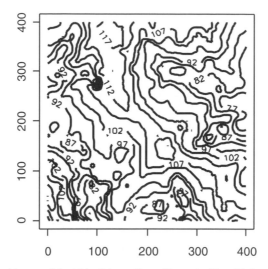

Fig. 35.2. Topographic map of the 16-ha Palanan Forest Dynamics Plot with 5-m contour intervals.

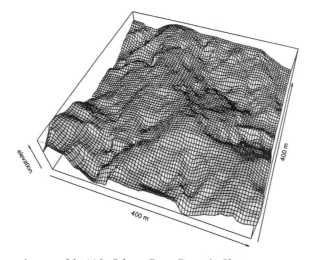

Fig. 35.3. Perspective map of the 16-ha Palanan Forest Dynamics Plot.

Forest Type and Characteristics

Palanan includes a rich mosaic of forest types largely influenced by the diverse underlying soils. The Palanan Forest Dynamics Plot itself encompasses evergreen dipterocarp forest dominated in basal area by species of *Shorea* (Dipterocarpaceae), *Dipterocarpus* (Dipterocarpaceae), and *Hopea* (Dipterocarpaceae).

Table 35.2. Palanan Plot Census History

Census	Dates	Number of Trees (≥ 1 cm dbh)	Number of Species (≥ 1 cm dbh)	Number of Trees (≥ 10 cm dbh)	Number of Species (≥ 10 cm dbh)
First	1994–1999	65,986	335	8593	262

Notes: An initial census was completed of the 8-ha Palanan Forest Dynamics Plot. A second census of the first 8 ha and a first census of a supplemental 8 ha was completed in 1999. Data from the second census of the first 8 ha are not yet available.

Table 35.3. Palanan Summary Tally

Size Class (cm dbh)	Average per Hectare							16-ha Plot				
	BA	N	S	G	F	H'	α	S	G	F	H'	α
≥ 1	39.8	4124	197	111	48	1.92	43.4	335	155	60	2.00	46.1
≥ 10	36.1	537	100	63	35	1.63	36.5	262	137	55	1.79	51.1
≥ 30	26.0	110	32	24	18	1.21	16.0	130	78	39	1.44	32.4
≥ 60	15.6	28	8	4	4	0.69	4.0	36	27	20	0.89	9.2

Notes: BA represents basal area in m^2, N is the total number of individual trees, S is the total number of species, G is the total number of genera, F is the total number of families, H' is the Shannon–Wiener diversity index using $\log_{10}$, and α is Fisher's α. Basal area includes all multiple stems for each individual. Individuals are counted using their largest stem. 3370 individuals were not identified to species or morphospecies. Data are from 16 ha of the first census.

Trees from the Dipterocarpaceae represent over half of all trees over 30 cm dbh. The next closest families (Tiliaceae and Meliaceae) are represented by 126 and 90 trees respectively. The Meliaceae is expected to be among the 10 most abundant and species-rich families in any Asian forest. In Palanan it is especially rich: ranking third in species number and first in total abundance. Also notable is the rich abundance of the genus *Leea* [Leeaceae]. While *Leea* is found in all Asian forests (and is listed among all of the Asian plots), it is only in the Palanan plot that it becomes a major component of the forest understory, represented by seven species and over 3400 trees, or nearly 5% of the total census. For census data and rankings, see tables 35.2–35.6.

Fauna

The Northern Sierra Madre Natural Park contains a rich diversity of faunal life. Of 291 species of birds thus far documented in the park, 83 are endemic to the Philippines. The park is also host to 44 species of terrestrial mammals, of which 21 are endemic to the country (NSMNP Management Plan 2001). Notable large mammal species are few, consisting of a macaque, *Macaca fascicularis,* two species of civets, *Paradoxurus hermaphroditus* and *Viverra tangalunga,* a pig, *Sus philippinensis,* and a deer, *Cervus mariannus* (Danielsen et al. 1994). A distinctive feature of the mammalian fauna of Luzon Island is the absence of squirrels, which

Table 35.4. Palanan Rankings by Family

Rank	Family	Basal Area (m²)	% BA	% Trees	Family	Trees	% Trees	Family	Species
1	Dipterocarpaceae	323.4	52.8	11.7	Meliaceae	7561	12.1	Euphorbiaceae	37
2	Euphorbiaceae	37.1	6.1	9.9	Dipterocarpaceae	7310	11.7	Myrtaceae	23
3	Tiliaceae	35.9	5.9	3.5	Euphorbiaceae	6200	9.9	Meliaceae	20
4	Meliaceae	33.2	5.4	12.1	Sapindaceae	5742	9.2	Lauraceae	18
5	Sapindaceae	28.4	4.6	9.2	Lauraceae	5154	8.2	Moraceae	16
6	Olacaceae	16.1	2.6	2.6	Annonaceae	3171	5.1	Rubiaceae	16
7	Lauraceae	15.4	2.5	8.2	Leeaceae	2941	4.7	Annonaceae	14
8	Myrtaceae	15.1	2.5	4.5	Myrtaceae	2823	4.5	Guttiferae	13
9	Leguminosae	14.1	2.3	1.7	Rubiaceae	2222	3.5	Sapindaceae	13
10	Moraceae	7.5	1.2	1.6	Tiliaceae	2177	3.5	Dipterocarpaceae	12
								Myristicaceae	12
								Sapotaceae	12

Notes: The top 10 families for trees ≥1 cm dbh ranked in terms of basal area, number of individual trees, and number of species, with the percentage of trees in the plot. Data are from 16 ha of the first census.

Table 35.5. Palanan Rankings by Genus

Rank	Genus	Basal Area (m²)	% BA	% Trees	Genus	Trees	% Trees	Genus	Species
1	*Shorea* (Dipterocarpaceae)	311.6	50.9	9.8	*Shorea* (Dipterocarpaceae)	6144	9.8	*Eugenia* (Myrtaceae)	23
2	*Diplodiscus* (Tiliaceae)	29.8	4.9	2.1	*Litsea* (Lauraceae)	3809	6.1	*Ficus* (Moraceae)	14
3	*Drypetes* (Euphorbiaceae)	29.3	4.8	4.9	*Nephelium* (Sapindaceae)	3525	5.6	*Palaquium* (Sapotaceae)	12
4	*Nephelium* (Sapindaceae)	23.5	3.8	5.6	*Drypetes* (Euphorbiaceae)	3038	4.9	*Dysoxylum* (Meliaceae)	8
5	*Strombosia* (Olacaceae)	16.1	2.6	2.6	*Leea* (Leeaceae)	2941	4.7	*Garcinia* (Guttiferae)	7
6	*Eugenia* (Myrtaceae)	15.1	2.5	4.5	*Eugenia* (Myrtaceae)	2823	4.5	*Glochidion* (Euphorbiaceae)	7
7	*Dysoxylum* (Meliaceae)	14.8	2.4	3.9	*Chisocheton* (Meliaceae)	2610	4.2	*Leea* (Leeaceae)	7
8	*Chisocheton* (Meliaceae)	12.5	2.0	4.2	*Dysoxylum* (Meliaceae)	2450	3.9	*Litsea* (Lauraceae)	7
9	*Dipterocarpus* (Dipterocarpaceae)	11.3	1.8	1.7	*Aglaia* (Meliaceae)	2263	3.6	*Aglaia* (Meliaceae)	6
10	*Litsea* (Lauraceae)	8.1	1.3	6.1	*Haplostichanthus* (Annonaceae)	2198	3.5	*Shorea* (Dipterocarpaceae)	6

Notes: The top 10 tree genera for trees ≥1 cm dbh ranked by basal area, number of individual trees, and number of species with the percentage of trees in the plot. Data are from 16 ha of the first census.

Table 35.6. Palanan Rankings by Species

Rank	Species	Number Trees	% Trees	Species	Basal Area (m^2)	% BA	% Trees
1	*Nephelium lappeceum* (Sapindaceae)	3525	5.6	*Shorea palosapis* (Dipterocarpaceae)	117.4	19.2	3.6
2	*Shorea palosapis* (Dipterocarpaceae)	2237	3.6	*Shorea negrosensis* (Dipterocarpaceae)	101.3	16.6	2.2
3	*Drypetes megacarpa* (Euphorbiaceae)	2179	3.5	*Shorea polysperma* (Dipterocarpaceae)	37.0	6.0	0.5
4	*Haplostichanthus sp.* (Annonaceae)	2020	3.2	*Shorea contorta* (Dipterocarpaceae)	30.5	5.0	2.8
5	*Shorea contorta* (Dipterocarpaceae)	1719	2.8	*Diplodiscus paniculatus* (Tiliaceae)	29.8	4.9	2.1
6	*Dysoxylum oppositifolium* (Meliaceae)	1667	2.7	*Drypetes megacarpa* (Euphorbiaceae)	25.2	4.1	3.5
7	*Praravinia sp.* (Rubiaceae)	1638	2.6	*Nephelium lappaceum* (Sapindaceae)	23.5	3.8	5.6
8	*Strombosia philippinensis* (Olacaceae)	1622	2.6	*Shorea assamica* ssp. *philippinensis* (Dipterocarpaceae)	18.0	3.0	0.5
9	*Leea congesta* (Leeaceae)	1482	2.4	*Strombosia philippinensis* (Olacaceae)	16.1	2.6	2.6
10	*Shorea negrosensis* (Dipterocarpaceae)	1363	2.2	*Dysoxylum oppositifolium* (Meliaceae)	9.2	1.5	2.7

Notes: The top 10 tree species for trees ≥1 cm dbh are ranked by number and percentage of trees and basal area. Data are from 16 ha of the first census.

are perhaps replaced ecologically by various forest rats. The largest colony of flying foxes in Luzon Island is also found in the park, most noteworthy of which are the golden crowned flying fox (*Acerodon jubatus*) and the mottled-winged flying fox (*Pteropus leucopterus*). Although no systematic survey has been conducted on amphibians or reptiles in the park, preliminary data indicate high levels of species endemicity. Of about 18 species of amphibians thus far known to occur in the park, 13 are endemics.

In 1997, a rapid survey of the terrestrial vertebrate fauna of the 16-ha Palanan Forest Dynamics Plot and immediate 3-km radius recorded 95 species including 7 amphibians, 5 reptiles, 67 birds, and 16 mammals. Rufous hornbill (*Buceros hydrocorax*), Gray's monitor lizard (*Varanus olivaceus*), Luzon pygmy fruit bat (*Otopteropus cartilagonodus*), and wild pigs (*Sus philippinensis*) are some of the more prominent endemic wildlife species that frequent the 16-ha plot (Duya et al. unpublished data).

Natural Disturbances

Palanan Forest Dynamics Plot is subjected to periodic severe typhoons that are strong enough to either knock over exposed trees or strip their canopy bare of leaves. The area is buffeted by three to eight typhoons per year, mostly of modest strength (class one). Powerful typhoons (class three and above) strike every few years. The exact frequency of these typhoons is not clear, but the living memory of Palanan residents suggests that typhoons destructive to the forest occur every 3 years or so.

From November to June, northeast winds off the Pacific dominate the weather of northern Luzon. These steady winds bring daily clouds and rain to east-facing slopes of the Sierra Madre, while leaving the western side dry. Thus, during the driest months of the year (January to May) when 4 months of rainless days are expected in the Cagayan Valley and Northwest Luzon, Palanan may be receiving daily storms.

From June to November, Philippine weather is dominated by the great cyclonic storms that rise up in the Pacific. They begin their life 1000 km away to the southeast and sweep north and west, typically either crossing the central Philippine islands and southern Luzon, or moving northward up the Luzon coast westward across the strait between Luzon and Taiwan. These latter storms are an annual feature of Palanan weather. Relatively mild storms occur every year, while highly destructive storms (wind gusts over 100 km/hr) occur at least every decade. Effects on the forest include a direct impact on the crown, branches, and form, producing a short, even stature, and an indirect impact through litter deposition and the subsequent greater chance of fires. There is no known impact of El Niño events.

Human Disturbance

Although Palanan is completely cut off via road network from the economically more developed western side of the Sierra Madre, it suffers from a variety of human disturbances. Individual trees of valued species, particularly narra, *Pterocarpus indicus* (Leguminosae), and ebonies, *Diospyros* spp. (Ebenaceae), have been routinely extracted over the past decades. Typically trees are cut and squared in the forest and the log is drawn out using the native buffalo. The logs are taken to the ocean side and picked up by large outrigger boats bound elsewhere in Luzon. Deep logging trails can be found in the plot itself, although the number of trees extracted from the plot appears to be low.

Some hunting of deer and pig continues as does the extraction of rattan. The greatest danger to the residual lowland forest is the continuing loss of forest land to small-scale, low-income farming, especially for corn, pineapple, and banana. No figures are available for the recent loss of forest cover.

Much of the high diversity of faunal life in the Northern Sierra Madre Natural Park is threatened because of past and current human activity throughout the region. Fifty-one endemic bird species and 12 of the endemic terrestrial mammal species are categorized as threatened or near-threatened species by the International Union for Conservation of Nature and Natural Resources (IUCN)

Red List of Threatened Species (IUCN 2002; NSMNP Management Plan 2001). The golden crowned flying fox and mottled-winged flying fox are categorized as endangered in the IUCN list, while two forest tree frogs, *Platymantis pygmaeus* and *P. sierramadrensis*, are categorized as vulnerable in the list. Among the threatened herpetofauna, Gray's monitor lizard, *Varanus olivaceus*, and the Philippine crocodile, *Crocodylus mindorensis* (both Philippine endemics), are respectively categorized as vulnerable and critically endangered (NORDECO-DENR 1998; NSMNP Conservation Project unpublished data; Duya personal communication).

The 16-ha plot is located within a multiple-use zone within the park, which allows limited and regulated extraction of timber and nontimber products and game hunting for domestic, noncommercial consumption. However, local communities around the plot do not engage in any open extractive activities within the plot or its associated field station. There is currently a move within the NSMNP Protected Area Management Board to declare the plot's perimeter as a special-use zone for scientific research and ecotourism.

Plot Size and Location

The initial census was 8 ha, 200 × 400 m; the long axis lies north and south. The second 8 ha, 200 × 400 m, is located to the east of the first 8 ha and also has a north-south orientation. The plot is located at approximately 17°02′36″ N, 122°22′58″ E.

Funding Sources

The Palanan Forest Dynamics Plot was initially funded by Conservation International, the U.S. Agency for International Development, Isabela State University, and Keidanran Nature Conservation Fund. The participation of the College of Forestry was also supported by a long-standing environmental program of Leiden University, Netherlands. In 1997, support for this work was greatly augmented by a grant from the Dutch government to PLAN International for a community-based conservation program. Additional funding for the topographic mapping, as well as for infrastructure development for the plot's field station, was provided by Siemens, RICOH-Japan, and the First Philippine Conservation, Inc., in 2000–02.

References

Ashton, P. S. 1997. Before the memory fades: Some notes on the indigenous forest of the Philippines. *Sandakania* 9:1–19.

Co, L. L., and B. C. Tan. 1992. Botanical Exploration in Palanan Wilderness, Isabela Province, The Philippines: First Report. *Flora Malesiana Bulletin* 11:49–53.

Danielsen, F., D. S. Balete, T. D. Christensen, M. Heegaard, O. F. Jakobsen, A. Jensen, T. Lund, M. K. Poulsen. 1994. *Conservation of Biological Diversity in the Sierra Madre Mountains of Isabela and Southern Cagayan Province, the Philippines.* Department of Environment and Natural Resources, Bird Life International, Zoological Museum of Copenhagen University, and Danish Ornithological Society, Copenhagen.

Department of Environment and Natural Resources (DENR), NGOs for Integrated Protected Areas (NIPA), Nordic Agency for Development and Ecology (NORDECO), Conservation International and Plan International. 2001. Management Plan, Northern Sierra Madre Natural Park, 204 pp.

International Union for the Conservation of Nature and Natural Resources (IUCN). *2002 IUCN Red List of Threatened Species.* http://www.redlist.org, downloaded 15 March 2003.

Mabberley, D. J. 1997. *The Plant-Book: A Portable Dictionary of the Vascular Plants.* Cambridge University Press, Cambridge, U.K.

Nordic Agency for Development & Ecology (NORDECO) and Department of Environment and Natural Resources (DENR). 1998. Technical Report. Integrating Conservation and Development in Protected Area Management in the Northern Sierra Madre Natural Park, the Philippines, 205 pp. NORDECO, Copenhagen and DENR, Manila.

36

Pasoh Forest Dynamics Plot, Peninsular Malaysia

N. Manokaran, Quah Eng Seng, Peter S. Ashton, James V. LaFrankie,
Nur Supardi Mohd. Noor, Wan Mohd Shukri Wan Ahmad, and
Toshinori Okuda

Site Location, Administration, and Scientific Infrastructure

The Pasoh Forest Reserve lies in the Jelebu District of the State of Negeri
Sembilan in peninsular Malaysia. When gazetted in 1917, Pasoh was the largest
forest reserve in the Malayan system, but over the years, portions have been
de-gazetted for agriculture. Today the entire reserve encompasses less than
11,000 ha. It is administered by the State of Negeri Sembilan as a Permanent
Forest Estate for goods and services such as timber. The southernmost 1200 ha
of the reserve is the Pasoh Research Forest, which is administered by the Forest Research Institute of Malaysia (FRIM), headquartered in Kepong, Kuala
Lumpur, on behalf of the State Forest Department. The State Forest Department, which retains ultimate responsibility for the research forest, has made a
commitment to total protection for the Pasoh Forest Reserve as long as research
continues.

Because Pasoh is one of the most intensively researched rainforests in Asia, it
is also one of the best understood (Okuda et al. 2003). Silvicultural regeneration
surveys were carried out in various parts of the original Pasoh Forest Reserve
beginning in the 1950s, but the first study in the research forest was undertaken
by Wong Yew Kwan of FRIM in 1961 when he established ten 1-acre, permanent
north-south-oriented strip plots. In these plots, all trees greater than 12 inches dbh
were censused (Wong and Whitmore 1970). In 1970, P. S. Ashton extended five
of these plots to 2-ha, 200 × 100 m blocks. In the early 1970s, the International
Biological Program (IBP), a joint Malaysian–Japanese–British project, selected
the Pasoh Forest Reserve as a site for leading international research (Ashton et al.
2003; Soepadmo 1978). In 1986, FRIM, Harvard University, and the Smithsonian
Tropical Research Institute initiated a 50-ha Forest Dynamics Plot, encompassing
two of the previously established 2-ha IBP plots. The 50-ha plot lies within a core
remnant of primary forest that is at least 600 ha and is located 1.6 km from the
forest edge (fig. 36.1). Oil palm plantation surrounds the western half of the forest
primarily. Rubber plantations are found on the south and southeast. In 1991, the
National Institute of Environmental Studies (NIES) of Japan established a forest

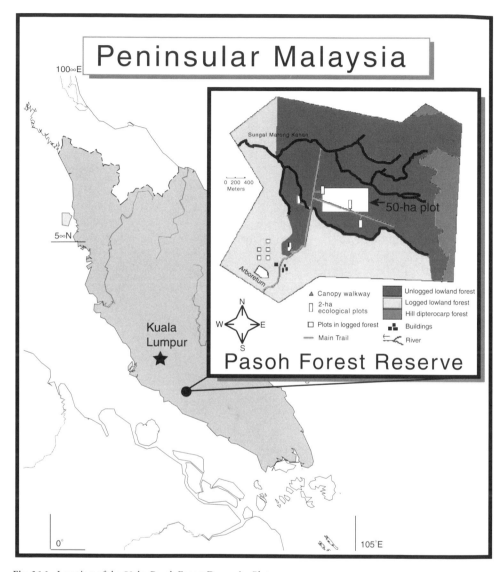

Fig. 36.1. Location of the 50-ha Pasoh Forest Dynamics Plot.

research program of rainforest research in collaboration with FRIM and Universiti Putra Malaysia (UPM), which partially utilized the 50-ha plot. Later NIES became a partner in the Forest Dynamics Plot research.

The research station in the Pasoh Research Forest includes housing, a variety of long-term ecological facilities, and canopy towers.

Table 36.1. Pasoh Climate Data

	Jan	Feb	Mar	Apr	May	Jun	Jul	Aug	Sep	Oct	Nov	Dec	Total/Averages
Rain (mm)	94	109	153	167	162	125	115	120	162	189	224	168	1788
ADTMx (°C)	32.0	33.1	33.9	34.6	34.4	33.8	32.8	33.3	33.2	33.7	32.0	31.6	33.2
ADTMn (°C)	21.9	22.2	22.6	23.1	23.4	23.1	22.6	22.8	22.8	22.9	22.7	22.8	22.7
Q (kW/m^2)	0.14	0.17	0.18	0.17	0.16	0.14	0.14	0.16	0.16	0.15	0.12	0.10	0.15
Wind (kph)	0.08	0.07	0.06	0.04	0.03	0.03	0.02	0.03	0.04	0.05	0.05	0.10	0.05

Notes: Data for mean monthly rainfall are for 1975–1998 and average daily temperatures (maximum and minimum) for 1991–1997 (Malaysian Meteorological Service 1977–2000; Nur Supardi 1999; Hydrology Unit FRIM, unpublished data). Data for solar radiation (Q) and wind speed are for 1991–1993 (Saifuddin et. al. 1994). Rainfall data between 1975 and 1998 excludes 3 years with incomplete records; data were collected by the Malaysian Meteorological Service from Pasoh Dua, a town 4 km to the south of Pasoh Forest Reserve. Temperature data from December 1991 to June 1995 were taken at an automatic weather station in the clearing of Pasoh Research Station. From July 1995 to December 1997, temperature data were taken from an automatic weather station positioned on a tower located about 600 m NE of the other weather station at Pasoh.

Climate

The climate of the Pasoh area is the driest and hottest of the southern Malay peninsula (see table 36.1). Mean annual rainfall recorded at Pasoh Forest Reserve from 1996 to 1999 was 1571 mm with a range of 1182 to 2065 (Tani et al. 2003). Long-term data from Pasoh Dua, an agricultural settlement 4 km from Pasoh, recorded a mean annual rainfall of 1788 mm over 24 years (Malaysian Meteorological Service 1977–2000). According to climate data collected at Kuala Pilah, 30 km away from the Pasoh Reserve, each year on record has at least one 20-day period without rain, typically in late January and/or late July. Average annual minimum and maximum temperatures are 22.7°C and 33.2°C, respectively (Nur Supardi 1999; Hydrology Unit FRIM unpublished data).

Topography and Soil

At approximately 80 m above sea level, the Pasoh Forest Reserve lies on a level plain of raised Pleistocene alluvium from which low undulating hills of Triassic sediments and granite arise (figs. 36.2 and 36.3). It is bordered to the east by a sharp north-south granite ridge that reaches a peak at Bukit Palong, 645 m above sea level. The 50-ha Pasoh Forest Dynamics Plot differs by only 24 m from high to low point. While the plot contains no permanent streams, a significant portion lies under standing water for more than 1 month, typically during November and December. Allbrook (1973) mapped Pasoh soils using the Malaysian agricultural

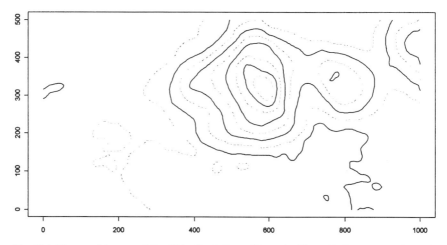

Fig. 36.2. Topographic map of the 50-ha Pasoh Forest Dynamics Plot with 5-m contour intervals (solid line) and intermediate 2.5 m contour intervals (dashed line).

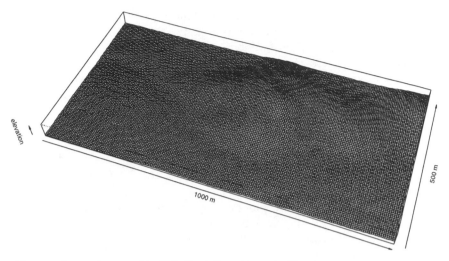

Fig. 36.3. Perspective map of the 50-ha Pasoh Forest Dynamics Plot.

soil classification system based on substrate and profile morphology. Dr. Amir Husni extended this work by mapping in detail soils of the Pasoh Forest Dynamics Plot (Amir Husni and Miller 1990; Amir Husni et al. 1991). Throughout the 50-ha Pasoh Forest Dynamics Plot, soils on the hills are predominantly well-drained Ultisols derived from alluvial sediment except in the northeast portion where granite is the parent material. Sandy, usually well-drained Entisols predominate on the plains. Poorly drained clay Entisols occur along the water courses.

Forest Type and Characteristics

Pasoh Forest Reserve includes one of the last remnants of the lowland mixed dipterocarp forests that once covered the south central peninsula. Following the classification of Wyatt-Smith (1987), they are sometimes described as south central red-meranti–keruing forest. An interesting and perhaps unique feature of Pasoh is the strong co-abundance of *Neobalanocarpus heimii* (Dipterocarpaceae) and *Dipterocarpus kunstleri* (Dipterocarpaceae). The closed canopy averages 35 m tall, with the emergent layer reaching 50–60 m. The upper canopy is noticeably dominated by the Dipterocarpaceae family: meranti (*Shorea* section Muticae, especially *S. leprosula* Miq., *S. acuminata* Dyer, and *S. macroptera* Dyer), keruing (*Dipterocarpus cornutus* Dyer), balau (*Shorea maxwelliana* King), and chengal (*Neobalanocarpus heimii* King). There are three main types of tree communities found in the plot: a low-lying swamp community in the north and northwestern corner, the hill community in the center and eastern portion, and the alluvium forest community in the remaining portions of the plot (Davies et al. 2003). This last type can be further broken down into sandy alluvium in the western portion of the plot, clay alluvium in the east, and an intermediate community in the northwest. There are strong species associations with each of these community types, including *Saraca thaipingensis* (Leguminosae), *Diospyros andamanica* (Ebenaceae), and *Iguanura wallichiana* (Palmae) in the swamp community; *Cleistanthus myrianthus* (Euphorbiaceae), *Pentace strychnoidea* (Tiliaceae), *Anisophyllea corneri* (Anisophylleaceae), and *Elateriospermum tapos* (Euphorbiaceae) in the hill community; and *Shorea maxwelliana* (Dipterocarpaceae), *Dipterocarpus crinitus* (Dipterocarpaceae), *Hopea mengarawan* (Dipterocarpaceae), and *Pavetta graciliflora* (Rubiaceae) in the alluvium communities. For census data and species rankings see tables 36.2–36.7.

Flowering and fruiting in the plot occur throughout the year, though supra-annual general mast flowering tends to occur from March to May (Yap and Chan

Table 36.2. Pasoh Plot Census History

Census	Dates	Number of Trees ($\geq$1 cm dbh)	Number of Species ($\geq$1 cm dbh)	Number of Trees ($\geq$10 cm dbh)	Number of Species ($\geq$10 cm dbh)
First	January 1986–September 1988	335,352	814	26,554	678
Second	January 1990–October 1990	323,237*	814	27,699	666
Third	January 1995–November 1996	320,808*	817*	29,288	674
Fourth	February 2000–July 2001	305,942[†]	816[†]	28,279	673

*In the Pasoh Forest Dynamics Plot dataset, during a recensus, trees whose main stems were broken or had disappeared and whose secondary stems were alive and larger than 1 cm dbh, were recorded as alive in the dataset, but their dbh measurements were not entered. These individuals have not been included in this table. However, if one were to include them in the dataset and assume that their measurement is between 1 and 10 cm dbh, then the following numbers (all in $\geq$1 cm dbh class) would be modified for the second the third censuses:

   Number of trees for second census: 326,797
   Number of trees for third census: 338,262
   Number of species for third census: 818

[†]Number of trees and species do not include the new recruits in the fourth census.

Note: Four censuses have been completed, the next census is expected to begin in January 2005.

Table 36.3. Pasoh Summary Tally

Size Class (cm dbh)	Average per Hectare							50-ha Plot				
	BA	N	S	G	F	H'	α	S	G	F	H'	α
$\geq$1	31.0	6707	495	210	68	2.31	123.9	814	288	82	2.45	100.3
$\geq$10	25.7	531	206	115	44	2.15	124.9	678	250	72	2.45	126.7
$\geq$30	15.9	76	47	35	21	1.57	53.5	375	163	56	2.13	103.2
$\geq$60	7.8	15	10	7	5	0.94	17.2*	103	58	30	1.61	32.7

*Based on 47 hectares.

Notes: BA represents basal area in m^2, N is the number of individual trees, S is the number of species, G is the number of genera, F is the number of families, H' is the Shannon–Wiener diversity index using $\log_{10}$, and α is Fisher's α. All individuals were identified. Data are from the first census.

Table 36.4. Pasoh Rankings by Family

Rank	Family	Basal Area (m²)	% BA	% Trees	Family	Trees	% Trees	Family	Species
1	Dipterocarpaceae	437.2	28.2	9.2	Euphorbiaceae	44,096	13.1	Euphorbiaceae	85
2	Leguminosae	132.3	8.5	3.3	Dipterocarpaceae	30,913	9.2	Lauraceae	48
3	Euphorbiaceae	106.7	6.9	13.1	Annonaceae	23,888	7.1	Myrtaceae	48
4	Burseraceae	96.5	6.2	5.4	Rubiaceae	19,395	5.8	Rubiaceae	45
5	Myrtaceae	52.6	3.4	3.1	Burseraceae	18,035	5.4	Meliaceae	43
6	Fagaceae	51.2	3.3	1.6	Sapindaceae	16,512	4.9	Annonaceae	42
7	Annonaceae	50.4	3.3	7.1	Myristicaceae	14,347	4.3	Guttiferae	34
8	Anacardiaceae	46.7	3.0	2.3	Ebenaceae	13,996	4.2	Anacardiaceae	32
9	Sapindaceae	42.4	2.7	4.9	Myrsinaceae	11,814	3.5	Myristicaceae	31
10	Myristicaceae	42.2	2.7	4.3	Guttiferae	11,072	3.3	Dipterocarpaceae	30

Notes: Top 10 tree families for basal area, number of trees ≥1 cm dbh, and number of species, and percentage of all trees in plot. Data are from the first census.

1990; Numata et al. 2003). Since 2001, 247 traps (0.5 m²) have been censused weekly for flowers, fruits, and seeds, following the same protocol used in the Barro Colorado Island and Yasuní Forest Dynamics Plots.

Fauna

In the Pasoh Forest Reserve, 89 mammal species, of which 12 are bats, are known to occur (Kemper 1988). Based on geographic distributions and habitat preferences, an additional 98 species may also reside in the forest reserve. Resident mammals within the research forest number 42 species, with perhaps four local extirpations. Mammal species include the long-tailed giant rat (*Leopoldamys sabanus*), lesser mouse deer (*Tragulus javanicus*), Malayan flying lemur (*Cynocephalus variegates*), slow loris (*Nycticebus coucang*), siamang (*Hylobates syndactylus*), red giant flying squirrel (*Petaurista petaurista*), spotted giant flying squirrel (*P. elegans*), and large black flying squirrel (*Aeromys tephromelas*) (Kemper 1988). Resident birds of the primary forest number approximately 166, with no known local extirpations, although populations of several species—especially the larger hornbills—have ostensibly declined from pre-1970 levels. The Pasoh forest also contains 489 ant species (Malsch et al. 2003). A survey on herpetofauna conducted from 1968 to 1991 in the Pasoh Forest Reserve recorded 75 species comprising 26 amphibians, 24 tortoises, turtles, and lizards, and 25 snakes (Lim et al. 2003).

Natural Disturbances

In general, the south central Malay peninsula lacks large-scale natural disturbances such as floods, droughts, fire, typhoons, and volcanoes. Lightning strikes

Table 36.5. Pasoh Rankings by Genus

Rank	Genus	Basal Area (m²)	% BA	% Trees	Genus	Trees	% Trees	Genus	Species
1	*Shorea* (Dipterocarpaceae)	301.6	19.5	6.4	*Shorea* (Dipterocarpaceae)	21,437	6.4	*Eugenia* (Myrtaceae)	45
2	*Dipterocarpus* (Dipterocarpaceae)	65.8	4.2	0.7	*Aporusa* (Euphorbiaceae)	17,000	5.1	*Diospyros* (Ebenaceae)	23
3	*Eugenia* (Myrtaceae)	50.7	3.3	3.0	*Diospyros* (Ebenaceae)	13,996	4.2	*Aglaia* (Meliaceae)	22
4	*Neobalanocarpus* (Dipterocarpaceae)	43.5	2.8	1.0	*Ardisia* (Myrsinaceae)	11,545	3.4	*Garcinia* (Guttiferae)	16
5	*Koompassia* (Leguminosae)	36.8	2.4	0.2	*Knema* (Myristicaceae)	10,697	3.2	*Litsea* (Lauraceae)	14
6	*Dacryodes* (Burseraceae)	35.1	2.3	3.2	*Dacryodes* (Burseraceae)	10,614	3.2	*Shorea* (Dipterocarpaceae)	14
7	*Ixonanthes* (Ixonanthaceae)	27.8	1.8	1.0	*Eugenia* (Myrtaceae)	10,044	3.0	*Aporusa* (Euphorbiaceae)	13
8	*Quercus* (Fagaceae)	26.0	1.7	0.4	*Xerospermum* (Sapindaceae)	8,969	2.7	*Knema* (Myristicaceae)	13
9	*Santiria* (Burseraceae)	25.9	1.7	0.8	*Rinorea* (Violaceae)	8,544	2.5	*Mangifera* (Anacardiaceae)	13
10	*Canarium* (Burseraceae)	23.8	1.5	1.3	*Anaxagorea* (Annonaceae)	7,076	2.1	*Memecylon* (Melastomataceae)	12

Notes: The top 10 tree genera for trees ≥1 cm dbh ranked by basal area, number of individual trees, and number of species with the percentage of trees in the plot. Data are from the first census.

Table 36.6. Pasoh Rankings by Species

Rank	Species	Number Trees	% Trees	Species	Basal Area (m²)	% BA	% Trees
1	*Xerospermum noronhianum* (Sapindaceae)	8961	2.7	*Shorea maxwelliana* (Dipterocarpaceae)	54.7	3.5	1.7
2	*Rinorea anguifera* (Violaceae)	8262	2.5	*Shorea leprosula* (Dipterocarpaceae)	53.0	3.4	0.9
3	*Ardisia crassa* (Myrsinaceae)	7641	2.3	*Neobalanocarpus heimii* (Dipterocarpaceae)	43.5	2.8	1.0
4	*Anaxagorea javanica* (Annonaceae)	7076	2.1	*Shorea lepidota* (Dipterocarpaceae)	41.4	2.7	0.4
5	*Aporusa microstachya* (Euphorbiaceae)	6509	1.9	*Shorea pauciflora* (Dipterocarpaceae)	40.5	2.6	0.7
6	*Shorea maxwelliana* (Dipterocarpaceae)	5676	1.7	*Koompassia malaccensis* (Leguminosae)	36.8	2.4	0.2
7	*Dacryodes rugosa* (Burseraceae)	5649	1.7	*Shorea acuminata* (Dipterocarpaceae)	36.4	2.4	0.7
8	*Knema laurina* (Myristicaceae)	4489	1.3	*Dipterocarpus cornutus* (Dipterocarpaceae)	34.6	2.2	0.4
9	*Gironniera parvifolia* (Ulmaceae)	3961	1.2	*Ixonanthes icosandra* (Ixonanthaceae)	27.8	1.8	1.0
10	*Barringtonia macrostachya* (Lecythidaceae)	3705	1.1	*Shorea parvifolia* (Dipterocarpaceae)	26.7	1.7	0.5

Notes: The top 10 tree species for trees ≥ 1 cm dbh are ranked by number and percentage of trees and basal area. Data are from the first census.

are extremely frequent in this region but are not known to ignite forest fires. Windthrows of less than 1-ha size are relatively common during April, especially in areas with impeded drainage and shallow rooting. Small-scale disturbances created by wild pigs (*Sus scrofa*) are also common. Pigs root through soil for food and build nests with hundreds of saplings each, significantly reducing plant recruitment. Termites (*Microcerotermes dubius*) are also known to destroy living trees in the area, causing forest gaps (Tho 1982).

Human Disturbance

Humans have lived in the Malay peninsula for perhaps 40,000 years and have traded extracted forest products since prehistoric times (Dunn 1975). Until recently, there has been no history of shifting cultivation in this area of predominantly proto-Malay Semelai and Mon-Khmer (Jakun) people. Jakun collectors had traditionally tapped various species of *Dipterocarpus* in the Pasoh forest for keruing oil. This practice continued until very recently through resin extraction of the highly productive *D. kerrii*. In 1970–71 the area immediately surrounding the forest to the north, west, and south sides of the reserve was cleared for oil palm (*Elaeis guinensis* [Palmae]) production. At this time, a large number of

Table 36.7. Pasoh Tree Demographic Dynamics

Size Class (cm dbh)	Growth Rate (mm/yr)			Mortality Rate (%/yr)			Recruitment Rate (%/yr)			BA Losses (m²/ha/yr)			BA Gains (m²/ha/yr)		
	86–90	90–95	95–00	86–90	90–95	95–00	86–90	90–95	95–00	86–90	90–95	95–00	86–90	90–95	95–00
1–9.9	1.36	0.63	0.42	1.16	1.42*	1.78*	0.65	2.37*	1.33*	0.07	0.10	0.11	0.40	0.25	0.15
10–29.9	2.33	2.13	1.38	1.19	1.49	1.98	3.39	3.05	1.84	0.13	0.19	0.21	0.41	0.39	0.26
≥30	3.37	3.79	2.45	1.11	1.76	2.05	2.37	2.39	1.90	0.18	0.32	0.37	0.33	0.34	0.26

*In the Pasoh Forest Dynamics Plot dataset, during a recensus, trees whose main stems were broken or had disappeared and whose secondary stems were alive and larger than 1 cm dbh, were recorded as alive in the dataset, but their dbh measurements were not entered. These individuals have not been included in these tables. However, if one were to include them in the dataset and assume that their measurement is between 1 and 10 cm dbh, then the following rates (all in the 1–9.9 cm dbh class) would be modified:

Mortality for 90–95: 1.47%/yr
Mortality for 95–00: 2.00%/yr
Recruitment for 90–95: 2.39%/yr
Recruitment for 95–00: 0.88%/yr

forest fragments were formed in areas of similar soil and topography to the reserve, within the watershed of the Pertang river. Between 1930 and 1960, logging with buffalo was carried out in many of the peripheral compartments of the Pasoh forest, but not within the core forest compartments that include the 50-ha plot.

Occasional hunting exists in the nature reserve as native pigs are abundant due to the additional food supply from surrounding oil palm plantations. Locally extirpated populations include the rhinoceros, tapir, elephant, and tiger, although all species—except the rhino—may still rarely wander into the Pasoh Forest Reserve. The only invasive plant species is the neotropical *Clidemia hirta* (Melastomataceae) which has become increasingly common in the forest during the last 10 years. Plans for an east-west road north of the research forest are currently underway. In general, the central human disturbance in Pasoh has been the indirect effects of fragmentation and size reduction of the forest.

In the 1950s, a buffer zone of 700 ha within the western and southern area of the reserve was logged under a regime called the Malayan Uniform System (MUS). Distinct differences were found between the primary (unlogged) and regenerating (logged) forest in canopy height, canopy structure, stand structure, and tree species compositions (Manokaran and Swaine 1994; Okuda et al. 2003), flower-visiting beetles (Fukuyama et al. 2003), termite communities (Takamura 2003), small mammal community (Yasuda et al. 2003; Lim et al. 2003), herpetofauna (Lim and Yaakob 2003), fungi (wood-decaying basidiomycetes) (Hattori and Lee 2003), and bird community (Styring and Ickes 2003).

Plot Size and Location

Pasoh is a 50-ha, 1000 × 500 m plot; the long axis lies east-west. The southwest corner of the plot (coordinates 0, 0) is located at 2°58′47″N, 102°18′29″E.

Funding Sources

The Pasoh Forest Dynamics Plot has been funded primarily by the Forest Research Institute Malaysia, U.S. National Science Foundation, Smithsonian Tropical Research Institute, and Center for Global Environmental Research at the National Institute for Environmental Studies.

References

Allbrook, R. F. 1973. The soils of Pasoh Forest Reserve, Negeri Sembilan. *Malaysian Forester* 36:22–33.

Amir Husni, M. S., and H. G. Miller. 1990. *Shorea leprosula* as an indicator species for site fertility evaluation in dipterocarp forests of Peninsular Malaysia. *Journal of Tropical Forest Science* 3:101–10.

Amir Husni, M. S., H. G. Miller, and S. Appanah. 1991. Soil fertility and tree species diversity in two Malaysian forests. *Journal of Tropical Forest Science* 3:318–31.

Appanah, S., A. H. Gentry, and J. V. LaFrankie. 1993. Liana diversity and species richness of Malaysian rain forests. *Journal of Tropical Forest Science* 6:116–23.

Appanah, S., and G. Weinland. 1993. *A Preliminary Analysis of the 50-Hectare Pasoh Demography Plot: I. Dipterocarpaceae.* FRIM Research Pamphlet No. 112. Forest Research Institute Malaysia, Kepong, Malaysia.

Ashton, P. S., T. Okuda, and N. Manokaran. 2003. Pasoh research, past and present. Pages 1–14 in T. Okuda, N. Manokaran, Y. Matsumoto, K. Niiyama, S. C. Thomas, and P. S. Ashton, editors. *Pasoh: Ecology of a Lowland Rain Forest in Southeast Asia.* Springer-Verlag, Tokyo.

Ashton, P. S. 1976. Mixed dipterocarp forest and its variation with habitat in the Malayan lowlands: A re-evaluation at Pasoh. *Malaysian Forester* 39:56–72.

Chan, H. T. 1981. Reproductive biology of some Malaysian dipterocarps: III. Breeding systems. *Malaysian Forester* 44:28–36.

Davies, S. J., Nur Supardi Md. Noor, J. V. LaFrankie, and P. S. Ashton. 2003. The trees of Pasoh Forest: Stand structure and floristics of the 50-ha Forest Research Plot. Pages 35–50 in T. Okuda, N. Manokaran, Y. Matsumoto, K. Niiyama, S. C. Thomas, and P. S. Ashton, editors. *Pasoh: Ecology of a Lowland Rain Forest in Southeast Asia.* Springer-Verlag, Tokyo.

Dunn, F. L. 1975. Rainforest collectors and traders, a study of resource utilization in modern and ancient Malaya. Monographs of the Malaysian Branch of the Royal Asiatic Society 5:1–151.

Francis, C. 1985. *A Field Guide to the Mammals of Borneo.* Kota Kinabalu, Sabah. The Sabah Society, Sabah, Malaysia.

Fukuyama, K., Maeto Kaoru, and Sajap S. Ahmad. 2003. Spatial distribution of flower visiting beetles in Pasoh forest reserve and its study technique. Pages 421–36 in T. Okuda, N. Manokaran, Y. Matsumoto, K. Niiyama, S. C. Thomas, and P. S. Ashton, editors. *Pasoh: Ecology of a Lowland Rain Forest in Southeast Asia.* Springer-Verlag, Tokyo.

Hattori, T., and S. Lee. 2003. Community structure of wood-decaying basidiomycetes in Pasoh. Pages 161–70 in T. Okuda, N. Manokaran, Y. Matsumoto, K. Niiyama, S. C. Thomas, and P. S. Ashton, editors. *Pasoh: Ecology of a Lowland Rain Forest in Southeast Asia.* Springer-Verlag, Tokyo.

Kemper, C. M. 1988. The mammals of Pasoh Forest Reserve, Peninsular Malaysia. *Malayan Nature Journal* 42:1–19.

Kira, T. 1978. Community architecture and organic matter dynamics in tropical lowland rain forests of southeast Asia with special reference to Pasoh Forest, West Malaysia. Pages 561–90 in P. B. Tomlinson and M. H. Zimmermann, editors. *Tropical Trees as Living Systems: The Proceedings of the Fourth Cabot Symposium, Harvard Forest, Petersham, Mass., April 26–30, 1976.* Cambridge University Press, New York.

Kochummen, K. M. 1997. *Tree Flora of Pasoh Forest.* Malayan Forest Records No. 44. Forest Research Institute of Malaysia, Kepong, Malaysia.

Kochummen, K. M., J. V. LaFrankie, and N. Manokaran. 1990. Floristic composition of Pasoh Forest Reserve, a lowland rain forest in Peninsular Malaysia. *Journal of Tropical Forest Science* 3:1–13.

Kochummen, K. M., J. V. LaFrankie, and N. Manokaran. 1992. Diversity of trees and shrubs in Malaya at regional and local levels. *Malayan Nature Journal* 45:545–54.

LaFrankie, J. V. 1996. Distribution and abundance of Malayan trees: Significance of family characteristics for conservation. *Gardens' Bulletin Singapore* 48:75–87.

Lee, S. S. 1995. *A Guide Book to Pasoh*. Forest Research Institute of Malaysia, Kepong, Malaysia.

Lim, B. L., and N. Yaakob. 2003. Herpetofauna diversity survey in Pasoh Forest Reserve, Negeri Sembilan, Peninsular Malaysia. Pages 395–402 in T. Okuda, N. Manokaran, Y. Matsumoto, K. Niiyama, S. C. Thomas, and P. S. Ashton, editors. *Pasoh: Ecology of a Lowland Rain Forest in Southeast Asia*. Springer-Verlag, Tokyo.

Lim, B. L., L. Ratnam, and N. Hussein. 2003. Small mammal diversity in Pasoh Forest Reserve, Negeri Sembilan, Peninsular Malaysia. Pages 403–11 in T. Okuda, N. Manokaran, Y. Matsumoto, K. Niiyama, S. C. Thomas, and P. S. Ashton, editors. *Pasoh: Ecology of a Lowland Rain Forest in Southeast Asia*. Springer-Verlag, Tokyo.

Mabberley, D. J. 1997. *The Plant-Book: A Portable Dictionary of the Vascular Plants*. Cambridge University Press, Cambridge, U.K.

Malaysian Meteorological Service. 1977–2000. *Annual Summary of Meteorological Observation 1975–1998*. Malaysian Meteorological Service, Kuala Lumpur, Malaysia.

Malsch, A. K. F., K. Rosciszewski, and U. Maschwitz. 2003. The ant species richness and diversity of a primary lowland rain forest, the Pasoh Forest Reserve, West-Malaysia. Pages 347–74 in T. Okuda, N. Manokaran, Y. Matsumoto, K. Niiyama, S. C. Thomas, and P. S. Ashton, editors. *Pasoh: Ecology of a Lowland Rain Forest in Southeast Asia*. Springer-Verlag, Tokyo.

Manokaran, N. 1998. Effects, 34 years later, of selective logging in the lowland dipterocarp forest at Pasoh, peninsular Malaysia, and implications on present day logging in the hill forests. Pages 41–60 in Lee, S. S., D. Y. May, I. D. Gauld, and J. Bishop, editors. *Conservation, Management and Development of Forest Resources*. Forest Research Institute Malaysia, Kepong, Malaysia.

Manokaran, N., and J. V. LaFrankie. 1990. Stand structure of Pasoh Forest Reserve, a lowland rain forest in Peninsular Malaysia. *Journal of Tropical Forest Science* 3:14–24.

Manokaran, N., and M. D. Swaine. 1994. *Population Dynamics of Trees in Dipterocarp Forests of Peninsular Malaysia*. Malayan Forest Records No. 40. Forest Research Institute Malaysia, Kepong, Malaysia.

Manokaran, N., J. V. LaFrankie, K. Kochummen, E. Quah, J. Klahn, P. S. Ashton, and S. P. Hubbell. 1990. *Methodology for the Fifty Hectare Research Plot at Pasoh Forest Reserve*. FRIM Pamphlet 104. Forest Research Institute Malaysia, Kepong, Malaysia.

————. 1992. *Stand Table and Distribution of Species in the 50-ha Research Plot at Pasoh Forest Reserve*. FRIM Research Data. Forest Research Institute Malaysia, Kepong, Malaysia.

Numata, S., M. Yasuda, T. Okuda, N. Kachi, and M. N. Supardi. 2003. Temporal and spatial patterns of mass flowerings on the Malay Peninsula. *American Journal of Botany* 90:1025–31.

Nur Supardi, M. N. 1999. *The Impact of Logging on the Diversity and Values Of Palms (Arecaceae) in the Lowland Dipterocarp Forest of Pasoh, Peninsular Malaysia*. Ph.D. thesis, University of Reading, Reading, England.

Okuda, T., N. Kachi, S. K. Yap, and N. Manokaran. 1997. Tree distribution pattern and fate of juveniles in a lowland tropical rain forest—Implications for regeneration and maintenance of species diversity. *Plant Ecology* 131:155–71.

Okuda, T., N. Manokaran, Y. Matsumoto, K. Niiyama, S. C. Thomas, and P. S. Ashton, editors. 2003. *Pasoh: Ecology of a Lowland Rain Forest in Southeast Asia*. Springer-Verlag, Tokyo.

Okuda, T., M. Suzuki, N. Adachi, E. S. Quah, H. Nor Azman, and N. Manokaran. 2003. Effect of selective logging on canopy and stand structure and tree species composition in a lowland dipterocarp forest in Peninsular Malaysia. *Forest Ecology and Management* 175:297–320.

Saifuddin S., A. R. Nik, and J. V. LaFrankie. 1994. *Pasoh Climatic Summary (1991–1993)*. FRIM Research Data No. 3. Forest Research Institute Malaysia, Kepong, Malaysia.

Saw, L. G., J. V. LaFrankie, K. M. Kochummen, and S. K. Yap. 1991. Fruit trees in a Malaysian rainforest. *Economic Botany* 45:120–36.

Soepadmo, E. 1978. Introduction to the Malaysian IBP Synthesis Meeting. *Malayan Nature Journal* 30:119–24.

Styring, R. A., and K. Ickes. 2003. Woodpeckers (Picidae) at Pasoh: Foraging ecology, flocking and the impacts of logging on abundance and diversity. Pages 547–58 in T. Okuda, N. Manokaran, Y. Matsumoto, K. Niiyama, S. C. Thomas, and P. S. Ashton, editors. *Pasoh: Ecology of a Lowland Rain Forest in Southeast Asia*. Springer-Verlag, Tokyo.

Takamura, K. 2003. Is the termite community disturbed by logging? Pages 521–31 in T. Okuda, N. Manokaran, Y. Matsumoto, K. Niiyama, S. C. Thomas, and P. S. Ashton, editors. *Pasoh: Ecology of a Lowland Rain Forest in Southeast Asia*. Springer-Verlag, Tokyo.

Tani, M., A. R. Nik, Y. Ohtani, Y. Yasuda, M. M. Sahat, B. Kasran, S. Takanashi, S. Noguchi, Z. Yusop, and T. Watanabe. 2003. Characteristics of energy exchange and surface conductance of a tropical rain forest in peninsular Malaysia. Pages 73–88 in T. Okuda, N. Manokaran, Y. Matsumoto, K. Niiyama, S. C. Thomas, and P. S. Ashton, editors. *Pasoh: Ecology of a Lowland Rain Forest in Southeast Asia*. Springer-Verlag, Tokyo.

Tho, Y. P. 1982. Gap formation by the termite *Microcerotermes dubius* in lowland forests of Peninsular Malaysia. *Malaysian Forester* 45:184–192.

Wells, D. R. 1978. Number and biomass of insectivorous birds in the understorey of rain forest at Pasoh Forest. *Malayan Nature Journal* 30:353–62.

Wong, M. 1983. Understory phenology of the virgin and regenerating habitats in Pasoh Forest Reserve, Negeri Sembilan, West Malaysia. *Malaysian Forester* 46:197–223.

Wong, M. 1985. Understorey birds as indicators of regeneration in a patch of selectively logged Malaysian rain forest. *ICBP Technical Publication* 4:249–63.

———. 1986. Trophic organization of understorey birds in a Malaysian dipterocarp forest. *Auk* 103:100–16.

Wong, Y. K., and T. C. Whitmore. 1970. On the influence of soil properties on species distribution in a Malayan lowland dipterocarp rainforest. *Malaysian Forester* 33:42–54.

Wyatt-Smith, J. 1961. A note on the fresh-water swamp, lowland and hill forest types of Malaya. *Malaysian Forester* 24:110–21.

———. 1964. A preliminary vegetation map of Malaya with description of the vegetation types. *Journal of Tropical Geography* 18:200–13.

———. 1987. *Manual of Malayan Silviculture for Inland Forests*. Part III. Chapter 7. Red Meranti-Keruing forest. FRIM Research Pamphlet No. 101. Forest Research Institute of Malaysia, Kepong, Malaysia.

Yap, S. K., and H. T. Chan. 1990. Phenological behaviour of some *Shorea* species in peninsular Malaysia. Pages 21–35 in K. S. Bawa and M. Hadley, editors. *Reproductive Ecology of Tropical Forest Plants*. Man and the Biosphere Series, Vol. 7. Parthenon Publications, Paris.

Yasuda, M., N. Ishii, T. Okuda, and N. A. Hussein. 2003. Small mammal community: habitat preference and effects after selective logging. Pages 533–46 in T. Okuda, N. Manokaran, Y. Matsumoto, K. Niiyama, S. C. Thomas, and P. S. Ashton, editors. *Pasoh: Ecology of a Lowland Rain Forest in Southeast Asia*. Springer-Verlag, Tokyo.

37

Sinharaja Forest Dynamics Plot, Sri Lanka

C. V. S. Gunatilleke, I. A. U. N. Gunatilleke, Peter S. Ashton, A. U. K. Ethugala, N. S. Weerasekera, and Shameema Esufali

Site Location, Administration, and Scientific Infrastructure

Located in the southwest region of Sri Lanka, the Sinharaja forest is the largest block of relatively undisturbed lowland evergreen rainforest on the island. Preservation of the Sinharaja forest goes back to 1907, when a 3200-ha area was demarcated as the result of a Forest Ordinance. In 1926, the ordinance was extended to cover 3724 ha. In 1978, after a short period of selective logging, the area was declared an 8800-ha International Man and Biosphere Reserve (de Zoysa and Raheem 1990). Subsequently, the forest, together with an eastern extension of 2450 ha, was declared a National Heritage Wilderness Area in 1988 and a UNESCO Natural World Heritage Site in 1989.

The 25-ha Sinharaja Forest Dynamics Plot was established in 1993 by the University of Peradeniya (Sri Lanka), the Forest Department of Sri Lanka, Harvard Institute of International Development (now the Center for International Development), and the Smithsonian Tropical Research Institute. The plot, which contains only undisturbed forest, is located in the southwestern portion of the reserve (fig. 37.1). The eastern, northern, and northwestern sides of the plot are surrounded by forests that are regenerating after they were selectively logged in the 1970s.

Between 1977 and 1981, 100 small plots, each 0.25-ha in extent, were set up in five different areas of the Sinharaja forest by I. A. U. N. Gunatilleke and C. V. S. Gunatilleke, to document the differences in floristic composition and community structure in this forest. In these plots, all trees above 10 cm dbh were enumerated. A recensus of these plots was conducted in 1999.

Site facilities include research vehicles, a building with accommodations for at least eight researchers, a basic field laboratory, shade houses, and facilities for cooking, dining, and bathing. The nearby Forest Department station at Kudawa can accommodate larger parties and student classes.

Climate

Over a 17-year period from 1984 to 1999, annual rainfall ranged from 4087 to 5907 mm, and averaged 5016 mm. There is no distinct dry season (average

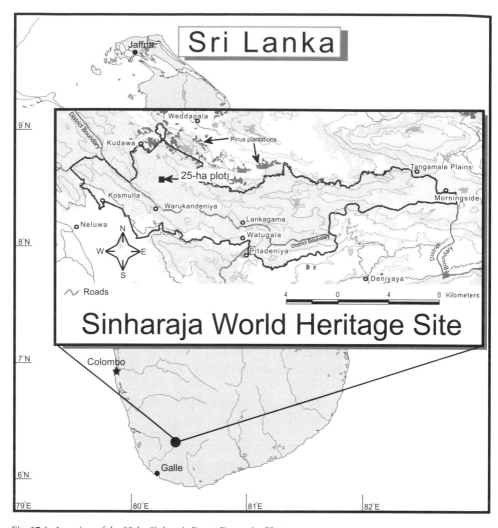

Fig. 37.1. Location of the 25-ha Sinharaja Forest Dynamics Plot.

monthly rainfall less than 100 mm). The forest receives rain from both the southwest monsoon, from May through July, and the northeast monsoon, from October through January (Munidasa et al. 2002). For climate data see table 37.1.

Topography and Soil

The Sinharaja Forest Dynamics Plot lies between 424 and 575 m above sea level (figs. 37.2 and 37.3). It has a central valley bounded by two slopes, the steeper

Table 37.1. Sinharaja Climate Data

	Jan	Feb	Mar	Apr	May	Jun	Jul	Aug	Sep	Oct	Nov	Dec	Total/ Averages
Rain (mm)	191	171	237	434	695	610	390	424	553	562	471	274	5016
ADTMx	24.4	25.5	26.8	26.1	25.0	24.0	23.7	23.9	24.1	23.9	24.1	24.5	24.7
ADTMn	19.3	19.3	19.6	20.8	21.7	21.4	21.1	20.8	20.8	20.4	20.1	19.5	20.4

Notes: Rainfall data were collected in a large opening in the forest and temperature data were collected in the adjacent forest near the field station in Sinharaja at 520 m above sea level. Values given are rainfall averages taken over the period 1984–2002 and temperature averages over the period 1992–1999 from the understory of the forest.

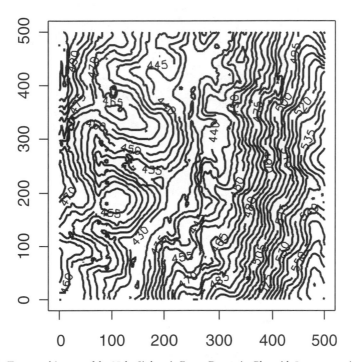

Fig. 37.2. Topographic map of the 25-ha Sinharaja Forest Dynamics Plot with 5-m contour intervals.

and higher slope faces the southwest, and the other faces the northeast. Seepage ways, spurs, and small hillocks cut across both slopes. Two perennial streams and several seasonal streamlets are present in the plot.

The geology of the area represents metamorphic rocks of Sri Lanka's Highland Series of the Precambrian Age (Cooray 1967). The more siliceous charnokite forms the prominent parallel crests; khondalite, rich in hornblende, underlies the fertile lower slopes and valleys. The soils overlying them are Ultisols (humults to udults) (Panabokke 1996). Soil physical–chemical characteristics for the Sinharaja forest in general are reported as follows: pH = 3.8–4.77; C = 1.44–4.73%; N =

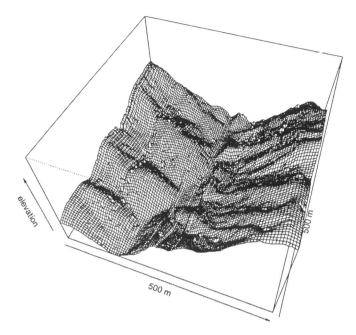

Fig. 37.3. Perspective map of the 25-ha Sinharaja Forest Dynamics Plot.

0.09–0.22%; P = 0.033–0.045%; K = 0.27–1.03%; Ca = 0.09–0.77%; Mg = 0.20–0.73% (Gunatilleke and Maheswaran 1988; Hafeel 1991; Gunatilleke et al. 1996).

Forest Type and Characteristics

Although many species in the Sinharaja Forest Dynamics Plot are endemic to Sri Lanka (69%), the generic composition is strikingly similar to the lowland mixed dipterocarp forest of western Malesia. It is therefore categorized as mixed dipterocarp forest, though the south Asian endemic generic element differentiates it as a distinct regional type. Unlike other mixed dipterocarp forests, those of Sri Lanka's southwestern hills do not have an emergent layer of trees, possibly due to strong winds. Average canopy height in the plot is about 30 m. The forest structure is skewed toward small individuals.

Forest composition in the plot tends to change along an elevational gradient. As one ascends from the lower slopes at 430–459 m above sea level to the ridgetop above 520 m, the number of trees per hectare increases from 7272 to 11,278, the basal area per hectare increases from 39.2 to 59.4 m², Fisher's alpha

Table 37.2. Sinharaja Plot Census History

	Dates	Number of Trees (≥ 1 cm dbh)	Number of Tree Species (≥ 1 cm dbh)	Number of Lianas (≥ 1 cm dbh)	Number of Liana Species (≥ 1 cm dbh)	Number of Trees (≥ 10 cm dbh)	Number of Tree Species (≥ 10 cm dbh)
First	August 1993–April 1996	205,373	205	1128	10*	16,937	167

* Number of liana species may increase as lianas are further identified.

Notes: One census has been completed. The second census was completed in 2002.

Table 37.3. Sinharaja Summary Tally

Size Class (cm dbh)	Average per Hectare							25-ha Plot				
	BA	N	S	G	F	H'	α	S	G	F	H'	α
≥ 1	45.6	8215	142	81	38	1.53	24.4	205	116	46	1.71	22.5
≥ 10	39.9	677	72	48	28	1.41	20.4	167	95	42	1.66	25.7
≥ 30	27.0	143	34	24	17	1.16	14.1	109	71	35	1.44	21.2
≥ 60	10.7	23	8	7	6	0.70	4.3	47	32	21	1.13	12.1

Notes: BA represents basal area in m^2, N is the number of individual trees, S is number of species, G is number of genera, F is number of families, H' is Shannon–Wiener diversity index using $\log_{10}$, and α is Fisher's α. Basal area includes all multiple stems for each individual. 9132 individuals were not identified to species or morphospecies. Data are from the first census.

Table 37.4. Sinharaja Rankings by Family

Rank	Family	Basal Area (m^2)	% BA	% Trees	Family	No. of Trees	% Trees	Family	No. of Species
1	Guttiferae	304.9	26.7	15.5	Guttiferae	31,825	15.5	Euphorbiaceae	21
2	Dipterocarpaceae	246.4	21.6	14.1	Euphorbiaceae	30,388	14.8	Rubiaceae	18
3	Bombacaceae	100.3	8.8	3.0	Dipterocarpaceae	28,984	14.1	Melastomataceae	16
4	Euphorbiaceae	67.6	5.9	14.8	Leguminosae	22,537	11.0	Myrtaceae	14
5	Anacardiaceae	44.7	3.9	3.1	Rubiaceae	16,169	7.9	Dipterocarpaceae	13
6	Sapotaceae	44.6	3.9	4.2	Sapotaceae	8,527	4.2	Lauraceae	11
7	Lauraceae	37.9	3.3	2.2	Anacardiaceae	6,465	3.1	Guttiferae	10
8	Myristicaceae	37.0	3.2	1.3	Bombacaceae	6,157	3.0	Sapotaceae	9
9	Myrtaceae	30.4	2.7	2.3	Myrtaceae	4,744	2.3	Anacardiaceae	8
10	Meliaceae	22.7	2.0	0.6	Lauraceae	4,606	2.2	Ebenaceae	8

Notes: The top 10 families for trees ≥ 1 cm dbh are ranked in terms of basal area, number of individual trees, and number of species, with the percentage of trees in the plot. Data are from the first census.

decreases from 25.38 to 23.30, and the dominant canopy species change from *Shorea megistophylla* (Dipterocarpaceae), *Mesua ferrea* (Guttiferae), and *Shorea trapezifolia* (Dipterocarpaceae) to *Mesua nagassarium* (Guttiferae) and *Shorea disticha* (Dipterocarpaceae). See chapter 10 and tables 37.2–37.6.

Table 37.5. Sinharaja Rankings by Genus

Rank	Genus	Basal Area (m²)	% BA	% Trees	Genus	No. of Trees	% Trees	Genus	No. of Species
1	Mesua (Guttiferae)	245.5	21.9	10.6	Shorea (Dipterocarpaceae)	27,824	14.2	Memecylon (Melastomataceae)	13
2	Shorea (Dipterocarpaceae)	233.0	20.8	14.2	Agrostistachys (Euphorbiaceae)	25,343	12.9	Syzygium (Myrtaceae)	12
3	Cullenia* (Bombacaceae)	99.8	8.9	3.1	Humboldtia (Leguminosae)	22,459	11.4	Diospyros (Ebenaceae)	8
4	Garcinia (Guttiferae)	51.8	4.6	4.6	Mesua (Guttiferae)	20,863	10.6	Shorea (Dipterocarpaceae)	8
5	Palaquium (Sapotaceae)	42.3	3.8	4.1	Garcinia (Guttiferae)	9,014	4.6	Palaquium (Sapotaceae)	6
6	Myristica (Myristicaceae)	35.6	3.2	1.4	Psychotria (Rubiaceae)	8,690	4.4	Garcinia (Guttiferae)	5
7	Chaetocarpus (Euphorbiaceae)	31.0	2.8	0.8	Palaquium (Sapotaceae)	8,069	4.1	Glochidion (Euphorbiaceae)	5
8	Syzygium (Myrtaceae)	30.2	2.7	2.3	Cullenia* (Bombacaceae)	6,154	3.1	Semecarpus (Anacardiaceae)	5
9	Litsea (Lauraceae)	29.0	2.6	0.9	Syzygium (Myrtaceae)	4,593	2.3	Symplocos (Symplocaceae)	5
10	Semecarpus (Anacardiaceae)	22.8	2.0	2.3	Semecarpus (Anacardiaceae)	4,530	2.3	Cinnamomum (Lauraceae)	4
11								Psychotria (Rubiaceae)	4

*Genus Cullenia has been changed to Durio according to Mabberley (1997), though Durio is not known to occur in Sri Lanka.

Notes: The top 10 tree genera for trees ≥1 cm dbh are ranked by basal area, number of individual trees, and number of species with the percentage of trees in the plot. Data are from the first census.

Each vertical stratum of the forest is dominated by one or two genera or species. In the canopy, 85% of the individuals are in the genera *Mesua* or *Shorea*, 49% of the understory trees are *Humboldtia laurifolia* (Leguminosae), and 41% of the treelets and shrubs are *Agrostistachys intramarginalis* (Euphorbiaceae) or *A. hookeri.*

Phenological studies on the *Shorea* species (Dipterocarpaceae) of section Doona present in Sinharaja have been monitored since 1984, in the selectively logged roadside areas of the forest where the tree crowns are easily visible. These *Shorea* spp. fall into two groups: the Thiniya-Dun and Beraliya. Members of the Thiniya-Dun group such as *Shorea trapezifolia, S. affinis,* and *S. congestiflora* have soft wood and very resinous small nonedible fruits. Species in this group bloom annually, and they flush almost all year long. In the Beraliya group, represented by *S. megistophylla, S. distica, S. worthingtonii* and *S. cordifolia,* trees possess medium hard wood and relatively less resinous, large edible fruits. They show sequential flowering with little overlap and synchronized fruiting. They flower supra-annually and flush during limited periods. All these *Shorea* species bloom at dawn, have fragrant flowers, and are pollinated by bees (Dayanandan

Table 37.6. Sinharaja Ranking by Species

Rank	Species	Number of Trees	% Trees	Species	Basal Area (m^2)	% BA	% Trees
1	*Humboldtia laurifolia* (Leguminosae)	22,459	10.9	*Mesua nagassarium* (Guttiferae)	230.9	20.24	7.3
2	*Agrostistachys intramarginalis* (Euphorbiaceae)	18,022	8.8	*Cullenia ceylanica** (Bombacaceae)	86.6	7.59	1.9
3	*Mesua nagassarium* (Guttiferae)	14,881	7.2	*Shorea trapezifolia* (Dipterocarpaceae)	68.2	5.98	1.3
4	*Garcinia hermonii* (Guttiferae)	8,133	4.0	*Garcinia hermonii** (Guttiferae)	49.6	4.34	4.0
5	*Shorea disticha* (Dipterocarpaceae)	7,397	3.6	*Shorea disticha* (Dipterocarpaceae)	42.8	3.75	3.6
6	*Agrostistachys hookeri* (Euphorbiaceae)	7,321	3.6	*Myristica dactyloides* (Myristicaceae)	35.6	3.12	1.3
7	*Psychotria nigra* (Rubiaceae)	6,087	3.0	*Shorea stipularis* (Dipterocarpaceae)	30.8	2.70	0.5
8	*Mesua ferrea* (Guttiferae)	5,982	2.9	*Litsea gardneri* (Lauraceae)	27.9	2.45	0.1
9	*Shorea worthingtonii* (Dipterocarpaceae)	4,628	2.3	*Shorea affinis* (Dipterocarpaceae)	22.3	1.96	2.0
10	*Shorea affinis* (Dipterocarpaceae)	4,193	2.0	*Shorea megistophylla* (Dipterocarpaceae)	21.5	1.88	1.6

*Genus *Cullenia* has been changed to *Durio* according to Mabberley (1997), though *Durio* is not known to occur in Sri Lanka.

Notes: The top 10 tree species for trees ≥1 cm dbh are ranked by number of trees and basal area. The percentage of the total population is also shown. Data are from the first census.

et al. 1990). Short-term studies on the phenology and reproductive biology have also been carried out on the palm *Caryota urens* (Palmae) (Ratnayake et al. 1991), the medicinal liana *Coscinium fenestratum* (Menispermaceae) (Senerath 1990), and the medicinal and spice herb *Elettaria cardamomum* (Zingiberaceae).

Sinharaja was included as one of 10 international sites for the Tropical Soil Biology and Fertility Programme (TSBF) of the International Union of Biological Sciences (IUBS) and UNESCO's Programme on Man and the Biosphere (MAB) under which comparative studies were carried out on the structure and functioning of natural and managed ecosystems (primary forest versus *Pinus* and *Hevea* plantations). A comprehensive site characterization study using standardized and calibrated methods selected and/or developed by TSBF was also carried out (Woomer and Swift 1994). During this study litter fall in the natural forest was found to be 832 g/m^2/year. Leaf litter in some species, such as *Palaquium petiolare* (Sapotaceae), *Shorea affinis*, and *Shorea disticha* (Dipterocarpaceae), showed a single distinct peak during the relatively dry period of March to May. Multiple peaks in litter fall were observed in *Mesua nagassarium* (Guttiferae), *Palaquium thwaitesii* (Sapotaceae), and *Strombosia nana* (Olacaceae) (Myers et al.1994).

Fauna

Sinharaja has a rich faunal diversity. Of the island's vertebrate species, 36% (262 species, including 112 endemics) have been recorded at Sinharaja. Among the mammals at Sinharaja are the leopard (*Panthera pardus*), the endemic purple-faced leaf monkey (*Trachypithecus vetulus*), loris (*Loris tardigradus*), several species of shrews—many endemic—and squirrels such as the giant squirrel (*Ratufa macroura*), small flying squirrel (*Petinomys fuscocapillus*), and flame-striped jungle squirrel (*Funambulus layardi*). Sinharaja is home to 147 bird species, 18 of which are endemic. Among these endemics are the rare green-billed coucal (*Centropus chlororynchus*), colorful Sri Lanka blue magpie (*Urocissa ornata*), Sri Lanka grey hornbill (*Tockus gingalensis*), and the yellow-fronted barbet (*Megalaima flavifrons*). Twenty amphibian species, half of them endemic, and 65 species of butterflies, including the rare endemic Ceylon rose (*Atrophaneura japhon*), have also been recorded (de Zoysa and Raheem 1990; Ministry of Forestry and Environment 1999).

Natural Disturbances

Some evidence exists that much of the forest may indeed be successional, implying occasional catastrophic disturbance presumably by wind (Ashton and Gunatilleke 1987). The mixed dipterocarp forests of southwestern Sri Lanka characteristically support a number of genera associated with forest succession elsewhere, notably *Dillenia* (Dilleniaceae) and *Vitex* (Labiatae), as well as the endemic *Schumacheria* (Dilleniaceae).

Human Disturbance

Although all of the forest in the Sinharaja 25-ha plot is primary, 2000 ha of the Sinharaja reserve were selectively logged in 1972–77. Today, there is no selective timber harvesting within the reserve. In the past, local small-scale shifting agriculture was carried out, mostly near the perimeter of the current reserve. Extraction of nontimber forest products does not occur in the plot, but illicit extraction of nontimber forest products takes place near the reserve boundaries, which has led to the depletion of fruit and seed resources for germination, especially in species where the whole plant has been extracted from the wild, as in the rattans and the medicinal vine *Coscinium fenestratum* (Menispermaceae). In the case of the fishtail palm *Caryota urens* (Palmae), the distal part of the immature inflorescence, long before the flowers even open, is cut and discarded, and the exudate emanating from the inflorescence stump is tapped and used for the production of sugar candy (jaggary), treacle, and an alcoholic beverage called toddy. Thus, as most of the fruit is prevented from ripening in this species, the terminal inflorescence is never tapped, leaving enough fruit to maintain the population (Gunatilleke and Gunatilleke 1993; Ratnayake et al. 1991). Due to human disturbance, only about 75% of the Sinharaja reserve consists of mature forest, 4% is secondary forest, and 21% is nonforested and covered by fern lands and grasslands (Banyard and Fernando 1988).

Plot Size and Location

Sinharaja is a 25-ha, 500 × 500 m plot. The plot is approximately located at 06°24'N and 80°24'E.

Funding Sources

The Sinharaja Forest Dynamics Plot has been funded by The John D. and Catherine T. MacArthur Foundation, the Rockefeller Foundation, Harvard Institute of International Development (now Center for International Development), Harvard University, USA, and National Institute for Environmental Studies, Japan.

References

Ashton, P. S., and C. V. S. Gunatilleke. 1987. New light on the plant geography of Ceylon I. Historical plant geography. *Journal of Biogeography* 14:249–85.

Banyard, S. G., and D. Fernando. 1988. Sinharaja forest: Monitoring changes by using aerial photographs of two different dates. *Sri Lanka Forester* 18:101–08.

Cooray, P. G. 1967. An introduction to the geology of Ceylon. *Spolia Zeylanica* 31:314.

Dayanandan, S., D. N. C. Attygala, A. W. W. L. Abeygunasekara, I. A. U. N. Gunatilleke, and C. V. S. Gunatilleke. 1990. Phenology and floral morphology in relation to pollination of some Sri Lankan Dipterocarps. Pages 103–133 in K. S. Bawa and M. Hadley, editors. *Reproductive Ecology of Tropical Forest Plants*. Man and the Biosphere Series, Vol. 20. Parthenon Publishing Group, Pearl River, NY.

de Zoysa, N., and R. Raheem. 1990. *Sinharaja: A Rain Forest in Sri Lanka*. Aitkin Spence & Co., Ltd., Colombo, Sri Lanka.

Gunatilleke, C. V. S., and P. S. Ashton. 1987. New light on the plant geography of Ceylon II. The ecological biogeography of the lowland endemic tree flora. *Journal of Biogeography* 14:295–327.

Gunatilleke, C. V. S., I. A .U. N. Gunatilleke, and P. M. S. Ashton. 1995. Rainforest research and conservation: The Sinharaja experience in Sri Lanka. *Sri Lanka Forester* 22:49–60.

Gunatilleke, C. V. S., G. A. D. Perera, P. M. S. Ashton, P. S. Ashton, and I. A. U. N. Gunatilleke. 1996. Seedling growth of *Shorea* section *Doona* (Dipterocarpaceae) in soils from topographically different sites of Sinharaja rain forest, Sri Lanka. Pages 245–65 in M. D. Swaine, editor. *The Ecology of Tropical Forest Tree Seedlings*. Man and the Biosphere Series, Vol. 16. Parthenon Publishing Group, Carnforth, U.K.

Gunatilleke, I. A. U. N., and C. V. S. Gunatilleke. 1993. Underutilized food plant resources of Sinharaja rain forest in Sri Lanka. Pages 183–98 in C. M. Hladik, A. Hladik, H. Pagzy, O. F. Linares, and M. Hadley, editors. *Food and Nutrition in the Tropical Rain Forest: Bicultural Interactions.* Man and the Biosphere Series, Vol. 15, UNESCO, Paris, and Parthenon Publishing, Carnforth, U.K.

Gunatilleke, I. A. U. N., and J. Maheswaran. 1988. Nutrient change during litter decomposition in a lowland rain forest and a deforested area in Sri Lanka. Pages 291–305 in F. S. P. Ng, editor. *Trees and Mycorrhiza: Proceedings of the Asian Seminar, 13–17 April 1987.* Forest Research Institute Malaysia, Kepong, Malaysia.

Hadley, M., and N. Ishwaran. 1997. Conservation, research and capacity building in the forest of the Lion King, Sri Lanka. Pages 89–102 in P. Gobel, editor. *Science and Technology in Asia and the Pacific. Co-operation for Development.* UNESCO, Paris.

Hafeel, K. M. 1991. *Soil Physico-Chemical and Endomycorrhizal Studies in Natural and Modified Sites of Sinharaja Rain Forest, Sri Lanka.* Masters thesis, University of Peradeniya, Peradeniya, Sri Lanka.

Mabberley, D. J. 1997. *The Plant-Book: A Portable Dictionary of the Vascular Plants.* Cambridge University Press, Cambridge, U.K.

Ministry of Forestry and Environment. 1999. *Biodiversity Conservation in Sri Lanka: A Framework for Action.* Ministry of Forestry and Environment, Sri Lanka.

Munidasa, B. K. H. C., C. V. S. Gunatilleke, and I. A. U. N. Gunatilleke. 2002. Climate of Sinharaja rain forest, Sri Lanka: An attempt to understand the El Niño and La Niña events. *Ceylon Journal of Science (Biological Sciences)* 30: 37–54.

Myers, R. J. K., C. A. Palm, E. Cuevas, I. U. N. Gunatilleke, and M. Brossard. 1994. The synchronisation of nutrient mineralization and plant nutrient demand. Pages 81–116 in P. L. Woomer and M. J. Swift, editors. *The Biological Management of Tropical Soil Fertility.* Wiley, New York.

Panabokke, C. R. 1996. *Soils and Agro-Ecological Environments of Sri Lanka.* Natural Resources, Energy and Science Authority, Sri Lanka.

Ratnayake, P. D. K. C., C. V. S. Gunatilleke, and I. A. U. N. Gunatilleke. 1991. *Caryota urens* L. (Palmae): An indigenous multipurpose tree species in the wet lowlands of Sri Lanka. Pages 77–88 in D. A. Taylor and K. Macdicken, editors. *Research on Multipurpose Tree Species in Asia.* Winrock International F/FRED, Bangkok.

Senerath, M. A. B. D. 1990. *Biological studies on Coscinium fenestratum Clobber (Menispermaceae).* Masters thesis, University of Peradeniya, Peradeniya, Sri Lanka.

Woomer, P. J., and M. J. Swift. 1994. *The Biological Management of Tropical Soil Fertility.* Wiley, New York.

38

Yasuní Forest Dynamics Plot, Ecuador

Renato Valencia, Richard Condit, Robin B. Foster, Katya Romoleroux,
Gorky Villa Muñoz, Jens-Christian Svenning, Else Magård, Margot Bass,
Elizabeth C. Losos, and Henrik Balslev

Site Location, Administration, and Scientific Infrastructure

The Yasuní Forest Dynamics Plot is located in mature terra firme forest in Yasuní
National Park and Biosphere Reserve. The park and adjacent Huaorani territory
comprise 1.6 million ha, representing the largest protected area of mature forest
in the Amazon region of Ecuador. The 50-ha Yasuní Forest Dynamics Plot is
located in the northwest corner of the park, on a ridge above the Tiputini River,
a tributary of the Napo River (fig. 38.1). It was initiated in 1995 by the Pontificia
Universidad Católica del Ecuador, the Smithsonian Tropical Research Institute,
and the University of Aarhus (Denmark). The enumeration of all trees ≥ 1 cm
dbh was completed for the first 25 ha on the western side of the plot in November
1999; the enumeration of all trees ≥ 10 cm dbh was completed for the second 25 ha
in April 2001. At present, the enumeration of all trees ≥ 1 cm dbh is underway
but not yet completed for the second 25 ha. In this chapter, we report on results
from the first 25 ha.

The Yasuní Biological Research Station, managed by the Pontificia Universidad Católica del Ecuador in Quito, is approximately 1 km from the Yasuní Forest
Dynamics Plot and 47 km south of the Napo River. The station is a permanent facility managed strictly for scientific research, with electricity, simple laboratories,
accommodations for 60 people, classroom and dining facilities, and a herbarium.

Climate

Over a 53-month period from 1995 to 1999, annual rainfall averaged 3081 mm
at the Yasuní Research Station. These figures agree closely with more extensive
records analyzed by Pitman (2000). Using data from eight long-term sites in lowland Amazonian Ecuador, he concluded that annual rainfall averaged 3214 mm.
Although the area had no mean monthly rainfall below 100 mm, the data did show
clear seasonal rainfall variation, with two peaks and two troughs per year and the
rainiest month averaging 72% more rain than the driest month. The wettest
months were April–May and October–November, with the earlier peak slightly
rainier than the latter one. Drier months were characterized by both fewer days

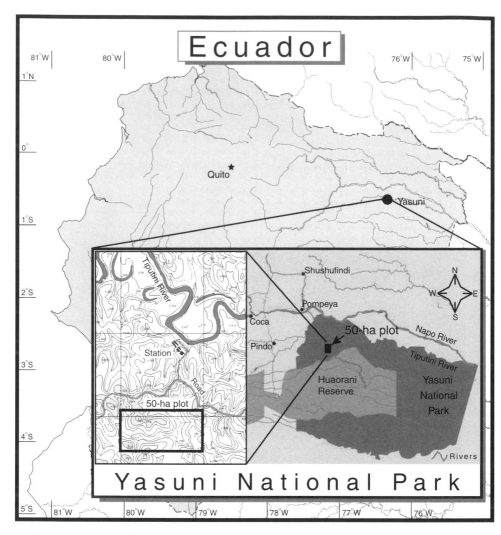

Fig. 38.1. Location of the 25-ha Yasuni Forest Dynamics Plot.

of rain and smaller daily maximums, and accounted for proportionally more of the annual solar radiation and potential evapotranspiration budgets (359.9 cal/cm^2/day [174 watts/m^2] and 1033 mm/year respectively; data from Tiputini Research Station, reported in Marengo 1998, cf. Pitman 2000). At all sites, relative humidity averaged 80–94% throughout the year. The average annual maximum

Table 38.1. Yasuní Climate Data

	Jan.	Feb.	Mar.	Apr.	May	Jun.	Jul.	Aug.	Sep.	Oct.	Nov.	Dec.	Total/ Averages
Rain (mm)	226	344	200	253	412	374	227	193	174	253	196	229	3081
ADTMx (°C)	33.7	35.4	34.8	34.8	33.5	33.9	34.2	35.2	36.6	36.6	36.1	35.1	35.0
ADTMn (°C)	21.7	23.4	22.0	21.7	21.4	21.3	21.3	21.2	21.2	21.4	21.5	21.7	21.7

Notes: Mean monthly rainfall and average daily temperature are based on climate data measured at the Yasuní Research Station from 1995 to 1999. Monthly records with incomplete data (i.e., <26 days; the great majority have 30 or 31 days) were excluded. Thus, some of the monthly temperature means and rainfall means, maxima, and minima were averaged from only 4 years during this period, with 7 months of data excluded. The weather station is in full light.

temperature from 1995 to 1999 at the Yasuní Biological Station is 35°C in the full sun. More recent climate data (January 2000 to May 2001) from the station revealed a cooler average annual maximum temperature by a couple of degrees, but a similar average annual minimum temperature (Garwood and Persson unpublished data). See table 38.1.

Topography and Soil

The 50-ha Yasuní Forest Dynamics Plot is located at 230 m above sea level and the difference between the plot's lowest and the highest points is 33.5 m (figures 38.2 and 38.3 illustrate the west 25-ha half of the 50-ha plot where all the woody plants ≥1 cm dbh were enumerated mapped and identified to date). The average slope of the plot is 13%. The plot is bound by two ridges, each dominated by red clays. The ridges are composed mostly of gently sloping hills, though steep slopes are found along erosion gullies. Bottomlands, characterized by brown or gray alluvium, separate the ridges. The bottomland includes several small permanent streams and a small swamp (frequently flooded) in the eastern half of the plot.

Detailed soil analyses have not been carried out in the Yasuní Forest Dynamics Plot. However, soil studies within the Yasuní National Park concluded that topographic variation is generally low. In upland areas, most soils are clayey, acidic, low in most cations while rich in aluminum and iron, and lacking rocks and pebbles (Korning et al. 1994; Pitman 2000). Most soils in Yasuní are classified as udult Ultisols. The remainder of soils in Yasuní are those influenced by flooding, either in swamps or floodplains, with that in the swamps classified as Histosols (Pitman 2000). However, in a wetland approximately 4 km away from the study plot, two layers of about 10 cm of volcanic ash have been found some 5 m below the surface (Athens 1997), providing evidence of some volcanic influence on soil formation. The terra firme hills in Yasuni are composed of sediments of the Curaray and Chambira formations (Lips and Duivenvoorden 2001).

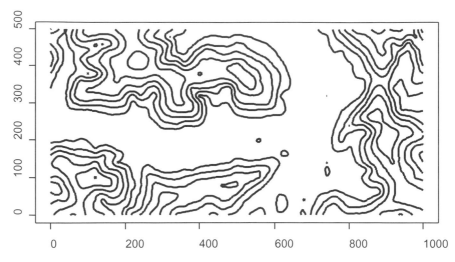

Fig. 38.2. Topographic map of the 50-ha Yasuni Forest Dynamics Plot with 5-m contour intervals.

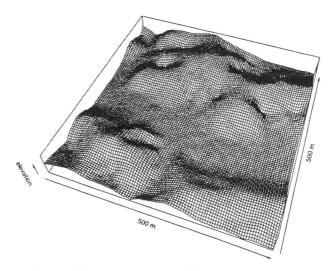

Fig. 38.3. Perspective map of the western-most 25 ha of the 50-ha Yasuni Forest Dynamics Plot, where census of trees ≥ 1 cm dbh has been completed.

Forest Type and Characteristics

The forest is evergreen lowland wet forest, with a canopy mostly 15–30 m tall and some emergent trees reaching 40 and rarely 50 m. The largest stem diameters are usually 2 m, frequently *Ceiba pentandra* (Bombacaceae) and the slightly smaller *Tessmannianthus heterostemon* (Melastomataceae). The forest around the plot

Table 38.2. Yasuní Plot Census History

Census	Dates	Number Trees (≥1 cm dbh)	Number Species (≥1 cm dbh)	Number Trees (≥10 cm dbh)	Number Species (≥10 cm dbh)
First	April 1995–November 1999	152,353	1104	17,546	820

Notes: One full census has been completed of the Yasuní Forest Dynamics Plot for trees ≥1 cm dbh in 25 ha; includes 4215 deaths before identification. The first recensus of the first 25 ha, not yet finalized, was initiated in September 2002.

Table 38.3. Yasuní Summary Tally

Size Class (cm dbh)	Average per Hectare							25-ha Plot				
	BA	N	S	G	F	H'	α	S	G	F	H'	α
≥1	33.0	6094	655	243	69	2.44	187.1	1104	328	81	2.57	161.1
≥10	27.3	702	251	132	47	2.11	141.7	820	274	69	2.37	178.3
≥30	13.4	81	55	41	24	1.65	80.8	397	180	57	2.25	147.2
≥60	4.1	8	6	6	5	0.73	20.5*	76	55	27	1.65	46.5

*Fisher's alpha based on 15 ha.

Notes: BA represents basal area in m^2, N is the number of individual trees, S is number of species, G is number of genera, F is number of families, H' is Shannon–Wiener diversity index using $\log_{10}$, and α is Fisher's α. Basal area includes all multiple stems for each individual. 6943 individuals were not identified to species or morphospecies. Data are from the first census.

appears to be maturing, undisturbed for several centuries and possibly much longer. See census data and rankings in tables 38.2–38.6.

Species composition changes slightly between the ridges and the bottomland (chap. 9). A comparison between the ridge and an adjacent bottomland revealed that many more midcanopy species grow exclusively on the ridge than on the bottomland. Among the dominant species that preferred the ridge forest are *Brownea grandiceps* (Leguminosae), *Macrolobium* sp. nov. (Leguminosae), *Tachigali formicarum* (Leguminosae), *Protium aracouchini* (Burseraceae), and *Ocotea javitensis* (Lauraceae), whereas *Bauhinia brachycalyx* (Leguminosae), *Coccoloba densifrons* (Polygonaceae), *Guarea grandifolia* (Meliaceae), *Guarea pubescens* (Meliaceae), *Talauma ovata* (Magnoliaceae), and *Astrocaryum murumuru* (Palmae) dominate in the bottomland. The most common species, such as *Iriartea deltoidea* (Palmae), *Matisia oblongifolia* (Bombacaceae), *Matisia malacocalyx* (Bombacaceae), and *Marmaroxylon basijugum* (Leguminosae), grow abundantly in both habitats.

The swampy area in the eastern half of the plot is most notably different. The palm *Mauritia flexuosa* (Palmae), a *Sapium* sp. (Euphorbiaceae), and several species of *Piper* (Piperaceae) are found only in the swamp. This small swamp, topographically recessed, contains water throughout the year.

Families and genera listed in tables 38.4 and 38.5 are typical of species-rich tracts of lowland neotropical forests (e.g., Gentry 1988; Valencia et al. 1994). Among the shrubby genera not well sampled in the Yasuní Forest Dynamics Plot

Table 38.4. Yasuni Rankings by Family

Rank	Family	Basal Area (m²)	% BA	% Trees	Family	Trees	% Trees	Family	Species
1	Leguminosae	112.1	14.9	13.0	Leguminosae	18,860	13.0	Leguminosae	108
2	Palmae	63.8	8.5	3.6	Bombacaceae	10,109	7.0	Lauraceae	81
3	Lecythidaceae	45.5	6.0	2.6	Lauraceae	8,055	5.5	Rubiaceae	80
4	Moraceae	41.7	5.5	4.2	Meliaceae	7,733	5.3	Melastomataceae	59
5	Cecropiaceae	41.3	5.5	2.2	Violaceae	7,239	5.0	Myrtaceae	56
6	Meliaceae	40.4	5.4	5.3	Euphorbiaceae	6,775	4.7	Sapotaceae	54
7	Euphorbiaceae	40.4	5.4	4.7	Rubiaceae	6,569	4.5	Moraceae	51
8	Lauraceae	39.1	5.2	5.5	Moraceae	6,136	4.2	Annonaceae	40
9	Bombacaceae	36.5	4.8	7.0	Myrtaceae	5,670	3.9	Meliaceae	39
10	Myristicaceae	35.5	4.7	1.8	Palmae	5,231	3.6	Euphorbiaceae	34

Notes: The top 10 families for trees ≥1 cm dbh are ranked in terms of basal area, number of individual trees, and number of species, with the percentage of trees in the plot. Data are from 25 ha of the first census.

Table 38.5. Yasuní Rankings by Genus

Rank	Genus	Basal Area (m²)	% BA	% Trees	Genus	Trees	% Trees	Genus	Species
1	Iriartea (Palmae)	49.3	6.7	1.7	Matisia (Bombacaceae)	9188	6.6	Miconia (Melastomataceae)	45
2	Inga (Leguminosae)	38.3	5.2	6.0	Inga (Leguminosae)	8380	6.0	Inga (Leguminosae)	43
3	Eschweilera (Lecythidaceae)	35.3	4.8	1.7	Rinorea (Violaceae)	6023	4.3	Pouteria (Sapotaceae)	30
4	Matisia (Bombacaceae)	25.8	3.5	6.6	Guarea (Meliaceae)	5351	3.8	Piper (Piperaceae)	22
5	Alchornea (Euphorbiaceae)	21.8	3.0	0.2	Miconia (Melastomataceae)	4128	2.9	Guarea (Meliaceae)	20
6	Pourouma (Cecropiaceae)	19.3	2.6	1.8	Zygia (Leguminosae)	4042	2.9	Neea (Nyctaginaceae)	20
7	Guarea (Meliaceae)	18.8	2.6	3.8	Piper (Piperaceae)	3766	2.7	Licania (Chrysobalanaceae)	17
8	Cecropia (Cecropiaceae)	18.6	2.5	0.5	Neea (Nyctaginaceae)	3480	2.5	Sloanea (Elaeocarpaceae)	17
9	Otoba (Myristicaceae)	16.8	2.3	0.3	Siparuna (Monimiaceae)	2619	1.9	Psychotria (Rubiaceae)	16
10	Virola (Myristicaceae)	15.9	2.2	0.7	Pourouma (Cecropiaceae)	2534	1.8	Trichilia (Meliaceae)	15

Notes: The top 10 tree genera for trees ≥ 1 cm dbh are ranked by basal area, number of individual trees, and number of species with the percentage of trees in the plot. Data are from 25 ha of the first census.

Table 38.6. Yasuní Rankings by Species

Rank	Species	Number Trees	% Trees	Species	Basal Area (m^2)	% BA	% Trees
1	*Matisia oblongifolia* (Bombacaceae)	4581	3.2	*Iriartea deltoidea* (Palmae)	49.3	6.5	1.6
2	*Rinorea lindeniana* (Violaceae)	3241	2.2	*Eschweilera coriacea* (Lecythidaceae)	24.3	3.2	0.9
3	*Matisia malacocalyx* (Bombacaceae)	2323	1.6	*Alchornea triplinervia* (Euphorbiaceae)	21.8	2.9	0.2
4	*Iriartea deltoidea* (Palmae)	2313	1.6	*Otoba glycycarpa* (Myristicaceae)	16.8	2.2	0.3
5	*Brownea grandiceps* (Leguminosae)	2156	1.5	*Cecropia sciadophylla* (Cecropiaceae)	14.8	2.0	0.3
6	*Memora cladotricha* (Bignoniaceae)	2075	1.4	*Cedrelinga cateniformis* (Leguminosae)	14.6	1.9	0.0
7	*Piper* sp. (Piperaceae)	2074	1.4	*Apeiba membranacea* (Tiliaceae)	13.2	1.8	0.3
8	*Zygia basijugum* (Leguminosae)	1913	1.3	*Matisia malacocalyx* (Bombacaceae)	11.0	1.5	1.6
9	*Zygia heteroneura* (Leguminosae)	1764	1.2	*Pourouma bicolor* (Cecropiaceae)	10.5	1.4	1.1
10	*Inga auristellae* (Leguminosae)	1701	1.2	*Cedrela fissilis* (Meliaceae)	7.8	1.0	0.1

Notes: The top 10 tree species for trees ≥1 cm dbh are ranked by number of trees and basal area. The percentage of the total population is also shown. Data are from 25 ha of the first census.

census due to their small size, *Miconia* (Melastomataceae) is the most species-rich. In the Yasuní plot, only four species of *Miconia* are midcanopy trees. Yet, with 89 species, this genus is the most species-rich genus of all flowering plants in Amazonian Ecuador (Jørgensen and León-Yánez 1999; see also Renner et al. 1990).

Tree phenology was studied from February 2000 to March 2001 using 200 traps (74 × 74 cm) scattered inside the 50-ha plot, with a minimum distance between then of 13.5 m, following the same protocol used in the Barro Colorado Island and Pasoh Forest Dynamics Plots. Flowering in trees and lianas was inversely related to rainfall. However, when trees were in peak flowering, lianas flowered less, and vice versa. Flowers of Leguminosae and Palmae were the most common in the traps (Aguilar 2002).

Fauna

Yasuní National Park, including the area around the research station, has an essentially intact vertebrate fauna, including the larger species of birds and mammals. The nest of a harpy eagle (*Harpia harpyja*) was found a few kilometers from the station, macaws (*Ara* spp.) and guans (*Aburria pipile*) are numerous, and pumas (*Felis concolor*) and giant anteaters (*Myrmecophaga tridactyla*) occur in the area. White-lipped peccaries (*Tayassu pecari*), tapirs (*Tapirus terrestris*), and jaguars (*Felis onca*) have been seen in the Yasuní Forest Dynamics Plot. There are 11 monkey species near the station, and many are bold and easy to observe. The ecology and behavior of woolly monkeys (*Lagothrix lagothricha*) and other species of monkeys have been investigated within a 350-ha plot close to the station. Inventories of other groups of vertebrates and invertebrates are underway. To date there have been 84 species of amphibians and 77 species of reptiles recorded around the Yasuní Research Station (S. Ron personal communication), whereas 60,000 species of insects are estimated to exist in a hectare of forest at Yasuní (T. Erwin personal communication).

Natural Disturbances

Most canopy disturbances are from small treefall gaps created when one or a few trees fall. The importance of large-scale wind storms is unknown. Occasionally significant blowdowns do cause major disturbance in the forest, as occurred in March 2002 when 96 trees over 10 cm dbh were downed by a wind storm. During that event, big trees such as *Cedrelinga cataeniformis* (>100 cm dbh) together with other Leguminosae species (especially *Inga*, *Lonchocarpus*, and *Parkia*) and Cecropiaceae (*Cecropia* and *Pourouma*) accounted for more than 40% of the total fallen trees. There is no indication that El Niño events have any impact in the region.

Human Disturbance

In 1995, an archeological survey was carried out on a hilltop near the northwest corner of the study plot. There were ceramic shards just 50 cm below the forest floor, estimated to be roughly 500–1000 years old (Netherly 1997). The artifacts may belong to the nomadic Huaorani Indian group, which formerly opened small clearings or used natural gaps for plantations of manioc and temporary home sites. Evidence of prehistoric burnings, presumably for agriculture or subsistence, were found in soil cores taken near the study site and preliminarily dated to 7700 years before the present (Athens 1997). There is evidence of Native American settlements in the area, but the existence of extensive clearings is unknown.

The most conspicuous present day human disturbances are concessions for oil exploitation and new settlements of indigenous groups along a new road. The large oil reserves in the national park were leased to various oil companies for prospecting and exploitation in the 1990s. From 1992 to 1995, the oil company Maxus established facilities for oil exploitation, including an underground pipeline and a 150-km road that crosses the northwestern part of the park. Although the oil company controls the road access to prevent human colonization, in the last several years Huaorani and Quichua settlements have appeared near the station. A new Huaorani settlement of about 20 people has arisen 4 km east of the NE corner of the plot, as well as a bigger settlement about 12 km to the west. All Huaorani communities have given up their nomadic lifestyle for permanent houses. Consequently, hunting intensity has increased along the road. The Huaorani occasionally hunt in the 50-ha Forest Dynamics Plot, which is only 100 m from an oil road. Recent evidence indicates that hunting along the road has already had an impact on mammal populations. Monkey densities are lower now than they were 5 years ago in a study plot near a road and only 10 km from the Yasuní Research Station. Sites further from the road appear to be unaffected (S. Suarez personal communication). The vast majority of Yasuní National Park, however, is largely inaccessible and its forests remain undisturbed.

A 1-ha area near the southwest corner of the plot—presently dominated by *Ceropia* species (Cecropiaceae)—was a heliport used for oil exploration before 1990. In addition, the exotic plant *Muntingia calabura* (Elaeocarpaceae) grows inside the plot.

Plot Size and Location

Yasuní is a 50-ha, 1000 × 500 m plot; its long axis lies east-west. The western 25 ha have been fully enumerated for all trees ≥1 cm dbh; in the eastern 25 ha, all trees ≥10 cm dbh have been enumerated. The southwest corner is at 00°41′14″ S and 76°23′72″ W. The southeast corner is at 00°40′84″ S and 76°23′72″ W. The northwest corner is at 00°41′14″ S and 76°24′20″ W. The northeast corner is at 00°40′45″ S and 76°23′08″ W.

Funding Sources

The Yasuní Forest Dynamics Plot has been funded by the Andrew W. Mellon Foundation, the U.S. National Science Foundation, the Diva Project (Ecuador-Denmark), the Tupper Family Foundation, and the Smithsonian Tropical Research Institute.

References

Aguilar, Z. 2002. *Tesis de Fenología en Yasuní.* Tesis de Licenciatura, Departamento de Biología, Pontificia Universidad Católica del Ecuador, Quito.

Athens, J. S. 1997. Paleoambiente del Oriente ecuatoriano: Resultados preliminares de columnas de sedimentos procedentes de humedales. *Fronteras de la Ciencia* 1:15–32.

Balslev, H., J. Luteyn, B. Ollgaard, and L. Holm-Nielsen. 1987. Composition and structure of adjacent unflooded and flooded forest in Amazonian Ecuador. *Opera Botanica* 92:37–57.

Cerón, C. E., and C. G. Montalvo. 1997. Composición y estructura de una hectárea de bosque en la Amazonía Ecuatoriana—Con información etnobotánica de los Huaorani. Pages 153–72 in R. Valencia and H. Balslev, editors. *Estudios sobre Diversidad y Ecología de Plantas*. Pontificia Universidad Católica del Ecuador, Quito.

———. 1997. Composición de una hectárea de bosque en la comunidad Huaorani de Quehueiri–Ono, zona de amortiguamiento del Parque Nacional Yasuní, Napo, Ecuador. Pages 279–98 in P. A. Mena, A. Soldi, R. Alarcón, C. Chiriboga, and L. Suárez, editors. *Estudios Biológicos para la Conservación. Diversidad, Ecología y Etnobiología*, EcoCiencia, Quito.

———. 1998. *Etnobotánica de los Huaorani de Quehueiri-Ono*. Ediciones Abya Yala, Quito.

Davis, E. W., and J. A. Yost. 1983. The ethnobotany of the Waorani of eastern Ecuador. *Harvard University, Botanical Museum Leaflets* 29:159–217.

Di Fiore, A., and P. S. Rodman. 2001. Time allocation patterns of lowland woolly monkeys (*Lagothrix lagotricha poeppigii*) in a neotropical terre firma forest. *International Journal of Primatology* 22:449–80.

Gentry, A. 1988. Changes in plant community diversity and floristic composition on environmental and geographical gradients. *Annals of the Missouri Botanical Garden* 75:1–34.

Herrera-MacBryde, O., and D. A. Neill. 1997. Yasuní National Park and the Waorani Ethnic Reserve, Ecuador. Pages 344–48 in S. D. Davis, V. H. Heywood, O. Herrera-MacBryde, J. Villa-Lobos, and A. C. Hamilton, editors. *Centers of Plant Diversity*, Vol. 3. World Wildlife Fund, Oxford, U.K.

Jørgensen, P. M., and S. León-Yánez. 1999. *Catalogue of the Vascular Plants of Ecuador*. Missouri Botanical Garden Press, St. Louis, MO.

Kjaer-Pedersen, N. 2000. *Composition, Diversity, and Abundance of Herbs and Palms in an Amazonian Rain Forest: Changes in Relation to Flooding and Topography*. Masters thesis, Botanical Institute University of Copenhagen, Denmark.

Korning, J. 1987. *Studies of Amazonian Tree and Understory Vegetation and Associated Soils in Añangu, East Ecuador*. Masters thesis. University of Aarhus, Denmark.

Korning, J., and H. Balslev. 1994a. Growth and mortality of trees in Amazonian tropical rain forest in Ecuador. *Journal of Vegetation Science* 4:77–86.

Korning, J., and H. Balslev. 1994b. Growth rates and mortality patterns of tropical lowland tree species and the relation to forest structure in Amazonian Ecuador. *Journal of Tropical Ecology* 10:151–66.

Korning, J., K. Thomsen, and B. Olgaard. 1991. Composition and structure of a species rich Amazonian rain forest obtained by two different sample methods. *Nordic Journal of Botany* 11:103–10.

Korning, J., K. Thomsen, K. Dalsgaard, and P. Nornberg. 1994. Characters of three udults and their relevance to the composition and structure of virgin forest of Amazonian Ecuador. *Geoderma* 63:145–64.

Lips, J., and J. Duivenvoorden. 2001. Caracterización ambiental. Pages 19–46 in J. Duivenvoorden, H. Balslev, J. Cavalier, C. Grandez, H. Tuomisto, and R. Valencia, editors. *Evaluación de Recursos Forestales Nomaderables en la Amazonía Noroccidental*. University of Amsterdam, Netherlands.

Mabberley, D. J. 1997. *The Plant-Book: A Portable Dictionary of the Vascular Plants*. Cambridge University Press, Cambridge, U.K.

Macía, M., H. Romero-Saltos, and R. Valencia. 2001. Patrones de uso en un bosque primario de la Amazonía ecuatoriana: comparación entre dos comunidades Huaorani. Pages 225–62 in J. Duivenvoorden, H. Balslev, J. Cavalier, C. Grandez, H. Tuomisto, and

R. Valencia, editors. *Evaluación de Recursos Forestales Nomaderables en la Amazonía Noroccidental.* University of Amsterdam, Netherlands.

Malo, J. E., and J. M. Olano. 1997. Predicción de la frecuencia de especies nemorales del bosque amazónico a partir de variables topográficas sencillas. Pages 279–89 in R. Valencia and H. Balslev, editors. *Estudios sobre Diversidad y Ecología de Plantas.* Pontificia Universidad Católica del Ecuador, Quito.

Marengo, J. A. 1998. Climatología de la zona de Iquitos. Pages 35–57 in R. Kalliola and P. Flores editors. *Geoecología y Desarrollo Amazónico.* Annales Universitatis Turkuensis Ser A II 114, University of Turku, Finland.

Netherly, P. 1997. Loma y ribera: Patrones de asentamiento prehistórico en la Amazonía ecuatoriana. *Fronteras de la Ciencia* 1:33–54.

Ojeda, P. 1994. Diagnóstico etnobotánico y comercialización del morete *Mauritia flexuosa* (Arecaceae), en la zona del Alto Napo, Ecuador. Pages 90–109 in R. Alarcón, P. Mena, and A. Soldi, editors. *Etnobotánica, Valoración Económica y Comercialización de Recursos Florísticos Silvestres en el Alto Napo, Ecuador.* EcoCiencia, Quito.

Ojeda, P. 1997. Floración y frutificación de *Mauritia flexuosa* L.f. en la Amazonía ecuatoriana. Pages 269–77 in P. Mena, A. Soldi, R. Alarcón, C. Chiriboga, and L. Suárez, editors. *Estudios Biológicos para la Conservación: Diversidad, Ecología y Etnobiología.* EcoCiencia, Quito.

Pitman, N. C. A. 2000. *A Large-Scale Inventory of Two Amazonian Tree Communities.* Ph.D. dissertation. Department of Botany, Duke University, Durham, NC.

Pitman, N. C. A., J. Terborgh, M.R. Silman, P. Núñez V., D. A. Neill, C. E. Cerón, W. A. Palacios, and M. Aulestia. 2001. Dominance and distribution of tree species in upper Amazonian terra firme forests. *Ecology* 82:2101–17.

Renner, S. S., H. Balslev, and L. B. Holm-Nielsen. 1990. Flowering plants of Amazonian Ecuador—A checklist. *AAU Reports* 24:1–41.

Romero-Saltos, H., R. Valencia, and M. Macía. 2001. Patrones de diversidad, distribución y rareza de plantas leñosas en el Parque Nacional Yasuní y la Reserva Étnica Huaorani, Amazonía ecuatoriana. Pages 131–62 in J. Duivenvoorden, H. Balslev, J. Cavalier, C. Grandez, H. Tuomisto, and R. Valencia, editors. *Evaluación de Recursos Forestales Nomaderables en la Amazonía Noroccidental.* University of Amsterdam, Netherlands.

Romeroleroux, K., R. Foster, R. Valencia, R. Condit, H. Balslev, and E. Losos. 1997. Especies leñosas (dap ≥ 1 cm) encontradas en dos hectáreas de un bosque de la Amazonía ecuatoriana. Pages 189–215 in R. Valencia and H. Balslev, editors. *Estudios Sobre Diversidad y Ecología de Plantas.* Pontificia Universidad Católica del Ecuador, Quito.

Svenning, J. C. 1999. Microhabitat specialization in a species-rich palm community in Amazonian Ecuador. *Journal of Ecology* 87:55–65.

Svenning, J. C. 1999. Recruitment of tall arborescent palms in the Yasuní National Park, Amazonian Ecuador: Are large treefall gaps important? *Journal of Tropical Ecology* 15: 355–66.

Valencia, R., H. Balslev, and G. Paz y Miño. 1994. High tree alpha-diversity in Amazonian Ecuador. *Biodiversity and Conservation* 3:21–28.

Contributors

Naoki Adachi
21st Century COE Program
Biodiversity and Ecosystem Restoration
The University of Tokyo
1-1-1, Yayoi, Bunkyo-ku
Tokyo 113-8657
Japan

Jorge A. Ahumada
U.S. Geological Survey
National Wildlife Health Center
6006 Schroeder Road
Madison, Wisconsin 53711

Roberto R. Araño
Kabang Kalikasanng Pilipinas
WWF-Philippines
23 Maalindog Street, UP Village
Diliman, Quezon City 1101
Philippines

P. Mark S. Ashton
School of Forestry and
    Environment
Yale University, Marsh Hall
360 Prospect Street
New Haven, Connecticut 06511

Peter S. Ashton
Harvard University Herbaria
22 Divinity Avenue
Cambridge, Massachusetts 02138

Patrick J. Baker
USDA Forest Service
Institute of Pacific Islands Forestry
23 East Kawili Street
Hilo, Hawaii 96720

Henrik Balslev
Department for Systematic Botany
University of Aarhus
Nordlandsvej 68
DK-8240 Risskov
Denmark

Nestor A. Bartolome
Conservation International
5 South Lawin Avenue
Philan Homes, Quezon City
Philippines

Margot Bass
Finding Species Foundation
PO Box 5717
Takoma Park, Maryland 20914

John P. Bautista
Conservation International -Philippines
5 South Lawin Avenue
Philan Homes, Quezon City
Philippines

Nicholas Brokaw
Institute for Tropical
Ecosystems Studies
University of Puerto Rico
PO Box 23341
San Juan, Puerto Rico 00931-3341

Sarayudh Bunyavejchewin
Royal Forest Department
Chatuchak, Bangkok 10900
Thailand

George B. Chuyong
Department of Life Sciences
University of Buea
PO Box 63
Buea
Cameroon

Leonardo L. Co
Herbarium, Institute of Biology
College of Science
University of the Philippines
Diliman, Quezon City 1101
Philippines

Richard Condit
Center for Tropical Forest Science
Smithsonian Tropical Research Institute
Unit 0948
APO AA 34002-0948 USA

James W. Dalling
University of Illinois, Urbana-Champaign
Plant Biology Department
265 Morrill Hall
505 South Goodwin Avenue
Urbana, Illinois 61801

Handanakere Shivaramiah Dattaraja
Center for Ecological Sciences
Indian Institute of Science
Bangalore 560 012
India

Stuart J. Davies
Center for Tropical Forest Science—AA
   Asia Program
Harvard University Herbaria
22 Divinity Avenue
Cambridge, Massachusetts 02138

Shameema Esufali
Department of Botany
Faculty of Science
University of Peradeniya
Peradeniya
Sri Lanka

A. U. K. Ethugala
Department of Botany
Faculty of Science
University of Peradeniya
Peradeniya
Sri Lanka

Edwin M. Everham III
Department of Environmental Science
Florida Gulf Coast University
19501 Treeline Avenue South
Fort Myers, Florida 33965-6565

Corneille Ewango
Centre de Formation et de Recherche en
   Conservation Forestiere (CEFRECOF)
Epulu, Ituri Forest, Reserve de Faune
   a Okapis
Democratic Republic of Congo
c/o J. & T. Hart
The Wildlife Conservation
   Society—African Program
2300 Southern Boulevard
Bronx, New York 10460

Robin B. Foster
Department of Botany
Field Museum of Natural History
Roosevelt Road at Lakeshore Drive
Chicago, Illinois 60605

Shawn Fraver
Department of Forest Ecosystem
   Science
University of Maine
5755 Nutting Hall
Omo, Maine 04469-5755

Hubert G. Garcia
College of Forestry
Isabela State University
Garita Heights, Cabagan
Isabela 3328
Philippines

Jason S. Grear
School of Forestry and
   Environment Studies
Yale University
205 Prospect Street
New Haven, Connecticut 06511

Edmundo C. Gumpal
College of Forestry and Environmental
    Management
Isabela State University
Garita Heights, Cabagan
Isabela 3328
Philippines

C. V. S. Gunatilleke
Department of Botany
Faculty of Science
University of Peradeniya
Peradeniya
Sri Lanka

I. A. U. N. Gunatilleke
Department of Botany
Faculty of Science
University of Peradeniya
Peradeniya
Sri Lanka

James L. Hamrick
Department of Botany
University of Georgia
2502 Plant Sciences
Athens, Georgia 30602

Masatoshi Hara
Chiba Natural History Museum and
    Institute
955-2 Aoba-cho
Chuo-ku
Chiba 260-8682
Japan

Kyle E. Harms
Department of Biological Sciences
Louisiana State University
Baton Rouge, Louisiana 70803

Rhett Harrison
Smithsonian Tropical Research Institute
Unit 0948
APO AA 34002-0948 USA

John A. Hart
International Programs, Building A
Wildlife Conservation Society
185th Street and Southern Boulevard
Bronx, New York 10460

Terese B. Hart
International Programs, Building A
Wildlife Conservation Society
185th Street and Southern Boulevard
Bronx, New York 10460

David E. Hibbs
Department of Forest Science
Oregon State University
Richardson Hall 321K
Corvallis, Oregon 97331-5752

Chang-Fu Hsieh
National Taiwan University
Department of Botany
1, Section 3
Roosevelt Road
Taipei
Taiwan 106, Republic of China

Stephen P. Hubbell
Department of Botany
University of Georgia
2502 Plant Sciences
Athens, Georgia 30602

Amir Husni Mohd. Shariff
Adabi Consumer Industries Sdn. Bhd.
Industrial Park, Malaysia,
48000, Rawang, Selangor D.E.
Malaysia

Nor Azman Hussein
Forest Research Institute Malaysia
Kepong
52109 Kuala Lumpur
Malaysia

Akira Itoh
Lab Plant Ecology
Faculty of Science
Osaka City University
3-3-138 Sugimoto
Sumiyoshi
Japan

Robert John
University of Illinois, Urbana-Champaign
Plant Biology Department
265 Morrill Hall
505 South Goodwin Avenue
Urbana, Illinois 61801

Niranjan V. Joshi
Center for Ecological
    Sciences
Indian Institute of Science
Bangalore 560 012
India

Mamoru Kanzaki
Tropical Forest Resources and
    Environments
Graduate School of Agriculture
Kyoto University
Oiwake Kitashirakawa
Kyoto 606-8502
Japan

David Kenfack
Department of Biology
University of Missouri
St. Louis, Missouri 63121

James V. LaFrankie
Center for Tropical Forest Science—AA
    Asia Program
National Institute of Education
Nanyang Technological University
1 Nanyang Walk
Singapore 637616
Singapore

Daniel A. Lagunzad
Institute of Biology
College Of Science
University of the Philippines
Diliman, Quezon City 1100
Philippines

Suzanne Loo de Lao
Center for Tropical Forest Science
Smithsonian Tropical Research Institute
Unit 0948
APO AA 34002-0948 USA

Sing Kong Lee
Biology Department
National Institute of Education
Nanyang Technological University
1 Nanyang Walk
Singapore 637616
Singapore

Hua Seng Lee
Sarawak Forest Department
Jalan Stadium, Petra Jaya
Kuching, Sarawak 93660
Malaysia

Egbert G. Leigh, Jr.
Smithsonian Tropical Research Institute
Unit 0948
APO AA 34002-0948 USA

Innocent Liengola
Centre de Formation et de Recherche en
    Conservation Forestiere (CEFRECOF)
Epulu, D. R. Congo
Via P.O Box 4930, Kampala
Uganda

Elizabeth C. Losos
Center for Tropical Forest Science
Smithsonian Tropical Research Institute
Smithsonian Institution, Suite 3123
1100 Jefferson Drive, Southwest
Washington DC 20560

Shawn K. Y. Lum
National Institute of Education
Nanyang Technological University
1 Nanyang Walk
Singapore 637616
Singapore

Else Magård
Herbarium AAU
University of Aarhus
Uni-parken, bygn. 137
DK-8000 Aarhus C.
Denmark

Jean-Remy Makana
Faculty of Forestry
University of Toronto
33 Willcocks Street
Toronto, Ontario M5S 3B3
Canada

N. Manokaran
Forest Research Institute Malaysia
Kepong
52109 Kuala Lumpur
Malaysia

Humberto Mendoza
Instituto Alexander von Humboldt
Claustro San Agustin
Villa de Leyva, Boyaca
Colombia

Jeanmaire E. Molina
Conservation International–
    Philippines
5 South Lawin Avenue
Philan Homes, Quezon City
Philippines

Sainge Nsanyi Moses
Korup Forest Dynamic Program
Center for Tropical Forest Science
Bioresouces Development and
    Conservation Programme–Cameroon

c/o Korup National Park
PO Box 36
Mundemba Southwest Province
Cameroon

Helene C. Muller-Landau
National Center for Ecological Analysis
    and Synthesis
735 State Street, Suite 300
Santa Barbara, California 93101-5504

Tatsuhiro Ohkubo
Department of Forest Science
Utsunomiya University
350 Minemachi
Utsonomiya 321-8505
Japan

Toshinori Okuda
Global Environment Division
National Institute for Environmental
    Studies
16-2 Onogawa
Tsukuba Ibaraki-ken 305-0053
Japan

Peter A. Palmiotto
Department of Environmental Studies
Antioch New England Graduate School
40 Avon Street
Keene, New Hampshire 03431-3516

Rolando Pérez
Center for Tropical Forest Science
Smithsonian Tropical Research Institute
Unit 0948
APO AA 34002-0948 USA

Naoki Rokujo
Laboratory of Plant Ecology
Osaka City University
Sugimoto, Sumiyoshi
Osaka 558-85858
Japan

Katya Romoleroux
Herbario QCA
Department de Ciencias Biologicas
Pontifica University Católica de Ecuador
Apartado 17-01-2184
Quito
Ecuador

Pongsak Sahunalu
Faculty of Forestry
Kasetsart University
Chatuchak
Bangkok 10900
Thailand

Cristián Samper
National Museum of Natural History
10th and Constitution Northwest
Smithsonian Institution
Washington DC 20560

Leng Guan Saw
Forest Research Institute Malaysia
Kepong
52109 Kuala Lumpur
Malaysia

Douglas A. Schaefer
Institute for Tropical Ecosystem Studies
University of Puerto Rico
PO Box 23341
San Juan Puerto Rico 00931-3341

Hardy Semui
Kolej Chermaijaya
Lot 295, Jalan Sultan Tengah
Petra Jaya, 93050 Kuching
Sarawak, Malaysia

Quah Eng Seng
Forest Research Institute Malaysia
Kepong
52109 Kuala Lumpur
Malaysia

Wan Mohd Shukri Wan Ahmad
Natural Forest Division
Forest Research Institute Malaysia (FRIM)
Kepong
52109 Kuala Lumpur
Malaysia

Nicholas C. Songwe
PO Box 4056
Upstation Bamenda Northwestern
   Province
Cameroon

Kriangsak Sri-ngernyuang
Maejo University
Chiang Mai Phrao Road
Kansai Chiang Mai 50290
Thailand

Elizabeth A. Stacy
Department of Biology
Concordia University
1455 de Maisonneuve Boulevard W.
Montreal QC H3G 1M8
Canada

Raman Sukumar
Centre for Ecological Sciences
Indian Institute of Science
Bangalore 560012
India

I-Fang Sun
Center for Tropical Forest Science–AA
   Asia Program
National Institute of Education
Nanyang Technological University
1 Nanyang Walk
Singapore 637616
Singapore

Nur Supardi Mohd. Noor
Forest Research Institute Malaysia
Kepong
52109 Kuala Lumpur
Malaysia

Hebbalalu Sathyanarayana Suresh
Center for Ecological Sciences
Indian Institute of Science
Bangalore 560012
India

Jens-Christian Svenning
Herbarium AAU
University of Aarhus
Uni-parken, bygn 137
DK-8000 Aarhus C
Denmark

Mariko Suzuki
Laboratory of Tropical Forest Ecology
National Institute of Environmental
    Studies
16-2 Onogawa
Tsukuba 305-0053
Japan

Minoru N. Tamura
Botanical Garden
Faculty of Science
Osaka City University
Kisaichi, Katano
Osaka 576-0004
Japan

Sylvester Tan
Sarawak Forest Department
Bangunan Wisma Sumber Alam
Jalan Stadium, Petra Jaya
Kuching Sarawak 93660
Malaysia

Sakhan Teejuntuk
Faculty of Forestry
Kasetsart University
Chatuchak, Bangkok 10900
Thailand

Duncan W. Thomas
Oregon State University

College of Forestry
37309 Kings Valley Highway
Philomath Oregon 973770

Sean C. Thomas
Faculty of Forestry
University of Toronto
33 Willcock Street
Toronto, Ontario M5S 3B3
Canada

Jill Thompson
El Verde Field Station
PO Box 1690
Luquillo, Puerto Rico 00773

Joel Tupac Otero
CSIRO Plant Industry
GPO Box 1600
Canberra ACT 2601
Australia

Renato Valencia
Departmento de Ciencias Biologicas
Pontificia Universidad Catolica del Ecuador
Apartado 17-01:2184
Quito
Ecuador

Martha Isabel Vallejo
Instituto Alexander von Humboldt
Carrera 13 # 28-01, Piso 7
Edficio Palma Real
Bogota DC
Colombia

Gorky Villa Muñoz
Departmento de Ciencias Biologicas
Pontificia Universidad Catolica del
    Ecuador
Apartado 17-01:2184
Quito
Ecuador

Daniel J. Vogt
College of Forest Resources
University of Washington
Seattle, Washington 98195

Kristina A. Vogt
College of Forest Resources
University of Washington
Seattle, Washington 98195

Robert B. Waide
Terrestrial Ecology Division
University of Puerto Rico
PO Box 23341
San Juan, Puerto Rico
　00931-3341

N. S. Weerasekara
Department of Botany
Faculty of Science
University of Peradeniya
Peradeniya
Sri Lanka

D. S. A. Wijesundara
Department of Botany
Faculty of Science
University of Peradeniya
Peradeniya
Sri Lanka

Christopher Wills
Department of Biology
University of California–San Diego
La Jolla, California 92093-0116

S. Joseph Wright
Smithsonian Tropical Research Institute
Unit 0948
APO AA 34002-0948 USA

Takuo Yamakura
Department of Biology
Faculty of Science
Osaka City University
3-3-138 Sugimoto, Sumiyoshi
Osaka 558-8585
Japan

Sandar L. Yap
Conservation International–Philippines
5 South Lawin Avenue
Philan Homes, Quezon City
Philippines

Jess K. Zimmerman
Terrestrial Ecology Division
University of Puerto Rico
PO Box 23341
San Juan, Puerto Rico 00931-3341

Index

Page numbers in italics refer to figures.

abiotic conditions, *32,* 198
abundance. *See also* rankings, tree abundance;
  BCI FDP, *409*; distributions, 87–88; by
  family, *96*; habitat specificity and, *151*; HKK
  FDP, 147–51, *148, 149*; Ituri FDPs, 165–66;
  life-history parameters and, 387–88;
  mortality rates and, 366; Sinharaja FDP, *134*;
  species, 81–82; tree mortality and, 373–75;
  Yasuní FDP, *115*
access, site selection, 435
Acer oblongum (Aceraceae), 485
administration: BCI FDP, Panama, 451; Bukit
  Timah FDP, 464; Doi Inthanon FDP, 474;
  HKK FDP, 482; Ituri FDPs, 492–93; Korup
  FDP, 506; La Planada FDP, 517; Lambir FDP,
  527; Luquillo FDP, 540; Mudumalai FDP, 551;
  Nanjenshan FDP, 564; Palanan FDP, 574;
  Pasoh FDP, 585–86; Sinharaja FDP, 599;
  Yasuní FDP, 609
aerial photographs, 226
African elephants *(Loxodonta africana cyclotis),*
  500, 512
African golden cat *(Profelis aurata),* 500
agamospermy, 286
aggregation, 81–82, 85
agoutis *(Dasyprocta punctata),* 457
agriculture. *See also* disturbances; BCI FDP, 459;
  Doi Inthanon FDP, 480; fragmentation and,
  290; HKK FDP and, 489; Ituri FDPs, 502;
  Korup FDP and, 514; La Planada FDP, 524;
  Luquillo FDP, 548; Nanjenshan FDP, 571–72;
  Yasuní FDP, 617
Agrostistachys spp. (Euphorbiaceae): *A. hookeri,*
  128, 130, 604; *A. intramarginalis,* 126, 128,
  130, 133, *137,* 604
Ailanthus triphysa (Simaroubaceae), 147
Alfisols, *63,* 64–65, 92, 553
allozyme diversity, 286
allozyme electrophoresis, 267, 271–72
Alouatta palliata (howler monkeys), 457
Alseis blackiana (Rubiaceae), 340
altitude above sea level, *32*
Amazonian climate cycles, 39
Anacardiaceae, *96, 98. See also specific plants*
Andean bears *(Tremarctos ornatus),* 523
andesitic cap, 453
Andigena laminirostris (mountain toucans), 522

Andisols, 66, 519–20
angiosperms, 37–39, 244. *See also specific plants*
Anisophyllea corneri (Anisophylleaceae), 590
Annonaceae. *See also specific plants;* abundance,
  96, 100; dominance, 93, *98*; HKK FDP, 147;
  species richness, *99*
Anogeissus latifolia (Combretaceae), 364, 367,
  374, 554, 555
anoxic conditions, 58–59
antelope, 500, 559. *See also specific antelope*
Anthocephalus cadamba (Rubiaceae), 147, 154
Anthrophyllum diversifolium (Araliaceae), 293
anthropogenic disturbances. *See* disturbances
antiherbivore defenses, 47
ants, 513, 591
apomixis, 285–86
Aporusa spp. (Euphorbiaceae), 192, 295, *296*;
  A. benthamiana, 295; *A. nigricans, 284*
aquatic genets *(Osbornictis piscivora),* 499
Ardisia spp. (Myrsinaceae); *A. blackiana,* 387;
  A. crassa, 234, 387
Areca catechu (betel palm, Palmae), 571–72
Aridisols, *63,* 64–65
army ants, 513
Artocarpus spp. (Moraceae), 467
Ashton, P.S., 585
Asian elephants *(Elephas maximus),* 471, 480,
  486, 487, 489, 559
Asiatic bear *(Ursus thibetanus),* 479
Astrocaryum spp. (Palmae); *A. murumuru,* 613;
  A. standleyanum, 251
Astronia ferruginea (Melastomataceae), 568
asymptotic heights, 292, 298
Attalea butyracea (Palmae), 251
average daily temperatures: BCI FDP, 451; Bukit
  Timah FDP, 446, *466*; Doi Inthanon FDP,
  476; HKK FDP, *484*; Ituri FDPs, *494*; Korup
  FDP, *508*; La Planada FDP, *519*; Lambir FDP,
  529; Luquillo FDP, 540, *542*; Mudumalai
  FDP, *553*; Nanjenshan FDP, 564–65, *566*;
  Palanan FDP, *576*; Pasoh FDP, *587*; Sinharaja
  FDP, *601*; Yasuní FDP, 610–11, *611*

Baccaurea brevipes (Euphorbiaceae), 295, 298
Baccaurea (Euphorbiaceae), 192, 295, *296*
Bactris spp. (Palmae), 15
Baird's tapir *(Tapirus bairdii),* 457

629